A POLICY on

GEOMETRIC DESIGN of

HIGHWAYS and STREETS

2001

Second Printing

American Association of State
Highway and Transportation Officials
444 North Capitol Street, N.W., Suite 249
Washington, D.C. 20001
(202) 624-5800
www.transportation.org

ISBN: 1-56051-156-7

American Association of State Highway and Transportation Officials Executive Committee 2000-2001

TASK FORCE ON GEOMETRIC DESIGN 2000
Members

Terry L. Abbott	California	1999 – 2000
Don T. Arkle	Alabama	1991 – Present
Ray Ballentine	Mississippi	1997 – 1999
Harold E. Bastin	National League of Cities	1993 – 1999
Paul Bercich	Wyoming	1995 – Present
James O. Brewer	Kansas	1986 – Present
Jerry Champa	California	1997 – 1999
Philip J. Clark	New York	1992 – Present
Susan Davis	Oklahoma	1994 – 1995
Alan Glenn	California	1992 – 1997
Charles A. Goessel	New Jersey	1986 – Present
Dennis A. Grylicki	National Association of County Engineers	1992 – 1999
Irving Harris	Mississippi	1992 – 1997
David Hutchison	National League of Cities	1999 – Present
John LaPlante	American Public Works Association	1989 – Present
Ken Lazar	Illinois	1990 – 2000
Donald A. Lyford	New Hampshire	1992 – Present
Mark A. Marek	Texas	l986 – Present
Terry H. Otterness	Arizona	1997 – Present
Steven R. Oxoby	Nevada	1993 – Present
Robert P. Parisi	Port Authority of New York and New Jersey	1992 – Present
Randy Peters	Nebraska	1993 – 1998
John Pickering	Mississippi	1999 – Present
William A. Prosser	FHWA, Secretary	1995 – Present
Norman H. Roush	West Virginia	1979 – Present
Joe Ruffer	National Association of County Engineers	1999 – Present
John Sacksteder	Kentucky	1991 – 2000
Larry Sutherland	Ohio	1991 – Present
Charlie V. Trujillo	New Mexico	1998 – Present
Robert L. Walters	Arkansas, Chairman	1982 – Present
Ted Watson	Nebraska	1998 – Present

TASK FORCE ON GEOMETRIC DESIGN 2002
Members

Reza Amini	Oklahoma
Don T. Arkle	Alabama
Paul Bercich	Wyoming
James O. Brewer	Kansas
Philip J. Clark	New York
Ron Erickson	Minnesota
Charles A. Goessel	New Jersey
David Hutchison	National League of Cities
Jeff C. Jones	Tennessee
Wayne Kinder	Nevada
John LaPlante	American Public Works Association
Donald A. Lyford	New Hampshire
Mark A. Marek	Texas
Robert P. Parisi	Port Authority of New York and New Jersey
John Pickering	Mississippi
William A. Prosser	Federal Highway Administration, Secretary
Norman H. Roush	West Virginia
Joe Ruffer	National Association of County Engineers
Larry Sutherland	Ohio
Karla Sutliff	California
Charlie V. Trujillo	New Mexico
Robert L. Walters	Arkansas, Chair
Ted Watson	Nebraska

AASHTO Highway Subcommittee on Design 2000-2001

Dr. Kam K. Movassaghi, LOUISIANA, Chair
Susan Martinovich, NEVADA, Vice-Chair
Dwight A. Horne, FHWA, Secretary

—

Ken F. Kobetsky, P.E., AASHTO, Staff Liaison

ALABAMA— Arkle, Don T.
Chief, Design Bureau
Alabama Department of Transportation
1409 Coliseum Boulevard
Montgomery, AL 36130-3050

ALABAMA— Walker, Steven E.
Assistant Chief Design Engineer
Alabama Department of Transportation
1409 Coliseum Boulevard
Montgomery, AL 36130-3050

ALASKA— Hogins, Gary
Chief of Design & Construction
Standards
Alaska Department of Transportation &
Public Facilities
3132 Channel Drive
Juneau, AK 99801-7898

ARIZONA— Louis, John L.
Assistant State Engineer, Roadway Group
Arizona Department of Transportation
205 South 17th Ave., Mail Drop 611E
Phoenix, AZ 85007-3213

ARKANSAS— Loe, Dale F.
Assistant Chief Engineer—Design
Arkansas State Highway & Transportation
Department
P.O. Box 2261, 10324 Interstate 30
State Highway Building
Little Rock, AR 72203-2261

ARKANSAS— McConnell, Phillip L.
Engineer of Roadway Design
Arkansas State Highway & Transportation
Department
P.O. Box 2261, 10324 Interstate 30
State Highway Building
Little Rock, AR 72203-2261

CALIFORNIA— Buckley, Robert L.
State and Local Project Development
Program Manager
California Department of Transportation
P.O. Box 942874, 1120 N Street
Sacramento, CA 94273

COLORADO— Harris, Timothy J.
Project Development Branch Manager
Colorado Department of Transportation
4201 East Arkansas Avenue, Room 406
Denver, CO 80222

CONNECTICUT— Bard, Carl F.
Principal Engineer
Connecticut Department of Transportation
P.O. Box 317546/2800 Berlin Turnpike
Newington, CT 06131-7546

CONNECTICUT— Byrnes, James F.
Chief Engineer
Connecticut Department of Transportation
P.O. Box 317546/2800 Berlin Turnpike
Newington, CT 06131-7546

CONNECTICUT— Smith, Bradley J.
Manager of State Design
Connecticut Department of Transportation
P.O. Box 317546/2800 Berlin Turnpike
Newington, CT 06131-7546

DELAWARE— Angelo, Michael A.
Assistant Director, Design Support
Delaware Department of Transportation
P.O. Box 778, Bay Road, Route 113
Dover, DE 19903-0778

DELAWARE— Canning, Kevin
Supervising Engineer—Road Design
Delaware Department of Transportation
P.O. Box 778, Bay Road, Route 113
Dover, DE 19903-0778

DELAWARE— Satterfield, Joe
Specifications Engineer
Delaware Department of Transportation
P.O. Box 778, Bay Road, Route 113
Dover, DE 19903-0778

DELAWARE— Simmons, Michael H.
Road Design Engineer
Delaware Department of Transportation
P.O. Box 778, Bay Road, Route 113
Dover, DE 19903-0778

FLORIDA— Hattaway, Billy L.
State Roadway Design Engineer
Florida Department of Transportation
605 Suwannee Street
Tallahassee, FL 32399-0450

FLORIDA— Mills, Jim
Roadway Design Engineer
Florida Department of Transportation
605 Suwannee Street
Tallahassee, FL 32399-0450

FLORIDA— Simmons, Freddie L.
State Highway Engineer
Florida Department of Transportation
605 Suwannee Street, MS 38
Tallahassee, FL 32311-0450

GEORGIA— Kennerly, James
State Road and Airport Design Engineer
Georgia Department of Transportation
2 Capitol Square, Room 444
Atlanta, GA 30334

GEORGIA— Palladi, Joseph
State Urban and Multi-Modal Design
    Engineer
Georgia Department of Transportation
No. 2 Capitol Square, Room 356
Atlanta, GA 30334

GEORGIA— Scott, Walker W.
Georgia Department of Transportation
2 Capitol Square
Atlanta, GA 30334

HAWAII— Abe, Casey
Engineer Program Manager, Design
    Branch, Highways Division
Hawaii Department of Transportation
601 Kamokila Boulevard, Room 688A
Kapolei, HI 96707

HAWAII— Fronda, Julius
Highway Design Section Head
Hawaii Department of Transportation
601 Kapolei Boulevard, Room 609
Kapolei, HI 96707

IDAHO— Hutchinson, Steven C.
Assistant Chief Engineer—Development
Idaho Transportation Department
P.O. Box 7129, 3311 W. State Street
Boise, ID 83707

IDAHO— Laragan, Gregory
Roadway Design Engineer
Idaho Transportation Department
P.O. Box 7129, 3311 W. State Street
Boise, ID 83707

ILLINOIS— Hine, Michael
Chief of Design and Environment
Illinois Department of Transportation
2300 S. Dirksen Parkway
Springfield, IL 62764

ILLINOIS— Seyfried, Robert
Northwestern University Center for
    Public Safety
405 Church Street
Evanston, IL 60204

INDIANA— Klika, Phelps H.
Director, Division of Design
Indiana Department of Transportation
100 N. Senate Avenue
Indianapolis, IN 46204-2217

IOWA— Dillavou, Mitch
Director, Office of Design
Iowa Department of Transportation
800 Lincoln Way
Ames, IA 50010

IOWA— Little, David
Deputy Director
Iowa Department of Transportation,
    Engineering Division
800 Lincoln Way
Ames, IA 50010

KANSAS— Adams, Richard G.
Road Design Engineer
Kansas Department of Transportation
915 Harrison Ave., 9th Floor
Topeka, KS 66612-1568

KANSAS— Brewer, James O.
Engineering Manager—State Road Office
Kansas Department of Transportation
Docking State Office Building, 9th Floor
Topeka, KS 66612-1568

KENTUCKY— Kratt, David
Location Branch Manager
Kentucky Transportation Cabinet,
    Division Of Highway Design
High and Clinton streets, 6th Floor
Frankfort, KY 40622

KENTUCKY— Sperry, Kenneth R.
Assistant State Highway Engineer
Kentucky Transportation Cabinet, State
    Highway Engineer's Office
501 High Street, State Office Building
Frankfort, KY 40622

LOUISIANA— Israel, N. Kent
Roadway Design Engineer Administrator
Louisiana Department of Transportation
    and Development
P.O. Box 94245, 1201 Capitol Access Road
Baton Rouge, LA 70804-9245

LOUISIANA— Kalivoda, Nicholas
Traffic and Geometrics Design Engineer
Louisiana Department of Transportation
    and Development
Trenton, LA 86250

LOUISIANA— Porta, Lloyd E.
Design Squad
Louisiana Department of Transportation
    and Development
P.O. Box 94245, 1201 Capitol Access Road
Baton Rouge, LA 70804-9245

MAINE— Casey, Jerry A.
Program Manager—Urban and Arterial
    Highways
Maine Department of Transportation
Transportation Building, State House
Station 16
Augusta, ME 04333-0016

MARYLAND— Douglass, Robert D.
Deputy Chief Engineer-Highway
    Development
Maryland Department of
    Transportation, State Highway
    Administration
707 N. Calvert Street, Mail Stop C102
Baltimore, MD 21202

MARYLAND— McClelland, Kirk G.
Highway Design Division Chief
Maryland Department of
    Transportation, State Highway
    Administration
707 N. Calvert Street
Baltimore, MD 21202

MASSACHUSETTS— Blundo, John
Deputy Chief Engineer, Highway
    Engineering
Massachusetts Highway Department
10 Park Plaza, Room 6340
Boston, MA 02116-3973

MASSACHUSETTS— Wood, Stanley
Highway Location and Design Engineer
Massachusetts Highway Department
10 Park Plaza
Boston, MA 02116

MICHIGAN— Miller, Paul F.
Engineer of Design
Michigan Department of Transportation,
    Design Division
State Transportation Building
425 W. Ottawa Street, P.O. Box 30050
Lansing, MI 48909

MINNESOTA— Gerdes, Delbert
Director, Technical Support
Minnesota Department of Transportation
Transportation Building, MS 675, 395
John Ireland Boulevard
St. Paul, MN 55155-1899

MISSISSIPPI— Pickering, John B.
Roadway Design Engineer
Mississippi Department of Transportation
P.O. Box 1850, 401 North West Street
Jackson, MS 39215-1850

MISSISSIPPI— Ruff, Wendel T.
Assistant Chief Engineer—Preconstruction
Mississippi Department of Transportation
P.O. Box 1850, 401 North West Street
Jackson, MS 39215-1850

MISSOURI— Nichols, David B.
Director of Project Development
Missouri Department of Transportation
P.O. Box 270
Jefferson City, MO 65102-0207

MISSOURI— Yarnell, William (Bill)
Division Engineer Design
Missouri Department of Transportation
105 West Capitol Avenue, P.O. Box 270
Jefferson City, MO 65102-0207

MONTANA— Peil, Carl S.
Preconstruction Engineer
Montana Department of Transportation
P.O. Box 201001, 2701 Prospect Avenue
Helena, MT 59620-1001

NEBRASKA— Poppe, Eldon D.
Engineer, Roadway Design Division
Nebraska Department of Roads
1500 Nebraska Highway 2
P.O. Box 94759
Lincoln, NE 68509-4759

NEVADA— Oxoby, Steve R.
Chief Road Design Engineer
Nevada Department of Transportation
1263 S. Stewart Street
Carson City, NV 89712

NEW HAMPSHIRE— Green, Craig A.
Administrator, Bureau of Highway Design
New Hampshire Department of
Transportation
John O. Morton Building, P.O. Box 483
1 Hazen Drive
Concord, NH 03301-0483

NEW JERSEY— Dunne, Richard W.
Director, Design Services
New Jersey Department of Transportation
1035 Parkway Avenue, CN 600
Trenton, NJ 08625-0600

NEW JERSEY— Eisdorfer, Arthur J.
Manager, Bureau of Civil Engineering
New Jersey Department of Transportation
1035 Parkway Avenue, CN 600
Trenton, NJ 08625-0600

NEW JERSEY— Miller, Charles
Executive Assistant, Office of the Director
New Jersey Department of Transportation,
Division Of Design Services
1035 Parkway Avenue, CN 600
Trenton, NJ 08625-0600

NEW MEXICO— Maestas, Roy
Chief, Internal Design Bureau
New Mexico State Highway and
Transportation Department
P.O. Box 1149, 1120 Cerrillos Road
Santa Fe, NM 87504-1149

NEW MEXICO— Trujillo, Charlie V.
Deputy Secretary of Transportation
Planning and Design
New Mexico State Highway and
Transportation Department
P.O. Box 1149, 1120 Cerrillos Road
Santa Fe, NM 87504-1149

NEW YORK— Bellair, Peter J.
Director of Design Quality Insurance
Bureau
New York Department of Transportation
Building 5, State Office Campus
1220 Washington Avenue
Albany, NY 12232-0750

NEW YORK— Clark, Phillip J.
Deputy Chief Engineer/Director, Design
Division
New York Department of Transportation
Building 5, State Office Campus
1220 Washington Avenue
Albany, NY 12232-0748

NEW YORK— D'Angelo, Daniel
Director, Design Quality Assurance Bureau
New York Department of Transportation
1220 Washington Ave.
Building 5, Room 410
Albany, NY 12232-0751

NORTH CAROLINA— Alford, John E.
State Roadway Design Unit
North Carolina Department of
Transportation
P.O. Box 25201, 1 South Wilmington Street
Raleigh, NC 27611-5201

NORTH CAROLINA— Barbour, Deborah M.
State Design Engineer
North Carolina Department of
Transportation
P.O. Box 25201, 1 South Wilmington Street
Raleigh, NC 27611-5201

NORTH CAROLINA— Hill, Len
Deputy Administrator, Pre-Construction
North Carolina Department of
Transportation
P.O. Box 25201, 1 South Wilmington Street
Raleigh, NC 27611-5201

NORTH CAROLINA— Morton, Don R.
Deputy Administrator—Preconstruction
North Carolina Department of
Transportation, Division of Highways
P.O. Box 25201, 1 South Wilmington Street
Raleigh, NC 27611-5201

NORTH DAKOTA— Birst, Kenneth E.
Design Engineer
North Dakota Department of Transportation
608 E. Boulevard Avenue
Bismarck, ND 58505-0700

OHIO— Misel, Cash
Assistant Director and Chief Engineer
Ohio Department of Transportation,
Planning and Production Management
1980 West Broad Street
Columbus, OH 43223-1102

OHIO— Sutherland, Larry F.
Deputy Director, Office of Roadway
    Engineering Services
Ohio Department of Transportation
1980 West Broad Street
Columbus, OH 43223

OKLAHOMA— Senkowski, Christine M.
Division Engineer, Roadway Design
Oklahoma Department of Transportation
200 N. E. 21st Street, Room 2c-2
Oklahoma City, OK 73105-3204

OKLAHOMA— Taylor, Bruce E.
Chief Engineer
Oklahoma Department of Transportation
200 N.E. 21st Street
Oklahoma City, OK 73105-3204

OREGON— Greenberg, Dave
Design Unit Manager
Oregon Department of Transportation
355 Capitol Street N.E., Room 200
Salem, OR 97310

OREGON— Nelson, Catherine
Manager, Roadway Engineering Section
Oregon Department of Transportation
200 Transportation Building
Salem, OR 97310

OREGON— Scheick, Jeff
Manager, Technical Services
Oregon Department of Transportation
Transportation Building, 355 Capitol Street
Salem, OR 97310

PENNSYLVANIA— Schreiber, Dean A.
Chief, Highway Quality Assurance Div.
Pennsylvania Department of Transportation
P.O. Box 3161
Harrisburg, PA 17105-3161

PUERTO RICO— Hernandez Borges, Jose E.
Director, Design Area
Puerto Rico Highway and Transportation
    Authority
P.O. Box 42007, Minillas Station
San Juan, PR 00940-2007

RHODE ISLAND— Bennett, J. Michael
Managing Engineer, Highway Design
Rhode Island Department of Transportation
State Office Building, 2 Capitol Hill
Providence, RI 02903-1124

SOUTH CAROLINA— Kneece, Rocque L.
C Fund Manager
South Carolina Department of
    Transportation
Silas N. Pearman Building, 955 Park Street
Box 191
Columbia, SC 29202-0191

SOUTH CAROLINA— Pratt, Robert I.
Project Development Engineer
South Carolina Department of
    Transportation
Silas N. Pearman Building, 955 Park Street
Box 191
Columbia, SC 29202-0191

SOUTH CAROLINA— Walsh, John V.
Program Development Engineer
South Carolina Department of
    Transportation
Silas N. Pearman Building, 955 Park Street
Box 191
Columbia, SC 29202-0191

SOUTH DAKOTA— Bjorneberg, Timothy
Chief Road Design Engineer
South Dakota Department of Transportation
700 East Broadway Avenue
Pierre, SD 57501-2586

SOUTH DAKOTA— Feller, Joe
Chief Materials and Surfacing Engineer
South Dakota Department of Transportation
700 East Broadway Avenue
Pierre, SD 57501-2586

TENNESSEE— Jones, Jeff C.
Civil Engineer Director, Design Division
Tennessee Department of Transportation
505 Deaderick Street, Suite 700
Nashville, TN 37243-0339

TENNESSEE— Zeigler, James
Director, Bureau of Planning and
    Development
Tennessee Department of Transportation
700 James K. Polk Building,
Fifth and Deaderick
Nashville, TN 37243-0339

TEXAS— Wilson, Robert L.
Director, Design
Texas Department of Transportation
125 East 11th Street
Austin, TX 78701-2483

UTAH— Mohanty, P. K.
Roadway Design Engineer
Utah Department of Transportation
4501 South 2700 West
Salt Lake City, UT 84119

VERMONT— Lathrop, Donald H.
Plan Support Engineer
Vermont Agency of Transportation
State Administration Building
133 State Street
Montpelier, VT 05633-5001

VERMONT— Shattuck, Robert F.
Roadway and Traffic Design Program
    Manager
Vermont Agency of Transportation
State Administration Building
133 State Street
Montpelier, VT 05633-5001

VIRGINIA— Harris, James T.
Assisant Division Administrator
Virginia Department of Transportation,
    Location and Design Division
1401 E. Broad Street
Richmond, VA 23219

VIRGINIA— Mills, Jimmy
Location and Testing Engineer
Virginia Department of Transportation
1401 East Broad Street
Richmond, VA 23219

VIRGINIA— Mirshahi, Mohammad
Assistant Division Administrator
Virginia Department of Transportation
1401 E. Broad Street
Richmond, VA 23219

WASHINGTON— Albin, Richard
Standards Engineer
Washington State Department of
    Transportation
Transportation Building
310 Maple Park, P.O. Box 47329
Olympia, WA 98504-7329

WASHINGTON— Ziegler, Brian J.
State Design Engineer
Washington State Department of
  Transportation
505 Deaderick Street, Suite 700
Olympia, WA 98504-7300

WEST VIRGINIA— Clevenger, David E.
Consultant Review Section Head
West Virginia Department of
  Transportation, Engineering Division
1900 Kanawha Boulevard East, Building 5
Charleston, WV 25305-0440

WEST VIRGINIA— Epperly, Randolph T.
Deputy State Highway Engineer-Project
  Development
West Virginia Department of
  Transportation
1900 Kanawha Boulevard East, Building 5
Charleston, WV 25305-0440

WEST VIRGINIA— Roush, Norman H.
Deputy Commissioner of Highways
West Virginia Department of
  Transportation
1900 Kanawha Boulevard East, Building 5
Charleston, WV 25305-0440

WISCONSIN— Haverberg, John E.
Director, Bureau of Highway Development
Wisconsin Department of Transportation
P.O. Box 7910
4802 Sheboygan Avenue
Madison, WI 53707-7910

WISCONSIN— Pfeiffer, Robert F.
Project Development Chief
Wisconsin Department of Transportation,
District 2, Waukesha
P.O. Box 7910
4802 Sheboygan Avenue
Madison, WI 53707-7910

WYOMING— Bercich, Paul
Project Development Engineer
Wyoming Department of Transportation
P.O. Box 1708, 5300 Bishop Boulevard
Cheyenne, WY 82003-1708

DISTRICT OF
COLUMBIA— Rice, John
Manager, Engineering and Specifications
  Division
Federal Aviation Administration
800 Independence Avenue, S.W.
Room 616C, AAS-200
Washington, D.C. 20591-0001

DISTRICT OF
COLUMBIA— Sandhu, Harbhajan S.
Chief, Design and Engineering Division
District of Columbia Department of Public
  Works
2000 14th Street, N.W., 5th Floor
Washington, D.C. 20009

BRITISH COLUMBIA,
CANADA— Voyer, Richard
Senior Standards and Design Engineer
British Columbia Ministry of
  Transportation and Highways
5B - 940 Blanshard Street
Victoria, British Columbia V8W 3E6

Preface

This Policy was developed as part of the continuing work of the Standing Committee on Highways. The Committee, then titled the Committee on Planning and Design Policies, was established in 1937 to formulate and recommend highway engineering policies. This Committee has developed *A Policy on Geometric Design of Rural Highways*, 1954 and 1965 editions; *A Policy on Arterial Highways in Urban Areas*, 1957; *A Policy on Design of Urban Highways and Arterial Streets*, 1973; *Geometric Design Standards for Highways Other Than Freeways*, 1969; *A Policy on Geometric Design of Highways and Streets*, 1984, 1990, and 1994; *A Policy on Design Standards—Interstate System*, 1956, 1967, and 1991; and a number of other AASHO and AASHTO policy and "guide" publications.

An AASHTO publication is typically developed through the following steps: (1) The Committee selects subjects and broad outlines of material to be covered. (2) The appropriate subcommittee and its task forces, in this case, the Subcommittee on Design and its Task Force on Geometric Design, assemble and analyze relevant data and prepare a tentative draft. Working meetings are held and revised drafts are prepared, as necessary, and reviewed by the Subcommittee, until agreement is reached. (3) The manuscript is then submitted for approval by the Standing Committee on Highways and then the Executive Committee. Standards and policies must be adopted by a two-thirds vote by the Member Departments before publication. During the developmental process, comments are sought and considered from all the states, the Federal Highway Administration, and representatives of the American Public Works Association, the National Association of County Engineers, the National League of Cities, and other interested parties.

Notice to User

This second printing of the Fourth Edition of AASHTO's *A Policy on Geometric Design of Highways and Streets* incorporates the revisions that were approved by the Task Force on Geometric Design during the meeting of July 2002. To view a complete list of the changes that were made, please visit AASHTO's online bookstore at www.transportation.org.

Table of Contents

CHAPTER TITLES

Chapter 1
HIGHWAY FUNCTIONS

Page

Chapter 3
ELEMENTS OF DESIGN

Chapter 4
CROSS SECTION ELEMENTS

Chapter 5
LOCAL ROADS AND STREETS

Chapter 6
COLLECTOR ROADS AND STREETS

Chapter 7
RURAL AND URBAN ARTERIALS

Chapter 8
FREEWAYS

Chapter 9
INTERSECTIONS

Chapter 10
GRADE SEPARATIONS AND INTERCHANGES

LIST OF EXHIBITS

Foreword

As highway designers, highway engineers strive to provide for the needs of highway users while maintaining the integrity of the environment. Unique combinations of design requirements that are often conflicting result in unique solutions to the design problems. The guidance supplied by this text, *A Policy on Geometric Design of Highways and Streets*, is based on established practices and is supplemented by recent research. This text is also intended to form a comprehensive reference manual for assistance in administrative, planning, and educational efforts pertaining to design formulation.

Design values are presented in this document in both metric and U.S. customary units and were developed independently within each system. The relationship between the metric and U.S. customary values is neither an exact (soft) conversion nor a completely rationalized (hard) conversion. The metric values are those that would have been used had the policy been presented exclusively in metric units; the U.S. customary values are those that would have been used if the policy had been presented exclusively in U.S. customary units. Therefore, the user is advised to work entirely in one system and not attempt to convert directly between the two.

The fact that new design values are presented herein does not imply that existing streets and highways are unsafe, nor does it mandate the initiation of improvement projects. This publication is not intended as a policy for resurfacing, restoration, or rehabilitation (3R) projects. For projects of this type, where major revisions to horizontal or vertical curvature are not necessary or practical, existing design values may be retained. Specific site investigations and crash history analysis often indicate that the existing design features are performing in a satisfactory manner. The cost of full reconstruction for these facilities, particularly where major realignment is not needed, will often not be justified. Resurfacing, restoration, and rehabilitation projects enable highway agencies to improve highway safety by selectively upgrading existing highway and roadside features without the cost of full reconstruction. When designing 3R projects, the designer should refer to TRB Special Report 214, *Designing Safer Roads: Practices for Resurfacing, Restoration, and Rehabilitation* and related publications for guidance.

The intent of this policy is to provide guidance to the designer by referencing a recommended range of values for critical dimensions. It is not intended to be a detailed design manual that could supercede the need for the application of sound principles by the knowledgeable design professional. Sufficient flexibility is permitted to encourage independent designs tailored to particular situations. Minimum values are either given or implied by the lower value in a given range of values. The larger values within the ranges will normally be used where the social, economic, and environmental (S.E.E.) impacts are not critical.

The highway, vehicle, and individual users are all integral parts of transportation safety and efficiency. While this document primarily addresses geometric design issues, a properly equipped and maintained vehicle and reasonable and prudent performance by the user are also necessary for safe and efficient operation of the transportation facility.

Emphasis has been placed on the joint use of transportation corridors by pedestrians, cyclists, and public transit vehicles. Designers should recognize the implications of this sharing of the transportation corridors and are encouraged to consider not only vehicular movement, but also movement of people, distribution of goods, and provision of essential services. A more comprehensive transportation program is thereby emphasized.

Cost-effective design is also emphasized. The traditional procedure of comparing highway-user benefits with costs has been expanded to reflect the needs of non-users and the environment. Although adding complexity to the analysis, this broader approach also takes into account both the need for a given project and the relative priorities among various projects. The results of this approach may need to be modified to meet the needs-versus-funds problems that highway administrators face. The goal of cost-effective design is not merely to give priority to the most beneficial individual projects but to provide the most benefits to the highway system of which each project is a part.

Most of the technical material that follows is detailed or descriptive design information. Design guidelines are included for freeways, arterials, collectors, and local roads, in both urban and rural locations, paralleling the functional classification used in highway planning. The book is organized into functional chapters to stress the relationship between highway design and highway function. An explanation of functional classification is included in Chapter 1.

These guidelines are intended to provide operation efficiency, comfort, safety, and convenience for the motorist. The design concepts presented herein were also developed with consideration for environmental quality. The effects of the various environmental impacts can and should be mitigated by thoughtful design processes. This principle, coupled with that of aesthetic consistency with the surrounding terrain and urban setting, is intended to produce highways that are safe and efficient for users, acceptable to non-users, and in harmony with the environment.

This publication supersedes the 1994 AASHTO publication of the same name. Because the concepts presented could not be completely covered in one book, references to additional literature are given at the end of each chapter.

CHAPTER 1
HIGHWAY FUNCTIONS
SYSTEMS AND CLASSIFICATIONS

The classification of highways into different operational systems, functional classes, or geometric types is necessary for communication among engineers, administrators, and the general public. Different classification schemes have been applied for different purposes in different rural and urban regions. Classification of highways by design types based on the major geometric features (e.g., freeways and conventional streets and highways) is the most helpful one for highway location and design procedures. Classification by route numbering (e.g., U.S., State, County) is the most helpful for traffic operations. Administrative classification (e.g., National Highway System or Non-National Highway System) is used to denote the levels of government responsible for, and the method of financing, highway facilities. Functional classification, the grouping of highways by the character of service they provide, was developed for transportation planning purposes. Comprehensive transportation planning, an integral part of total economic and social development, uses functional classification as an important planning tool. The emergence of functional classification as the predominant method of grouping highways is consistent with the policies contained in this publication.

THE CONCEPT OF FUNCTIONAL CLASSIFICATION

This section introduces the basic concepts needed for understanding the functional classification of highway facilities and systems.

Hierarchies of Movements and Components

A complete functional design system provides a series of distinct travel movements. The six recognizable stages in most trips include main movement, transition, distribution, collection, access, and termination. For example, Exhibit 1-1 shows a hypothetical highway trip using a freeway, where the main movement of vehicles is uninterrupted, high-speed flow. When approaching destinations from the freeway, vehicles reduce speed on freeway ramps, which act as transition roadways. The vehicles then enter moderate-speed arterials (distributor facilities) that bring them nearer to the vicinity of their destination neighborhoods. They next enter collector roads that penetrate neighborhoods. The vehicles finally enter local access roads that provide direct approaches to individual residences or other terminations. At their destinations the vehicles are parked at an appropriate terminal facility.

Each of the six stages of a typical trip is handled by a separate facility designed specifically for its function. Because the movement hierarchy is based on the total amount of traffic volume, freeway travel is generally highest in the movement hierarchy, followed by distributor arterial travel, which is in turn higher in the movement hierarchy than travel on collectors and local access routes.

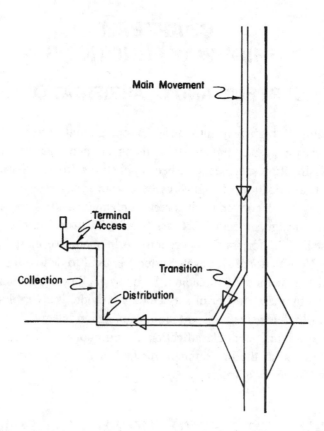

Exhibit 1-1. Hierarchy of Movement

Although many trips can be subdivided into all of the six recognizable stages, intermediate facilities are not always needed. The complete hierarchy of circulation facilities relates especially to conditions of low-density suburban development, where traffic flows are cumulative on successive elements of the system. However, it sometimes is desirable to reduce the number of components in the chain. For instance, a large single traffic generator may fill one or more lanes of a freeway during certain periods. In this situation, it is expedient to lead traffic directly onto a freeway ramp without introducing arterial facilities that unnecessarily mix already-concentrated traffic flows with additional vehicles. This deletion of intermediate facilities does not eliminate the functional need for the remaining parts of the flow hierarchy or the functional design components, although it may change their physical characters. The order of movement is still identifiable.

The failure to recognize and accommodate by suitable design each of the different trip stages of the movement hierarchy is a prominent cause of highway obsolescence. Conflicts and congestion occur at interfaces between public highways and private traffic-generating facilities when the functional transitions are inadequate. Examples are commercial driveways that lead directly from a relatively high-speed arterial into a parking aisle without intermediate provisions for transition deceleration and arterial distribution or, more seriously, freeway ramps that lead directly into or from large traffic generators such as major shopping centers.

Inadequate acceptance capacity of the distributor arterial or internal circulation deficiencies within the traffic absorber may lead to traffic backing up onto the freeway. Successful internal design that provides facilities to accommodate all the intermediate functions between the high-speed freeway and the terminal parking facility will alleviate such a situation.

In the case of the freeway leading to a large traffic generator, deceleration from rapid movement on the freeway occurs on the exit ramp. Distribution to various parking areas is then accomplished by primary distribution-type roads or lanes within the parking facility. These roads or lanes supplant the distributor arterial function. Collector-type roads or lanes within the parking facility may then deliver segments of the entering flow to the parking bays. The parking aisle, in leading to individual parking space terminals, then becomes the equivalent of an access street. Thus, the principal functions within the hierarchical movement system are recognizable. In addition, each functional category also is related to a range of vehicle speeds.

The same principles of design are also relevant to terminal facilities that adjoin distributor arterials or collectors. The functional design of the facility includes each movement stage, with internal circulation in the terminal design to accommodate the order of movement. The need to design for all stages of the movement hierarchy varies with the size of the traffic generator. For relatively small generators, two or more stages may be accommodated on the same internal facility. For larger traffic generators, each movement stage should have a separate functional facility.

To determine the number of design components needed, the customary volumes of traffic handled by public streets of different functional categories can be compared. The volume range on private internal facilities can be related to the comparable range on public streets. These volumes may not be directly comparable, inasmuch as the physical space available within a private facility is smaller and the operational criteria are necessarily quite different. However, the same principles of flow specialization and movement hierarchy can be applied.

Some further examples may demonstrate how the principles of movement hierarchy are related to a logical system of classification of traffic generation intensity. At the highest practical level of traffic generation, a single generator fills an entire freeway, and for this condition, intermediate public streets could not be inserted between the generator and the freeway, so the various movement stages should be accommodated internally with appropriate design features. At the next level of traffic generation a single traffic generator could fill a single freeway lane. It is then appropriate to construct a freeway ramp for the exclusive use of the generator without intervening public streets. At still smaller volumes it becomes desirable to combine the traffic from several generators with additional traffic before the flow arrives at a freeway entrance ramp. The road performing this function then becomes a collector facility, accumulating these small flows until a traffic volume that will fill the freeway ramp is reached.

Similar principles can be applied at the distributor arterial level of service. If a given traffic generator is of sufficient size, an exclusive intersection driveway for that generator is justified. In other cases an intermediate collector street should combine smaller traffic flows until they reach a volume that warrants an intersection along the distributor. The same theory can be applied with regard to the criteria for direct access to the collector street. A moderately sized traffic generator usually warrants a direct connection to the collector without an intermediate access street; however, in a district of single-family residences a local access street should assemble the traffic from a group of residences and lead it into a collector street at a single point of access. In practice, direct access to arterials and collectors should be provided from commercial and residential properties, particularly in established neighborhoods.

In short, each element of the functional hierarchy can serve as a collecting facility for the next higher element, but an element should be present only where the intermediate collection is needed to satisfy the spacing needs and traffic volume demands of the next higher facility. By defining the spacing needs and traffic volume demands for a system element, it is possible to determine which cases should use the full system and in which cases intermediate elements may be bypassed.

Functional Relationships

Functional classification thus groups streets and highways according to the character of service they are intended to provide. This classification recognizes that individual roads and streets do not serve travel independently. Rather, most travel involves movement through networks of roads and can be categorized relative to such networks in a logical and efficient manner. Thus, functional classification of roads and streets is also consistent with categorization of travel.

A schematic illustration of this basic idea is shown in Exhibit 1-2. In Exhibit 1-2A, lines of travel desire are straight lines connecting trip origins and destinations (circles). The relative widths of the lines indicate the relative amounts of travel desire. The relative sizes of the circles indicate the relative trip generating and attracting power of the places shown. Because it is impractical to provide direct-line connections for every desire line, trips should be channelized on

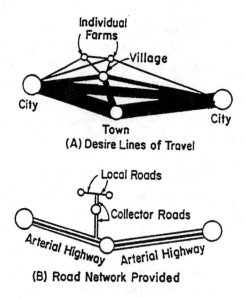

Exhibit 1-2. Channelization of Trips

a limited road network in the manner shown in Exhibit 1-2B. Heavy travel movements are directly served or nearly so the smaller movements are channeled into somewhat indirect paths. The facilities in Exhibit 1-2 are labeled local access, collector, and arterial, which are terms that describe their functional relationships. In this scheme the functional hierarchy is also seen to be related to the hierarchy of trip distances served by the network.

A more complete illustration of a functionally classified rural network is shown in Exhibit 1-3. The arterial highways generally provide direct service between cities and larger towns, which generate and attract a large proportion of the relatively longer trips. Roads of the intermediate functional category (collectors) serve small towns directly, connecting them to the arterial network. Roads of this category collect traffic from the local roads, which serve individual farms and other rural land uses or distribute traffic to these local roads from the arterials.

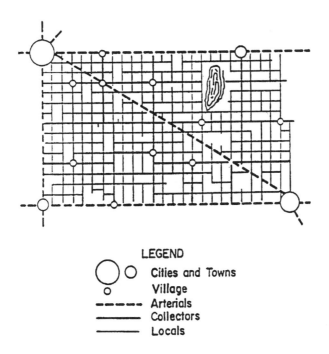

LEGEND

◯ ○ Cities and Towns
 ○ Village
- - - - - Arterials
――――― Collectors
――――― Locals

Exhibit 1-3. Schematic Illustration of a Functionally Classified Rural Highway Network

Although this example has a rural setting, the same basic concepts also apply in urban and suburban areas. A similar hierarchy of systems can be defined; however, because of the high intensity of land use and travel, specific travel generation centers are more difficult to identify. In urban and suburban areas additional considerations, such as the spacing of intersections, become more important in defining a logical and efficient network. A schematic illustration of a functionally classified suburban street network is shown in Exhibit 1-4.

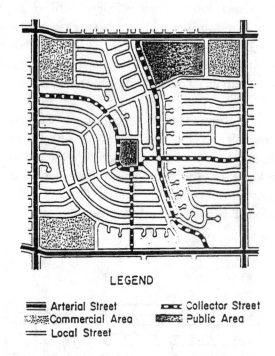

LEGEND

▬▬ Arterial Street   ▭▭ Collector Street
▨▨ Commercial Area   ▨▨ Public Area
═══ Local Street

Exhibit 1-4. Schematic Illustration of a Portion of a Suburban Street Network

Access Needs and Controls

The two major considerations in classifying highway and street networks functionally are access and mobility. The conflict between serving through movement and providing access to a dispersed pattern of trip origins and destinations necessitates the differences and gradations in the various functional types. Regulated limitation of access is needed on arterials to enhance their primary function of mobility.

Conversely, the primary function of local roads and streets is to provide access (implementation of which causes a limitation of mobility). The extent and degree of access control is thus a significant factor in defining the functional category of a street or highway.

Allied to the idea of traffic categorization is the dual role that the highway and street network plays in providing (1) access to property and (2) travel mobility. Access is a fixed need for every area served by the highway system. Mobility is provided at varying levels of service. Mobility can incorporate several qualitative elements, such as riding comfort and absence of speed changes, but the most basic factor is operating speed or trip travel time.

Exhibit 1-2 shows that the concept of traffic categorization leads logically not only to a functional hierarchy of road classes but also to a similar hierarchy of relative travel distances served by these road classes. The hierarchy of travel distances can be related logically to functional specialization in meeting the property access and travel mobility needs. Local rural facilities emphasize the land access function. Arterials for main movement or distribution emphasize the high level of mobility for through movement. Collectors offer approximately balanced service for both functions. This scheme is illustrated conceptually in Exhibit 1-5.

Further discussion of the various degrees of access control appropriate to street and highway development is provided in the section on "Access Control and Access Management" in Chapter 2.

PROPORTION OF SERVICE

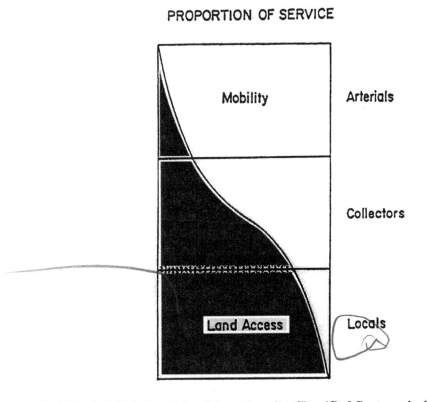

Exhibit 1-5. Relationship of Functionally Classified Systems in Serving Traffic Mobility and Land Access

FUNCTIONAL SYSTEM CHARACTERISTICS

This section contains definitions and characteristics of highway facilities in urban and rural settings based on their functional classifications. It presents information, in revised form, from the Federal Highway Administration publication *Highway Functional Classification: Concepts, Criteria, and Procedures* (**1**).

Definitions of Urban and Rural Areas

Urban and rural areas have fundamentally different characteristics with regard to density and types of land use, density of street and highway networks, nature of travel patterns, and the way in which these elements are related. Consequently, urban and rural functional systems are classified separately.

Urban areas are those places within boundaries set by the responsible State and local officials having a population of 5,000 or more. Urban areas are further subdivided into *urbanized areas* (population of 50,000 and over) and *small urban areas* (population between 5,000 and 50,000). For design purposes, the population forecast for the design year should be used. (For legal definition of urban areas, see Section 101 of Title 23, U.S. Code.)

Rural areas are those areas outside the boundaries of urban areas.

Functional Categories

The roads making up the functional systems differ for urban and rural areas. The hierarchy of the functional systems consists of principal arterials (for main movement), minor arterials (distributors), collectors, and local roads and streets; however, in urban areas there are relatively more arterials with further functional subdivisions of the arterial category whereas in rural areas there are relatively more collectors with further functional subdivisions of the collector category.

Functional Systems for Rural Areas

Rural roads consist of facilities outside of urban areas. The names provided for the recognizable systems are principal arterials (roads), minor arterials (roads), major and minor collectors (roads), and local roads.

Rural Principal Arterial System

The rural principal arterial system consists of a network of routes with the following service characteristics:

1. Corridor movement with trip length and density suitable for substantial statewide or interstate travel.

2. Movements between all, or virtually all, urban areas with populations over 50,000 and a large majority of those with populations over 25,000.

3. Integrated movement without stub connections except where unusual geographic or traffic flow conditions dictate otherwise (e.g., international boundary connections or connections to coastal cities).

In the more densely populated states, this class of highway includes most (but not all) heavily traveled routes that might warrant multilane improvements in the majority of states; the principal arterial system includes most (if not all) existing rural freeways.

The principal arterial system is stratified into the following two design types: (1) freeways and (2) other principal arterials.

Rural Minor Arterial System

The rural minor arterial road system, in conjunction with the rural principal arterial system, forms a network with the following service characteristics:

1. Linkage of cities, larger towns, and other traffic generators (such as major resort areas) that are capable of attracting travel over similarly long distances.

2. Integrated interstate and intercounty service.

3. Internal spacing consistent with population density, so that all developed areas of the state are within reasonable distances of arterial highways.

4. Corridor movements consistent with items (1) through (3) with trip lengths and travel densities greater than those predominantly served by rural collector or local systems.

Minor arterials therefore constitute routes, the design of which should be expected to provide for relatively high travel speeds and minimum interference to through movement.

Rural Collector System

The rural collector routes generally serve travel of primarily intracounty rather than statewide importance and constitute those routes on which (regardless of traffic volume) predominant travel distances are shorter than on arterial routes. Consequently, more moderate speeds may be typical. To define rural collectors more clearly, this system is subclassified according to the following criteria:

- Major Collector Roads. These routes (1) serve county seats not on arterial routes, larger towns not directly served by the higher systems, and other traffic generators of equivalent intracounty importance, such as consolidated schools, shipping points, county parks, and important mining and agricultural areas; (2) link these places with nearby larger towns or cities, or with routes of higher classifications; and (3) serve the more important intracounty travel corridors.

- Minor Collector Roads. These routes should (1) be spaced at intervals consistent with population density to accumulate traffic from local roads and bring all developed areas within reasonable distances of collector roads; (2) provide service to the remaining smaller communities; and (3) link the locally important traffic generators with their rural hinterland.

Rural Local Road System

The rural local road system, in comparison to collectors and arterial systems, primarily provides access to land adjacent to the collector network and serves travel over relatively short distances. The local road system constitutes all rural roads not classified as principal arterials, minor arterials, or collector roads.

Extent of Rural Systems

The functional criteria for road systems have been expressed herein primarily in qualitative rather than quantitative terms. Because of varying geographic conditions (e.g., population densities, spacing between and sizes of cities, and densities and patterns of road networks), criteria on sizes of population centers, trip lengths, traffic volumes, and route spacings do not apply to all systems in all States. However, the results of classification studies conducted in many States show considerable consistency (when expressed in percentages of the total length of rural roads) in the relative extents of the functional systems.

Highway systems developed by using these criteria are generally expected, in all States except Alaska and Hawaii, to fall within the percentage ranges shown in Exhibit 1-6. The higher values of the ranges given in Exhibit 1-6 apply to States having less extensive total road networks relative to the population density. In States having more extensive total road networks relative to the population density, the lower values are applicable. The range of percentages of rural collectors represents the total length of both major and minor collector roads and applies to the statewide rural roadway totals the percentages in particular counties may vary considerably from the statewide average. Areas having an extensive regular grid pattern of roads usually have a smaller percentage of collectors than areas within which geographic conditions have imposed a restricted or less regular pattern of road development.

Systems	Percentage of Total Rural Road Length
Principal arterial system	2–4%
Principal arterial plus minor arterial system	6–12%, with most States falling in 7–10% range
Collector road	20–25%
Local road system	65–75%

Exhibit 1-6. Typical Distribution of Rural Functional Systems

Functional Highway Systems in Urbanized Areas

The four functional highway systems for urbanized areas are urban principal arterials (streets), minor arterials (streets), collectors (streets), and local streets. The differences in the nature and intensity of development in rural and urban areas warrant corresponding differences in urban system characteristics relative to the correspondingly named rural systems.

Urban Principal Arterial System

In every urban environment, one system of streets and highways can be identified as unusually significant in terms of the nature and composition of travel it serves. In small urban areas (population under 50,000), these facilities may be very limited in number and extent, and their importance may be derived primarily from the service provided to through travel. In urbanized areas, their importance also derives from service to rurally oriented traffic, but equally or even more importantly, from service for major circulation movements within these urbanized areas.

The urban principal arterial system serves the major centers of activity of urbanized areas, the highest traffic volume corridors, and the longest trip desires and carries a high proportion of the total urban area travel even though it constitutes a relatively small percentage of the total roadway network. The system should be integrated both internally and between major rural connections.

The principal arterial system carries most of the trips entering and leaving the urban area, as well as most of the through movements bypassing the central city. In addition, significant intra-area travel, such as between central business districts and outlying residential areas, between major inner-city communities, and between major suburban centers, is served by this class of facility. Frequently, the principal arterial system carries important intra-urban as well as intercity bus routes. Finally, in urbanized areas, this system provides continuity for all rural arterials that intercept the urban boundary.

Because of the nature of the travel served by the principal arterial system, almost all fully and partially controlled access facilities are usually part of this functional class. However, this system is not restricted to controlled-access routes. To preserve the identification of controlled-access facilities, the principal arterial system should be stratified as follows: (1) interstate, (2) other freeways, and (3) other principal arterials (with partial or no control of access).

The spacing of urban principal arterials is closely related to the trip-end density characteristics of particular portions of the urban areas. Although no firm spacing rule applies in all or even in most circumstances, the spacing between principal arterials (in larger urban areas) may vary from less than 1.6 km [1 mi] in the highly developed central business areas to 8 km [5 mi] or more in the sparsely developed urban fringes.

For principal arterials, service to abutting land is subordinate to travel service to major traffic movements. Only facilities within the subclass of other principal arterials are capable of providing any direct access to land, and such service should be purely incidental to the primary functional responsibility of this class of roads.

Urban Minor Arterial Street System

The minor arterial street system interconnects with and augments the urban principal arterial system. It accommodates trips of moderate length at a somewhat lower level of travel mobility than principal arterials do. This system distributes travel to geographic areas smaller than those identified with the higher system.

The minor arterial street system includes all arterials not classified as principal. This system places more emphasis on land access than the higher system does and offers lower traffic mobility. Such a facility may carry local bus routes and provide intracommunity continuity but ideally does not penetrate identifiable neighborhoods. This system includes urban connections to rural collector roads where such connections have not been classified as urban principal arterials for internal reasons.

The spacing of minor arterial streets may vary from 0.2 to 1.0 km [0.1 to 0.5 mi] in the central business district to 3 to 5 km [2 to 3 mi] in the suburban fringes but is normally not more than 2 km [1 mi] in fully developed areas.

Urban Collector Street System

The collector street system provides both land access service and traffic circulation within residential neighborhoods and commercial and industrial areas. It differs from the arterial system in that facilities on the collector system may penetrate residential neighborhoods, distributing trips from the arterials through the area to their ultimate destinations. Conversely, the collector street also collects traffic from local streets in residential neighborhoods and channels it into the arterial system. In the central business district, and in other areas of similar development and traffic density, the collector system may include the entire street grid. The collector street system may also carry local bus routes.

Urban Local Street System

The local street system comprises all facilities not in one of the higher systems. It primarily permits direct access to abutting lands and connections to the higher order systems. It offers the lowest level of mobility and usually contains no bus routes. Service to through-traffic movement usually is deliberately discouraged.

Length of Roadway and Travel on Urban Systems

Exhibit 1-7 contains the typical distribution of travel volume and length of roadway of the functional systems for urbanized areas. Systems developed for urbanized areas using the criteria herein usually fall within the percentage ranges shown.

Systems	Range	
	Travel volume (%)	Length (%)
Principal arterial system	40–65	5–10
Principal arterial plus minor arterial street system	65–80	15–25
Collector road	5–10	5–10
Local road system	10–30	65–80

Exhibit 1-7. Typical Distribution of Urban Functional Systems

Functional Classification as a Design Type

This text has utilized the functional classification system as a design type of highway. Two major difficulties arise from this usage. The first major problem involves freeways. A freeway is not a functional class in itself but is normally classified as a principal arterial. It does, however, have unique geometric criteria that demand a separate design designation apart from other arterials. Therefore, a separate chapter on freeways has been included along with chapters on arterials, collectors, and local roads and streets. The addition of the universally familiar term "freeway" to the basic functional classes seems preferable to the adoption of a complete separate system of design types.

The second major difficulty is that, in the past, geometric design criteria and capacity levels have traditionally been based on a classification of traffic volume ranges. Under such a system, highways with comparable traffic volumes are constructed to the same criteria and provide identical levels of service, although there may be considerable difference in the functions they serve.

Under a functional classification system, design criteria and level of service vary according to the function of the highway facility. Volumes serve to further refine the design criteria for each class.

Arterials are expected to provide a high degree of mobility for the longer trip length. Therefore, they should provide a high operating speed and level of service. Since access to abutting property is not their major function, some degree of access control is desirable to enhance mobility. The collectors serve a dual function in accommodating the shorter trip and feeding the arterials. They should provide some degree of mobility and also serve abutting property. Thus, an intermediate design speed and level of service is appropriate. Local roads and streets have relatively short trip lengths, and, because property access is their main function, there is little need for mobility or high operating speeds. This function is reflected by use of a lower design speed and level of service.

The functional concept is important to the designer. Even though many of the geometric design values could be determined without reference to the functional classification, the designer should keep in mind the overall purpose that the street or highway is intended to serve. This concept is consistent with a systematic approach to highway planning and design.

The first step in the design process is to define the function that the facility is to serve. The level of service needed to fulfill this function for the anticipated volume and composition of traffic provides a rational and cost-effective basis for the selection of design speed and geometric criteria within the ranges of values available to the designer. The use of functional classification as a design type should appropriately integrate the highway planning and design process.

REFERENCES

1. U.S. Department of Transportation, Federal Highway Administration. *Highway Functional Classification: Concepts, Criteria, and Procedures,* Washington, D.C.: 1989.

2. U.S. Department of Transportation, Federal Highway Administration, Office of Information Management. *Our Nation's Highways—Selected Facts and Figures*, Report No. FHWA-PL-98-015, Washington, D.C.: 1998.

CHAPTER 2
DESIGN CONTROLS AND CRITERIA

INTRODUCTION

This chapter discusses those characteristics of vehicles, pedestrians, and traffic that act as criteria for the optimization or improvement in design of the various highway and street functional classes.

DESIGN VEHICLES

General Characteristics

Key controls in geometric highway design are the physical characteristics and the proportions of vehicles of various sizes using the highway. Therefore, it is appropriate to examine all vehicle types, establish general class groupings, and select vehicles of representative size within each class for design use. These selected vehicles, with representative weight, dimensions, and operating characteristics, used to establish highway design controls for accommodating vehicles of designated classes, are known as design vehicles. For purposes of geometric design, each design vehicle has larger physical dimensions and a larger minimum turning radius than most vehicles in its class. The largest design vehicles are usually accommodated in freeway design.

Four general classes of design vehicles have been established: (1) passenger cars, (2) buses, (3) trucks, and (4) recreational vehicles. The passenger-car class includes passenger cars of all sizes, sport/utility vehicles, minivans, vans, and pick-up trucks. Buses include inter-city (motor coaches), city transit, school, and articulated buses. The truck class includes single-unit trucks, truck tractor-semitrailer combinations, and truck tractors with semitrailers in combination with full trailers. Recreational vehicles include motor homes, cars with camper trailers, cars with boat trailers, motor homes with boat trailers, and motor homes pulling cars. In addition, the bicycle should also be considered as a design vehicle where bicycle use is allowed on a highway.

Dimensions for 19 design vehicles representing vehicles within these general classes are given in Exhibit 2-1. In the design of any highway facility, the designer should consider the largest design vehicle likely to use that facility with considerable frequency or a design vehicle with special characteristics appropriate to a particular intersection in determining the design of such critical features as radii at intersections and radii of turning roadways. In addition, as a general guide, the following may be considered when selecting a design vehicle:

- A passenger car may be selected when the main traffic generator is a parking lot or series of parking lots.
- A single-unit truck may be used for intersection design of residential streets and park roads.

Metric

Design Vehicle Type	Symbol	Dimensions (m)											
		Overall			Overhang		WB$_1$	WB$_2$	S	T	WB$_3$	WB$_4$	Typical Kingpin to Center of Rear Axle
		Height	Width	Length	Front	Rear							
Passenger Car	P	1.3	2.1	5.8	0.9	1.5	3.4	–	–	–	–	–	–
Single Unit Truck	SU	3.4-4.1	2.4	9.2	1.2	1.8	6.1	–	–	–	–	–	–
Buses													
Inter-city Bus (Motor Coaches)	BUS-12	3.7	2.6	12.2	1.8	1.9^a	7.3	1.1	–	–	–	–	8.4
	BUS-14	3.7	2.6	13.7	1.8	2.6^a	8.1	1.2	–	–	–	–	11.4
City Transit Bus	CITY-BUS	3.2	2.6	12.2	2.1	2.4	7.6	–	–	–	–	–	13.0
Conventional School Bus (65 pass.)	S-BUS 11	3.2	2.4	10.9	0.8	3.7	6.5	–	–	–	–	–	–
Large School Bus (84 pass.)	S-BUS 12	3.2	2.4	12.2	2.1	4.0	6.1	–	–	–	–	–	–
Articulated Bus	A-BUS	3.4	2.6	18.3	2.6	3.1	6.7	5.9	1.9^b	4.0^b	–	–	–
Trucks													
Intermediate Semitrailer	WB-12	4.1	2.4	13.9	0.9	0.8^a	3.8	8.4	–	–	–	–	8.4
Intermediate Semitrailer	WB-15	4.1	2.6	16.8	0.9	0.6^a	4.5	10.8	–	–	–	–	11.4
Interstate Semitrailer	WB-19*	4.1	2.6	20.9	1.2	0.8^a	6.6	12.3	–	–	–	–	13.0
Interstate Semitrailer	WB-20**	4.1	2.6	22.4	1.2	1.4-0.8^a	6.6	13.2-13.8	–	–	–	–	13.9-14.5
"Double-Bottom"-Semitrailer/Trailer	WB-20D	4.1	2.6	22.4	0.7	0.9	3.4	7.0	0.9^c	2.1^c	7.0	–	7.0
Triple-Semitrailer/ Trailers	WB-30T	4.1	2.6	32.0	0.7	0.9	3.4	6.9	0.9^d	2.1^d	7.0	7.0	7.0
Turnpike Double-Semitrailer/Trailer	WB-33D*	4.1	2.6	34.8	0.7	0.8^a	4.4	12.2	0.8^e	3.1^e	13.6	–	13.0
Recreational Vehicles													
Motor Home	MH	3.7	2.4	9.2	1.2	1.8	6.1	–	–	–	–	–	–
Car and Camper Trailer	P/T	3.1	2.4	14.8	0.9	3.1	3.4	–	1.5	5.8	–	–	–
Car and Boat Trailer	P/B	–	2.4	12.8	0.9	2.4	3.4	–	1.5	4.6	–	–	–
Motor Home and Boat Trailer	MH/B	3.7	2.4	16.2	1.2	2.4	6.1	–	1.8	4.6	–	–	–
Farm Tractor^f	TR	3.1	2.4-3.1	4.9^g	–	–	3.1	2.7	0.9	2.0	–	–	–

Note: Since vehicles are manufactured in U.S. Customary dimensions and to provide only one physical size for each design vehicle, the values shown in the design vehicle drawings have been soft converted from numbers listed in feet, and then the numbers in this table have been rounded to the nearest tenth of a meter.

* = Design vehicle with 14.63 m trailer as adopted in 1982 Surface Transportation Assistance Act (STAA).

** = Design vehicle with 16.16 m trailer as grandfathered in with 1982 Surface Transportation Assistance Act (STAA).

a = This is overhang from the back axle of the tandem axle assembly.

b = Combined dimension is 5.91 m and articulating section is 1.22 m wide.

c = Combined dimension is typically 3.05 m.

d = Combined dimension is typically 3.05 m.

e = Combined dimension is typically 3.81 m.

f = Dimensions are for a 150–200 hp tractor excluding any wagon length.

g = To obtain the total length of tractor and one wagon, add 5.64 m to tractor length. Wagon length is measured from front of drawbar to rear of wagon, and drawbar is 1.98 m long.

• WB$_1$, WB$_2$, and WB$_3$ and WB$_4$ are the effective vehicle wheelbases, or distances between axle groups, starting at the front and working towards the back of each unit.

• S is the distance from the rear effective axle to the hitch point or point of articulation.

• T is the distance from the hitch point or point of articulation measured back to the center of the next axle or center of tandem axle assembly.

Exhibit 2-1. Design Vehicle Dimensions

US Customary

Design Vehicle Type	Symbol	Overall			Overhang		Dimensions (ft)						Typical Kingpin to Center of Rear Axle
		Height	Width	Length	Front	Rear	WB$_1$	WB$_2$	S	T	WB$_3$	WB$_4$	
Passenger Car	P	4.25	7	19	3	5	11	-	-	-	-	-	-
Single Unit Truck	SU	11-13.5	8.0	30	4	6	20	-	-	-	-	-	-
Buses													
Inter-city Bus (Motor Coaches)	BUS-40	12.0	8.5	40	6	6.3[a]	24	3.7	-	-	-	-	27.5
	BUS-45	12.0	8.5	45	6	8.5[a]	26.5	4.0	-	-	-	-	37.5
City Transit Bus	CITY-BUS	10.5	8.5	40	7	8	25	-	-	-	-	-	42.5
Conventional School Bus (65 pass.)	S-BUS 36	10.5	8.0	35.8	2.5	12	21.3	-	-	-	-	-	-
Large School Bus (84 pass.)	S-BUS 40	10.5	8.0	40	7	13	20	-	-	-	-	-	-
Articulated Bus	A-BUS	11.0	8.5	60	8.6	10	22.0	19.4	6.2[b]	13.2[b]	-	-	-
Trucks													
Intermediate Semitrailer	WB-40	13.5	8.0	45.5	3	2.5[a]	12.5	27.5	-	-	-	-	27.5
Intermediate Semitrailer	WB-50	13.5	8.5	55	3	2[a]	14.6	35.4	-	-	-	-	37.5
Interstate Semitrailer	WB-62*	13.5	8.5	68.5	4	2.5[a]	21.6	40.4	-	-	-	-	42.5
Interstate Semitrailer	WB-65** or WB-67	13.5	8.5	73.5	4	4.5-2.5[a]	21.6	43.4-45.4	-	-	-	-	45.5-47.5
"Double-Bottom"-Semitrailer/Trailer	WB-67D	13.5	8.5	73.3	2.33	3	11.0	23.0	3.0[c]	7.0[c]	23.0	-	23.0
Triple-Semitrailer/ Trailers	WB-100T	13.5	8.5	104.8	2.33	3	11.0	22.5	3.0[d]	7.0[d]	23.0	23.0	23.0
Turnpike Double-Semitrailer/Trailer	WB-109D*	13.5	8.5	114	2.33	2.5[e]	14.3	39.9	2.5[e]	10.0[e]	44.5	-	42.5
Recreational Vehicles													
Motor Home	MH	12	8	30	4	6	20	-	-	-	-	-	-
Car and Camper Trailer	P/T	10	8	48.7	3	10	11	-	5	19	-	-	-
Car and Boat Trailer	P/B	-	8	42	3	8	11	-	5	15	-	-	-
Motor Home and Boat Trailer	MH/B	12	8	53	4	8	20	-	6	15	-	-	-
Farm Tractor	TR	10	8-10	16[g]	-	-	10	9	3	6.5	-	-	-

* = Design vehicle with 48 ft trailer as adopted in 1982 Surface Transportation Assistance Act (STAA).
** = Design vehicle with 53 ft trailer as grandfathered in with 1982 Surface Transportation Assistance Act (STAA).
a = This is overhang from the back axle of the tandem axle assembly.
b = Combined dimension is 19.4 ft and articulating section is 4 ft wide.
c = Combined dimension is typically 10.0 ft.
d = Combined dimension is typically 10.0 ft.
e = Combined dimension is typically 12.5 ft.
f = Dimensions are for a 150-200 hp tractor excluding any wagon length.
g = To obtain the total length of tractor and one wagon, add 18.5 ft to tractor length. Wagon length is measured from front of drawbar to rear of wagon, and drawbar is 6.5 ft long.

- WB$_1$, WB$_2$, and WB$_4$ are the effective vehicle wheelbases, or distances between axle groups, starting at the front and working towards the back of each unit.
- S is the distance from the rear effective axle to the hitch point or point of articulation.
- T is the distance from the hitch point or point of articulation measured back to the center of the next axle or center of tandem axle assembly.

Exhibit 2-1. Design Vehicle Dimensions (Continued)

- A city transit bus may be used in the design of state highway intersections with city streets that are designated bus routes and that have relatively few large trucks using them.

- Depending on expected usage, a large school bus (84 passengers) or a conventional school bus (65 passengers) may be used for the design of intersections of highways with low-volume county highways and township/local roads under 400 ADT. The school bus may also be appropriate for the design of some subdivision street intersections.

- The WB-20 [WB-65 or 67] truck should generally be the minimum size design vehicle considered for intersections of freeway ramp terminals with arterial crossroads and for other intersections on state highways and industrialized streets that carry high volumes of traffic and/or that provide local access for large trucks.

In addition to the 19 design vehicles, dimensions for a typical farm tractor are shown in Exhibit 2-1, and the minimum turning radius for a farm tractor with one wagon is shown in Exhibit 2-2. Turning paths of design vehicles can be determined from the dimensions shown in Exhibit 2-1 and 2-2 and through the use of commercially available computer programs.

Minimum Turning Paths of Design Vehicles

Exhibits 2-3 through 2-23 present the minimum turning paths for 19 typical design vehicles. The principal dimensions affecting design are the minimum centerline turning radius (CTR), the out-to-out track width, the wheelbase, and the path of the inner rear tire. Effects of driver characteristics (such as the speed at which the driver makes a turn) and of the slip angles of wheels are minimized by assuming that the speed of the vehicle for the minimum turning radius is less than 15 km/h [10 mph].

The boundaries of the turning paths of each design vehicle for its sharpest turns are established by the outer trace of the front overhang and the path of the inner rear wheel. This turn assumes that the outer front wheel follows the circular arc defining the minimum centerline turning radius as determined by the vehicle steering mechanism. The minimum radii of the outside and inside wheel paths and the centerline turning radii (CTR) for specific design vehicles are given in Exhibit 2-2.

Trucks and buses generally require more generous geometric designs than do passenger vehicles. This is largely because trucks and buses are wider and have longer wheelbases and greater minimum turning radii, which are the principal vehicle dimensions affecting horizontal alignment and cross section. Single-unit trucks and buses have smaller minimum turning radii than most combination vehicles, but because of their greater offtracking, the longer combination vehicles need greater turning path widths. Exhibit 2-11 defines the turning characteristics of a typical tractor/semitrailer combination. Exhibit 2-12 defines the lengths of tractors commonly used in tractor/semitrailer combinations.

A combination truck is a single-unit truck with a full trailer, a truck tractor with a semitrailer, or a truck tractor with a semitrailer and one or more full trailers. Because combination truck sizes and turning characteristics vary widely, there are several combination truck design

Metric

Design Vehicle Type	Pas-senger Car	Single Unit Truck	Inter-city Bus (Motor Coach)		City Transit Bus	Conven-tional School Bus (65 pass.)	Large[2] School Bus (84 pass.)	Articu-lated Bus	Intermed-iate Semi-trailer	Intermed-iate Semi-trailer
Symbol	P	SU	BUS-12	BUS-14	CITY-BUS	S-BUS11	S-BUS12	A-BUS	WB-12	WB-15
Minimum Design Turning Radius (m)	7.3	12.8	13.7	13.7	12.8	11.9	12.0	12.1	12.2	13.7
Center-line[1] Turning Radius (CTR) (m)	6.4	11.6	12.4	12.4	11.5	10.6	10.8	10.8	11.0	12.5
Minimum Inside Radius (m)	4.4	8.6	8.4	7.8	7.5	7.3	7.7	6.5	5.9	5.2

Design Vehicle Type	Interstate Semi-trailer		"Double Bottom" Combina-tion	Triple Semi-trailer/ trailers	Turnpike Double Semi-trailer/ trailer	Motor Home	Car and Camper Trailer	Car and Boat Trailer	Motor Home and Boat Trailer	Farm Tractor w/One Wagon
Symbol	WB-19*	WB-20**	WB-20D	WB-30T	WB-33D*	MH	P/T	P/B	MH/B	TR/W
Minimum Design Turning Radius (m)	13.7	13.7	13.7	13.7	18.3	12.2	10.1	7.3	15.2	5.5
Center-line[1] Turning Radius (CTR) (m)	12.5	12.5	12.5	12.5	17.1	11.0	9.1	6.4	14.0	4.3
Minimum Inside Radius (m)	2.4	1.3	5.9	3.0	4.5	7.9	5.3	2.8	10.7	3.2

Note: Numbers in table have been rounded to the nearest tenth of a meter.

* = Design vehicle with 14.63 m trailer as adopted in 1982 Surface Transportation Assistance Act (STAA).

** = Design vehicle with 16.16 m trailer as grandfathered in with 1982 Surface Transportation Assistance Act (STAA).

1 = The turning radius assumed by a designer when investigating possible turning paths and is set at the centerline of the front axle of a vehicle. If the minimum turning path is assumed, the CTR approximately equals the minimum design turning radius minus one-half the front width of the vehicle.

2 = School buses are manufactured from 42 passenger to 84 passenger sizes. This corresponds to wheelbase lengths of 3,350 mm to 6,020 mm, respectively. For these different sizes, the minimum design turning radii vary from 8.78 m to 12.01 m and the minimum inside radii vary from 4.27 m to 7.74 m.

3 = Turning radius is for 150–200 hp tractor with one 5.64 m long wagon attached to hitch point. Front wheel drive is disengaged and without brakes being applied.

Exhibit 2-2. Minimum Turning Radii of Design Vehicles

US Customary

Design Vehicle Type	Pas-senger Car	Single Unit Truck	Inter-city Bus (Motor Coach)		City Transit Bus	Conven-tional School Bus (65 pass.)	Large[2] School Bus (84 pass.)	Articu-lated Bus	Intermed-iate Semi-trailer	Intermed-iate Semi-trailer
Symbol	P	SU	BUS-40	BUS-45	CITY-BUS	S-BUS36	S-BUS40	A-BUS	WB-40	WB-50
Minimum Design Turning Radius (ft)	24	42	45	45	42.0	38.9	39.4	39.8	40	45
Center-line[1] Turning Radius (CTR) (ft)	21	38	40.8	40.8	37.8	34.9	35.4	35.5	36	41
Minimum Inside Radius (ft)	14.4	28.3	27.6	25.5	24.5	23.8	25.4	21.3	19.3	17.0

Design Vehicle Type	Interstate Semi-trailer		"Double Bottom" Combina-tion	Triple Semi-trailer/ trailers	Turnpike Double Semi-trailer/ trailer	Motor Home	Car and Camper Trailer	Car and Boat Trailer	Motor Home and Boat Trailer	Farm[3] Tractor w/One Wagon
Symbol	WB-62*	WB-65** or WB-67	WB-67D	WB-100T	WB-109D*	MH	P/T	P/B	MH/B	TR/W
Minimum Design Turning Radius (ft)	45	45	45	45	60	40	33	24	50	18
Center-line[1] Turning Radius (CTR) (ft)	41	41	41	41	56	36	30	21	46	14
Minimum Inside Radius (ft)	7.9	4.4	19.3	9.9	14.9	25.9	17.4	8.0	35.1	10.5

* = Design vehicle with 48 ft trailer as adopted in 1982 Surface Transportation Assistance Act (STAA).

** = Design vehicle with 53 ft trailer as grandfathered in with 1982 Surface Transportation Assistance Act (STAA).

1 = The turning radius assumed by a designer when investigating possible turning paths and is set at the centerline of the front axle of a vehicle. If the minimum turning path is assumed, the CTR approximately equals the minimum design turning radius minus one-half the front width of the vehicle.

2 = School buses are manufactured from 42 passenger to 84 passenger sizes. This corresponds to wheelbase lengths of 132 in to 237 in, respectively. For these different sizes, the minimum design turning radii vary from 28.8 ft to 39.4 ft and the minimum inside radii vary from 14.0 ft to 25.4 ft.

3 = Turning radius is for 150–200 hp tractor with one 18.5 ft long wagon attached to hitch point. Front wheel drive is disengaged and without brakes being applied.

Exhibit 2-2. Minimum Turning Radii of Design Vehicles (Continued)

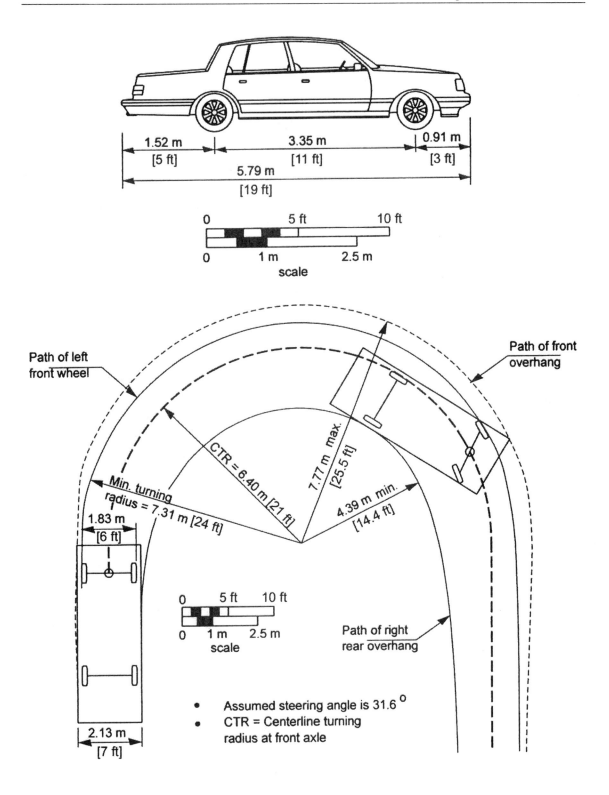

Exhibit 2-3. Minimum Turning Path for Passenger Car (P) Design Vehicle

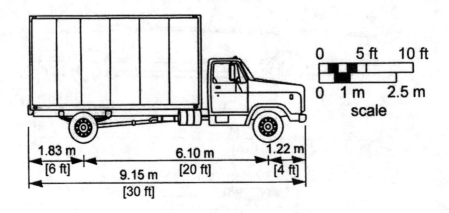

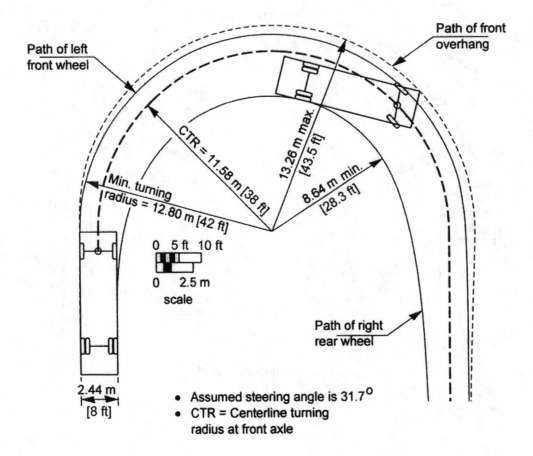

Exhibit 2-4. Minimum Turning Path for Single-Unit (SU) Truck Design Vehicle

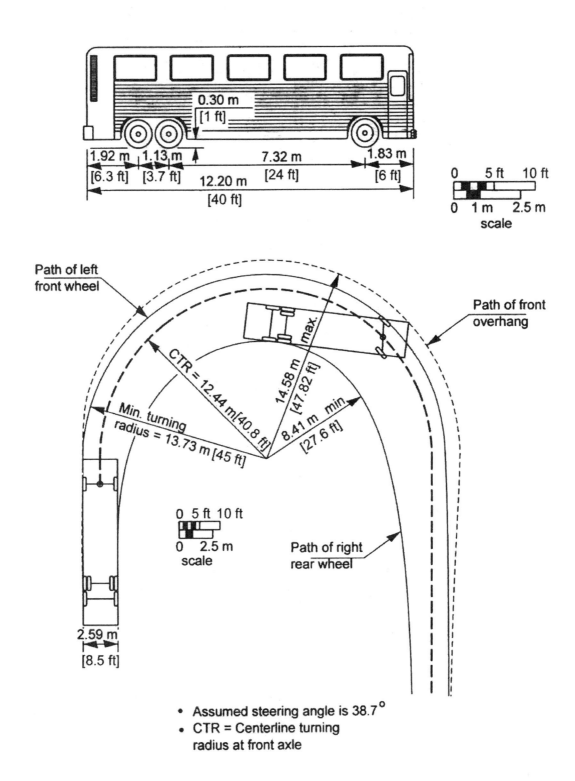

Exhibit 2-5. Minimum Turning Path for Intercity Bus (BUS-12 [BUS-40]) Design Vehicle

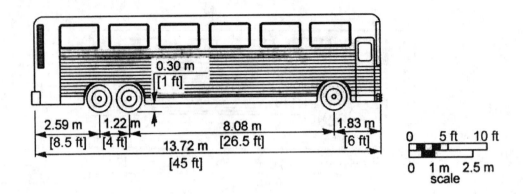

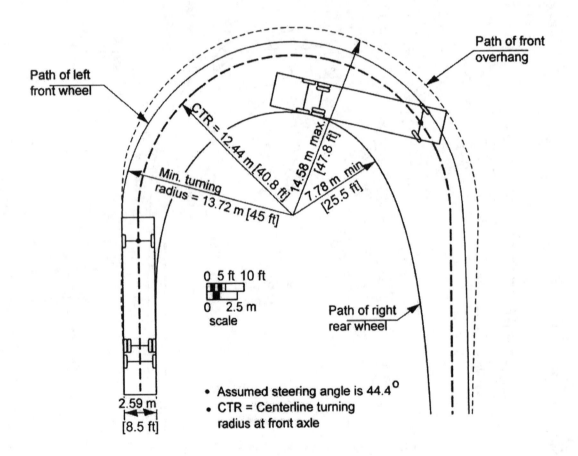

Exhibit 2-6. Minimum Turning Path for Intercity Bus (BUS-14 [BUS-45]) Design Vehicle

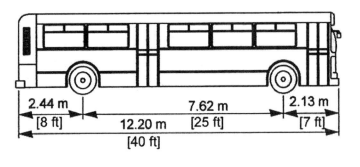

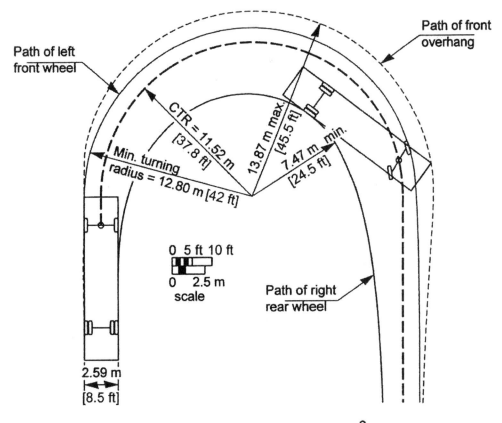

Exhibit 2-7. **Minimum Turning Path for City Transit Bus (CITY-BUS) Design Vehicle**

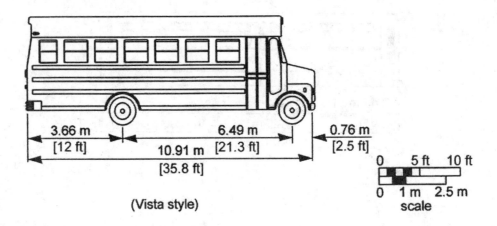

(Vista style)

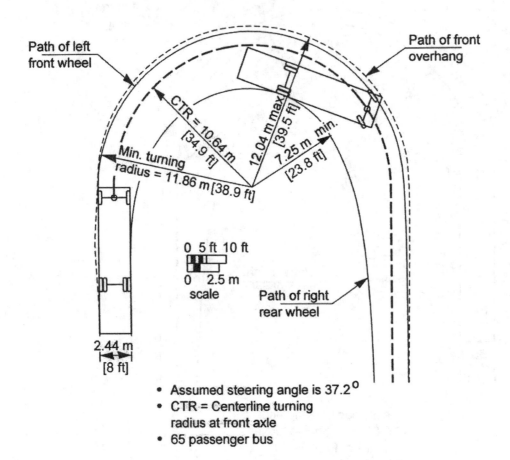

- Assumed steering angle is 37.2°
- CTR = Centerline turning radius at front axle
- 65 passenger bus

**Exhibit 2-8. Minimum Turning Path for Conventional School Bus (S-BUS-11 [S-BUS-36])
Design Vehicle**

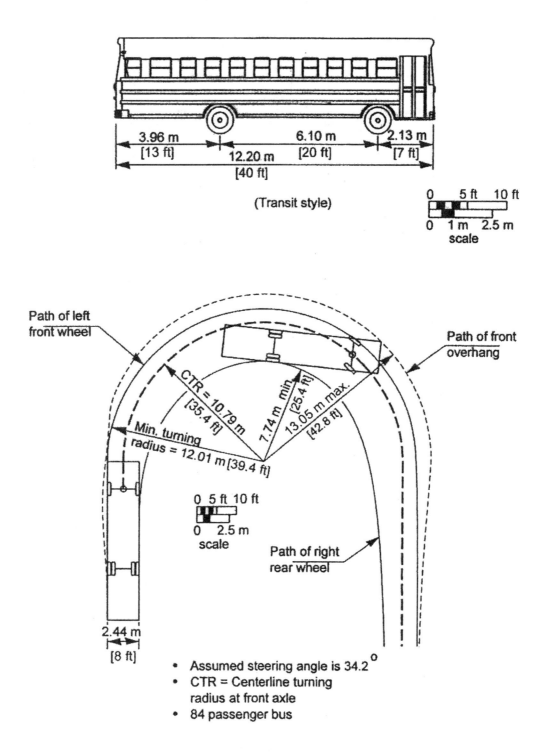

(Transit style)

• Assumed steering angle is 34.2°
• CTR = Centerline turning radius at front axle
• 84 passenger bus

**Exhibit 2-9. Minimum Turning Path for Large School Bus (S-BUS-12 [S-BUS-40])
Design Vehicle**

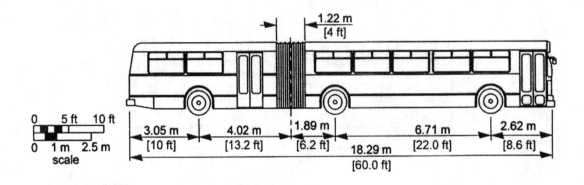

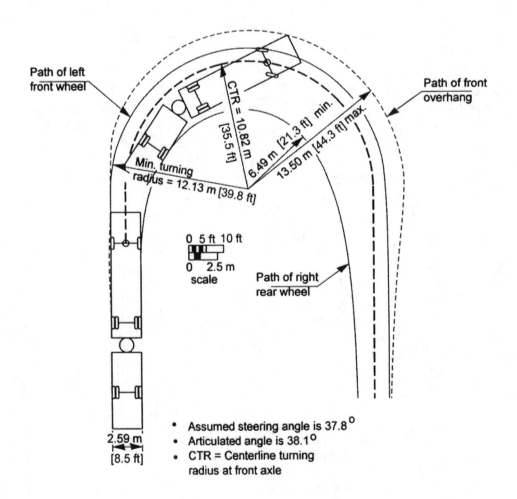

Exhibit 2-10. Minimum Turning Path for Articulated Bus (A-BUS) Design Vehicle

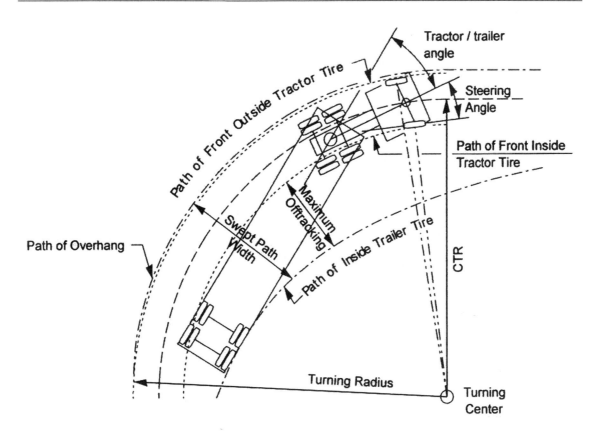

Definitions:

1. Turning radius—The circular arc formed by the turning path radius of the front outside tire of a vehicle. This radius is also described by vehicle manufacturers as the "turning curb radius."

2. CTR—The turning radius of the centerline of the front axle of a vehicle.

3. Offtracking—The difference in the paths of the front and rear wheels of a tractor/semitrailer as it negotiates a turn. The path of the rear tires of a turning truck does not coincide with that of the front tires, and this effect is shown in the drawing above.

4. Swept path width—The amount of roadway width that a truck covers in negotiating a turn and is equal to the amount of offtracking plus the width of the tractor unit. The most significant dimension affecting the swept path width of a tractor/semitrailer is the distance from the kingpin to the rear trailer axle or axles. The greater this distance is, the greater the swept path width.

5. Steering angle—The maximum angle of turn built into the steering mechanism of the front wheels of a vehicle. This maximum angle controls the minimum turning radius of the vehicle.

6. Tractor/trailer angle—The angle between adjoining units of a tractor/semitrailer when the combination unit is placed into a turn; this angle is measured between the longitudinal axes of the tractor and trailer as the vehicle turns. The maximum tractor/trailer angle occurs when a vehicle makes a 180° turn at the minimum turning radius; this angle is reached slightly beyond the point where maximum swept path width is achieved.

Exhibit 2-11. Turning Characteristics of a Typical Tractor-Semitrailer Combination Truck

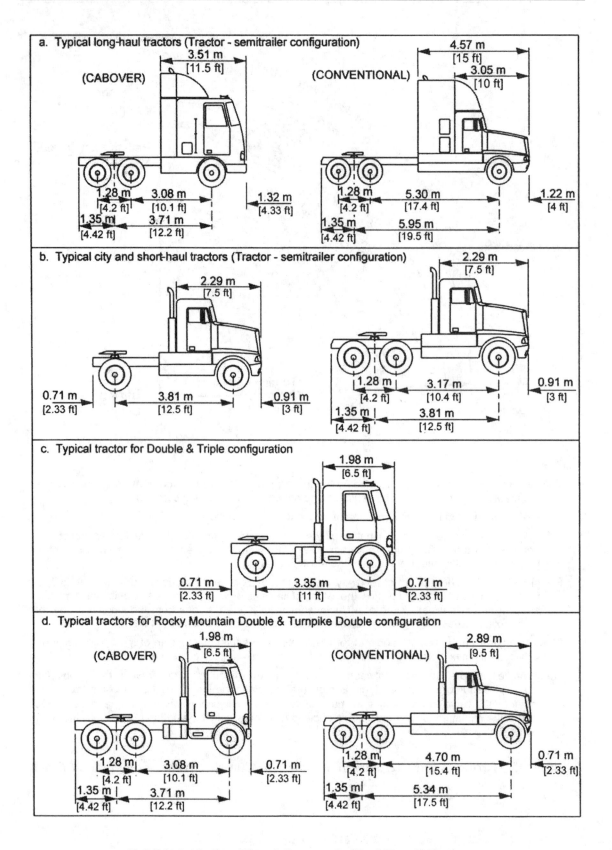

Exhibit 2-12. Lengths of Commonly Used Truck Tractors

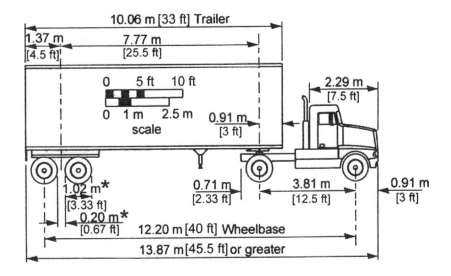

* Typical tire size and space between
tires applies to all trailers.

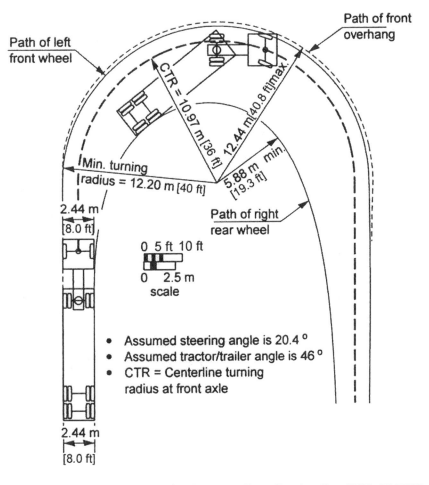

- Assumed steering angle is 20.4 °
- Assumed tractor/trailer angle is 46 °
- CTR = Centerline turning
  radius at front axle

**Exhibit 2-13. Minimum Turning Path for Intermediate Semitrailer (WB-12 [WB-40])
Design Vehicle**

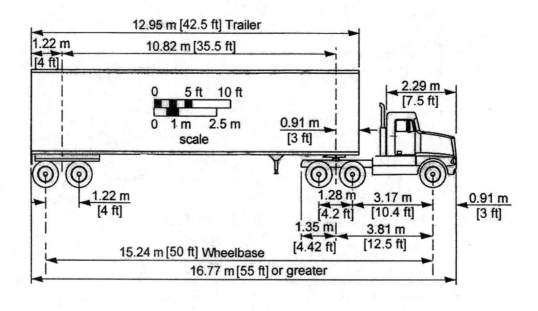

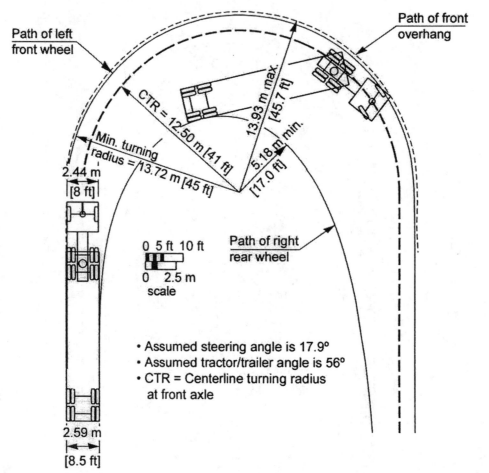

**Exhibit 2-14. Minimum Turning Path for Intermediate Semitrailer (WB-15 [WB-50])
Design Vehicle**

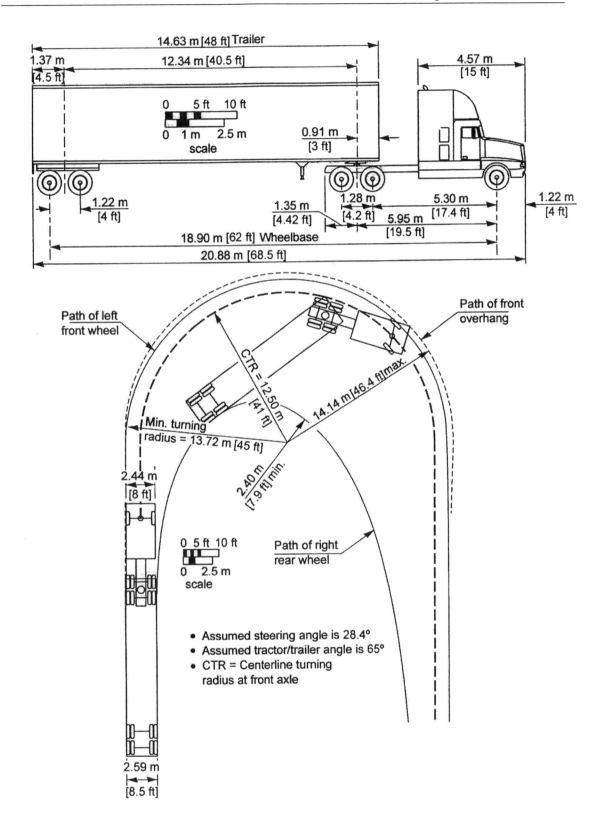

**Exhibit 2-15. Minimum Turning Path for Interstate Semitrailer (WB-19 [WB-62])
Design Vehicle**

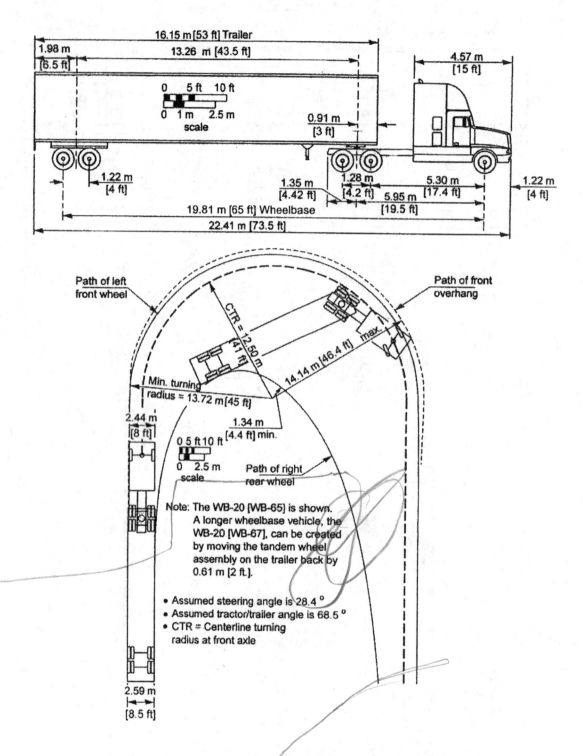

Exhibit 2-16. Minimum Turning Path for Interstate Semitrailer (WB-20 [WB-65 and WB-67]) Design Vehicle

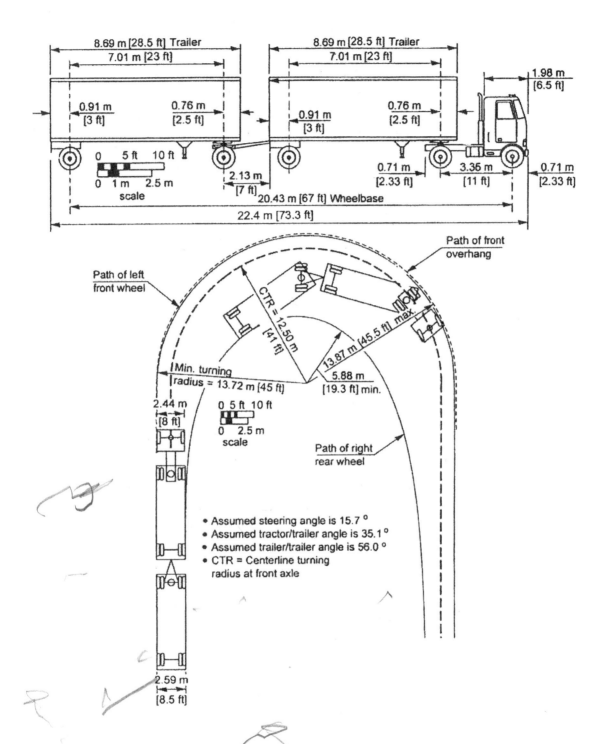

Exhibit 2-17. Minimum Turning Path for Double-Trailer Combination (WB-20D [WB-67D]) Design Vehicle

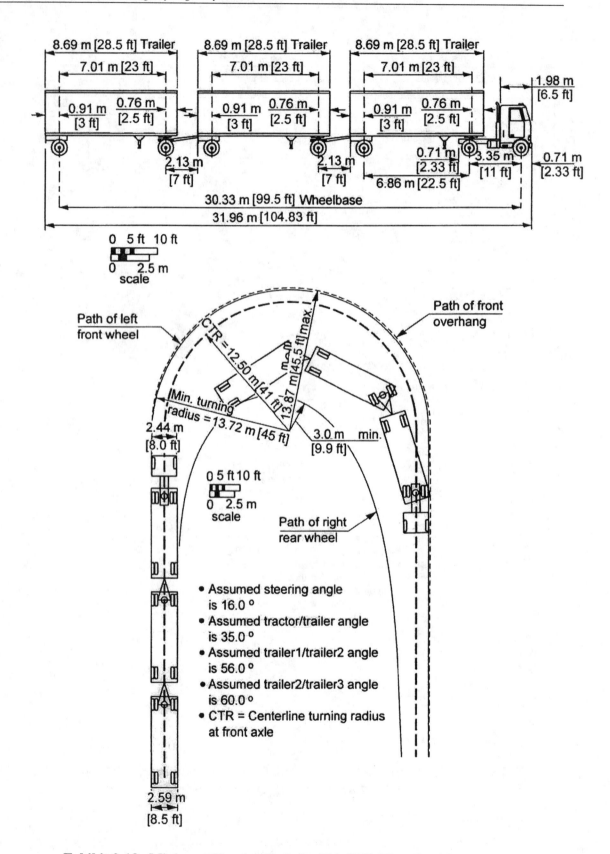

Exhibit 2-18. Minimum Turning Path for Triple-Trailer Combination (WB-30T [WB-100T]) Design Vehicle

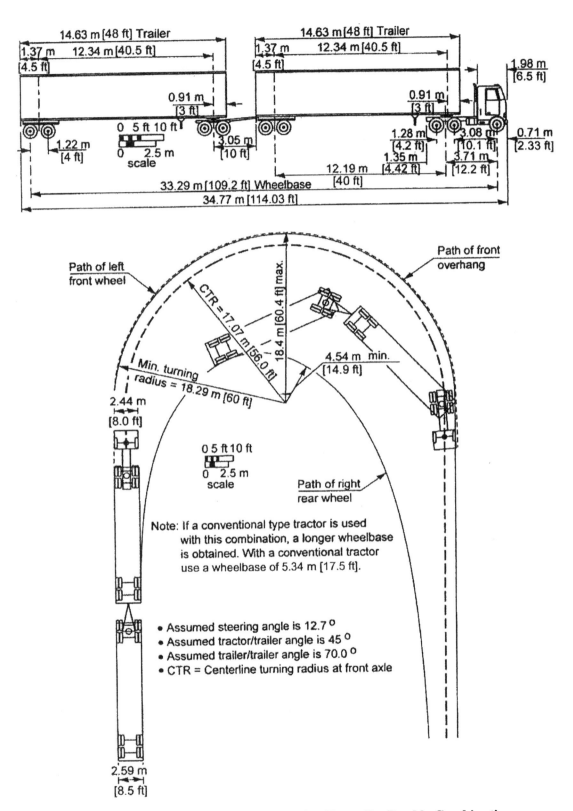

Exhibit 2-19. Minimum Turning Path for Turnpike-Double Combination (WB-33D [WB-109D]) Design Vehicle

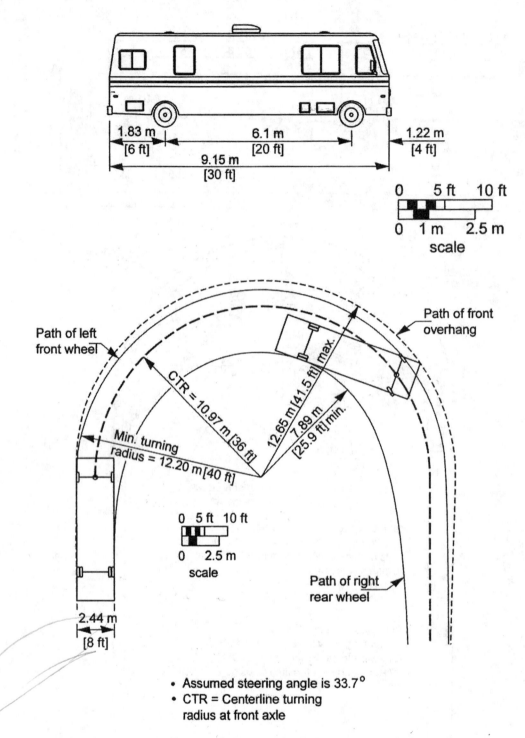

- Assumed steering angle is 33.7°
- CTR = Centerline turning radius at front axle

Exhibit 2-20. Minimum Turning Path for Motor Home (MH) Design Vehicle

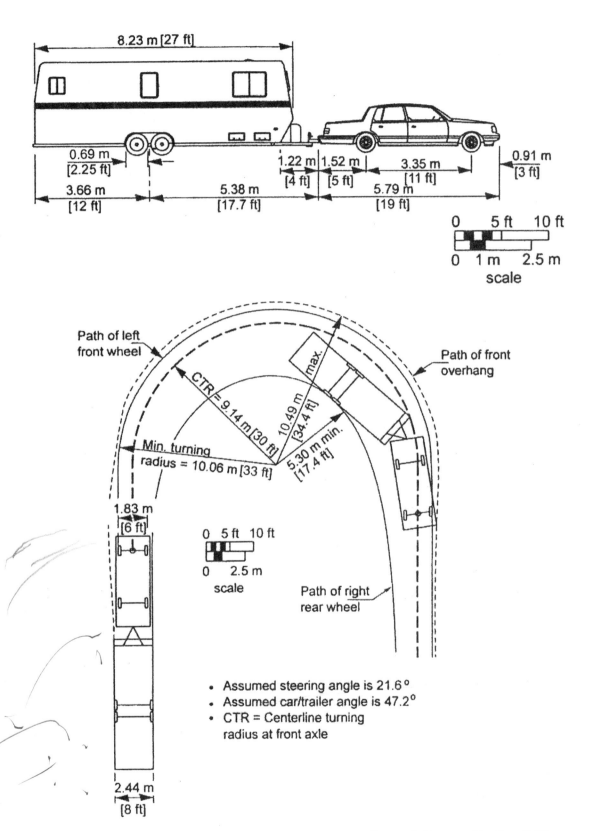

Exhibit 2-21. Minimum Turning Path for Passenger Car and Camper Trailer (P/T) Design Vehicle

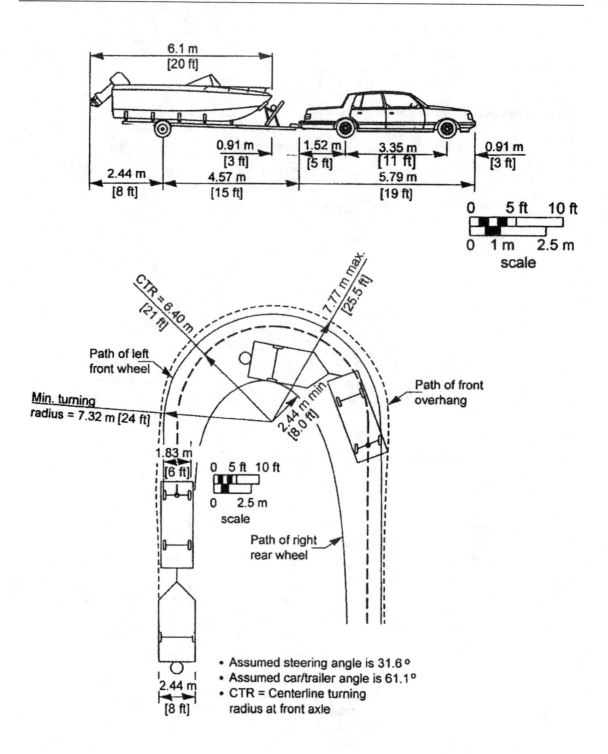

Exhibit 2-22. Minimum Turning Path for Passenger Car and Boat Trailer (P/B) Design Vehicle

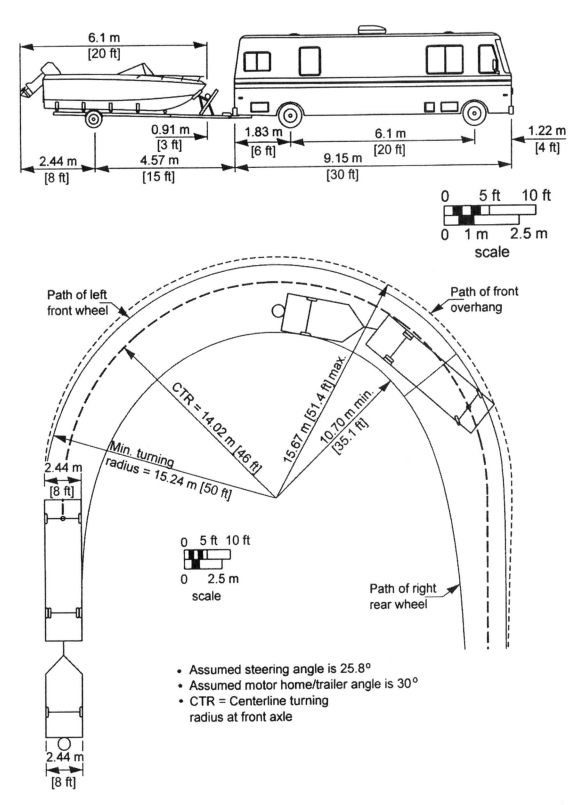

Exhibit 2-23. Minimum Turning Path for Motor Home and Boat Trailer (MH/B) Design Vehicle

vehicles. These combination trucks are identified by the designation WB, together with the wheel base or another length dimension in both metric and U.S. customary units. The combination truck design vehicles are: (1) the WB-12 [WB-40] design vehicle representative of intermediate size tractor-semitrailer combinations, (2) the WB-15 [WB-50] design vehicle representative of a slightly larger intermediate size tractor-semitrailer combination commonly in use, (3) the WB-19 [WB-62] design vehicle representative of larger tractor semitrailer combinations allowed on selected highways by the Surface Transportation Assistance Act of 1982, (4) the WB-20 [WB-65 or WB-67] design vehicle representative of a larger tractor-semitrailer allowed to operate on selected highways by "grandfather" rights under the Surface Transportation Assistance Act of 1982, (5) the WB-20D [WB-67D] design vehicle representative of a tractor-semitrailer/full trailer (doubles or twin trailer) combination commonly in use, (6) the WB-30T [WB-100T] design vehicle representative of tractor-semitrailer/full trailer/full trailer combinations (triples) selectively in use, and (7) the WB-33D [WB-109D] design vehicle representative of larger tractor-semitrailer/full trailer combinations (turnpike double) selectively in use. Although turnpike doubles and triple trailers are not permitted on many highways, their occurrence does warrant inclusion in this publication.

The minimum turning radii and transition lengths shown in the exhibits are for turns at less than 15 km/h [10 mph]. Longer transition curves and larger curve radii are needed for roadways with higher speeds. The radii shown are considered appropriate minimum values for use in design, although skilled drivers might be able to turn with a slightly smaller radius.

The dimensions of the design vehicles take into account recent trends in motor vehicle sizes manufactured in the United States and represent a composite of vehicles currently in operation. However, the design vehicle dimensions are intended to represent vehicle sizes that are critical to geometric design and thus are larger than nearly all vehicles belonging to their corresponding vehicle classes.

The turning paths shown in Exhibits 2-3 through 2-10 and Exhibits 2-13 through 2-23 were derived by using commercially available computer programs.

The P design vehicle, with the dimensions and turning characteristics shown in Exhibit 2-3, represents a larger passenger car.

The SU design vehicle represents a larger single-unit truck. The control dimensions indicate the minimum turning path for most single-unit trucks now in operation (see Exhibit 2-4). On long-distance facilities serving large over-the-road truck traffic or inter-city buses (motor coaches), the design vehicle should generally be either a combination truck or an inter-city bus (see Exhibit 2-5 or Exhibit 2-6).

For intra-city or city transit buses, a design vehicle designated as CITY-BUS is shown in Exhibit 2-7. This design vehicle has a wheel base of 7.62 m [25 ft] and an overall length of 12.20 m [40 ft]. Buses serving particular urban areas may not conform to the dimensions shown in Exhibit 2-7. For example, articulated buses, which are now used in certain cities, are longer than a conventional bus, with a permanent hinge near the vehicle's center that allows more maneuverability. Exhibit 2-10 displays the critical dimensions for the A-BUS design vehicle.

Also, due to the importance of school buses, two design vehicles designated as S-BUS 11 [S-BUS 36] and S-BUS 12 [S-BUS 40] are shown in Exhibits 2-8 and 2-9, respectively. The larger design vehicle is an 84-passenger bus and the smaller design vehicle is a 65-passenger bus. The highway designer should also be aware that for certain buses the combination of ground clearance, overhang, and vertical curvature of the roadway may present problems in hilly areas.

Exhibits 2-13 through 2-19 show dimensions and the minimum turning paths of the design vehicles that represent various combination trucks. For local roads and streets, the WB-15 [WB-50] or WB-12 [WB-40] is often considered an appropriate design vehicle. The larger combination trucks are appropriate for design of facilities that serve over-the-road trucks.

Exhibits 2-20 through 2-23 indicate minimum turning paths for typical recreational vehicles.

In addition to the vehicles shown in Exhibits 2-3 through 2-10 and Exhibits 2-13 through 2-23, other vehicles may be used for selected design applications, as appropriate. With the advent of computer programs that can derive turning path plots, the designer can determine the path characteristics of any selected vehicle if it differs from those shown (**1**).

Vehicle Performance

Acceleration and deceleration rates of vehicles are often critical parameters in determining highway design. These rates often govern the dimensions of such design features as intersections, freeway ramps, climbing or passing lanes, and turnout bays for buses. The following data are not meant to depict average performance for specific vehicle classes but rather lower performance vehicles suitable for design application, such as a low-powered (compact) car and a loaded truck or bus.

Based on its acceleration and deceleration performance, the passenger car seldom controls design. From Exhibits 2-24 and 2-25, it is obvious that relatively rapid accelerations and decelerations are possible, although they may be uncomfortable for the vehicle's passengers. Also, due to the rapid changes being made in vehicle operating characteristics, current data on acceleration and deceleration may soon become outdated. In addition, refer to the NCHRP Report 400, *Determination of Stopping Sight Distances* (**2**). Exhibit 2-24 is based on NCHRP Report 270 (**3**).

When a highway is located in a recreational area, the performance characteristics of recreational vehicles should be considered.

Vehicular Pollution

Pollutants emitted from motor vehicles and their impact on land uses adjacent to highways are factors affecting the highway design process. As each vehicle travels along the highway, it emits pollutants into the atmosphere and transmits noise to the surrounding area. The highway

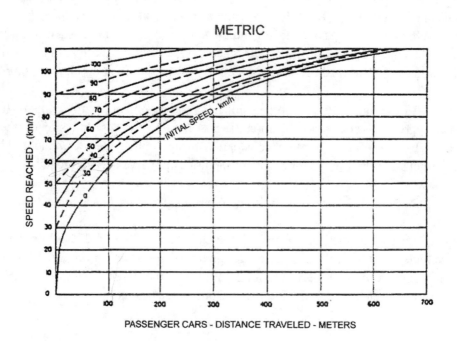

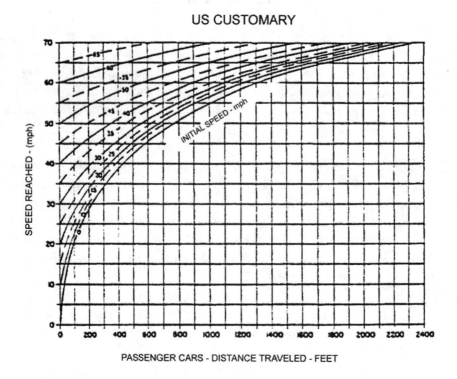

Exhibit 2-24. Acceleration of Passenger Cars, Level Conditions

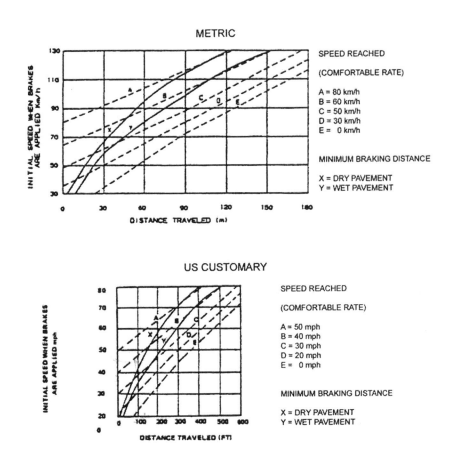

Exhibit 2-25. Deceleration Distances for Passenger Vehicles Approaching Intersections

designer should recognize these impacts and evaluate them in selecting appropriate transportation alternatives. Many factors affect the rate of pollutant emission from vehicles, including vehicle mix, vehicle speed, ambient air temperature, vehicle age distribution, and percentage of vehicles operating in a cold mode.

In addition to air pollution, the highway designer should also consider noise pollution. Noise is unwanted sound, a subjective result of sounds that intrude on or interfere with activities such as conversation, thinking, reading, or sleeping. Thus, sound can exist without people—noise cannot.

Motor vehicle noise is generated by the mechanical operation of the vehicle and its equipment, by its aerodynamics, by the action of its tires on the pavement, and, in metropolitan areas, by the short-duration sounds of brake squeal, exhaust backfires, horns, and, in the case of emergency vehicles, sirens.

Trucks and passenger cars are the major noise-producing vehicles on the nation's highways. Motorcycles are also a factor to be considered because of the rapid increase in their numbers in recent years. Modern passenger cars are relatively quiet, particularly at the lower cruising speeds, but exist in such numbers as to make their total noise contribution significant. While noise produced by passenger cars increases dramatically with speed, steep grades have little influence on passenger car noise.

For passenger cars, noise produced under normal operating conditions is primarily from the engine exhaust system and the tire-roadway interaction. Constant highway speeds provide much the same noise reading whether or not a car's engine is operating, because the noise is principally produced by the tire-roadway interaction with some added wind noise. For conditions of maximum acceleration, the engine system noise is predominant.

Trucks, particularly heavy diesel-powered trucks, present the most difficult noise problem on the highway, with the development of more powerful engines generally increasing noise. Truck noise levels are not greatly influenced by speed, however, because the factors (including acceleration noise) that are not directly affected by speed usually make up a major portion of the total noise. In contrast, steep grades can cause an increase in noise levels for large trucks.

Truck noise has several principal components originating from such sources as exhaust, engine gears, fans, and air intake. At higher speeds, tire-roadway interaction and wind noise add to the problem. As in passenger cars, the noise produced by large diesel trucks is primarily from the engine exhaust system and the tire-roadway interaction. For trucks, however, engine exhaust noise tends to dominate tire-roadway interaction for most operating conditions, particularly during acceleration.

The quality of noise varies with the number and operating conditions of the vehicles while the directionality and amplitude of the noise vary with highway design features. The highway designer should therefore be concerned with how highway location and design influence the vehicle noise perceived by persons residing or working nearby. The perceived noise level decreases as the distance to the highway from a residence or workplace increases.

DRIVER PERFORMANCE

Introduction

An appreciation of driver performance is essential to proper highway design and operation. The suitability of a design rests as much on how safely and efficiently drivers are able to use the highway as on any other criterion. When drivers use a highway designed to be compatible with their capabilities and limitations, their performance is aided. When a design is incompatible with the capabilities of drivers, the chance for driver errors increase, and crashes or inefficient operation may result.

This section provides information about driver performance useful to highway engineers in designing and operating highways. It describes drivers in terms of their performance—how they interact with the highway and its information system and why they make errors.

The material draws extensively from *A User's Guide to Positive Guidance* (4), which contains information on driver attributes, driving tasks, and information handling by the driver.

Where positive guidance is applied to design, competent drivers, using well-designed highways with appropriate information displays, can perform safely and efficiently. Properly designed and operated highways, in turn, provide positive guidance to drivers. In addition, Transportation Research Record 1281 entitled *Human Factors and Safety Research Related to Highway Design and Operations* (**5**), provides background information.

Older Drivers

At the start of the 20th century, approximately 4 percent of America's population was 65 years of age or older. This group, which accounted for 15 percent of the driving population in 1986, and is expected to increase to 22 percent by the year 2030.

Older drivers and pedestrians are a significant and rapidly growing segment of the highway user population with a variety of age-related diminished capabilities. As a group, they have the potential to adversely affect the highway system's safety and efficiency. There is agreement that older road users require mobility and that they should be accommodated by the design and operational characteristics of a highway to the extent practical.

Older drivers have special needs that should be considered in highway design and traffic control. For example, for every decade after age 25, drivers need twice the brightness at night to receive visual information. Hence, by age 75, some drivers may need 32 times the brightness they did at age 25.

Research findings show that enhancements to the highway system to improve its usability for older drivers and pedestrians can also improve the system for all users. Thus, designers and engineers should be aware of the capabilities and needs of older road users and consider appropriate measures to aid their performance. A Federal Highway Administration report, entitled *Older Driver Highway Design Handbook: Recommendations and Guidelines* (**6**), provides information on how geometric design elements and traffic control devices can be modified to better meet the needs and capabilities of older road users.

The Driving Task

The driving task depends on drivers receiving and using information correctly. The information received by drivers as they travel is compared with the information they already possess. Decisions are then made by drivers based on the information available to them and appropriate control actions are taken.

Driving encompasses a number of discrete and interrelated activities. When grouped by performance, the components of the driving task fall into three levels: control, guidance, and navigation. These activities are ordered on scales of complexity of task and importance for safety. Simple steering and speed control are at one end of the scale (control). Road-following and safe-path maintenance in response to road and traffic conditions are at midlevel of the scale (guidance). At the other end of the scale are trip planning and route following (navigation).

47

The driving task may be complex and demanding, and several individual activities may need to be performed simultaneously, requiring smooth and efficient processing and integration of information. Driving often occurs at high speeds, under time pressure, in unfamiliar locations, and under adverse environmental conditions. The driving task may at other times be so simple and undemanding that a driver becomes inattentive. The key to safe, efficient driver performance in this broad range of driving situations is error-free information handling.

Driver errors result from many driver, vehicle, roadway, and traffic factors. Some driver errors occur because drivers may not always recognize what particular roadway traffic situations are require of them, because situations may lead to task overload or inattentiveness, and because deficient or inconsistent designs or information displays may cause confusion. Driver errors may also result from pressures of time, complexity of decisions, or profusion of information. Control and guidance errors by drivers may also contribute directly to crashes. In addition, navigational errors resulting in delay contribute to inefficient operations and may lead indirectly to crashes.

The Guidance Task

Of the three major components of the driving task, highway design and traffic operations have the greatest effect on guidance. An appreciation of the guidance component of the driving task is needed by the highway designer to aid driver performance.

Lane Placement and Road Following

Lane placement and road-following decisions, including steering and speed control judgments, are basic to vehicle guidance. Drivers use a feedback process to follow alignment and grade within the constraints of road and environmental conditions. Obstacle-avoidance decisions are integrated into lane placement and road-following activities. This portion of the guidance task level is continually performed both when no other traffic is present (singularly) or when it is shared with other activities (integrated).

Car Following

Car following is the process by which drivers guide their vehicles when following another vehicle. Car-following decisions are more complex than road-following decisions because they involve speed-control modifications. In car following, drivers need to constantly modify their speed to maintain safe gaps between vehicles. To proceed safely, they have to assess the speed of the lead vehicle and the speed and position of other vehicles in the traffic stream and continually detect, assess, and respond to changes.

Passing Maneuvers

The driver decision to initiate, continue, or complete a passing maneuver is even more complex than the decisions involved in lane placement or car following. Passing decisions require modifications in road- and car-following and in speed control. In passing, drivers must judge the speed and acceleration potential of their own vehicle, the speed of the lead vehicle, the speed and rate of closure of the approached vehicle, and the presence of an acceptable gap in the traffic stream.

Other Guidance Activities

Other guidance activities include merging, lane changing, avoidance of pedestrians, and response to traffic control devices. These activities also require complex decisions, judgments, and predictions.

The Information System

Each element that provides information to drivers is part of the information system of the highway. Formal sources of information are the traffic control devices specifically designed to display information to drivers. Informal sources include such elements as roadway and roadside design features, pavement joints, tree lines, and traffic. Together, the formal and informal sources provide the information drivers need to drive safely and efficiently. Formal and informal sources of information are interrelated and must reinforce and augment each other to be most useful.

Traffic Control Devices

Traffic control devices provide guidance and navigation information that often is not otherwise available or apparent. Such devices include regulatory, warning, and guide signs, and other route guidance information. Other traffic control devices, such as markings and delineation, display additional information that augments particular roadway or environmental features. These devices help drivers perceive information that might otherwise be overlooked or difficult to recognize. Information on the appropriate use of traffic control devices is presented in the *Manual on Uniform Traffic Control Devices* (7).

The Roadway and its Environment

Selection of speeds and paths is dependent on drivers being able to see the road ahead. Drivers must see the road directly in front of their vehicles and far enough in advance to perceive with a high degree of accuracy the alignment, profile gradeline, and related aspects of the roadway. The view of the road also includes the environment immediately adjacent to the roadway. Such appurtenances as shoulders and roadside obstacles (including sign supports, bridge piers, abutments, guardrail, and median barriers) affect driving behavior and, therefore, should be clearly visible to the driver.

Information Handling

Drivers use many of their senses to gather information. Most information is received visually by drivers from their view of the roadway alignment, markings, and signs. However, drivers also detect changes in vehicle handling through instinct. They do so, for example, by feeling road surface texture through vibrations in the steering wheel and hearing emergency vehicle sirens.

Throughout the driving task, drivers perform several functions almost simultaneously. They look at information sources, make numerous decisions, and perform necessary control actions. Sources of information (some needed, others not) compete for their attention. Needed information should be in the driver's field of view, available when and where needed, available in a usable form, and capable of capturing the driver's attention.

Because drivers can only attend to one visual information source at a time, they integrate the various information inputs and maintain an awareness of the changing environment through an attention-sharing process. Drivers sample visual information obtained in short-duration glances, shifting their attention from one source to another. They make some decisions immediately, and delay others, through reliance on judgment, estimation, and prediction to fill in gaps in available information.

Reaction Time

Information takes time to process. Drivers' reaction times increase as a function of decision complexity and the amount of information to be processed. Furthermore, the longer the reaction time, the greater the chance for error. Johannson and Rumar (**8**) measured brake reaction time for expected and unexpected events. Their results show that when an event is expected, reaction time averages about 0.6 s, with a few drivers taking as long as 2 s. With unexpected events, reaction times increased by 35 percent. Thus, for a simple, unexpected decision and action, some drivers may take as long as 2.7 s to respond. A complex decision with several alternatives may take several seconds longer than a simple decision. Exhibit 2-26 shows this relationship for median-case drivers, whereas Exhibit 2-27 shows this relationship for 85th-percentile drivers. The figures quantify the amount of information to be processed in bits. Long processing times decrease the time available to attend to other tasks and increase the chance for error.

Highway designs should take reaction times into account. It should be recognized that drivers vary in their responses to particular events and take longer to respond when decisions are complex or events are unexpected. Clear sight lines and adequate decision sight distance provide a margin for error.

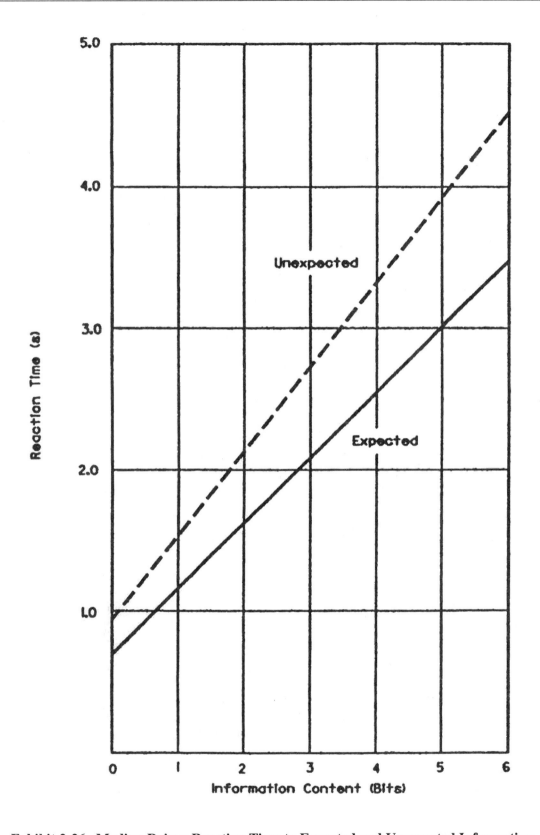

Exhibit 2-26. Median Driver Reaction Time to Expected and Unexpected Information

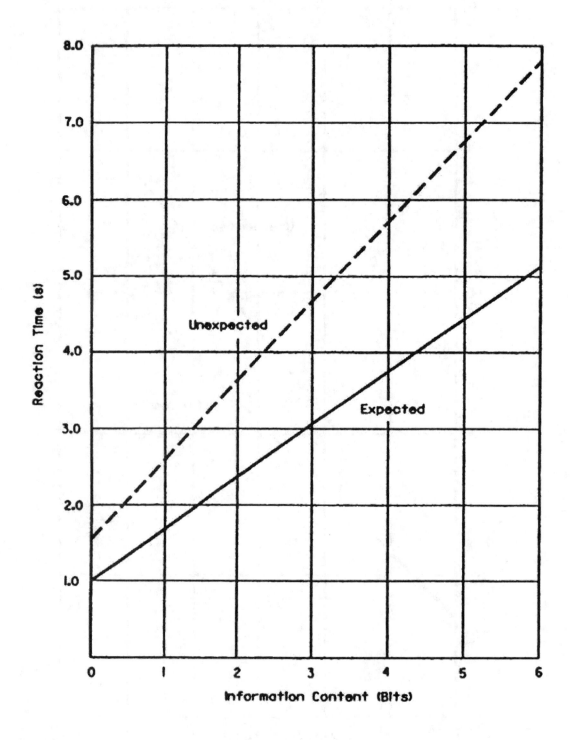

Exhibit 2-27. 85th-Percentile Driver Reaction Time to Expected and Unexpected Information

Primacy

Primacy relates the relative importance to safety of competing information. Control and guidance information is important because the related errors may contribute directly to crashes. Navigation information has a lower primacy because errors may lead to inefficient traffic flow, but are less likely to lead to crashes. Accordingly, the design should focus the drivers' attention on the safety-critical design elements and high-priority information sources. This goal may be achieved by providing clear sight lines and good visual quality.

Expectancy

Driver expectancies are formed by the experience and training of drivers. Situations that generally occur in the same way, and successful responses to these situations, are incorporated into each driver's store of knowledge. Expectancy relates to the likelihood that a driver will respond to common situations in predictable ways that the driver has found successful in the past. Expectancy affects how drivers perceive and handle information and modify the speed and nature of their responses.

Reinforced expectancies help drivers respond rapidly and correctly. Unusual, unique, or uncommon situations that violate driver expectancies may cause longer response times, inappropriate responses, or errors.

Most highway design features are sufficiently similar to create driver expectancies related to common geometric, operational, and route characteristics. For example, because most freeway interchanges have exits on the right side of the road, drivers generally expect to exit from the right. This aids performance by enabling rapid and correct responses when exits on the right are to be negotiated. There are, however, instances where expectancies are violated. For example, if an exit ramp is on the left, then the right-exit expectancy is incorrect, and response times may be lengthened or errors committed.

One of the most important ways to aid driver performance is to develop designs in accordance with prevalent driver expectancies. Unusual design features should be avoided, and design elements should be applied consistently throughout a highway segment. Care should also be taken to maintain consistency from one segment to another. When drivers obtain the information they expect from the highway and its traffic control devices, their performance tends to be error free. Where they do not get what they expect, or get what they do not expect, errors may result.

Driver Error

A common characteristic of many high-crash locations is that they place large or unusual demands on the information-processing capabilities of drivers. Inefficient operation and crashes usually occur where the driver's chances for information-handling errors are high. At locations where information-processing demands on the driver are high, the possibility of error and inappropriate driver performance increases.

Errors Due to Driver Deficiencies

Many driving errors are caused by deficiencies in a driver's capabilities or temporary states, which, in conjunction with inappropriate designs or difficult traffic situations, may produce a failure in judgment. For example, insufficient experience and training may contribute to a driver's inability to recover from a skid. Similarly, inappropriate risk taking may lead to errors in gap acceptance while passing (**9**). In addition, poor glare recovery may cause older drivers to miss information at night (**10**).

Adverse psychophysiological states also lead to driver failures. These include decreased performance caused by alcohol and drugs, for which a link to crashes has been clearly established. The effects of fatigue, caused by sleep deprivation from extended periods of driving without rest or prolonged exposure to monotonous environments, or both, also contribute to crashes (**11**).

It is not generally possible for a design or an operational procedure to reduce errors caused by innate driver deficiencies. However, designs should be as forgiving as practical to lessen the consequences of such failures. Errors committed by competent drivers can be reduced by proper design and operation. Most individuals possess the attributes and skills to drive properly and are neither drunk, drugged, nor fatigued at the start of their trips. When drivers overextend themselves, fail to take proper rest breaks, or drive for prolonged periods, they ultimately reach a less-than-competent state. Fatigued drivers represent a sizable portion of the long-trip driving population and should therefore be considered in freeway design.

Although opinions among experts are not unanimous, there is general agreement that advancing age has a deleterious effect on an individual's perceptual, mental, and motor skills. These skills are critical factors in vehicular operation. Therefore, it is important for the road designer to be aware of the needs of the older driver, and where appropriate, to consider these needs in the roadway design.

Some of the more important information and observations from recent research studies concerning older drivers is summarized below:

1. **Characteristics of the Older Driver**. In comparison to younger drivers, older drivers often exhibit the following operational deficiencies:

    *   slower information processing
    *   slower reaction times
    *   slower decision making
    *   visual deterioration
    *   hearing deterioration
    *   decline in ability to judge time, speed, and distance

- limited depth perception
- limited physical mobility
- side effects from prescription drugs

2. **Crash Frequency**. Older drivers are involved in a disproportionate number of crashes where there is a higher-than-average demand imposed on driving skills. The driving maneuvers that most often precipitate higher crash frequencies among older drivers include:

- making left turns across traffic
- merging with high-speed traffic
- changing lanes on congested streets in order to make a turn
- crossing a high-volume intersection
- stopping quickly for queued traffic
- parking

3. **Countermeasures**. The following countermeasures may help to alleviate the potential problems of the older driver:

- assess all guidelines to consider the practicality of designing for the 95th- or 99th-percentile driver, as appropriate, to represent the performance abilities of an older driver
- improve sight distance by modifying designs and removing obstructions, particularly at intersections and interchanges
- assess sight triangles for adequacy of sight distance
- provide decision sight distances
- simplify and redesign intersections and interchanges that require multiple information reception and processing
- consider alternate designs to reduce conflicts
- increase use of protected left-turn signal phases
- increase vehicular clearance times at signalized intersections
- provide increased walk times for pedestrians
- provide wider and brighter pavement markings
- provide larger and brighter signs
- reduce sign clutter
- provide more redundant information such as advance guide signs for street name, indications of upcoming turn lanes, and right-angle arrows ahead of an intersection where a route turns or where directional information is needed
- enforce speed limits
- increase driver education

In roadway design, perhaps the most practical measure related to better accommodate older drivers is an increase in sight distance, which may be accomplished through increased use of decision sight distance. The gradual aging of the driver population suggests that increased use of decision sight distance may help to reduce future crash frequencies for older drivers. Where provision of decision sight distance is impractical, increased use of advance warning or guide signs may be appropriate.

Errors Due to Situation Demands

Drivers often commit errors when they have to perform several highly complex tasks simultaneously under extreme time pressure (**12**). Errors of this type usually occur at urban locations with closely spaced decision points, intensive land use, complex design features, and heavy traffic. Information-processing demands beyond the drivers' capabilities may cause information overload or confuse drivers, resulting in an inadequate understanding of the driving situation.

Other locations present the opposite situations and are associated with different types of driver errors. Typically these are rural locations where there may be widely spaced decision points, sparse land use, smooth alignment, and light traffic. Information demands are thus minimal, and rather than being overloaded with information, the lack of information and decision-making demands may result in inattentiveness by drivers. Driving errors may be caused by a state of decreased vigilance in which drivers fail to detect, recognize, or respond to new, infrequently encountered, or unexpected design elements or information sources.

Speed and Design

Speed reduces the visual field, restricts peripheral vision, and limits the time available for drivers to receive and process information. Highways built to accommodate high speeds help compensate for these limitations by simplifying control and guidance activities, by aiding drivers with appropriate information, by placing this information within the cone of clear vision, by eliminating much of the need for peripheral vision, and by simplifying the decisions required and spacing them farther apart to decrease information-processing demands.

Current freeway designs have nearly reached the goal of allowing drivers to operate at high speeds in comfort and safety. Control of access to the traveled way reduces the potential for conflicts by giving drivers a clear path. Clear roadsides have been provided by eliminating obstructions or designing them to be more forgiving. The modern freeway provides an alignment and profile that, together with other factors, encourages high operating speeds.

Although improved design has produced significant benefits, it has also created potential problems. For example, driving at night at high speeds may lead to reduced forward vision because of the inability of headlights to illuminate objects in the driver's path in sufficient time for some drivers to respond (**13**). In addition, the severity of crashes is generally greater with increased speed.

Finally, the very fact that freeways succeed in providing safe, efficient transportation can lead to difficulties. The Institute of Traffic Engineers (**14**) indicated that "Freeways encourage drivers to extend the customary length and duration of their trips. This results in driver fatigue and slower reaction as well as a reduction in attention and vigilance."

Thus, extended periods of high-speed driving on highways with low demand for information processing may not always be conducive to proper information handling by drivers and may therefore lead to driver fatigue. Highway design should take these possible adverse effects into account and seek to lessen their consequences. For example, long sections of flat, tangent roadway should be avoided and flat, curving alignment that follows the natural contours of the terrain should be used whenever practical. Rest areas spaced at intervals of approximately one hour or less of driving time have also proved beneficial.

Design Assessment

The preceding sections of this chapter have described the way drivers use information provided by the highway and its appurtenances. This discussion has shown the interdependence between design and information display. Both should be assessed in the design of highway projects. Because drivers "read" the road and the adjacent environment and make decisions based on what they see (even if traffic control devices making up the formal information system indicate inconsistencies with the driver's view), a highway segment that is inappropriately designed may not operate safely and efficiently. Conversely, an adequately designed highway may not operate properly without the appropriate complement of traffic control devices.

Designers should consider how the highway will fit into the existing landscape, how the highway should be signed, and the extent to which the information system will complement and augment the proposed design. The view of the road is very important, especially to the unfamiliar driver. Therefore, consideration should be given to the visual qualities of the road. This can be accomplished through the use of 3-D computer visualization programs.

Locations with potential for information overload should be identified and corrected. The adequacy of the sight lines and sight distances should be assessed, and it should be determined whether unusual vehicle maneuvers are required and whether likely driver expectancies may be violated.

Potential driver problems can be anticipated before a facility is built by using information about the driving tasks and possible driver errors to assess the design. When trade-offs are appropriate, they should be made with the drivers' capabilities in mind to ensure that the resultant design is compatible with those capabilities. Properly designed highways that provide positive guidance to drivers can operate at a high level of safety and efficiency; therefore, designers should seek to incorporate these principles in highway design.

TRAFFIC CHARACTERISTICS

General Considerations

The design of a highway and its features should be based upon explicit consideration of the traffic volumes and characteristics to be served. All information should be considered jointly. Financing, quality of foundations, availability of materials, cost of right-of-way, and other factors all have important bearing on the design; however, traffic volumes indicate the need for the improvement and directly affect the geometric design features, such as number of lanes, widths, alignments, and grades. It is no more rational to design a highway without traffic information than it is to design a bridge without knowledge of the weights and numbers of vehicles it is intended to support. Information on traffic volumes serves to establish the loads for the geometric highway design.

Traffic data for a road or section of road are generally available or can be obtained from field studies. The data collected by State or local agencies include traffic volumes for days of the year and time of the day, as well as the distribution of vehicles by type and weight. The data also include information on trends from which the designer may estimate the traffic to be expected in the future.

Volume

Average Daily Traffic

The most basic measure of the traffic demand for a highway is the average daily traffic (ADT) volume. The ADT is defined as the total volume during a given time period (in whole days), greater than one day and less than one year, divided by the number of days in that time period. The current ADT volume for a highway can be readily determined when continuous traffic counts are available. When only periodic counts are taken, the ADT volume can be estimated by adjusting the periodic counts according to such factors as the season, month, or day of week.

Knowledge of the ADT volume is important for many purposes, such as determining annual highway usage as justification for proposed expenditures or designing the structural elements of a highway. However, the direct use of ADT volume in the geometric design of highways is not appropriate except for local and collector roads with relatively low volumes because it does not indicate traffic volume variations occurring during the various months of the year, days of the week, and hours of the day. The amount by which the volume of an average day is exceeded on certain days is appreciable and varied. At typical rural locations, the volume on certain days may be significantly higher than the ADT. Thus, a highway designed for the traffic on an average day would be required to carry a volume greater than the design volume for a considerable portion of the year, and on many days the volume carried would be much greater than the design volume.

Peak-Hour Traffic

Traffic volumes for an interval of time shorter than a day more appropriately reflect the operating conditions that should be used for design. The brief, but frequently repeated, rush-hour periods are significant in this regard. In nearly all cases, a practical and adequate time period is one hour.

The traffic pattern on any highway shows considerable variation in traffic volumes during the various hours of the day and in hourly volumes throughout the year. A key design decision involves determining which of these hourly traffic volumes should be used as the basis for design. While it would be wasteful to predicate the design on the maximum peak-hour traffic that occurs during the year, the use of the average hourly traffic would result in an inadequate design. The hourly traffic volume used in design should not be exceeded very often or by very much. On the other hand, it should not be so high that traffic would rarely be sufficient to make full use of the resulting facility. One guide in determining the hourly traffic volume that is best suited for use in design is a curve showing variation in hourly traffic volumes during the year.

Exhibit 2-28 shows the relationship between the highest hourly volumes and ADT on rural arterials. This figure was produced from an analysis of traffic count data covering a wide range of volumes and geographic conditions. The curves in the chart were prepared by arranging all of the hourly volumes for one year, expressed as a percentage of ADT, in a descending order of magnitude. The middle curve is the average for all locations studied and represents a highway with average fluctuation in traffic flow.

Based on a review of these curves, it is recommended that the hourly traffic volume that should generally be used in design is the 30th highest hourly volume of the year, abbreviated as 30 HV. The reasonableness of 30 HV as a design control is indicated by the changes that result from choosing a somewhat higher or lower volume. The curve in Exhibit 2-28 steepens quickly to the left of the point showing the 30th highest hour volume and indicates only a few more hours with higher volumes. The curve flattens to the right of the 30th highest hour and indicates many hours in which the volume is not much less than the 30 HV.

On rural roads with average fluctuation in traffic flow, the 30 HV is typically about 15 percent of the ADT. Whether or not this hourly volume is too low to be appropriate for design can be judged by the 29 hours during the year when it is exceeded. The maximum hourly volume, which is approximately 25 percent of the ADT on the graph, exceeds 30 HV by about 67 percent.

Whether the 30 HV is too high for practical economy in design can be judged by the trend in the hourly volumes lower than the 30th highest hour. The middle curve in Exhibit 2-28 indicates that the traffic volume exceeds 11.5 percent of the ADT during 170 hours of the year. The lowest of this range of hourly volumes is about 23 percent less than the 30 HV.

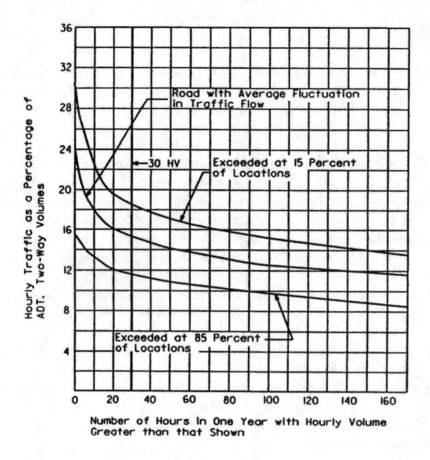

Exhibit 2-28. Relation Between Peak-Hour and Average Daily Traffic Volumes on Rural Arterials

Another fortunate characteristic of 30 HV is that, as a percentage of ADT, it generally varies only slightly from year to year even though the ADT may change substantially. Increased ADT generally results in a slight decrease in the percentage of ADT during the 30 HV. Thus, the percentage of ADT used for determining the 30 HV from current traffic data for a given facility can generally be used with confidence in computing the 30 HV from an ADT volume determined for some future year. This consistency between current and future may not apply where there is a radical change in the use of the land area served by the highway. In cases where the character and magnitude of future development can be foreseen, the relationship of 30 HV to ADT may be based on experience with other highways serving areas with similar land-use characteristics.

For highway design purposes, the variation in hourly traffic volumes should be measured and the percentage of ADT during the 30th highest hour determined. Where such measurements are impractical and only the ADT is known, the 30 HV should be estimated from 30th-hour percentage factors for similar highways in the same locality, operating under similar conditions.

On a typical rural arterial, the 30 HV is about 15 percent of ADT, and the maximum hourly volume is about 25 percent of ADT. As indicated in Exhibit 2-28, the 30 HV at 70 percent of all locations, except those having unusually high or low fluctuation in traffic flow, is in the range of 12 to 18 percent of the ADT. Likewise the range in maximum hourly volumes for the same groups of roads varies approximately from 16 to 32 percent of the ADT. These criteria for design apply to most rural highways. There are highways, however, for which there are unusual or highly seasonal fluctuations in traffic flow, such as resort roads on which weekend traffic during a few months of the year far exceeds the traffic during the rest of the year. Seasonal fluctuations result in high peak-hour volumes relative to ADT, high percentages for high-volume hours, and low percentages for low-volume hours.

Because the percentage represented by the 30 HV for a road with large seasonal fluctuations may not be much different from the percentage represented by the 30 HV on most rural roads, the 30 HV criterion may not be appropriate for such roads. A design that results in somewhat less satisfactory traffic operation during seasonal peaks than on rural roads with normal traffic fluctuations, will generally be accepted by the public. On the other hand, design should not be so economical that severe congestion results during peak hours. It may be desirable, therefore, to choose an hourly volume for design, which is about 50 percent of the volumes expected to occur during a few highest hours of the design year, whether or not that volume is equal to 30 HV. Some congestion would be experienced by traffic during peak hours but the capacity would not be exceeded. A check should be made to ensure that the expected maximum hourly traffic does not exceed the capacity.

The design hourly volume (DHV) for rural highways, therefore, should generally be the 30 HV of the future year chosen for design. Exceptions may be made on roads with high seasonal traffic fluctuation, where a different hourly volume may need to be used. The 30-HV criterion also applies in general to urban areas; however, where the fluctuation in traffic flow is markedly different from that on rural highways, other hours of the year should be considered as the basis for design.

In urban areas, an appropriate DHV may be determined from the study of traffic during the normal daily peak periods. Because of the recurring morning and afternoon peak traffic flow, there is usually little difference between the 30th and the 200th highest hourly volume. For typical urban conditions, the highest hourly volume is found during the afternoon work-to-home travel peak. One approach for determining a suitable DHV is to select the highest afternoon peak traffic flow for each week and then average these values for the 52 weeks of the year. If the morning peak-hour volumes for each week of the year are all less than the afternoon peak volumes, the average of the 52 weekly afternoon peak-hour volumes would have about the same value as the 26th highest hourly volume of the year. If the morning peaks are equal to the afternoon peaks, the average of the afternoon peaks would be about equal to the 50th highest hourly volume.

The volumes represented by the 26th and 50th highest hours of the year are not sufficiently different from the 30 HV value to affect design. Therefore, in urban design, the 30th highest hourly volume can also be assumed to be a reasonable representation of daily peak hours during the year. Exceptions may be appropriate in those areas or locations where recreational or other travel is concentrated during particular seasons. At such locations, a distribution of traffic volume where the hourly volumes are much greater than the 30 HV may result; the 30 HV in such cases may be inappropriate as the DHV and a higher value should be considered in design. Specific measurements of traffic volumes should be made and evaluated to determine the appropriate DHV.

Traffic estimates used for the design of urban streets and highways are usually expressed as ADT volumes derived from the urban transportation planning process. In recent years, however, consideration has been given to the development of DHVs by making peak-hour traffic assignments in lieu of ADT assignments. The availability of the 1980 and 1990 census journey-to-work information has had a major influence on this latter approach.

In the usual case, future travel demand is determined from the urban transportation planning process in terms of total daily trips that are assigned to the transportation system. Consideration of the split between public and private transportation is also incorporated into this process. These assigned trips constitute the traffic volumes on links of the future street and highway network.

In some instances, these volumes (ADT) are provided directly to highway designers. In others, they are converted by the operational transportation study staff to directional volumes for the design hour. From a practical standpoint, the latter approach may be the more desirable because the transportation study staff is often in a better position to evaluate the effects that the assumptions inherent in the planning process have on the resulting design volumes.

Two-way DHVs (i.e., the 30 HV, or its equivalent) may be determined by applying a representative percentage (usually 8 to 12 percent in urban areas) to the ADT. In many cases this percentage, based on data obtained in a traffic count program, is developed and applied system-wide; in other cases, factors may be developed for different facility classes or different areas of an urban region, or both. At least one highway agency has developed regression equations representing the relationship between peak flow and ADT; different equations are applied, depending on the number of lanes and the range of the ADT volumes.

Directional Distribution

For two-lane rural highways, the DHV is the total traffic in both directions of travel. In the design of highways with more than two lanes and on two-lane roads where important intersections are encountered or where additional lanes are to be provided later, knowledge of the hourly traffic volume for each direction of travel is essential.

A multilane highway with a high percentage of traffic in one direction during the peak hours may need more lanes than a highway having the same ADT but with a lesser percentage of directional traffic. During peak hours on most rural highways, from 55 to 70 percent of the traffic is traveling in the peak direction, with up to as much as 80 percent occasionally. Directional distributions of traffic vary enough between sites that two multilane highways carrying equal traffic may have peak direction volumes that differ by as much as 60 percent. For example, consider a rural road with a design volume of 4,000 vehicles per hour (vph) for both directions of travel combined. If during the design hour, the directional distribution is equally split, or 2,000 vph is one direction, two lanes in each direction may be adequate. If 80 percent of the DHV is in one direction, at least three lanes in each direction would be needed for the 3,200 vph; and if a 1,000-vehicles-per-lane criterion is applied, four lanes in each direction would be needed.

The peak-hour traffic distribution by direction of travel is generally consistent from day to day and from year to year on a given rural road, except on some highways serving recreational areas. Except for urban highways, the directional distribution of traffic measured for current conditions may generally be assumed to apply to the DHV for the future year for which the facility is designed.

The directional distribution of traffic on multilane facilities during the design hour (DDHV) should be determined by making field measurements on the facility under consideration or on parallel and similar facilities. In the latter case, the parallel facilities should preferably be those from which traffic, for the most part, would be diverted to the new highway. The DDHV applicable for use on multilane facilities may be computed by multiplying the ADT by the percentage that 30 HV is of the ADT, and then by the percentage of traffic in the peak direction during the design hour. Thus, if the DHV is 15 percent of the ADT and the directional distribution at the peak hour is 60:40, the DDHV is 0.15 x 0.60 x ADT, or 9 percent of the ADT. If the directional ADT is known for only one direction, the ADT is nearly always twice the directional ADT.

In designing intersections and interchanges, the volumes of all movements occurring during the design hour should be known. This information is needed for both the morning and evening peak periods because the traffic pattern may change significantly from one peak hour to the other. Normally, a design is based on the DHV, which is to be accommodated during the morning rush hour in one direction and during the evening rush hour in the other direction. Total (two-way) volumes may be the same during both of these peaks, but the percentage of traffic in the two directions of travel is reversed. At intersections, the percentage of approaching traffic that turns to the right and to the left on each intersection leg should be determined separately for the morning and evening peak periods. This information should be determined from actual counts, from origin and destination data, or both.

Composition of Traffic

Vehicles of different sizes and weights have different operating characteristics that should be considered in highway design. Besides being heavier, trucks are generally slower and occupy more roadway space. Consequently, trucks have a greater individual effect on highway traffic operation than do passenger vehicles. The effect on traffic operation of one truck is often equivalent to several passenger cars. The number of equivalent passenger cars equaling the effect of one truck is dependent on the roadway gradient and, for two-lane highways, on the available passing sight distance. Thus, the larger the proportion of trucks in a traffic stream, the greater the equivalent traffic demand and the greater the highway capacity needed.

For uninterrupted traffic flow, as typically found in rural areas, the various sizes and weights of vehicles, as they affect traffic operation, can be grouped into two general classes:

- Passenger cars—all passenger cars, including mini-vans, vans, pick-up trucks, and sport/utility vehicles
- Trucks—all buses, single-unit trucks, combination trucks, and recreation vehicles

For traffic-classification purposes, trucks are normally defined as those vehicles having manufacturer's gross vehicle weight (GVW) ratings of 4,000 kg [9,000 lb] or more and having dual tires on at least one rear axle.

In the passenger-car class, as defined above, most of the vehicles have similar operating characteristics. In the truck class, operating characteristics vary considerably, particularly in size and weight/power ratio. Despite this variation in the operating characteristics of trucks, the average effect of all trucks in a traffic stream is similar on most highways under comparable conditions. Accordingly, for the geometric design of a highway, it is essential to have traffic data on vehicles in the truck class. These data generally indicate the major types of trucks and buses as percentages of all traffic expected to use the highway.

For design purposes, the percentage of truck traffic during the peak hours should be determined. In rural areas, comprehensive data usually are not available on the distribution of traffic by vehicle types during the peak hours; however, the percentage of truck traffic during the peak hours is generally less than the percentage for a 24-hour period. As the peak hour approaches, the volume of passenger-car traffic generally increases at a greater rate than does the volume of truck traffic. Most trucks operate steadily throughout the day, and much over-the-road hauling is done at night and during early morning hours. In the vicinity of major truck and bus terminals, the scheduling of regular truck and bus runs may result in the concentration of trucks during certain hours of the day. However, because of the delays caused by other traffic during peak hours, such schedules generally are made to avoid these hours.

For design of a particular highway, data on traffic composition should be determined by traffic studies. Truck traffic should be expressed as a percentage of total traffic during the design hour (in the case of a two-lane highway, as a percentage of total two-way traffic, and in the case of a multilane highway, as a percentage of total traffic in the peak direction of travel).

Under urban interrupted-flow conditions, the criteria for determining traffic composition differ from those used elsewhere. At important intersections, the percentage of trucks during the morning and evening peak hours should be determined separately. Variations in truck traffic between the various traffic movements at intersections may be substantial and may influence the appropriate geometric layout. The percentage of trucks may also vary considerably during a particular hour of the day. Therefore, it is advisable to count trucks for the several peak hours that are considered representative of the 30th highest or design hour. A convenient value, that appears appropriate for design use, is the average of the percentages of truck traffic percentages for a number of weekly peak hours. For highway-capacity analysis purposes, local city-transit buses should be considered separately from other trucks and buses.

Projection of Future Traffic Demands

Geometric design of new highways or improvements to existing highways should not usually be based on current traffic volumes alone, but should consider future traffic volumes expected to use the facility. A highway should be designed to accommodate the traffic volume that is likely to occur within the design life of the facility.

It is difficult to define the life of a highway because major segments may have different lengths of physical life. Each segment is subject to variations in estimated life expectancy for reasons not readily subject to analysis, such as obsolescence or unexpected radical changes in land use, with the resulting changes in traffic volumes, patterns, and demands. Right-of-way and grading may be considered to have a physical life expectancy of 100 years; minor drainage structures and base courses, 50 years; bridges, 25 to 100 years; resurfacing, 10 years; and pavement structure, 20 to 30 years, assuming adequate maintenance and no allowance for obsolescence. Bridge life may vary depending on the cumulative frequency of heavy loads. Pavement life can vary widely, depending largely on initial expenditures and the repetition of heavy axle loads.

The assumption of no allowance for functional obsolescence is open to serious debate. The principal causes of obsolescence are increases in the number of intersections and driveways, and increases in traffic demand beyond the design capacity. On non-freeway highways, obsolescence due to addition of intersections and driveways is much more difficult to forestall; this occurs particularly in urban and suburban areas, but may occur in rural areas as well.

It is a moot question whether the design capacity of a highway should be based on its life expectancy. The decision is greatly influenced by economics. For example, a highway might be designed for traffic volumes 50 years hence with the expectation that the pavement structure would be restored in 20 to 25 years. However, if the added cost of a 50-year design over a design with a 25-year life expectancy is appreciable, it may be imprudent to make a further investment providing capacity that will not be needed for at least 25 years. The construction cost savings could be used to construct another currently needed highway project. Furthermore, the cost of increased maintenance for the larger highway would be avoided for at least 25 years. Also, most highways are capable of handling higher traffic volumes than their design volume indicates, but this may cause more inconvenience, such as a reduction in speed and less maneuverability.

For example, a four-lane divided highway with a design ADT of 10,000 or 15,000 vehicles per day could handle two or three times that design volume depending on several factors discussed later. Thus, the four-lane divided highway could adequately serve traffic long after the design year and, in many cases, indefinitely.

In a practical sense, the design volume should be a value that can be estimated with reasonable accuracy. Many highway engineers believe the maximum design period is in the range of 15 to 24 years. Therefore, a period of 20 years is widely used as a basis for design. Traffic cannot usually be forecast accurately beyond this period on a specific facility because of probable changes in the general regional economy, population, and land development along the highway, which cannot be predicted with any degree of assurance.

Estimating traffic volumes for a 20-year design period may not be appropriate for many reconstruction or rehabilitation projects. These projects may be developed on the basis of a shorter design period (5 to 10 years) because of the uncertainties of predicting traffic and funding constraints.

Speed

Speed is one of the most important factors considered by travelers in selecting alternative routes or transportation modes. Travelers assess the value of a transportation facility in moving people and goods by its convenience and economy, which are directly related to its speed. The attractiveness of a public transportation system or a new highway are each weighed by the travelers in terms of time, convenience, and money saved. Hence, the desirability of rapid transit may well rest with how rapid it actually is. The speed of vehicles on a road or highway depends, in addition to capabilities of the drivers and their vehicles, upon five general conditions: the physical characteristics of the highway, the amount of roadside interference, the weather, the presence of other vehicles, and the speed limitations (established either by law or by traffic control devices). Although any one of these factors may govern travel speed, the effect of these general conditions is usually interrelated.

The objective in design of any engineered facility used by the public is to satisfy the public's demand for service in a safe and economical manner. The facility should, therefore, accommodate nearly all demands with reasonable adequacy and also should not fail under severe or extreme traffic demands. Therefore, highways should be designed to operate at a speed that satisfies nearly all drivers. Because only a small percentage of drivers travel at extremely high speed, it is not economically practical to design for them. They can use the highway, of course, but will be constrained to travel at speeds less than they consider desirable. On the other hand, the speed chosen for design should not be that used by drivers under unfavorable conditions, such as inclement weather, because the highway would then be inefficient, and possibly unsafe, for drivers under favorable conditions, and would not satisfy reasonable public expectations for the facility.

Operating Speed

Operating speed is the speed at which drivers are observed operating their vehicles during free-flow conditions. The 85th percentile of the distribution of observed speeds is the most frequently used measure of the operating speed associated with a particular location or geometric feature.

Running Speed

The speed at which an individual vehicle travels over a highway section is known as its running speed. The running speed is the length of the highway section divided by the running time required for the vehicle to travel through the section. The average running speed of all vehicles is the most appropriate speed measure for evaluating level of service and road user costs. The average running speed is the sum of the distances traveled by vehicles on a highway section during a specified time period divided by the sum of their running times.

One means of estimating the average running speed for an existing facility where flow is reasonably continuous is to measure the spot speed at one or more locations. The average spot speed is the arithmetic mean of the speeds of all traffic as measured at a specified point on the roadway. For short sections of highway, on which speeds do not vary materially, the average spot speed at one location may be considered an approximation of the average running speed. On longer stretches of rural highway, average spot speeds measured at several points, where each point represents the speed characteristics of a selected segment of highway, may be averaged (taking relative lengths of the highway segments into account) to provide a better approximation of the average running speed.

The average running speed on a given highway varies somewhat during the day, depending primarily on the traffic volume. Therefore, when reference is made to a running speed, it should be clearly stated whether this speed represents peak hours, off-peak hours, or an average for the day. Peak and off-peak running speeds are used in design and operation; average running speeds for an entire day are used in economic analyses.

The effect of traffic volume on average running speed can be determined using the procedures of the *Highway Capacity Manual* (HCM) (**15**). The HCM shows that:

- for freeways and multilane highways, there is a substantial range of flow rates over which speed is relatively insensitive to the flow rate; this range extends to fairly high flow rates. Then, as the flow rate per lane approaches capacity, speed decreases substantially with increasing flow rate.
- for two-lane highways, speed decreases linearly with increasing flow rate over the entire range of flow rates between zero and capacity.

Design Speed

Design speed is a selected speed used to determine the various geometric design features of the roadway. The assumed design speed should be a logical one with respect to the topography, anticipated operating speed, the adjacent land use, and the functional classification of highway. Except for local streets where speed controls are frequently included intentionally, every effort should be made to use as high a design speed as practical to attain a desired degree of safety, mobility, and efficiency within the constraints of environmental quality, economics, aesthetics, and social or political impacts. Once the design speed is selected, all of the pertinent highway features should be related to it to obtain a balanced design. Above-minimum design values should be used, where practical. Some design features, such as curvature, superelevation, and sight distance, are directly related to, and vary appreciably with, design speed. Other features, such as widths of lanes and shoulders and clearances to walls and rails, are not directly related to design speed, but they do affect vehicle speeds. Therefore, wider lanes, shoulders, and clearances should be considered for higher design speeds. Thus, when a change is made in design speed, many elements of the highway design will change accordingly.

The selected design speed should be consistent with the speeds that drivers are likely to expect on a given highway facility. Where a reason for limiting speed is obvious, drivers are more apt to accept lower speed operation than where there is no apparent reason. A highway of higher functional classification may justify a higher design speed than a lesser classified facility in similar topography, particularly where the savings in vehicle operation and other operating costs are sufficient to offset the increased costs of right-of-way and construction. A low design speed, however, should not be selected where the topography is such that drivers are likely to travel at high speeds. Drivers do not adjust their speeds to the importance of the highway, but to their perception of the physical limitations of the highway and its traffic.

The selected design speed should fit the travel desires and habits of nearly all drivers expected to use a particular facility. Where traffic and roadway conditions are such that drivers can travel at their desired speed, there is always a wide range in the speeds at which various individuals will choose to operate their vehicles. A cumulative distribution of free-flow vehicle speeds typically has an S-shape when plotted as the percentage of vehicles versus observed speed. The selected design speed should be a high-percentile value in this speed distribution curve (i.e., inclusive of nearly all of the desired speeds of drivers, wherever practical).

Speed distribution curves illustrate the range of speeds that should be considered in selecting an appropriate design speed. A design speed of 110 km/h [70 mph] should be used for freeways, expressways, and other arterial highways in rural areas.

It is desirable that the running speed of a large proportion of drivers be lower than the design speed. Experience indicates that deviations from this desired goal are most evident and problematic on sharper horizontal curves. In particular, curves with low design speeds (relative to driver expectation) are frequently overdriven and tend to have poor safety records. Therefore, it is important that the design speed used for horizontal curve design be a conservative reflection of the expected speed on the constructed facility.

Where the physical features of the highway are the principal speed controls and where most drivers choose to operate near the speed limit, a design speed of 120 km/h [75 mph] would serve a very high percentage of drivers. On a highway designed for this speed, only a small percentage of drivers might operate at higher speeds when volume is low and all other conditions are favorable. However, for a design speed of 80 km/h [50 mph], satisfactory performance could be expected only on certain types of highways. When a low speed design is selected, it may be important to have the speed limit enforced during off-peak hours.

On many freeways, particularly in suburban and rural areas, a design speed of 100 km/h [60 mph] or higher can be provided with little additional cost above that required for a design speed of 80 km/h [50 mph]. If the freeway alignment is relatively straight and the character and location of interchanges permit design for high-speed operation, a design speed of 110 km/h [70 mph] is desirable.

Generally, there is no distinction in design speed between ground-level, elevated, and depressed freeways. However, the operating characteristics of elevated freeways differ from those on depressed freeways. On a depressed freeway, traffic exits the freeway on upgrade ramps and enters the freeway on downgrade ramps, which encourages good operation. By contrast, on an elevated freeway, traffic exits the freeway on downgrade ramps and enters the freeway on upgrade ramps, which is less desirable because vehicles entering the elevated freeway on an ascending grade, particularly loaded trucks, require long distances to reach the running speed of the freeway. Furthermore, vehicles leaving the elevated freeway on a descending grade need additional braking distance before reaching the arterial street, and therefore, may tend to slow down in the through-traffic lanes in advance of the ramp terminal. Parallel deceleration lanes or longer ramp lengths and lesser grades are frequently used on elevated freeways to reduce the likelihood that vehicles will slow in the main lanes. Nevertheless, running speeds on elevated freeways are apt to be slightly lower than those on similar depressed freeways, especially when access points are closely spaced. In northern climates, elevated structures are subject to rapid freezing of precipitation as a result of their exposure; the use of lower superelevation rates may be appropriate under such conditions. Although speeds on viaducts are less than those on comparable depressed sections, the difference probably is small. Therefore, design speeds of 80 to 110 km/h [50 to 70 mph] apply to both elevated and depressed freeways.

Given an overall range in design speeds of 20 to 120 km/h [15 to 75 mph] used in geometric design, it is desirable to select design speeds in increments of 10 km/h [5 mph]. Smaller increments would result in little distinction in the dimensions of design elements between one design speed and the next higher design speed; larger increments of 20 to 30 km/h [15 to 20 mph] would result in too large a difference in the dimensions of design features between any two design speeds. In some instances, however, there may be an advantage in using intermediate increments to effect changes in the design speed. Increments in design speed of 10 km/h [5 mph] should also be used in the design of turning roadways, ramps, and low-speed roads.

Exhibit 2-29 shows the corresponding design speeds in metric and U.S. customary units in 10-km/h [5-mph] increments. This table should be used in converting the units of measurement of design speeds.

Although the selected design speed establishes the limiting values of curve radius and minimum sight distance that should be used in design, there should be no restriction on the use of flatter horizontal curves or greater sight distances where such improvements can be provided as a part of an economical design. Even in rugged terrain, an occasional tangent or flat curve may be desirable. Isolated features designed for higher speeds would not necessarily encourage drivers to speed up, although a succession of such features might. In such cases, the entire section of highway should be designed for a higher speed. A substantial length of tangent between sections of curved alignment is also likely to encourage high-speed operation. In such situations, a higher design speed should be selected for all geometric features, particularly sight distance on crest vertical curves and across the inside of horizontal curves.

Metric	US Customary
Design speed (km/h)	Corresponding design speed (mph)
20	15
30	20
40	25
50	30
60	40
70	45
80	50
90	55
100	60
110	70
120	75
130	80

Exhibit 2-29. Corresponding Design Speeds in Metric and US Customary Units

A pertinent consideration in selecting design speeds is the average trip length. The longer the trip, the greater the driver's desire to use higher speeds. In the design of a substantial length of highway, it is desirable to select a uniform design speed. However, changes in terrain and other physical controls may dictate a change in design speed on certain sections. If so, the introduction of a lower design speed should not be done abruptly but should be effected over sufficient distance to permit drivers to gradually change speed before reaching the highway section with the lower design speed.

Where it is appropriate to reduce horizontal and vertical alignment features, many drivers may not perceive the lower speed condition ahead, and therefore, it is important that they be warned well in advance. The changing condition should be indicated by such controls as speed-zone and curve-speed signs.

On rural highways and on high-type urban facilities, a percentage of vehicles is usually able to travel at near the free-flow speed governed by geometric design elements; therefore, the selection of an appropriate design speed is particularly important. However, in many arterial streets, vehicle speeds during several hours of the day are limited or regulated more by the presence of large volumes of vehicles and by traffic control devices, rather than by the physical characteristics of the street. In such cases, the selection of a design speed is less critical to safe and efficient operation.

During periods of low-to-moderate volume, speeds on arterial streets are governed by such factors as posted speed limits, midblock turns into and out of driveways, intersectional turns, traffic signal spacing, and signal timing for progression. When arterial street improvements are being planned, factors such as future posted speed limits, physical and economic constraints, and running speeds likely to be attained during off-peak hours should be considered. All of these factors should influence the selection of an appropriate design speed.

Horizontal alignment generally is not the governing factor in restricting speeds on arterial streets. Proposed improvements generally are patterned to the existing street system, and minor horizontal alignment changes are commonly made at intersections. The effect of these alignment changes is usually small because operation through the intersection is regulated by the type of traffic controls needed to handle the volume of cross and turning traffic. Superelevation may be provided at curves on urban arterial streets, but the amount of superelevation needed is determined in a different manner than for open-road rural conditions. Wide pavement areas, proximity of adjacent development, control of cross slope and profile for drainage, and the frequency of cross streets and entrances all contribute to the need for lower superelevation rates on urban arterial streets. The width of lanes, offset to curbs, proximity of poles and trees to the traveled way, presence of pedestrians within the right-of-way, and nearness of business or residential buildings, individually or in combination, often limit speeds even on highways with good alignment and flat profiles. Despite these factors, designers should strive for good alignment and flat profiles in the design of urban arterial streets, since safety and operating characteristics can be improved, particularly during off-peak periods. Chapter 3 provides guidance on horizontal alignment design for low-speed urban conditions.

Topography can materially affect the choice of design speed on arterial streets. Many cities were developed along watercourses and include areas varying from gently rolling to mountainous terrain. Streets may have been constructed originally with only minor grading to fit the topography. Because an arterial street is usually developed to fit the alignment of an existing street, both through business and residential areas, it generally follows a varying vertical profile. Once the design speed is selected, appropriate sight distance should be provided at all crests and across the inside of horizontal curves. Profiles with long, continuous grades should be designed with proper consideration for the speeds of mass transit and commercial vehicles. Extra lanes on the upgrades may be needed so that the grade can match other portions of the facility in capacity and enable vehicles that can proceed at a reasonable speed to pass slower moving vehicles.

Urban arterial streets should be designed and control devices regulated, where practical, to permit running speeds of 30 to 75 km/h [20 to 45 mph]. Speeds in the lower portion of this range are applicable to local and collector streets through residential areas and to arterial streets through more crowded business areas, while the speeds in the higher portion of the range apply to high-type arterials in outlying suburban areas. For arterial streets through crowded business areas, coordinated signal control through successive intersections is generally needed to permit attainment of even the lower speeds. Many cities have substantial lengths of signal controlled streets that operate at speeds of 20 to 40 km/h [15 to 25 mph].

Under less crowded conditions in suburban areas, it is common on preferred streets to adopt some form of speed zoning or control to limit high operating speeds. In such areas, pedestrians along the arterial or vehicles on cross streets, although relatively infrequent, may be exposed to potential collisions with through drivers. Such through drivers may gradually gain speed as urban restrictions are left behind or may retain their open-road speeds as they enter the city. Thus, although through traffic should be expedited to the extent practical, it may be equally important to limit speeds to reduce the risk of crashes and to serve local traffic.

Posted speed limits, as a matter of policy, are not the highest speeds that might be used by drivers. Instead, such limits are usually set to approximate the 85th percentile speed of traffic as determined by measuring the speeds of a sizable sample of vehicles. The 85th-percentile speed is usually within the "pace" or the 15-km/h [10-mph] speed range used by most drivers. Speed zones cannot be made to operate properly if the posted speed limit is determined arbitrarily. In addition, speed zones should be determined from traffic engineering studies, should be consistent with prevailing conditions along the street and with the cross section of the street, and should be capable of reasonable enforcement.

Urban arterial streets and highways generally have running speeds of 30 to 70 km/h [20 to 45 mph]. It follows that the appropriate design speeds for arterials should range from 50 to 100 km/h [30 to 60 mph]. The design speed selected for an urban arterial should depend largely on the spacing of signalized intersections, the selected type of median cross section, the presence or absence of curb and gutter along the outside edges of the traveled way, and the amount and type of access to the street. Reconstructed urban arterial highways should generally be designed for an operating speed of at least 50 km/h [30 mph].

The preceding discussion describes the considerations in selecting an appropriate design speed. From this discussion, it should be evident that there are important differences between the design criteria applicable to low- and high-speed designs. Because of these distinct differences, the upper limit for low-speed design is 70 km/h [45 mph] and the lower limit for high-speed design is 80 km/h [50 mph].

Traffic Flow Relationships

Traffic flow conditions on roadways can be characterized by the volume flow rate expressed in vehicles per hour, the average speed in kilometers per hour [miles per hour], and the traffic density in vehicles per kilometer [vehicles per mile]. These three variables—volume, speed, and density—are interrelated and have predictable relationships. The generalized relationships between volume, speed, and density for uninterrupted flow facilities, as presented in the HCM (**15**) are shown in Exhibit 2-30. The relationships shown in the exhibit are conceptual in nature and do not necessarily correspond to the actual relationships used in specific HCM procedures. For example, the HCM procedures for freeways and multilane highways show that speed does not vary with volume through most of the low and intermediate volume range, as shown in the exhibit. The HCM procedures for two-lane highways show that speed varies linearly with volume throughout the entire volume range from zero to capacity.

Density, the number of vehicles per unit length of roadway, increases as vehicles crowd closer together. As Exhibit 2-30 shows, when speeds decrease, increased crowding can occur and drivers can comfortably follow more closely behind other vehicles. Density is used in the HCM as the measure of quality of traffic service for freeways and multilane highways.

Traffic volumes also vary with density from zero to maximum flow rate, as shown in Exhibit 2-30. The two points of zero flow in the exhibit represent either no vehicles at all or so many vehicles on the roadway that flow has stopped. The maximum flow is reached at the point of maximum density.

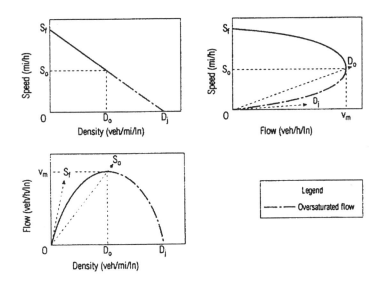

Exhibit 2-30. Generalized Speed-Volume-Density Relationships (15)

Interference to traffic flow causes speeds to be reduced, vehicles to travel closer together, and density to increase. Interference may be caused by weather conditions, cross traffic, disabled vehicles, crashes, or other conditions. As these conditions cause more interference, the flow rates within certain limits can still be maintained but with reduced speed, closer vehicle spacing, and greater density. When interference becomes so great (despite closer vehicle spacing and greater density) that the average speed drops below that necessary to maintain stable flow, there is a rapid decrease in speed and traffic flow, and severe congestion occurs.

When traffic on a highway encounters interference that limits or reduces the roadway capacity in a single area, the result is a "bottleneck." If the flow entering this bottleneck does not exceed its capacity, flow remains stable and no problems arise. However, when the upstream section carries more vehicles than the bottleneck can accommodate, a breakdown in traffic flow results. Speeds are reduced to a crawl and vehicles begin to queue upstream until incoming flow again falls below the outflow capacity. To avoid bottleneck situations, care should be taken to design roadways with consistent volume-carrying capacity. The level-of-service concept discussed in the next section helps in obtaining this consistency.

An intersection is often an unavoidable bottleneck. This reduction in capacity becomes acute when the intersection is controlled by stop signs or traffic signals. At a traffic signal, vehicles that arrive during the red phase encounter a zero-capacity bottleneck. These vehicles form a queue until the green phase begins, removing the restraint, and discharging the queue. If the incoming volume is too high, not all vehicles in the queue can be discharged during the green phase, and there is a continuing buildup of the queue.

Arrivals at the intersection are generally predictable in urban areas where the approaching vehicles are platooned by upstream signals. In suburban or rural locations, vehicle arrivals are often random. This random arrival pattern should be recognized in the design of appropriate cycle times, turn-lane storage lengths, and approach capacity.

At bottlenecks where the traffic must slow down or stop, each vehicle and its occupants incur a certain delay. Delays increase fuel consumption and air pollution, which create undesirable economic and environmental effects.

HIGHWAY CAPACITY

General Characteristics

The term "capacity" is used to express the maximum hourly rate at which persons or vehicles can reasonably be expected to traverse a point (i.e., a uniform section of a lane or a roadway) during a given time period under prevailing roadway and traffic conditions. The range of traffic flow on a highway can vary from very light volumes to volumes that equal the capacity of the facility as defined above. In the generic sense, the term also encompasses broader relations between highway characteristics and conditions, traffic composition and flow patterns, and the relative degree of congestion at various traffic volumes. Highway capacity issues in this broad sense are discussed below.

The following sections provide a brief overview of the principles and major factors concerning highway design capacity. To determine the capacity for a particular highway design, the designer should refer to the most recent edition of the *Highway Capacity Manual* (HCM) (**15**) for guidance. The HCM is used as the basic reference for the following discussion.

Application

Highway capacity analysis serves three general purposes, including:

- **Transportation planning studies**—Highway capacity analysis is used in these studies to assess the adequacy or sufficiency of existing highway networks to service current traffic. In addition, it is used to estimate the time in the future when traffic growth may overtake the capacity of a highway or perhaps reach a level of congestion below capacity that is considered undesirable.
- **Highway design**—A knowledge of highway capacity is essential to properly fit a planned highway to traffic demands. Highway capacity analysis is used both to select the highway type and to determine dimensions such as the number of lanes and the minimum lengths for weaving sections.
- **Traffic operational analyses**—Highway capacity analysis is used in these analyses for many purposes, but especially for identifying bottleneck locations (either existing or potential). It is also used in preparing estimates of operational improvements that may

be expected to result from prospective traffic control measures or from spot alterations in the highway geometry.

The traffic data for these uses varies with the degree of accuracy needed. For traffic-operational analyses, in which the success of minor improvements may be measured in terms of a few vehicles per hour, a high degree of precision is desirable. For highway design, a much lower order of precision suffices because the traffic data are frequently estimated for a period 10 to 20 years in the future and involve not only approximations of traffic volumes but also approximations of such factors as traffic composition and movement patterns. The discussion below shows the appropriate level of detail to ensure a reasonable balance between the design of the highway and the estimated future traffic. Such an analysis ensures that future operating conditions will not fall below an acceptable level. If a greater accuracy than is available from the suggested procedures is needed, refer to the HCM and other reports on traffic operational analysis.

Capacity as a Design Control

Design Service Flow Rate Versus Design Volume

The design volume is the volume of traffic projected to use a particular facility during the design year, which is usually 10 to 20 years in the future. Design volumes are estimated in the planning process and are often expressed as the expected traffic volume during a specified design hour. The derivation of the DHV has been discussed earlier in this chapter in the section on "Traffic Characteristics."

Design service flow rate is the maximum hourly flow rate of traffic that a highway with particular design features would be able to serve without the degree of congestion falling below a pre-selected level, as described below.

A major objective in designing a highway is to create a facility with dimensions and alignment that can serve the design service flow rate, which should be at least as great as the flow rate during the peak 15-minute period of the design hour, but not so great as to represent an extravagance in the design. Where this objective is accomplished, a well-balanced, economical highway facility will result.

Measures of Congestion

Three key considerations in geometric design are the roadway design, the traffic using the roadway, and the degree of congestion on the roadway. The first two considerations can be measured in exact units. For example, the roadway either is or is not a highway with full control of access, its cross-section dimensions can be expressed in meters [feet], and the steepnesses of its grades can be expressed as a percentage. Likewise, traffic flow can be expressed as the number of vehicles per unit of time, traffic composition can be expressed as the percentage of vehicles of each class, and the peaking characteristics and directional distribution of traffic can also be quantified.

A scale of values for expressing the degree of congestion is, however, a much more elusive measure. Numerous measures of the overall service provided by a roadway section have been suggested, including safety, freedom to maneuver, the ratio of traffic volume to capacity (v/c), operating speed, average running speed, and others. In the case of signalized intersections, the stopped delay encountered by motorists is a commonly used measure of congestion.

For uninterrupted traffic flow (i.e., flow not influenced by signalized intersections), traffic operational conditions are defined by using three primary measures: speed, volume (or rate of flow), and density. Density describes the proximity of vehicles to one another and reflects the freedom to maneuver within the traffic stream. It is a critical parameter describing traffic operations with uninterrupted flow. As density increases from zero, the rate of flow also increases because more vehicles are on the roadway. While this is happening, speed begins to decline (due to the vehicle interactions). This decline is virtually negligible at low densities and flow rates. However, as density continues to increase, a point is reached at which speed declines noticeably. A maximum rate of flow is eventually reached at which the high density of traffic results in markedly decreased speeds and a reduced flow rate. This maximum rate of flow for any given facility is defined as its capacity. As capacity is approached, flow becomes more unstable because available gaps in the traffic stream become fewer and fewer. At capacity, there are no usable gaps in the traffic stream, and any conflict from vehicles entering or leaving the facility, or from internal lane changing maneuvers, creates a disturbance that cannot be effectively damped or dissipated. Thus, operation at or near capacity is difficult to maintain for long periods of time without the formation of upstream queues, and forced or breakdown flow becomes almost unavoidable. For this reason, most facilities are designed to operate at volumes less than their capacity.

For interrupted flow, such as that occurring on streets where traffic is controlled by signals, the highway user is not as concerned with attaining a high travel speed as with avoiding lengthy stops at intersections or a succession of stops at several intersections. Average stopped-time delay is the principal measure of effectiveness used in evaluating signalized intersections. Stopped-time delay, which is used because it is reasonably easy to measure and is conceptually simple, is a characteristic of intersection operations that is closely related to motorist perceptions of quality of traffic flow.

Relation Between Congestion and Traffic Flow Rate

Congestion does not necessarily signify a complete stoppage of traffic flow. Rather it can be thought of as a restriction or interference to normal free flow. For any given class of highway, congestion increases with an increase in flow rate until the flow rate is almost equal to the facility's capacity, at which point congestion becomes acute. The gradual increase in congestion with increase in flow rate is apparent no matter what measure is used as an index of congestion.

The relationship between running speed and traffic flow rate for freeways, multilane highways, and two-lane highways has been discussed earlier in this chapter in the section on "Running Speed." As the traffic flow rate approaches a facility's capacity, as defined in the HCM (**15**), any minor disruption in the free flow of traffic may cause traffic on a roadway to operate on a stop-and-go basis, with a resulting decrease in traffic flow rate that can be served.

Highway sections where the paths of traffic must merge and diverge within relatively short distances are called "weaving sections." Average running speed, and hence the degree of congestion, is a function not only of the volume of traffic involved in the weaving (crossing) movements but also of the distance within which the weaving maneuvers must be completed. (Weaving is addressed under a separate subsection later in this chapter.)

On arterial streets within the urban environment, average running speed varies only slightly with changes in traffic flow rate. However, delay at signalized intersections may increase dramatically as flow rates approach capacity. Therefore, greater degrees of congestion occur, and this results in reduced overall travel speeds, higher average travel times, and traffic spill-backs into upstream intersections.

Acceptable Degrees of Congestion

From the standpoint of the highway user, it would be preferable for each user to have an exclusive right to the highway at the time the motorist finds occasion or need to use it. Moreover, a motorist would prefer that all highways be of types that would permit speeds far in excess of those normally afforded by urban surface streets. However, users recognize that if others are to share in the costs of transportation facilities, they are also entitled to share in their use. Therefore, they will readily accept a moderate amount of congestion. Just what degree of congestion the motoring public is willing to accept as reasonable remains a matter of conjecture, but it is known to vary with a number of factors.

The average motorist understands in a general sense that corrective measures to alleviate congestion may be more costly in some instances than in others. As a result, motorists will generally accept a higher degree of congestion in those areas where improvements can be made only at a substantial cost. Also, motorists are more willing to accept a higher degree of restraint in short trips than they are in long trips, but motorists are generally not satisfied with the type of operation that occurs when the volume of traffic approaches the facility's capacity.

From a highway administrator's point of view, the degree of congestion that highway users experience is geared to the availability of resources. Historically, funds have never been sufficient to meet all needs, causing severe strain in improving highways rapidly enough to prevent the traffic demand from exceeding the capacity of the facility.

The appropriate degree of congestion that should be used in planning and designing highway improvements is determined by weighing the desires of the motorists against the resources available for satisfying these desires. The degree of congestion that should not be exceeded during the design year on a proposed highway can be realistically assessed by: (1) determining the operating conditions that the majority of motorists will accept as satisfactory, (2) determining the most extensive highway improvement that the governmental jurisdiction considers practical, and (3) reconciling the demands of the motorist and the general public with the finances available to meet those demands.

This reconciliation of desires with available resources is an administrative process of high importance. The decision should first be made as to the degree of congestion that should not be exceeded during the design period. The appropriate design for a particular facility (such as number of lanes) can then be estimated from the concepts discussed in the following sections.

Principles for Acceptable Degrees of Congestion

No scientific method exists for deciding the maximum degree of congestion that might be accepted as a basis for design. This decision lends itself neither to a modeling technique nor to the insertion of coefficients into a computer program. Nevertheless, some principles or guidelines that should aid in arriving at such decisions are itemized and discussed on the following pages.

1. *The highway should be so designed that, when it is carrying the design volume, the traffic demand will not exceed the capacity of the facility even during short intervals of time.*

Conditions can become intolerable for the motorist when the traffic demand exceeds the capacity of the street or highway. Moreover, when stop-and-go traffic develops due to congestion on highways (other than those controlled by signals), the flow rate that can be served by the highway is drastically reduced. Stoppages will occur if the capacity is exceeded even for short intervals of time. Because traffic does not flow uniformly throughout a full hour, allowance should be made for peaking within the hour. This allowance is made in the HCM procedures with an adjustment known as the "peak hour factor," which is discussed later in this chapter.

Where traffic is controlled by signals at intersections, the relationship between delay and capacity may be extremely complex. It is possible to have unacceptably large delays and long queues where traffic demand approaches 75 to 85 percent of capacity. The reverse is also possible—an intersection approach where traffic demand equals capacity may have low delays if the signal cycle is short and/or if signal progression is possible.

2. *The design volume per lane should not exceed the rate at which traffic can dissipate from a standing queue.*

This principle is applicable primarily to freeways and other high-type multilane highways. For example, if traffic on a freeway lane is stopped even momentarily, it cannot recover at a rate equal to the capacity of a freely flowing lane. If the traffic demand exceeds the rate at which cars can depart from the head of a standing queue, the queue will increase in length rather than dissipate, even after the cause of the stoppage is removed. The rate at which vehicles can depart from a standing queue is estimated by various authorities as being within the range of 1,500 to 1,800 passenger cars per lane per hour.

3. Drivers should be afforded some choice of speed. The latitude in choice of speed should be related to the length of trip.

This principle is applicable to all types of streets and highways. The degree of freedom that should be afforded is a subjective determination. On congested freeways with average speeds of about 100 km/h [60 mph], for example, the range of speeds between the slowest and fastest driver would typically be about 25 km/h [15 mph]. This may be satisfactory for short trips.

For longer trips, higher average speeds may be warranted, perhaps 10 km/h [5 mph] higher than for short trips in densely developed areas. An average speed of 110 km/h [70 mph] or more can be achieved on freeways with low to moderate traffic volumes. However, the high cost of construction of urban freeways and the impact on the surrounding neighborhoods usually works against achieving an operating speed this high except in suburban areas, as discussed further under Principle 6 below.

4. Operating conditions should be such that they provide a degree of freedom from driver tension that is related to or consistent with the length and duration of the trip.

This principle may appear to be a corollary of the previous principle. However, Principle 3 represents tensions stemming from impatience, whereas this one deals with tensions that develop from driving in a dense traffic stream at speeds that an individual driver may consider to be too fast for comfort but over which that driver is powerless to exercise control. If the driver reduces speed, this induces others to pass and cut in front of them, thereby reducing the gap that the driver was seeking to enlarge. Freeway travel at speeds of 65 to 100 km/h [40 to 60 mph] under very high-density conditions is a rather tense experience to many and is one that should not be endured if avoidable. Presently, no research data exist to support any recommendations as to the maximum length of time that drivers can or should endure travel under high-density conditions, but it is commonly accepted that tensions build up with continued exposure.

Driver tensions associated with freeway densities of 26 passenger cars per kilometer per lane [42 passenger cars per mile per lane] or less are generally considered acceptable for trips within most metropolitan areas. For long trips the mental concentration that is required and the tensions that develop while driving in such heavy traffic are excessive; consequently, lower volumes should be used for designing freeways that serve relatively long trips.

5. There are practical limitations that preclude the design of an ideal freeway.

An ideal section of freeway would have wide lanes and shoulders on tangent alignment with no restrictions in lateral clearance. Such a freeway would be capable of carrying the capacity specified for basic freeway segments in the HCM (**15**). More often than not, it is necessary to compromise design features to fit the freeway (or other arterial) within attainable right-of-way, to economize on certain features such as curvature or lengths of speed-change lanes, or to locate interchanges closer to each other than would be desirable. It is usually not practical to design a section of freeway with uniform capacity throughout its length.

6. *The attitude of motorists toward adverse operating conditions is influenced by their awareness of the construction and right-of-way costs that might be necessary to provide better service.*

Highway users will accept poor operating conditions if they perceive that the highway is the best design that can be reasonably provided at the particular location. They recognize in a general way that highways are extremely costly in densely developed areas with high land values, in difficult terrain, and at major obstacles to be crossed, such as navigable streams or harbors. Consequently, they will accept poorer operating conditions where highway costs are high than where there is no apparent reason for deficiencies that can be corrected at moderate expense. Because construction costs are frequently much higher in large cities than in small cities, the net result is that this principle tends to offset Principle 3 insofar as the effect of trip length within densely developed areas is concerned.

Reconciliation of Principles for Acceptable Degrees of Congestion

As noted above, the capacities for the base conditions presented in the HCM (**15**) may not be obtainable or desirable on specific highway facilities, depending on the design and desired use of the highway. These principles point to the broad general conclusions that are summarized below.

Freeways. For short trips, tolerance to congestion is governed to a considerable extent by driving tensions. Loss of travel time is of secondary importance, except that complete stoppages, or stop-and-go driving, may be intolerable. These considerations suggest that the density of traffic on urban freeways preferably should not exceed 26 passenger cars per kilometer per lane [42 passenger cars per mile per lane]. Furthermore, if density does not exceed this level, little difficulty from momentary stoppages will result, and minor design restrictions will have no noticeable adverse effect on operating conditions.

For longer trips in metropolitan areas, travel time becomes more important to the user. Driver tensions associated with densities of 26 passenger cars per kilometer per lane [42 passenger cars per mile per lane], while not unbearable, are decidedly unpleasant. No criteria are available for fixing upon any definite value, but indications point to 20 passenger cars per kilometer per lane [30 passenger cars per mile per lane] as resulting in an acceptable degree of congestion.

For rural freeways, travel speed is the dominant consideration. On the basis of past experience, a density of 13 passenger cars per kilometer per lane [20 passenger cars per mile per lane] will permit desirable operations in rural areas.

Other Multilane Highways. Except where traffic is controlled by signals, measures of congestion on other multilane highways are similar to those for freeways. Where the interference with traffic from marginal development is slight, the traffic densities that result in acceptable degrees of congestion on freeways may also be served by other multilane highways. This situation is notably true in rural areas. In urban areas, the traffic volumes that can be served on other multilane highways, at acceptable levels of congestion, are generally somewhat lower than those for freeways, as will be discussed subsequently in this chapter.

Factors Other Than Traffic Volume That Affect Operating Conditions

The ability of a highway to serve traffic efficiently and effectively is influenced by the characteristics of the traffic and by the design features of the highway.

Highway Factors

Few highways have ideal designs. Although most modern freeways have adequate cross-sectional dimensions, many are not ideal with respect to design speed, weaving section design, and ramp terminal design. Inadequacies in these features will result in inefficient use of the remaining portions of the freeway.

On other classes of multilane highways, intersections, even though unsignalized, often interfere with the free-flow operation of traffic. Development adjacent to the highway with attendant driveways and interference from traffic entering and leaving the through-traffic lanes cause a loss in efficiency and lead to congestion and safety problems at relatively low volumes. The adverse effect, although readily apparent, can be difficult to quantify (16). Sharp curves and steep grades cannot always be avoided, and it is sometimes appropriate to compromise on cross-sectional dimensions. All of these conditions combine to cause the effects of congestion to be felt at lower traffic volumes than would be the case for highways designed with ideal features and protected by full access control or by access management.

For urban streets with signalized intersections at relatively close intervals, the traffic volumes that could otherwise be served are reduced because a portion of each signal cycle must be assigned exclusively to the crossing highway.

For a highway that is deficient in some of its characteristics and where the traffic stream is composed of a mixture of vehicle classes rather than passenger cars only, compensatory adjustment factors need to be applied to the traffic flow rates used as design values for ideal highway conditions. These adjustments are necessary to determine the volume of mixed traffic that can be served under minimum acceptable operating conditions on the highway under consideration.

The HCM (15) identifies significant highway features that may have an adverse effect on operating conditions. The HCM provides factors and outlines procedures for determining the traffic volumes that can be served by highways that are not ideal in all respects. Features that could result in a highway being less than ideal in its operational characteristics include narrow lanes and shoulders, steep grades, low design speed, and the presence of intersections, ramp terminals, and weaving sections. The HCM should be referred to for a discussion of these features and their effects on operating conditions. However, the HCM discussion concerning horizontal alignment, weaving sections, and ramp terminals is supplemented and amplified below.

Alignment

For traffic traveling at any given speed, the better the roadway alignment, the more traffic it can carry. It follows that congestion will generally be perceived at lower volumes if the design speed is low than if the design speed is high. The highway should be subdivided into sections of consistent geometric design characteristics for analysis using the HCM techniques. A single limiting curve or steep grade in an otherwise gentle alignment will thus be identified as the critical feature limiting roadway capacity.

Weaving Sections

Weaving sections are highway segments where the pattern of traffic entering and leaving at contiguous points of access results in vehicle paths crossing each other. Where the distance in which the crossing is accomplished is relatively short in relation to the volume of weaving traffic, operations within the highway section will be congested. Some reduction in operating efficiency through weaving sections can be tolerated by highway users if the reduction is minor and the frequency of occurrence is not high. It is generally accepted that a reduction in operating speed of about 10 km/h [5 mph] below that for which the highway as a whole operates can be considered a tolerable degree of congestion for weaving sections.

Operating conditions within weaving sections are affected by both the length and width of the section as well as by the volume of traffic in the several movements. These relationships are discussed later in this chapter and in the HCM.

Ramp Terminals

Ramps and ramp terminals are features that can adversely influence operating conditions on freeways if the demand for their use is excessive or if their design is deficient. When congestion develops at freeway ramp junctions, some through vehicles avoid the outside lane of the freeway, thereby adding to the congestion in the remaining lanes. Thus, if there are only two lanes in one direction, the efficiency per lane is not as high on the average as that for three or more lanes in one direction.

The loss in efficiency is a function of the volume of traffic entering or leaving ramps, the distance between points of entry and exit, and the geometric layout of the terminals. Too little is known of these separate variables to permit a quantitative assessment of their effect when taken individually. Their combined effect is accounted for by levying a uniform assessment against the outside lane, regardless of the causes or extent of interference at individual locations.

Apart from the effect on through traffic, traffic that uses ramps is exposed to a different form of congestion, which does not lend itself to measurement in terms of travel speed, delay, or driver tension. The degree of congestion for a ramp is related to the total volume of traffic in the outside lane of the freeway in the vicinity of the ramp junction (i.e., the combined volume of through traffic using the outside lane and the volume of traffic using the ramp).

The HCM provides procedures for estimating volumes of through traffic in the outside lane of a freeway just upstream of an entrance or an exit ramp for various combinations of highway and traffic conditions.

Traffic Factors

Traffic streams are usually composed of a mixture of vehicles: passenger cars, trucks, buses, and, occasionally, recreational vehicles and bicycles. Furthermore, traffic does not flow at a uniform rate throughout the hour, day, season, or year. Consideration should be given to these two variables, composition of traffic and fluctuations in flow, in deciding upon volumes of traffic that will result in acceptable degrees of congestion (see the subsequent discussion on "Levels of Service") and also upon the period of time over which the flow should extend.

The effect of trucks and buses on highway congestion is discussed in the HCM (**15**). Detailed procedures are provided for converting volumes of mixed traffic to equivalent volumes of passenger cars. These passenger-car equivalency (PCE) factors used in the HCM differ substantially between facility types.

Peak Hour Factor

The accepted unit of time for expressing flow rate is a 1-hour period. It is customary to design highways with a sufficient number of lanes and with other features that will enable the highway to accommodate the forecasted DHV for the design year, which is frequently 20 years from the date of construction.

Because flow is not uniform throughout an hour, there are certain periods within an hour during which congestion is worse than at other times. The HCM considers operating conditions prevailing during the most congested 15-minute period of the hour to establish the service level for the hour as a whole. Accordingly, the total hourly volume that can be served without exceeding a specified degree of congestion is equal to or less than four times the maximum 15-minute count.

The factor used to convert the rate of flow during the highest 15-minute period to the total hourly volume is the peak hour factor (PHF). The PHF may be described as the ratio of the total hourly volume to the number of vehicles during the highest 15-minute period multiplied by 4. The PHF is never greater than 1.00 and is normally within the range of 0.75 to 0.95. Thus, for example, if the maximum flow rate that can be served by a certain freeway without excessive congestion is 4,200 vehicles per hour during the peak 15-minute period, and further, if the PHF is 0.80, the total hourly volume that can be accommodated at that service level is 3,360 vehicles, or 80 percent of the traffic flow rate, during the most congested 15-minute period.

Levels of Service

Techniques and procedures for adjusting operational and highway factors to compensate for conditions that are other than ideal are found in the HCM (**15**). It is desirable that the results of these procedures be made adaptable to highway design.

The HCM defines the quality of traffic service provided by specific highway facilities under specific traffic demands by means of a level of service. The level of service characterizes the operating conditions on the facility in terms of traffic performance measures related to speed and travel time, freedom to maneuver, traffic interruptions, and comfort and convenience. The levels of service range from level-of-service A (least congested) to level-of-service F (most congested). Exhibit 2-31 shows the general definitions of these levels of service. The specific definitions of level of service differ by facility type. The HCM presents a more thorough discussion of the level-of-service concept.

Level of service	General operating conditions
A	Free flow
B	Reasonably free flow
C	Stable flow
D	Approaching unstable flow
E	Unstable flow
F	Forced or breakdown flow

NOTE: Specific definitions of levels-of-service A through F vary by facility type and are presented in the HCM (**15**).

Exhibit 2-31. General Definitions of Levels of Service

The division points between levels-of-service A through F were determined subjectively. Furthermore, the HCM contains no recommendations for the applicability of the levels of service in highway design. Choice of an appropriate level of service for design is properly left to the highway designer. The guidance in the preceding discussion should enable the designer to link the appropriate degrees of congestion to specific levels of service. The relationship between highway type and location and the level of service appropriate for design is summarized in Exhibit 2-32. This relationship is derived from the criteria for acceptable degrees of congestion, as outlined earlier in this discussion.

As may be fitting to the conditions, highway agencies should strive to provide the highest level of service practical. For example, in heavily developed sections of metropolitan areas, conditions may make the use of level-of-service D appropriate for freeways and arterials; however, this level should be used sparingly and at least level-of-service C should be sought.

Functional class	Appropriate level of service for specified combinations of area and terrain type			
	Rural level	Rural rolling	Rural mountainous	Urban and suburban
Freeway	B	B	C	C
Arterial	B	B	C	C
Collector	C	C	D	D
Local	D	D	D	D

Exhibit 2-32. Guidelines for Selection of Design Levels of Service

Design Service Flow Rates

The traffic flow rates that can be served at each level of service are termed "service flow rates." Once a particular level of service has been identified as applicable for design, the corresponding service flow rate logically becomes the design service flow rate, implying that if the traffic flow rate using the facility exceeds that value, operating conditions will fall below the level of service for which the facility was designed.

Once a level of service has been selected, it is desirable that all elements of the roadway are designed consistent to this level. This consistency of design service flow rate results in near-constant freedom of traffic movement and operating speed, and flow interruptions due to bottlenecks can be avoided.

The HCM supplies the analytical base for design calculations and decisions, but the designer should use his or her judgment to select the appropriate level of service. Exhibit 2-32 provides guidance that may be used by designers in selecting an appropriate level of service. For certain recreational routes or for environmental or land use planning reasons, the designer may possibly select a design service flow rate less than the anticipated demand.

Whether designing an intersection, interchange, arterial, or freeway, the selection of the desired level of service should be carefully weighed because the traffic operational adequacy of the roadway is dependent on this choice.

Weaving Sections

Weaving sections occur where one-way traffic streams cross by merging and diverging maneuvers. The principal types of weaving sections are illustrated in Exhibit 2-33. Weaving sections are designed, checked, and adjusted so that the level of service is consistent with the remaining highway. The design level of service of a weaving section is dependent on its length, number of lanes, acceptable degree of congestion, and relative volumes of individual movements. Large-volume weaving movements usually result in considerable friction and reduction in speed of all traffic. Further, there is a definite limit to the amount of traffic that can be handled on a given weaving section without undue congestion. This limiting volume is a function of the distribution of traffic between the weaving movements, the length of weaving section, and the number of lanes.

Weaving sections may be considered as simple or multiple. Exhibit 2-34A shows a simple weaving section in which a single entrance is followed by a single exit. A multiple-weaving section consists of two or more overlapping weaving sections. A multiple weave may also be defined as that portion of a one-way roadway that has two consecutive entrances followed closely by one or more exits, or one entrance followed closely by two or more exits, as shown in Exhibit 2-34B. Multiple weaving sections occur frequently in urban areas where there is need for collection and distribution of high concentrations of traffic. For further information concerning the operation and analysis of simple and multiple weaving sections, refer to the HCM.

The weaving section should have a length and number of lanes based on the appropriate level of service, as given in Exhibit 2-32. The HCM presents an equation for predicting the average running speed of weaving and non-weaving traffic based on roadway and traffic conditions. Level-of-service criteria for weaving sections are based on these average running speeds.

Multilane Highways Without Access Control

Multilane highways may be treated as similar to freeways if major crossroads are infrequent, or if many of the crossroads are grade separated, and if adjacent development is sparse so as to generate little interference. Even on those highways where such interference is currently only marginal, the designer should consider the possibility that by the design year the interference may be extensive unless access to the highway is well managed. In most cases, the designer should assume that extensive crossroad and business improvements are likely over the design life of the facility.

Where there are major crossroads or where adjacent development results in more than slight interference, the facility should be treated as a multilane highway without access control.

Arterial Streets and Urban Highways

It is often difficult to establish design service flow rates for arterial streets and urban highways because the level of service provided by such facilities does not remain stable with the passage of time and tends to deteriorate in an unpredictable manner. However, if the principles of access management are applied initially to the street or highway, a high level of operations can be maintained over time (**16, 17, 18**). The capacity of an arterial is generally dominated by the capacity of its individual signalized intersections. The level of service for a section of an arterial is defined by the average overall travel speed for the section.

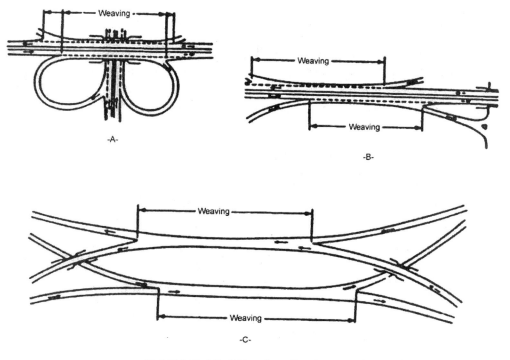

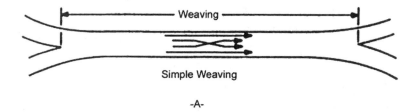

Exhibit 2-33. Weaving Sections

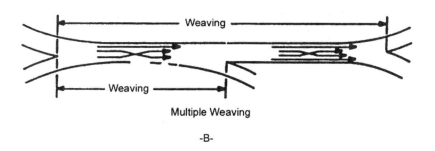

Exhibit 2-34. Simple and Multiple Weaving Sections

Intersections

Design capacities of intersections are affected by a very large number of variables. To the extent that these variables can be predicted for the design year, design capacities can be estimated by procedures for signalized and unsignalized intersections given in the HCM. The design and spacing of signalized intersections should also be coordinated with traffic signal design and phasing.

Pedestrians and Bicycles

The level of service for pedestrian and bicycle facilities can be evaluated using procedures presented in the HCM.

ACCESS CONTROL AND ACCESS MANAGEMENT

General Conditions

Regulating access is called "access control." It is achieved through the regulation of public access rights to and from properties abutting the highway facilities. These regulations generally are categorized as full control of access, partial control of access, access management, and driveway/entrance regulations. The principal advantages of controlling access are the preservation or improvement of service and safety.

The functional advantage of providing access control on a street or highway is the management of the interference with through traffic. This interference is created by vehicles or pedestrians entering, leaving, and crossing the highway. Where access to a highway is managed, entrances and exits are located at points best suited to fit traffic and land-use needs and are designed to enable vehicles to enter and leave safely with minimum interference from through traffic. Vehicles are prevented from entering or leaving elsewhere so that, regardless of the type and intensity of development of the roadside areas, a high quality of service is preserved and crash potential is lessened. Conversely, on streets or highways where there is no access management and roadside businesses are allowed to develop haphazardly, interference from the roadside can become a major factor in reducing the capacity, increasing the crash potential, and eroding the mobility function of the facility.

Access control techniques can be implemented with two basic legal powers: police power and eminent domain. This first power allows a state to restrict individual actions for the public welfare. Police power provides sufficient authority for most access control techniques associated with highway operations, driveway location, driveway design, and access denials. The second power allows a state to take property for public use provided an owner is compensated for his loss. A State may need to use eminent domain when building local service roads, buying abutting property, acquiring additional right-of-way, and taking access rights. However, an agency usually has the power to deny direct access through the use of police power when reasonable alternative access is available.

Generally, States have adequate power to manage access to a highway as long as reasonable access is provided to abutting property. However, providing reasonable access does not necessarily mean providing direct access to the state highway system. Coordinating access policies into a clear and definitive regulation facilitates the use of police power. Because authority and interpretations vary from state to state, each State should evaluate its particular legal powers for controlling access. Certain techniques may not be legally feasible in a state that has neither the policy nor precedent to uphold them.

Full control of access means that preference is given to through traffic by providing access connections by means of ramps with only selected public roads and by prohibiting crossings at grade and direct private driveway connections.

With partial control of access, preference is given to through traffic to a degree. Access connections, which may be at-grade or grade-separated, are provided with selected public roads, and private driveways. Generally, full or partial access control is accomplished by legally obtaining the access rights from the abutting property owners (usually at the time of purchase of the right-of-way) or by the use of frontage roads.

Access management involves providing (or managing) access to land development while simultaneously preserving the flow of traffic on the surrounding road system in terms of safety, capacity, and speed (**17**). Access management applies to all types of roads and streets. It calls for setting access policies for various types of roadway, keying designs to these policies, having the access policies incorporated into legislation, and having the legislation upheld in the courts.

Access management views the highway and its surrounding activities as part of a single system. Individual parts of the system include the activity center and its circulation systems, access to and from the center, the availability of public transportation, and the roads serving the center. All parts are important and interact with each other. The goal is to coordinate the planning and design of each activity center to preserve the capacity of the overall system and to allow efficient access to and from the activities.

Access management extends traffic engineering principles to the location, design, and operation of access roads that serve activities along streets and highways. It also includes evaluating the suitability of a site for different types of development from an access standpoint and is, in a sense, a new element of roadway design.

Driveway/entrance regulations may be applied even though no control of access is obtained. Each abutting property is permitted access to the street or highway; however, the location, number, and geometric design of the access points are governed by the regulations.

Access management addresses the basic questions of when, where, and how access should be provided or denied, and what legal or institutional changes are needed to enforce these decisions. In a broad context, access management is resource management, since it is a way to anticipate and prevent congestion and to improve traffic flow.

Key elements of access management include: defining the allowable access and access spacings for various classes of highways, providing a mechanism for granting variances when reasonable access cannot otherwise be provided, and establishing means of enforcing policies and decisions. These key elements, along with appropriate design policies, should be implemented through a legal code that provides a systematic and supportable basis for making access decisions. The code should provide a common basis for decisions for both the public and private sectors.

Basic Principles of Access Management

The following principles define access management techniques:

- *Classify the road system by the primary function of each roadway.* Freeways emphasize movement and provide complete control of access. Local streets emphasize property access rather than traffic movement. Arterial and collector roads must serve a combination of both property access and traffic movement.
- *Limit direct access to roads with higher functional classifications.* Direct property access should be denied or limited along higher class roadways, whenever *reasonable* access can be provided to a lower class roadway.
- *Locate traffic signals to emphasize through traffic movements.* Signalized access points should fit into the overall signal coordination plan for traffic progression.
- *Locate driveways and major entrances to minimize interference with traffic operations.* Driveways and entrances should be located away from other intersections to minimize crashes, to reduce traffic interference, and to provide for adequate storage lengths for vehicles turning into entrances.
- *Use curbed medians and locate median openings to manage access movements and minimize conflicts.*

The extent of access management depends upon the location, type and density of development, and the nature of the highway system. Access management actions involve both the planning and design of new roads and the retrofitting of existing roads and driveways.

Access Classifications

Access classification is the foundation of a comprehensive access management program. It defines when, where, and how access can be provided between public highways and private driveways or entrances. Access classification relates the allowable access to each type of highway in conjunction with its purpose, importance, and functional characteristics.

The functional classification system provides the starting point in assigning highways to different access categories. Modifying factors include existing land development, driveway density, and geometric design features, such as the presence or absence of a raised-curb median.

An access classification system defines the type and spacing of allowable access for each class of road. Direct access may be denied, limited to right turns in and out, or allowed for all or most movements depending upon the specific class and type of road. Spacing of signals in terms of distance between signals or through band width (progression speed) is also specified. Examples of access classification schemes are presented in NCHRP Report 348, *Access Management Guidelines for Activity Centers* (**17**).

Methods of Controlling Access

Public agencies can manage and control access by means of statutes, land-use ordinances, geometric design policies, and driveway regulations.

- *Control by the transportation agency:* Every State and local transportation agency has the basic statutory authority to control all aspects of highway design to protect public safety, health, and welfare. The extent to which an agency can apply specific policies for driveways/entrances, traffic signal locations, land use controls, and denial of direct access is specifically addressed by legislation and, to some degree, by the State courts.
- *Land-use ordinances:* Land-use control is normally administered by local governments. Local zoning ordinances and subdivision requirements can specify site design, setback distances, type of access, parking restrictions, and other elements that influence the type, volume, and location of generated traffic.
- *Geometric design:* Geometric design features, such as the use of raised-curb medians, the spacing of median openings, use of frontage roads, closure of median openings, and raised-curb channelization at intersections, all assist in controlling access.
- *Driveway regulations:* Agencies may develop detailed access and driveway/entrance policies by guidelines, regulations, or ordinances, provided specific statutory authority exists. Guidelines usually need no specific authority, but are weak legally. Cities can pass ordinances implementing access management policies. Likewise, state agencies may develop regulations when authorized by legislation. Regulations can deny direct access to a road if reasonable, alternative access is provided, but they cannot "take away" access rights.

Benefits of Controlling Access

Highways with full access control consistently experience only 25 to 50% of the crash rates observed on roadways without access control. These rates are defined in terms of crashes per million vehicle kilometers [miles] of travel. Freeways limit the number and variety of events to which drivers must respond and thus lower crash rates result.

The safety and operating benefits of controlling access to a highway have long been recognized and well documented. As access density increases, there is a corresponding increase in crashes and travel times.

A study on congestion by the Texas Transportation Institute has reported a 5- to 8-km/h reduction in speed for every added signal per kilometer [2- to 3-mph speed reduction for every added signal per mile] (**19**). A research study on the impact of access management found that through vehicles in the curb or right lane approximated 20% of the right turns desiring to enter a development (**18**).

As the number of driveways along a highway increases, the crash rate also increases. The effect of driveway and business frequency on crash rates is shown in Exhibit 2-35 through 2-37. As the number of business and access points increases along a roadway, there is a corresponding increase in crash rates. This contrasts sharply with freeway crash rates that remain the same or even decrease slightly over time.

The generalized effects of access spacing on traffic crashes were derived from a literature synthesis and an analysis of 37,500 crashes (**18**). This study's analysis shows the relative increase in crash rates that can be expected as the total driveway density increases. Increasing the access frequency from 10 to 30 access points per kilometer [20 to 50 access points per mile] will result in almost a doubling of the crash rate. Each additional access point per kilometer increases the crash rate about 5 percent; thus, each additional access point per mile increases the crash rate about 3 percent.

Exhibits 2-35 and 2-36 show crash rates by access frequency and type of median for urban/suburban and rural roads, respectively. Crash rates rise for each type of median treatment with an increase in access frequency. Non-traversable medians generally have a lower crash rate than two-way left-turn lanes and undivided roadway sections for all access densities. However, as discussed in Chapter 7, provision of non-traversable medians will eliminate left-turn movements at some intersections and driveways, but may increase U-turn volumes at other locations on the same road or may divert some traffic to other roads. The safety consequences of increased U-turn volumes or diverted traffic may not be reflected in Exhibits 2-35 and 2-36.

For urban/suburban roads, representative crash rates for combinations of signalized and unsignalized access density are shown in Exhibit 2-37. This figure indicates that crash rates rise with increases in either unsignalized or signalized access density.

In summary, some degree of access control or access management should be included in the development of any street or highway, particularly on a new facility where the likelihood of commercial development exists. The type of street or highway to be built should be coordinated with the local land-use plan to ensure that the desired type of access can be maintained through local zoning ordinances or subdivision regulations. The control of access may range from minimal driveway regulations to full control of access. Thus, the extent and degree of access management that is practical is a significant factor in defining the type of street or highway.

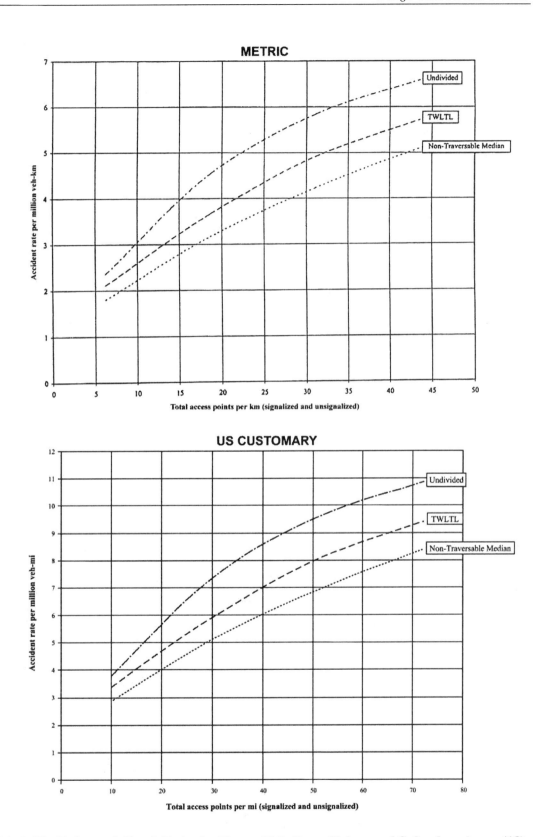

Exhibit 2-35. Estimated Crash Rates by Type of Median—Urban and Suburban Areas (18)

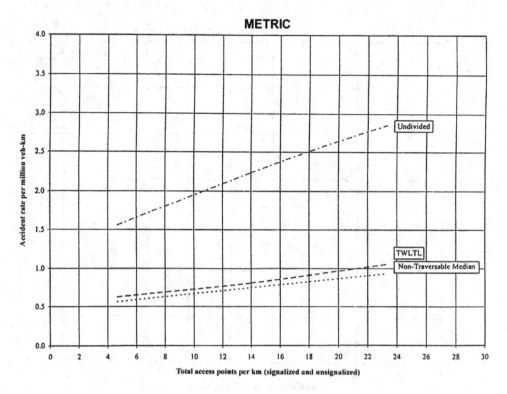

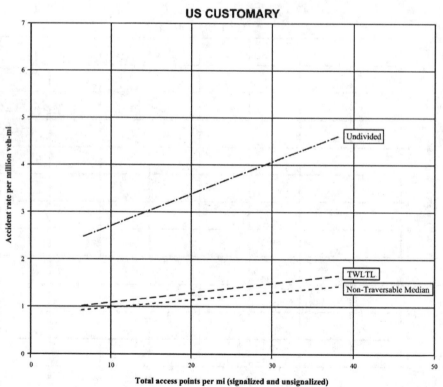

Exhibit 2-36. Estimated Crash Rates by Type of Median—Rural Areas (18)

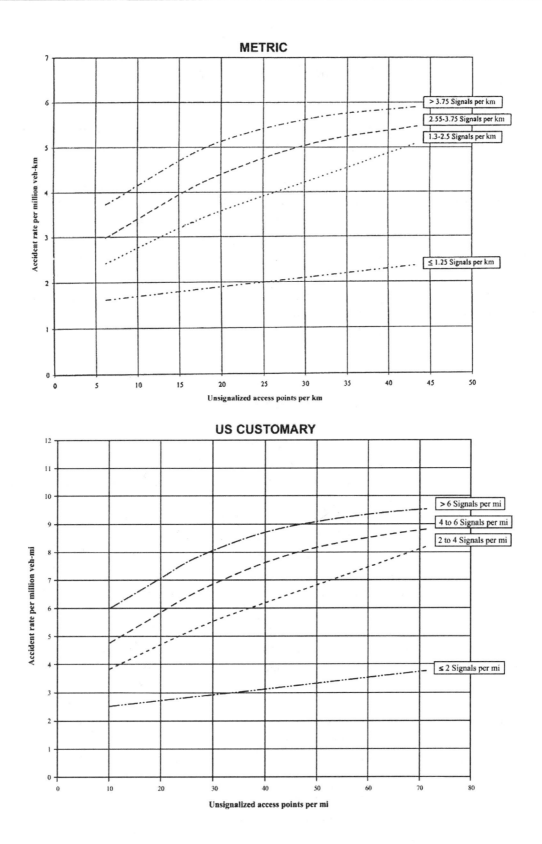

Exhibit 2-37. Estimated Crash Rates by Unsignalized and Signalized Access Density—Urban and Suburban Areas (18)

THE PEDESTRIAN

General Considerations

Interactions of pedestrians with traffic are a major consideration in highway planning and design. Pedestrians are a part of every roadway environment, and attention should be paid to their presence in rural as well as urban areas. The urban pedestrian, being far more prevalent, more often influences roadway design features than the rural pedestrian does. Because of the demands of vehicular traffic in congested urban areas, it is often very difficult to make adequate provisions for pedestrians. Yet provisions should be made, because pedestrians are the lifeblood of our urban areas, especially in the downtown and other retail areas. In general, the most successful shopping sections are those that provide the most comfort and pleasure for pedestrians. Pedestrian facilities include sidewalks, crosswalks, traffic control features, and curb cuts (depressed curbs and ramped sidewalks) and ramps for the older walkers and persons with mobility impairments. Pedestrian facilities also include bus stops or other loading areas, sidewalks on grade separations, and the stairs, escalators, or elevators related to these facilities. The *Americans with Disabilities Act Accessibility Guidelines* (ADAAG) (**23**) must be considered when designing roadways where pedestrian traffic is expected.

General Characteristics

To effectively plan and design pedestrian facilities, it is necessary to understand the typical pedestrian. The pedestrian most likely will not walk over 1.5 km [1 mi] to work or over 1.0 km [0.5 mi] to catch a bus, and about 80 percent of the distances traveled by the pedestrian will be less than 1.0 km [0.5 mi] (**24**). The typical pedestrian is a shopper about 50 percent of the time that he or she is a pedestrian and a commuter only about 11 percent of the time. As a consequence, pedestrian volumes peak at about noon rather than at the peak commuter times. Pedestrian volumes are influenced by such transient conditions as weather or, in specific locations, advertised sales. Hourly fluctuations in pedestrian volumes on a city street can be found in the AASHTO *Guide for the Planning, Design, and Operation of Pedestrian Facilities* (**25**).

Pedestrian actions are less predictable than those of motorists. Many pedestrians consider themselves outside the law in traffic matters, and in many cases, pedestrian regulations are not fully enforced. This makes it difficult to design a facility for safe and orderly pedestrian movements.

Pedestrians tend to walk in a path representing the shortest distance between two points. Therefore, crossings in addition to those at corners and signalized intersections may be appropriate at particular locations.

Pedestrians also have a basic resistance to changes in grade or elevation when crossing roadways and tend to avoid using special underpass or overpass pedestrian facilities. Also, pedestrian underpasses may be potential crime areas, lessening their usage.

A pedestrian's age is an important factor that may explain behavior that leads to collisions between motor vehicles and pedestrians. Very young pedestrians are often careless in traffic from either ignorance or exuberance, whereas older pedestrians may be affected by limitations in sensory, perceptual, cognitive, or motor skills. Pedestrian collisions can also be related to the lack of sidewalks, which may force pedestrians to share the traveled way with motorists. Therefore, sidewalk construction should be considered as part of any urban/suburban street improvement.

The following have been suggested as measures with the potential to aid older pedestrians and road users:

- Use simple designs that minimize crossing widths and minimize the use of more complex elements such as channelization and separate turning lanes. Where these features are appropriate, assess alternative designs that will assist older pedestrians, such as 3.3-m [11-ft] lane widths.
- Assume lower walking speeds.
- Provide median refuge islands of sufficient width at wide intersections.
- Provide lighting and eliminate glare sources at locations that demand multiple information gathering and processing.
- Consider the traffic control system in the context of the geometric design to assure compatibility and to provide adequate advance warning or guide signs for situations that could surprise or adversely affect the safety of older drivers or pedestrians.
- Use enhanced traffic control devices.
- Provide oversized, retroreflective signs with suitable legibility.
- Consider increasing sign letter size and retroreflectivity to accommodate individuals with decreased visual acuity.
- Use properly located signals with large signal indications.
- Provide enhanced markings and delineation.
- Use repetition and redundancy in design and in signing.

For further information on older pedestrians and drivers, refer to the FHWA publication entitled *Older Driver Highway Design Handbook: (Recommendations and Guidelines)* (**6**).

Walking Speeds

Because pedestrians have a broad range of walking speeds, the speeds at which they may cross a street is significant in design. Average pedestrian walking speeds range from approximately 0.8 to 1.8 m/s [2.5 to 6.0 ft/sec]. The *Manual on Uniform Traffic Control Devices (MUTCD)* (**7**) uses a normal walking speed of 1.2 m/s [4.0 ft/s]. Older people will generally walk at speeds in the lower end of this range.

Walking speeds are faster at midblock locations than at intersections, are faster for men than for women, and are affected by steep grades. Air temperature, time of day, trip purpose, and ice and snow all affect pedestrian walking speeds. Age is the most common cause of slower walking speeds, and in areas where there are many older people, a speed of 0.9 m/s [2.8 ft/s] should be considered for use in design.

Walkway Capacities

Walking speeds decrease as the pedestrian density of the walkway increases. As with roadway capacities, there is an optimum speed and density under which the walkway will carry the largest volume. The effective width used for walkway calculations should be reduced where parking meters, hydrants, newsstands, litter barrels, utility poles, or similar obstructions preclude the use of the full walkway. Walkway capacity calculations for sidewalks, stairs, and the effect of traffic signals involve differing procedures as discussed below. For a more detailed analysis of walkway design and capacities, see the AASHTO *Guide for the Planning, Design, and Operation of Pedestrian Facilities* (**25**) and the current edition of the *Highway Capacity Manual* (**15**).

Sidewalks

Levels of service have been developed to quantify the relative mobility of the pedestrian and his or her conflicts with other pedestrians that influence walking speed, maneuvering room, and the feeling of comfort (**26**). As in the level-of-service concept for motor vehicle traffic discussed earlier in this chapter and in the HCM (**15**), levels of service (A to F) reflect increasing crowding and decreasing freedom of movement. These levels of service are based on the available area per person and are defined as follows (**27**):

Level-of-service A allows each person to choose a desired walking speed and to avoid conflicts with other pedestrians.

At *Level-of-service B*, pedestrians begin to be aware of other pedestrians.

Level-of-service C requires minor adjustments to speed and direction by pedestrians to avoid conflicts.

At *Level-of-service D*, freedom to select individual walking speed and bypass other pedestrians is restricted. Frequent changes in speed and position are required.

Level-of-service E provides for very crowded walking, at times reduced to shuffling, making reverse or cross-traffic flow very difficult. The speed of virtually all pedestrians is reduced.

At *Level-of-service F,* a person is likely to be standing stationary in a waiting area or is able to walk only by shuffling. There is frequent, unavoidable contact with other pedestrians.

Computations of walkway capacity should use walkway widths that are reduced about 500 mm [18 in] if there are adjacent walls, with an additional 500 mm [18 in] if window shoppers are expected. Street hardware such as parking meters and poles also reduces the available walkway width.

Intersections

When pedestrians encounter an intersection, there is a major interruption in pedestrian flow. The sidewalk should provide sufficient storage area for those waiting to cross as well as an area for pedestrian cross traffic to pass.

Once pedestrians are given the walk indication, the crosswalk width and length become important. Crosswalks should be wide enough to accommodate the pedestrian flow in both directions within the duration of the pedestrian signal phase. The wider the street, the longer it takes a pedestrian to cross and proportionately less green signal time will be available for the primary street movements. Additionally, the longer the pedestrian crossing time, the longer the exposure to potential pedestrian/vehicular conflicts.

If the intersection is not signal controlled or if stop signs do not control the through motor vehicular traffic, pedestrians must wait for suitable gaps in the traffic to cross. The wider the street, the longer the gaps must be to afford safe pedestrian crossing times. Under urban conditions, pedestrian crossing times may be reduced by using narrower lanes or by providing raised-curb medians. However, traffic safety and reasonable roadway and intersection capacity requirements should still be met when considering reduced crossing times.

Reducing Pedestrian-Vehicular Conflicts

When designing urban highways with substantial pedestrian-vehicular conflicts, the following are some measures that could be considered to help reduce these conflicts and may increase the efficient operation of the roadway: (1) eliminate left and/or right turns, (2) prohibit free-flow right-turn movements, (3) prohibit right turn on red, (4) convert from two-way to one-way street operation, (5) provide separate signal phases for pedestrians, (6) eliminate selected crosswalks, and (7) provide for pedestrian grade separations. These and other pedestrian considerations are detailed in subsequent chapters and in the AASHTO *Guide for the Planning, Design, and Operation of Pedestrian Facilities* (**25**).

Characteristics of Persons With Disabilities

Consideration of persons with disabilities in highway design can greatly enhance the mobility of this sector of our society. To adequately provide for persons with disabilities, the designer must be aware of the range of disabilities to expect so that the design can appropriately accommodate them. The designer is cautioned to adequately review all local and national guidelines to assure proper compliance with applicable rules and regulations (**26**). For further details, see the section on "Sidewalk Curb Ramps" in Chapter 4, as well as the AASHTO *Guide for the Planning, Design, and Operation of Pedestrian Facilities* (**25**) and the ADAAG (**23**).

Mobility Impairments

Ambulatory difficulties include persons who walk without assistive devices, but with difficulty, to persons who require aid from braces, canes, or crutches, to persons who use wheelchairs. Stairs, curbs, and raised channelizing islands are the major roadway obstructions to these pedestrians. Design modifications should provide ramps rather than stairs or curbs. The front wheels of a wheelchair are very sensitive to obstacles; any bump may impair the progress of a wheelchair and may increase the possibility that a user will be propelled out of the wheelchair.

Visual Impairments

Pedestrians with visual impairments need special consideration. Intersections are the most complicated transportation element for visually impaired people. Complicated crossings such as those at channelized intersections can be improved by installing guide strips. Sidewalk curb cuts for wheelchairs make it difficult for visually impaired pedestrians to locate the curb line. Adding a 600-mm [2-ft] detectable warning strip at the bottom of the sidewalk ramp that meets the design specifications of the ADAAG (23) will benefit people with visual impairments. Because the visually impaired often rely on the sound of traffic when crossing intersections, caution should be used when considering exclusive turn phases or other unusual traffic movements.

Developmental Impairments

Many people with developmental impairments are unable to drive and, therefore, often travel as pedestrians. To help ensure correct responses from these pedestrians, including young children, pedestrian signals or other pedestrian-related facilities should be simple, straightforward, and consistent in their meaning.

BICYCLE FACILITIES

The bicycle has become an important element for consideration in the highway design process. Fortunately, the existing street and highway system provides most of the mileage needed for bicycle travel. While many highway agencies allow bicycles on partially access controlled facilities, most highway agencies do not allow bicycles on fully access controlled facilities unless no other alternative route is available.

Improvements such as the following, which generally are of low to moderate cost, can considerably enhance the safety of a street or highway and provide for bicycle traffic:

- paved shoulders
- wider outside traffic lanes (4.2 m [14 ft] minimum), if no shoulders exist
- bicycle-safe drainage grates
- adjusting manhole covers to the grade
- maintaining a smooth, clean riding surface

At certain locations or in certain corridors, it is appropriate to further supplement the existing highway system by providing specifically designated bikeways (for either exclusive or non-exclusive bicycle use). To provide adequately for bicycle traffic, the designer should be familiar with bicycle dimensions, operating characteristics, and needs. These factors determine acceptable turning radii, grades, and sight distance. In many instances, design features of separate bike facilities are controlled by the adjoining roadway and by the design of the highway itself. For further guidance, refer to the latest edition of the AASHTO *Guide for Development of Bicycle Facilities* (**28**) and other current research (**29**).

SAFETY

Attention to highway safety has been emphasized by the Congress of the United States as well as other national committees concerned with safety. In July 1973, after hearings on highway safety, design, and operations were conducted by subcommittees of the House Committee on Public Works, the following mandate was published by the Committee:

> Whose responsibility is it to see that maximum safety is incorporated into our motor vehicle transportation system? On this, the subcommittee is adamant. It is the responsibility of Government and specifically those agencies that, by law, have been given that mandate. This responsibility begins with the Congress and flows through the Department of Transportation, its Federal Highway Administration, the State Highway Departments and safety agencies, and the street and highway units of counties, townships, cities, and towns. There is no retreating from this mandate, either in letter or in spirit (**30**).

This emphasis by Congress on safety has also been evidenced by passage of the Highway Safety Act of 1966, and from the Federal Highway Administration (FHWA) by adoption of the AASHTO publications, *Highway Design and Operational Practices Related to Highway Safety* (**30**) and *Highway Safety Design and Operations Guide* (**31**). Other safety resources include the report entitled *Enhancing Highway Safety in an Age of Limited Resources* (**32**), which resulted from the TRB-conducted symposium sponsored by AASHTO and others in 1981.

Crashes seldom result from a single cause—usually several influences affect the situation at any given time. These influences can be separated into three groups: the human element, the vehicle element, and the highway element. Although this policy is primarily concerned with highway characteristics and design, the role of psychological factors is ever present. An error in perception or judgment or a faulty action on the driver's part can easily lead to a crash.

Highways should be designed to minimize driver decisions and to reduce unexpected situations. The number of crashes increases with an increase in the number of decisions required of the driver. Uniformity in highway design features and traffic control devices plays an important role in reducing the number of required decisions, and by this means, the driver becomes aware of what to expect on a certain type of highway.

The most significant design factor contributing to safety is the provision of full access control. Full access control reduces the number, frequency, and variety of events to which drivers must respond. The beneficial effect of this element has been documented in reports of a cooperative research study (**33**) of the FHWA and 39 state highway agencies. One of the principal findings of this study is that highways without access control invariably had higher crash rates than those with access control. This study showed that crash, injury, and fatality rates on Interstate highways are between 30 and 76 percent of comparable rates of conventional highways that existed before the Interstate highways were opened to traffic. No other single design element can claim comparable reductions.

Research has demonstrated a relationship between crashes and number of access points on a roadway (**18, 19, 34**). Relationships of this type have been illustrated in Exhibits 2-35 through 2-37.

The principle of full access control is invaluable as a means for preserving the capacity of arterial highways and of minimizing crash potential; however, this principle does not have universal application. Highways without control of access are essential as land service facilities, and the design features and operating characteristics of these highways need to be carefully planned so that they will reduce conflicts and minimize the interference between vehicles and still meet the needs of highway users.

Speed is often a contributing factor in crashes, but its role must be related to actual conditions at a crash site to be understood. It is improper to conclude that any given speed is safer than another for all combinations of the many kinds of drivers, vehicles, highways, and local conditions. For a highway with particularly adverse roadway conditions, a relatively low speed may result in fewer crashes than a high speed, but this does not necessarily mean that all potential crashes can be eliminated by low speeds. Likewise, vehicles traveling on good roads at relatively high speed may have lower crash involvement rates than vehicles traveling at lower speeds, but it does not necessarily follow that yet a higher speed would be even safer.

The safest speed for any highway depends on design features, road conditions, traffic volumes, weather conditions, roadside development, spacing of intersecting roads, cross-traffic volumes, and other factors. Crashes are not related as much to speed as to the range in speeds from the highest to the lowest. Regardless of the average speed on a main rural highway, the greater a driver's deviation from this average speed, either lower or higher, the greater the probability that the driver will be involved in crashes. Thus, design features that reduce the variance in speed of vehicles (such as flat grades, speed-change lanes, grade separations, and good signing and marking) contribute to highway safety. Normally, crashes involving vehicles traveling at high speed are more severe than those at low speed.

When designing a highway, consideration should be given to the type and characteristics of the drivers expected to use the highway. Trip purposes (such as recreation, commuting to work, and through travel) are factors affecting the design to some extent. Trip purposes are related to the mix of vehicle types likely to use the highway, ranging from all passenger vehicles to a high percentage of heavy commercial vehicles. Where trips of one type predominate, the facility should be designed to fit the specific needs of that type of trip.

A study on the effect of the Interstate highway system on crashes found a lower crash rate on four-lane divided highways than on four-lane undivided highways (**35**). This study was developed from data for highways within Interstate highway corridors during periods before and after opening new sections of Interstate highways to traffic.

A highway with a median width of 15 m [50 ft] or more has a very low incidence of head-on collisions caused by vehicles crossing the median. A median width of 23 to 30 m [75 to 100 ft] on freeways is very desirable as a means of reducing cross-median collisions. On a divided highway with partial access control (i.e., an expressway) or where no access control exists, the width of median should also take into account the operation of at-grade intersections.

With narrower medians, median barriers will eliminate head-on collisions, but at the cost of some increase in same-direction crashes because recovery space is decreased. Properly designed median barriers minimize vehicle damage and lessen the crash severity. However, if a narrow median with a median barrier is proposed on a high-speed highway, the design should include adequate shoulder widths in the median for emergency stops and emergency vehicle use.

Another study relating crashes to shoulder width, alignment, and grade found that crash rates on sections with curves or grades were much higher than on level tangent highway sections. This study also found that crash rates were highest on roads having combinations of sharp curves and steep grades (**36**). This study, which was limited to rural two-lane roads, lends strong support to the postulate that straight, level rural roads without intersections or significant numbers of private driveways are the safest highways within their general class. The few crashes that occur on straight, level, rural roads without intersections do not represent a stable source of information, however. There are wide variations between similar roadway sections and between different years for the same section. An apparent correlation between a geometric design element such as shoulder width, for example, and crash rates is almost certain to be clouded by random variations of the crash pattern.

Zegeer (**37**) developed relationships between the geometric design of horizontal curves and their safety performance. This research addresses the relationship to safety of both length and radius of horizontal curves.

Crashes are likely to occur where drivers are called upon to make decisions under circumstances where their vehicles are unable to respond properly, for example, where a truck is descending a grade. It would be logical to expect more crashes on grades and curves than on level tangent highways where driver decisions are needed less frequently and vehicles are fully responsive. However, design with tangent alignment can be overdone.

On extremely long tangents, drivers have a tendency to completely relax, especially after driving on a congested highway before entering a freeway. On some freeways, there has been concern over the number of crashes that occur when the driver apparently goes to sleep. It is considered highly desirable to provide gentle curvature and to avoid a fixed cross section for long tangent sections of roadway. This can be achieved by varying the median width, using independent roadway alignments, and taking advantage of the terrain, wherever practical. In addition, rumble strips can be added to shoulders to reduce run-off-the-road crashes caused by drivers falling asleep at the wheel.

As the design of alignment, grade, and traveled-way cross section has improved, roadside design has also become increasingly important. Crashes involving single vehicles running off the road constitute more than one-half of all fatal crashes on freeways.

When a vehicle leaves the roadway, the driver no longer has the ability to fully control the vehicle. Any object in or near the path of the vehicle becomes a potential contributing factor to crash severity. The concept of the safer or forgiving roadside should not be viewed as a by-product of the application of safety criteria to each element but as a planned segment of the total engineering for the highway. The AASHTO publication *Highway Safety Design and Operation Guide* (**31**) presents an overview of the AASHTO policies in this area; these policies are reflected throughout this book in the criteria for specific geometric design elements.

Basic to the concept of the forgiving roadside is the provision of a clear recovery area. Studies have indicated that on high-speed highways, a relatively level traversable width of approximately 9 m [30 ft] from the edge of the traveled way permits about 80 percent of the vehicles leaving the highway to safely stop or return to the roadway. Even though the 9-m [30-ft] width is not a "magic number" and the application of engineering judgment is necessary, the 9-m [30-ft] width has been used extensively as a guide for recovery zones.

In roadside design, two major elements should be controlled by the designer: roadside slopes and unyielding obstacles. NCHRP Report 247 (**38**) discusses the effectiveness of clear recovery areas. The AASHTO *Roadside Design Guide* (**39**) also discusses the effects that slope and other topographic features have on the effectiveness of recovery areas. On existing highways, AASHTO recommends the following priorities for treatment of roadside obstacles:

- Remove the obstacle or redesign it so it can be safely traversed.
- Relocate the obstacle to a point where it is less likely to be struck.
- Reduce severity of impacts with the obstacle by using an appropriate breakaway device.
- Redirect a vehicle by shielding the obstacle with a longitudinal traffic barrier and/or crash cushion.
- Delineate the obstacle if the above alternatives are not appropriate.

The design of guardrails and barrier systems has become a subject of considerable research. AASHTO *Roadside Design Guide* (**39**) and NCHRP Report 350 (**40**) are some of many published reports that deal with this subject. These publications note that the treatment of end sections on guardrail or a barrier is of particular concern.

Highway designers should recognize the dynamic developments *currently* under way in the entire area of roadside design. Although this publication has attempted to present the most current information available on roadside design, ongoing research and implementation projects will undoubtedly offer newer and better results in the future. Highway designers should endeavor to use the most current acceptable information in their designs.

Communication with the motorist is probably one of the most complex problems for the designer. One of the best available tools concerning motorist communication is the MUTCD (7), which presents national criteria for uniform application of signing, signalization, painted channelization, and pavement markings for all highways in the United States. A primary message of the MUTCD is the importance of uniformity.

Highway users are dependent on traffic control devices (signs, markings, and signals) for information, warning, and guidance. So great is the dependence of highway users on such information that uniform, high-quality traffic control devices are necessary for safe, efficient use and public acceptance of any highway regardless of its excellence in width, alignment, and structural design.

All traffic control devices should have the following characteristics: (1) fulfill an important need, (2) command attention, (3) convey a clear, simple meaning, (4) command respect of road users, and (5) provide adequate response time. In addition, devices that control or regulate traffic must be sanctioned by law.

Four basic attributes of traffic control devices are essential to ensure that these devices are effective: design, placement, maintenance, and uniformity. Consideration should be given to these attributes during the design of a highway to ensure that the required number of devices can be kept to a minimum and that those that are needed can be properly placed.

The operation of a motor vehicle takes considerable concentration, particularly in congested areas. A driver should be able to operate his or her vehicle with minimum distractions. Advertising or other roadside signs should not be placed where they would interfere with or confuse the meaning of standard traffic control devices. Advertising signs with bright colors or flashing lights are especially objectionable in this respect. Lights shining toward a driver can be blinding, partially or fully, for varying periods of time, depending on individual eye capability. Bright lights, in effect, can form a curtain hiding what is ahead and can thus put motorists and pedestrians at risk.

A large proportion of crashes on rural highways occur at intersections. Several studies have been made at intersections with varying conditions, and the results vary according to conditions studied. Factors to be considered in designing an intersection are total traffic volume, amount of cross traffic, turning movements, type of highway, type of traffic control needed, design of the crossroad sight distance, and the utilization of islands and channelization.

Various studies indicate improvements in safety at intersections can be accomplished by channelizing intersections, providing appropriate sight distances (including stopping, decision, and intersection sight distance), and providing safety refuge islands and sidewalks for pedestrians, lighting, signing, and traffic control devices. These concepts have been incorporated in the geometric design guidelines presented in this policy.

A viable safety evaluation and improvement program is a vital part of the overall highway improvement program. The identification of potential safety problems, the evaluation of the effectiveness of alternative solutions, and the programming of available funds for the most effective improvements are of primary importance. The safety of the traveling public should be reflected throughout the highway program: in spot safety projects, in rehabilitation projects, in the construction of new highways, and elsewhere. *Highway Design and Operational Practices Related to Highway Safety* (**30**) and *Highway Safety Design and Operations Guide* (**31**) provide a number of important recommendations on safety as part of a total highway program.

ENVIRONMENT

A highway necessarily has wide-ranging effects in addition to providing traffic service to users. It is essential that the highway be considered as an element of the total environment. The term *"environment,"* as used here refers to the totality of humankind's surroundings: social, physical, natural, and synthetic. It includes the human, animal, and plant communities and the forces that act on all three. The highway can and should be located and designed to complement its environment and serve as a catalyst to environmental improvement.

The area surrounding a proposed highway is an interrelated system of natural, synthetic, and sociologic variables. Changes in one variable within this system cannot be made without some effect on other variables. The consequences of some of these effects may be negligible, but others may have a strong and lasting impact on the environment, including the sustenance and quality of human life. Because highway location and design decisions have an effect on the development of adjacent areas, it is important that environmental variables be given full consideration. Also, care should be exercised to ensure that applicable local, state, and federal environmental requirements are met.

ECONOMIC ANALYSIS

Highway economics is concerned with the cost of a proposed improvement and the benefits resulting from it. The AASHTO *Manual on User Benefit Analysis of Highway and Bus-Transit Improvements* (**41**) may be used to perform economic analyses of proposed highway improvements.

REFERENCES

1. Fong, K. T., and D.C. Chenu. "Simulation of Truck Turns With a Computer Model," *Transportation Research Record 1100*, Transportation Research Board, 1985: 20-29.
2. Fambro, D. B., K. Fitzpatrick, and R. J. Koppa. *Determination of Stopping Sight Distances*, NCHRP Report 400, Washington, D.C.: Transportation Research Board, 1997.
3. Olson, P. L., D. E. Cleveland, P. S. Fancher, L. P. Kostyniuk, and L. W. Schneider. *Parameters Affecting Stopping Sight Distance*, NCHRP Report 270, Washington, D.C.: Transportation Research Board, 1984.
4. Alexander, G. H., and H. Lunenfeld. *A User's Guide to Positive Guidance* (3rd Edition), Report No. FHWA/SA-90/017, Washington, D.C.: U.S. Department of Transportation, Federal Highway Administration, 1990.

5. "Human Factors and Safety Research Related to Highway Design and Operations," *Transportation Research Record 1281*, Transportation Research Board, 1990.

6. Staplin, L., K. Lococo, and S. Byington. *Older Driver Highway Design Handbook*, Report No. FHWA-RD-97-135, McLean, Virginia: U.S. Department of Transportation, Federal Highway Administration, December 1998.

7. U.S. Department of Transportation, Federal Highway Administration. *Manual on Uniform Traffic Control Devices for Streets and Highways*, Washington, D.C.: 1988 or most current edition.

8. Johannson, C., and K. Rumar. "Driver's Brake Reaction Time," *Human Factors*, Vol. 13, No. 1, 1971: 22-27.

9. Fell, J. C. *A Motor Vehicle Accident Causal System. The Human Element*, Report No. DOT-HS-801-214, Washington, D.C.: U.S. Department of Transportation, National Highway Traffic Safety Administration, July 1974.

10. Schmidt, I., and P. D. Connolly. "Visual Considerations of Man, the Vehicle and the Highways," *Paper No. SP-279-SAE*, New York: Society of Automotive Engineers, 1966.

11. Tilley, D. H., C. W. Erwin, and D. T. Gianturco. "Drowsiness and Driving; Preliminary Report of a Population Survey," *Paper No. 730121-SAE*, New York: Society of Automotive Engineers, 1973.

12. Alexander, G. J., and H. Lunenfeld. *Driver Expectancy in Highway Design and Operations*, Report No. FHWA-TO-86-1, Washington, D.C.: U.S. Department of Transportation, Federal Highway Administration, May 1986.

13. Adler, B., and H. Lunenfeld. *Three Beam Headlight Evaluation*, Report No. DOT/HS-800-844, Washington, D.C.: U.S. Department of Transportation, National Highway Traffic Safety Administration, April 1973.

14. Institute of Traffic Engineers. *Freeway Operations*, Washington, D.C.: Institute of Traffic Engineers, 1961.

15. Transportation Research Board. *Highway Capacity Manual,* HCM2000, Washington, D.C.: Transportation Research Board, 2000 or most current edition.

16. Bonneson, J. A., and P. T. McCoy. *Capacity and Operational Effects of Midblock Left-Turn Lanes*, NCHRP Report 395, Washington, D.C.: Transportation Research Board, 1997.

17. Koepke, F. J. and H. S. Levinson. *Access Management Guidelines for Activity Centers*, NCHRP Report 348, Washington, D.C.: Transportation Research Board, 1992.

18. Gluck, J., H. S. Levinson, and V. Stover. *Impacts of Access Management Techniques*, NCHRP Report 420, Washington, D.C.: Transportation Research Board, 1999.

19. Lomax T., S. Turner, H. S. Levinson, R. Pratt, P. Bay, and T. Douglas. *Quantifying Congestion*, NCHRP Report 398, Washington, D.C.: Transportation Research Board, March 1997.

20. State of Colorado Access Control Demonstration Project. Colorado Department of Highways, June 1985.

21. Cribbins, P. D., J. W. Horn, F. V. Beeson, and R. D. Taylor. "Median Openings on Divided Highways: Their Effect on Accident Rates and Level of Service," *Highway Research Record 188*, Highway Research Board, 1967.

22. Glennon, J. C., J. J. Valenta, B. A. Thorson, and J. A. Azzeh. Technical Guidelines for the Control of Direct Access to Arterial Highways, Volumes 1 and 2, Report Nos. FHWA-RD-76-87 and -88, McLean, Virginia: U.S. Department of Transportation, Federal Highway Administration, August 1975.

23. Architectural and Transportation Barriers Compliance Board (Access Board). *Americans With Disabilities Act Accessibility Guidelines* (ADAAG), Washington, D.C.: July 1994 or most current edition.

24. Maring, G. E. "Pedestrian Travel Characteristics," *Highway Research Record 406*, Highway Research Board 1972: 14-20.

25. AASHTO. *Guide for the Planning, Design, and Operation of Pedestrian Facilities*, Washington, D.C.: AASHTO, forthcoming.

26. Older, S. J. "Movement of Pedestrians on Footways in Shopping Streets," *Traffic Engineering and Control*, August 1968: 160-163.

27. Fruin, J. J. "Designing for Pedestrians: A Level-of-Service Concept," *Highway Research Record 355*, Highway Research Board, 1971: 1-15.

28. AASHTO. *Guide for the Development of Bicycle Facilities*, Washington, D.C.: AASHTO, 1999.

29. Wilkinson, III, W. C., A. Clarke, B. Epperson, and R. L. Knoblauch. *Selecting Roadway Design Treatments to Accommodate Bicycles*, Report No. FHWA-RD-92-073, McLean, Virginia: U.S. Department of Transportation, Federal Highway Administration, January 1994.

30. AASHTO. *Highway Design and Operational Practices Related to Highway Safety*, Washington, D.C.: AASHTO, 1974.

31. AASHTO. *Highway Safety Design and Operations Guide*, Washington, D.C.: AASHTO, 1997.

32. AASHTO, et al. *Enhancing Highway Safety in an Age of Limited Resources*, A report resulting from a symposium conducted by the Transportation Research Board, unpublished, November 1981.

33. Fee, J. A., et al. *Interstate System Accident Research Study 1*, Washington, D.C.: U.S. Department of Transportation, Federal Highway Administration, October 1970.

34. Dart, Jr., O. K. and L. Mann, Jr. "Relationship of Rural Highway Geometry to Accident Rates in Louisiana," *Highway Research Record 312*, Highway Research Board, 1970.

35. Byington, S. R. "Interstate System Accident Research," *Public Roads*, Vol. 32, December 1963.

36. Billion, C. E., and W. R. Stohner. "A Detailed Study of Accidents as Related to Highway Shoulders in New York State," *Proceedings of HRB*, Vol. 36, Highway Research Board, 1957: 497-508.

37. Zegeer, C. V., J. R. Stewart, F. M. Council, D. W. Reinfurt, and E. Hamilton. "Safety Effects of Geometric Improvements on Horizontal Curves," *Transportation Research Board 1356*, Transportation Research Board, 1992.

38. Graham, J. L., and D. W. Harwood. *Effectiveness of Clear Recovery Zones*, NCHRP Report 247, Washington, D.C.: Transportation Research Board, 1982.

39. AASHTO. *Roadside Design Guide*, Washington, D.C.: AASHTO, 1996.

40. Ross, H. E., D. L. Sicking, R. A. Zimmer, and J. D. Michie. *Recommended Procedures for the Safety Performance Evaluation of Highway Features*, NCHRP Report 350, Washington, D.C.: Transportation Research Board, 1993.

41. AASHTO. *A Manual on User Benefit Analysis of Highway and Bus-Transit Improvements*, Washington, D.C.: AASHTO, 1977.

CHAPTER 3
ELEMENTS OF DESIGN

INTRODUCTION

The alignment of a highway or street produces a great impact on the environment, the fabric of the community, and the highway user. The alignment is comprised of a variety of elements joined together to create a facility that serves the traffic in a safe and efficient manner, consistent with the facility's intended function. Each alignment element should complement others to produce a consistent, safe, and efficient design.

The design of highways and streets within particular functional classes is treated separately in later chapters. Common to all classes of highways and streets are several principal elements of design. These include sight distance, superelevation, traveled way widening, grades, horizontal and vertical alignments, and other elements of geometric design. These alignment elements are discussed in this chapter, and, as appropriate, in the later chapters pertaining to specific highway functional classes.

SIGHT DISTANCE

General Considerations

A driver's ability to see ahead is of the utmost importance in the safe and efficient operation of a vehicle on a highway. For example, on a railroad, trains are confined to a fixed path, yet a block signal system and trained operators are needed for safe operation. On the other hand, the path and speed of motor vehicles on highways and streets are subject to the control of drivers whose ability, training, and experience are quite varied. For safety on highways, the designer should provide sight distance of sufficient length that drivers can control the operation of their vehicles to avoid striking an unexpected object in the traveled way. Certain two-lane highways should also have sufficient sight distance to enable drivers to occupy the opposing traffic lane for passing other vehicles without risk of a crash. Two-lane rural highways should generally provide such passing sight distance at frequent intervals and for substantial portions of their length. By contrast, it is normally of little practical value to provide passing sight distance on two-lane urban streets or arterials. The proportion of a highway's length with sufficient sight distance to pass another vehicle and interval between passing opportunities should be compatible with the design criteria established in the subsequent chapter pertaining to the functional classification of the specific highway or street.

Four aspects of sight distance are discussed below: (1) the sight distances needed for stopping, which are applicable on all highways; (2) the sight distances needed for the passing of overtaken vehicles, applicable only on two-lane highways; (3) the sight distances needed for decisions at complex locations; and (4) the criteria for measuring these sight distances for use in design. The design of alignment and profile to provide sight distances and that satisfy the applicable design criteria are described later in this chapter. The special conditions related to sight distances at intersections are discussed in Chapter 9.

Stopping Sight Distance

Sight distance is the length of the roadway ahead that is visible to the driver. The available sight distance on a roadway should be sufficiently long to enable a vehicle traveling at or near the design speed to stop before reaching a stationary object in its path. Although greater lengths of visible roadway are desirable, the sight distance at every point along a roadway should be at least that needed for a below-average driver or vehicle to stop.

Stopping sight distance is the sum of two distances: (1) the distance traversed by the vehicle from the instant the driver sights an object necessitating a stop to the instant the brakes are applied; and (2) the distance needed to stop the vehicle from the instant brake application begins. These are referred to as brake reaction distance and braking distance, respectively.

Brake Reaction Time

Brake reaction time is the interval from the instant that the driver recognizes the existence of an obstacle on the roadway ahead that necessitates braking to the instant that the driver actually applies the brakes. Under certain conditions, such as emergency situations denoted by flares or flashing lights, drivers accomplish these tasks almost instantly. Under most other conditions, the driver must not only see the object but must also recognize it as a stationary or slowly moving object against the background of the roadway and other objects, such as walls, fences, trees, poles, or bridges. Such determinations take time, and the amount of time needed varies considerably with the distance to the object, the visual acuity of the driver, the natural rapidity with which the driver reacts, the atmospheric visibility, the type and the condition of the roadway, and nature of the obstacle. Vehicle speed and roadway environment probably also influence reaction time. Normally, a driver traveling at or near the design speed is more alert than one traveling at a lesser speed. A driver on an urban street confronted by innumerable potential conflicts with parked vehicles, driveways, and cross streets is also likely to be more alert than the same driver on a limited-access facility where such conditions should be almost nonexistent.

The study of reaction times by Johansson and Rumar (**1**) referred to in Chapter 2 was based on data from 321 drivers who expected to apply their brakes. The median reaction-time value for these drivers was 0.66 s, with 10 percent using 1.5 s or longer. These findings correlate with those of earlier studies in which alerted drivers were also evaluated. Another study (**2**) found 0.64 s as the average reaction time, while 5 percent of the drivers needed over 1 s. In a third study (**3**), the values of brake reaction time ranged from 0.4 to 1.7 s. In the Johansson and Rumar study (**1**), when the event that required application of the brakes was unexpected, the drivers' response times were found to increase by approximately 1 s or more; some reaction times were greater than 1.5 s. This increase in reaction time substantiated earlier laboratory and road tests in which the conclusion was drawn that a driver who needed 0.2 to 0.3 s of reaction time under alerted conditions would need 1.5 s of reaction time under normal conditions.

Minimum brake reaction times for drivers could thus be at least 1.64 s and 0.64 s for alerted drivers as well as 1 s for the unexpected event. Because the studies discussed above used simple prearranged signals, they represent the least complex of roadway conditions. Even under these simple conditions, it was found that some drivers took over 3.5 s to respond. Because actual conditions on the highway are generally more complex than those of the studies, and because there is wide variation in driver reaction times, it is evident that the criterion adopted for use should be greater than 1.64 s. The brake reaction time used in design should be large enough to include the reaction times needed by nearly all drivers under most highway conditions. Both recent research (**4**) and the studies documented in the literature (**1, 2, 3**) show that a 2.5-s brake reaction time for stopping sight situations encompasses the capabilities of most drivers, including those of older drivers. The recommended design criterion of 2.5 s for brake reaction time exceeds the 90th percentile of reaction time for all drivers and has been used in the development of Exhibit 3-1.

A brake reaction time of 2.5 s is considered adequate for conditions that are more complex than the simple conditions used in laboratory and road tests, but it is not adequate for the most complex conditions encountered in actual driving. The need for greater reaction time in the most complex conditions encountered on the roadway, such as those found at multiphase at-grade intersections and at ramp terminals on through roadways, can be found later in this chapter in the section on "Decision Sight Distance."

$$d = Vt = 0.278\,Vt$$

Braking Distance

The approximate braking distance of a vehicle on a level roadway traveling at the design speed of the roadway may be determined from the following equation:

Metric	US Customary	
$d = 0.039\,\dfrac{V^2}{a}$	$d = 1.075\,\dfrac{V^2}{a}$	(**3-1**)
where: d = braking distance, m; V = design speed, km/h; a = deceleration rate, m/s^2	where: d = braking distance, ft; V = design speed, mph; a = deceleration rate, ft/s^2	

Studies documented in the literature (**4**) show that most drivers decelerate at a rate greater than 4.5 m/s^2 [14.8 ft/s^2] when confronted with the need to stop for an unexpected object in the roadway. Approximately 90 percent of all drivers decelerate at rates greater than 3.4 m/s^2 [11.2 ft/s^2]. Such decelerations are within the driver's capability to stay within his or her lane and maintain steering control during the braking maneuver on wet surfaces. Therefore, 3.4 m/s^2 [11.2 ft/s^2] (a comfortable deceleration for most drivers) is recommended as the deceleration

Metric					US Customary				
Design speed (km/h)	Brake reaction distance (m)	Braking distance on level (m)	Stopping sight distance Calculated (m)	Design (m)	Design speed (mph)	Brake reaction distance (ft)	Braking distance on level (ft)	Stopping sight distance Calculated (ft)	Design (ft)
20	13.9	4.6	18.5	20	15	55.1	21.6	76.7	80
30	20.9	10.3	31.2	35	20	73.5	38.4	111.9	115
40	27.8	18.4	46.2	50	25	91.9	60.0	151.9	155
50	34.8	28.7	63.5	65	30	110.3	86.4	196.7	200
60	41.7	41.3	83.0	85	35	128.6	117.6	246.2	250
70	48.7	56.2	104.9	105	40	147.0	153.6	300.6	305
80	55.6	73.4	129.0	130	45	165.4	194.4	359.8	360
90	62.6	92.9	155.5	160	50	183.8	240.0	423.8	425
100	69.5	114.7	184.2	185	55	202.1	290.3	492.4	495
110	76.5	138.8	215.3	220	60	220.5	345.5	566.0	570
120	83.4	165.2	248.6	250	65	238.9	405.5	644.4	645
130	90.4	193.8	284.2	285	70	257.3	470.3	727.6	730
					75	275.6	539.9	815.5	820
					80	294.0	614.3	908.3	910

Note: Brake reaction distance predicated on a time of 2.5 s; deceleration rate of 3.4 m/s^2 [11.2 ft/s^2] used to determine calculated sight distance.

Exhibit 3-1. Stopping Sight Distance

threshold for determining stopping sight distance. Implicit in the choice of this deceleration threshold is the assessment that most vehicle braking systems and the tire-pavement friction levels of most roadways are capable of providing a deceleration of at least 3.4 m/s² [11.2 ft/s²]. The friction available on most wet pavement surfaces and the capabilities of most vehicle braking systems can provide braking friction that exceeds this deceleration rate.

Design Values

The sum of the distance traversed during the brake reaction time and the distance to brake the vehicle to a stop is the stopping sight distance. The computed distances for various speeds at the assumed conditions are shown in Exhibit 3-1 and were developed from the following equation:

Metric	US Customary	
$$d = 0.278Vt + 0.039\frac{V^2}{a}$$	$$d = 1.47Vt + 1.075\frac{V^2}{a}$$	(3-2)
where:	where:	
t = brake reaction time, 2.5 s; V = design speed, km/h; a = deceleration rate, m/s²	t = brake reaction time, 2.5 s; V = design speed, mph; a = deceleration rate, ft/s²	

Stopping sight distances exceeding those shown in Exhibit 3-1 should be used as the basis for design wherever practical. Use of longer stopping sight distances increases the margin of safety for all drivers and, in particular, for those who operate at or near the design speed. To ensure that new pavements will have initially, and will retain, friction coefficients comparable to the deceleration rates used to develop Exhibit 3-1, pavement designs should meet the criteria established in the AASHTO *Guidelines for Skid Resistant Pavement Design* (**5**).

In computing and measuring stopping sight distances, the height of the driver's eye is estimated to be 1,080 mm [3.5 ft] and the height of the object to be seen by the driver is 600 mm [2.0 ft], equivalent to the taillight height of a passenger car. The application of these eye-height and object-height criteria is discussed further in the section on "Vertical Alignment" in this chapter.

Effect of Grade on Stopping

When a highway is on a grade, the equation for braking distance should be modified as follows:

Metric	US Customary	
$$d = \dfrac{V^2}{254\left(\left(\dfrac{a}{9.81}\right) \pm G\right)}$$	$$d = \dfrac{V^2}{30\left(\left(\dfrac{a}{32.2}\right) \pm G\right)}$$	(3-3)

In this equation, G is the percent of grade divided by 100, and the other terms are as previously stated. The stopping distances needed on upgrades are shorter than on level roadways; those on downgrades are longer. The stopping sight distances for various grades are shown in Exhibit 3-2. These adjusted sight distance values are computed for wet-pavement conditions using the same design speeds and brake reaction times used for level roadways in Exhibit 3-1.

On nearly all roads and streets, the grade is traversed by traffic in both directions of travel, but the sight distance at any point on the highway generally is different in each direction, particularly on straight roads in rolling terrain. As a general rule, the sight distance available on downgrades is larger than on upgrades, more or less automatically providing the appropriate corrections for grade. This may explain why designers do not adjust stopping sight distance because of grade. Exceptions are one-way roads or streets, as on divided highways with independent design profiles for the two roadways. For these separate roadways, adjustments for grade may be needed.

Variation for Trucks

The recommended stopping sight distances are based on passenger car operation and do not explicitly consider design for truck operation. Trucks as a whole, especially the larger and heavier units, need longer stopping distances from a given speed than passenger vehicles. However, there is one factor that tends to balance the additional braking lengths for trucks with those for passenger cars. The truck driver is able to see substantially farther beyond vertical sight obstructions because of the higher position of the seat in the vehicle. Separate stopping sight distances for trucks and passenger cars, therefore, are not generally used in highway design.

There is one situation in which every effort should be made to provide stopping sight distances greater than the design values in Exhibit 3-1. Where horizontal sight restrictions occur on downgrades, particularly at the ends of long downgrades where truck speeds closely approach or exceed those of passenger cars, the greater height of eye of the truck driver is of little value, even when the horizontal sight obstruction is a cut slope. Although the average truck driver tends to be more experienced than the average passenger car driver and quicker to recognize potential risks, it is desirable under such conditions to provide stopping sight distance that exceeds the values in Exhibits 3-1 or 3-2.

Metric							US Customary						
Design speed (km/h)	Stopping sight distance (m)						Design speed (mph)	Stopping sight distance (ft)					
	Downgrades			Upgrades				Downgrades			Upgrades		
	3%	6%	9%	3%	6%	9%		3%	6%	9%	3%	6%	9%
20	20	20	20	19	18	18	15	80	82	85	75	74	73
30	32	35	35	31	30	29	20	116	120	126	109	107	104
40	50	50	53	45	44	43	25	158	165	173	147	143	140
50	66	70	74	61	59	58	30	205	215	227	200	184	179
60	87	92	97	80	77	75	35	257	271	287	237	229	222
70	110	116	124	100	97	93	40	315	333	354	289	278	269
80	136	144	154	123	118	114	45	378	400	427	344	331	320
90	164	174	187	148	141	136	50	446	474	507	405	388	375
100	194	207	223	174	167	160	55	520	553	593	469	450	433
110	227	243	262	203	194	186	60	598	638	686	538	515	495
120	263	281	304	234	223	214	65	682	728	785	612	584	561
130	302	323	350	267	254	243	70	771	825	891	690	658	631
							75	866	927	1003	772	736	704
							80	965	1035	1121	859	817	782

Exhibit 3-2. Stopping Sight Distance on Grades

Decision Sight Distance

Stopping sight distances are usually sufficient to allow reasonably competent and alert drivers to come to a hurried stop under ordinary circumstances. However, these distances are often inadequate when drivers must make complex or instantaneous decisions, when information is difficult to perceive or when unexpected or unusual maneuvers are required. Limiting sight distances to those needed for stopping may preclude drivers from performing evasive maneuvers, which often involve less risk and are otherwise preferable to stopping. Even with an appropriate complement of standard traffic control devices in accordance with the MUTCD (**6**), stopping sight distances may not provide sufficient visibility distances for drivers to corroborate advance warning and to perform the appropriate maneuvers. It is evident that there are many locations where it would be prudent to provide longer sight distances. In these circumstances, decision sight distance provides the greater visibility distance that drivers need.

Decision sight distance is the distance needed for a driver to detect an unexpected or otherwise difficult-to-perceive information source or condition in a roadway environment that may be visually cluttered, recognize the condition or its potential threat, select an appropriate speed and path, and initiate and complete the maneuver safely and efficiently (**7**). Because decision sight distance offers drivers additional margin for error and affords them sufficient length to maneuver their vehicles at the same or reduced speed, rather than to just stop, its values are substantially greater than stopping sight distance.

Drivers need decision sight distances whenever there is a likelihood for error in either information reception, decision-making, or control actions (**8**). Examples of critical locations where these kinds of errors are likely to occur, and where it is desirable to provide decision sight distance include interchange and intersection locations where unusual or unexpected maneuvers are required, changes in cross section such as toll plazas and lane drops, and areas of concentrated demand where there is apt to be "visual noise" from competing sources of information, such as roadway elements, traffic, traffic control devices, and advertising signs.

The decision sight distances in Exhibit 3-3: (1) provide values for sight distances that may be appropriate at critical locations and (2) serve as criteria in evaluating the suitability of the available sight distances at these locations. Because of the additional safety and maneuvering space provided, it is recommended that decision sight distances be provided at critical locations or that critical decision points be moved to locations where sufficient decision sight distance is available. If it is not practical to provide decision sight distance because of horizontal or vertical curvature or if relocation of decision points is not practical, special attention should be given to the use of suitable traffic control devices for providing advance warning of the conditions that are likely to be encountered.

Metric						US Customary					
Design speed (km/h)	Decision sight distance (m)					Design speed (mph)	Decision sight distance (ft)				
	Avoidance maneuver						Avoidance maneuver				
	A	B	C	D	E		A	B	C	D	E
50	70	155	145	170	195	30	220	490	450	535	620
60	95	195	170	205	235	35	275	590	525	625	720
70	115	235	200	235	275	40	330	690	600	715	825
80	140	280	230	270	315	45	395	800	675	800	930
90	170	325	270	315	360	50	465	910	750	890	1030
100	200	370	315	355	400	55	535	1030	865	980	1135
110	235	420	330	380	430	60	610	1150	990	1125	1280
120	265	470	360	415	470	65	695	1275	1050	1220	1365
130	305	525	390	450	510	70	780	1410	1105	1275	1445
						75	875	1545	1180	1365	1545
						80	970	1685	1260	1455	1650

Avoidance Maneuver A: Stop on rural road—t = 3.0 s
Avoidance Maneuver B: Stop on urban road—t = 9.1 s
Avoidance Maneuver C: Speed/path/direction change on rural road—t varies between 10.2 and 11.2 s
Avoidance Maneuver D: Speed/path/direction change on suburban road—t varies between 12.1 and 12.9 s
Avoidance Maneuver E: Speed/path/direction change on urban road—t varies between 14.0 and 14.5 s

Exhibit 3-3. Decision Sight Distance

Decision sight distance criteria that are applicable to most situations have been developed from empirical data. The decision sight distances vary depending on whether the location is on a rural or urban road and on the type of avoidance maneuver required to negotiate the location properly. Exhibit 3-3 shows decision sight distance values for various situations rounded for design. As can be seen in the exhibit, shorter distances are generally needed for rural roads and for locations where a stop is the appropriate maneuver.

For the avoidance maneuvers identified in Exhibit 3-3, the pre-maneuver time is increased above the brake reaction time for stopping sight distance to allow the driver additional time to detect and recognize the roadway or traffic situation, identify alternative maneuvers, and initiate a response at critical locations on the highway (**9**). The pre-maneuver component of decision sight distance uses a value ranging between 3.0 and 9.1 s (**10**).

The braking distance from the design speed is added to the pre-maneuver component for avoidance maneuvers A and B as shown in Equation (3-4). The braking component is replaced in avoidance maneuvers C, D, and E with a maneuver distance based on maneuver times between 3.5 and 4.5 s, that decrease with increasing speed (**9**) in accordance with Equation (3-5).

The decision sight distances for avoidance maneuvers A and B are determined as:

Metric	**US Customary**	
$d = 0.278Vt + 0.039\dfrac{V^2}{a}$	$d = 1.47Vt + 1.075\dfrac{V^2}{a}$	(**3-4**)
where: t = pre-maneuver time, s (see notes in Exhibit 3-3); V = design speed, km/h; a = driver deceleration, m/s^2	where: t = pre-maneuver time, s (see notes in Exhibit 3-3); V = design speed, mph; a = driver deceleration, ft/s^2	

The decision sight distances for avoidance maneuvers C, D, and E are determined as:

Metric	**US Customary**	
$d = 0.278Vt$	$d = 1.47Vt$	(**3-5**)
where: t = total pre-maneuver and maneuver time, s (see notes in Exhibit 3-3); V = design speed, km/h	where: t = total pre-maneuver and maneuver time, s (see notes in Exhibit 3-3); V = design speed, mph	

In computing and measuring decision sight distances, the same 1,080-mm [3.5-ft] eye-height and 600-mm [2.0-ft] object-height criteria used for stopping sight distance have been adopted. Although drivers may have to be able to see the entire roadway situation, including the road surface, the rationale for the 600-mm [2.0-ft] object height is as applicable to decision sight distance as it is to stopping sight distance.

Passing Sight Distance for Two-Lane Highways

Criteria for Design

Most roads and many streets are two-lane, two-way highways on which vehicles frequently overtake slower moving vehicles. Passing maneuvers in which faster vehicles move ahead of slower vehicles must be accomplished on lanes regularly used by opposing traffic. If passing is to be accomplished safely, the passing driver should be able to see a sufficient distance ahead, clear of traffic, to complete the passing maneuver without cutting off the passed vehicle before meeting an opposing vehicle that appears during the maneuver. When appropriate, the driver can return to the right lane without completing the pass if he or she sees opposing traffic is too close when the maneuver is only partially completed. Many passing maneuvers are accomplished without the driver being able to see any potentially conflicting vehicle at the beginning of the maneuver, but design should not be based on such maneuvers. Because many cautious drivers would not attempt to pass under such conditions, design on this basis would reduce the usefulness of the highway. An alternative to providing passing sight distance is found later in this chapter in the section on "Passing Lanes."

Passing sight distance for use in design should be determined on the basis of the length needed to complete normal passing maneuvers in which the passing driver can determine that there are no potentially conflicting vehicles ahead before beginning the maneuver. While there may be occasions to consider multiple passings, where two or more vehicles pass or are passed, it is not practical to assume such conditions in developing minimum design criteria. Instead, sight distance should be determined for a single vehicle passing a single vehicle. Longer sight distances occur in design and such locations can accommodate an occasional multiple passing.

Minimum passing sight distances for design of two-lane highways incorporate certain assumptions about driver behavior. Actual driver behavior in passing maneuvers varies widely. To accommodate these variations in driver behavior, the design criteria for passing sight distance should accommodate the behavior of a high percentage of drivers, rather than just the average driver. The following assumptions are made concerning driver behavior in passing maneuvers:

1. The overtaken vehicle travels at uniform speed.
2. The passing vehicle has reduced speed and trails the overtaken vehicle as it enters a passing section.
3. When the passing section is reached, the passing driver needs a short period of time to perceive the clear passing section and to react to start his or her maneuver.
4. Passing is accomplished under what may be termed a delayed start and a hurried return in the face of opposing traffic. The passing vehicle accelerates during the maneuver, and its average speed during the occupancy of the left lane is 15 km/h [10 mph] higher than that of the overtaken vehicle.
5. When the passing vehicle returns to its lane, there is a suitable clearance length between it and an oncoming vehicle in the other lane.

Some drivers accelerate at the beginning of a passing maneuver to an appreciably higher speed and then continue at a uniform speed until the maneuver is completed. Many drivers accelerate at a fairly high rate until just beyond the vehicle being passed and then complete the maneuver either without further acceleration or at reduced speed. For simplicity, such extraordinary maneuvers are ignored and passing distances are developed with the use of observed speeds and times that fit the practices of a high percentage of drivers.

The minimum passing sight distance for two-lane highways is determined as the sum of the following four distances (shown in Exhibit 3-4):

- d_1—Distance traversed during perception and reaction time and during the initial acceleration to the point of encroachment on the left lane.
- d_2—Distance traveled while the passing vehicle occupies the left lane.
- d_3—Distance between the passing vehicle at the end of its maneuver and the opposing vehicle.
- d_4—Distance traversed by an opposing vehicle for two-thirds of the time the passing vehicle occupies the left lane, or 2/3 of d_2 above.

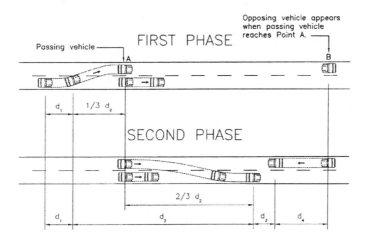

Exhibit 3-4. Elements of Passing Sight Distance for Two-Lane Highways

Various distances for the components of passing maneuvers, based on extensive field observations of driver behavior (**11**) are presented for four passing speed groups in Exhibit 3-5. Time and distance values were determined in relation to the average speed of the passing vehicle. The speeds of the overtaken vehicles were approximately 15 km/h [10 mph] less than the speeds of the passing vehicles.

Very little change was noted in the passing practices of drivers in a restudy of three of the original sections despite increased vehicle performance capabilities. A later study (**12**) of vehicle passing performance on two-lane highways produced a different set of passing sight distance values. These values were subsequently reviewed (**13**) to evaluate minimum passing sight distances. This evaluation reported the total passing sight distances as seen in Exhibit 3-5 are greater than those determined in subsequent studies for all speeds except 110 km/h [70 mph]. Thus, the minimum passing sight distances presented in Exhibit 3-7 are generally conservative for modern vehicles and are used below.

Component of passing maneuver	Metric				US Customary			
	Speed range (km/h)				Speed range (mph)			
	50-65	66-80	81-95	96-110	30-40	40-50	50-60	60-70
	Average passing speed (km/h)				Average passing speed (mph)			
	56.2	70.0	84.5	99.8	34.9	43.8	52.6	62.0
Initial maneuver:								
a = average acceleration[a]	2.25	2.30	2.37	2.41	1.40	1.43	1.47	1.50
t_1 = time (sec)[a]	3.6	4.0	4.3	4.5	3.6	4.0	4.3	4.5
d_1 = distance traveled	45	66	89	113	145	216	289	366
Occupation of left lane:								
t_2 = time (sec)[a]	9.3	10.0	10.7	11.3	9.3	10.0	10.7	11.3
d_2 = distance traveled	145	195	251	314	477	643	827	1030
Clearance length:								
d_3 = distance traveled[a]	30	55	75	90	100	180	250	300
Opposing vehicle:								
d_4 = distance traveled	97	130	168	209	318	429	552	687
Total distance, $d_1 + d_2 + d_3 + d_4$	317	446	583	726	1040	1468	1918	2383

[a] For consistent speed relation, observed values adjusted slightly.

Note: In the metric portion of the table, speed values are in km/h, acceleration rates in km/h/s, and distances are in meters. In the U.S. customary portion of the table, speed values are in mph, acceleration rates in mph/sec, and distances are in feet.

Exhibit 3-5. Elements of Safe Passing Sight Distance for Design of Two-Lane Highways

Initial maneuver distance (d_1). The initial maneuver period has two components, a time for perception and reaction, and an interval during which the driver brings the vehicle from the trailing speed to the point of encroachment on the left or passing lane. To a great extent these two periods overlap. As a passing section of highway comes into the view of a driver desiring to pass, the driver may begin to accelerate and maneuver his or her vehicle toward the centerline of the highway while deciding whether or not to pass. Studies show that the average passing vehicle accelerates at less than its maximum potential, indicating that the initial maneuver period contains an element of time for perception and reaction. However, some drivers may remain in normal lane position while deciding to pass. The exact position of the vehicle during initial maneuver is unimportant because the differences in resulting passing distances are insignificant.

The acceleration rate obtained from the passing study data in the first three speed groups during the initial maneuver period varied from 2.25 to 2.37 km/h/s [1.41 to 1.47 mph/s]; the average time varied from 3.7 to 4.3 s; and the average passing speeds were 56.2, 70.0, and 84.5 km/h [34.9, 43.8, and 52.6 mph]. For the 96 to 100 km/h [60 to 70 mph] group, on the basis of extrapolated data, the average acceleration was assumed to be 2.41 km/h/s [1.50 mph/s]; the maneuver time, 4.5 s; and the average speed, 99.8 km/h [62.0 mph].

The distance d_1 traveled during the initial maneuver period is computed with the following equation:

Metric	US Customary	
$$d_1 = 0.278\, t_i \left(v - m + \frac{at_i}{2} \right)$$	$$d_1 = 1.47\, t_i \left(v - m + \frac{at_i}{2} \right)$$	(3-6)
where:	where:	
t_i = time of initial maneuver, s; a = average acceleration, km/h/s; v = average speed of passing vehicle, km/h; m = difference in speed of passed vehicle and passing vehicle, km/h	t_i = time of initial maneuver, s; a = average acceleration, mph/s; v = average speed of passing vehicle, mph; m = difference in speed of passed vehicle and passing vehicle, mph	

The acceleration, time, and distance traveled during the initial maneuver periods in passing are given in Exhibit 3-5. The d_1 line in Exhibit 3-6 shows the distance plotted against the average speed of the passing vehicle.

Distance while passing vehicle occupies left lane (d_2). Passing vehicles were found in the study to occupy the left lane from 9.3 to 10.4 s. The distance d_2 traveled in the left lane by the passing vehicle is computed with the following equation:

Metric	US Customary	
$d_2 = 0.278vt_2$	$d_2 = 1.47vt_2$	(3-7)
where:	where:	
t_2 = time passing vehicle occupies the left lane, s; v = average speed of passing vehicle, km/h	t_2 = time passing vehicle occupies the left lane, s; v = average speed of passing vehicle, mph	

The time and distance traveled while the passing vehicle occupies the left lane are given in Exhibit 3-5. Distances are plotted against average passing speeds as curve d_2 in Exhibit 3-6.

Clearance length (d_3). The clearance length between the opposing and passing vehicles at the end of the passing maneuvers was found in the passing study to vary from 30 to 75 m [100 to 250 ft]. This length, adjusted somewhat for practical consistency, is shown as the clearance length d_3 in Exhibits 3-5 and 3-6.

Distance traversed by an opposing vehicle (d_4). Passing sight distance includes the distance traversed by an opposing vehicle during the passing maneuver, to minimize the chance that a passing vehicle will meet an opposing vehicle while in the left lane. Conservatively, this distance should be the distance traversed by an opposing vehicle during the entire time it takes to pass or during the time the passing vehicle is in the left lane, but such distance is questionably long. During the first phase of the passing maneuver, the passing vehicle has not yet pulled abreast of the vehicle being passed, and even though the passing vehicle occupies the left lane, its driver can return to the right lane if an opposing vehicle is seen. It is unnecessary to include this trailing time interval in computing the distance traversed by an opposing vehicle. This time interval, which can be computed from the relative positions of passing and passed vehicle, is about one-third the time the passing vehicle occupies the left lane, so that the passing sight distance element for the opposing vehicle is the distance it traverses during two-thirds of the time the passing vehicle occupies the left lane. The opposing vehicle is assumed to be traveling at the same speed as the passing vehicle, so $d_4 = 2d_2/3$. The distance d_4 is shown in Exhibits 3-5 and 3-6.

Design Values

The "total" curve in Exhibit 3-6 is determined by the sum of the elements d_1 through d_4. For each passing speed, this total curve indicates the minimum passing sight distance for a vehicle to pass another vehicle traveling 15 km/h [10 mph] slower, in the face of an opposing vehicle traveling at the same speed as the passing vehicle. On determination of a likely and logical relation between average passing speed and the highway design speed, these distances can be used to express the minimum passing sight distance needed for design purposes.

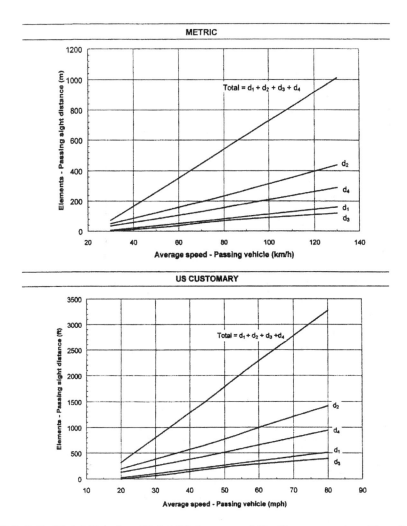

Exhibit 3-6. Total Passing Sight Distance and Its Components—Two-Lane Highways

The ranges of speeds of the passed and passing vehicles are affected by traffic volume. When traffic volume is low (level-of-service A), there are few vehicles that need to be passed, but as the volume increases (level-of-service D or lower) there are few, if any, passing opportunities. The speed of the passed vehicle has been assumed to be the average running speed at a traffic volume near capacity. The speed of the passing vehicle is assumed to be 15 km/h [10 mph] greater. The assumed speeds for passing vehicles in Exhibit 3-7 represent the likely passing speeds on two-lane highways. Passing sight distances for these passing speeds would accommodate a majority of the desired passing maneuvers and correspond to the total curve in Exhibit 3-6. The values in the last column of Exhibit 3-7 are design values for minimum passing sight distance. In designing a highway these distances should be exceeded as much as practical, and passing sections should be provided as often as can be done at reasonable cost to provide as many passing opportunities as practical.

These minimum passing sight distances for design should not be confused with other distances used as the warrants for placing no-passing zone pavement markings on completed highways. Such values as shown in the MUTCD (**6**) are substantially less than design distances

Metric

Design speed (km/h)	Assumed speeds (km/h)		Passing sight distance (m)	
	Passed vehicle	Passing vehicle	From Exhibit 3-6	Rounded for design
30	29	44	200	200
40	36	51	266	270
50	44	59	341	345
60	51	66	407	410
70	59	74	482	485
80	65	80	538	540
90	73	88	613	615
100	79	94	670	670
110	85	100	727	730
120	90	105	774	775
130	94	109	812	815

US Customary

Design speed (mph)	Assumed speeds (mph)		Passing sight distance (ft)	
	Passed vehicle	Passing vehicle	From Exhibit 3-6	Rounded for design
20	18	28	706	710
25	22	32	897	900
30	26	36	1088	1090
35	30	40	1279	1280
40	34	44	1470	1470
45	37	47	1625	1625
50	41	51	1832	1835
55	44	54	1984	1985
60	47	57	2133	2135
65	50	60	2281	2285
70	54	64	2479	2480
75	56	66	2578	2580
80	58	68	2677	2680

Exhibit 3-7. Passing Sight Distance for Design of Two-Lane Highways

and are derived for traffic operating-control needs that are based on different assumptions from those for highway design.

Effect of Grade on Passing Sight Distance

Appreciable grades affect the sight distance needed for passing. Passing is easier for the vehicle traveling downgrade because the overtaking vehicle can accelerate more rapidly than on the level and thus can reduce the time of passing. However, the overtaken vehicle can also accelerate easily so that a situation akin to a racing contest may result.

The sight distances needed to permit vehicles traveling upgrade to pass safely are greater than those needed on level roads because of reduced acceleration of the passing vehicle (which increases the time of passing) and the likelihood that opposing traffic may speed up (which increases the distance traveled by an opposing vehicle during the passing maneuver). Compensating for this somewhat are the factors that the passed vehicle frequently is a truck that usually loses some speed on appreciable upgrades and that many drivers are aware of the greater distances needed for passing upgrade compared with level conditions.

If passing maneuvers are to be performed on upgrades under the same assumptions about the behavior of the passing and passed vehicles discussed above, the passing sight distance should be greater than the derived design values. Specific adjustments for design use are unavailable, but the designer should recognize the desirability of exceeding the values shown in Exhibit 3-7.

Frequency and Length of Passing Sections

Sight distance adequate for passing should be encountered frequently on two-lane highways. Each passing section along a length of roadway with sight distance ahead equal to or greater than the minimum passing sight distance should be as long as practical. The frequency and length of passing sections for highways depend, principally on the topography, the design speed of highway, and the cost; for streets, the spacing of intersections is the principal consideration.

It is not practical to directly indicate the frequency with which passing sections should be provided on two-lane highways due to the physical and cost limitations. During the course of normal design, passing sections are provided on almost all highways and selected streets, but the designer's appreciation of their importance and a studied attempt to provide them can usually ensure others at little or no additional cost. In steep mountainous terrain, it may be more economical to build intermittent four-lane sections or passing lanes with stopping sight distance on some two-lane highways, in lieu of two-lane sections with passing sight distance. Alternatives are discussed later in this chapter in the section on "Passing Lanes."

The passing sight distances shown in Exhibit 3-7 are sufficient for a single or isolated pass only. Designs with infrequent passing sections will not assure that opportunities for passing are available. Even on low-volume roadways, a driver desiring to pass may, on reaching the passing section, find vehicles in the opposing lane and thus be unable to use the passing section or at least may not be able to begin to pass at once.

The importance of frequent passing sections is illustrated by their effect on the level of service of a two-lane, two-way highway. The procedures in the *Highway Capacity Manual* (**14**) to analyze two-lane, two-way highways base the level-of-service criteria on two measures of effectiveness—percent time spent following and average travel speed. Both of these criteria are affected by the lack of passing opportunities. The HCM procedures show, for example, up to a 19 percent increase in the percent time spent following when the directional split is 50/50 and no-passing zones comprise 40 percent of the analysis length compared to a highway with similar traffic volumes and no sight restrictions. The effect of restricted passing sight distance is even more severe for unbalanced flow and where the no-passing zones comprise more than 40 percent of the length.

There is a similar effect on the average travel speed. As the percent of no-passing zones increases, there is an increased reduction in the average travel speed for the same demand flow rate. For example, a demand flow rate of 800 passenger cars per hour incurs a reduction of 3.1 km/h (1.9 mph) when no-passing zones comprise 40 percent of the analysis length compared to no reduction in speed on a route with unrestricted passing.

The HCM procedures indicate another possible criterion for passing sight distance design on two-lane highways that are several miles or more in length. The available passing sight distances along this length can be summarized to show the percentage of length with greater-than-minimum passing sight distance. Analysis of capacity related to this percentage would indicate whether or not alignment and profile adjustments are needed to accommodate the design hourly volume (DHV). When highway sight distances are analyzed over the whole range of lengths within which passing maneuvers are made, a new design criterion may be evaluated. Where high traffic volumes are expected on a highway and a high level of service is to be maintained, frequent or nearly continuous passing sight distances should be provided.

Sight Distance for Multilane Highways

It is not necessary to consider passing sight distance on highways or streets that have two or more traffic lanes in each direction of travel. Passing maneuvers on multilane roadways are expected to occur within the limits of the traveled way for each direction of travel. Thus, passing maneuvers that involve crossing the centerline of four-lane undivided roadways or crossing the median of four-lane roadways should be prohibited.

Multilane roadways should have continuously adequate stopping sight distance, with greater-than-design sight distances preferred. Design criteria for stopping sight distance vary with vehicle speed and are discussed in detail at the beginning of this chapter.

Criteria for Measuring Sight Distance

Sight distance is the distance along a roadway throughout which an object of specified height is continuously visible to the driver. This distance is dependent on the height of the driver's eye above the road surface, the specified object height above the road surface, and the height and lateral position of sight obstructions within the driver's line of sight.

Height of Driver's Eye

For sight distance calculations for passenger vehicles, the height of the driver's eye is considered to be 1,080 mm [3.5 ft] above the road surface. This value is based on a study (**4**) found that average vehicle heights have decreased to 1,300 mm [4.25 ft] with a comparable decrease in average eye heights to 1,080 mm [3.5 ft]. Because of various factors that appear to place practical limits on further decreases in passenger car heights and the relatively small increases in the lengths of vertical curves that would result from further changes that do occur, 1,080 mm [3.5 ft] is considered to be the appropriate height of driver's eye for measuring both stopping and passing sight distances. For large trucks, the driver eye height ranges from 1,800 to 2,400 mm [5.9 to 7.9 ft]. The recommended value of truck driver eye height for design is 2,330 mm [7.6 ft] above the roadway surface.

Height of Object

For stopping sight distance calculations, the height of object is considered to be 600 mm [2.0 ft] above the road surface. For passing sight distance calculations, the height of object is considered to be 1,080 mm [3.5 ft] above the road surface.

Stopping sight distance object. The basis for selection of a 600-mm [2.0-ft] object height was largely an arbitrary rationalization of the size of object that might potentially be encountered in the road and of a driver's ability to perceive and react to such situations. It is considered that an object 600 mm [2.0 ft] high is representative of an object that involves risk to drivers and can be recognized by a driver in time to stop before reaching it. Using object heights of less than 600 mm [2.0 ft] for stopping sight distance calculations would result in longer crest vertical curves without documented safety benefits (**4**). Object height of less than 600 mm [2.0 ft] could substantially increase construction costs because additional excavation would be needed to provide the longer crest vertical curves. It is also doubtful that the driver's ability to perceive situations involving risk of collisions would be increased because recommended stopping sight distances for high-speed design are beyond most drivers' capabilities to detect small objects (**4**).

Passing sight distance object. An object height of 1,080 mm [3.5 ft] is adopted for passing sight distance. This object height is based on a vehicle height of 1,330 mm [4.35 ft], which represents the 15th percentile of vehicle heights in the current passenger car population, less an allowance of 250 mm [0.82 ft], which represents a near-maximum value for the portion of the vehicle height that needs to be visible for another driver to recognize a vehicle as such (**15**). Passing sight distances calculated on this basis are also considered adequate for night conditions because headlight beams of an opposing vehicle generally can be seen from a greater distance than a vehicle can be recognized in the daytime. The choice of an object height equal to the driver eye height makes passing sight distance design reciprocal (i.e., when the driver of the passing vehicle can see the opposing vehicle, the driver of the opposing vehicle can also see the passing vehicle).

Sight Obstructions

On a tangent roadway, the obstruction that limits the driver's sight distance is the road surface at some point on a crest vertical curve. On horizontal curves, the obstruction that limits the driver's sight distance may be the road surface at some point on a crest vertical curve, or it may be some physical feature outside of the traveled way, such as a longitudinal barrier, a bridge-approach fill slope, a tree, foliage, or the backslope of a cut section. Accordingly, all highway construction plans should be checked in both the vertical and horizontal plane for sight distance obstructions.

Measuring and Recording Sight Distance on Plans

The design of horizontal alignment and vertical profile using sight distance and other criteria is addressed later in this chapter, including the detailed design of horizontal and vertical curves. Sight distance should be considered in the preliminary stages of design when both the horizontal and vertical alignment are still subject to adjustment. By determining the available sight distances graphically on the plans and recording them at frequent intervals, the designer can appraise the overall layout and effect a more balanced design by minor adjustments in the plan or profile. Methods for scaling sight distances on plans are demonstrated in Exhibit 3-8, which also shows a typical sight distance record that would be shown on the final plans.

Because the view of the highway ahead may change rapidly in a short distance, it is desirable to measure and record sight distance for both directions of travel at each station. Both horizontal and vertical sight distances should be measured and the shorter lengths recorded. In the case of a two-lane highway, passing sight distance should be measured and recorded in addition to stopping sight distance.

Sight distance charts such as those in Exhibit 3-74 and 3-77 may be used to establish minimum lengths of vertical curves. Charts similar to Exhibit 3-57 are useful for determining the radius of horizontal curve or the lateral offset from the traveled way needed to provide the design sight distance. Once the horizontal and vertical alignments are tentatively established, the most practical means of examining sight distances along the proposed highway is by direct scaling on the plans.

Horizontal sight distance on the inside of a curve is limited by obstructions such as buildings, hedges, wooded areas, high ground, or other topographic features. These are generally plotted on the plans. Horizontal sight is measured with a straightedge, as indicated in the upper left portion of Exhibit 3-8. The cut slope obstruction is shown on the worksheets by a line

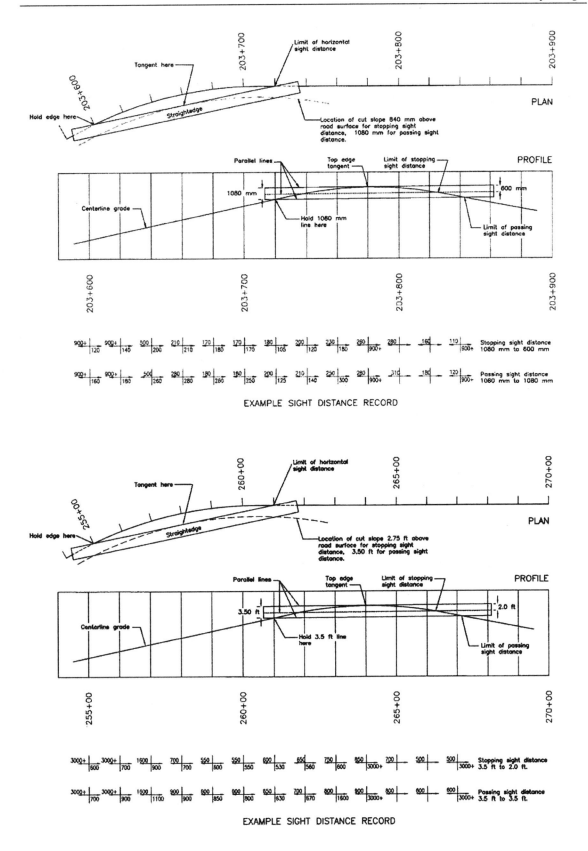

Exhibit 3-8. Scaling and Recording Sight Distances on Plans

representing the proposed excavation slope at a point 840 mm [2.75 ft] above the road surface (i.e., the approximate average of 1,080 mm and 600 mm [3.5 ft and 2.0 ft]) for stopping sight distance and at a point about 1,080 mm [3.5 ft] above the road surface for passing sight distance. The position of this line with respect to the centerline may be scaled from the plotted highway cross sections. Preferably, the stopping sight distance should be measured between points on one traffic lane, and passing sight distance from the middle of one lane to the middle of the other lane.

Such refinement on two-lane highways generally is not necessary and measurement of sight distance along the centerline or traveled way edge is suitable. Where there are changes of grade coincident with horizontal curves that have sight-limiting cut slopes on the inside, the line-of-sight intercepts the slope at a level either lower or higher than the assumed average height. In measuring sight distance, the error in the use of the assumed 840 or 1,080 mm [2.75 or 3.5 ft] height usually can be ignored.

Vertical sight distance may be scaled from a plotted profile by the method illustrated at the right center of Exhibit 3-8. A transparent strip with parallel edges 1,080 mm [3.5 ft] apart and with a scratched line 600 mm [2.0 ft] from the upper edge, in accordance with the vertical scale, is a useful tool. The lower edge of the strip is placed on the station from which the vertical sight distance is desired, and the strip is pivoted about this point until the upper edge is tangent to the profile. The distance between the initial station and the station on the profile intersected by the 600 mm [2.0 ft] line is the stopping sight distance. The distance between the initial station and the station on the profile intersected by the lower edge of the strip is the passing sight distance.

A simple sight distance record is shown in the lower part of Exhibit 3-8. Sight distances in both directions are indicated by arrows and figures at each station on the plan and profile sheet of the proposed highway. To avoid the extra work of measuring unusually long sight distances that may occasionally be found, a selected maximum value may be recorded. In the example shown, all sight distances of more than 1,000 m [3,000 ft] are recorded as 1,000 m+ [3,000 ft+], and where this occurs for several consecutive stations, the intermediate values are omitted. Sight distances less than 500 m [1,500 ft] may be scaled to the nearest 10 m [50 ft] and those greater than 500 m [1,500 ft] to the nearest 50 m [100 ft]. The available sight distances along a proposed highway also may be shown by other methods. Several States use a sight distance graph, plotted in conjunction with the plan and profile of the highway, as a means of demonstrating sight distances. Sight distances can also be easily determined where plans and profiles are drawn using computer-aided design and drafting (CADD) systems.

Sight distance records for two-lane highways may be used effectively to tentatively determine the marking of no-passing zones in accordance with criteria given in the MUTCD (6). Marking of such zones is an operational rather than a design problem. No-passing zones thus established serve as a guide for markings when the highway is completed. The zones so determined should be checked and adjusted by field measurements before actual markings are placed.

Sight distance records also are useful on two-lane highways for determining the percentage of length of highway on which sight distance is restricted to less than the passing minimum, which is important in evaluating capacity. With recorded sight distances, as in the lower part of

Exhibit 3-8, it is a simple process to determine the percentage of length of highway with a given sight distance or greater.

HORIZONTAL ALIGNMENT

Theoretical Considerations

For balance in highway design all geometric elements should, as far as economically practical, be designed to provide safe, continuous operation at a speed likely to be observed under the normal conditions for that roadway. For the most part, this can be achieved through the use of design speed as an overall design control. The design of roadway curves should be based on an appropriate relationship between design speed and curvature and on their joint relationships with superelevation and side friction. Although these relationships stem from the laws of mechanics, the actual values for use in design depend on practical limits and factors determined more or less empirically over the range of variables involved. These limits and factors are explained in the following discussion, as they relate to the determination of logical controls for roadway curve design.

When a vehicle moves in a circular path, it undergoes a centripetal acceleration that acts toward the center of curvature. This acceleration is sustained by a component of the vehicle's weight related to the roadway superelevation, by the side friction developed between the vehicle's tires and the pavement surface, or by a combination of the two. As a matter of conceptual convenience, centripetal acceleration is sometimes equated to centrifugal force. However, this is an imaginary force that motorists believe is pushing them outward while cornering when, in fact, they are truly feeling the vehicle being accelerated in an inward direction. The term "centripetal acceleration" and its equivalent in horizontal curve design, "lateral acceleration," are used in this policy as they are fundamentally correct.

From the laws of mechanics, the basic formula that governs vehicle operation on a curve is:

Metric	US Customary
$$\frac{0.01e+f}{1-0.01ef} = \frac{v^2}{gR} = \frac{0.0079V^2}{R} = \frac{V^2}{127R}$$	$$\frac{0.01e+f}{1-0.01ef} = \frac{v^2}{gR} = \frac{0.067V^2}{R} = \frac{V^2}{15R} \quad (3\text{-}8)$$

where:
 	
e = rate of roadway superelevation, percent;
f = side friction (demand) factor;
v = vehicle speed, m/s;
g = gravitational constant, 9.81 m/s²;
V = vehicle speed, km/h;
R = radius of curve, m
	where:
	
	e = rate of roadway superelevation, percent;
	f = side friction (demand) factor;
	v = vehicle speed, ft/s;
	g = gravitational constant, 32.2 ft/s²;
	V = vehicle speed, mph;
	R = radius of curve, ft

Equation (3-8), which models the moving vehicle as a point mass, is often referred to as the basic curve formula.

When a vehicle travels at constant speed on a curve superelevated so that the f value is zero, the centripetal acceleration is sustained by a component of the vehicle's weight and, theoretically, no steering force is needed. A vehicle traveling faster or slower than the balance speed develops tire friction as steering effort is applied to prevent movement to the outside or to the inside of the curve. On nonsuperelevated curves, travel at different speeds is also possible by utilizing appropriate amounts of side friction to sustain the varying centripetal acceleration.

General Considerations

From accumulated research and experience, limiting values for superelevation rate (e_{max}) and side friction demand (f_{max}) have been established for curve design. Using these established limiting values in the basic curve formula permits determination of a minimum curve radius for various design speeds. Use of curves with radii larger than this minimum allows superelevation, side friction, or both to have values below their respective limits. The amount by which each factor is below its respective limit is chosen to provide an equitable contribution of each factor toward sustaining the resultant centripetal acceleration. The methods used to achieve this equity for different design situations are discussed below.

Superelevation

There are practical upper limits to the rate of superelevation on a horizontal curve. These limits relate to considerations of climate, constructability, adjacent land use, and the frequency of slow-moving vehicles. Where snow and ice are a factor, the rate of superelevation should not exceed the rate on which vehicles standing or traveling slowly would slide toward the center of the curve when the pavement is icy. At higher speeds, the phenomenon of partial hydroplaning can occur on curves with poor drainage that allows water to build up on the pavement surface. Skidding occurs, usually at the rear wheels, when the lubricating effect of the water film reduces the available lateral friction below the friction demand for cornering. When travelling slowly around a curve with high superelevation, negative lateral forces develop and the vehicle is held in the proper path only when the driver steers up the slope or against the direction of the horizontal curve. Steering in this direction seems unnatural to the driver and may explain the difficulty of driving on roads where the superelevation is in excess of that needed for travel at normal speeds. Such high rates of superelevation are undesirable on high-volume roads, as in urban and suburban areas, where there are numerous occasions when vehicle speeds may be considerably reduced because of the volume of traffic or other conditions.

Some vehicles have high centers of gravity and some passenger cars are loosely suspended on their axles. When these vehicles travel slowly on steep cross slopes, a high percentage of their weight is carried by the inner tires. A vehicle can roll over if this condition becomes extreme.

A discussion of these considerations and the rationale used to establish an appropriate maximum rate of superelevation for design of horizontal curves is provided in the subsequent section on "Maximum Superelevation Rates."

Side Friction Factor

The side friction factor represents the vehicle's need for side friction, also called the side friction demand; it also represents the lateral acceleration a_f that acts on the vehicle. This acceleration can be computed as the product of the side friction demand factor f and the gravitational constant g (i.e., $a_f = fg$). It should be noted that the lateral acceleration actually experienced by vehicle occupants tends to be slightly larger than predicted by the product fg due to vehicle body roll angle.

With the wide variation in vehicle speeds on curves, there usually is an unbalanced force whether the curve is superelevated or not. This force results in tire side thrust, which is counterbalanced by friction between the tires and the pavement surface. This frictional counterforce is developed by distortion of the contact area of the tire.

The coefficient of friction f is the friction force divided by the component of the weight perpendicular to the pavement surface and is expressed as a simplification of the basic curve formula shown as Equation (3-8). The value of the product ef in this formula is always small. As a result, the 1-0.01ef term is nearly equal to 1.0 and is normally omitted in roadway design. Omission of this term yields the following basic side friction equation:

Metric	US Customary	
$f = \dfrac{V^2}{127R} - 0.01e$	$f = \dfrac{V^2}{15R} - 0.01e$	(3-9)

This equation is referred to as the simplified curve formula and yields slightly larger (and, thus, more conservative) estimates of friction demand than would be obtained using the basic curve formula.

The coefficient f has been called lateral ratio, cornering ratio, unbalanced centrifugal ratio, friction factor, and side friction factor. Because of its widespread use, the last term is used in this discussion. The upper limit of the side friction factor is the point at which the tire would begin to skid; this is known as the point of impending skid. Because highway curves are designed to avoid skidding conditions with a margin of safety, the f values used in design should be substantially less than the coefficient of friction at impending skid.

The side friction factor at impending skid depends on a number of other factors, among which the most important are the speed of the vehicle, the type and condition of the roadway surface, and the type and condition of the vehicle tires. Different observers have recorded different maximum side friction factors at the same speeds for pavements of similar composition, and logically so, because of the inherent variability in pavement texture, weather conditions, and tire condition. In general, studies show that the maximum side friction factors developed between new tires and wet concrete pavements range from about 0.5 at 30 km/h [20 mph] to approximately 0.35 at 100 km/h [60 mph]. For normal wet concrete pavements and smooth tires the maximum side friction factor at impeding skid is about 0.35 at 70 km/h [45 mph]. In all cases, the studies show a decrease in friction values as speeds increase (**16, 17, 18**).

Horizontal curves should not be designed directly on the basis of the maximum available side friction factor. Rather, the maximum side friction factor used in design should be that portion of the maximum available side friction that can be used with comfort and safety by the vast majority of drivers. Side friction levels that represent pavements that are glazed, bleeding, or otherwise lacking in reasonable skid-resistant properties should not control design because such conditions are avoidable and geometric design should be based on acceptable surface conditions attainable at reasonable cost.

A key consideration in selecting maximum side friction factors for use in design is the level of centripetal or lateral acceleration that is sufficient to cause drivers to experience a feeling of discomfort and to react instinctively to avoid higher speed. The speed on a curve at which discomfort due to the lateral acceleration is evident to drivers has been accepted as a design control for the maximum side friction factor. At lower nonuniform running speeds, which are typical in urban areas, drivers are more tolerant of discomfort, thus permitting employment of an increased amount of side friction for use in design of horizontal curves.

The ball-bank indicator has been widely used by research groups, local agencies, and highway departments as a uniform measure of lateral acceleration to set speeds on curves that avoid driver discomfort. It consists of a steel ball in a sealed glass tube; except for the damping effect of the liquid in the tube, the ball is free to roll. Its simplicity of construction and operation has led to widespread acceptance as a guide for determination of appropriate curve speeds. With such a device mounted in a vehicle in motion, the ball-bank reading at any time is indicative of the combined effect of body roll, lateral acceleration angle, and superelevation as shown in Exhibit 3-9.

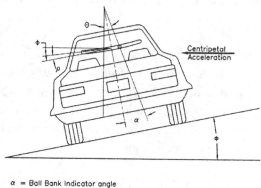

α = Ball Bank Indicator angle
ρ = Body roll angle
φ = Superelevation angle
Θ = Centripetal acceleration angle

Exhibit 3-9. Geometry for Ball-Bank Indicator

The centripetal acceleration developed as a vehicle travels at uniform speed on a curve causes the ball to roll out to a fixed angle position as shown in Exhibit 3-9. A correction should be made for that portion of the force taken up in the small body roll angle. The indicated side force perceived by the vehicle occupants is thus on the order of $F \approx \tan(\alpha - \rho)$.

In a series of definitive tests (**18**), it was concluded that speeds on curves that avoid driver discomfort are indicated by ball-bank readings of 14 degrees for speeds of 30 km/h [20 mph] or less, 12 degrees for speeds of 40 and 50 km/h [25 and 30 mph], and 10 degrees for speeds of 55 through 80 km/h [35 through 50 mph]. These ball-bank readings are indicative of side friction factors of 0.21, 0.18, and 0.15, respectively, for the test body roll angles and provide ample margin of safety against skidding.

From other tests (**19**), a maximum side friction factor of 0.16 for speeds up to 100 km/h [60 mph] was recommended. For higher speeds, the incremental reduction of this factor was recommended. Speed studies on the Pennsylvania Turnpike (**17**) led to a conclusion that the side friction factor should not exceed 0.10 for design speeds of 110 km/h [70 mph] and higher. A recent study (**20**) re-examined previously published findings and analyzed new data collected at numerous horizontal curves. The side friction demand factors developed in that study are generally consistent with the side friction factors reported above.

An electronic accelerometer provides an alternative to the ball-bank indicator for use in determining advisory speeds for horizontal curves and ramps. An accelerometer is a gravity-sensitive electronic device that can measure the lateral forces and accelerations that drivers experience while traversing a highway curve (**65**).

It should be recognized that other factors influence driver speed choice under conditions of high friction demand. Swerving becomes perceptible, drift angle increases, and increased steering effort is needed to avoid involuntary lane line violations. Under these conditions, the cone of vision narrows and is accompanied by an increasing sense of concentration and intensity considered undesirable by most drivers. These factors are more apparent to a driver under open-road conditions.

Where practical, the maximum side friction factors used in design should be conservative for dry pavements and should provide an ample margin of safety against skidding on pavements that are wet as well as ice or snow covered. The need to provide skid-resistant pavement surfacing for these conditions cannot be overemphasized because superimposed on the frictional demands resulting from roadway geometry are those that result from driving maneuvers such as braking, sudden lane changes, and minor changes in direction within a lane. In these short-term maneuvers, high friction demand can exist but the discomfort threshold may not be perceived in time for the driver to take corrective action.

Exhibit 3-10 summarizes the findings of the cited tests relating to side friction factors recommended for curve design. Although some variation in the test results is noted, all are in agreement that the side friction factor should be lower for high-speed design than for low-speed design. The subsequent sections in this chapter on "Design for Rural Highways, Urban Freeways, and High-Speed Urban Streets" and "Design for Low-Speed Urban Streets" should be referred to for the values of the side friction factor recommended for use in horizontal curve design. Exhibit 3-11 compares the friction factors assumed for the three different types of highway facilities for which different friction factors are assumed herein: (1) rural highways and high-speed urban streets, (2) low-speed urban streets, and (3) turning roadways.

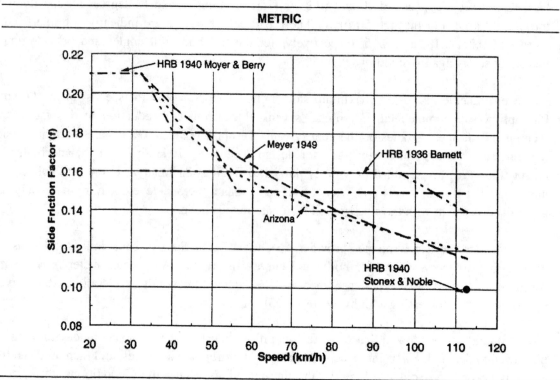

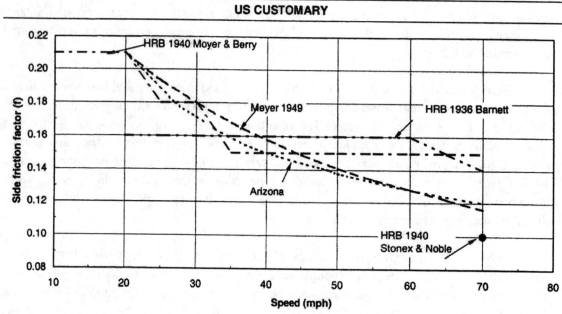

Exhibit 3-10. Side Friction Factors

METRIC

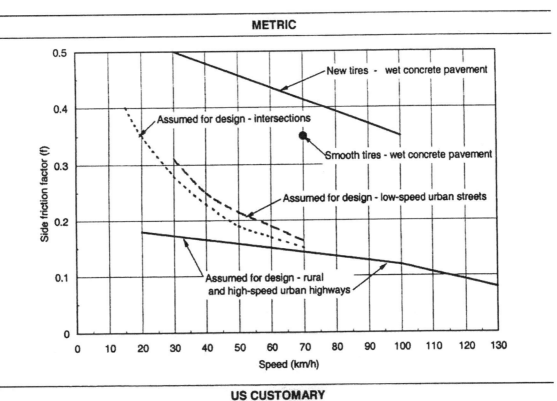

US CUSTOMARY

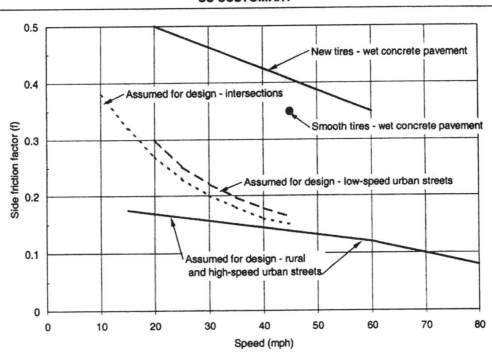

Exhibit 3-11. Comparison of Side Friction Factors Assumed for Design of Different Types of Facilities

Distribution of e and f Over a Range of Curves

For a given design speed there are five methods for sustaining centripetal acceleration on curves by use of e or f, or both. These methods are discussed below, and the resulting relationships are illustrated in Exhibit 3-12:

- Method 1—Superelevation and side friction are directly proportional to the inverse of the radius (i.e., a straight-line relation exists between $1/R = 0$ and $1/R = 1/R_{min}$).
- Method 2—Side friction is such that a vehicle traveling at design speed has all lateral acceleration sustained by side friction on curves up to those requiring f_{max}. For sharper curves, f remains equal to f_{max} and superelevation is then used to sustain lateral acceleration until e reaches e_{max}. In this method, first f and then e are increased in inverse proportion to the radius of curvature.
- Method 3—Superelevation is such that a vehicle traveling at the design speed has all lateral acceleration sustained by superelevation on curves up to that requiring e_{max}. For sharper curves, e remains at e_{max} and side friction is then used to sustain lateral acceleration until f reaches f_{max}. In this method, first e and then f are increased in inverse proportion to the radius of curvature.
- Method 4—This method is the same as method 3, except that it is based on average running speed instead of design speed.
- Method 5—Superelevation and side friction are in a curvilinear relation with the inverse of the radius of the curve, with values between those of methods 1 and 3.

Exhibit 3-12A compares the relationship between superelevation and the inverse of the radius of the curve for these five methods. Exhibit 3-12B shows the corresponding value of side friction for a vehicle traveling at design speed, and Exhibit 3-12C for a vehicle traveling at the corresponding average running speed.

The straight-line relationship between superelevation and the inverse of the radius of the curve in method 1 results in a similar relationship between side friction and the radius for vehicles traveling at either the design or average running speed. This method has considerable merit and logic in addition to its simplicity. On any particular highway, the horizontal alignment consists of tangents and curves of varying radius greater than or equal to the minimum radius appropriate for the design speed (R_{min}). Application of superelevation in amounts directly proportional to the inverse of the radius would, for vehicles traveling at uniform speed, result in side friction factors with a straight-line variation from zero on tangents (ignoring cross slope) to the maximum side friction at the minimum radius. This method might appear to be an ideal means of distributing the side friction factor, but its appropriateness depends on travel at a constant speed by each vehicle in the traffic stream, regardless of whether travel is on a tangent, a curve of intermediate degree, or a curve with the minimum radius for that design speed. While uniform speed is the aim of most drivers, and can be obtained on well-designed highways when volumes are not heavy, there is a tendency for some drivers to travel faster on tangents and the flatter curves than on the sharper curves, particularly after being delayed by inability to pass slower moving vehicles. This tendency points to the desirability of providing superelevation rates for intermediate curves in excess of those that result from use of method 1.

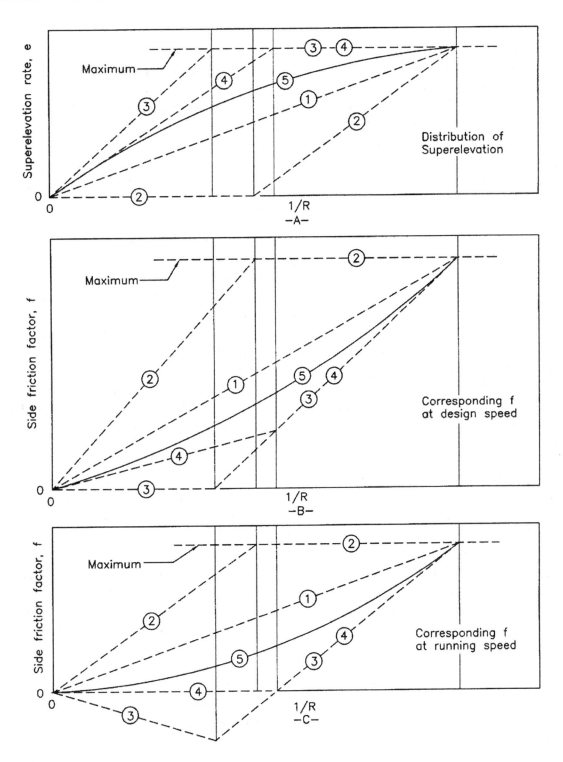

KEY: ◯ Method of distributing e and f, refer to text for explanation.

Exhibit 3-12. Methods of Distributing Superelevation and Side Friction

Method 2 uses side friction to sustain all lateral acceleration up to the curvature corresponding to the maximum side friction factor, and this maximum side friction factor is available on all sharper curves. In this method, superelevation is introduced only after the maximum side friction has been used. Therefore, no superelevation is needed on flatter curves that need less than maximum side friction for vehicles traveling at the design speed (see curve 2 in Exhibit 3-12A). When superelevation is needed, it increases rapidly as curves with maximum side friction grow sharper. Because this method is completely dependent on available side friction, its use is generally limited to locations where travel speed is not uniform, such as on urban streets. This method is particularly advantageous on low-speed urban streets where, because of various constraints, superelevation frequently cannot be provided.

In method 3, which was practiced many years ago, superelevation to sustain all lateral acceleration for a vehicle traveling at the design speed is provided on all curves up to that needing maximum practical superelevation, and this maximum superelevation is provided on all sharper curves. Under this method, no side friction is provided on flat curves with less than maximum superelevation for vehicles traveling at the design speed, as shown by curve 3 in Exhibit 3-12B, and the appropriate side friction increases rapidly as curves with maximum superelevation grow sharper. Further, as shown by curve 3 in Exhibit 3-12C, for vehicles traveling at average running speed, this superelevation method results in negative friction for curves from very flat radii to about the middle of the range of curve radii; beyond this point, as curves become sharper, the side friction increases rapidly up to a maximum corresponding to the minimum radius of curvature. This marked difference in side friction for different curves is not logical and may result in erratic driving, either at the design or average running speed.

Method 4 is intended to overcome the deficiencies of method 3 by using superelevation at speeds lower than the design speed. This method has been widely used with an average running speed for which all lateral acceleration is sustained by superelevation of curves flatter than that needing the maximum rate of superelevation. This average running speed was an approximation that, as presented in Exhibit 3-26, varies from 78 to 100 [80 to 100] percent of design speed. Curve 4 in Exhibit 3-12A shows that in using this method the maximum superelevation is reached near the middle of the curvature range. Exhibit 3-12C shows that at average running speed no side friction is needed up to this curvature, and side friction increases rapidly and in direct proportion for sharper curves. This method has the same disadvantages as method 3, but they apply to a smaller degree.

To accommodate overdriving that is likely to occur on flat to intermediate curves, it is desirable that the superelevation approximate that obtained by method 4. Overdriving on such curves involves very little risk that a driver will lose control of the vehicle because superelevation sustains nearly all lateral acceleration at the average running speed, and considerable side friction is available for greater speeds. On the other hand, method 1, which avoids use of maximum superelevation for a substantial part of the range of curve radii, is also desirable. In method 5, a curved line (curve 5, as shown within the triangular working range between curves 1 and 4 in Exhibit 3-12A) represents a superelevation and side friction distribution reasonably retaining the advantages of both methods 1 and 4. Curve 5 has an unsymmetrical parabolic form and represents a practical distribution for superelevation over the range of curvature.

Design Considerations

Superelevation rates that are applicable over the range of curvature for each design speed have been determined for use in highway design. One extreme of this range is the maximum superelevation rate established by practical considerations and used to determine the maximum curvature for each design speed. The maximum superelevation may be different for different highway conditions. At the other extreme, no superelevation is needed for tangent highways or highways with extremely long-radius curves. For curvature between these extremes and for a given design speed, the superelevation should be chosen in such manner that there is a logical relation between the side friction factor and the applied superelevation rate.

Maximum Superelevation Rates

The maximum rates of superelevation used on highways are controlled by four factors: climate conditions (i.e., frequency and amount of snow and ice); terrain conditions (i.e., flat, rolling, or mountainous); type of area (i.e., rural or urban); and frequency of very slow-moving vehicles whose operation might be affected by high superelevation rates. Consideration of these factors jointly leads to the conclusion that no single maximum superelevation rate is universally applicable and that a range of values should be used. However, using only one maximum superelevation rate within a region of similar climate and land use is desirable, as such a practice promotes design consistency.

Design consistency relates to the uniformity of the highway alignment and its associated design element dimensions. This uniformity allows drivers to improve their perception-reaction skills by developing expectancies. Design elements that are not uniform for similar types of roadways may be counter to a driver's expectancy and result in an increase in driver workload. Logically, there is an inherent relationship between design consistency, driver workload, and motorist safety with "consistent" designs being associated with lower workloads and safer highways.

The highest superelevation rate for highways in common use is 10 percent, although 12 percent is used in some cases. Superelevation rates above 8 percent are only used in areas without snow and ice. Although higher superelevation rates offer an advantage to those drivers traveling at high speeds, current practice considers that rates in excess of 12 percent are beyond practical limits. This practice recognizes the combined effects of construction processes, maintenance difficulties, and operation of vehicles at low speeds.

Thus, a superelevation rate of 12 percent appears to represent a practical maximum value where snow and ice do not exist. A superelevation rate of 12 percent may be used on low-volume gravel-surfaced roads to facilitate cross drainage; however, superelevation rates of this magnitude can cause higher speeds, which are conducive to rutting and displacement of gravel. Generally, 8 percent is recognized as a reasonable maximum value for superelevation rate.

Where snow and ice are factors, tests and experience show that a superelevation rate of about 8 percent is a logical maximum to minimize slipping across a highway by stopped vehicles or vehicles attempting to start slowly from a stopped position. One series of tests (**16**) found coefficients of friction for ice ranging from 0.050 to 0.200, depending on the condition of the ice (i.e., wet, dry, clean, smooth, or rough). Tests on loose or packed snow show coefficients of friction ranging from 0.200 to 0.400. Other tests (**21**) have corroborated these values. The lower extreme of this range of coefficients of friction probably occurs only under thin film "quick freeze" conditions at a temperature of about –1°C [30°F] in the presence of water on the pavement. Similar low friction values may occur with thin layers of mud on the pavement surface, with oil or flushed spots, and with high speeds and a sufficient depth of water on the pavement surface to permit hydroplaning. For these reasons some highway agencies have adopted a maximum superelevation rate of 8 percent. Such agencies believe that 8 percent represents a logical maximum superelevation rate, regardless of snow or ice conditions. Such a limit tends to reduce the likelihood that slow drivers will experience negative side friction, which can result in excessive steering effort and erratic operation.

Where traffic congestion or extensive marginal development acts to restrict top speeds, it is common practice to utilize a low maximum rate of superelevation, usually 4 to 6 percent. Similarly, either a low maximum rate of superelevation or no superelevation is employed within important intersection areas or where there is a tendency to drive slowly because of turning and crossing movements, warning devices, and signals. In these areas it is difficult to warp crossing pavements for drainage without providing negative superelevation for some turning movements.

In summary, it is recommended that (1) several rates, rather than a single rate, of maximum superelevation should be recognized in establishing design controls for highway curves, (2) a rate of 12 percent should not be exceeded, (3) a rate of 4 or 6 percent is applicable for urban design in areas with little or no constraints, and (4) superelevation may be omitted on low-speed urban streets where severe constraints are present. Accordingly, five maximum superelevation rates— 4, 6, 8, 10, and 12 percent—are used below. For each of these rates the maximum curvature and actual superelevation rates for flatter curves are determined. In actual design practice, an agency will generally use different superelevation rates within the normal range of rates described above for different road systems.

Minimum Radius

The minimum radius is a limiting value of curvature for a given design speed and is determined from the maximum rate of superelevation and the maximum side friction factor selected for design (limiting value of f). Use of sharper curvature for that design speed would call for superelevation beyond the limit considered practical or for operation with tire friction and lateral acceleration beyond what is considered comfortable by many drivers, or both. Although based on a threshold of driver comfort, rather than safety, the minimum radius of curvature is a significant value in alignment design. The minimum radius of curvature is also an important control value for determination of superelevation rates for flatter curves.

The minimum radius of curvature, R_{min}, can be calculated directly from the simplified curve formula introduced above in the section on the "Side Friction Factor." This formula can be recast to determine R_{min} as follows:

Metric	US Customary	
$R_{min} = \dfrac{V^2}{127(0.01e_{max} + f_{max})}$	$R_{min} = \dfrac{V^2}{15(0.01e_{max} + f_{max})}$	(3-10)

Design for Rural Highways, Urban Freeways, and High-Speed Urban Streets

On rural highways, on urban freeways, and on urban streets where speed is relatively high and relatively uniform, horizontal curves are generally superelevated and successive curves are generally balanced to provide a smooth-riding transition from one curve to the next. A balanced design for a series of curves of varying radii is provided by the appropriate distribution of e and f values, as discussed above, to select an appropriate superelevation rate in the range from the normal cross slope to maximum superelevation.

Exhibit 3-13 shows the recommended values of the side friction factor for rural highways, urban freeways, and high-speed urban streets as a solid line superimposed on the analysis curves from Exhibit 3-10. These recommended side friction factors provide a reasonable margin of safety at high speeds and lead to somewhat lower superelevation rates for low design speeds than do some of the other curves. The lower superelevation rates at the low speeds provide a greater margin of safety to offset the tendency of many motorists to overdrive highways with low design speeds.

For the reasons discussed above, it is recommended that maximum side friction factors for design of rural highways, urban freeways, and high-speed urban streets should be those represented by the solid line in Exhibit 3-13. These maximum side friction factors vary directly with design speed from 0.17 at 30 km/h [20 mph] to 0.14 at 80 km/h [50 mph] and then directly with design speed from 0.14 at 80 km/h [50 mph] to 0.08 at 130 km/h [80 mph]. The research report *Side Friction for Superelevation on Horizontal Curves* (22) confirms the appropriateness of these design values.

Based on the maximum allowable side friction factors from Exhibit 3-13, Exhibit 3-14 gives the minimum radius for each of the five maximum superelevation rates for design speeds from 20 to 130 km/h [15 to 80 mph].

Method 5, described previously, is recommended for the distribution of e and f for all curves with radii greater than the minimum radius of curvature on rural highways, urban freeways, and high-speed urban streets. Use of method 5 is discussed in the following text and exhibits.

METRIC

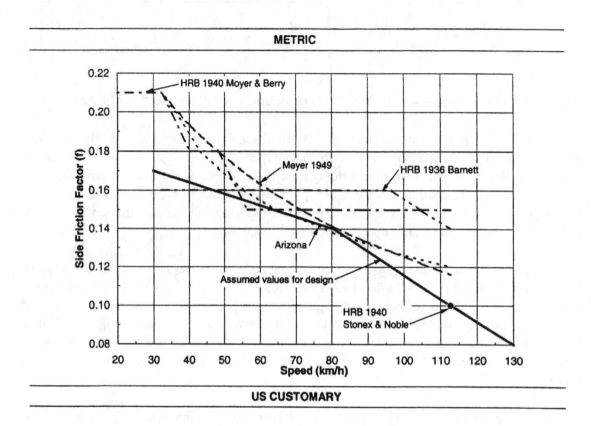

US CUSTOMARY

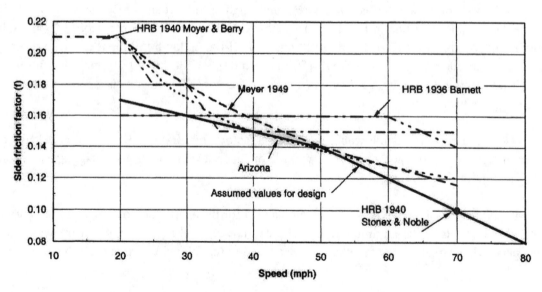

Exhibit 3-13. Side Friction Factors for Rural Highways and High-Speed Urban Streets

Metric

Design Speed (km/h)	Maximum e (%)	Limiting Values of f	Total (e/100 + f)	Calculated Radius (m)	Rounded Radius (m)
20	4.0	0.18	0.22	14.3	15
30	4.0	0.17	0.21	33.7	35
40	4.0	0.17	0.21	60.0	60
50	4.0	0.16	0.20	98.4	100
60	4.0	0.15	0.19	149.1	150
70	4.0	0.14	0.18	214.2	215
80	4.0	0.14	0.18	279.8	280
90	4.0	0.13	0.17	375.0	375
100	4.0	0.12	0.16	491.9	490
20	6.0	0.18	0.24	13.1	15
30	6.0	0.17	0.23	30.8	30
40	6.0	0.17	0.23	54.7	55
50	6.0	0.16	0.22	89.4	90
60	6.0	0.15	0.21	134.9	135
70	6.0	0.14	0.20	192.8	195
80	6.0	0.14	0.20	251.8	250
90	6.0	0.13	0.19	335.5	335
100	6.0	0.12	0.18	437.2	435
110	6.0	0.11	0.17	560.2	560
120	6.0	0.09	0.15	755.5	755
130	6.0	0.08	0.14	950.0	950
20	8.0	0.18	0.26	12.1	10
30	8.0	0.17	0.25	28.3	30
40	8.0	0.17	0.25	50.4	50
50	8.0	0.16	0.24	82.0	80
60	8.0	0.15	0.23	123.2	125
70	8.0	0.14	0.22	175.3	175
80	8.0	0.14	0.22	228.9	230
90	8.0	0.13	0.21	303.6	305
100	8.0	0.12	0.20	393.5	395
110	8.0	0.11	0.19	501.2	500
120	8.0	0.09	0.17	666.6	665
130	8.0	0.08	0.16	831.3	830
20	10.0	0.18	0.28	11.2	10
30	10.0	0.17	0.27	26.2	25
40	10.0	0.17	0.27	46.6	45
50	10.0	0.16	0.26	75.7	75
60	10.0	0.15	0.25	113.3	115
70	10.0	0.14	0.24	160.7	160
80	10.0	0.14	0.24	209.9	210
90	10.0	0.13	0.23	277.2	275
100	10.0	0.12	0.22	357.7	360
110	10.0	0.11	0.21	453.5	455
120	10.0	0.09	0.19	596.5	595
130	10.0	0.08	0.18	738.9	740
20	12.0	0.18	0.30	10.5	10
30	12.0	0.17	0.29	24.4	25
40	12.0	0.17	0.29	43.4	45
50	12.0	0.16	0.28	70.3	70
60	12.0	0.15	0.27	104.9	105
70	12.0	0.14	0.26	148.3	150
80	12.0	0.14	0.26	193.7	195
90	12.0	0.13	0.25	255.0	255
100	12.0	0.12	0.24	327.9	330
110	12.0	0.11	0.23	414.0	415
120	12.0	0.09	0.21	539.7	540
130	12.0	0.08	0.20	665.0	665

US Customary

Design Speed (mph)	Maximum e (%)	Limiting Values of f	Total (e/100 + f)	Calculated Radius (ft)	Rounded Radius (ft)
15	4.0	0.175	0.215	70.0	70
20	4.0	0.170	0.210	127.4	125
25	4.0	0.165	0.205	203.9	205
30	4.0	0.160	0.200	301.0	300
35	4.0	0.155	0.195	420.2	420
40	4.0	0.150	0.190	563.3	565
45	4.0	0.145	0.185	732.2	730
50	4.0	0.140	0.180	929.0	930
55	4.0	0.130	0.170	1190.2	1190
60	4.0	0.120	0.160	1505.0	1505
15	6.0	0.175	0.235	64.0	65
20	6.0	0.170	0.230	116.3	115
25	6.0	0.165	0.225	185.8	185
30	6.0	0.160	0.220	273.6	275
35	6.0	0.155	0.215	381.1	380
40	6.0	0.150	0.210	509.6	510
45	6.0	0.145	0.205	660.7	660
50	6.0	0.140	0.200	836.1	835
55	6.0	0.130	0.190	1065.0	1065
60	6.0	0.120	0.180	1337.8	1340
65	6.0	0.110	0.170	1662.4	1660
70	6.0	0.100	0.160	2048.5	2050
75	6.0	0.090	0.150	2508.4	2510
80	6.0	0.080	0.140	3057.8	3060
15	8.0	0.175	0.255	59.0	60
20	8.0	0.170	0.250	107.0	105
25	8.0	0.165	0.245	170.8	170
30	8.0	0.160	0.240	250.8	250
35	8.0	0.155	0.235	348.7	350
40	8.0	0.150	0.230	465.3	465
45	8.0	0.145	0.225	602.0	600
50	8.0	0.140	0.220	760.1	760
55	8.0	0.130	0.210	963.5	965
60	8.0	0.120	0.200	1204.0	1205
65	8.0	0.110	0.190	1487.4	1485
70	8.0	0.100	0.180	1820.9	1820
75	8.0	0.090	0.170	2213.3	2215
80	8.0	0.080	0.160	2675.6	2675
15	10.0	0.175	0.275	54.7	55
20	10.0	0.170	0.270	99.1	100
25	10.0	0.165	0.265	157.8	160
30	10.0	0.160	0.260	231.5	230
35	10.0	0.155	0.255	321.3	320
40	10.0	0.150	0.250	428.1	430
45	10.0	0.145	0.245	552.9	555
50	10.0	0.140	0.240	696.8	695
55	10.0	0.130	0.230	879.7	880
60	10.0	0.120	0.220	1094.6	1095
65	10.0	0.110	0.210	1345.8	1345
70	10.0	0.100	0.200	1638.8	1640
75	10.0	0.090	0.190	1980.3	1980
80	10.0	0.080	0.180	2378.3	2380
15	12.0	0.175	0.295	51.0	50
20	12.0	0.170	0.290	92.3	90
25	12.0	0.165	0.285	146.7	145
30	12.0	0.160	0.280	215.0	215
35	12.0	0.155	0.275	298.0	300
40	12.0	0.150	0.270	396.4	395
45	12.0	0.145	0.265	511.1	510
50	12.0	0.140	0.260	643.2	645
55	12.0	0.130	0.250	809.4	810
60	12.0	0.120	0.240	1003.4	1005
65	12.0	0.110	0.230	1228.7	1230
70	12.0	0.100	0.220	1489.8	1490
75	12.0	0.090	0.210	1791.7	1790
80	12.0	0.080	0.200	2140.5	2140

Note: In recognition of safety considerations, use of e_{max} = 4.0% should be limited to urban conditions.

Exhibit 3-14. Minimum Radius for Design of Rural Highways, Urban Freeways, and High-Speed Urban Streets Using Limiting Values of e and f

Procedure for Development of Finalized e Distribution

The side friction factors shown as the solid line on Exhibit 3-13 represent the maximum f values selected for design for each speed. When these values are used in conjunction with the recommended method 5, they determine the f distribution curves for the various speeds. Subtracting these computed f values from the computed value of e/100 + f at the design speed, the finalized e distribution is thus obtained (see Exhibit 3-15). The finalized e distribution curves resulting from this approach, based on method 5 and used below, are shown in Exhibits 3-16 to 3-20.

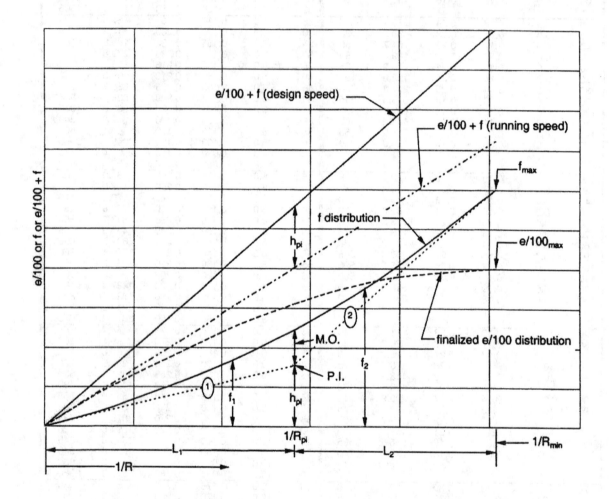

Exhibit 3-15. Method 5 Procedure for Development of the Finalized e Distribution

METRIC

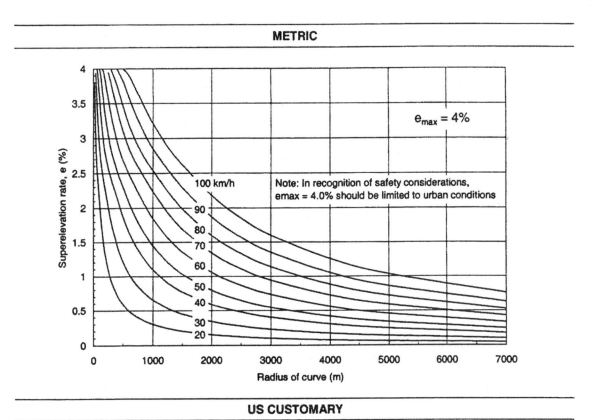

US CUSTOMARY

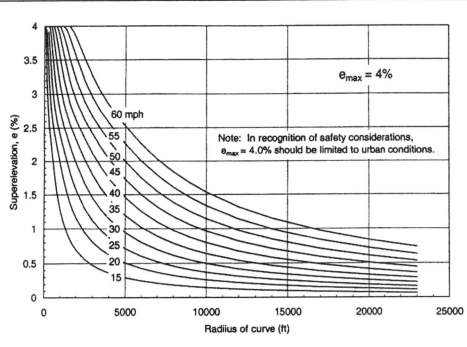

Exhibit 3-16. Design Superelevation Rates for Maximum Superelevation Rate of 4 Percent

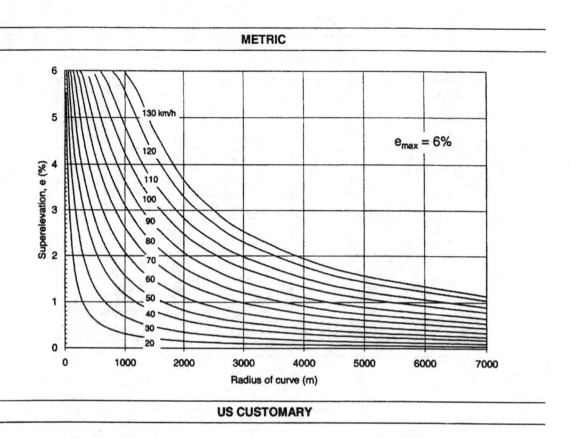

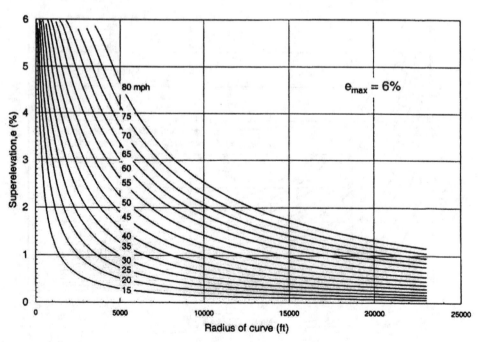

Exhibit 3-17. Design Superelevation Rates for Maximum Superelevation Rate of 6 Percent

METRIC

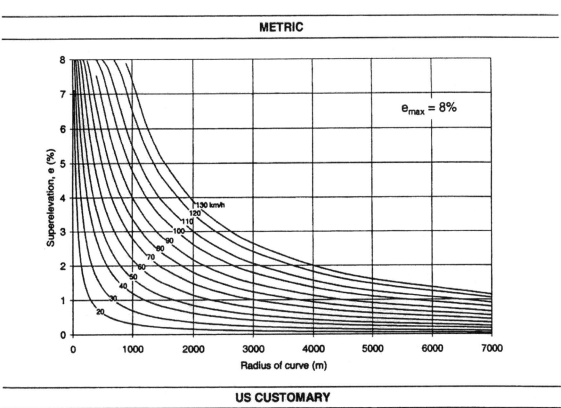

US CUSTOMARY

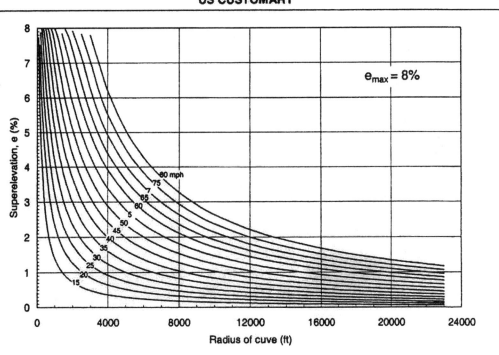

Exhibit 3-18. Design Superelevation Rates for Maximum Superelevation Rate of 8 Percent

METRIC

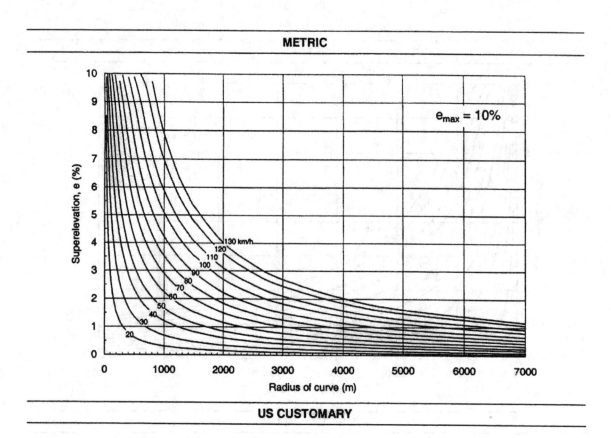

US CUSTOMARY

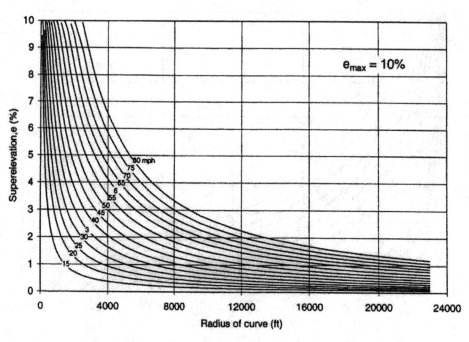

Exhibit 3-19. Design Superelevation Rates for Maximum Superelevation Rate of 10 Percent

METRIC

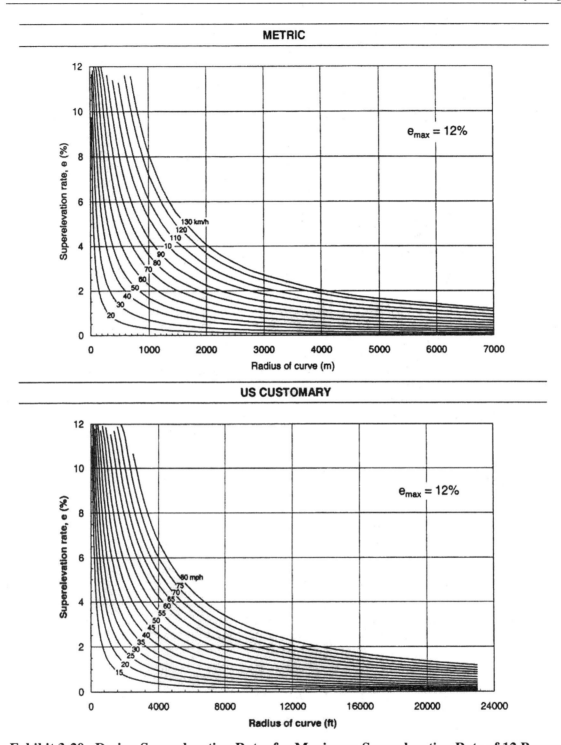

US CUSTOMARY

Exhibit 3-20. Design Superelevation Rates for Maximum Superelevation Rate of 12 Percent

As Exhibit 3-15 illustrates, the f distribution curve at the design speed, using method 5, results in an unsymmetrical parabolic curve with legs 1 and 2. These legs correspond to curves 4 and 3-4, respectively, in Exhibit 3-12B. The terms used in the derivation of the equations used to compute the f and finalized e distributions are illustrated in Exhibit 3-15.

The e and f distributions for method 5 may be derived using the basic curve formula, neglecting the (1 - 0.01ef) term, using the following sequence of equations:

Metric	US Customary	
$$0.01e + f = \frac{0.0079V^2}{R}$$	$$0.01e + f = \frac{0.067V^2}{R}$$	(3-11)

Metric	US Customary
where:	where:
V_D = V = design speed, km/h;	V_D = V = design speed, mph;
e_{max} = e = maximum superelevation, percent;	e_{max} = e = maximum superelevation, percent;
f_{max} = f = maximum allowable side friction factor;	f_{max} = f = maximum allowable side friction factor;
R_{min} = R = minimum radius, meters;	R_{min} = R = minimum radius, feet;
R_{PI} = R = radius at the point of intersection, PI, of legs (1) and (2) of the f distribution parabolic curve (= R at the point of intersection of $0.01e_{max}$ and $(0.01e + f)_R$);	R_{PI} = R = radius at the point of intersection, PI, of legs (1) and (2) of the f distribution parabolic curve (= R at the point of intersection of $0.01e_{max}$ and $(0.01e + f)_R$);
V_R = running speed, km/h	V_R = running speed, mph

Metric	US Customary	
then:	then:	(3-12)
$$R_{min} = \frac{0.0079V_D^2}{0.01e_{max} + f_{max}}$$	$$R_{min} = \frac{0.067V_D^2}{0.01e_{max} + f_{max}}$$	
and	and	(3-13)
$$R_{PI} = \frac{0.0079V_R^2}{0.01e_{max}}$$	$$R_{PI} = \frac{0.067V_R^2}{0.01e_{max}}$$	

Metric	US Customary	
Because $(0.01e + f)_D - (0.01e + f)_R = h$, at point R_{PI} the equations reduce to the following:	Because $(0.01e + f)_D - (0.01e + f)_R = h$, at point R_{PI} the equations reduce to the following:	(3-14)
$$h_{PI} = \left(\frac{(0.01e_{max})V_D^2}{V_R^2} \right) - 0.01e_{max}$$	$$h_{PI} = \left(\frac{(0.01e_{max})V_D^2}{V_R^2} \right) - 0.01e_{max}$$	
where h_{PI} = PI offset from the 1/R axis.	where h_{PI} = PI offset from the 1/R axis.	
Also,	Also,	(3-15)
$$S_1 = h_{PI}(R_{PI})$$	$$S_1 = \frac{h_{PI}(R_{PI})}{5729.58}$$	

Metric	US Customary	
where S_1 = slope of leg 1 and	where S_1 = slope of leg 1 and	
		(3-16)
$$S_2 = \frac{f_{max} - h_{PI}}{\dfrac{1}{R_{min}} - \dfrac{1}{R_{PI}}}$$	$$S_2 = \frac{f_{max} - h_{PI}}{5729.58 \left(\dfrac{1}{R_{min}} - \dfrac{1}{R_{PI}} \right)}$$	
where S_2 = slope of leg 2.	where S_2 = slope of leg 2.	
		(3-17)
The equation for the middle ordinate (MO) of an unsymmetrical vertical curve is the following:	The equation for the middle ordinate (MO) of an unsymmetrical vertical curve is the following:	
$$MO = \frac{L_1 L_2 (S_2 - S_1)}{2(L_1 + L_2)}$$	$$MO = \frac{L_1 L_2 (S_2 - S_1)}{2(L_1 + L_2)}$$	
where: $L_1 = 1/R_{PI}$ and $L_2 = 1/R_{min} - 1/R_{PI}$. It follows that:	where: $L_1 = 5729.58/R_{PI}$ and $L_2 = 5729.58(1/R_{min} - 1/R_{PI})$. It follows that:	
		(3-18)
$$MO = \frac{1}{R_{PI}} \left(\frac{1}{R_{min}} - \frac{1}{R_{PI}} \right) \left(\frac{S_2 - S_1}{2} \right) R_{min}$$	$$MO = \frac{5729.58}{R_{PI}} \left(\frac{1}{R_{min}} - \frac{1}{R_{PI}} \right) \left(\frac{S_2 - S_1}{2} \right) (R_{min})$$	
where MO = middle ordinate of the f distribution curve, and	where MO = middle ordinate of the f distribution curve, and	
		(3-19)
$$(0.01e + f)_D = \frac{(0.01e_{max} + f_{max})R_{min}}{R}$$	$$(0.01e + f)_D = \frac{(0.01e_{max} + f_{max})R_{min}}{R}$$	
in which R = radius at any point.	in which R = radius at any point.	

Metric	US Customary	
Use the general vertical curve equation:	Use the general vertical curve equation:	(3-20)
$$\frac{Y}{MO} = \left(\frac{x}{L}\right)^2$$	$$\frac{Y}{MO} = \left(\frac{x}{L}\right)^2$$	
with 1/R measured from the vertical axis.	with 1/R measured from the vertical axis.	
with 1/R ≤ 1/R$_{PI}$,	with 1/R ≤ 1/R$_{PI}$,	
		(3-21)
$$f_1 = MO\left(\frac{R_{PI}}{R}\right)^2 + \frac{S_1}{R}$$	$$f_1 = MO\left(\frac{R_{PI}}{R}\right)^2 + \frac{5729.58(S_1)}{R}$$	
where: f_1 = f distribution at any point 1/R ≤ 1/R$_{PI}$; and	where: f_1 = f distribution at any point 1/R ≤ 1/R$_{PI}$; and	
		(3-22)
$$0.01e_1 = (0.01e + f)_D - f_1$$	$$0.01e_1 = (0.01e + f)_D - f_1$$	
where: $0.01e_1$ = 0.01e distribution at any point 1/R ≤ 1/R$_{PI}$.	where: $0.01e_1$ = 0.01e distribution at any point 1/R ≤ 1/R$_{PI}$.	
For 1/R > 1/R$_{PI}$,	For 1/R > 1/R$_{PI}$,	(3-23)
$$f_2 = MO\left(\frac{\frac{1}{R_{min}} - \frac{1}{R}}{\frac{1}{R_{min}} - \frac{1}{R_{PI}}}\right)^2 + h_{PI} + S_2\left(\frac{1}{R} - \frac{1}{R_{PI}}\right)$$	$$f_2 = MO\left(\frac{\frac{1}{R_{min}} - \frac{1}{R}}{\frac{1}{R_{min}} - \frac{1}{R_{PI}}}\right)^2 + h_{PI} + 5729.58\left(S_2\right)\left(\frac{1}{R} - \frac{1}{R_{PI}}\right)$$	
where: f_2 = f distribution at any point 1/R > 1/R$_{PI}$; and	where: f_2 = f distribution at any point 1/R > 1/R$_{PI}$; and	
$$0.01e_2 = (0.01e + f)_D - f_2$$	$$0.01e_2 = (0.01e + f)_D - f_2$$	(3-24)
where: $0.01e_2$ = 0.01e distribution at any point 1/R >1/R$_{PI}$.	where: $0.01e_2$ = 0.01e distribution at any point 1/R >1/R$_{PI}$.	

Exhibit 3-15 is a typical layout illustrating the method 5 procedure for development of the finalized e distribution. The figure depicts how the f value is determined for 1/R and then subtracted from the value of (e/100 + f) to determine e/100.

An example of the procedure to calculate e for a design speed of 80 km/h [50 mph] and an e_{max} of 10 percent is shown below:

Example	
Metric	**US Customary**

Determine e given: V_D = 80 km/h
                      e_{max} = 10 percent
	Determine e given: V_D = 50 mph
	                      e_{max} = 10 percent

From Exhibit 3-26: V_R = 70 km/h
From Exhibit 3-14:   f = 0.14
                      (maximum allowable
                      side friction factor)
	From Exhibit 3-26: V_R = 44 mph
	From Exhibit 3-14: f = 0.14
	                      (maximum allowable
	                      side friction factor)

Using the appropriate equations yields:	Using the appropriate equations yields:
R_{min} = 210.7, R_{PI} = 387.1, and h_{PI} = 0.031	R_{min} = 697.9, R_{PI} = 1297.12, and h_{PI} = 0.029
S_1 = 11.95 and S_2 = 50.23	S_1 = 0.0066 and S_2 = 0.0293
Substituting, the middle ordinate becomes 0.022.	Substituting, the middle ordinate becomes 0.0231.
The e distribution value for any radius is found by taking the $(0.01e + f)_D$ value minus the f_1 or f_2 value (refer to Exhibit 3-15). Thus, the e distribution value for an R = R_{PI} would be $(0.01e + f)_D = 0.0079(V_D)^2/R = 0.131$ minus an f_1 = 0.053, which results in 0.078. This value multiplied by 100 to convert to percent corresponds to the e value, which can be interpolated for R = 386 m at the 80 km/h design speed in Exhibit 3-24.	The e distribution value for any radius is found by taking the $(0.01e + f)_D$ value minus the f_1 or f_2 value (refer to Exhibit 3-15). Thus, the e distribution value for an R = R_{PI} would be $(0.01e + f)_D = 0.067(V_D)^2/R = 0.129$ minus an f_1 = 0.052, which results in 0.077. This value multiplied by 100 to convert to percent corresponds to the e value, which can be interpolated for R = 1,298 ft at the 50 mph design speed in Exhibit 3-24.

Design Superelevation Tables

Exhibits 3-21 to 3-25 show, in addition to length of runoff or transition discussed later in this chapter, values of R and the resulting superelevation for different design speeds for each of five values of maximum superelevation rate (i.e., for a full range of common design conditions). The minimum radii for each of the five maximum superelevation rates were calculated from the simplified curve formula, with the use of f values from Exhibit 3-13. Method 5 was used to distribute e and f in calculating the appropriate superelevation rates for the remainder of the range of curvature. Under all but extreme weather conditions, vehicles can travel safely at speeds higher than the design speed on horizontal curves with the superelevation rates indicated in the tables. This is due to the development of a radius/superelevation relationship that uses friction factors that are generally considerably less than can be achieved. This is illustrated in Exhibit 3-11, which compares the friction factors used in design of various types of highway facilities and the maximum side friction factors available on certain wet and dry concrete pavements.

METRIC

Values are tabulated for design speeds V_d = 20, 30, 40, 50, 60, 70, 80, 90, and 100 km/h. For each speed the columns give e (%) and the minimum length of runoff L (m) for 2 Lns and 4 Lns.

R (m)	20 e(%)	20 L 2Lns	20 L 4Lns	30 e(%)	30 L 2Lns	30 L 4Lns	40 e(%)	40 L 2Lns	40 L 4Lns	50 e(%)	50 L 2Lns	50 L 4Lns	60 e(%)	60 L 2Lns	60 L 4Lns	70 e(%)	70 L 2Lns	70 L 4Lns	80 e(%)	80 L 2Lns	80 L 4Lns	90 e(%)	90 L 2Lns	90 L 4Lns	100 e(%)	100 L 2Lns	100 L 4Lns
7000	NC	0	0	NC	0	0	NC	0	0	NC	0	0	NC	0	0	NC	0	0	NC	0	0	NC	0	0	NC	0	0
5000	NC	0	0	NC	0	0	NC	0	0	NC	0	0	NC	0	0	NC	0	0	NC	0	0	NC	0	0	NC	0	0
3000	NC	0	0	NC	0	0	NC	0	0	NC	0	0	NC	0	0	NC	0	0	NC	0	0	NC	0	0	RC	16	25
2500	NC	0	0	NC	0	0	NC	0	0	NC	0	0	NC	0	0	NC	0	0	NC	0	0	RC	15	23	RC	16	25
2000	NC	0	0	NC	0	0	NC	0	0	NC	0	0	NC	0	0	NC	0	0	RC	14	22	RC	15	23	2.2	18	27
1500	NC	0	0	NC	0	0	NC	0	0	NC	0	0	NC	0	0	RC	13	20	RC	14	22	2.3	18	26	2.6	21	32
1400	NC	0	0	NC	0	0	NC	0	0	NC	0	0	NC	0	0	RC	13	20	2.1	15	23	2.4	18	28	2.7	22	33
1300	NC	0	0	NC	0	0	NC	0	0	NC	0	0	NC	0	0	RC	13	20	2.2	16	24	2.5	19	29	2.8	23	34
1200	NC	0	0	NC	0	0	NC	0	0	NC	0	0	RC	12	18	RC	13	20	2.3	17	25	2.6	20	30	2.9	24	36
1000	NC	0	0	NC	0	0	NC	0	0	NC	0	0	RC	12	18	2.2	14	22	2.5	18	27	2.8	21	32	3.2	26	39
900	NC	0	0	NC	0	0	NC	0	0	RC	11	17	RC	12	18	2.4	16	24	2.7	19	29	3.0	23	34	3.4	28	42
800	NC	0	0	NC	0	0	NC	0	0	RC	11	17	RC	12	19	2.5	16	25	2.8	20	30	3.2	25	37	3.5	29	43
700	NC	0	0	NC	0	0	NC	0	0	RC	11	17	2.1	13	19	2.7	18	27	3.0	22	32	3.4	26	39	3.7	30	45
600	NC	0	0	NC	0	0	RC	10	15	2.1	11	17	2.3	14	21	2.9	19	28	3.2	23	35	3.6	28	41	3.8	32	48
500	NC	0	0	NC	0	0	RC	10	15	2.3	12	19	2.5	15	23	3.1	20	30	3.5	25	38	3.8	29	44	4.0	33	49
400	NC	0	0	NC	0	0	2.1	11	16	2.5	14	21	2.7	16	24	3.4	22	33	3.7	27	40	4.0	31	46			
300	RC	9	14	RC	10	14	2.4	12	19	2.8	16	23	3.0	18	27	3.8	25	37	4.0	29	43						
250	RC	9	14	RC	10	14	2.6	13	20	3.0	17	25	3.3	20	30	3.9	26	38									
200	RC	9	14	2.3	11	17	2.8	14	22	3.3	18	27	3.6	22	32												
175	RC	9	14	2.4	12	17	2.9	15	22	3.5	19	29	3.8	23	34												
150	RC	9	14	2.5	12	18	3.1	16	24	3.7	20	31	3.9	23	35												
140	RC	9	14	2.6	12	19	3.2	16	25	3.8	21	32	4.0	24	36												
130	RC	9	14	2.6	12	19	3.3	17	25	3.8	21	32															
120	RC	9	14	2.7	13	19	3.4	17	26	3.9	22	32															
110	RC	9	14	2.8	13	20	3.5	18	27	4.0	22	33															
100	2.1	9	14	2.9	14	21	3.6	19	28	4.0	22	33															
90	2.2	10	15	3.0	14	22	3.7	19	29																		
80	2.4	11	16	3.2	15	23	3.8	20	29																		
70	2.5	11	17	3.3	16	24	3.9	20	30																		
60	2.6	12	18	3.5	17	25	4.0	21	31																		
50	2.8	13	19	3.7	18	27																					
40	3.0	14	20	3.9	19	28																					
30	3.3	15	22																								
20	3.8	17	26																								
R_{min}	15			35			60			100			150			215			280			375			490		

Legend:

e_{max}	=	4%
R	=	radius of curve
V_d	=	assumed design speed
e	=	rate of superelevation
L	=	minimum length of runoff (does not include tangent runout) as discussed in "Tangent-to-Curve Transition" section
NC	=	normal crown section
RC	=	remove adverse crown, superelevate at normal crown slope

Use of e_{max} = 4% should be limited to urban conditions

Exhibit 3-21. Values for Design Elements Related to Design Speed and Horizontal Curvature

US CUSTOMARY

R (ft)	V=15 mph e (%)	15 L 2Lns	15 L 4Lns	V=20 mph e (%)	20 L 2Lns	20 L 4Lns	V=25 mph e (%)	25 L 2Lns	25 L 4Lns	V=30 mph e (%)	30 L 2Lns	30 L 4Lns	V=35 mph e (%)	35 L 2Lns	35 L 4Lns	V=40 mph e (%)	40 L 2Lns	40 L 4Lns	V=45 mph e (%)	45 L 2Lns	45 L 4Lns	V=50 mph e (%)	50 L 2Lns	50 L 4Lns	V=55 mph e (%)	55 L 2Lns	55 L 4Lns	V=60 mph e (%)	60 L 2Lns	60 L 4Lns
23000	NC	0	0	NC	0	0	NC	0	0	NC	0	0	NC	0	0	NC	0	0	NC	0	0	NC	0	0	NC	0	0	NC	0	0
20000	NC	0	0	NC	0	0	NC	0	0	NC	0	0	NC	0	0	NC	0	0	NC	0	0	NC	0	0	NC	0	0	NC	0	0
17000	NC	0	0	NC	0	0	NC	0	0	NC	0	0	NC	0	0	NC	0	0	NC	0	0	NC	0	0	NC	0	0	NC	0	0
14000	NC	0	0	NC	0	0	NC	0	0	NC	0	0	NC	0	0	NC	0	0	NC	0	0	NC	0	0	NC	0	0	NC	0	0
12000	NC	0	0	NC	0	0	NC	0	0	NC	0	0	NC	0	0	NC	0	0	NC	0	0	NC	0	0	NC	0	0	NC	0	0
10000	NC	0	0	NC	0	0	NC	0	0	NC	0	0	NC	0	0	NC	0	0	NC	0	0	NC	0	0	NC	0	0	NC	0	0
8000	NC	0	0	NC	0	0	NC	0	0	NC	0	0	NC	0	0	NC	0	0	NC	0	0	NC	0	0	RC	51	77	RC	53	80
6000	NC	0	0	NC	0	0	NC	0	0	NC	0	0	NC	0	0	NC	0	0	NC	0	0	RC	48	72	RC	51	77	2.3	61	92
5000	NC	0	0	NC	0	0	NC	0	0	NC	0	0	NC	0	0	NC	0	0	RC	44	67	RC	46	72	2.3	59	88	2.5	67	100
4000	NC	0	0	NC	0	0	NC	0	0	NC	0	0	NC	0	0	RC	41	62	RC	44	67	2.3	55	83	2.6	66	100	2.8	75	112
3500	NC	0	0	NC	0	0	NC	0	0	NC	0	0	RC	39	58	RC	41	62	2.2	49	73	2.5	60	90	2.7	69	103	3.0	80	120
3000	NC	0	0	NC	0	0	NC	0	0	NC	0	0	RC	39	58	2.1	43	65	2.4	53	80	2.7	65	97	2.9	74	111	3.3	88	132
2500	NC	0	0	NC	0	0	NC	0	0	RC	36	55	RC	39	58	2.4	50	74	2.6	58	87	2.9	70	104	3.2	82	123	3.5	93	140
2000	NC	0	0	NC	0	0	RC	34	51	RC	36	55	2.3	45	67	2.6	54	81	2.9	64	97	3.2	77	115	3.5	89	134	3.8	101	152
1800	NC	0	0	NC	0	0	RC	34	51	2.1	38	57	2.4	46	70	2.7	56	84	3.0	67	100	3.3	79	119	3.7	94	142	3.9	104	156
1600	NC	0	0	NC	0	0	RC	34	51	2.2	40	60	2.6	50	75	2.9	60	90	3.2	71	107	3.5	84	126	3.8	97	146	4.0	107	160
1400	NC	0	0	NC	0	0	RC	34	51	2.4	44	65	2.7	52	78	3.0	62	93	3.4	76	113	3.7	89	133	3.9	100	149			
1200	NC	0	0	RC	32	49	2.2	38	57	2.5	45	68	2.9	56	84	3.2	66	99	3.6	80	120	3.9	94	140	4.0	102	153			
1000	NC	0	0	RC	32	49	2.4	41	62	2.7	49	74	3.1	60	90	3.5	72	109	3.8	84	127	4.0	96	144						
900	NC	0	0	2.1	34	51	2.5	43	64	2.9	53	79	3.2	62	93	3.6	74	112	3.9	87	130									
800	NC	0	0	2.2	36	54	2.6	45	67	3.0	55	82	3.4	66	99	3.8	79	118	4.0	89	133									
700	RC	31	46	2.3	37	56	2.7	46	69	3.2	58	87	3.6	70	105	3.9	81	121												
600	RC	31	46	2.5	41	61	2.9	50	75	3.4	62	93	3.8	74	110	4.0	83	124												
500	2.1	32	48	2.6	42	63	3.1	53	80	3.6	65	98	3.9	75	113															
450	2.2	34	51	2.7	44	66	3.2	55	82	3.7	67	101	4.0	77	116															
400	2.3	35	53	2.9	47	71	3.4	58	87	3.8	69	104																		
350	2.4	37	55	3.0	49	73	3.6	62	93	3.9	71	106																		
300	2.6	40	60	3.2	52	78	3.7	63	95	4.0	73	109																		
250	2.7	42	62	3.4	55	83	3.9	67	100																					
200	3.0	46	69	3.7	60	90																								
150	3.3	51	76	3.9	63	95																								
100	3.8	58	88																											
75	4.0	62	92																											
R_min	70			125			205			300			420			565			730			930			1190			1505		

Legend:

e_{max}	=	4%
R	=	radius of curve
V_d	=	assumed design speed
e	=	rate of superelevation
L	=	minimum length of runoff (does not include tangent runout) as discussed in "Tangent-to-Curve Transition" section
NC	=	normal crown section
RC	=	remove adverse crown, superelevate at normal crown slope

Use of e_{max} = 4% should be limited to urban conditions

Exhibit 3-21. **Values for Design Elements Related to Design Speed and Horizontal Curvature (Continued)**

METRIC

Exhibit 3-22 — Values for Design Elements Related to Design Speed and Horizontal Curvature

Legend:

e_{max}	= 6%
R	= radius of curve
V_d	= assumed design speed
e	= rate of superelevation
L	= minimum length of runoff (does not include tangent runout) as discussed in "Tangent-to-Curve Transition" section
NC	= normal crown section
RC	= remove adverse crown, superelevate at normal crown slope

R (m)	V_d=20 e (%)	2 Lns	4 Lns	V_d=30 e (%)	2 Lns	4 Lns	V_d=40 e (%)	2 Lns	4 Lns	V_d=50 e (%)	2 Lns	4 Lns	V_d=60 e (%)	2 Lns	4 Lns	V_d=70 e (%)	2 Lns	4 Lns	V_d=80 e (%)	2 Lns	4 Lns	V_d=90 e (%)	2 Lns	4 Lns	V_d=100 e (%)	2 Lns	4 Lns	V_d=110 e (%)	2 Lns	4 Lns	V_d=120 e (%)	2 Lns	4 Lns	V_d=130 e (%)	2 Lns	4 Lns
7000	NC	0	0	NC	0	0	NC	0	0	NC	0	0	NC	0	0	NC	0	0	NC	0	0	NC	0	0	NC	0	0	NC	0	0	NC	0	0	NC	0	0
5000	NC	0	0	NC	0	0	NC	0	0	NC	0	0	NC	0	0	NC	0	0	NC	0	0	NC	0	0	NC	0	0	NC	0	0	NC	0	0	NC	0	0
3000	NC	0	0	NC	0	0	NC	0	0	NC	0	0	NC	0	0	NC	0	0	NC	0	0	NC	0	0	RC	0	0	RC	0	0	2.3	22	33	2.5	26	39
2500	NC	0	0	NC	0	0	NC	0	0	NC	0	0	NC	0	0	RC	0	0	RC	0	0	RC	0	0	RC	0	0	2.3	18	26	2.7	26	38	3.0	31	46
2000	NC	0	0	NC	0	0	NC	0	0	NC	0	0	RC	0	0	RC	0	0	2.2	14	22	2.1	15	23	2.5	20	31	2.8	25	37	3.3	31	47	3.7	38	57
1500	NC	0	0	NC	0	0	NC	0	0	NC	0	0	RC	0	0	2.1	13	20	2.4	16	24	2.7	18	24	3.1	25	38	3.6	32	47	4.2	40	60	4.7	48	73
1400	NC	0	0	NC	0	0	NC	0	0	NC	0	0	2.1	12	18	2.2	13	20	2.5	17	26	2.8	21	31	3.3	27	41	3.8	33	50	4.4	42	63	5.0	51	77
1300	NC	0	0	NC	0	0	NC	0	0	NC	0	0	2.3	12	18	2.4	14	21	2.7	18	27	3.0	21	32	3.5	29	43	4.0	35	53	4.7	45	67	5.3	55	82
1200	NC	0	0	NC	0	0	NC	0	0	NC	0	0	2.5	13	19	2.6	14	22	3.1	19	33	3.2	23	34	3.7	30	45	4.2	37	55	5.0	47	71	5.6	58	86
1000	NC	0	0	NC	0	0	NC	0	0	NC	0	0	2.8	14	21	2.8	17	26	3.4	24	37	3.6	25	37	4.2	34	52	4.8	45	67	5.6	53	80	6.0	62	93
900	NC	0	0	NC	0	0	NC	0	0	NC	0	0	3.1	15	23	3.1	18	27	3.6	26	39	3.9	30	45	4.5	37	55	5.1	47	71	5.8	55	82			
800	NC	0	0	NC	0	0	NC	0	0	RC	0	0	3.5	17	25	3.4	20	30	4.0	29	43	4.2	32	48	4.9	40	60	5.4	51	76	6.0	57	85			
700	NC	0	0	NC	0	0	NC	0	0	RC	0	0	4.0	19	28	3.8	22	33	4.3	31	46	4.6	35	53	5.2	43	64	5.8	53	79						
600	NC	0	0	NC	0	0	NC	0	0	RC	0	0	4.6	21	33	4.2	27	41	4.8	35	52	5.0	38	57	5.6	46	69									
500	NC	0	0	NC	0	0	NC	0	0	2.1	11	17	5.0	23	36	4.7	31	46	5.3	42	57	5.4	41	62	5.9	48	72									
400	NC	0	0	NC	0	0	RC	0	0	2.4	13	20	5.5	28	41	5.4	35	53	5.9	48	64	5.9	45	68												
300	RC	0	0	RC	0	0	RC	0	0	2.8	16	23	5.8	30	45	5.8	38	57	6.0	50	67															
250	RC	0	0	RC	0	0	2.1	10	15	3.3	18	27	6.0	33	52	6.0	39	59	6.0	43	65															
200	RC	0	0	2.3	10	14	2.5	10	15	3.9	22	32	6.0	36	54	6.0	41	59																		
175	2.1	9	14	2.8	10	14	3.1	11	16	4.2	23	35																								
150	2.2	10	14	3.0	11	17	3.5	13	19	4.7	26	39	5.8	35	53																					
140	2.4	10	15	3.3	13	20	3.9	16	24	5.0	28	44	6.0	36	54																					
130	2.1	9	14	3.6	14	22	4.1	17	26	5.3	31	47																								
120	2.2	9	14	3.8	16	24	4.4	20	30	5.7	32	48																								
110	2.4	10	14	3.9	17	26	4.5	21	32	5.8	33	50																								
100	2.5	10	15	4.1	18	27	4.8	23	34	6.0	33	50																								
90	2.7	11	16	4.2	18	28	5.2	25	38																											
80	3.0	12	18	4.5	20	30	5.4	28	42																											
70	3.2	14	20	4.7	22	32	5.6	30	45																											
60	3.5	16	24	5.0	23	34	6.0	31	46																											
50	3.8	17	26	5.4	24	36																														
40	4.2	19	29	5.8	28	42																														
30	4.7	21	32	6.0	29	43																														
20	5.5	25	37																																	
R_{min}	15			30			55			90			135			195			250			335			435			660			755			950		

Exhibit 3-22. Values for Design Elements Related to Design Speed and Horizontal Curvature

US CUSTOMARY

R (ft)	15 mph e(%)	15 L₂	15 L₄	20 mph e(%)	20 L₂	20 L₄	25 mph e(%)	25 L₂	25 L₄	30 mph e(%)	30 L₂	30 L₄	35 mph e(%)	35 L₂	35 L₄	40 mph e(%)	40 L₂	40 L₄	45 mph e(%)	45 L₂	45 L₄	50 mph e(%)	50 L₂	50 L₄	55 mph e(%)	55 L₂	55 L₄	60 mph e(%)	60 L₂	60 L₄	65 mph e(%)	65 L₂	65 L₄	70 mph e(%)	70 L₂	70 L₄	75 mph e(%)	75 L₂	75 L₄	80 mph e(%)	80 L₂	80 L₄
23000	NC	0	0	NC	0	0	NC	0	0	NC	0	0	NC	0	0	NC	0	0	NC	0	0	NC	0	0	NC	0	0	NC	0	0	NC	0	0	NC	0	0	NC	0	0	NC	0	0
20000	NC	0	0	NC	0	0	NC	0	0	NC	0	0	NC	0	0	NC	0	0	NC	0	0	NC	0	0	NC	0	0	NC	0	0	NC	0	0	NC	0	0	NC	0	0	NC	0	0
17000	NC	0	0	NC	0	0	NC	0	0	NC	0	0	NC	0	0	NC	0	0	NC	0	0	NC	0	0	NC	0	0	NC	0	0	NC	0	0	NC	0	0	NC	0	0	NC	0	0
14000	NC	0	0	NC	0	0	NC	0	0	NC	0	0	NC	0	0	NC	0	0	NC	0	0	NC	0	0	NC	0	0	NC	0	0	RC	56	84	RC	60	90	RC	63	95	RC	69	103
12000	NC	0	0	NC	0	0	NC	0	0	NC	0	0	NC	0	0	NC	0	0	NC	0	0	NC	0	0	NC	0	0	RC	53	80	RC	56	84	RC	60	90	RC	63	95	2.1	72	108
10000	NC	0	0	NC	0	0	NC	0	0	NC	0	0	NC	0	0	NC	0	0	NC	0	0	NC	0	0	RC	51	77	RC	53	80	2.9	64	96	2.1	63	95	2.3	73	109	2.5	86	129
8000	NC	0	0	NC	0	0	NC	0	0	NC	0	0	NC	0	0	NC	0	0	NC	0	0	RC	48	72	RC	51	77	2.6	69	104	3.4	81	121	2.5	75	113	2.8	88	133	3.1	106	159
6000	NC	0	0	NC	0	0	NC	0	0	NC	0	0	NC	0	0	NC	0	0	RC	44	67	RC	48	72	2.2	66	100	3.0	80	120	4.0	95	142	3.7	111	167	3.6	114	171	4.0	137	206
5000	NC	0	0	NC	0	0	NC	0	0	NC	0	0	NC	0	0	RC	41	62	RC	44	67	2.2	53	79	2.6	79	119	3.6	96	144	4.4	112	167	4.4	132	198	4.2	133	199	4.7	161	242
4000	NC	0	0	NC	0	0	NC	0	0	NC	0	0	RC	39	58	RC	41	62	2.3	51	77	2.7	65	97	3.1	89	134	3.9	104	156	4.8	123	184	4.9	147	221	4.9	165	232	5.5	189	283
3500	NC	0	0	NC	0	0	NC	0	0	NC	0	0	RC	39	58	2.1	43	65	2.6	58	87	3.0	72	108	3.5	100	149	4.3	115	172	5.3	134	201	5.3	159	236	5.4	171	256	5.9	202	303
3000	NC	0	0	NC	0	0	NC	0	0	RC	36	55	2.3	45	67	2.4	50	74	2.9	64	97	3.4	82	122	3.9	110	165	4.8	128	192	5.6	148	222	5.8	174	261	5.8	183	275	**Rₘᵢₙ = 3060**		
2500	NC	0	0	NC	0	0	NC	0	0	RC	36	55	2.8	54	81	2.8	58	87	3.3	73	110	3.8	91	137	4.3	125	188	5.4	144	216	6.0	162	243									
2000	NC	0	0	NC	0	0	RC	34	51	2.2	40	60	3.0	58	87	3.3	68	102	3.8	84	127	4.3	103	155	4.9	130	195	5.6	149	224	6.0	167	251									
1800	NC	0	0	NC	0	0	RC	34	51	2.4	44	65	3.3	64	96	3.6	74	112	4.1	91	137	4.6	110	166	5.1	138	207	5.9	157	236												
1600	NC	0	0	NC	0	0	2.1	36	54	2.7	49	74	3.6	70	105	3.8	79	118	4.4	98	147	4.9	116	176	5.4	146	218	6.0	160	240												
1400	NC	0	0	NC	0	0	2.3	39	59	2.9	53	79	3.9	75	113	4.1	85	127	4.7	104	157	5.2	125	187	5.7	151	226															
1200	NC	0	0	RC	32	49	2.6	45	67	3.3	60	90	4.3	83	125	4.5	93	140	5.0	111	167	5.6	134	202	5.9															
1000	NC	0	0	RC	32	49	2.9	51	77	3.7	67	101	4.5	87	131	4.9	101	152	5.5	122	183	5.9	142	212	**Rₘᵢₙ = 1065**														
900	NC	0	0	2.2	36	54	3.0	55	82	3.9	71	106	4.8	93	139	5.1	106	158	5.7	127	190	6.0	144	216															
800	RC	31	46	2.4	39	58	3.2	58	87	4.1	76	112	5.1	99	148	5.4	112	168	5.9	131	197	**Rₘᵢₙ = 835**															
700	RC	31	46	2.7	44	66	3.4	63	95	4.4	80	120	5.4	105	157	5.7	118	177	6.0	133	200															
600	2.1	32	48	2.9	47	71	3.7	69	103	4.7	85	128	5.7	110	166	5.9	122	183	**Rₘᵢₙ = 660**															
500	2.4	37	55	3.2	52	78	4.0	74	111	5.1	93	139	5.9	114	171	**Rₘᵢₙ = 510**															
450	2.7	42	62	3.6	58	88	4.3	77	116	5.3	96	145	6.0	116	174															
400	2.9	45	67	3.8	62	92	4.5	82	123	5.6	102	153	**Rₘᵢₙ = 380**															
350	3.2	49	74	4.0	65	97	4.8	87	131	5.8	105	158															
300	3.5	54	81	4.2	68	102	5.1	93	139	6.0	109	164															
250	3.8	58	88	4.5	73	109	5.7	98	147	**Rₘᵢₙ = 275**															
200	4.1	63	95	4.8	78	117	6.0	103	154															
150	4.7	72	108	5.3	86	129	**Rₘᵢₙ = 185**															
100	5.5	85	127	5.8	94	141															
75	5.9	91	136	**Rₘᵢₙ = 115**															
	Rₘᵢₙ = 65																																									

Legend:

e_max	=	6%
R	=	radius of curve
V_d	=	assumed design speed
e	=	rate of superelevation
L	=	minimum length of runoff (does not include tangent runout) as discussed in "Tangent-to-Curve Transition" section
NC	=	normal crown section
RC	=	remove adverse crown, superelevate at normal crown slope

Exhibit 3-22. Values for Design Elements Related to Design Speed and Horizontal Curvature (Continued)

METRIC

R (m)	Vd=20 e(%)	Vd=20 L 2Lns	Vd=20 L 4Lns	Vd=30 e(%)	Vd=30 L 2Lns	Vd=30 L 4Lns	Vd=40 e(%)	Vd=40 L 2Lns	Vd=40 L 4Lns	Vd=50 e(%)	Vd=50 L 2Lns	Vd=50 L 4Lns	Vd=60 e(%)	Vd=60 L 2Lns	Vd=60 L 4Lns	Vd=70 e(%)	Vd=70 L 2Lns	Vd=70 L 4Lns	Vd=80 e(%)	Vd=80 L 2Lns	Vd=80 L 4Lns	Vd=90 e(%)	Vd=90 L 2Lns	Vd=90 L 4Lns	Vd=100 e(%)	Vd=100 L 2Lns	Vd=100 L 4Lns	Vd=110 e(%)	Vd=110 L 2Lns	Vd=110 L 4Lns	Vd=120 e(%)	Vd=120 L 2Lns	Vd=120 L 4Lns	Vd=130 e(%)	Vd=130 L 2Lns	Vd=130 L 4Lns
7000	NC	0	0	NC	0	0	NC	0	0	NC	0	0	NC	0	0	NC	0	0	NC	0	0	NC	0	0	NC	0	0	NC	0	0	NC	0	0	NC	0	0
5000	NC	0	0	NC	0	0	NC	0	0	NC	0	0	NC	0	0	NC	0	0	NC	0	0	NC	0	0	NC	0	0	RC	18	28	RC	23	34	RC	21	31
3000	NC	0	0	NC	0	0	NC	0	0	NC	0	0	NC	0	0	NC	0	0	RC	14	22	RC	15	23	RC	17	26	2.1	21	32	2.4	23	34	2.6	27	40
2500	NC	0	0	NC	0	0	NC	0	0	NC	0	0	NC	0	0	RC	13	20	RC	17	26	2.2	17	26	2.6	21	32	2.4	26	39	2.9	27	41	3.1	32	48
2000	NC	0	0	NC	0	0	NC	0	0	NC	0	0	RC	12	18	RC	14	21	2.5	18	27	2.8	21	32	3.4	28	42	3.0	32	47	3.5	41	50	3.9	40	60
1500	NC	0	0	NC	0	0	NC	0	0	RC	11	17	RC	16	24	3.3	22	33	3.4	25	37	3.6	28	42	4.6	44	69	4.1	44	65	4.9	46	70	5.1	52	79
1400	NC	0	0	NC	0	0	NC	0	0	RC	11	17	2.1	18	27	3.6	25	37	3.5	28	42	4.1	31	46	4.9	47	70	4.4	47	70	5.2	49	74	5.4	56	83
1300	NC	0	0	NC	0	0	RC	10	14	RC	12	18	2.2	18	27	3.8	28	42	3.8	37	55	4.6	37	55	5.0	50	77	4.7	50	77	5.6	53	80	5.8	60	89
1200	NC	0	0	NC	0	0	NC	0	0	RC	12	18	2.4	18	27	4.0	32	48	4.0	42	63	4.8	42	64	5.7	59	89	5.5	59	87	6.0	62	92	6.3	65	97
1000	NC	0	0	NC	0	0	NC	0	0	RC	13	20	2.7	22	33	4.4	34	51	4.4	44	66	5.3	55	77	6.3	64	101	6.0	66	101	6.5	67	101	7.4	76	114
900	NC	0	0	NC	0	0	NC	0	0	RC	14	21	3.0	24	37	4.8	37	55	4.8	48	71	5.7	61	87	6.9	77	95	6.6	68	101	7.1	72	108	7.9	81	122
800	NC	0	0	NC	0	0	RC	0	0	RC	17	17	3.3	27	40	5.1	41	61	5.3	55	77	6.7	63	93	7.5	85	101	7.2	72	108	8.0	76	114	R_min=830
700	NC	0	0	NC	0	0	RC	0	0	2.2	18	18	3.6	30	42	5.7	47	70	6.3	61	85	7.6	68	101	8.0	93	105	7.8	76	108	R_min=665		
600	NC	0	0	NC	0	0	RC	0	0	2.6	14	22	3.9	32	48	6.3	52	78	6.9	66	93	8.0	77	105	R_min=500			8.0	70	R_min=500						
500	NC	0	0	NC	0	0	2.2	11	17	3.0	17	25	4.7	37	56	6.6	55	82	7.6	77	105	R_min=395															
400	RC	10	14	RC	0	0	2.7	14	21	3.8	20	30	5.6	42	63	7.4	61	79	7.9	85	R_min=305																	
300	2.1	10	16	RC	0	0	3.4	17	26	4.5	25	37	6.2	50	73	R_min=175																				
250	2.5	12	18	RC	0	0	4.0	21	31	5.1	28	42	7.0	56	78																					
200	3.0	14	22	RC	0	0	4.6	24	35	5.8	32	48	7.4	63	R_min=125																					
175	3.4	16	24	RC	0	0	5.0	26	39	6.2	34	52	7.8	67																						
150	3.8	18	27	RC	0	0	5.4	28	42	6.7	37	56	8.0	70																						
140	4.0	19	29	RC	0	0	5.6	29	43	6.9	38	57																								
130	4.2	20	30	2.2	20	30	5.8	30	45	7.1	39	59																								
120	4.4	21	32	2.3	21	32	6.0	31	46	7.4	41	61																								
110	4.7	23	34	2.5	23	34	6.3	32	49	7.6	42	63																								
100	5.0	24	36	2.7	24	36	6.6	34	51	7.8	43	65																								
90	5.2	25	37	3.0	25	37	6.9	35	53	7.9	44	66																								
80	5.5	26	40	3.3	26	40	7.2	37	56	8.0	44	66	R_min=80																							
70	5.9	28	42	3.6	28	42	7.5	39	58	R_min=80																									
60	6.4	31	46	4.1	31	46	7.8	40	60																										
50	6.9	33	50	4.6	33	50	8.0	41	62	R_min=50																									
40	7.5	36	54	5.2	35	50	R_min=50																												
30	8.0	38	58	5.9	38	58	R_min=30																												
20	7.1	32	48	R_min=30																															
R_min	=10			=30			=50			=80			=125			=175			=230			=305			=395			=500			=665			=830		

Legend

e_max	=	8%
R	=	radius of curve
V_d	=	assumed design speed
e	=	rate of superelevation
L	=	minimum length of runoff (does not include tangent runout) as discussed in "Tangent-to-Curve Transition" section
NC	=	normal crown section
RC	=	remove adverse crown, superelevate at normal crown slope

Exhibit 3-23. Values for Design Elements Related to Design Speed and Horizontal Curvature

US CUSTOMARY

R (ft)	Vd=15 e(%)	15 L 2Lns	15 L 4Lns	Vd=20 e(%)	20 L 2Lns	20 L 4Lns	Vd=25 e(%)	25 L 2Lns	25 L 4Lns	Vd=30 e(%)	30 L 2Lns	30 L 4Lns	Vd=35 e(%)	35 L 2Lns	35 L 4Lns	Vd=40 e(%)	40 L 2Lns	40 L 4Lns	Vd=45 e(%)	45 L 2Lns	45 L 4Lns	Vd=50 e(%)	50 L 2Lns	50 L 4Lns	Vd=55 e(%)	55 L 2Lns	55 L 4Lns	Vd=60 e(%)	60 L 2Lns	60 L 4Lns	Vd=65 e(%)	65 L 2Lns	65 L 4Lns	Vd=70 e(%)	70 L 2Lns	70 L 4Lns	Vd=75 e(%)	75 L 2Lns	75 L 4Lns	Vd=80 e(%)	80 L 2Lns	80 L 4Lns
23000	NC	0	0	NC	0	0	NC	0	0	NC	0	0	NC	0	0	NC	0	0	NC	0	0	NC	0	0	NC	0	0	NC	0	0	NC	0	0	NC	0	0	NC	0	0	NC	0	0
20000	NC	0	0	NC	0	0	NC	0	0	NC	0	0	NC	0	0	NC	0	0	NC	0	0	NC	0	0	NC	0	0	NC	0	0	NC	0	0	NC	0	0	NC	0	0	NC	0	0
17000	NC	0	0	NC	0	0	NC	0	0	NC	0	0	NC	0	0	NC	0	0	NC	0	0	NC	0	0	NC	0	0	NC	0	0	RC	53	80	RC	60	90	RC	63	95	RC	69	103
14000	NC	0	0	NC	0	0	NC	0	0	NC	0	0	NC	0	0	NC	0	0	NC	0	0	NC	0	0	NC	0	0	RC	56	84	RC	60	90	RC	63	95	RC	69	103
12000	NC	0	0	NC	0	0	NC	0	0	NC	0	0	NC	0	0	NC	0	0	NC	0	0	RC	48	72	RC	51	77	2.1	56	84	2.4	67	100	2.6	78	117	2.4	76	114	2.2	75	113
10000	NC	0	0	NC	0	0	NC	0	0	NC	0	0	NC	0	0	NC	0	0	RC	44	67	RC	53	80	2.4	61	92	2.7	72	108	3.1	87	130	3.4	102	153	2.9	92	137	2.6	89	134
8000	NC	0	0	NC	0	0	NC	0	0	NC	0	0	NC	0	0	RC	41	62	RC	44	67	2.4	58	86	2.8	71	107	3.2	85	128	3.6	100	151	4.1	123	184	3.8	120	180	3.3	113	170
6000	NC	0	0	NC	0	0	NC	0	0	NC	0	0	RC	39	58	RC	41	62	2.4	53	80	2.9	70	104	3.4	87	130	3.9	104	156	4.4	123	184	4.9	147	221	4.5	142	213	4.3	147	221
5000	NC	0	0	NC	0	0	NC	0	0	RC	36	55	RC	39	58	2.3	48	71	2.7	60	90	3.2	77	115	3.8	97	146	4.4	117	176	4.9	137	205	5.5	165	248	5.5	174	261	5.1	175	262
4000	NC	0	0	NC	0	0	RC	36	55	2.1	41	61	2.6	54	81	3.0	62	93	3.4	81	121	3.9	103	155	4.5	128	191	5.0	152	228	5.6	179	268	6.3	216	324	6.2	196	294	6.2	213	319
3500	NC	0	0	RC	36	55	2.4	39	63	2.6	48	73	3.0	61	87	3.4	73	110	3.7	98	147	4.3	103	155	5.0	151	228	5.7	176	264	6.4	207	310	7.2	237	356	7.0	221	332	7.0	240	360
3000	NC	0	0	RC	36	55	2.6	44	65	3.0	55	82	3.3	64	96	3.7	77	115	4.4	104	157	4.7	132	198	5.9	181	241	6.6	176	284	7.4	264		7.9	310		7.8	246	369	7.8	267	401
2500	NC	0	0	RC	34	44	3.0	51	77	3.3	65	97	3.6	70	105	4.0	91	137	4.7	116	173	5.5	142	212	6.3	181	257	7.2	207	300	8.0	215	312				7.8					
2000	NC	0	0	RC	34	51	3.3	58	71	3.6	79	87	4.0	96	116	4.4	124	149	5.2	137	187	5.9	157	230	6.7	226	276	7.5	223	335				7.9	237	356						
1800	NC	0	0	RC	32	49	2.9	53	79	3.6	79	87	4.0	105	116	4.8	137	149	5.5	173	212	6.4	187	230	7.1	189	284	8.0	223	335												
1600	RC	31	46	RC	32	49	3.2	58	87	4.0	87	98	4.5	116	131	5.4	149	167	6.2	187	207	7.1	212	230	7.5	200	300												
1400	RC	31	46	2.4	41	62	3.6	65	98	4.5	98	116	5.1	131	148	6.0	168	186	6.8	207	227	7.6	224	252	7.8	208	312												
1200	RC	31	46	2.4	41	62	4.2	76	115	5.1	115	132	5.6	148	160	6.4	186	211	7.6	240	253	7.8	252	274															
1000	NC	0	0	2.8	48	72	4.5	90	100	5.5	106	123	6.0	160	181	7.2	199	253	7.8	187	281	8.0	204	306															
900	RC	31	46	3.3	57	85	4.9	100	111	5.9	134	145	6.3	171	223	7.6	211	263	8.0	263																			
800	RC	31	46	3.5	60	90	5.3	111	123	6.3	145	158	7.2	183	240	7.9	223	281	8.0	192	288																		
700	RC	31	46	3.9	67	100	5.8	123	132	6.8	158	171	7.6	197		8.0	236																					
600	2.2	34	51	4.3	74	111	6.4	116	144	7.4	175	197	8.0	216	263	8.0	176	263																				
500	2.6	40	60	4.8	82	123	6.7	122	175	7.7	183	216	8.0	224																							
450	2.9	45	67	5.3	91	136	7.1	129	183	7.9	194	224	8.0	165	248																						
400	3.2	49	74	5.6	96	144	7.5	136	194	8.0	205	232																								
350	3.5	54	81	6.0	103	154	7.6	142	205	8.0	155	232																								
300	3.9	60	90	6.4	110	165	7.8	145	213																											
250	4.5	69	104	6.8	117	176	8.0	145	218																											
200	5.1	78	118	7.4	127	190																														
150	5.9	91	136	7.5	122	182																														
100	7.0	108	162																																	
75	7.7	118	178																																	
Rmin	60			105			170			250			350			465			600			760			965			1205			1485			1820			2215			2675		

Legend:

e_max	=	8%
R	=	radius of curve
V_d	=	assumed design speed
e	=	rate of superelevation
L	=	minimum length of runoff (does not include tangent runout) as discussed in "Tangent-to-Curve Transition" section
NC	=	normal crown section
RC	=	remove adverse crown, superelevate at normal crown slope

Exhibit 3-23. Values for Design Elements Related to Design Speed and Horizontal Curvature (Continued)

161

METRIC

R (m)	V_d = 20 km/h			V_d = 30 km/h			V_d = 40 km/h			V_d = 50 km/h			V_d = 60 km/h			V_d = 70 km/h			V_d = 80 km/h			V_d = 90 km/h			V_d = 100 km/h			V_d = 110 km/h			V_d = 120 km/h			V_d = 130 km/h		
	e (%)	L (m) 2 Lns	4 Lns	e (%)	2 Lns	4 Lns	e (%)	2 Lns	4 Lns	e (%)	2 Lns	4 Lns	e (%)	2 Lns	4 Lns	e (%)	2 Lns	4 Lns	e (%)	2 Lns	4 Lns	e (%)	2 Lns	4 Lns	e (%)	2 Lns	4 Lns	e (%)	2 Lns	4 Lns	e (%)	2 Lns	4 Lns	e (%)	2 Lns	4 Lns
7000	NC	0	0	NC	0	0	NC	0	0	NC	0	0	NC	0	0	NC	0	0	NC	0	0	NC	0	0	NC	0	0	NC	0	0	NC	0	0	NC	0	0
5000	NC	0	0	NC	0	0	NC	0	0	NC	0	0	NC	0	0	NC	0	0	NC	0	0	NC	0	0	RC	16	25	RC	18	28	RC	24	36	RC	21	31
3000	NC	0	0	NC	0	0	NC	0	0	NC	0	0	NC	0	0	NC	0	0	NC	0	0	RC	15	22	2.2	18	27	2.1	22	33	2.5	27	41	2.7	28	42
2500	NC	0	0	NC	0	0	NC	0	0	NC	0	0	NC	0	0	RC	13	20	RC	14	21	2.2	17	24	2.7	22	33	2.5	27	41	2.9	34	51	3.2	33	49
2000	NC	0	0	NC	0	0	NC	0	0	NC	0	0	NC	0	0	RC	14	21	2.4	17	26	2.9	22	33	3.5	29	43	3.1	27	41	3.6	34	51	4.0	41	62
1500	NC	0	0	NC	0	0	NC	0	0	NC	0	0	RC	12	18	2.1	16	24	2.6	20	28	3.1	24	36	3.8	31	47	4.1	38	54	4.8	45	68	5.3	55	82
1400	NC	0	0	NC	0	0	NC	0	0	NC	0	0	RC	12	18	2.1	16	24	2.6	20	30	3.3	24	36	4.0	33	49	4.3	38	57	5.1	48	72	5.7	59	88
1300	NC	0	0	NC	0	0	NC	0	0	NC	0	0	RC	12	18	2.3	18	26	2.8	22	32	3.3	26	38	4.3	35	53	4.6	44	66	5.5	52	78	6.1	63	94
1200	NC	0	0	NC	0	0	NC	0	0	RC	11	17	2.2	13	20	2.4	19	28	3.0	25	38	3.6	28	42	4.6	42	63	5.0	52	66	5.9	56	84	6.6	68	102
1000	NC	0	0	NC	0	0	RC	10	15	2.3	13	19	2.6	15	23	2.9	21	31	3.5	28	42	4.2	32	48	5.1	46	69	5.9	55	83	7.0	66	99	7.9	81	122
900	NC	0	0	NC	0	0	RC	10	15	2.7	15	22	2.7	18	24	3.2	23	34	3.9	31	46	4.6	35	50	5.6	46	69	6.4	58	86	7.7	73	109	8.7	89	134
800	NC	0	0	RC	10	14	2.3	12	18	3.1	17	26	3.1	19	28	3.6	26	39	4.5	35	52	5.1	39	58	6.2	51	76	7.1	62	94	8.5	81	121	9.7	100	150
700	NC	0	0	RC	11	16	2.8	14	22	3.8	21	32	3.6	22	32	4.0	29	44	4.9	37	55	5.8	44	66	6.9	56	85	8.0	70	105	9.5	90	135			
600	NC	0	0	2.2	12	19	3.6	19	28	4.8	27	40	4.2	25	38	4.5	35	52	5.3	43	64	6.5	50	75	7.8	64	95	9.0	79	119	10.0	95	142			
500	RC	10	14	2.6	14	22	4.2	22	32	5.6	31	47	5.0	30	45	5.3	41	62	6.4	52	77	7.6	58	87	8.9	73	109	9.9	87	130						
400	RC	10	14	3.1	16	25	5.0	28	39	6.6	37	55	6.3	38	57	6.3	51	77	7.5	54	81	8.8	67	101	9.8	80	120									
300	2.2	12	18	3.8	19	29	5.6	29	43	7.6	39	59	7.1	43	64	7.8	57	85	9.0	65	97	9.9	76	114												
250	2.6	14	22	4.2	22	32	6.6	35	55	8.1	45	67	8.2	49	74	8.7	63	94	9.7	70	105															
200	3.1	17	26	5.0	28	39	7.8	43	59	8.8	47	71	8.8	63	79	9.6	65	97																		
175	RC	9	14	5.6	29	43	7.8	43	59	9.1	49	73	9.4	56	85																					
150	RC	9	14	6.2	32	48	8.1	45	65	9.5	50	76	9.7	58	87																					
140	2.1	9	14	6.4	33	49	8.5	47	67				9.8	59	88																					
130	2.2	10	14	6.7	34	52	8.8	49	71																										
120	2.4	10	15	7.0	36	54	9.1	53	73																										
110	2.6	11	16	7.4	38	57	9.5	53	76																										
100	2.8	12	18	7.7	40	59	9.8	54	79																										
90	3.1	13	19	8.2	42	63																													
80	3.4	14	21	8.6	44	66																													
70	3.8	15	23	9.1	47	70																													
60	4.4	17	26	9.6	49	74																													
50	5.0	23	34																																
40	5.9	27	40																																
30	7.0	31	44																																
20	8.5	38	57																																

R_{min} values: V_d=20 → 10; V_d=30 → 25; V_d=40 → 45; V_d=50 → 75; V_d=60 → 115; V_d=70 → 160; V_d=80 → 210; V_d=90 → 275; V_d=100 → 360; V_d=110 → 455; V_d=120 → 595; V_d=130 → 740.

Legend:

e_{max}	=	10%
R	=	radius of curve
V_d	=	assumed design speed
e	=	rate of superelevation
L	=	minimum length of runoff (does not include tangent runout) as discussed in "Tangent-to-Curve Transition" section
NC	=	normal crown section
RC	=	remove adverse crown, superelevate at normal crown slope

Exhibit 3-24. Values for Design Elements Related to Design Speed and Horizontal Curvature

US CUSTOMARY

The following table gives, for each design speed V_d, the rate of superelevation e (%) and the minimum length of runoff L (ft) for 2-lane (2 Lns) and 4-lane (4 Lns) conditions, as a function of radius of curve R (ft).

R (ft)	15 mph e(%)	15 L 2	15 L 4	20 mph e(%)	20 L 2	20 L 4	25 mph e(%)	25 L 2	25 L 4	30 mph e(%)	30 L 2	30 L 4	35 mph e(%)	35 L 2	35 L 4	40 mph e(%)	40 L 2	40 L 4	45 mph e(%)	45 L 2	45 L 4	50 mph e(%)	50 L 2	50 L 4	55 mph e(%)	55 L 2	55 L 4	60 mph e(%)	60 L 2	60 L 4	65 mph e(%)	65 L 2	65 L 4	70 mph e(%)	70 L 2	70 L 4	75 mph e(%)	75 L 2	75 L 4	80 mph e(%)	80 L 2	80 L 4
23000	NC	0	0	NC	0	0	NC	0	0	NC	0	0	NC	0	0	NC	0	0	NC	0	0	NC	0	0	NC	0	0	NC	0	0	NC	0	0	NC	0	0	NC	0	0	NC	0	0
20000	NC	0	0	NC	0	0	NC	0	0	NC	0	0	NC	0	0	NC	0	0	NC	0	0	NC	0	0	NC	0	0	NC	0	0	NC	0	0	NC	0	0	NC	0	0	NC	0	0
17000	NC	0	0	NC	0	0	NC	0	0	NC	0	0	NC	0	0	NC	0	0	NC	0	0	NC	0	0	NC	0	0	NC	0	0	NC	0	0	NC	0	0	NC	0	0	RC	69	103
14000	NC	0	0	NC	0	0	NC	0	0	NC	0	0	NC	0	0	NC	0	0	NC	0	0	NC	0	0	NC	0	0	NC	0	0	NC	0	0	NC	0	0	NC	0	0	RC	69	103
12000	NC	0	0	NC	0	0	NC	0	0	NC	0	0	NC	0	0	NC	0	0	NC	0	0	NC	0	0	NC	0	0	NC	0	0	NC	0	0	NC	0	0	RC	63	95	2.2	75	113
10000	NC	0	0	NC	0	0	NC	0	0	NC	0	0	NC	0	0	NC	0	0	NC	0	0	NC	0	0	NC	0	0	RC	53	80	RC	56	84	RC	60	90	RC	63	95	2.7	93	139
8000	NC	0	0	NC	0	0	NC	0	0	NC	0	0	NC	0	0	NC	0	0	NC	0	0	RC	48	72	RC	51	77	2.2	59	88	2.4	67	100	2.2	66	99	2.4	76	114	3.3	113	170
6000	NC	0	0	NC	0	0	NC	0	0	NC	0	0	NC	0	0	RC	41	62	RC	44	67	2.1	50	76	2.4	61	92	2.8	75	112	3.2	89	134	2.7	81	122	3.0	95	142	4.4	151	226
5000	NC	0	0	NC	0	0	NC	0	0	NC	0	0	NC	0	0	RC	41	62	RC	44	67	2.5	58	86	2.9	74	111	3.4	91	136	3.8	106	159	3.6	108	162	4.0	126	189	5.3	182	273
4000	NC	0	0	NC	0	0	NC	0	0	NC	0	0	RC	39	58	2.3	48	71	2.5	56	83	3.0	72	108	3.5	89	134	4.1	109	164	4.6	128	193	4.2	126	189	4.7	148	223	6.5	223	334
3500	NC	0	0	NC	0	0	NC	0	0	RC	36	55	RC	39	58	2.7	56	84	2.8	62	93	3.4	82	122	4.0	102	153	4.6	123	184	5.2	145	218	5.2	156	234	5.8	183	276	7.4	254	381
3000	NC	0	0	NC	0	0	RC	36	55	RC	36	55	2.1	41	61	3.2	66	99	3.3	73	110	3.9	94	140	4.6	117	176	5.3	141	212	6.0	167	251	5.9	177	266	6.6	208	313	8.6	295	442
2500	NC	0	0	RC	32	49	2.2	38	57	2.4	45	65	2.5	48	73	3.9	81	121	3.8	84	127	4.5	110	166	5.3	135	203	6.2	165	248	7.0	195	293	6.7	201	302	7.6	240	360	9.9	339	509
2000	RC	31	46	RC	32	49	2.7	44	65	3.0	55	82	3.1	60	90	4.2	87	130	4.7	104	157	5.5	132	198	6.4	163	245	7.4	197	296	8.1	216	324	7.9	237	356	8.9	281	422			
1800	RC	31	46	2.2	38	57	3.0	49	74	3.4	62	93	3.4	66	99	4.7	97	146	5.1	113	170	6.0	144	216	7.0	179	268	8.1	216	324	8.7	232	348	9.4	282	423	10.0	316	474			
1600	RC	31	46	2.5	43	64	3.4	57	82	3.8	74	110	3.8	74	110	5.2	108	161	5.6	124	187	6.6	158	238	7.6	194	291	8.7	232	348	9.6	268	402	9.9	297	446						
1400	2.1	34	51	2.9	49	73	3.8	67	93	4.3	83	125	4.3	83	125	5.9	122	183	6.2	138	209	7.3	175	263	8.4	214	322	9.4	251	376	10.0	279	419									
1200	2.6	41	61	3.4	58	87	4.5	79	118	4.8	104	139	4.8	93	139	6.8	141	211	7.0	156	233	8.1	194	292	9.2	235	352	9.9	264	396												
1000	3.4	44	66	3.7	63	95	4.9	87	130	5.6	120	180	5.6	108	163	7.3	151	227	7.9	176	263	9.0	216	324	9.8	250	375															
900	RC	46	69	4.1	70	105	5.4	98	147	6.1	134	177	6.1	118	177	7.9	163	245	8.5	189	283	9.4	226	338	10.0	255	383															
800	RC	51	76	4.6	79	118	5.9	107	161	6.6	147	192	6.6	128	192	8.5	176	264	9.0	200	300	9.8	235	353																		
700	RC	57	85	5.2	89	134	6.6	120	180	7.2	161	209	7.2	139	209	9.2	190	285	9.5	211	317	10.0	240	360																		
600	2.3	35	53	5.9	104	155	7.4	135	202	8.0	180	215	8.0	155	232	9.8	203	304	9.9	220	330																					
500	2.7	42	62	6.8	120	181	7.9	144	215	8.8	202	255	8.8	170	255	10.0	207	310																								
450	3.0	46	69	7.4	130	195	8.4	153	229	9.2	215	232	9.2	178	267																										
400	3.3	51	76	8.0	141	208	9.0	164	245	9.6	229	250	9.6	186	279																										
350	3.7	57	85	8.6	151	226	9.5	173	260	9.9	245	271	9.9	192	287																										
300	4.2	65	97	9.6	165	247	10.0	181	271																															
250	4.9	75	113	10.0	181	271																																	
200	5.7	88	132																																				
150	6.8	105	157																																				
100	8.4	129	194																																				
75	9.4	145	217																																				

R_{min} values: 15 mph = 55; 20 mph = 100; 25 mph = 160; 30 mph = 230; 35 mph = 320; 40 mph = 430; 45 mph = 555; 50 mph = 695; 55 mph = 880; 60 mph = 1095; 65 mph = 1345; 70 mph = 1640; 75 mph = 1980; 80 mph = 2380.

Legend:

e_{max} = 10%
R = radius of curve
V_d = assumed design speed
e = rate of superelevation
L = minimum length of runoff (does not include tangent runout) as discussed in "Tangent-to-Curve Transition" section
NC = normal crown section
RC = remove adverse crown, superelevate at normal crown slope

Exhibit 3-24. Values for Design Elements Related to Design Speed and Horizontal Curvature (Continued)

METRIC

Exhibit 3-25. Values for Design Elements Related to Design Speed and Horizontal Curvature

In the exhibit, each design speed V_d block contains three sub-columns: **e (%)**, and **L (m)** for 2 lanes (Lns) and 4 lanes (Lns). NC = normal crown section; RC = remove adverse crown, superelevate at normal crown slope. The shared left column is **R (m)**.

V_d = 20 km/h

R (m)	e (%)	L (m) 2 Lns	L (m) 4 Lns
7000–175	NC	0	0
150	RC	9	9
140	2.1	9	14
130	2.3	10	14
120	2.5	11	17
110	2.7	12	18
100	2.9	13	20
90	3.2	14	21
80	3.5	16	24
70	4.0	18	27
60	4.6	20	31
50	5.3	24	36
40	6.3	28	43
30	7.7	35	52
20	9.7	44	65

R_{min} = 10

V_d = 30 km/h

R (m)	e (%)	L (m) 2 Lns	L (m) 4 Lns
7000–500	NC	0	0
400	RC	10	14
300	2.2	11	16
250	2.6	12	19
200	3.2	15	23
175	3.6	17	26
150	4.2	20	30
140	4.4	21	32
130	4.7	23	34
120	5.1	24	37
110	5.4	26	39
100	5.9	28	42
90	6.4	31	46
80	6.9	33	50
70	7.6	36	55
60	8.4	40	60
50	9.3	45	67
40	10.4	50	75
30	11.6	55	84

R_{min} = 25

V_d = 40 km/h

R (m)	e (%)	L (m) 2 Lns	L (m) 4 Lns
7000–700	NC	0	0
600	RC	10	15
500	2.4	10	15
400	2.9	12	19
300	3.8	15	23
250	4.4	20	29
200	5.3	23	34
175	5.9	27	41
150	6.7	30	46
140	7.0	34	52
130	7.4	36	54
120	7.8	38	57
110	8.2	40	60
100	8.7	42	63
90	9.3	45	67
80	9.9	48	72
70	10.5	54	81
60	11.2	58	86
50	11.8	61	91

R_{min} = 45

V_d = 50 km/h

R (m)	e (%)	L (m) 2 Lns	L (m) 4 Lns
7000–1000	NC	0	0
900	RC	11	17
800	2.1	11	17
700	2.4	12	17
600	2.7	13	20
500	3.2	15	22
400	3.9	18	27
300	5.1	22	33
250	5.9	27	39
200	7.1	33	49
175	7.8	39	59
150	8.7	43	65
140	9.1	48	72
130	9.5	50	76
120	10.0	53	79
110	10.5	58	83
100	11.0	61	91
90	11.4	63	95
80	11.8	65	98
70	12.0	66	100

R_{min} = 70

V_d = 60 km/h

R (m)	e (%)	L (m) 2 Lns	L (m) 4 Lns
7000–1500	NC	0	0
1400	RC	12	18
1300	RC	12	18
1200	RC	12	18
1000	2.3	14	21
900	2.5	15	23
800	2.8	17	25
700	3.2	19	29
600	3.7	22	33
500	4.3	26	39
400	5.3	32	48
300	6.7	40	60
250	7.7	46	69
200	9.1	55	82
175	10.0	60	90
150	10.9	65	95
140	11.2	67	101
130	11.5	69	104
120	11.8	71	106
110	12.0	72	108

R_{min} = 105

V_d = 70 km/h

R (m)	e (%)	L (m) 2 Lns	L (m) 4 Lns
7000–2000	NC	0	0
1500	RC	13	20
1400	2.1	14	21
1300	2.3	15	23
1200	2.5	16	26
1000	2.9	19	28
900	3.3	22	32
800	3.6	24	35
700	4.1	27	40
600	4.7	31	46
500	5.5	36	54
400	6.7	44	66
300	8.5	58	83
250	9.7	66	95
200	11.1	73	109
175	11.7	77	115
150	12.0	79	118

R_{min} = 150

V_d = 80 km/h

R (m)	e (%)	L (m) 2 Lns	L (m) 4 Lns
7000–2500	NC	0	0
2000	RC	14	22
1500	2.5	18	27
1400	2.6	19	28
1300	2.8	20	30
1200	3.0	22	32
1000	3.6	26	39
900	4.0	29	43
800	4.4	32	48
700	5.0	36	54
600	5.7	41	62
500	6.7	48	72
400	8.1	58	87
300	10.1	73	109
250	11.2	81	121
200	12.0	85	130

R_{min} = 195

V_d = 90 km/h

R (m)	e (%)	L (m) 2 Lns	L (m) 4 Lns
7000–3000	NC	0	0
2500	RC	15	23
2000	2.3	18	27
1500	3.0	23	34
1400	3.2	25	37
1300	3.4	26	39
1200	3.7	28	43
1000	4.4	34	51
900	4.8	37	55
800	5.3	41	62
700	6.0	46	69
600	6.9	53	78
500	8.1	62	93
400	9.7	74	111
300	11.6	89	133

R_{min} = 255

V_d = 100 km/h

R (m)	e (%)	L (m) 2 Lns	L (m) 4 Lns
7000–5000	NC	0	0
3000	RC	16	23
2500	2.2	18	26
2000	2.7	22	33
1500	3.6	29	44
1400	3.9	32	48
1300	4.1	34	51
1200	4.5	37	55
1000	5.3	43	65
900	5.8	47	71
800	6.5	53	80
700	7.3	60	90
600	8.3	68	102
500	9.7	79	119
400	11.4	93	133

R_{min} = 330

V_d = 110 km/h

R (m)	e (%)	L (m) 2 Lns	L (m) 4 Lns
7000	NC	0	0
5000	RC	18	28
3000	2.1	22	33
2500	2.5	28	42
2000	3.2	37	55
1500	4.2	39	58
1400	4.4	42	63
1300	4.8	45	67
1200	5.1	54	80
1000	6.1	59	88
900	6.7	66	99
800	7.5	75	112
700	8.5	85	128
600	9.7	99	149
500	11.3	99	149

R_{min} = 415

V_d = 120 km/h

R (m)	e (%)	L (m) 2 Lns	L (m) 4 Lns
7000	NC	0	0
5000	NC	0	0
3000	RC	24	36
2500	2.5	28	43
2000	3.0	35	53
1500	3.7	46	70
1400	4.9	49	74
1300	5.2	53	80
1200	5.6	58	87
1000	6.1	68	102
900	7.2	76	114
800	8.0	84	126
700	8.9	96	144
600	11.6	110	185

R_{min} = 540

V_d = 130 km/h

R (m)	e (%)	L (m) 2 Lns	L (m) 4 Lns
7000	NC	0	0
5000	RC	21	31
3000	2.7	28	42
2500	3.3	34	51
2000	4.1	42	63
1500	5.4	56	83
1400	5.8	60	89
1300	6.3	65	97
1200	6.8	70	105
1000	8.1	83	125
900	9.0	93	139
800	10.1	104	156
700	11.6	119	179

R_{min} = 665

Legend

e_{max}	=	12%
R	=	radius of curve
V_d	=	assumed design speed
e	=	rate of superelevation
L	=	minimum length of runoff (does not include tangent runout) as discussed in "Tangent-to-Curve Transition" section
NC	=	normal crown section
RC	=	remove adverse crown, superelevate at normal crown slope

Exhibit 3-25. Values for Design Elements Related to Design Speed and Horizontal Curvature

US CUSTOMARY

R (ft)	V_d=15 mph e(%)	L(ft) 2 Lns	L(ft) 4 Lns	V_d=20 mph e(%)	L 2	L 4	V_d=25 mph e(%)	L 2	L 4	V_d=30 mph e(%)	L 2	L 4	V_d=35 mph e(%)	L 2	L 4	V_d=40 mph e(%)	L 2	L 4	V_d=45 mph e(%)	L 2	L 4	V_d=50 mph e(%)	L 2	L 4	V_d=55 mph e(%)	L 2	L 4	V_d=60 mph e(%)	L 2	L 4	V_d=65 mph e(%)	L 2	L 4	V_d=70 mph e(%)	L 2	L 4	V_d=75 mph e(%)	L 2	L 4	V_d=80 mph e(%)	L 2	L 4
23000	NC	0	0	NC	0	0	NC	0	0	NC	0	0	NC	0	0	NC	0	0	NC	0	0	NC	0	0	NC	0	0	NC	0	0	NC	0	0	NC	0	0	NC	0	0	NC	0	0
20000	NC	0	0	NC	0	0	NC	0	0	NC	0	0	NC	0	0	NC	0	0	NC	0	0	NC	0	0	NC	0	0	NC	0	0	NC	0	0	NC	0	0	NC	0	0	NC	0	0
17000	NC	0	0	NC	0	0	NC	0	0	NC	0	0	NC	0	0	NC	0	0	NC	0	0	NC	0	0	NC	0	0	NC	0	0	NC	0	0	NC	0	0	NC	0	0	RC	69	103
14000	NC	0	0	NC	0	0	NC	0	0	NC	0	0	NC	0	0	NC	0	0	NC	0	0	NC	0	0	NC	0	0	NC	0	0	NC	0	0	RC	60	90	RC	63	95	RC	69	103
12000	NC	0	0	NC	0	0	NC	0	0	NC	0	0	NC	0	0	NC	0	0	NC	0	0	NC	0	0	NC	0	0	RC	53	80	RC	55	84	RC	60	90	RC	66	99	2.3	79	118
10000	NC	0	0	NC	0	0	NC	0	0	NC	0	0	NC	0	0	NC	0	0	NC	0	0	NC	0	0	RC	51	77	RC	56	84	RC	56	84	2.2	60	90	2.1	66	99	2.7	93	139
8000	NC	0	0	NC	0	0	NC	0	0	NC	0	0	NC	0	0	NC	0	0	RC	44	67	RC	48	72	RC	64	96	2.2	59	88	2.5	70	105	2.7	81	122	2.5	79	118	3.4	117	175
6000	NC	0	0	NC	0	0	NC	0	0	NC	0	0	NC	0	0	RC	41	62	2.1	47	70	2.1	50	76	2.5	74	111	2.9	77	116	3.2	89	134	3.6	108	162	3.1	98	147	4.5	154	231
5000	NC	0	0	NC	0	0	NC	0	0	NC	0	0	RC	39	58	2.1	43	65	2.5	55	83	2.5	60	90	2.9	74	111	3.4	91	136	3.9	109	163	4.3	129	193	4.1	129	194	5.4	185	278
4000	NC	0	0	NC	0	0	NC	0	0	RC	36	55	2.2	39	58	2.4	50	74	2.9	64	97	3.1	74	112	3.6	92	138	4.2	112	168	4.8	134	201	5.4	162	243	4.8	152	227	6.7	230	345
3500	NC	0	0	NC	0	0	NC	0	0	2.2	38	55	2.6	43	64	2.7	56	84	3.3	73	110	3.5	84	126	4.1	105	157	4.8	128	192	5.4	151	226	6.1	183	274	6.0	189	284	7.7	264	396
3000	NC	0	0	NC	0	0	RC	34	51	2.6	45	68	3.2	50	75	3.2	66	99	4.0	89	133	4.0	96	144	4.7	120	180	5.5	147	220	6.2	173	260	7.0	210	315	6.8	215	322	8.9	305	458
2500	NC	0	0	NC	0	0	2.3	39	59	3.2	49	74	3.9	62	93	4.0	83	124	4.8	107	160	4.7	113	169	5.8	143	214	6.5	173	260	7.4	207	310	8.3	249	374	7.9	249	374	10.6	363	545
2000	NC	0	0	RC	32	49	2.6	45	67	3.9	56	85	4.4	68	102	4.4	91	137	5.3	118	177	5.8	139	209	6.8	174	260	7.9	211	316	9.0	251	377	10.2	306	459	9.4	297	445	R_{min} = 2140		
1800	NC	0	0	RC	32	49	3.1	51	77	4.4	62	93	5.1	75	113	4.9	101	152	5.9	131	197	6.3	151	227	7.4	189	283	8.6	229	344	10.2	306	459	11.1	333	499	11.5	363	545			
1600	NC	0	0	2.1	34	51	3.4	57	85	5.1	73	109	5.9	85	128	5.5	114	171	6.6	147	220	7.0	168	252	8.2	209	314	9.5	253	380	10.8	301	452	11.9	357	536	12.0	379	568			
1400	NC	0	0	2.6	41	61	3.9	67	100	5.6	85	128	6.5	99	148	6.2	128	192	7.5	167	250	7.8	187	281	9.1	232	349	10.6	283	424	11.7	327	490	R_{min} = 1490								
1200	NC	0	0	2.8	45	68	4.0	73	109	6.3	102	153	7.1	114	171	7.3	151	227	8.7	193	290	8.8	211	317	10.3	263	394	11.5	307	460	R_{min} = 1230											
1000	NC	0	0	3.1	50	75	4.7	82	123	7.1	115	172	7.9	128	189	7.9	163	245	9.4	209	313	10.1	242	364	11.4	291	437	R_{min} = 1005													
900	RC	31	46	3.5	57	85	5.1	94	141	7.9	129	194	8.6	137	206	8.6	178	267	10.2	227	340	10.8	259	389	11.8	301	452	R_{min} = 810													
800	RC	31	46	3.9	67	100	5.6	111	166	8.8	149	224	9.5	153	229	9.5	197	295	11.0	244	367	11.4	274	410																	
700	2.4	37	56	4.3	74	111	6.3	123	184	10.0	160	240	10.6	170	255	10.5	217	326	11.7	260	390	11.9	286	428																	
600	2.8	43	65	4.8	82	123	7.1	141	211	10.6	173	259	11.2	184	276	11.4	236	354	R_{min} = 510			R_{min} = 645																				
500	3.1	48	72	5.5	94	141	8.2	165	248	11.2	185	278	11.7	194	290	11.8	244	366																							
450	3.4	52	78	6.4	110	165	8.9	177	266	11.7	194	291	12.0	205	308	12.0	248	372																							
400	3.8	58	88	6.9	118	177	9.5	193	290	12.0	200	300	12.0	217	325	R_{min} = 395																								
350	4.4	68	102	7.5	129	193	10.2	213	320	11.2	217	325	12.0	226	340																									
300	5.1	78	118	8.3	142	213	11.0	232	348	11.7	226	340	12.0	232	348																									
250	6.1	94	141	9.1	156	234	11.7	213?	260	12.0	213	319	R_{min} = 300																										
200	7.5	115	173	10.1	173	260	12.0	192	288	R_{min} = 215																													
150	9.6	146	222	11.2	192	288	12.0	206	309																													
100	11.0	169	254	11.9	193	289	R_{min} = 145																															
75	R_{min} = 50			R_{min} = 90																																					

Legend:

e_{max}	=	12%
R	=	radius of curve
V_d	=	assumed design speed
e	=	rate of superelevation
L	=	minimum length of runoff (does not include tangent runout) as discussed in "Tangent-to-Curve Transition" section
NC	=	normal crown section
RC	=	remove adverse crown, superelevate at normal crown slope

Exhibit 3-25. Values for Design Elements Related to Design Speed and Horizontal Curvature (Continued)

The term "normal cross slope" (NC) designates curves that are so flat that the elimination of adverse cross slope is not considered necessary, and thus the normal cross slope sections can be used. The term "remove cross slope" (RC) designates curves where it is adequate to eliminate the adverse cross slope by superelevating the entire roadway at the normal cross slope.

Sharpest Curve Without Superelevation

The minimum rate of cross slope applicable to the traveled way is determined by drainage needs. Consistent with the type of highway and amount of rainfall, snow, and ice, the usually accepted minimum values for cross slope range from 1.5 percent for high-type surfaces to approximately 2.0 percent for low-type surfaces (for further information, see the section on "Cross Slope" in Chapter 4). For discussion purposes, a value of 1.5 percent is used below as a single value representative of the cross slope for high-type, uncurbed pavements. Steeper cross slopes are generally needed where curbs are used to minimize ponding of water on the outside through lane.

The shape or form of the normal cross slope varies. Some States and many municipalities use a curved traveled way cross section for two-lane roadways, usually parabolic in form. Others employ a straight-line section for each lane.

Very flat horizontal curves need no superelevation. Traffic entering a curve to the right has some superelevation provided by the normal cross slope. Traffic entering a curve to the left has an adverse or negative superelevation resulting from the normal cross slope, but with flat curves the side friction needed to sustain the lateral acceleration and counteract the negative superelevation is small. However, on successively sharper curves for the same speed, a point is reached where the combination of lateral acceleration and negative superelevation overcomes the allowable side friction, and a positive slope across the entire roadway is desirable to help sustain the lateral acceleration. Thus, an important part of superelevation design policy is a criterion for the minimum curvature for which superelevation is needed, or conversely, the maximum curvature for which a normal roadway cross section is appropriate.

Many highway agencies express this criterion as a single radius of curvature applicable to all design speeds. When using this method, care should be taken to ensure that the cross slope is sufficient to provide surface drainage and reduce the potential for vehicle hydroplaning and wet weather skidding, especially on flat, high-speed curves. Some agencies use a different criterion for maximum superelevation without curvature for each design speed. The latter method is more realistic and conforms to the previously discussed superelevation-speed-curvature relationships. The maximum curvature for sections without superelevation for each design speed should be determined by setting consistently low values of side friction factor, considering the effect of normal cross slope and both directions of travel. The result is an increasing radius for successively higher design speeds.

For an average rate of cross slope of 1.5 percent and the superelevation curves of Exhibit 3-19 (e_{max} = 10 percent), the corresponding minimum radius for each design speed is shown in the third column of Exhibit 3-26. These are curvatures calling for superelevation equal to the normal cross slope, and therefore indicate the limit of curvature with normal cross slopes. The side friction factors developed because of adverse cross slope at both the design speed and the average running speed are shown in the right columns. It is evident from their uniform and low values over the range of design speeds that these radii are logical limiting values for sections with normal cross slopes.

For a limited range of horizontal curves sharper than those shown in Exhibit 3-26, a practical superelevation adjustment can be obtained by retaining the shape of the normal traveled way cross section but rotating it around the edge or centerline of the traveled way. This adjustment makes it unnecessary to change the screeds used in constructing rigid pavements, and the construction procedures are the same on such curves as on tangent sections except that the side forms should be set to the proper difference in elevation. This method of eliminating adverse slope results in a steeper slope at the lower edge of traveled way than would otherwise be obtained, which may be desirable for drainage. However, traffic operating on the higher side of the traveled way does not receive as much benefit as it does when the normal section is changed to a plane section for the full width of the roadway.

On a curve sharp enough to need a superelevation rate in excess of about 2.0 percent, a plane slope across the whole traveled way should be used. A transition from the normal to a straight-line cross slope is needed. For short lengths of highway needing cross-slope reversals, the difficulties and extra costs involved in constructing such transitions may supersede the desirable design refinement. This practical limit of 2.0 percent corresponds to curves with radii ranging from 700 m [2,290 ft] for 50-km/h [30-mph] design speeds to about 3,500 m [11,500 ft] for 110-km/h [70-mph] design speeds. For curves between these values and those in Exhibit 3-26, the superelevation adjustment can be made by rotation of the normal traveled way cross section or, preferably, by change to a plane slope across the whole traveled way.

Effects of Grades

On long or fairly steep grades, drivers tend to travel faster in the downgrade than in the upgrade direction. In a refined design this tendency should be recognized, and some adjustment in superelevation rates should be made. In the case of a divided highway with each roadway independently superelevated, or on a one-way ramp, such an adjustment can be readily made. In the simplest practical form, values from Exhibits 3-21 to 3-25 can be used directly by assuming a slightly higher design speed for the downgrade and a slightly lower design speed for the upgrade. The appropriate variation in design speed depends on the particular conditions, especially the rate and length of grade and the magnitude of the curve radius in comparison to other curves on the approach highway section.

It is questionable whether similar adjustments should be made on two-lane and multilane undivided roadways. In one respect the two directions of traffic tend to balance each other, and adjustment of superelevation is not needed. However, the downgrade speed is the most critical, and adjustment for it may be desirable in some cases. Although not common practice, lanes can be constructed at different cross slopes in the same direction. More practical would be an adjustment for the whole traveled way as determined by the downgrade speed, because the extra cross slope would not significantly affect upgrade travel, with the possible exception of heavy trucks on long grades. The desirability of avoiding minor changes in design speed should also be considered. In general, it is advisable to follow the common practice of not making such superelevation adjustments on undivided roadways.

Metric					US Customary				
Design speed (km/h)	Average running speed (km/h)	Minimum curve radius (m)	Resulting side friction factor, f, with adverse cross slope		Design speed (mph)	Average running speed (mph)	Minimum curve radius (ft)	Resulting side friction factor, f, with adverse cross slope	
			at design speed	at running speed				at design speed	at running speed
20	20	200	0.031	0.031	15	15	960	0.031	0.031
30	30	450	0.031	0.031	20	20	1700	0.031	0.031
40	40	800	0.031	0.031	25	24	2460	0.032	0.031
50	47	1110	0.033	0.031	30	28	3350	0.033	0.031
60	55	1530	0.034	0.031	35	32	4390	0.034	0.031
70	63	2020	0.034	0.031	40	36	5570	0.034	0.031
80	70	2500	0.035	0.030	45	40	6880	0.035	0.031
90	77	3030	0.036	0.030	50	44	8350	0.035	0.031
100	85	3700	0.036	0.030	55	48	9960	0.035	0.031
110	91	4270	0.037	0.030	60	52	11720	0.036	0.030
120	98	4990	0.038	0.030	65	55	13180	0.036	0.030
130	102	5450	0.040	0.030	70	58	14730	0.037	0.030
					75	61	16380	0.038	0.030
					80	64	18130	0.039	0.030

Exhibit 3-26. Minimum Curve Radius for Section with Normal Cross Slopes ($e_{max} = 10\%$)

Transition Design Controls

General Considerations

The design of transition sections includes consideration of transitions in the roadway cross slope and possible transition curves incorporated in the horizontal alignment. The former consideration is referred to as superelevation transition and the latter is referred to as alignment transition. Where both transition components are used, they occur together over a common section of roadway at the beginning and end of the mainline circular curves.

The superelevation transition section consists of the superelevation runoff and tangent runout sections. The superelevation runoff section consists of the length of roadway needed to accomplish a change in outside-lane cross slope from zero (flat) to full superelevation, or vice versa. The tangent runout section consists of the length of roadway needed to accomplish a change in outside-lane cross slope from the normal cross slope rate to zero (flat), or vice versa. For reasons of safety and comfort, the pavement rotation in the superelevation transition section should be effected over a length that is sufficient to make such rotation imperceptible to drivers. To be pleasing in appearance, the pavement edges should not appear distorted to the driver.

In the alignment transition section, a spiral or compound transition curve may be used to introduce the main circular curve in a natural manner (i.e., one that is consistent with the driver's steered path). Such transition curvature consists of one or more curves aligned and located to provide a gradual change in alignment radius. As a result, an alignment transition introduces the lateral acceleration associated with the curve in a gentle manner. While such a gradual change in path and lateral acceleration is appealing, there is no definitive evidence that transition curves are essential to the safe operation of the roadway and, as a result, they are not used by many agencies.

When a transition curve is not used, the roadway tangent directly adjoins the main circular curve. This type of transition design is referred to below as the "tangent-to-curve" transition.

Some agencies employ spiral curves and use their length to make the appropriate superelevation transition. A spiral curve approximates the natural turning path of a vehicle. One agency believes that the length of spiral should be based on a 4-s minimum maneuver time at the design speed of the highway. Other agencies do not employ spiral curves but empirically designate proportional lengths of tangent and circular curve for the same purpose. In either case, as far as can be determined, the length of roadway to effect the superelevation runoff should be the same for the same rate of superelevation and radius of curvature.

Review of current design practice indicates that the length of a superelevation runoff section is largely governed by its appearance. Spiral transition curve lengths as determined otherwise often are shorter than those determined for general appearance, so that theoretically derived spiral lengths are replaced with longer empirically derived runoff lengths. A number of agencies have established one or more control runoff lengths within a range of about 30 to 200 m [100 to 650 ft], but there is no universally accepted empirical basis for determining runoff length, considering all likely traveled way widths. In one widely used empirical expression, the runoff length is determined as a function of the slope of the outside edge of the traveled way relative to the centerline profile.

Tangent-to-Curve Transition

Minimum length of superelevation runoff. For appearance and comfort, the length of superelevation runoff should be based on a maximum acceptable difference between the longitudinal grades of the axis of rotation and the edge of pavement. The axis of rotation is generally represented by the alignment centerline for undivided roadways; however, other pavement reference lines can be used. These lines and the rationale for their use is discussed below in the section on "Methods of Attaining Superelevation."

Current practice is to limit the grade difference, referred to as the relative gradient, to a maximum value of 0.50 percent or a longitudinal slope of 1:200 at 80 km/h [50 mph]. In one source (**23**), this same 1:200 slope is used for a design speed of 80 km/h [50 mph] and higher. Where design speeds are less than 80 km/h [50 mph], greater relative slopes are used. To reflect the importance of the higher design speed and to harmonize with the flatter curving elements, both horizontal and vertical, it appears logical to extrapolate the relative slopes for the higher design speeds.

The maximum relative gradient is varied with design speed to provide longer runoff lengths at higher speeds and shorter lengths at lower speeds. Experience indicates that relative gradients of 0.80 and 0.35 percent [0.78 and 0.35 percent] provide acceptable runoff lengths for design speeds of 20 and 130 km/h [15 and 80 mph], respectively.

Interpolation between these values provides the maximum relative gradients shown in Exhibit 3-27. The maximum relative gradient between profiles of the edges of two-lane traveled ways should be double those given in the exhibit. Runoff lengths determined on this basis are directly proportional to the total superelevation, which is the product of the lane width and superelevation rate.

Previous editions of this policy have suggested that runoff lengths should be at least equal to the distance traveled in 2.0 s at the design speed. This criterion tended to determine the runoff lengths of curves with small superelevation rates, high speed, or both. Experience with the 2.0-s criterion indicates that the improvement in appearance is outweighed by a tendency to aggravate problems associated with pavement drainage in the transition section. In fact, it is noted that some agencies do not use this control. From this evidence, it is concluded that a comfortable and aesthetically pleasing runoff design can be attained through the exclusive use of the maximum relative gradient criterion.

Metric			US Customary		
Design speed (km/h)	Maximum relative gradient (%)	Equivalent maximum relative slope	Design speed (mph)	Maximum relative gradient (%)	Equivalent maximum relative slope
20	0.80	1:125	15	0.78	1:128
30	0.75	1:133	20	0.74	1:135
40	0.70	1:143	25	0.70	1:143
50	0.65	1:154	30	0.66	1:152
60	0.60	1:167	35	0.62	1:161
70	0.55	1:182	40	0.58	1:172
80	0.50	1:200	45	0.54	1:185
90	0.47	1:213	50	0.50	1:200
100	0.44	1:227	55	0.47	1:213
110	0.41	1:244	60	0.45	1:222
120	0.38	1:263	65	0.43	1:233
130	0.35	1:286	70	0.40	1:250
			75	0.38	1:263
			80	0.35	1:286

Exhibit 3-27. Maximum Relative Gradients

On the basis of the preceding discussion, the minimum length of runoff should be determined as:

Metric	US Customary
$$L_r = \dfrac{(wn_1)\,e_d}{\Delta}\,(b_w)$$	$$L_r = \dfrac{(wn_1)\,e_d}{\Delta}\,(b_w) \qquad (\,3\text{-}25\,)$$
where:	where:
L_r = minimum length of superelevation runoff, m; Δ = maximum relative gradient, percent; n_1 = number of lanes rotated; b_w = adjustment factor for number of lanes rotated; w = width of one traffic lane, m (typically 3.6 m); e_d = design superelevation rate, percent	L_r = minimum length of superelevation runoff, ft; Δ = maximum relative gradient, percent; n_1 = number of lanes rotated; b_w = adjustment factor for number of lanes rotated; w = width of one traffic lane, ft (typically 12 ft); e_d = design superelevation rate, percent

Equation (3-25) can be used directly for undivided streets or highways where the cross section is rotated about the highway centerline and n_1 is equal to one-half the number of lanes in the cross section. More generally, Equation (3-25) can be used for rotation about any pavement reference line provided that the rotated width (wn_1) has a common superelevation rate and is rotated as a plane.

A strict application of the maximum relative gradient criterion provides runoff lengths for four-lane undivided roadways that are double those for two-lane roadways; those for six-lane undivided roadways would be tripled. While lengths of this order may be considered desirable, it is often not practical to provide such lengths in design. On a purely empirical basis, it is recommended that minimum superelevation runoff lengths be adjusted downward to avoid excessive lengths for multilane roadways. The recommended adjusted factors are presented in Exhibit 3-28.

The adjustment factors listed in Exhibit 3-28 are directly applicable to undivided streets and highways. Development of runoff for divided highways is discussed in more detail in the later section on "Axis of Rotation with a Median." The topic of runoff superelevation for turning roadway designs at intersections and through interchanges is discussed in Chapters 9 and 10, respectively.

Typical minimum superelevation runoff lengths are presented in Exhibit 3-29. The lengths shown represent cases where one or two lanes are rotated about a pavement edge. The former case is found on two-lane roadways where the pavement is rotated about the centerline or on one-lane interchange ramps where the pavement rotation is about an edge line. The latter case is found on multilane undivided roadways where each direction is separately rotated about an edge line.

METRIC			US CUSTOMARY		
Number of Lanes Rotated, n_l	Adjustment Factor, b_w [a]	Length Increase Relative to One-lane Rotated ($=n_l b_w$)	Number of Lanes Rotated, n_l	Adjustment Factor, b_w [a]	Length Increase Relative to One-lane Rotated ($=n_l b_w$)
1	1.00	1.0	1	1.00	1.0
1.5	0.83	1.25	1.5	0.83	1.25
2	0.75	1.5	2	0.75	1.5
2.5	0.70	1.75	2.5	0.70	1.75
3	0.67	2.0	3	0.67	2.0
3.5	0.64	2.25	3.5	0.64	2.25

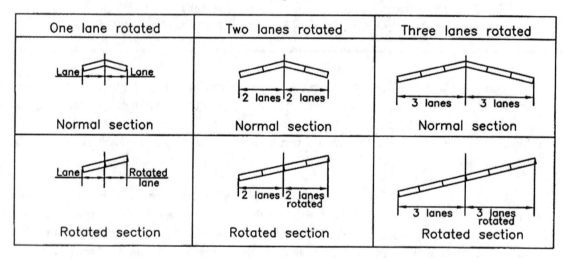

Exhibit 3-28. Adjustment Factor for Number of Lanes Rotated

Elimination of the 2.0-s travel-time criterion discussed above results in shorter runoff lengths for smaller superelevation rates and higher speeds. However, even the shortest runoff lengths (corresponding to a superelevation rate of 2.0 percent) correspond to travel times of 0.6 s, which is sufficient to provide a smooth edge-of-pavement profile.

For high-type alignments, superelevation runoff lengths longer than those shown in Exhibit 3-28 may be desirable. In this case, drainage needs or the desire for smoothness in the traveled way edge profiles may call for a small increase in runoff length.

The superelevation runoff lengths given in Exhibit 3-28 are based on 3.6-m [12-ft] lanes. For other lane widths, the appropriate runoff length should vary in proportion to the ratio of the actual lane width to 3.6 m [12 ft]. Shorter lengths could be applied for designs with 3.0- and 3.3-m [10- and 11-ft] lanes, but considerations of consistency and practicality suggest that the runoff lengths for 3.6-m [12-ft] lanes should be used in all cases.

Minimum length of tangent runout. The length of tangent runout is determined by the amount of adverse cross slope to be removed and the rate at which it is removed. To effect a smooth edge of pavement profile, the rate of removal should equal the relative gradient used to define the superelevation runoff length. Based on this rationale, the following equation should be used to compute the minimum tangent runout length:

Metric	US Customary
$$L_t = \frac{e_{NC}}{e_d} L_r$$	$$L_t = \frac{e_{NC}}{e_d} L_r \qquad (3\text{-}26)$$
where: L_t = minimum length of tangent runout, m; e_{NC} = normal cross slope rate, percent; e_d = design superelevation rate, percent; L_r = minimum length of superelevation runoff, m	where: L_t = minimum length of tangent runout, ft; e_{NC} = normal cross slope rate, percent; e_d = design superelevation rate, percent; L_r = minimum length of superelevation runoff, ft

The tangent runout lengths determined with Equation (3-26) are listed in Exhibit 3-29.

Location with respect to end of curve. In the tangent-to-curve design, the location of the superelevation runoff length with respect to the point of curvature (PC) must be determined. Normal practice is to divide the runoff length between the tangent and curved sections and to avoid placing the entire runoff length on either the tangent or the curve. With full superelevation attained at the PC, the runoff lies entirely on the approach tangent, where theoretically no superelevation is needed. At the other extreme, placement of the runoff entirely on the circular curve results in the initial portion of the curve having less than the desired amount of superelevation. Both of these extremes tend to be associated with a large peak lateral acceleration.

Experience indicates that locating a portion of the runoff on the tangent, in advance of the PC, is preferable, since this tends to minimize the peak lateral acceleration and the resulting side friction demand. The magnitude of side friction demand incurred during travel through the runoff can vary with the actual vehicle travel path. Observations indicate that a spiral path results from a driver's natural steering behavior during curve entry or exit. This natural spiral usually begins on the tangent and ends beyond the beginning of the circular curve. Most evidence indicates that the length of this natural spiral ranges from 2- to 4-s travel time; however, its length may also be affected by lane width and the presence of other vehicles.

Based on the preceding discussion, locating a portion of the runoff on the tangent is consistent with the natural spiral path adopted by the driver during curve entry. In this manner, the gradual introduction of superelevation prior to the curve compensates for the gradual increase in lateral acceleration associated with the spiral path. As a result, the peak lateral acceleration incurred at the PC should theoretically be about equal to 50 percent of the lateral acceleration associated with the circular curve.

Metric

Minimum runoff and runout length (m)

Design speed (km/h)	Runoff — Superelevation						Runout
	2	4	6	8	10	12	any
One lane rotated							
20	9	18	27	36	45	54	9
30	10	19	29	38	48	57	10
40	10	21	31	41	51	62	10
50	11	22	32	43	54	65	11
60	12	24	36	48	60	72	12
70	13	26	39	52	66	79	13
80	14	29	43	58	72	86	14
90	15	31	46	61	77	92	15
100	16	33	49	65	82	98	16
110	18	35	53	70	88	105	18
120	19	38	57	76	95	114	19
130	21	41	62	82	103	124	21
Two lanes rotated							
20	14	27	41	54	68	81	14
30	14	29	43	57	72	86	14
40	15	31	46	62	77	93	15
50	16	32	49	65	81	97	16
60	18	36	54	72	90	108	18
70	20	39	59	79	98	118	20
80	22	43	65	86	108	130	22
90	23	46	69	92	115	138	23
100	25	49	74	98	123	147	25
110	26	53	79	105	132	158	26
120	28	57	85	114	142	170	28
130	31	62	93	124	154	185	31

US Customary

Minimum runoff and runout length (ft)

Design speed (mph)	Runoff — Superelevation						Runout
	2	4	6	8	10	12	any
One lane rotated							
15	31	61	92	123	154	184	31
20	32	65	97	130	162	194	32
25	34	69	103	137	172	206	34
30	36	73	109	146	182	219	36
35	39	77	116	155	193	232	39
40	41	83	124	165	206	248	41
45	44	89	133	178	222	266	44
50	48	96	144	192	240	288	48
55	51	102	153	204	256	307	51
60	53	107	160	213	266	320	53
65	56	112	168	224	280	336	56
70	60	120	180	240	300	360	60
75	63	126	189	252	316	379	63
80	69	137	206	275	343	412	69
Two lanes rotated							
15	46	92	138	184	230	276	46
20	49	97	146	194	243	292	49
25	51	103	154	206	257	309	51
30	55	109	164	219	274	328	55
35	58	116	174	232	290	348	58
40	62	124	186	248	310	372	62
45	67	133	200	266	333	400	67
50	72	144	216	288	360	432	72
55	77	153	230	307	383	460	77
60	80	160	240	320	400	480	80
65	84	168	252	336	419	503	84
70	90	180	270	360	450	540	90
75	95	189	284	379	473	568	95
80	103	206	309	412	515	618	103

Note: Based on 3.6 m [12 ft] lanes and on a 2.0% normal cross slope.

Exhibit 3-29. Minimum Superelevation Runoff and Tangent Runout Lengths

To achieve this balance in lateral acceleration, most agencies locate a portion of the runoff length on the tangent prior to the curve. The proportion of runoff length placed on the tangent varies from 0.6 to 0.8 (i.e., 60 to 80 percent) with a large majority of agencies using 0.67 (i.e., 67 percent). Most agencies consistently use a single value of this proportion for all street and highway curves.

Theoretical considerations confirm the desirability of placing a larger portion of the runoff length on the approach tangent rather than on the circular curve. Such considerations are based on analysis of the acceleration acting laterally on the vehicle while it travels through the transition section. This lateral acceleration can induce a lateral velocity and lane shift that could lead to operational problems. Specifically, a lateral velocity in an outward direction (relative to the curve) requires a driver to make a corrective steer maneuver that produces a path radius sharper than that of the roadway curve. Such a critical radius produces an undesirable increase in peak side friction demand. Moreover, a lateral velocity of sufficient magnitude to shift the vehicle into an adjacent lane (without corrective steering) is also undesirable for safety reasons.

Analysis of the aforementioned theoretical considerations has led to the conclusion that an appropriate allocation of runoff length between the tangent and the curve can minimize the aforementioned operational problems (**24**). The values obtained from the analysis are listed in Exhibit 3-30. If used in design, the values listed in Exhibit 3-30 should minimize lateral acceleration and the vehicle's lateral motion. Values smaller than those listed tend to be associated with larger outward lateral velocities. Values larger than those listed tend to be associated with larger lateral shifts.

Metric					US Customary				
Design speed (km/h)	Portion of runoff located prior to the curve				Design speed (mph)	Portion of runoff located prior to the curve			
	No. of lanes rotated					No. of lanes rotated			
	1.0	1.5	2.0-2.5	3.0-3.5		1.0	1.5	2.0-2.5	3.0-3.5
20-70	0.80	0.85	0.90	0.90	15-45	0.80	0.85	0.90	0.90
80-130	0.70	0.75	0.80	0.85	50-80	0.70	0.75	0.80	0.85

Exhibit 3-30. Runoff Locations that Minimize the Vehicle's Lateral Motion

Theoretical considerations indicate that values for the proportion of runoff length on the tangent in the range of 0.7 to 0.9 (i.e., 70 to 90 percent) offer the best operating conditions; the specific value in this range should be dependent on design speed and rotated width. Experience obtained from existing practice indicates that deviation from the values in Exhibit 3-30 by 10 percent should not lead to measurable operational problems. In this regard, use of a single value for the proportion of runoff length on the tangent in the range of 0.6 to 0.9 (60 to 90 percent) for all speeds and rotated widths is considered acceptable. However, refinement of this value, based on the trends shown in Exhibit 3-30, is desirable when conditions allow.

Location with respect to end of curve. In alignment design with spirals, the superelevation runoff is effected over the whole of the transition curve. The length of the superelevation runoff should be equal to the spiral length for the tangent-to-spiral (TS) transition at the beginning and the spiral-to-curve (SC) transition at the end of the circular curve. The change in cross slope begins by removing the adverse cross slope from the lane or lanes on the outside of the curve on a length of tangent just ahead of TS (the tangent runout) (see Exhibit 3-37). Between the TS and SC, the spiral curve and the superelevation runoff are coincident and the traveled way is rotated to reach the full superelevation at the SC. This arrangement is reversed on leaving the curve. In this design, the whole of the circular curve has full superelevation.

Limiting superelevation rates. Theoretical considerations indicate that, when a vehicle is traveling through a tangent-to-curve transition, large superelevation rates are associated with large shifts in the vehicle's lateral position. In general, such shifts in lateral position can be minimized by the proper location of the superelevation runoff section, as described above. However, excessively large lateral shifts must be checked by the driver through steering action.

In recognition of the potential adverse effect that large shifts in lateral position may have on safety, the threshold superelevation rates associated with a lateral shift of 1.0 m [3.0 ft] are identified in Exhibit 3-31. These limiting superelevation rates do not apply for speeds of 80 km/h [50 mph] or more when combined with superelevation rates of 12 percent or less.

Metric		US Customary	
Design speed (km/h)	Limiting superelevation rate (%)	Design speed (mph)	Limiting superelevation rate (%)
20	8	15	8
30	8	20	8
40	10	25	10
50	11	30	11
60	11	35	11
70	12	40	11
		45	12

Exhibit 3-31. Limiting Superelevation Rates

Designs that incorporate superelevation in excess of the limiting rates may be associated with excessive lateral shift. Therefore, it is recommended that such superelevation rates be avoided. However, if they are used, consideration should be given to increasing the width of the traveled way along the curve to reduce the potential for vehicle encroachment into the adjacent lane.

Spiral Curve Transitions

General. Any motor vehicle follows a transition path as it enters or leaves a circular horizontal curve. The steering change and the consequent gain or loss of lateral force cannot be effected instantly. For most curves, the average driver can follow a suitable transition path within the limits of normal lane width. However, combinations of high speed and sharp curvature lead to longer transition paths, which can result in shifts in lateral position and sometimes actual encroachment on adjoining lanes. In such instances, incorporation of transition curves between the tangent and the sharp circular curve, as well as between circular curves of substantially different radii, may be appropriate in order to make it easier for a driver to keep his or her vehicle within its own lane.

The principal advantages of transition curves in horizontal alignment are the following:

1. A properly designed transition curve provides a natural, easy-to-follow path for drivers, such that the lateral force increases and decreases gradually as a vehicle enters and leaves a circular curve. Transition curves minimize encroachment on adjoining traffic lanes and tend to promote uniformity in speed. A spiral transition curve simulates the natural turning path of a vehicle.

2. The transition curve length provides a suitable location for the superelevation runoff. The transition from the normal pavement cross slope on the tangent to the fully superelevated section on the curve can be accomplished along the length of the transition curve in a manner that closely fits the speed-radius relationship for vehicles traversing the transition. Where superelevation runoff is introduced without a transition curve, usually partly on the curve and partly on the tangent, the driver approaching the curve may have to steer opposite to the direction of the approaching curve when on the superelevated tangent portion in order to keep the vehicle within its lane.

3. A spiral transition curve also facilitates the transition in width where the traveled way is widened on a circular curve. Use of spiral transitions provides flexibility in accomplishing the widening of sharp curves.

4. The appearance of the highway or street is enhanced by the application of spiral transition curves. The use of spiral transitions avoids noticeable breaks in the alignment as perceived by drivers at the beginning and end of circular curves. Exhibit 3-32 illustrates such breaks, which are made more prominent by the presence of superelevation runoff.

Length of Spiral

Length of spiral. Generally, the Euler spiral, which is also known as the clothoid, is used in the design of spiral transition curves. The radius varies from infinity at the tangent end of the spiral to the radius of the circular arc at the end that adjoins that circular arc. By definition, the radius of curvature at any point on an Euler spiral varies inversely with the distance measured along the spiral. In the case of a spiral transition that connects two circular curves having different radii, there is an initial radius rather than an infinite value.

The following equation, developed in 1909 by Shortt (**25**) for gradual attainment of lateral acceleration on railroad track curves, is the basic expression used by some highway agencies for computing minimum length of a spiral transition curve:

Metric	US Customary	
$$L = \frac{0.0214V^3}{RC}$$	$$L = \frac{3.15V^3}{RC}$$	(3-27)
where:	where:	
L = minimum length of spiral, m; V = speed, km/h; R = curve radius, m; C = rate of increase of lateral acceleration, m/s^3	L = minimum length of spiral, ft; V = speed, mph; R = curve radius, ft; C = rate of increase of lateral acceleration, ft/s^3	

Exhibit 3-32. Transition Spirals (23)

The factor C is an empirical value representing the comfort and safety levels provided by the spiral curve. The value of C = 0.3 m/s^3 [1 ft/s^3] is generally accepted for railroad operation, but values ranging from 0.3 to 0.9 m/s^3 [1 to 3 ft/s^3] have been used for highways. This equation is sometimes modified to take into account the effect of superelevation, which results in much shorter spiral curve lengths. Highways do not appear to need as much precision as is obtained from computing the length of spiral by this equation or its modified form. A more practical control for the length of spiral is that it should equal the length needed for superelevation runoff.

Maximum radius for use of a spiral. A review of guidance on the use of spiral curve transitions indicates a general lack of consistency among highway agencies. In general, much of this guidance suggests that an upper limit on curve radius can be established such that only radii below this maximum are likely to obtain safety and operational benefits from the use of spiral transition curves. Such a limiting radius has been established by several agencies based on a minimum lateral acceleration rate. Such minimum rates have been found to vary from 0.4 to 1.3 m/s² [1.3 to 4.25 ft/s²]. The upper end of this range of rates corresponds to the maximum curve radius for which some reduction in crash potential has also been noted. For these reasons, it is recommended that the maximum radius for use of a spiral should be based on a minimum lateral acceleration rate of 1.3 m/s² [4.25 ft/s²] (**20**). These radii are listed in Exhibit 3-33.

The radii listed in Exhibit 3-33 are intended for use by those highway agencies that desire to use spiral curve transitions. Exhibit 3-33 is not intended to define radii that require the use of a spiral.

Metric		US Customary	
Design speed (km/h)	Maximum radius (m)	Design speed (mph)	Maximum radius (ft)
20	24	15	114
30	54	20	203
40	95	25	317
50	148	30	456
60	213	35	620
70	290	40	810
80	379	45	1025
90	480	50	1265
100	592	55	1531
110	716	60	1822
120	852	65	2138
130	1000	70	2479
		75	2846
		80	3238
Note: The safety benefits of spiral curve transitions are likely to be negligible for larger radii.			

Exhibit 3-33. Maximum Radius for Use of a Spiral Curve Transition

Minimum length of spiral. Several agencies define a minimum length of spiral based on consideration of driver comfort and shifts in the lateral position of vehicles. Criteria based on driver comfort are intended to provide a spiral length that allows for a comfortable increase in lateral acceleration as a vehicle enters a curve. The criteria based on lateral shift are intended to ensure that a spiral curve is sufficiently long to provide a shift in a vehicle's lateral position within its lane that is consistent with that produced by the vehicle's natural spiral path. It is recommended that these two criteria be used together to determine the minimum length of spiral. Thus, the minimum spiral length can be computed as:

Metric	US Customary	
$L_{s, min}$ should be the larger of: $$L_{s,min} = \sqrt{24(p_{min})R}$$ or	$L_{s, min}$ should be the larger of: $$L_{s,min} = \sqrt{24(p_{min})R}$$ or	(3-28)
$$L_{s,min} = 0.0214\frac{V^3}{RC}$$	$$L_{s,min} = 3.15\frac{V^3}{RC}$$	(3-29)
where: $L_{s,min}$ = minimum length of spiral, m; p_{min} = minimum lateral offset between the tangent and circular curve (0.20 m); R = radius of circular curve, m; V = design speed, km/h; C = maximum rate of change in lateral acceleration (1.2 m/s³)	where: $L_{s,min}$ = minimum length of spiral, ft; p_{min} = minimum lateral offset between the tangent and circular curve (0.66 ft); R = radius of circular curve, ft; V = design speed, mph; C = maximum rate of change in lateral acceleration (4 ft/s³)	

A value of 0.20 m [0.66 ft] is recommended for p_{min}. This value is consistent with the minimum lateral shift that occurs as a result of the natural steering behavior of most drivers. The recommended minimum value for C is 1.2 m/s³ [4.0 ft/s³]. The use of lower values will yield longer, more "comfortable" spiral curve lengths; however, such lengths would not represent the minimum length consistent with driver comfort.

Maximum length of spiral. International experience indicates that there is a need to limit the length of spiral transition curves. Safety problems have been found to occur on spiral curves that are long (relative to the length of the circular curve). Such problems occur when the spiral is so long as to mislead the driver about the sharpness of the approaching curve. A conservative maximum length of spiral that should minimize the likelihood of such problems can be computed as:

Metric	US Customary	
$$L_{s,max} = \sqrt{24(p_{max})R}$$	$$L_{s,max} = \sqrt{24(p_{max})R}$$	(3-30)
where: $L_{s,max}$ = maximum length of spiral, m; p_{max} = maximum lateral offset between the tangent and circular curve (1.0 m); R = radius of circular curve, m	where: $L_{s,max}$ = maximum length of spiral, ft; p_{max} = maximum lateral offset between the tangent and circular curve (3.3 ft); R = radius of circular curve, ft	

A value of 1.0 m [3.3 ft] is recommended for p_{max}. This value is consistent with the maximum lateral shift that occurs as a result of the natural steering behavior of most drivers. It also provides a reasonable balance between spiral length and curve radius.

Desirable length of spiral. A recent study of the operational effects of spiral curve transitions (**20**) found that spiral length is an important design control. Specifically, the most desirable operating conditions were noted when the spiral curve length was approximately equal to the length of the natural spiral path adopted by drivers. Differences between these two lengths resulted in operational problems associated with large lateral velocities or shifts in lateral position at the end of the transition curve. Specifically, a large lateral velocity in an outward direction (relative to the curve) requires the driver to make a corrective steering maneuver that results in a path radius sharper than the radius of the circular curve. Such a critical radius produces an undesirable increase in peak side friction demand. Moreover, lateral velocities of sufficient magnitude to shift a vehicle into an adjacent lane (without corrective steering) are also undesirable for safety reasons.

Based on these considerations, desirable lengths of spiral transition curves are shown in Exhibit 3-34. These lengths correspond to 2.0 s of travel time at the design speed of the roadway. This travel time has been found to be representative of the natural spiral path for most drivers (**20**).

The spiral lengths listed in Exhibit 3-34 are recommended as desirable values for street and highway design. Theoretical considerations suggest that significant deviations from these lengths tend to increase the shifts in the lateral position of vehicles within a lane that may precipitate encroachment on an adjacent lane or shoulder. The use of longer spiral curve lengths that are less than $L_{s,max}$ is acceptable. However, where such longer spiral curve lengths are used, consideration should be given to increasing the width of the traveled way on the curve to minimize the potential for encroachments into the adjacent lanes.

Metric		US Customary	
Design speed (km/h)	Spiral length (m)	Design speed (mph)	Spiral length (ft)
20	11	15	44
30	17	20	59
40	22	25	74
50	28	30	88
60	33	35	103
70	39	40	117
80	44	45	132
90	50	50	147
100	56	55	161
110	61	60	176
120	67	65	191
130	72	70	205
		75	220
		80	235

Exhibit 3-34. Desirable Length of Spiral Curve Transition

Spiral curve lengths longer than those shown in Exhibit 3-34 may be needed at turning roadway terminals to adequately develop the desired superelevation. Specifically, spirals twice as long as those shown in Exhibit 3-34 may be needed in such situations. The resulting shift in lateral position may exceed 1.0 m [3.3 ft]; however, such a shift is consistent with driver expectancy at a turning roadway terminal and can be accommodated by the additional lane width typically provided on such turning roadways.

Finally, if the desirable spiral curve length shown in Exhibit 3-34 is less than the minimum spiral curve length determined from Equations (3-28) and (3-29), the minimum spiral curve length should be used in design.

Length of superelevation runoff. In transition design with a spiral curve, it is recommended that the superelevation runoff be accomplished over the length of spiral. For the most part the calculated values for length of spiral and length of runoff do not differ materially. However, in view of the empirical nature of both, an adjustment in one to avoid having two separate sets of design criteria is desirable. The length of runoff is applicable to all superelevated curves, and it is recommended that this value should be used for minimum lengths of spiral. In this manner, the length of spiral should be set equal to the length of runoff. The change in cross slope begins by introducing a tangent runout section just in advance of the spiral curve. Full attainment of superelevation is then accomplished over the length of the spiral. In such a design, the whole of the circular curve has full superelevation.

Limiting superelevation rates. One consequence of equating runoff length to spiral length is that the resulting relative gradient of the pavement edge may exceed the values listed in Exhibit 3-27. However, small increases in gradient have not been found to have an adverse effect on comfort or appearance. In this regard, the adjustment factors listed in Exhibit 3-28 effectively allow for a 50 percent increase in the maximum relative gradient when three lanes are rotated.

The superelevation rates that are associated with a maximum relative gradient that is 50 percent larger than the values in Exhibit 3-27 are listed in Exhibit 3-35. If the superelevation rate used in design exceeds the rate listed in this table, the maximum relative gradient will be at least 50 percent larger than the maximum relative gradient allowed for a tangent-to-curve design. In this situation, special consideration should be given to the transition's appearance and the abruptness of its edge-of-pavement profile.

Metric				US Customary			
Design speed (km/h)	Number of lanes rotated			Design speed (mph)	Number of lanes rotated		
	1	2	3		1	2	3
20	3.7	1.9	1.3	15	4.3	2.2	1.5
30	5.2	2.6	1.7	20	5.5	2.8	1.9
40	6.5	3.2	2.2	25	6.5	3.3	2.2
50	7.5	3.8	2.5	30	7.3	3.7	2.5
60	8.3	4.2	2.8	35	8.0	4.0	2.7
70	8.9	4.5	3.0	40	8.5	4.3	2.9
80	9.3	4.6	3.1	45	8.9	4.5	3.0
90	9.8	4.9	3.3	50	9.2	4.6	3.1
100	10.2	5.1	3.4	55	9.5	4.8	3.2
110	10.4	5.2	3.5	60	9.9	5.0	3.3
120	10.6	5.3	3.5	65	10.3	5.2	3.4
130	10.6	5.3	3.5	70	10.3	5.2	3.5
				75	10.5	5.3	3.5
				80	10.5	5.3	3.5
Note: Based on desirable length of spiral curve transition from Exhibit 3-34.							

Exhibit 3-35. Superelevation Rates Associated With Large Relative Gradients

Length of tangent runout. The tangent runout length for a spiral curve transition design is based on the same approach used for the tangent-to-curve transition design. Specifically, a smooth edge of pavement profile is desired such that a common edge slope gradient is maintained throughout the superelevation runout and runoff sections. Based on this rationale, the following equation can be used to compute the tangent runout length:

Metric	US Customary	
$$L_t = \frac{e_{NC}}{e_d} L_S$$	$$L_t = \frac{e_{NC}}{e_d} L_S$$	(3-31)
where:	where:	
L_t = length of tangent runout, m;	L_t = length of tangent runout, ft;	
L_S = length of spiral, m;	L_S = length of spiral, ft;	
e_d = design superelevation rate, percent;	e_d = design superelevation rate, percent;	
e_{NC} = normal cross slope rate, percent	e_{NC} = normal cross slope rate, percent	

The tangent runout lengths obtained from Equation (3-31) are presented in Exhibit 3-36. The lengths in this table tend to be longer than desirable for combinations of low superelevation rate and high speed. Such long lengths may present safety problems when there is insufficient profile grade to provide adequate pavement surface drainage. Such problems can be avoided when the profile grade criteria described in the section on "Minimum Transition Grades" are applied to the spiral curve transition.

Metric						US Customary					
Design speed (km/h)	Tangent runout length (m) Superelevation rate					Design speed (mph)	Tangent runout length (ft) Superelevation rate				
	2	4	6	8	10		2	4	6	8	10
20	11	–	–	–	–	15	44	–	–	–	–
30	17	8	–	–	–	20	59	30	–	–	–
40	22	11	7	–	–	25	74	37	25	–	–
50	28	14	9	–	–	30	88	44	29	–	–
60	33	17	11	8	–	35	103	52	34	26	–
70	39	19	13	10	–	40	117	59	39	29	–
80	44	22	15	11	–	45	132	66	44	33	–
90	50	25	17	13	10	50	147	74	49	37	–
100	56	28	19	14	11	55	161	81	54	40	–
110	61	31	20	15	12	60	176	88	59	44	–
120	67	33	22	17	13	65	191	96	64	48	38
130	72	36	24	18	14	70	205	103	68	51	41
						75	220	110	73	55	44
						80	235	118	78	59	47

Notes: 1. Based on 2.0% normal cross slope.
2. Superelevation rates above 10% and cells with "–" coincide with a pavement edge grade that exceeds the maximum relative gradient in Exhibit 3-27 by 50% or more. These limits apply to roads where one lane is rotated; lower limits apply when more lanes are rotated (see Exhibit 3-28).

Exhibit 3-36. Tangent Runout Length for Spiral Curve Transition Design

Compound Curve Transition

In general, compound curve transitions are most commonly considered for application to low-speed turning roadways at intersections. In contrast, tangent-to-curve or spiral curve transition designs are more commonly used on street and highway curves.

Guidance concerning compound curve transition design for turning roadways is provided in Chapters 9 and 10. The guidance in Chapter 9 applies to low-speed turning roadway terminals at intersections, while the guidance in Chapter 10 applies to interchange ramp terminals.

Methods of Attaining Superelevation

Four methods are used to transition the pavement to a superelevated cross section. These methods include: (1) revolving a traveled way with normal cross slopes about the centerline profile, (2) revolving a traveled way with normal cross slopes about the inside-edge profile, (3) revolving a traveled way with normal cross slopes about the outside-edge profile, and (4) revolving a straight cross-slope traveled way about the outside-edge profile. Exhibit 3-37 illustrates these four methods. The methods of changing cross slope are most conveniently shown in the exhibit in terms of straight line relationships, but it is emphasized that the angular breaks between the straight-line profiles are to be rounded in the finished design, as shown in the exhibit.

The profile reference line controls for the roadway's vertical alignment through the horizontal curve. Although shown as a horizontal line in Exhibit 3-37, the profile reference line may correspond to a tangent, a vertical curve, or a combination of the two. In Exhibit 3-37A, the profile reference line corresponds to the centerline profile. In Exhibits 3-37B and 3-37C, the profile reference line is represented as a "theoretical" centerline profile as it does not coincide with the axis of rotation. In Exhibit 3-37D, the profile reference line corresponds to the outside edge of traveled way. The cross sections at the bottom of each diagram in Exhibit 3-37 indicate the traveled way cross slope condition at the lettered points.

The first method, as shown in Exhibit 3-37A, revolves the traveled way about the centerline profile. This method is the most widely used because the change in elevation of the edge of the traveled way is made with less distortion than with the other methods. In this regard, one-half of the change in elevation is made at each edge.

The second method, as shown in Exhibit 3-37B, revolves the traveled way about the inside-edge profile. In this case, the inside-edge profile is determined as a line parallel to the profile reference line. One-half of the change in elevation is made by raising the actual centerline profile with respect to the inside-edge profile and the other half by raising the outside-edge profile an equal amount with respect to the actual centerline profile.

The third method, as shown in Exhibit 3-37C, revolves the traveled way about the outside-edge profile. This method is similar to that shown in Exhibit 3-37B except that the elevation change is accomplished below the outside-edge profile instead of above the inside-edge profile.

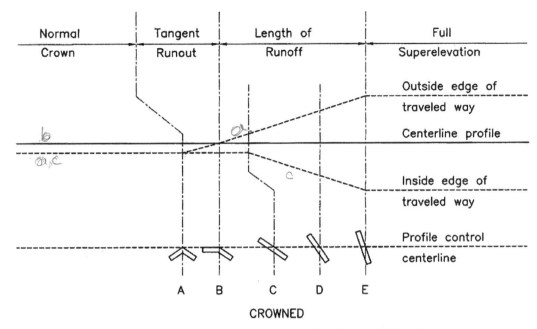

CROWNED

TRAVELED WAY REVOLVED ABOUT CENTERLINE

−A−

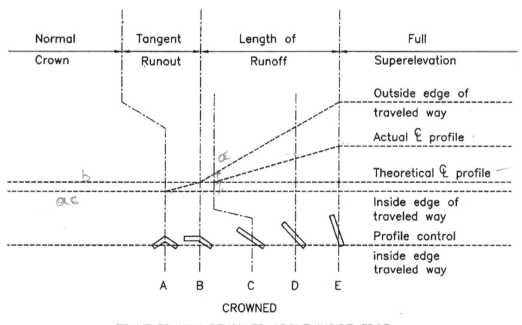

CROWNED

TRAVELED WAY REVOLVED ABOUT INSIDE EDGE

−B−

Exhibit 3-37. Diagrammatic Profiles Showing Methods of Attaining Superelevation for a Curve to the Right

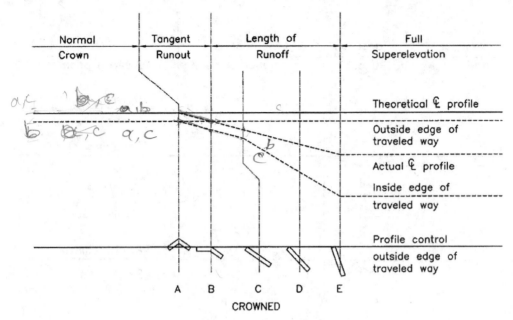

CROWNED

TRAVELED WAY REVOLVED ABOUT OUTSIDE EDGE

−C−

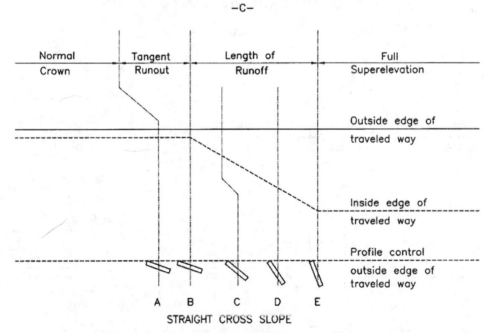

STRAIGHT CROSS SLOPE

TRAVELED WAY REVOLVED ABOUT OUTSIDE EDGE

−D−

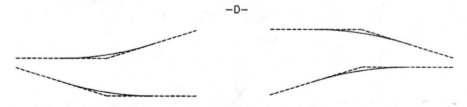

NOTE: ANGULAR BREAKS TO BE APPROPRIATELY ROUNDED AS SHOWN. (SEE TEXT)

Exhibit 3-37. Diagrammatic Profiles Showing Methods of Attaining Superelevation for a Curve to the Right (Continued)

The fourth method, as shown in Exhibit 3-37D, revolves the traveled way (having a straight cross-slope) about the outside-edge profile. This method is often used for two-lane one-way roadways where the axis of rotation coincides with the edge of the traveled way adjacent to the highway median.

The methods for attaining superelevation are nearly the same for all four methods. Cross section A at one end of the tangent runout is a normal (or straight) cross-slope section. At cross section B, the other end of the tangent runout and the beginning of the superelevation runoff, the lane or lanes on the outside of the curve are made horizontal (or level) with the actual centerline profile for Exhibits 3-37A, 3-37B, and 3-37C; there is no change in cross slope for Exhibit 3-37D.

At cross section C the traveled way is a plane, superelevated at the normal cross slope rate. Between cross sections B and C for Exhibits 3-37A, 3-37B, and 3-37C, the outside lane or lanes change from a level condition to one of superelevation at the normal cross slope rate and normal cross slope is retained on the inner lanes. There is no change between cross sections B and C for Exhibit 3-37D. Between cross sections C and E the pavement section is revolved to the full rate of superelevation. The rate of cross slope at an intermediate point (e.g., cross section D) is proportional to the distance from cross section C.

In an overall sense, the method of rotation about the centerline shown in Exhibit 3-37A is usually the most adaptable. On the other hand, the method shown in Exhibit 3-37B is preferable where the lower edge profile is a major control, as for drainage. With uniform profile conditions, its use results in the greatest distortion of the upper edge profile. Where the overall appearance is to be emphasized, the methods of Exhibits 3-37C and 3-37D are advantageous in that the upper-edge profile—the edge most noticeable to drivers—retains the smoothness of the control profile. Thus, the shape and direction of the centerline profile may determine the preferred method for attaining superelevation.

Considering the infinite number of profile arrangements that are possible and in recognition of such specific problems as drainage, avoidance of critical grades, aesthetics, and fitting the roadway to the adjacent topography, no general recommendation for the adoption of any particular axis of rotation can be made. To obtain the most pleasing and functional results, each superelevation transition section should be considered individually. In practice, any pavement reference line used for the axis of rotation may be best suited for the problem at hand.

Design of Smooth Profiles for Traveled Way Edges

In the diagrammatic profiles shown in Exhibit 3-37, the tangent profile control lines result in angular breaks at cross sections A, C, and E. For general appearance and safety, these breaks should be rounded in final design by insertion of vertical curves. Even when the maximum relative gradient is used to define runoff length, the length of vertical curve needed to conform to the 0.65 [0.66] percent break at the 50-km/h [30-mph] design speed (see Exhibit 3-27) and 0.38 [0.38] percent break at the 120-km/h [75 mph] design speed need not be great. Where the traveled way is revolved about an edge, these grade breaks are doubled to 1.30 [1.32] percent for the

50-km/h [30-mph] design speed and to 0.76 [0.76] percent for the 120-km/h [75-mph] design speed. Greater lengths of vertical curve are obviously needed in these cases. Specific criteria for the lengths of vertical curves at the breaks in the diagrammatic profiles have not been established. For an approximate guide, however, the minimum vertical curve length in meters [feet] can be used as numerically equal to the design speed in kilometers per hour [equal to the design speed in miles per hour]. Greater lengths should be used where practical, as the general profile condition may determine.

Several methods are available for the development of smooth-edge profiles in superelevation transition sections. One method defines the edge profiles on a straight-line basis, as shown in Exhibit 3-37, and then develops the profile details based on inserting parabolic vertical curves at each edge break. In such cases, the minimum vertical curve length is often set equal to a travel time at the design speed of about 0.7 s. This method is laborious when the edge vertical curves are superimposed on a centerline vertical curve. However, it does provide an essential control for the designer and should yield uniformity of results.

A second method uses a graphical approach to define the edge profile. The method essentially is one of spline-line development. In this method the centerline or other base profile, which usually is computed, is plotted on an appropriate vertical scale. Superelevation control points are in the form of the break points shown in Exhibit 3-37. Then by means of a spline, curve template, ship curve, or circular curve, smooth-flowing lines are drawn to approximate the straight-line controls. The natural bending of the spline nearly always satisfies the need for minimum smoothing. Once the edge profiles are drawn in the proper relation to one another, elevations can be read at the appropriate intervals (as needed for construction control).

An important advantage of the graphical or spline-line method is the infinite study alternatives it affords the designer. Alternate profile solutions can be developed with a minimum expenditure of time. The net result is a design that is well suited to the particular control conditions. The engineering design labor needed for this procedure is minimal. These several advantages make this method preferable to the other methods of developing profile details for runoff sections.

Divided highways warrant a greater refinement in design and greater attention to appearance than do two-lane highways because divided highways usually serve much greater traffic volumes. Moreover, the cost of such refinements is insignificant compared with the construction cost of the divided highway. Accordingly, there should be greater emphasis on the development of smooth-flowing traveled way edge profiles for divided highways.

Axis of Rotation with a Median

In the design of divided highways, streets, and parkways, the inclusion of a median in the cross section influences the superelevation transition design. This influence stems from the several possible locations for the axis of rotation. The most appropriate location for this axis depends on the width of the median and its cross section. Common combinations of these factors and the appropriate corresponding axis location are described in the following three cases:

Case I—The whole of the traveled way, including the median, is superelevated as a plane section. Case I should necessarily be limited to narrow medians and moderate superelevation rates to avoid substantial differences in elevation of the extreme edges of the traveled way arising from the median tilt. Specifically, Case I should be applied only to medians with widths of 4 m [15 ft] or less. Superelevation can be attained using a method similar to that shown in Exhibit 3-37A except for the two median edges, which will appear as profiles only slightly removed from the centerline.

Case II—The median is held in a horizontal plane and the two traveled ways are rotated separately around the median edges. Case II can be applied to any width of median but is most appropriate for medians with widths between 4 and 18 m [15 and 60 ft]. By holding the median edges level, the difference in elevation between the extreme traveled way edges can be limited to that needed to superelevate the roadway. Superelevation transition design for Case II usually has the median-edge profiles as the control. One traveled way is rotated about its lower edge and the other about its higher edge. Superelevation can be attained using any of the methods shown in Exhibits 3-37B, 3-37C, and 3-37D, with the profile reference line being the same for both traveled ways.

Case III—The two traveled ways are treated separately for runoff with a resulting variable difference in elevations at the median edges. Case III design can be used with wide medians (i.e., those having a width of 18 m [60 ft] or more). For this case, the differences in elevation of the extreme edges of the traveled way are minimized by a compensating slope across the median. With a wide median, it is possible to design the profiles and superelevation transition separately for the two roadways. Accordingly, superelevation can be attained by the method otherwise considered appropriate (i.e., any of the methods in Exhibit 3-37 can be used).

Superelevation runoff lengths vary for each of the three cases. For Case I designs, the length of runoff should be based on the total rotated width (including the median width). Runoff lengths for Case II designs should be the same as those for undivided highways with a similar number of lanes. Finally, runoff lengths for Case III designs are based on the needs of the separate one-way roadways, as defined by their superelevation rates and rotated widths.

Superelevation runoff lengths for four- and six-lane undivided highways have been shown in Exhibit 3-28 as 1.5 and 2 times, respectively, the lengths for two-lane highways. For Case I designs of divided highways the length of runoff should properly be increased in the proportion to the total width, including the median. Because Case I applies primarily to narrow medians, the added length usually will be insignificant. With medians of the order of 1 to 3 m [3 to 10 ft] wide, any increase in runoff length may well be ignored.

Under Case II conditions with narrow medians in a horizontal plane, the runoff lengths should be the same as those for undivided highways as shown in Exhibits 3-21 through 3-25 for four-lane highways. This length applies to highways with medians about 4 m [15 ft] or less in width. However, with medians about 12 m [40 ft] or more in width, the two-lane values should be used for the one-way roadways because the extreme traveled way edges are at least 24 m [80 ft] apart and are independent of each other. Values for the one-way roadways of six-lane highways when separated by a wide median should be 1.2 times the two-lane values of Exhibits 3-21 through 3-25. The one-way traveled ways of highways with medians between 4 and 12 m [15 and 40 ft] might be designed on the basis of either the suggested two-lane or multilane runoff lengths.

With Case III cross sections, the median generally will be 12 m [40 ft] or more in width, and the two-lane values for length of runoff are applicable for one-way roadways of four-lane divided highways. The values for the one-way roadways of six-lane divided highways should be somewhat greater. In situations where the median width is less than about 12 m [40 ft], the runoff length should be determined in the same manner as for Case II.

Divided highways warrant a greater refinement in design and greater attention to appearance than two-lane highways because they serve much greater traffic volumes and because the cost of such refinements is insignificant compared with the cost of construction. Accordingly, the values for length of runoff indicated above should be considered minimums, and the use of yet longer values should be considered. Likewise, there should be emphasis on the development of smooth-flowing traveled way edge profiles of the type obtained by spline-line design methods.

Minimum Transition Grades

Two potential pavement surface drainage problems are of concern in the superelevation transition section. One problem relates to the potential lack of adequate longitudinal grade. This problem generally occurs when the grade axis of rotation is equal to, but opposite in sign to, the effective relative gradient. It results in the edge of pavement having negligible longitudinal grade, which can lead to poor pavement surface drainage, especially on curbed cross sections.

The other potential drainage problem relates to inadequate lateral drainage due to negligible cross slope during pavement rotation. This problem occurs in the transition section where the cross slope of the outside lane varies from an adverse slope at the normal cross slope rate to a superelevated slope at the normal cross slope rate. This length of the transition section includes the tangent runout section and an equal length of the runoff section. Within this length, the pavement cross slope may not be sufficient to adequately drain the pavement laterally.

Two techniques can be used to alleviate these two potential drainage problems. One technique is to provide a minimum profile grade in the transition section. The second technique is to provide a minimum edge of pavement grade in the transition section. Both techniques can be incorporated in the design by use of the following grade criteria:

1. Maintain minimum profile grade of 0.5 percent through the transition section.

2. Maintain minimum edge of pavement grade of 0.2 percent (0.5 percent for curbed streets) through the transition section.

The second grade criterion is equivalent to the following series of equations relating profile grade and effective maximum relative gradient:

Metric		US Customary		
Uncurbed	Curbed	Uncurbed	Curbed	
G≤−Δ*−0.2	G≤−Δ*−0.5	G≤−Δ*−0.2	G≤−Δ*−0.5	
G≥−Δ*+0.2	G≥−Δ*+0.5	G≥−Δ*+0.2	G≥−Δ*+0.5	
G≤Δ*−0.2	G≤Δ*−0.5	G≤Δ*−0.2	G≤Δ*−0.5	
G≥Δ*+0.2	G≥Δ*+0.5	G≥Δ*+0.2	G≥Δ*+0.5	
with,		with,		
$$\Delta* = \frac{(wn_l)\,e_d}{L_r}$$		$$\Delta* = \frac{(wn_l)\,e_d}{L_r}$$		(3-32)

with, on both sides:

$$\Delta* = \frac{(wn_l)\,e_d}{L_r}$$

where:

G = profile grade, percent;

$\Delta*$ = effective maximum relative gradient, percent;

L_r = length of superelevation runoff, m;

n_l = number of lanes rotated, lanes;

w = width of one traffic lane, m (typically 3.6 m);

e_d = design superelevation rate, percent.

where:

G = profile grade, percent;

$\Delta*$ = effective maximum relative gradient, percent;

L_r = length of superelevation runoff, ft;

n_l = number of lanes rotated, lanes;

w = width of one traffic lane, ft (typically 12 ft);

e_d = design superelevation rate, percent.

The value of 0.2 in the grade control (G) equation represents the minimum edge of pavement grade for uncurbed roadways (expressed as a percentage). If this equation is applied to curbed streets, the value 0.2 should be replaced with 0.5.

To illustrate the combined use of the two grade criteria, consider an uncurbed roadway curve having an effective maximum relative gradient of 0.65 percent in the transition section. The first criterion would exclude grades between −0.50 and +0.50 percent. The second grade criterion would exclude grades in the range of −0.85 to −0.45 percent (via the first two components of the equation) and those in the range of 0.45 to 0.85 percent (via the last two components of the equation). Given the overlap between the ranges for Controls 1 and 2, the profile grade within the transition would have to be outside of the range of −0.85 to +0.85 percent in order to satisfy both criteria and provide adequate pavement surface drainage.

Turning Roadway Design

Turning roadways can be categorized as interchange ramps, roadways, or intersection curves for right-turning vehicles. Loop or diamond configurations for turning roadways are used at interchanges and consist of combinations of tangents and curves. At intersections, turning roadways have a diamond configuration and consist of curves (often compound curves). Turning roadway design does not apply to minimum edge-of-traveled-way design for turns at intersections. Here it is a matter of closely fitting compound curves to the inside edge of the design vehicle's swept path (as described in Chapter 9).

When the design speed of the turning roadway is 70 km/h [45 mph] or less, compound curvature can be used to form the entire alignment of the turning roadway. When the design speed exceeds 70 km/h [45 mph], the exclusive use of compound curves is often impractical, as it tends to need a large amount of right-of-way. Thus, high-speed turning roadways follow the interchange ramp design guidelines in Chapter 10 and include a mix of tangents and curves. By this approach, the design can be more sensitive to right-of-way impacts as well as to driver comfort and safety.

For compound curves at intersections, it is preferable that the ratio of the flatter radius to the sharper radius not exceed 2:1. This ratio results in a reduction of approximately 10 km/h [6 mph] in average running speeds for the two curves.

For compound curves at interchanges, it is preferable that the ratio of the flatter radius to the sharper radius not exceed 1.75:1. However, general observations on ramps having differences in radii with a ratio of 2:1 indicate that both operation and appearance are satisfactory.

Curves that are compounded should not be too short or their effect in enabling a change in speed from the tangent or flat curve to the sharp curve is lost. In a series of curves of decreasing radii, each curve should be long enough to enable the driver to decelerate at a reasonable rate. At intersections, a maximum deceleration rate of 5 km/h/s [3 mph/s] may be used (although 3 km/h/s [2 mph/s] is desirable). The desirable rate represents very light braking, because deceleration in gear alone generally results in overall rates between 1.5 and 2.5 km/h/s [1 and 1.5 mph/s]. Minimum compound curve lengths based on these criteria are presented in Exhibit 3-38.

The compound curve lengths in Exhibit 3-38 are developed on the premise that travel is in the direction of sharper curvature. For the acceleration condition, the 2:1 ratio is not as critical and may be exceeded.

Metric			US Customary		
	Minimum length of circular arc (m)			Minimum length of circular arc (ft)	
Radius (m)	Acceptable	Desirable	Radius (ft)	Acceptable	Desirable
30	12	20	100	40	60
50	15	20	150	50	70
60	20	30	200	60	90
75	25	35	250	80	120
100	30	45	300	100	140
125	35	55	400	120	180
150 or more	45	60	500 or more	140	200

Exhibit 3-38. Lengths of Circular Arcs for Different Compound Curve Radii

Design for Low-Speed Urban Streets

As previously discussed, the maximum allowable side friction factor for use in the design of horizontal curves is the point at which the lateral force causes the driver to experience a feeling of discomfort when driving a curve at a particular design speed. Exhibit 3-10 has summarized the test results of side friction factors developed on curves at these apparent limits of comfort. Use of the solid line in Exhibit 3-13 and method 5 was recommended for distributing e and f in the design of rural highways and high-speed urban streets. Method 2 is recommended for the design of horizontal curves on low-speed urban streets where, through conditioning, drivers have developed a higher threshold of discomfort. By this method, none of the lateral force is counteracted by superelevation so long as the side friction factor is less than the specified maximum for the radius of the curve and the design speed.

For sharper curves, f remains at the maximum and e is used in direct proportion to the continued increase in curvature until e reaches e_{max}. The recommended design values for f that are applicable to low-speed urban streets are shown in Exhibit 3-39 as a solid line superimposed on the analysis curves from Exhibit 3-10. They are based on a tolerable degree of discomfort and provide a reasonable margin of safety against skidding under normal driving conditions in the urban environment. These values vary with the design speed from 0.32 at 30 km/h [0.30 at 20 mph] to about 0.165 at 70 km/h [0.165 at 45 mph], with 70 km/h [45 mph] being the upper limit for low speed established in the design speed discussion of Chapter 2.

Although superelevation is advantageous for traffic operations, various factors often combine to make its use impractical in many built-up areas. These factors include wide pavement areas, the need to meet the grade of adjacent property, surface drainage considerations, and frequency of cross streets, alleys and driveways. Therefore, horizontal curves on low-speed streets in urban areas are frequently designed without superelevation, sustaining the lateral force solely with side friction. On these curves for traffic entering a curve to the left the normal cross slope is an adverse or negative superelevation, but with flat curves the resultant friction needed to sustain the lateral force, even given the negative superelevation, is small.

However, on successively sharper curves for the same design speed, the minimum radius or sharpest curve without superelevation is reached when the side friction factor developed to sustain the lateral force, given the adverse cross slope, reaches the maximum allowable value based on driver comfort considerations. The maximum allowable side friction factor based on driver comfort also provides an appropriate margin of safety against skidding. For travel on sharper curves, superelevation is needed.

The maximum superelevation rate of zero in Exhibit 3-41 establishes the minimum radius for each speed below which superelevation is not provided on local streets in residential and commercial areas but should be considered in industrial areas or other streets where operating speeds are higher. A maximum superelevation rate of 4.0 or 6.0 percent is commonly used. The maximum curvature for a given design speed is defined for low-speed urban streets when both the maximum superelevation rate and the maximum allowable side friction factors are utilized.

METRIC

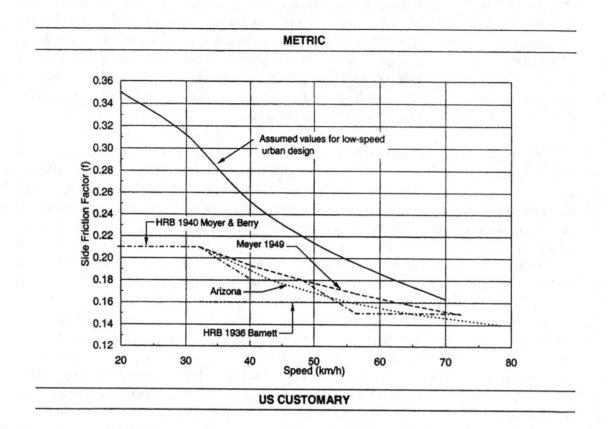

US CUSTOMARY

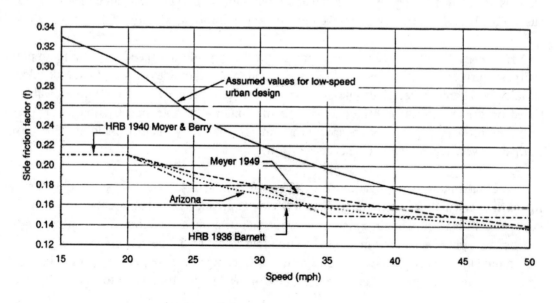

Exhibit 3-39. Side Friction Factors for Low-Speed Urban Streets

Maximum Comfortable Speed on Horizontal Curves

Exhibit 3-40 and 3-41, for low-speed urban streets, are derived from the simplified curve formula:

Metric	US Customary	
$\dfrac{e}{100} + f_{\max} = \dfrac{V^2}{127R}$	$\dfrac{e}{100} + f_{\max} = \dfrac{V^2}{15R}$	(3-33)

Exhibit 3-40 has been prepared by using the recommended values of *f* for low-speed urban streets from Exhibit 3-39 and by varying the rates of superelevation from –5.0 to +6.0 percent. This exhibit may be used for determining the maximum speeds for horizontal curves on low-speed urban streets. By interpolating, the exhibit may also be used to determine the minimum superelevation needed where a curve is to be provided with a radius greater than the minimum, but less than the radius for a cross section with normal cross slopes. However, it is desirable to provide the maximum superelevation for curves with these intermediate radii as well because of the tendency for drivers to overdrive curves with lower design speeds.

Minimum Superelevation Runoff Length

The following equation for deriving the minimum superelevation runoff length is based on the maximum allowable side friction factor, where *C* is the rate of change of the side friction factor obtained from Exhibit 3-41.

Metric	US Customary	
$L = \dfrac{2.72 f V_D}{C}$	$L = \dfrac{47.2 f V_D}{C}$	(3-34)
where: L = length of superelevation runoff, m; f = side friction factor; V_D = design speed, km/h; and C = rate of change of f, m/s^3.	where: L = length of superelevation runoff, ft; f = side friction factor; V_D = design speed, mph; and C = rate of change of f, ft/s3.	

METRIC

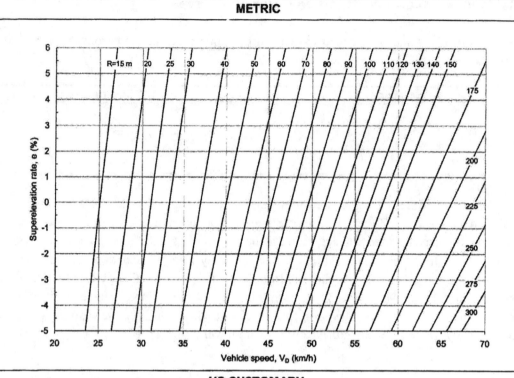

US CUSTOMARY

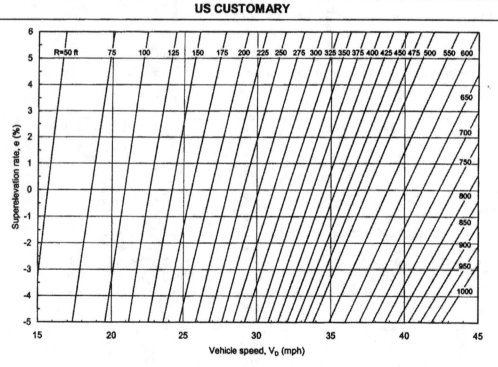

Exhibit 3-40. Relationship of Radius Superelevation, Cross Slope Rate, and Design Speed for Low-Speed Urban Street Design

Metric

Design speed (km/h)	Max e/100	Max f	Total (e/100 + f)	Min R (m)	Max e/100	Max f	Total (e/100 + f)	Min R (m)	Max e/100	Max f	Total (e/100 + f)	Min R (m)	C	Min L (m)
20	0.06	0.350	0.410	10	0.04	0.350	0.390	10	0.00	0.350	0.350	10	1.25	15
30	0.06	0.312	0.372	20	0.04	0.312	0.352	20	0.00	0.312	0.312	25	1.20	20
40	0.06	0.252	0.312	40	0.04	0.252	0.292	45	0.00	0.252	0.252	50	1.15	25
50	0.06	0.214	0.274	70	0.04	0.214	0.254	80	0.00	0.214	0.214	90	1.10	25
60	0.06	0.186	0.246	115	0.04	0.186	0.226	125	0.00	0.186	0.186	150	1.05	30
70	0.06	0.163	0.223	175	0.04	0.163	0.203	190	0.00	0.163	0.163	235	1.00	30

US Customary

Design speed (mph)	Max e/100	Max f	Total (e/100 + f)	Min R (ft)	Max e/100	Max f	Total (e/100 + f)	Min R (ft)	Max e/100	Max f	Total (e/100 + f)	Min R (ft)	C	Min L (ft)
15	0.06	0.330	0.390	40	0.04	0.330	0.370	40	0.00	0.330	0.330	45	4.25	55
20	0.06	0.300	0.360	75	0.04	0.300	0.340	80	0.00	0.300	0.300	90	4.00	75
25	0.06	0.252	0.312	135	0.04	0.252	0.292	145	0.00	0.252	0.252	165	3.75	80
30	0.06	0.221	0.281	215	0.04	0.221	0.261	230	0.00	0.221	0.221	275	3.50	90
35	0.06	0.197	0.257	320	0.04	0.197	0.231	345	0.00	0.197	0.197	415	3.25	100
40	0.06	0.178	0.238	450	0.04	0.178	0.218	490	0.00	0.178	0.178	600	3.00	115
45	0.06	0.163	0.223	605	0.04	0.163	0.203	665	0.00	0.163	0.163	830	2.75	125

Exhibit 3-41. Minimum Radii and Minimum Lengths of Superelevation Runoff for Limiting Values of e and f (Low-Speed Urban Streets)

For design on low-speed urban streets the value of *C* varies from 1.25 m/s^3 at 20 km/h [4.25 ft/s^3 at 15 mph] to 1.0 m/s^3 at 70 km/h [2.75 ft/s^3 at 45 mph]. The minimum lengths of superelevation runoff for limiting values of superelevation and side friction are shown in Exhibit 3-41. The formula for length of superelevation runoff is based on revolving the traveled way about the centerline of the street. On flat grades, revolving the traveled way about the centerline may result in low spots on the inner edge of the traveled way. To avoid this condition, the traveled way should be revolved about the inside edge and the length of superelevation runoff shown on Exhibit 3-41 should be doubled.

Minimum Radii and Minimum Lengths of Superelevation Runoff for Limiting Values of *e* and *f*

Exhibit 3-41 presents the minimum radii for three rates of superelevation: 0.0, 4.0, and 6.0 percent. The 6.0 percent rate is considered to be the desirable maximum superelevation for low-speed urban street design. In addition, the exhibit contains the minimum lengths of superelevation runoff for each superelevation rate. From this table, the minimum desirable tangent length between two reversing curves of minimum radii can be calculated. The superelevation rate of zero is included in the exhibit because an intervening length of tangent is needed between reversing curves even if neither is superelevated. As stated previously, the portion of the maximum superelevation provided at the PC and PT of the curves can range between 60 and 90 percent. The sum of the superelevation runoff lengths outside the PC or PT of the curves is the minimum intervening length of tangent.

Curvature of Turning Roadways and Curvature at Intersections

Curvature for through roads and streets has been discussed previously. Curvature of turning roadways and curvature at high-speed intersections are special cases discussed in the following four sections.

Minimum Radius for Turning Speed

As further discussed in Chapter 9, vehicles turning at intersections designed for minimum-radius turns have to operate at low speed, perhaps less than 15 km/h [10 mph]. While it is desirable and often practical to design for turning vehicles operating at higher speeds, it is often appropriate for safety and economy to use lower turning speeds at most intersections. The speeds for which these intersection curves should be designed depend on vehicle speeds on the approach highways, the type of intersection, and the volumes of through and turning traffic. Generally, a desirable turning speed for design is the average running speed of traffic on the highway approaching the turn. Designs at such speeds offer little hindrance to smooth flow of traffic and may be justified for some interchange ramps or, at intersections, for certain movements involving little or no conflict with pedestrians or other vehicular traffic.

Curves at intersections need not be considered in the same category as curves on the open highways because the various warnings provided and the anticipation of more critical conditions at an intersection permit the use of less liberal design factors. Drivers generally operate at higher speeds in relation to the radii on intersection curves than on open highway curves. This increased speed is accomplished by the drivers' acceptance and use of higher side friction factors in operating on curves at intersections than the side friction factors accepted and used on the high-speed highways.

Several studies (**26, 27, 28**) have been conducted to determine lateral vehicle placement and distribution of speeds on intersection curves. Results of these studies pertinent to speed-curvature relationships are plotted in Exhibit 3-42. In the analyses of these data the 95-percentile speed of traffic was assumed to be that closely representing the design speed, which generally corresponds to the speed adopted by the faster group of drivers. Side friction factors (taking superelevation into account) actually developed by drivers negotiating the curves at the 95-percentile speed are indicated for 34 locations in Exhibit 3-42. The dashed line at the upper left shows the side friction factors used for design of curves on rural highways and high-speed urban streets (Exhibit 3-13). Use of this control limit for high speeds, and a friction factor of about 0.5 that could be developed at a low speed as the other limit, gives an average or representative curve through the plottings of individual observations—a relation between design (95-percentile) speed and side friction factor that is considered appropriate for rural and high-speed urban curve design for at-grade intersections.

With this relation established and with logical assumptions for the superelevation rate that can be developed on intersection curves, minimum radii for various design speeds are derived from the simplified curve formula (see Equation (3-9) in the preceding section on "Theoretical Considerations"). Obviously, different rates of superelevation would produce somewhat different radii for a given design speed and side friction factor. For design of intersection curves it is desirable to establish a single minimum radius for each design speed. This is done by assuming a likely minimum rate of superelevation (a conservative value) that can nearly always be obtained for certain radii. If more superelevation than this minimum is actually provided, drivers will either be able to drive the curves a little faster or drive them more comfortably because of less friction.

In selecting a minimum rate of superelevation it is recognized that the sharper the curve the shorter its length and the less of an opportunity for developing a large rate of superelevation. This condition applies particularly to intersections where the turning roadway is often close to the intersection proper, where much of its area is adjacent to the through traveled way, and where the complete turn is made through a total angle of about 90 degrees. Assuming the more critical conditions and considering the lengths likely to be available for developing superelevation on curves of various radii, the minimum rate of superelevation for derivation purposes is taken as that varying from zero at 15 km/h to 9.0 percent at 70 km/h [zero at 10 mph to 10.0 percent at 45 mph]. By using these rates and the side friction factors of Exhibit 3-42 in the simplified curve formula, the minimum radii for intersection curves for operation at design speed are derived as shown in Exhibit 3-43.

METRIC

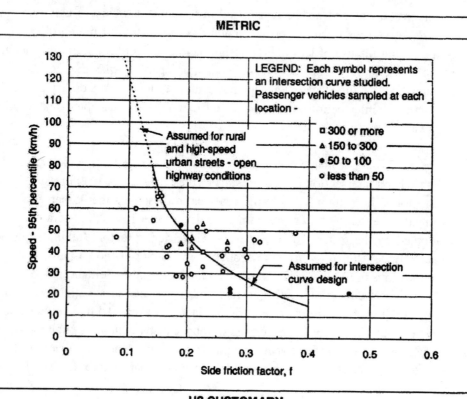

US CUSTOMARY

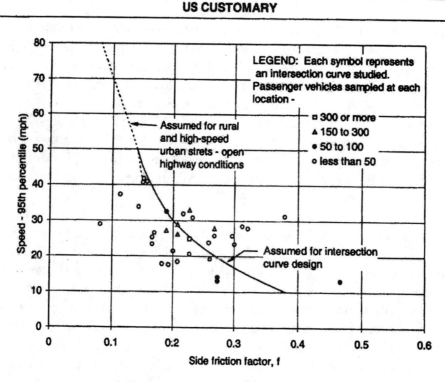

Exhibit 3-42. Relation Between Speed and Side Friction Factor on Curves at Intersections

Metric							
Design (turning) speed, V (km/h)	15	20	30	40	50	60	70
Side friction factor, f	0.40	0.35	0.28	0.23	0.19	0.17	0.15
Assumed minimum superelevation, e/100	0.00	0.00	0.02	0.04	0.06	0.08	0.09
Total e/100 + f	0.40	0.35	0.30	0.27	0.25	0.25	0.24
Calculated minimum radius, R (m)	5	9	24	47	79	113	161
Suggested minimum radius curve for design (m)	7	10	25	50	80	115	160
Average running speed (km/h)	15	20	28	35	42	51	57

Note: For design speeds greater than 70 km/h, use values for open highway conditions.

US Customary								
Design (turning) speed, V (mph)	10	15	20	25	30	35	40	45
Side friction factor, f	0.38	0.32	0.27	0.23	0.20	0.18	0.16	0.15
Assumed minimum superelevation, e/100	0.00	0.00	0.02	0.04	0.06	0.08	0.09	0.10
Total e/100 + f	0.38	0.32	0.29	0.27	0.26	0.26	0.25	0.25
Calculated minimum radius, R (ft)	18	47	92	154	231	314	426	540
Suggested minimum radius curve for design (ft)	25	50	90	150	230	310	430	540
Average running speed (mph)	10	14	18	22	26	30	34	36

Note: For design speeds greater than 45 mph, use values for open highway conditions.

Exhibit 3-43. Minimum Radii for Intersection Curves

The minimum radii of Exhibit 3-43 are represented by the solid line at the left in Exhibit 3-44. The solid line at the upper right shows the relation between design speed and minimum radius for open highway conditions, as derived with *e* values shown at the upper left. The joining of the two lines indicates that open-highway conditions on intersection curves are approached when the curvature is sufficiently flat to permit operation between 60 and 80 km/h [40 and 50 mph]. Thus, in design of intersection curves for design speeds of above 70 km/h [45 mph], open highway conditions should be assumed and the design based on Exhibit 3-21 to 3-25. The square points in Exhibit 3-44 are the observed 95-percentile speeds from the same studies represented in Exhibit 3-42 for the locations where more than 50 vehicles were sampled. The plotted line fits these points closely, indicating further that the assumptions made in the derivation of minimum radii in Exhibit 3-44 are appropriate and that a group of the higher speed drivers will use the design speed assumed.

In addition to the design speed, the average running speed is also used in consideration of certain elements of intersection design. The points indicated by crosses in Exhibit 3-44 are actual average speeds observed on the same intersection curves referred to. The long dashed line through these points is assumed to represent the average running speed for intersection curves. At the right this curve crosses the short dashed line, which indicates the average speed for open highways. For a given design speed on the solid (upper) curve in Exhibit 3-44, the assumed running speed lies vertically below on the dashed curve. These running speeds are shown in the last line of Exhibit 3-43.

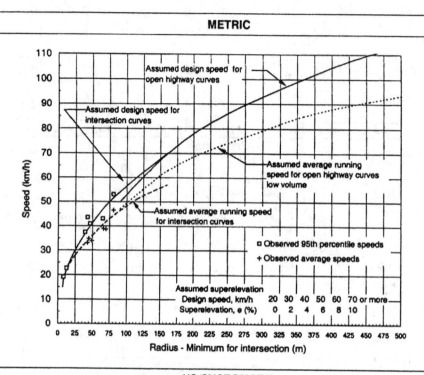

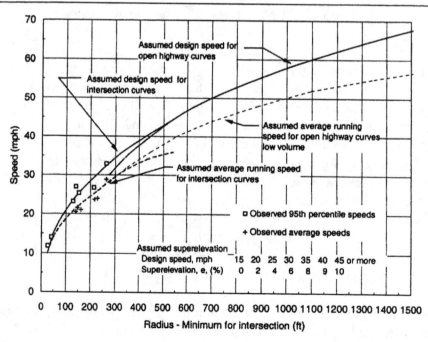

Exhibit 3-44. Minimum Radii for Curves at Intersections

The minimum radii established above should be used for design preferably on the inner edge of the traveled way rather than on the middle of the vehicle path or the centerline of the traveled way. In all cases, as much superelevation as practical up to the appropriate maximum value should be developed. For the suggested radii shown in Exhibit 3-43, a superelevation rate of at least 8.0 percent is desirable at all locations, and a rate of 8.0 to 10.0 percent for locations where snow or ice is not a factor. On those intersection legs where all traffic comes to a stop, as at stop signs, a lesser amount of superelevation is usually appropriate. Also where large trucks will be using an intersection, use of superelevation may need to be limited because these larger trucks may have trouble negotiating intersection curves with superelevation. This is particularly true where trucks cross over from a roadway or ramp sloping in one direction to one sloping the other way. Where there are a significant number of large trucks for each design speed, flatter curves and less superelevation should be provided. Superelevation for curves at intersections is further discussed under that heading in Chapter 9.

Transitions and Compound Curves

Drivers turning at intersections and at interchange ramp terminals naturally follow transitional travel paths just as they do at higher speeds on the open highway. If facilities are not provided for driving in this natural manner, many drivers may deviate from the intended path and develop their own transition, sometimes to the extent of encroaching on other lanes or on the shoulder. Natural travel paths can best be provided by the use of transition or spiral curves that may be inserted between a tangent and a circular arc or between two circular arcs of different radii. Practical designs that follow transitional paths may also be developed by the use of compound circular curves. Transitioned roadways have the added advantage of providing a practical means for changing from a normal to a superelevated cross section.

Length of Spiral

Lengths of spirals for use at intersections are determined in the same manner as they are for open highways. On intersection curves, lengths of spirals may be shorter than they are on the open highway curves, because drivers accept a more rapid change in direction of travel under intersection conditions. In other words, C (the rate of change of lateral acceleration on intersection curves) may be higher on intersection curves than on open highway curves, where values of C ranging from 0.3 to 1.0 m/s^3 [1 to 3 ft/sec^3] generally are accepted. Rates for curves at intersections are assumed to vary from 0.75 m/s^3 [2.5 ft/s^3] for a turnout speed of 80 km/h [50 mph] to 1.2 m/s^3 [4.0 ft/s^3] for 30 km/h [20 mph]. With the use of these values in the Shortt formula (**25**), lengths of spirals for intersection curves are developed in Exhibit 3-45. The minimum lengths of spirals shown are for minimum-radius curves as governed by the design speed. Somewhat lesser spiral lengths are suitable for above-minimum radii.

		Metric						US Customary					
Design speed	(km/h)	30	40	50	60	70	(mph)	20	25	30	35	40	45
Minimum radius	(m)	25	50	80	125	160	(ft)	90	150	230	310	430	550
Assumed C	(m/s³)	1.2	1.1	1.0	0.9	0.8	(ft/s³)	4.0	3.75	3.5	3.25	3.0	2.75
Calculated length of spiral	(m)	19	25	33	41	57	(ft)	70	87	105	134	156	190
Suggested minimum length of spiral	(m)	20	25	35	45	60	(ft)	70	90	110	130	160	200

Exhibit 3-45. Minimum Lengths of Spiral for Intersection Curves

Spirals also may be advantageous between two circular arcs of widely different radii. In this case, the length of spiral can be obtained from Exhibit 3-45 by using a radius that is the difference in the radii of the two arcs. For example, two curves to be connected by a spiral have radii of 250 and 80 m [820 and 262 ft]. This difference of 170 m [558 ft] is very close to the minimum radius of 160 m [550 ft] in Exhibit 3-45 for which the suggested minimum length is about 60 m [200 ft].

Compound curves at intersections for which the radius of one curve is more than twice the radius of the other should have either a spiral or a circular curve of intermediate radius inserted between the two. If, in such instances, the calculated length of spiral is less than 30 m [100 ft], it is suggested that a length of at least 30 m [100 ft] be used.

Compound Circular Curves

Compound circular curves are advantageous in effecting desirable shapes of turning roadways for at-grade intersections and for interchange ramps. Where circular arcs of widely different radii are joined, however, the alignment appears abrupt or forced, and the travel paths of vehicles need considerable steering effort.

On compound curves for open highways, it is generally accepted that the ratio of the flatter radius to the sharper radius should not exceed 1.5:1. For compound curves at intersections where drivers accept more rapid changes in direction and speed, the radius of the flatter arc can be as much as 100 percent greater than the radius of the sharper arc, a ratio of 2:1. The ratio of 2:1 for the sharper curves used at intersections results in approximately the same difference (about 10 km/h [6 mph]) in average running speeds for the two curves. These curves are compounded as for a ratio of 1.5:1 on the flatter curves used on the open highway. General observations on ramps having differences in radii with a ratio of 2:1 indicate that both operation and appearance normally are satisfactory.

Where practical, a smaller difference in radii should be used. A desirable maximum ratio is 1.75:1. Where the ratio is greater than 2:1, a suitable length of spiral or a circular arc of intermediate radius should be inserted between the two curves. In the case of very sharp curves designed to accommodate minimum turning paths of vehicles, it is not practical to apply this ratio control. In this case, compound curves should be developed that fit closely to the path of the design vehicle to be accommodated, for which higher ratios may be needed as shown in Chapter 9.

Curves that are compounded should not be too short or their effectiveness in enabling smooth transitions from tangent or flat-curve to sharp-curve operation may be lost. In a series of curves of decreasing radii, each curve should be long enough to enable the driver to decelerate at a reasonable rate, which at intersections is assumed to be not more than 5 km/h/s [3 mph/s], although 3 km/h/s [2 mph/s] is desirable. Minimum curve lengths that meet these criteria based on the running speeds shown in Exhibit 3-44, are indicated in Exhibit 3-46. They are based on a deceleration of 5 km/h/s [3 mph/s], and a desirable minimum deceleration of 3 km/h/s [2 mph/s].

The latter deceleration rate indicates very light braking, because deceleration in gear alone generally results in overall rates between 1.5 and 2.5 km/h/s [1 and 1.5 mph/s].

Metric			US Customary		
	Length of circular arc (m)			Length of circular arc (ft)	
Radius (m)	Minimum	Desirable	Radius (ft)	Minimum	Desirable
30	12	20	100	40	60
50	15	20	150	50	70
60	20	30	200	60	90
75	25	35	250	80	120
100	30	45	300	100	140
125	35	55	400	120	180
150 or more	45	60	500 or more	140	200

Exhibit 3-46. Length of Circular Arc for a Compound Intersection Curve When Followed by a Curve of One-Half Radius or Preceded by a Curve of Double Radius

These design guidelines for compound curves are developed on the premise that travel is in the direction of sharper curvature. For the acceleration condition, the 2:1 ratio is not as critical and may be exceeded.

Offtracking

Offtracking is the characteristic, common to all vehicles, although much more pronounced with the larger design vehicles, in which the rear wheels do not follow precisely the same path as the front wheels when the vehicle negotiates a horizontal curve or makes a turn. When a vehicle traverses a curve without superelevation at low speed, the rear wheels track inside the front wheels. When a vehicle traverses a superelevated curve, the rear wheels may track inside the front wheels more or less than the amount computed on the above basis. This is because of the slip angle assumed by the tires with respect to the direction of travel, which results from the side friction developed between the pavement and rolling tires. The relative position of the wheel tracks depends on the speed and the amount of friction developed to sustain the lateral force not sustained by superelevation or, when traveling slowly, by the friction developed to counteract the effect of superelevation not compensated by lateral force. At higher speeds, the rear wheels may even track outside the front wheels.

Derivation of Design Values for Widening on Horizontal Curves

In each case, the amount of offtracking, and therefore the amount of widening needed on horizontal curves, depends jointly on the length and other characteristics of the design vehicle and the radius of curvature negotiated. Selection of the design vehicle is based on the size and frequency of the various vehicle types at the location in question. The amount of widening needed increases with the size of the design vehicle (for single-unit vehicles or vehicles with the same number of trailers or semitrailers) and decreases with increasing radius of curvature. The width elements of the design vehicle used in determining the appropriate roadway widening on curves include the track width of the design vehicles that may meet or pass on the curve, U; the lateral clearance per vehicle, C; the width of front overhang of the vehicle occupying the inner lane or lanes, F_A; the width of rear overhang, F_B; and a width allowance for the difficulty of driving on curves, Z.

The track width (U) for a vehicle following a curve or making a turn, also known as the swept path width, is the sum of the track width on tangent (u) (2.44 or 2.59 m [8.0 or 8.5 ft] depending on the design vehicle) and the amount of offtracking. The offtracking depends on the radius of the curve or turn, the number and location of articulation points, and the lengths of the wheelbases between axles. The track width on a curve (U) is calculated using the equation:

Metric	US Customary	
$$U = u + R - \sqrt{R^2 - \sum L_i^2}$$	$$U = u + R - \sqrt{R^2 - \sum L_i^2}$$	(3-35)
where: U = track width on curve, m; u = track width on tangent (out-to-out of tires), m; R = radius of curve or turn, m; and L_i = wheelbase of design vehicle between consecutive axles (or sets of tandem axles) and articulation points, m.	where: U = track width on curve, ft; u = track width on tangent (out-to-out of tires), ft; R = radius of curve or turn, ft; and Li = wheelbase of design vehicle between consecutive axles (or sets of tandem axles) and articulation points, ft.	

This equation can be used for any combination of radius and number and length of wheelbases. The radius for open highway curves is the path of the midpoint of the front axle; however, for most design purposes on two-lane highways, the radius of the curve at the centerline of the highway may be used for simplicity of calculations. For turning roadways, the radius is the path of the outer front wheel (**31**). The wheelbases (L_i) used in the calculations include the distances between each axle and articulation point on the vehicle. For a single-unit truck only the distance between the front axle and the drive wheels is considered. For an articulated vehicle, each of the articulation points is used to determine U. For example, a tractor/semitrailer combination truck has three L_i values that are considered in determining offtracking: (1) the distance from the front axle to the tractor drive axle(s), (2) the distance from the drive axle(s) to the fifth wheel pivot, and (3) the distance from the fifth wheel pivot to the rear axle(s). In the summation process, some terms may be negative, rather than positive, if the articulation point is in front of, rather than behind, the drive axle(s) (**29**) or if there is a rear-axle overhang. Rear-axle overhang is the distance between the rear axle(s) and the pintle hook of a towing vehicle (**30**) in a multi-trailer combination truck. Representative values for the track width of design vehicles are shown in Exhibit 3-47 to illustrate the differences in relative widths between groups of design vehicles.

The lateral clearance allowance, C, provides for the clearance between the edge of the traveled way and nearest wheel path and for the body clearance between vehicles passing or meeting. Lateral clearance per vehicle is assumed to be 0.6, 0.75, and 0.9 m [2.0, 2.5, and 3.0 ft] for tangent lane widths, W_n, equal to 6.0, 6.6, and 7.2 m [20, 22, and 24 ft], respectively.

METRIC

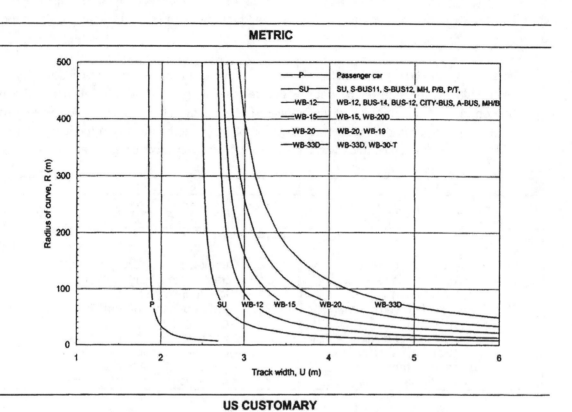

US CUSTOMARY

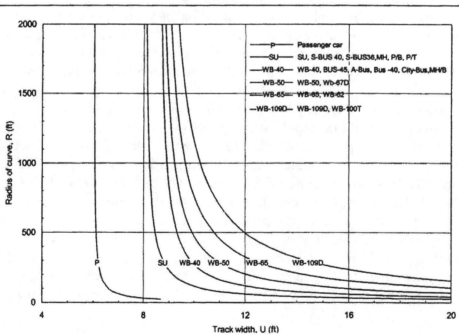

Exhibit 3-47. Track Width for Widening of Traveled Way on Curves

The width of the front overhang (F_A) is the radial distance between the outer edge of the tire path of the outer front wheel and the path of the outer front edge of the vehicle body. For curves and turning roadways, F_A depends on the radius of the curve, the extent of the front overhang of the design vehicle, and the wheelbase of the unit itself. In the case of tractor-trailer combinations, only the wheelbase of the tractor unit is used. Exhibit 3-48 illustrates relative overhang width values for F_A determined from:

Metric	US Customary	
$F_A = \sqrt{R^2 + A(2L + A)} - R$	$F_A = \sqrt{R^2 + A(2L + A)} - R$	(3-36)
where:	where:	
A = front overhang of inner lane vehicle, m; L = wheelbase of single unit or tractor, m.	A = front overhang of inner lane vehicle, ft; L = wheelbase of single unit or tractor, ft.	

The width of the rear overhang (F_B) is the radial distance between the outer edge of the tire path of the inner rear wheel and the inside edge of the vehicle body. For the passenger car (P) design vehicle, the width of the body is 0.3 m [1 ft] greater than the width of out-to-out width of the rear wheels, making $F_B = 0.15$ m [0.5 ft]. In the truck design vehicles, the width of body is the same as the width out-to-out of the rear wheels, and $F_B = 0$.

The extra width allowance (Z) is an additional radial width of pavement to allow for the difficulty of maneuvering on a curve and the variation in driver operation. This additional width is an empirical value that varies with the speed of traffic and the radius of the curve. The additional width allowance is expressed as:

Metric	US Customary	
$Z = 0.1\left(V/\sqrt{R}\right)$	$Z = V/\sqrt{R}$	(3-37)
where:	where:	
V = design speed of the highway, km/h.	V = design speed of the highway, mph.	

This expression, used primarily for widening of the traveled way on open highways, is also applicable to intersection curves. Exhibit 3-49 illustrates the computed values for Z for speeds between 20 and 100 km/h [15 and 60 mph]. For the normal range of curve radii at intersections, Z resolves into a nearly constant value of 0.6 m [2 ft] by using the speed-curvature relations in Exhibit 3-44 for radii in the range of 15 to 150 m [50 to 500 ft]. This added width, as shown diagrammatically in Exhibits 3-50 and 3-53, should be assumed to be evenly distributed over the traveled way width to allow for the inaccuracy in steering on curved paths.

METRIC

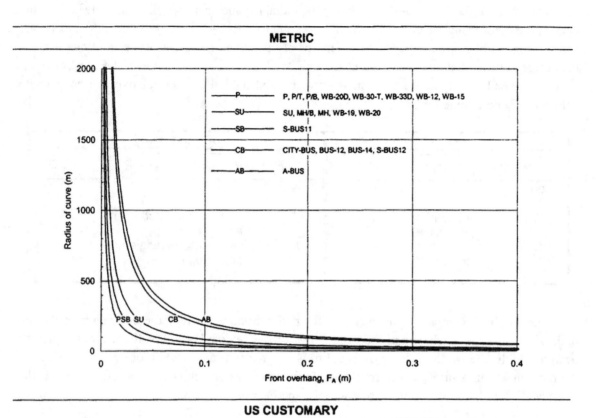

US CUSTOMARY

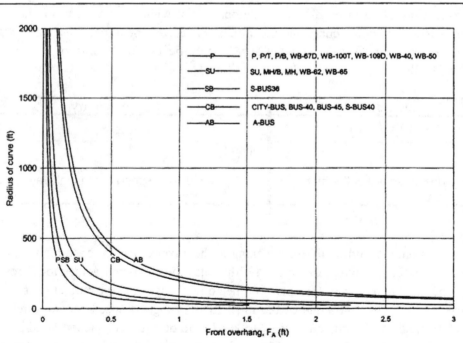

Exhibit 3-48. Front Overhang for Widening of Traveled Way on Curves

METRIC

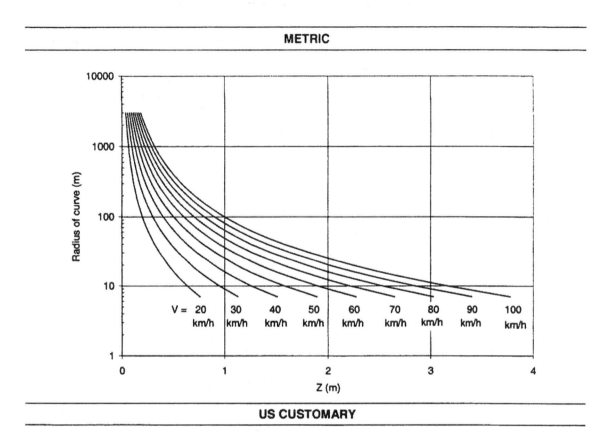

US CUSTOMARY

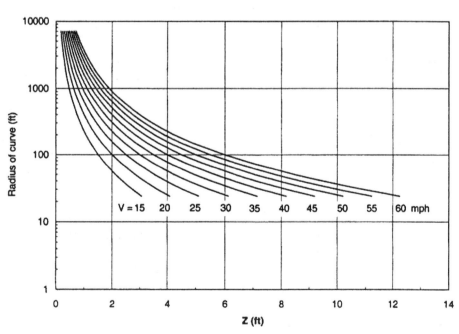

Exhibit 3-49. Extra Width Allowance for Difficulty of Driving on Traveled Way on Curves

Traveled Way Widening on Horizontal Curves

The traveled way on horizontal curves is sometimes widened to make operating conditions on curves comparable to those on tangents. On earlier highways with narrow lanes and sharp curves, there was considerable need for widening on curves, even though speeds were generally low. On modern highways and streets with 3.6 m [12 ft] lanes and high-type alignment, the need for widening has lessened considerably in spite of high speeds, but for some conditions of speed, curvature, and width it remains appropriate to widen traveled ways.

Widening is needed on certain curves for one of the following reasons: (1) the design vehicle occupies a greater width because the rear wheels generally track inside front wheels (offtracking) in negotiating curves, or (2) drivers experience difficulty in steering their vehicles in the center of the lane. The added width occupied by the vehicle as it traverses the curve as compared with the width of the traveled way on tangent can be computed by geometry for any combination of radius and wheelbase. The effect of variation in lateral placement of the rear wheels with respect to the front wheels and the resultant difficulty of steering should be accommodated by widening on curves, but the appropriate amount of widening cannot be determined as positively as that for simple offtracking.

The amount of widening of the traveled way on a horizontal curve is the difference between the width needed on the curve and the width used on a tangent:

Metric	US Customary	
$w = W_c - W_n$	$w = W_c - W_n$	(3-38)
where:	where:	
w = widening of traveled way on curve, m; W_c = width of traveled way on curve, m; W_n = width of traveled way on tangent, m	w = widening of traveled way on curve, ft; W_c = width of traveled way on curve, ft; W_n = width of traveled way on tangent, ft	

The traveled way width needed on a curve (W_c) has several components related to operation on curves, including: the track width of each vehicle meeting or passing, U; the lateral clearance for each vehicle, C; width of front overhang of the vehicle occupying the inner lane or lanes, F_A; and a width allowance for the difficulty of driving on curves, Z. The application of these components is illustrated in Exhibit 3-50. Each of these components is derived in the section on "Derivation of Design Values for Widening on Horizontal Curves," earlier in this chapter.

To determine width W_c, it is necessary to select an appropriate design vehicle. The design vehicle should usually be a truck because offtracking is much greater for trucks than for passenger cars. The WB-15 [WB-50] design vehicle is considered representative for two-lane open-highway conditions. Other design vehicles may be selected however, when representative of the actual traffic on a particular facility.

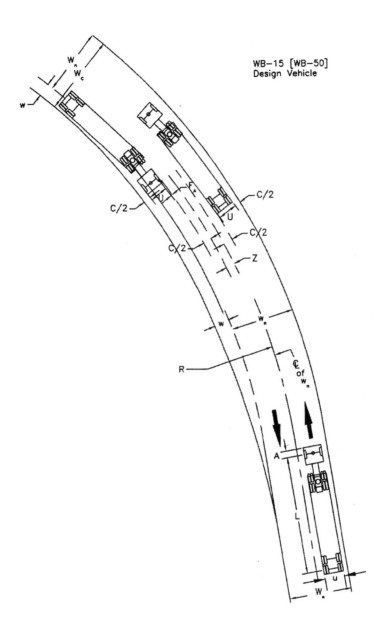

Exhibit 3-50. Widening Components on Open Highway Curves (Two-Lane Highways, One-Way or Two-Way)

The width W_c is calculated by the equation:

Metric	US Customary	
$W_c = N(U+C)+(N-1)F_A + Z$	$W_c = N(U+C)+(N-1)F_A + Z$	(3-39)
where: N = number of lanes; U = track width of design vehicle (out-to-out tires), m; C = lateral clearance, m; FA = width of front overhang of inner-lane vehicle, m; Z = extra width allowance, m	where: N = number of lanes; U = track width of design vehicle (out-to-out tires), ft; C = lateral clearance, ft; FA = width of front overhang of inner-lane vehicle, ft; Z = extra width allowance, ft	

The traveled way widening values for the assumed design condition for a WB-15 [WB-50] vehicle on a two-lane highway are presented in Exhibit 3-51. The differences in track widths of the SU, WB-12, WB-19, WB-20, WB-20D, WB-30T, and WB-33D [SU, WB-40, WB-62, WB-65, WB-67D, WB-100T, and WB-109D] design trucks are substantial for the sharp curves associated with intersections, but for open highways on which radii are usually larger than 200 m [650 ft], with design speeds over 60 km/h [30 mph], the differences are insignificant (see Exhibit 3-47). Where both sharper curves (as for a 50 km/h [30 mph] design speed) and large truck combinations are prevalent, the derived widening values for the WB-15 [WB-50] truck should be adjusted in accordance with Exhibit 3-52. The suggested increases of the tabular values for two ranges of radius of curvature are general and will not necessarily result in a full lateral clearance C or an extra width allowance Z, as shown in Exhibit 3-49 for the shorter radii. With the lower speeds and volumes on roads with such curvature, however, slightly smaller clearances may be tolerable.

Design Values for Traveled Way Widening

Widening is costly and very little is actually gained from a small amount of widening. It is suggested that a minimum widening of 0.6 m [2.0 ft] be used and that lower values in Exhibit 3-51 be disregarded. Note that the values in Exhibit 3-51 are for a WB-15 [WB-50] design vehicle. For other design vehicles, an adjustment from Exhibit 3-52 should be applied. Values in Exhibit 3-51 also are applicable to two-lane, one-way traveled ways (i.e., to each roadway of a divided highway or street). Studies show that on tangent alignment somewhat smaller clearances between vehicles are used in passing vehicles traveling in the same direction as compared with meeting vehicles traveling in opposite directions. There is no evidence that these smaller clearances are obtained on curved alignment on one-way roads. Moreover, drivers are not in position to judge clearances as well when passing vehicles as when meeting opposing vehicles on a curved two-way highway. For this reason and because all geometric elements on a divided highway are generally well maintained, widening on a two-lane, one-way traveled way of a divided highway should be the same as that on a two-lane, two-way highway, as noted in Exhibit 3-51.

Metric

Radius of curve (m)	Roadway width = 7.2 m Design Speed (km/h)						Roadway width = 6.6 m Design Speed (km/h)						Roadway width = 6.0 m Design Speed (km/h)					
	50	60	70	80	90	100	50	60	70	80	90	100	50	60	70	80	90	100
3000	0.0	0.0	0.0	0.0	0.0	0.0	0.2	0.2	0.3	0.3	0.3	0.3	0.5	0.5	0.6	0.6	0.6	0.6
2500	0.0	0.0	0.0	0.0	0.0	0.0	0.2	0.3	0.3	0.3	0.3	0.3	0.5	0.6	0.6	0.6	0.6	0.6
2000	0.0	0.0	0.0	0.0	0.1	0.1	0.3	0.3	0.3	0.3	0.3	0.4	0.6	0.6	0.6	0.6	0.6	0.7
1500	0.0	0.0	0.1	0.1	0.1	0.1	0.3	0.3	0.4	0.4	0.4	0.4	0.6	0.6	0.7	0.7	0.7	0.7
1000	0.1	0.1	0.1	0.2	0.2	0.2	0.4	0.4	0.4	0.5	0.5	0.5	0.7	0.7	0.7	0.8	0.8	0.8
900	0.1	0.1	0.2	0.2	0.2	0.3	0.4	0.4	0.5	0.5	0.5	0.6	0.7	0.7	0.8	0.8	0.8	0.9
800	0.1	0.2	0.2	0.2	0.3	0.3	0.4	0.5	0.5	0.5	0.6	0.6	0.7	0.8	0.8	0.8	0.9	0.9
700	0.2	0.2	0.2	0.3	0.3	0.3	0.5	0.5	0.5	0.6	0.6	0.6	0.8	0.8	0.8	0.9	0.9	1.0
600	0.2	0.3	0.3	0.3	0.4	0.4	0.5	0.5	0.6	0.6	0.7	0.7	0.8	0.9	0.9	0.9	1.0	1.0
500	0.3	0.3	0.4	0.4	0.5	0.5	0.6	0.6	0.7	0.7	0.8	0.8	0.9	0.9	1.0	1.0	1.1	1.1
400	0.4	0.4	0.5	0.5	0.6	0.6	0.7	0.7	0.8	0.8	0.9	0.9	1.0	1.0	1.1	1.1	1.2	1.2
300	0.5	0.6	0.6	0.7	0.8	0.8	0.8	0.9	0.9	1.0	1.1	1.1	1.1	1.2	1.2	1.3	1.4	1.4
250	0.6	0.7	0.8	0.8	0.9	0.8	0.9	1.0	1.1	1.1	1.2	1.1	1.2	1.3	1.4	1.4	1.5	1.4
200	0.8	0.9	1.0	1.0			1.1	1.2	1.3	1.3			1.4	1.5	1.6	1.6		
150	1.1	1.2	1.3	1.3			1.4	1.5	1.6	1.6			1.7	1.8	1.9	1.9		
140	1.2	1.3					1.5	1.6					1.8	1.9				
130	1.3	1.4					1.6	1.7					1.9	2.0				
120	1.4	1.5					1.7	1.8					2.0	2.1				
110	1.5	1.6					1.8	1.9					2.1	2.2				
100	1.6	1.7					1.9	2.0					2.2	2.3				
90	1.8						2.1						2.4					
80	2.0						2.3						2.6					
70	2.3						2.6						2.9					

Notes: Values shown are for WB-15 design vehicle and represent widening in meters. For other design vehicles, use adjustments in Exhibit 3-52.
Values less than 0.6 m may be disregarded.
For 3-lane roadways, multiply above values by 1.5.
For 4-lane roadways, multiply above values by 2.

Exhibit 3-51. Calculated and Design Values for Traveled Way Widening on Open Highway Curves (Two-Lane Highways, One-Way or Two-Way)

US Customary

Radius of curve (ft)	Roadway width = 24 ft Design Speed (mph)							Roadway width = 22 ft Design Speed (mph)							Roadway width = 20 ft Design Speed (mph)						
	30	35	40	45	50	55	60	30	35	40	45	50	55	60	30	35	40	45	50	55	60
7000	0.0	0.0	0.0	0.0	0.0	0.0	0.0	0.6	0.6	0.7	0.7	0.8	0.9	0.9	1.6	1.6	1.7	1.7	1.8	1.9	1.9
6500	0.0	0.0	0.0	0.0	0.0	0.0	0.0	0.6	0.7	0.7	0.8	0.8	0.9	1.0	1.6	1.7	1.7	1.8	1.8	1.9	2.0
6000	0.0	0.0	0.0	0.0	0.0	0.0	0.0	0.6	0.7	0.8	0.8	0.9	1.0	1.0	1.6	1.7	1.8	1.8	1.9	2.0	2.0
5500	0.0	0.0	0.0	0.0	0.0	0.0	0.1	0.7	0.7	0.8	0.9	0.9	1.0	1.1	1.7	1.7	1.8	1.9	1.9	2.0	2.1
5000	0.0	0.0	0.0	0.0	0.0	0.1	0.1	0.7	0.8	0.9	0.9	1.0	1.1	1.1	1.7	1.8	1.9	1.9	2.0	2.1	2.1
4500	0.0	0.0	0.0	0.0	0.1	0.1	0.2	0.8	0.8	0.9	1.0	1.1	1.1	1.2	1.8	1.8	1.9	2.0	2.1	2.1	2.2
4000	0.0	0.0	0.0	0.1	0.2	0.2	0.3	0.8	0.9	1.0	1.1	1.2	1.2	1.3	1.8	1.9	2.0	2.1	2.2	2.2	2.3
3500	0.0	0.0	0.1	0.2	0.3	0.3	0.4	0.9	1.0	1.1	1.2	1.3	1.3	1.4	1.9	2.0	2.1	2.2	2.3	2.3	2.4
3000	0.0	0.1	0.2	0.3	0.4	0.5	0.6	1.0	1.1	1.2	1.3	1.4	1.5	1.6	2.0	2.1	2.2	2.3	2.4	2.5	2.6
2500	0.2	0.3	0.4	0.5	0.6	0.7	0.8	1.2	1.3	1.4	1.5	1.6	1.7	1.8	2.2	2.3	2.4	2.5	2.6	2.7	2.8
2000	0.4	0.5	0.6	0.7	0.8	1.0	1.1	1.4	1.5	1.6	1.7	1.8	2.0	2.1	2.4	2.5	2.6	2.7	2.8	3.0	3.1
1800	0.5	0.6	0.8	0.9	1.0	1.1	1.2	1.5	1.6	1.8	1.9	2.0	2.1	2.2	2.5	2.6	2.8	2.9	3.0	3.1	3.2
1600	0.7	0.8	0.9	1.0	1.2	1.3	1.4	1.7	1.8	1.9	2.0	2.2	2.3	2.4	2.7	2.8	2.9	3.0	3.2	3.3	3.4
1400	0.8	1.0	1.1	1.2	1.4	1.5	1.6	1.8	2.0	2.1	2.2	2.4	2.5	2.6	2.8	3.0	3.1	3.2	3.4	3.5	3.6
1200	1.1	1.2	1.4	1.5	1.7	1.8	1.9	2.1	2.2	2.4	2.5	2.7	2.8	2.9	3.1	3.2	3.4	3.5	3.7	3.8	3.9
1000	1.4	1.6	1.7	1.9	2.0	2.2	2.4	2.4	2.6	2.7	2.9	3.0	3.2	3.4	3.4	3.6	3.7	3.9	4.0	4.2	4.4
900	1.6	1.8	2.0	2.1	2.3	2.5		2.6	2.8	3.0	3.1	3.3	3.5		3.6	3.8	4.0	4.1	4.3	4.5	
800	1.9	2.1	2.2	2.4	2.6	2.8		2.9	3.1	3.2	3.4	3.6	3.8		3.9	4.1	4.2	4.4	4.6	4.8	
700	2.2	2.4	2.6	2.8	3.0			3.2	3.4	3.6	3.8	4.0			4.2	4.4	4.6	4.8	5.0		
600	2.7	2.9	3.1	3.3	3.5			3.7	3.9	4.1	4.3	4.5			4.7	4.9	5.1	5.3	5.5		
500	3.3	3.5	3.7	3.9				4.3	4.5	4.7	4.9				5.3	5.5	5.7	5.9			
450	3.7	3.9	4.1					4.7	4.9	5.1					5.7	5.9	6.1				
400	4.2	4.4	4.7					5.2	5.4	5.7					6.2	6.4	6.7				
350	4.8	5.1	5.3					5.8	6.1	6.3					6.8	7.1	7.3				
300	5.6	5.9						6.6	6.9						7.6	7.9					
250	6.8							7.8							8.8						
200	8.5							9.5							10.5						

Notes: Values shown are for WB-50 design vehicle and represent widening in feet. For other design vehicles, use adjustments in Exhibit 3-52.
Values less than 2.0 ft may be disregarded.
For 3-lane roadways, multiply above values by 1.5.
For 4-lane roadways, multiply above values by 2.

Exhibit 3-51. Calculated and Design Values for Traveled Way Widening on Open Highway Curves (Two-Lane Highways, One-Way or Two-Way) (Continued)

Metric

Radius of curve (m)	Design vehicle						
	SU	WB-12	WB-19	WB-20	WB-20D	WB-30T	WB-33D
3000	-0.3	-0.3	0.0	0.0	0.0	0.0	0.1
2500	-0.3	-0.3	0.0	0.0	0.0	0.0	0.1
2000	-0.3	-0.3	0.0	0.0	0.0	0.0	0.1
1500	-0.4	-0.3	0.0	0.1	0.0	0.0	0.1
1000	-0.4	-0.4	0.1	0.1	0.0	0.0	0.2
900	-0.4	-0.4	0.1	0.1	0.0	0.0	0.2
800	-0.4	-0.4	0.1	0.1	0.0	0.0	0.2
700	-0.4	-0.4	0.1	0.1	0.0	0.0	0.3
600	-0.5	-0.4	0.1	0.2	0.0	0.1	0.3
500	-0.5	-0.4	0.2	0.2	0.0	0.1	0.3
400	-0.5	-0.4	0.2	0.2	0.0	0.1	0.4
300	-0.6	-0.5	0.2	0.3	-0.1	0.1	0.5
250	-0.7	-0.5	0.2	0.3	-0.1	0.1	0.6
200	-0.8	-0.6	0.3	0.4	-0.1	0.2	0.8
150	-0.9	-0.7	0.4	0.6	-0.1	0.2	1.0
140	-0.9	-0.7	0.4	0.6	-0.1	0.2	1.3
130	-1.0	-0.7	0.5	0.6	-0.2	0.2	1.4
120	-1.1	-0.8	0.5	0.7	-0.2	0.3	1.5
110	-1.1	-0.8	0.6	0.8	-0.2	0.3	1.6
100	-1.2	-0.9	0.6	0.8	-0.2	0.3	1.7
90	-1.3	-0.9	0.7	0.9	-0.2	0.3	1.9
80	-1.4	-1.0	0.8	1.1	-0.2	0.4	2.1
70	-1.6	-1.1	0.9	1.2	-0.3	0.5	2.4

US Customary

Radius of curve (ft)	Design vehicle						
	SU	WB-40	WB-62	WB-65	WB-67D	WB-100T	WB-109D
7000	-1.1	-1.1	0.1	0.1	0.0	0.0	0.3
6500	-1.1	-1.1	0.1	0.1	0.0	0.1	0.3
6000	-1.2	-1.1	0.1	0.2	0.0	0.1	0.3
5500	-1.2	-1.1	0.1	0.2	0.0	0.1	0.4
5000	-1.2	-1.1	0.1	0.2	0.0	0.1	0.4
4500	-1.2	-1.1	0.1	0.2	0.0	0.1	0.5
4000	-1.2	-1.2	0.2	0.2	-0.1	0.1	0.5
3500	-1.3	-1.2	0.2	0.3	-0.1	0.1	0.6
3000	-1.3	-1.2	0.2	0.3	-0.1	0.1	0.7
2500	-1.4	-1.2	0.3	0.4	-0.1	0.2	0.8
2000	-1.5	-1.3	0.3	0.5	-0.1	0.2	1.0
1800	-1.5	-1.3	0.3	0.5	-0.1	0.2	1.1
1600	-1.6	-1.4	0.4	0.6	-0.1	0.2	1.3
1400	-1.7	-1.4	0.4	0.6	-0.2	0.2	1.5
1200	-1.8	-1.5	0.5	0.8	-0.2	0.3	1.7
1000	-2.0	-1.6	0.5	0.9	-0.2	0.3	2.0
900	-2.1	-1.7	0.6	1.0	-0.2	0.4	2.3
800	-2.2	-1.8	0.7	1.1	-0.3	0.4	2.6
700	-2.4	-1.9	0.8	1.3	-0.3	0.5	2.9
600	-2.6	-2.0	0.9	1.5	-0.4	0.6	3.4
500	-2.9	-2.2	1.1	1.8	-0.4	0.7	4.1
450	-3.2	-2.4	1.3	2.0	-0.5	0.7	4.6
400	-3.4	-2.5	1.4	2.3	-0.5	0.8	5.1
350	-3.8	-2.8	1.6	2.6	-0.6	1.0	5.9
300	-4.3	-3.0	1.9	3.0	-0.7	1.1	6.9
250	-4.9	-3.5	2.2	3.7	-0.9	1.4	8.3
200	-5.9	-4.1	2.6	4.6	-1.1	1.7	10.5

Notes: Adjustments are applied by adding to or subtracting from the values in Exhibit 3-51.
Adjustments depend only on radius and design vehicle; they are independent of roadway width and design speed.
For 3-lane roadways, multiply values by 1.5.
For 4-lane roadways, multiply values by 2.0.

Exhibit 3-52. Adjustments for Traveled Way Widening Values on Open Highway Curves (Two-Lane Highways, One-Way or Two-Way)

On four-lane undivided highways or streets the widening of the traveled way should be double the design values indicated in Exhibit 3-51. This means that some values below 0.6 m [2 ft] in Exhibit 3-51, which were disregarded for two-lane highways, may now be used because, when doubled for undivided four-lane highways, they will be greater than the minimum.

The above values are applicable to open-highway curves. For intersection conditions, with generally smaller radii on turning roadways, the criteria for design widths are somewhat different. These criteria are presented in the section "Widths for Turning Roadways at Intersections" in this chapter, and design values are given in Exhibit 3-53.

Application of Widening on Curves

Widening should transition gradually on the approaches to the curve to ensure a reasonably smooth alignment of the edge of the traveled way and to fit the paths of vehicles entering or leaving the curve. The principal points of concern in the design of curve widening, which apply to both ends of highway curves, are presented below:

- On simple (unspiraled) curves, widening should be applied on the inside edge of the traveled way only. On curves designed with spirals, widening may be applied on the inside edge or divided equally on either side of the centerline. In the latter method, extension of the outer-edge tangent avoids a slight reverse curve on the outer edge. In either case, the final marked centerline, and desirably any central longitudinal joint, should be placed midway between the edges of the widened traveled way.
- Curve widening should transition gradually over a length sufficient to make the whole of the traveled way fully usable. Although a long transition is desirable for traffic operation, it may result in narrow pavement slivers that are difficult and expensive to construct. Preferably, widening should transition over the superelevation runoff length, but shorter lengths are sometimes used. Changes in width normally should be effected over a distance of 30 to 60 m [100 to 200 ft].
- From the standpoints of usefulness and appearance, the edge of the traveled way through the widening transition should be a smooth, graceful curve. A tangent transition edge should be avoided. On minor highways or in cases where plan details are not available, a curved transition staked by eye generally is satisfactory and better than a tangent transition. In any event, the transition ends should avoid an angular break at the pavement edge.
- On highway alignment without spirals, smooth and fitting alignment results from attaining widening with one-half to two-thirds of the transition length along the tangent and the balance along the curve. This is consistent with a common method for attaining superelevation. The inside edge of the traveled way may be designed as a modified spiral, with control points determined by the width/length ratio of a triangular wedge, by calculated values based on a parabolic or cubic curve, or by a larger radius (compound) curve. Otherwise, it may be aligned by eye in the field. On highway alignment with spiral curves, the increase in width is usually distributed along the length of the spiral.

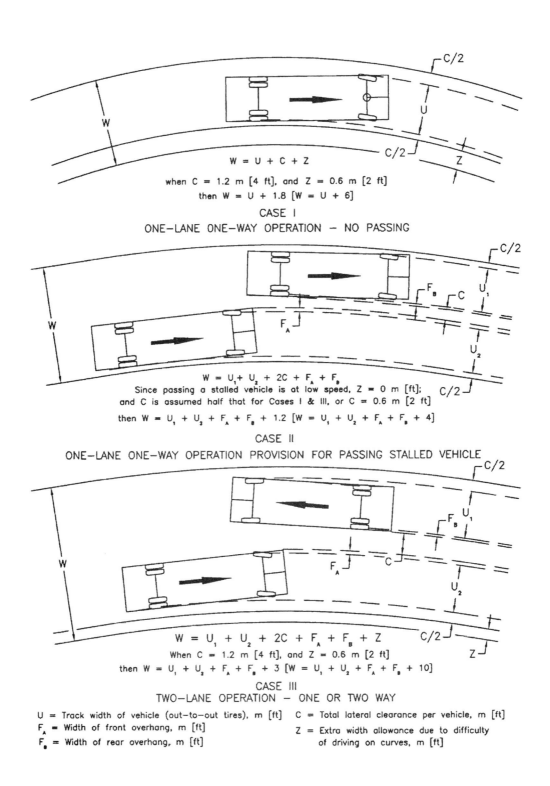

$$W = U + C + Z$$
when C = 1.2 m [4 ft], and Z = 0.6 m [2 ft]
then W = U + 1.8 [W = U + 6]
CASE I
ONE-LANE ONE-WAY OPERATION – NO PASSING

$$W = U_1 + U_2 + 2C + F_A + F_B$$
Since passing a stalled vehicle is at low speed, Z = 0 m [ft];
and C is assumed half that for Cases I & III, or C = 0.6 m [2 ft]
then W = U_1 + U_2 + F_A + F_B + 1.2 [W = U_1 + U_2 + F_A + F_B + 4]

CASE II
ONE-LANE ONE-WAY OPERATION PROVISION FOR PASSING STALLED VEHICLE

$$W = U_1 + U_2 + 2C + F_A + F_B + Z$$
When C = 1.2 m [4 ft], and Z = 0.6 m [2 ft]
then W = U_1 + U_2 + F_A + F_B + 3 [W = U_1 + U_2 + F_A + F_B + 10]

CASE III
TWO-LANE OPERATION – ONE OR TWO WAY

U = Track width of vehicle (out-to-out tires), m [ft]   C = Total lateral clearance per vehicle, m [ft]
F_A = Width of front overhang, m [ft]
F_B = Width of rear overhang, m [ft]   Z = Extra width allowance due to difficulty
of driving on curves, m [ft]

Exhibit 3-53. Derivation of Turning Roadway Widths on Curves at Intersections

- Widening areas can be fully detailed on construction plans. Alternatively, general controls can be cited on construction or standard plans with final details left to the field engineer.

Widths for Turning Roadways at Intersections

The widths of turning roadways at intersections are governed by the types of vehicles to be accommodated, the radius of curvature, and the expected speed. Turning roadways may be designed for one- or two-way operation, depending on the geometric pattern of the intersection.

Selection of an appropriate design vehicle should be based on the size and frequency of vehicle types using or expected to use the facility. The radius of curvature in combination with the track width of the design vehicle determine the width of a turning roadway. The width elements for the turning vehicle, shown diagrammatically in Exhibit 3-53, are explained in the section on "Derivation of Design Values for Widening on Horizontal Curves," presented earlier in this chapter. They ignore the effects of insufficient superelevation and of surfaces with low friction resistance that tend to cause the rear wheels of vehicles traveling at other than low speed to swing outward, developing the appropriate slip angles.

Turning roadways are classified for operational purposes as one-lane operation, with or without opportunity for passing a stalled vehicle, and two-lane operation, either one-way or two-way. Three cases are commonly considered in design:

Case I—One-lane, one-way operation with no provision for passing a stalled vehicle is usually appropriate for minor turning movements and moderate turning volumes where the connecting roadway is relatively short. Under these conditions, the chance of a vehicle breakdown is remote but one of the edges of the traveled way should preferably have a sloping curb or be flush with the shoulder.

Case II—One-lane, one-way operation with provision for passing a stalled vehicle is used to allow operation at low speed and with sufficient clearance so that other vehicles can pass a stalled vehicle. These widths are applicable to all turning movements of moderate to heavy traffic volumes that do not exceed the capacity of a single-lane connection. In the event of a breakdown, traffic flow can be maintained at a somewhat reduced speed. Many ramps and connections at channelized intersections are in this category. However, for Case II, the widths needed for the longer vehicles are very large as shown in Exhibit 3-54. Case I widths for these longer vehicles, including the WB-19, WB-20, WB-30T, and WB-33D [WB-62, WB-65, WB-100T, and WB-109D] design vehicles, may have to be used as the minimum values where they are present in sufficient numbers to be considered the appropriate design vehicle.

Case III—Two-lane operation, either one- or two-way, is applicable where operation is two way or where operation is one way, but two lanes are needed to handle the traffic volume.

METRIC

Case I, One-Lane, One-Way Operation, No Provision for Passing a Stalled Vehicle

Radius on Inner Edge of Pavement R (m)	P	SU	BUS-12	BUS-14	CITY-BUS	S-BUS11	S-BUS12	A-BUS	WB-12	WB-15	WB-19	WB-20	WB-20D	WB-30T	WB-33D	MH	P/T	P/B	MH/B
15	4.0	5.5	6.6	7.2	6.5	5.7	5.5	6.7	7.0	9.7	13.3	15.7	8.8	11.6	–	5.5	5.7	5.4	6.5
25	3.9	5.0	5.7	5.9	5.6	5.1	5.0	5.7	5.8	7.2	8.5	9.0	6.8	7.9	12.0	5.0	5.1	4.9	5.5
30	3.8	4.9	5.4	5.7	5.4	5.0	4.9	5.5	5.5	6.7	7.7	8.1	6.3	7.3	10.4	4.9	5.0	4.8	5.3
50	3.7	4.6	5.0	5.2	5.0	4.7	4.6	5.0	5.0	5.7	6.3	6.5	5.5	6.1	7.7	4.6	4.7	4.6	4.9
75	3.7	4.5	4.8	4.9	4.8	4.5	4.5	4.9	4.8	5.3	5.7	5.9	5.2	5.6	6.7	4.5	4.5	4.5	4.7
100	3.7	4.5	4.8	4.9	4.8	4.5	4.5	4.9	4.8	5.3	5.7	5.9	5.2	5.6	6.7	4.5	4.5	4.5	4.7
125	3.7	4.5	4.8	4.9	4.8	4.5	4.5	4.9	4.8	5.3	5.7	5.9	5.2	5.6	6.7	4.5	4.5	4.5	4.7
150	3.7	4.5	4.8	4.9	4.8	4.5	4.5	4.9	4.8	5.3	5.7	5.9	5.2	5.6	6.7	4.5	4.5	4.5	4.7
Tangent	3.6	4.2	4.4	4.4	4.4	4.2	4.2	4.4	4.2	4.4	4.4	4.4	4.4	4.4	4.4	4.2	4.2	4.2	4.2

Case II, One-Lane, One-Way Operation, with Provision for Passing a Stalled Vehicle by Another of the Same Type

Radius on Inner Edge of Pavement R (m)	P	SU	BUS-12	BUS-14	CITY-BUS	S-BUS11	S-BUS12	A-BUS	WB-12	WB-15	WB-19	WB-20	WB-20D	WB-30T	WB-33D	MH	P/T	P/B	MH/B
15	6.0	9.2	11.9	13.1	11.7	9.4	9.7	12.4	11.8	17.3	24.7	29.5	15.4	20.9	–	9.2	9.3	8.7	11.0
25	5.6	7.9	9.6	10.2	9.5	8.0	8.2	9.9	9.3	12.1	14.9	16.0	11.2	13.5	21.7	7.9	7.9	7.6	8.9
30	5.5	7.6	9.0	9.5	9.0	7.7	7.8	9.3	8.8	11.1	13.3	14.2	10.4	12.2	18.4	7.6	7.6	7.4	8.4
50	5.3	7.0	8.0	8.3	7.9	7.0	7.1	8.1	7.7	9.1	10.4	10.9	8.7	9.8	13.1	7.0	7.0	6.8	7.5
75	5.2	6.7	7.4	7.6	7.4	6.7	6.8	7.5	7.1	8.2	9.0	9.3	7.9	8.6	10.8	6.7	6.7	6.6	7.0
100	5.2	6.5	7.2	7.3	7.1	6.6	6.6	7.2	6.9	7.7	8.3	8.6	7.5	8.1	9.7	6.5	6.5	6.5	6.8
125	5.1	6.4	7.0	7.1	7.0	6.5	6.5	7.1	6.7	7.5	8.0	8.1	7.3	7.7	9.0	6.4	6.4	6.4	6.6
150	5.1	6.4	6.9	7.0	6.9	6.4	6.4	7.0	6.6	7.3	7.7	7.8	7.2	7.5	8.6	6.4	6.4	6.3	6.5
Tangent	5.0	6.1	6.4	6.4	6.4	6.1	6.1	6.4	6.1	6.4	6.4	6.4	6.4	6.4	6.4	6.1	6.1	6.1	6.1

Case III, Two-Lane Operation, Either One or Two-Way (Same Type Vehicle in Both Lanes)

Radius on Inner Edge of Pavement R (m)	P	SU	BUS-12	BUS-14	CITY-BUS	S-BUS11	S-BUS12	A-BUS	WB-12	WB-15	WB-19	WB-20	WB-20D	WB-30T	WB-33D	MH	P/T	P/B	MH/B
15	7.8	11.0	13.7	14.9	13.5	11.2	11.5	14.2	13.6	19.1	26.5	31.3	17.2	22.7	–	11.0	11.1	10.5	12.8
25	7.4	9.7	11.4	12.0	11.3	9.8	10.0	11.7	11.1	13.9	16.7	17.8	13.0	15.3	23.5	9.7	9.7	9.4	10.7
30	7.3	9.4	10.8	11.3	10.8	9.5	9.6	11.1	10.6	12.9	15.1	16.0	12.2	14.0	20.2	9.4	9.4	9.2	10.2
50	7.1	8.8	9.8	10.1	9.7	8.8	8.9	9.9	9.5	10.9	12.2	12.7	10.5	11.6	14.9	8.8	8.8	8.6	9.3
75	7.0	8.5	9.2	9.4	9.2	8.5	8.6	9.3	8.9	10.0	10.8	11.1	9.7	10.4	12.6	8.5	8.5	8.4	8.8
100	7.0	8.3	9.0	9.1	8.9	8.4	8.4	9.0	8.7	9.5	10.1	10.4	9.3	9.9	11.5	8.3	8.3	8.3	8.6
125	6.9	8.2	8.8	8.9	8.8	8.3	8.3	8.9	8.5	9.3	9.8	9.9	9.1	9.5	10.8	8.2	8.2	8.2	8.4
150	6.9	8.2	8.7	8.8	8.7	8.2	8.2	8.8	8.4	9.1	9.5	9.6	9.0	9.3	10.4	8.2	8.2	8.1	8.3
Tangent	6.8	7.9	8.2	8.2	8.2	7.9	7.9	7.9	7.9	8.2	8.2	8.2	8.2	8.2	8.2	7.9	7.9	7.9	7.9

Exhibit 3-54. Derived Pavement Widths for Turning Roadways for Different Design Vehicles

US CUSTOMARY

Case I, One-Lane, One-Way Operation, No Provision for Passing a Stalled Vehicle

Radius on Inner Edge of Pavement R (ft)	P	SU	BUS-40	BUS-45	CITY-BUS	S-BUS36	S-BUS40	A-BUS	WB-40	WB-50	WB-62	WB-65	WB-67D	WB-100T	WB-109D	MH	P/T	P/B	MH/B
50	13	18	22	23	21	19	18	22	23	32	43	49	29	37	--	18	19	18	21
75	13	17	19	20	19	17	17	19	20	25	29	32	23	27	43	17	17	17	19
100	13	16	18	19	18	16	16	18	18	22	25	27	21	24	34	16	16	16	17
150	12	15	17	17	17	16	15	17	17	19	21	22	19	21	27	15	16	15	16
200	12	15	16	17	16	15	15	16	16	18	20	20	18	19	23	15	15	15	16
300	12	15	16	16	16	15	15	16	15	17	18	18	17	17	20	15	15	15	15
400	12	15	16	16	16	15	15	16	15	17	18	18	17	17	20	15	15	15	15
500	12	15	16	16	16	15	15	16	15	17	18	18	17	17	20	15	15	15	15
Tangent	12	14	15	15	15	14	14	15	14	15	15	15	15	15	15	14	14	14	14

Case II, One-Lane, One-Way Operation with Provision for Passing a Stalled Vehicle by Another of the Same Type

Radius R (ft)	P	SU	BUS-40	BUS-45	CITY-BUS	S-BUS36	S-BUS40	A-BUS	WB-40	WB-50	WB-62	WB-65	WB-67D	WB-100T	WB-109D	MH	P/T	P/B	MH/B
50	20	30	39	42	38	31	32	40	39	56	79	93	50	67	--	30	30	28	36
75	19	27	32	35	32	27	28	34	32	42	52	56	39	47	79	27	27	26	30
100	18	25	30	31	29	25	26	30	29	36	43	46	34	40	60	25	25	24	28
150	18	23	27	28	27	24	24	27	26	31	35	37	29	33	45	23	23	23	25
200	17	22	25	26	25	23	23	26	24	28	32	33	27	30	39	22	22	22	24
300	17	22	24	24	24	22	22	24	23	26	28	29	25	27	33	22	22	21	23
400	17	21	23	24	23	21	21	23	22	25	26	27	24	25	30	21	21	21	22
500	17	21	23	23	23	21	21	23	22	24	25	26	23	25	28	21	21	21	21
Tangent	17	20	21	21	21	20	20	21	20	21	21	21	21	21	21	20	20	20	20

Case III, Two-Lane Operation, Either One or Two-Way (Same Type Vehicle in Both Lanes)

Radius R (ft)	P	SU	BUS-40	BUS-45	CITY-BUS	S-BUS36	S-BUS40	A-BUS	WB-40	WB-50	WB-62	WB-65	WB-67D	WB-100T	WB-109D	MH	P/T	P/B	MH/B
50	26	36	45	48	44	37	38	46	45	62	85	99	56	73	--	36	36	34	42
75	25	33	38	41	38	33	34	40	38	48	58	62	45	53	85	33	33	32	36
100	24	31	36	37	35	31	32	36	35	42	49	52	40	46	66	31	31	30	34
150	24	29	33	34	33	29	30	33	32	37	41	43	35	39	51	29	29	29	31
200	23	28	31	32	31	29	29	32	30	34	38	39	33	36	45	28	28	28	30
300	23	28	30	30	30	28	28	30	29	32	34	35	31	33	39	28	28	28	29
400	23	27	29	30	29	27	28	29	28	31	32	33	30	31	36	27	27	27	28
500	23	27	29	29	29	27	27	29	28	30	31	32	29	31	34	27	27	27	27
Tangent	23	26	27	27	27	26	26	27	26	27	27	27	27	27	27	26	26	26	26

Exhibit 3-54. Derived Pavement Widths for Turning Roadways for Different Design Vehicles (Continued)

Design Values

The total width, W, for separate turning roadways at intersections is derived by the summation of the proper width elements. The separate formulas for width and values for lateral clearance, C, and the allowance for difficulty of driving on curves, Z, for each case are shown in Exhibit 3-53. Values for track width, U, are obtained from Exhibit 3-47 and values for front overhang, F_A, from Exhibit 3-48. Values of U and F_A are read from the exhibit for the turning radius, R_T, which is closely approximated by adding the track width and proper clearances to the radius of the inner edge of the turning roadway.

When determining the width for Case I, a lateral clearance, C, of 1.2 m [4 ft] is considered appropriate. The allowance for difficulty of driving curves, Z, is constant, equal to about 0.6 m [2 ft] for all radii of 150m [500 ft] or less. In this case, the front overhang, F_A, need not be considered because no passing of another vehicle is involved.

For Case II, the width involves U and C for the stopped vehicle and the U and C for the passing vehicle. To this is added extra width for the front overhang, F_A, of one vehicle and the rear overhang, F_B, (if any) of the other vehicle. The width of rear overhang for a passenger car is considered to be 0.15 m [0.5 ft]. F_B for truck design vehicles is 0. A total clearance of one-half the value of C in the other two cases is assumed (i.e., 0.6 m [2 ft] for the stopped vehicle and 0.6 m [2 ft] for the passing vehicle). Because passing the stalled vehicle is accomplished at low speeds, the extra width allowance, Z, is omitted.

All the width elements apply for Case III. To the values of U and F_A obtained from Exhibits 3-47 and 3-48, respectively, the lateral clearance, C, of 1.2 m [4 ft], F_B of 0.15 m [0.5 ft] for passenger cars, and Z of 0.6 m [2 ft] is added to determine the total width.

The derived widths for various radii for each design vehicle are given in Exhibit 3-54. For general design use, the recommended widths given in Exhibit 3-54 seldom apply directly, because the turning roadways usually accommodate more than one type of vehicle. Even parkways designed primarily for P vehicles are used by buses and maintenance trucks. At the other extreme, few if any public highways are designed to fully accommodate the WB-15 [WB-50] or longer design vehicles. Widths needed for some combination of separate design vehicles become the practical design guide for intersection roadways. Such design widths are given in Exhibit 3-55 for three logical conditions of mixed traffic, which are defined below. However, where the larger design vehicles such as the WB-19 or WB-33D [WB-62 or WB-109D] will be using a turning roadway or ramp, the facility should accommodate their turning paths for at least the Case I condition. Therefore, Case I widths for the appropriate design vehicle and radius shown in Exhibit 3-54 should be checked to determine whether they exceed widths shown in Exhibit 3-55. If they do, consideration should be given to using the widths for Case I shown in Exhibit 3-54 as the minimum widths for the turning roadway or ramp.

Traffic conditions for defining turning roadway widths are described in broad terms because data concerning the traffic volume, or the percentage of the total volume, for each type of vehicle are not available to define these traffic conditions with precision in relation to width.

Metric

Pavement width (m) — Design traffic conditions

Radius on inner edge of pavement R (m)	Case I One-lane, one-way operation—no provision for passing a stalled vehicle			Case II One-lane, one-way operation—with provision for passing a stalled vehicle			Case III Two-lane operation—either one-way or two-way		
	A	B	C	A	B	C	A	B	C
15	5.4	5.5	7.0	6.0	7.8	9.2	9.4	11.0	13.6
25	4.8	5.0	5.8	5.6	6.9	7.9	8.6	9.7	11.1
30	4.5	4.9	5.5	5.5	6.7	7.6	8.4	9.4	10.6
50	4.2	4.6	5.0	5.3	6.3	7.0	7.9	8.8	9.5
75	3.9	4.5	4.8	5.2	6.1	6.7	7.7	8.5	8.9
100	3.9	4.5	4.8	5.2	5.9	6.5	7.6	8.3	8.7
125	3.9	4.5	4.8	5.1	5.9	6.4	7.6	8.2	8.5
150	3.6	4.5	4.5	5.1	5.8	6.4	7.5	8.2	8.4
Tangent	3.6	4.2	4.2	5.0	5.5	6.1	7.3	7.9	7.9

Width modification regarding edge treatment

	Case I	Case II	Case III
No stabilized shoulder	None	None	None
Sloping curb	None	None	None
Vertical curb: one side	Add 0.3 m	None	Add 0.3 m
two sides	Add 0.6 m	Add 0.3 m	Add 0.6 m
Stabilized shoulder, one or both sides	Lane width for conditions B & C on tangent may be reduced to 3.6 m where shoulder is 1.2 m or wider	Deduct shoulder width; minimum width as under Case I	Deduct 0.6 where shoulder is 1.2 m or wider

US Customary

Pavement width (ft) — Design traffic conditions

Radius on inner edge of pavement R (ft)	Case I One-lane, one-way operation—no provision for passing a stalled vehicle			Case II One-lane, one-way operation—with provision for passing a stalled vehicle			Case III Two-lane operation—either one-way or two-way		
	A	B	C	A	B	C	A	B	C
50	18	18	23	20	26	30	31	36	45
75	16	17	20	19	23	27	29	33	38
100	15	16	18	18	22	25	28	31	35
150	14	15	17	18	21	23	26	29	32
200	13	15	16	17	20	22	26	28	30
300	13	15	15	17	20	22	25	28	29
400	13	15	15	17	19	21	25	27	28
500	12	15	15	17	19	21	25	27	28
Tangent	12	14	14	17	18	20	24	26	26

Width modification regarding edge treatment

	Case I	Case II	Case III
No stabilized shoulder	None	None	None
Sloping curb	None	None	None
Vertical curb: one side	Add 1 ft	None	Add 1 ft
two sides	Add 2 ft	Add 1 ft	Add 2 ft
Stabilized shoulder, one or both sides	Lane width for conditions B & C on tangent may be reduced to 12 ft where shoulder is 4 ft or wider	Deduct shoulder width; minimum pavement width as under Case I	Deduct 2 ft where shoulder is 4 ft or wider

Note: A = predominantly P vehicles, but some consideration for SU trucks.
 B = sufficient SU vehicles to govern design, but some consideration for semitrailer combination trucks.
 C = sufficient bus and combination-trucks to govern design.

Exhibit 3-55. Design Widths of Pavements for Turning Roadways

Traffic Condition A. This traffic condition consists predominantly of P vehicles, but some consideration is also given to SU trucks; the values in Exhibit 3-55 are somewhat higher than those for P vehicles in Exhibit 3-54.

Traffic Condition B. This traffic condition includes sufficient SU trucks to govern design, but some consideration is also given to tractor-semitrailer combination trucks; values in Exhibit 3-55 for Cases I and III are those for SU vehicles in Exhibit 3-54. For Case II, values are reduced as explained later in this section.

Traffic Condition C. This traffic condition includes sufficient tractor-semitrailer combination trucks, WB-12 or WB-15 [WB-40 or WB-50], to govern design; the values in Exhibit 3-55 for Cases I and III are those for the WB-12 [WB-40] truck in Exhibit 3-54. For Case II, values are reduced.

In general, Traffic Condition A may be assumed to have a small volume of trucks or only an occasional large truck; Traffic Condition B, a moderate volume of trucks (e.g., in the range of 5 to 10 percent of the total traffic); and Traffic Condition C, more and larger trucks.

In Exhibit 3-55, smaller vehicles in combination are assumed for deriving Case II widths than for deriving Case III widths, because passing of stalled vehicles in the former is apt to be very infrequent. Moreover, full offtracking need not be assumed for both the stalled and the passing vehicles. Often the stalled vehicles can be drifted adjacent to the inner edge of roadway, thereby providing additional clearance for the passing vehicle.

The design vehicles or combinations of different design vehicles used in determination of values given in Exhibit 3-55 for the three traffic conditions, assuming full clearance for the design vehicles indicated, are:

	Metric			US Customary		
	Design Traffic Condition			Design Traffic Condition		
Case	A	B	C	A	B	C
I	P	SU	WB-12	P	SU	WB-40
II	P-P	P-SU	SU-SU	P-P	P-SU	SU-SU
III	P-SU	SU-SU	WB-12-WB-12	P-SU	SU-SU	WB-40-WB-40

The combination of letters, such as P-SU for Case II, means that the design width in this example allows a P design vehicle to pass a stalled SU design truck or vice versa. In assuming full clearance, allowance was made for the values of C as discussed.

In negotiating roadways designed for smaller vehicles, larger vehicles will have less clearance and will need to use lower speeds and will demand more caution and skill by drivers, but there is a limit to the size of vehicles that can be operated on these narrower roadways. The larger vehicles that can be operated on turning roadways of the widths shown in Exhibit 3-55, but with partial clearance varying from about one-half the total values of C, as discussed for the sharper curves, to nearly full values for the flatter curves, are:

	Metric			US Customary		
Case	Design Traffic Condition			Design Traffic Condition		
	A	B	C	A	B	C
I	WB-12	WB-12	WB-15	WB-40	WB-40	WB-50
II	P-SU	P-WB-12	SU-WB-12	P-SU	P-WB-40	SU-WB-40
III	SU-WB-12	WB-12-WB-12	WB-15-WB-15	SU-WB-40	WB-40-WB-40	WB-50-WB-50

The widths in Exhibit 3-55 are subject to some modification with respect to the treatment at the edge, as shown at the bottom of the table. An occasional large vehicle can pass another on a roadway designed for small vehicles if there is space and stability outside the roadway and there is no barrier to prevent its occasional use. In such cases the width can be a little narrower than the tabulated dimension. Vertical curbs along the edge of a lane give drivers a sense of restriction, and occasional large vehicles have no additional space in which to maneuver; for this reason, such roadways should be a little wider than the values shown in Exhibit 3-55.

When there is an adjacent stabilized shoulder, the widths for Cases II and III and under certain conditions for Case I on roadways on tangent may be reduced. Case II values may be reduced by the additional width of stabilized shoulder but not below the widths for Case I. Similarly, Case III values may be reduced by 0.6 m [2 ft]. Case I values for the individual design vehicles are recommended minimums and further reduction is not in order, even with a usable shoulder, except on tangents. When vertical curbs are used on both sides, the tabulated widths should be increased by 0.6 m [2 ft] for Cases I and III, or by 0.3 m [1 ft] for Case II, because stalled vehicles are passed at low speed. Where such a curb is on only one side of the roadway, the added width may be only 0.3 m [1 ft] for Cases I and III, and no added width is needed for Case II.

The use of Exhibit 3-55 in design is illustrated by the following example. Assume that the geometric layout and traffic volume for a specific turning movement are such that one-lane, one-way operation with provision for passing a stalled vehicle is called for (Case II), and that the traffic volume includes 10 to 12 percent trucks with an occasional large semitrailer combination for which traffic condition C is deemed applicable. Then, with a radius of 50 m [165 ft] for the inner edge of the traveled way, the width tabulated in Exhibit 3-55 is 7.0 m [23 ft]. With a 1.2-m [4-ft] stabilized shoulder, the turning roadway width may be reduced to 5.8 [19 ft] (see lower part of Exhibit 3-55). With a vertical curb on each side, the turning roadway width should be not less than 7.3 m [24 ft].

Widths Outside Traveled Way

The roadway width for a turning roadway includes the shoulders or equivalent lateral clearance outside the traveled way. Over the whole range of intersections, the appropriate shoulder width varies from none, or minimal, on curbed urban streets to the width of an open-highway cross section. The more general cases are discussed in the following paragraphs.

Within a channelized intersection, shoulders for turning roadways are usually unnecessary. The lanes may be defined by curbs, pavement markings, or islands. The islands may be curbed and the general dimensional controls for islands provide the appropriate lateral clearances outside the edges of the turning roadway. In most instances, the turning roadways are relatively short, and shoulder sections are not needed for the temporary storage of vehicles. A discussion of island dimensions can be found in Chapter 9.

Where there is a separate roadway for right turns, its left edge defines one side of the triangular island. If the island is small or especially important in directing movements, it may be defined by both curbs or pavement markings. On the other hand, where the turning radius is large, the side of the island may be defined by guideposts, by delineators, or simply by pavement markings and the edge of the pavement of the turning roadway. In any case, a developed left shoulder is normally unnecessary. However, there should be either an offset, if curbs are used, or a fairly level section of sufficient width on the left to avoid affecting the lateral placement of vehicles.

A shoulder usually is provided on the right side of a right-turning roadway in rural areas. In cross section and general treatment, the right shoulder should be essentially the same as the shoulder of the adjacent open-highway section, possibly somewhat reduced in width because of conditions at the intersections. Because turning vehicles have a tendency to encroach on the shoulder, consideration should be given to providing heavy-duty right shoulders to accommodate the associated wheel loads. Although a curb on the right side might be advantageous in reducing maintenance operations that result from vehicles hugging the inside of the curve and causing edge depressions or raveling, the introduction of curbing adjacent to high-speed highways should be discouraged. For low-speed urban conditions, curbing of the right edge of a turning roadway is normal practice. Curbs are discussed in greater detail in Chapter 4.

On large-scale channelized layouts and at interchanges, there may be turning roadways of sufficient curvature and length to be well removed from other roadways. Such turning roadways should have a shoulder on both sides. Curbs, when used, should be located at the outside edge of the shoulder and should be sloping.

Some turning roadways, particularly ramps, pass over drainage structures, pass over or under other roadways, or pass adjacent to walls or rock cuts on one or both sides. For such locations, the minimum clearances for structures, as established in later chapters and in the current edition of the AASHTO bridge specifications (**31**), apply directly. In addition, the design should be evaluated for adequate sight distance, as the sharp curve may need above-minimum lateral clearance.

Exhibit 3-56 is a summary of the range of design values for the general turning roadway conditions described above. On roadways without curbs or with sloping curbs, the adjacent shoulder should be of the same type and cross section as that on the approach highway. The widths shown are for usable shoulders. Where roadside barriers are provided, the width indicated should be measured to the face of the barrier, and the graded width should be about 0.6 m [2.0 ft] greater. For other than low-volume conditions, it is desirable that right shoulders be surfaced, surface treated, or otherwise stabilized for a width of 1.2 m [4.0 ft] or more.

	Metric		US Customary	
	Shoulder width or lateral clearance outside of traveled way edge (m)		Shoulder width or lateral clearance outside of traveled way edge (ft)	
Turning roadway condition	Left	Right	Left	Right
Short length, usually within channelized intersection	0.6 to 1.2	0.6 to 1.2	2 to 4	2 to 4
Intermediate to long length or in cut or on fill	1.2 to 3.0	1.8 to 3.6	4 to 10	6 to 12
Note: All dimensions should be increased, where necessary, for sight distance.				

Exhibit 3-56. Range of Usable Shoulder Widths or Equivalent Lateral Clearances Outside of Turning Roadways, Not on Structure

Sight Distance on Horizontal Curves

Another element of horizontal alignment is the sight distance across the inside of curves. Where there are sight obstructions (such as walls, cut slopes, buildings, and longitudinal barriers) on the inside of curves or the inside of the median lane on divided highways, a design may need adjustment in the normal highway cross section or change in the alignment if removal of the obstruction is impractical to provide adequate sight distance. Because of the many variables in alignment, in cross section, and in the number, type, and location of potential obstructions, specific study is usually needed for each individual curve. With sight distance for the design speed as a control, the designer should check the actual conditions on each curve and make the appropriate adjustments to provide adequate sight distance.

Stopping Sight Distance

For general use in design of a horizontal curve, the sight line is a chord of the curve, and the stopping sight distance is measured along the centerline of the inside lane around the curve. Exhibit 3-57 is a design chart showing the middle ordinates needed for clear sight areas that satisfy stopping sight distance criteria presented in Exhibit 3-1 for horizontal curves of various radii. Exhibit 3-57 includes radii for all superelevation rates to a maximum of 12 percent.

The middle-ordinate values in Exhibit 3-57 are derived from geometry for the several dimensions, as indicated in the diagrammatic sketch in Exhibit 3-58 and in Equation (3-40). The equation applies only to circular curves longer than the sight distance for the pertinent design speed. The relationships between R, M, and V in this chart can be quickly checked. For example, with an 80 km/h [50 mph] design speed and a curve with a 350 m [1,150 ft] radius, a clear sight area with a middle ordinate of approximately 6.0 m [20 ft] is needed for stopping sight distance. As another example, for a sight obstruction at a distance M equal to 6.0 m [20 ft] from the centerline of the inside lane on a curve with a 175-m [575-ft] radius, the sight distance needed is approximately at the upper end of the range for a speed of approximately 60 km/h [40 mph].

METRIC

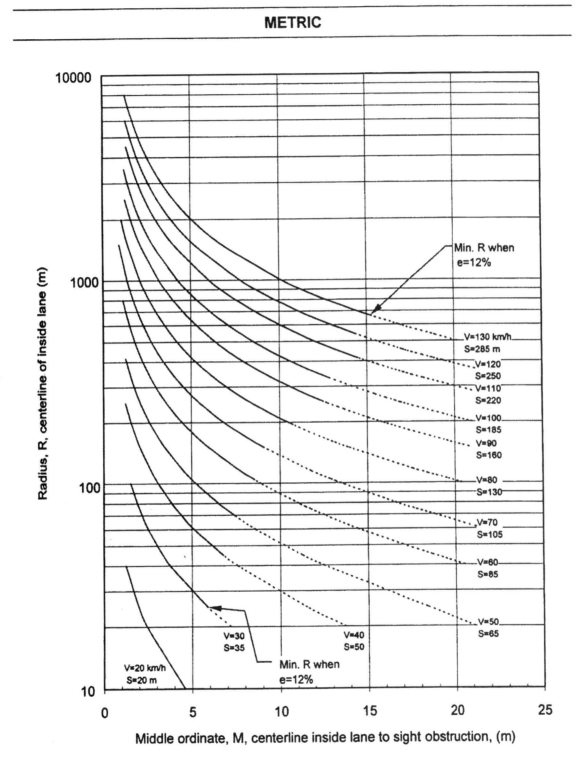

Exhibit 3-57. Design Controls for Stopping Sight Distance on Horizontal Curves

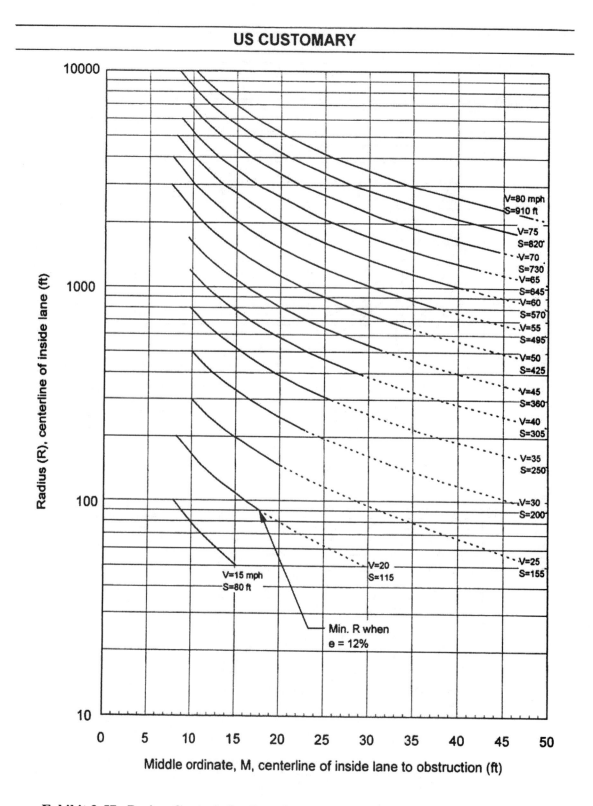

Exhibit 3-57. Design Controls for Stopping Sight Distance on Horizontal Curves (Continued)

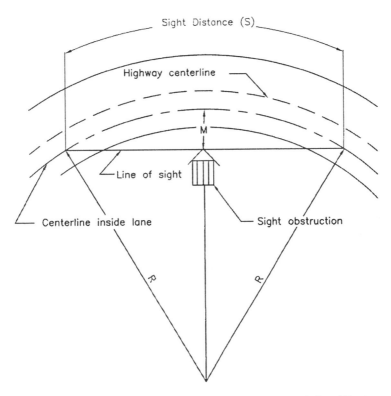

Exhibit 3-58. Diagram Illustrating Components for Determining Horizontal Sight Distance

Metric	US Customary	
$M = R\left[\left(1 - \cos\dfrac{28.65S}{R}\right)\right]$	$M = R\left[\left(1 - \cos\dfrac{28.65S}{R}\right)\right]$	(3-40)
where: S = Stopping sight distance, m; R = Radius of curve, m; M = Middle ordinate, m	where: S = Stopping sight distance, ft; R = Radius of curve, ft; M = Middle ordinate, ft	

Horizontal sight restrictions may occur where there is a cut slope on the inside of the curve. For the 1,080-mm [3.5-ft] eye height and the 600-mm [2.0-ft] object height used for stopping sight distance, a height of 840 mm [2.75 ft] may be used as the midpoint of the sight line where the cut slope usually obstructs sight. This assumes that there is little or no vertical curvature. For a highway with a 6.6-m [22-ft] traveled way, 1.2-m [4-ft] shoulders, an allowance of 1.2 m [4 ft] for a ditch section, and 1V:2H (1 m or 1 ft vertically for each, 2 m or 2 ft horizontally) cut slopes, the sight obstruction is about 5.75 m [19 ft] outside the centerline of the inside lane. This is sufficient for adequate sight distance at 50 km/h [30 mph] when curves have a radius of about 90 m [275 ft] or more and at 80 km/h [50 mph] when curves have a radius of about 375 m [1,230 ft] or more. Curves sharper than these would need flatter slopes, benching, or other adjustments. At the other extreme, highways with normal lateral dimensions of more than 16 m [52 ft] provide adequate stopping sight distances for horizontal curves over the entire range of design speeds and curves.

In some instances, retaining walls, concrete median safety barriers, and other similar features constructed on the inside of curves may be sight obstructions and should be checked for stopping sight distance. As an example, an obstruction of this type, located 1.2 m [4 ft] from the inside edge of a 7.2-m [24-ft] traveled way, has a middle ordinate of about 3.0 m [10 ft]. At 80 km/h [50 mph], this provides sufficient sight distance when a curve has a radius of about 700 m [2,300 ft] or more. If the obstruction is moved an additional 0.3 m [1 ft] away from the roadway creating a middle ordinate of 3.3 m [11 ft], a curve with a radius of 625 m [2,000 ft] or more provides sufficient sight distance at the same 80 km/h [50 mph] speed. The same finding would be applicable to existing buildings or similar sight obstructions on the inside of curves.

Where sufficient stopping sight distance is not available because a railing or a longitudinal barrier constitutes a sight obstruction, alternative designs should be considered for both safety and economic reasons. The alternatives are: (1) increase the offset to the obstruction, (2) increase the radius, or (3) reduce the design speed. However, the alternative selected should not incorporate shoulder widths on the inside of the curve in excess of 3.6 m [12 ft] because of the concern that drivers will use wider shoulders as a passing or travel lane.

As can be seen from Exhibit 3-58, the method presented is only exact when both the vehicle and the sight obstruction are located within the limits of the simple horizontal curve. When either the vehicle or the sight obstruction is situated beyond the limits of the simple curve, the values obtained are only approximate. The same is true if either the vehicle, the sight obstruction, or both is situated within the limits of a spiral or a compound curve. In these instances, the value obtained would result in middle ordinate values slightly larger than those needed to satisfy the desired stopping sight distance. In many instances, the resulting additional clearance will not be significant. Whenever Exhibit 3-57 is not applicable, the design should be checked either by utilizing graphical procedures or by utilizing a computational method. Reference (**32**) provides a computational method for making such checks.

Passing Sight Distance

The minimum passing sight distance for a two-lane road or street is about four times as great as the minimum stopping sight distance at the same design speed. To conform to those greater sight distances, clear sight areas on the inside of curves should have widths greatly in excess of those discussed. Equation (3-40) is directly applicable to passing sight distance but are of limited practical value except on long curves. A chart demonstration using this equation would be of value primarily in reaching negative conclusions—that it would be difficult to maintain passing sight distance on other than very flat curves.

Passing sight distance is measured between an eye height of 1,080 mm [3.5 ft] and an object height of 1,080 mm [3.5 ft]. The sight line near the center of the area inside a curve is about 240 mm [0.75 ft] higher than for stopping sight distance. The resultant lateral dimension for normal highway cross sections (1V:2H to 1V:6H backslopes) in cut between the centerline of the inside lane and the midpoint of the sight line is from 0.5 to 1.5 m [1.5 to 4.5 ft] greater than that for stopping sight distance. It is obvious that for many cut sections, design for passing sight distance should, for practical reasons, be limited to tangents and very flat curves. Even in level terrain, provision of passing sight distance would need a clear area inside each curve that would, in some instances, extend beyond the normal right-of-way line.

In general, the designer should use graphical methods to check sight distance on horizontal curves. This method is presented in Exhibit 3-8 and described in the accompanying discussion.

General Controls for Horizontal Alignment

In addition to the specific design elements for horizontal alignment discussed under previous headings, a number of general controls are recognized in practice. These controls are not subject to theoretical derivation, but they are important for efficient and smooth-flowing highways. Excessive curvature or poor combinations of curvature limit capacity, cause economic losses because of increased travel time and operating costs, and detract from a pleasing appearance. To avoid such poor design practices, the general controls that follow should be used where practical:

- Alignment should be as directional as practical, but should be consistent with the topography and with preserving developed properties and community values. A flowing line that conforms generally to the natural contours is preferable to one with long tangents that slashes through the terrain. With curvilinear alignment, construction scars can be kept to a minimum and natural slopes and growth can be preserved. Such design is desirable from a construction and maintenance standpoint. In general, the number of short curves should be kept to a minimum. Winding alignment composed of short curves should be avoided because it usually leads to erratic operation. Although the aesthetic qualities of curving alignment are important, long tangents are needed on two-lane highways so that sufficient passing sight distance is available on as great a percentage of the highway length as practical.

- In alignment developed for a given design speed, the minimum radius of curvature for that speed should be avoided wherever practical. The designer should attempt to use generally flat curves, saving the minimum radius for the most critical conditions. In general, the central angle of each curve should be as small as the physical conditions permit, so that the highway will be as directional as practical. This central angle should be absorbed in the longest practical curve, but on two-lane highways the exception noted in the preceding paragraph applies.

- Consistent alignment should always be sought. Sharp curves should not be introduced at the ends of long tangents. Sudden changes from areas of flat curvature to areas of sharp curvature should be avoided. Where sharp curvature is introduced, it should be approached, where practical, by a series of successively sharper curves.

- For small deflection angles, curves should be sufficiently long to avoid the appearance of a kink. Curves should be at least 150 m [500 ft] long for a central angle of 5 degrees, and the minimum length should be increased 30 m [100 ft] for each 1-degree decrease in the central angle. The minimum length for horizontal curves on main highways, $L_{c\,min}$, should be about three times the design speed expressed in km/h [15 times the design speed expressed in mph], or $L_{c\ min}=3V$ [15V]. On high speed controlled-access facilities that use flat curvature, for aesthetic reasons, the desirable minimum length for curves should be about double the minimum length described above, or $L_{c\ des}= 6V$ [30V].

- Sharp curvature should be avoided on long, high fills. In the absence of cut slopes, shrubs, and trees that extend above the level of the roadway, it is difficult for drivers to perceive the extent of curvature and adjust their operation accordingly.

- Caution should be exercised in the use of compound circular curves. While the use of compound curves affords flexibility in fitting the highway to the terrain and other ground controls, the ease with which such curves can be used may tempt the designer to use them without restraint. Preferably their use should be avoided where curves are sharp. Compound curves with large differences in radius introduce the same problems that arise at tangent approaches to circular curves. Where topography or right-of-way restrictions make their use appropriate, the radius of the flatter circular arc, R_1, should not be more than 50 percent greater than the radius of the sharper circular arc, R_2 (i.e., R_1 should not exceed 1.5 R_2). A multiple compound curve (i.e., several curves in sequence) may be suitable as a transition to sharp curves as discussed in the previous section on "Compound Circular Curves." A spiral transition between flat curves and sharp curves may be desirable. On one-way roads, such as ramps, the difference in radii of compound curves is not so important if the second curve is flatter than the first. However, the use of compound curves on ramps, with a flat curve between two sharper curves, is not good practice.

- Abrupt reversals in alignment should be avoided. Such changes in alignment make it difficult for drivers to keep within their own lane. It is also difficult to superelevate both curves adequately, and erratic operation may result. The distance between reverse curves should be the sum of the superelevation runoff lengths and the tangent runout lengths or, preferably, an equivalent length with spiral curves, as defined in the section on "Transition Design Controls" in this chapter. If sufficient distance (i.e., more than 100 m [300 ft]) is not available to permit the tangent runout lengths or preferably an equivalent length with spiral to return to a normal crown section, there may be a long length where the centerline and the edges of roadway are at the same elevation and poor transverse drainage can be expected. In this case, the superelevation runoff lengths should be increased until they adjoin, thus providing one instantaneous level section. For traveled ways with straight cross slopes, there is less difficulty in returning the edges of roadway to a normal section and the 100-m [300-ft] guideline discussed above may be decreased.

- The "broken-back" or "flat-back" arrangement of curves (with a short tangent between two curves in the same direction) should be avoided except where very unusual topographical or right-of-way conditions make other alternatives impractical. Except on circumferential highways, most drivers do not expect successive curves to be in the same direction; the preponderance of successive curves in opposite directions may develop a subconscious expectation among drivers that makes successive curves in the same direction unexpected. Broken-back alignments are also not pleasing in appearance. Use of spiral transitions or compound curve alignments, in which there is some degree of continuous superelevation, is preferable for such situations. The term "broken-back" usually is not applied when the connecting tangent is of considerable length. Even in this case, the alignment may be unpleasant in appearance when both curves are clearly visible for some distance ahead.

- To avoid the appearance of inconsistent distortion, the horizontal alignment should be coordinated carefully with the profile design. General controls for this coordination are discussed in the section of this chapter on "Combination of Horizontal and Vertical Alignment."

VERTICAL ALIGNMENT

Terrain

The topography of the land traversed has an influence on the alignment of roads and streets. Topography affects horizontal alignment, but has an even more pronounced effect on vertical alignment. To characterize variations in topography, engineers generally separate it into three classifications according to terrain.

In level terrain, highway sight distances, as governed by both horizontal and vertical restrictions, are generally long or can be made to be so without construction difficulty or major expense.

In rolling terrain, natural slopes consistently rise above and fall below the road or street grade, and occasional steep slopes offer some restriction to normal horizontal and vertical roadway alignment.

In mountainous terrain, longitudinal and transverse changes in the elevation of the ground with respect to the road or street are abrupt, and benching and side hill excavation are frequently needed to obtain acceptable horizontal and vertical alignment.

Terrain classifications pertain to the general character of a specific route corridor. Routes in valleys, passes, or mountainous areas that have all the characteristics of roads or streets traversing level or rolling terrain should be classified as level or rolling. In general, rolling terrain generates steeper grades than level terrain, causing trucks to reduce speeds below those of passenger cars; mountainous terrain has even greater effects, causing some trucks to operate at crawl speeds.

Grades

Roads and streets should be designed to encourage uniform operation throughout. As discussed earlier in this chapter, design speeds are used as a means toward this end by correlation of various geometric features of the road or street. Design criteria have been determined for many highway features, but few conclusions have been reached on the appropriate relationship of roadway grades to design speed. Vehicle operating characteristics on grades are discussed and established relationships of grades and their lengths to design speed are developed below.

Vehicle Operating Characteristics on Grades

Passenger cars. The practices of passenger car drivers on grades vary greatly, but it is generally accepted that nearly all passenger cars can readily negotiate grades as steep as 4 to 5 percent without an appreciable loss in speed below that normally maintained on level roadways, except for cars with high weight/power ratios, including some compact and subcompact cars.

Studies show that, under uncongested conditions, operation on a 3-percent upgrade, has only a slight effect on passenger car speeds compared to operations on the level. On steeper upgrades, speeds decrease progressively with increases in the grade. On downgrades, passenger car speeds generally are slightly higher than on level sections, but local conditions govern.

Trucks. The effect of grades on truck speeds is much more pronounced than on speeds of passenger cars. The average speed of trucks on level sections of highway approximates the average speed of passenger cars. Trucks generally increase speed by up to about 5 percent on downgrades and decrease speed by 7 percent or more on upgrades as compared to their operation on the level. On upgrades, the maximum speed that can be maintained by a truck is dependent primarily on the length and steepness of the grade and the truck's weight/power ratio, which is the gross vehicle weight divided by the net engine power. Other factors that affect the average truck speed on a grade are the entering speed, the aerodynamic resistance, and skill of the driver. The last two factors cause only minor variations in the average speed.

Extensive studies of truck performance have been conducted to determine the separate and combined effects of roadway grade, tractive effort, and gross vehicle weight (**33, 34, 35, 36, 37, 38, 39**).

The effect of rate and length of grade on the speed of a typical heavy truck is shown in Exhibits 3-59 and 3-60. From Exhibit 3-59 it can be determined how far a truck, starting its climb from any speed up to approximately 120 km/h [70 mph], travels up various grades or combinations of grades before a certain or uniform speed is reached. For instance, with an entering speed of approximately 110 km/h [70 mph], the truck travels about 950 m [2,700 ft] up a 6 percent grade before its speed is reduced to 60 km/h [35 mph]. If the entering speed is 60 km/h [35 mph], the speed at the end of a 300-m [1,000-ft] climb is about 43 km/h [26 mph]. This is determined by starting on the curve for a 6 percent grade corresponding to 60 km/h [35 mph] for which the distance is 750 m [2,500 ft], and proceeding along it to the point where the distance is 300 m [1,000 ft] more, or 1,050 m [3,500 ft], for which the speed is about 43 km/h [26 mph]. Exhibit 3-60 shows the performance on grade for a truck that approaches the grade at or below crawl speed. The truck is able to accelerate to a speed of 40 km/h [25 mph] or more only on grades of less than 3.5 percent. These data serve as a valuable guide for design in appraising the effect of trucks on traffic operation for a given set of profile conditions.

Travel time (and, therefore, speed) of trucks on grades is directly related to the weight/power ratio. Trucks of the same weight/power ratio typically have similar operating characteristics. Hence, this ratio is of considerable assistance in anticipating the performance of trucks. Normally,

METRIC

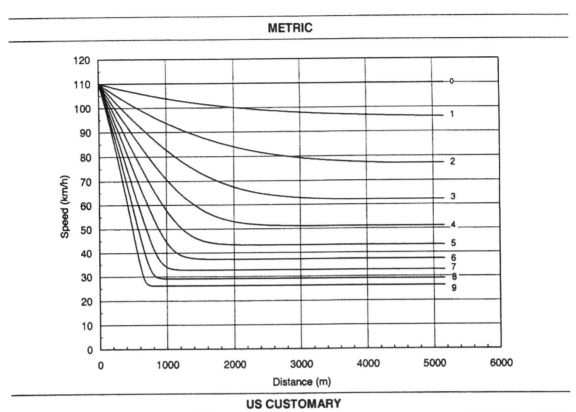

US CUSTOMARY

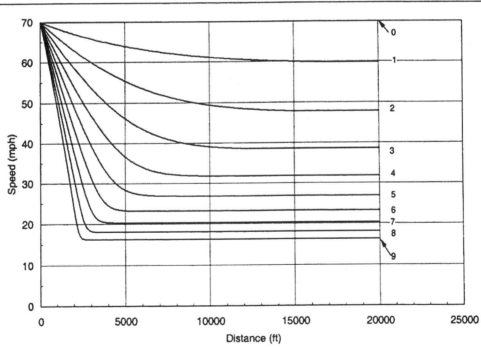

Exhibit 3-59. Speed-Distance Curves for a Typical Heavy Truck of 120 kg/kW [200 lb/hp] for Deceleration on Upgrades

METRIC

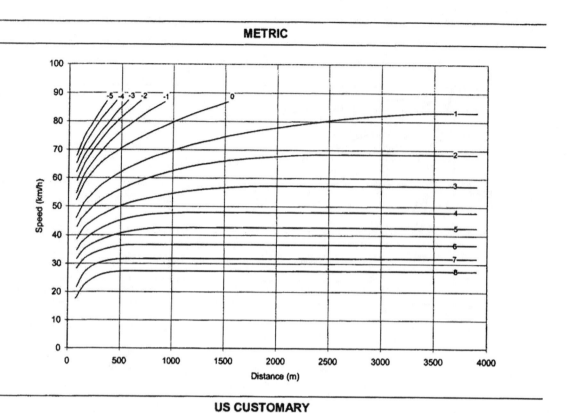

US CUSTOMARY

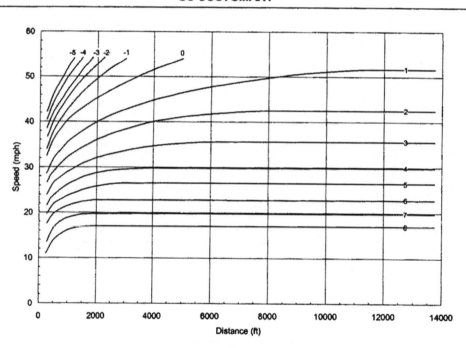

Exhibit 3-60. Speed-Distance Curves for Acceleration of a Typical Heavy Truck of 120 kg/kW [200 lb/hp] on Upgrades and Downgrades

the weight/power ratio is expressed in terms of gross weight and net power, in units of kg/kW [wt/hp]; while the metric unit kg is a unit of mass, rather than weight, it is commonly used to represent the weight of object. It has been found that trucks with weight/power ratios of about 120 kg/kW [200 lb/hp] have acceptable operating characteristics from the standpoint of the highway user. Such a weight/power ratio assures a minimum speed of about 60 km/h [35 mph] on a 3 percent upgrade. There is evidence that the automotive industry would find a weight/power ratio of this magnitude acceptable as a minimum goal in the design of commercial vehicles. There is also evidence that carrier operators are voluntarily recognizing this ratio as the minimum performance control in the loads placed on trucks of different power, the overall result being that weight/power ratio of trucks on highways has improved in recent years. Ratios developed from information obtained in conjunction with the nationwide brake performance studies conducted between 1949 and 1985 show, for example, that for a gross vehicle weight of 18,000 kg [40,000 lb], the average weight/power ratio decreased from about 220 kg/kW [360 lb/hp] in 1949, to about 130 kg/kW [210 lb/hp] in 1975; the weight/power ratio continued to fall to about 80 kg/kW [130 lb/hp] in 1985. This decreased weight/power ratio means greater power and better climbing ability for trucks on upgrades.

There is a trend toward larger and heavier trucks with as many as three trailer units allowed on certain highways in some States. Studies indicate that as the number of axles increases, the weight/power ratio increases. Taking all factors into account, it appears conservative to use a weight/power ratio of 120 kg/kW [200 lb/hp] in determining critical length of grade. However, there are locations where a weight/power ratio as high as 120 kg/kW [200 lb/hp] is not appropriate. Where this occurs, designers are encouraged to utilize either a more representative weight/power ratio or an alternate method that more closely fits the conditions.

Recreational vehicles. Consideration of recreational vehicles on grades is not as critical as consideration of trucks. However, on certain routes such as designated recreational routes, where a low percentage of trucks may not warrant a truck climbing lane, sufficient recreational vehicle traffic may indicate a need for an additional lane. This can be evaluated by using the design charts in Exhibit 3-61 in the same manner as for trucks described in the preceding section of this chapter. Recreational vehicles include self-contained motor homes, pickup campers, and towed trailers of numerous sizes. Because the characteristics of recreational vehicles vary so much, it is difficult to establish a single design vehicle. However, a recent study on the speed of vehicles on grades included recreational vehicles (**40**). The critical vehicle was considered to be a vehicle pulling a travel trailer, and the charts in Exhibits 3-61 for a typical recreational vehicle is based on that assumption.

Control Grades for Design

Maximum grades. On the basis of the data in Exhibits 3-59 through 3-62, and according to the grade controls now in use in a large number of States, reasonable guidelines for maximum grades for use in design can be established. Maximum grades of about 5 percent are considered

METRIC

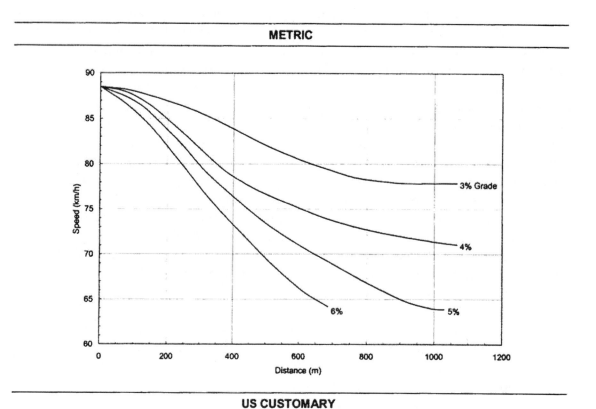

US CUSTOMARY

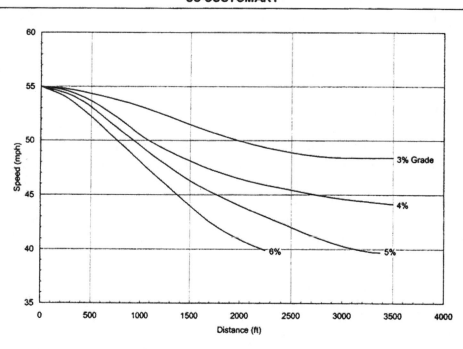

Exhibit 3-61. Speed-Distance Curves for a Typical Recreational Vehicle on the Selected Upgrades (40)

METRIC

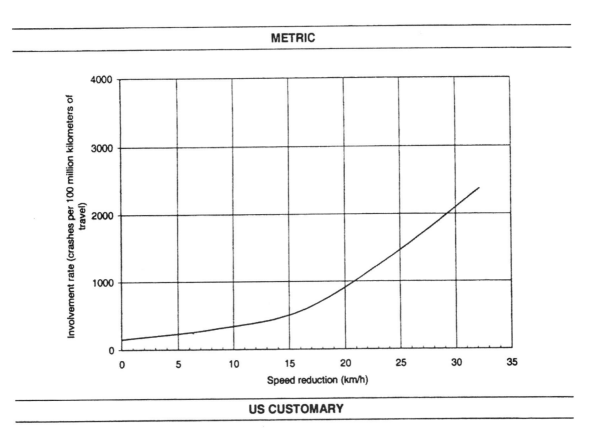

US CUSTOMARY

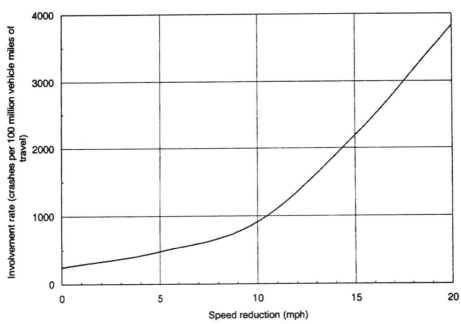

Exhibit 3-62. Crash Involvement Rate of Trucks for Which Running Speeds Are Reduced Below Average Running Speed of All Traffic (41)

appropriate for a design speed of 110 km/h [70 mph]. For a design speed of 50 km/h [30 mph], maximum grades generally are in the range of 7 to 12 percent, depending on terrain. If only the more important highways are considered, it appears that maximum grades of 7 or 8 percent are representative of current design practice for a 50-km/h [30-mph] design speed. Control grades for design speeds from 60 to 100 km/h [40 to 60 mph] fall between the above extremes. Maximum grade controls for each functional class of highway and street are presented in Chapters 5 through 8.

The maximum design grade should be used only infrequently; in most cases, grades should be less than the maximum design grade. At the other extreme, for short grades less than 150 m [500 ft] in length and for one-way downgrades, the maximum grade may be about 1 percent steeper than other locations; for low-volume rural highways, the maximum grade may be 2 percent steeper.

Minimum grades. Flat grades can typically be used without problem on uncurbed highways where the cross slope is adequate to drain the pavement surface laterally. With curbed highways or streets, longitudinal grades should be provided to facilitate surface drainage. An appropriate minimum grade is typically 0.5 percent, but grades of 0.30 percent may be used where there is a high-type pavement accurately sloped and supported on firm subgrade. Use of even flatter grades may be justified in special cases as discussed in subsequent chapters. Particular attention should be given to the design of storm water inlets and their spacing to keep the spread of water on the traveled way within tolerable limits. Roadside channels and median swales frequently need grades steeper than the roadway profile for adequate drainage. Drainage channels are discussed in Chapter 4.

Critical Lengths of Grade for Design

Maximum grade in itself is not a complete design control. It is also appropriate to consider the length of a particular grade in relation to desirable vehicle operation. The term "critical length of grade" is used to indicate the maximum length of a designated upgrade on which a loaded truck can operate without an unreasonable reduction in speed. For a given grade, lengths less than critical result in acceptable operation in the desired range of speeds. If the desired freedom of operation is to be maintained on grades longer than critical, design adjustments such as changes in location to reduce grades or addition of extra lanes should be considered. The data for critical lengths of grade should be used with other pertinent factors (such as traffic volume in relation to capacity) to determine where added lanes are warranted.

To establish design values for critical lengths of grade for which gradeability of trucks is the determining factor, data or assumptions are needed for the following:

1. Size and power of a representative truck or truck combination to be used as a design vehicle along with the gradeability data for this vehicle:
    A loaded truck, powered so that the weight/power ratio is about 120 kg/kW [200 lb/hp], is representative of the size and type of vehicle normally used for

design control for main highways. Data in Exhibits 3-59 and 3-60 apply to such a vehicle.

2. Speed at entrance to critical length of grade:

    The average running speed as related to design speed can be used to approximate the speed of vehicles beginning an uphill climb. This estimate is, of course, subject to adjustment as approach conditions may determine. Where vehicles approach on nearly level grades, the running speed can be used directly. For a downhill approach it should be increased somewhat, and for an uphill approach it should be decreased.

3. Minimum speed on the grade below in which interference to following vehicles is considered unreasonable:

    No specific data are available on which to base minimum tolerable speeds of trucks on upgrades. It is logical to assume that such minimum speeds should be in direct relation to the design speed. Minimum truck speeds of about 40 to 60 km/h [25 to 40 mph] for the majority of highways (on which design speeds are about 60 to 100 km/h [40 to 60 mph]) probably are not unreasonably annoying to following drivers unable to pass on two-lane roads, if the time interval during which they are unable to pass is not too long. The time interval is not likely to be annoying on two-lane roads with volumes well below their capacities, whereas it is likely to be annoying on two-lane roads with volumes near capacity. Lower minimum truck speeds can probably be tolerated on multilane highways rather than on two-lane roads because there is more opportunity for and less difficulty in passing. Highways should be designed so that the speeds of trucks will not be reduced enough to cause intolerable conditions for following drivers.

Studies show that, regardless of the average speed on the highway, the more a vehicle deviates from the average speed, the greater its chances of becoming involved in a crash. One such study (**41**) used the speed distribution of vehicles traveling on highways in one state, and related it to the crash involvement rate to obtain the rate for trucks of four or more axles operating on level grades. The crash involvement rates for truck speed reductions of 10, 15, 25, and 30 km/h [5, 10, 15, and 20 mph] were developed assuming the reduction in the average speed for all vehicles on a grade was 30 percent of the truck speed reduction on the same grade. The results of this analysis are shown in Exhibit 3-62.

A common basis for determining critical length of grade is based on a reduction in speed of trucks below the average running speed of traffic. The ideal would be for all traffic to operate at the average speed. This, however, is not practical. In the past, the general practice has been to use a reduction in truck speed of 25 km/h [15 mph] below the average running speed of all traffic to identify the critical length of grade. As shown in Exhibit 3-62, the crash involvement rate increases significantly when the truck speed reduction exceeds 15 km/h [10 mph] with the involvement rate being 2.4 times greater for a 25-km/h [15-mph] reduction than for a 15-km/h [10-mph] reduction. On the basis of these relationships, it is recommended that a 15-km/h [10-mph] reduction criterion be used as the general guide for determining critical lengths of grade.

The length of any given grade that will cause the speed of a representative truck (120 kg/kW [200 lb/hp]) entering the grade at 110 km/h [70 mph] to be reduced by various amounts below the average running speed of all traffic is shown graphically in Exhibit 3-63, which is based on the truck performance data presented in Exhibit 3-59. The curve showing a 15-km/h [10-mph] speed reduction is used as the general design guide for determining the critical lengths of grade. Similar information on the critical length of grade for recreational vehicles may be found in Exhibit 3-64, which is based on the recreational vehicle performance data presented in Exhibit 3-61.

Where the entering speed is less than 110 km/h [70 mph], as may be the case where the approach is on an upgrade, the speed reductions shown in Exhibits 3-63 and 3-64 will occur over shorter lengths of grade. Conversely, where the approach is on a downgrade, the probable approach speed is greater than 110 km/h [70 mph] and the truck or recreational vehicle will ascend a greater length of grade than shown in the exhibits before the speed is reduced to the values shown.

The method of using Exhibit 3-63 to determine critical lengths of grade is demonstrated in the following examples.

Assume that a highway is being designed for 100 km/h [60 mph] and has a fairly level approach to a 4 percent upgrade. The 15-km/h [10-mph] speed reduction curve in Exhibit 3-63 shows the critical length of grade to be 350 m [1,200 ft]. If, instead, the design speed was 60 km/h [40 mph], the initial and minimum tolerable speeds on the grade would be different, but for the same permissible speed reduction the critical length would still be 350 m [1,200 ft].

In another instance, the critical length of a 5 percent upgrade approached by a 500-m [1,650-ft] length of 2 percent upgrade is unknown. Exhibit 3-63 shows that a 2 percent upgrade of 500 m [1,650 ft] in length would result in a speed reduction of about 9 km/h [6 mph]. The chart further shows that the remaining tolerable speed reduction of 6 km/h [4 mph] would occur on 100 m [325 ft] of the 5 percent upgrade.

Where an upgrade is approached on a momentum grade, heavy trucks often increase speed, sometimes to a considerable degree in order to make the climb in the upgrade at as high a speed as practical. This factor can be recognized in design by increasing the tolerable speed reduction. It remains for the designer to judge to what extent the speed of trucks would increase at the bottom of the momentum grade above that generally found on level approaches. It appears that a speed increase of about 10 km/h [5 mph] can be considered for moderate downgrades and a speed increase of 15 km/h [10 mph] for steeper grades of moderate length or longer. On this basis, the tolerable speed reduction with momentum grades would be 25 or 30 km/h [15 or 20 mph]. For example, where there is a moderate length of 4 percent downgrade in advance of a 6 percent upgrade, a tolerable speed reduction of 25 km/h [15 mph] can be assumed. For this case, the critical length of the 6 percent upgrade is about 370 m [1,250 ft].

The critical length of grade in Exhibit 3-63 is derived as the length of tangent grade. Where a vertical curve is part of a critical length of grade, an approximate equivalent tangent grade length should be used. Where the condition involves vertical curves of Types II and IV shown later in this chapter in Exhibit 3-73 and the algebraic difference in grades is not too great, the measurement of critical length of grade may be made between the vertical points of intersection

METRIC

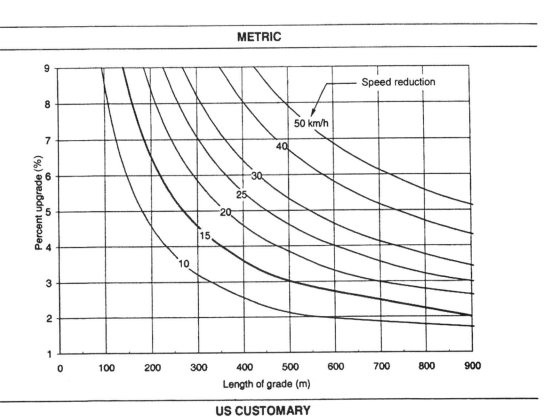

US CUSTOMARY

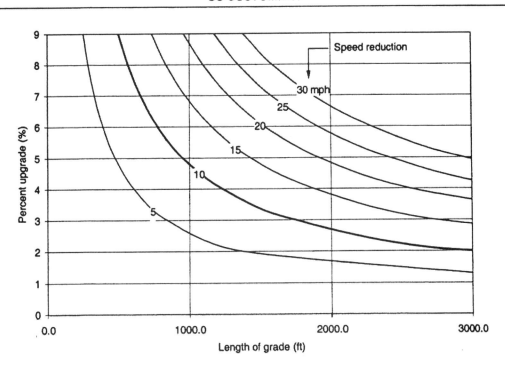

Exhibit 3-63. Critical Lengths of Grade for Design, Assumed Typical Heavy Truck of 120 kg/kW [200 lb/hp], Entering Speed = 110 km/h [70 mph]

METRIC

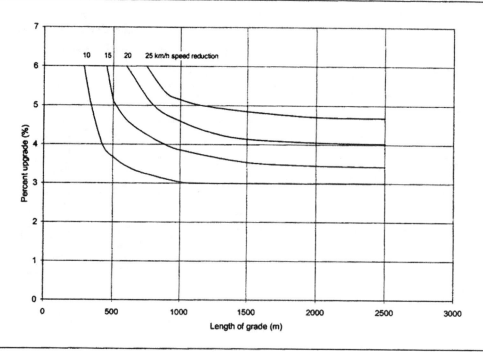

US CUSTOMARY

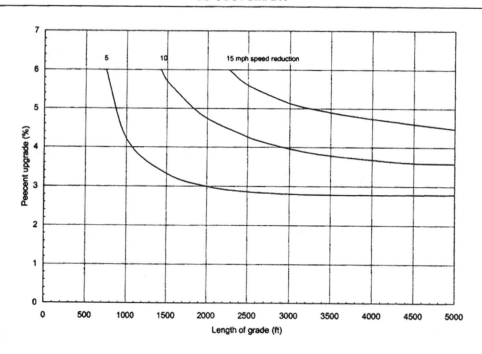

Exhibit 3-64. Critical Lengths of Grade Using an Approach Speed of 90 km/h [55 mph] for Typical Recreational Vehicle (40)

(VPI). Where vertical curves of Types I and III in Exhibit 3-73 are involved, about one-quarter of the vertical curve length should be considered as part of the grade under consideration.

Steep downhill grades can also have a detrimental effect on the capacity and safety of facilities with high traffic volumes and numerous heavy trucks. Some downgrades are long and steep enough that some heavy vehicles travel at crawl speeds to avoid loss of control on the grade. Slow-moving vehicles of this type may impede other vehicles. Therefore, there are instances where consideration should be given to providing a truck lane for downhill traffic. Procedures have been developed in the HCM (**14**) to analyze this situation.

The suggested design criterion for determining the critical length of grade is not intended as a strict control but as a guideline. In some instances, the terrain or other physical controls may preclude shortening or flattening grades to meet these controls. Where a speed reduction greater than the suggested design guide cannot be avoided, undesirable type of operation may result on roads with numerous trucks, particularly on two-lane roads with volumes approaching capacity and in some instances on multilane highways. Where the length of critical grade is exceeded, consideration should be given to providing an added uphill lane for slow-moving vehicles, particularly where volume is at or near capacity and the truck volume is high. Data in Exhibit 3-63 can be used along with other pertinent considerations, particularly volume data in relation to capacity and volume data for trucks, to determine where such added lanes are warranted.

Climbing Lanes

Climbing Lanes for Two-Lane Highways

General. Freedom and safety of operation on two-lane highways, besides being influenced by the extent and frequency of passing sections, are adversely affected by heavily loaded vehicle traffic operating on grades of sufficient length to result in speeds that could impede following vehicles. In the past, provision of added climbing lanes to improve operations on upgrades has been rather limited because of the additional construction costs involved. However, because of the increasing amount of delay and the number of serious crashes occurring on grades, such lanes are now more commonly included in original construction plans and additional lanes on existing highways are being considered as safety improvement projects. The crash potential created by this condition is illustrated in Exhibit 3-62.

A highway section with a climbing lane is not considered a three-lane highway, but a two-lane highway with an added lane for vehicles moving slowly uphill so that other vehicles using the normal lane to the right of the centerline are not delayed. These faster vehicles pass the slower vehicles moving upgrade, but not in the lane for opposing traffic, as on a conventional two-lane road. A separate climbing lane exclusively for slow-moving vehicles is preferred to the addition of an extra lane carrying mixed traffic. Designs of two-lane highways with climbing lanes are illustrated in Exhibits 3-65A and 3-65B. Climbing lanes are designed for each direction independently of the other. Depending on the alignment and profile conditions, they may not overlap, as in Exhibit 3-65A, or they may overlap, as in Exhibit 3-65B, where there is a crest with a long grade on each side.

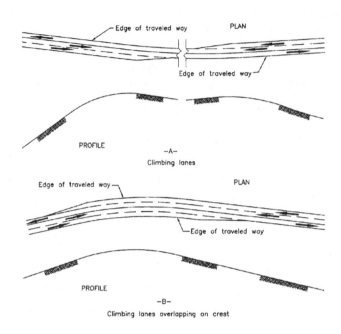

Exhibit 3-65. Climbing Lanes on Two-Lane Highways

It is desirable to provide a climbing lane, as an added lane for the upgrade direction of a two-lane highway where the grade, traffic volume, and heavy vehicle volume combine to degrade traffic operations from those on the approach to the grade. Where climbing lanes are provided there has been a high degree of compliance in their use by truck drivers.

On highways with low volumes, only an occasional car is delayed, and climbing lanes, although desirable, may not be justified economically even where the critical length of grade is exceeded. For such cases, slow-moving vehicle turnouts should be considered to reduce delay to occasional passenger cars from slow-moving vehicles. Turnouts are discussed in the section on "Methods for Increasing Passing Opportunities on Two-Lane Roads" in this chapter.

The following three criteria, reflecting economic considerations, should be satisfied to justify a climbing lane:

1. Upgrade traffic flow rate in excess of 200 vehicles per hour.

2. Upgrade truck flow rate in excess of 20 vehicles per hour.

3. One of the following conditions exists:
    *   A 15 km/h [10 mph] or greater speed reduction is expected for a typical heavy truck.
    *   Level-of-service E or F exists on the grade.

• A reduction of two or more levels of service is experienced when moving from the approach segment to the grade.

In addition, safety considerations may justify the addition of a climbing lane regardless of grade or traffic volumes.

The upgrade flow rate is determined by multiplying the predicted or existing design hour volume by the directional distribution factor for the upgrade direction and dividing the result by the peak hour factor (the peak hour and directional distribution factors are discussed in Chapter 2). The number of upgrade trucks is obtained by multiplying the upgrade flow rate by the percentage of trucks in the upgrade direction.

Trucks. As indicated in the preceding section, only one of the three conditions specified in criterion 3 must be met. The critical length of grade to effect a 15 km/h [10 mph] truck speed reduction is found using Exhibit 3-63. This critical length is compared with the length of the particular grade being evaluated. If the critical length of grade is less than the length of the grade being studied, criterion 3 is satisfied. This evaluation should be done first because, where the critical length of grade is exceeded, no further evaluations under criterion 3 will be needed.

Justification for climbing lanes where the critical length of grade is not exceeded should be considered from the standpoint of highway capacity. The procedures used are those from the HCM (**14**) for analysis of specific grades on two-lane highways. The remaining conditions in criterion 3 are evaluated using these HCM procedures. The effect of trucks on capacity is primarily a function of the difference between the average speed of the trucks and the average running speed of the passenger cars on the highway. Physical dimensions of heavy trucks and their poorer acceleration characteristics also have a bearing on the space they need in the traffic stream.

On individual grades the effect of trucks is more severe than their average effect over a longer section of highway. Thus, for a given volume of mixed traffic and a fixed roadway cross section, a higher degree of congestion is experienced on individual grades than for the average operation over longer sections that include downgrades as well as upgrades. Determination of the design service volume on individual grades should use truck factors derived from the geometrics of the grade and the level of service selected by the highway agency for use as the basis for design of the highway under consideration.

If there is no 15-km/h [10-mph] reduction in speed (i.e., if the critical length of grade is not exceeded), the level of service on the grade should be examined to determine if level-of-service E or F exists. This is done by calculating the limiting service flow rate for level-of-service D and comparing this rate to the actual flow rate on the grade. The actual flow rate is determined by dividing the hourly volume of traffic by the peak hour factor. If the actual flow rate exceeds the service flow rate at level-of-service D, criterion 3 is satisfied. When the actual flow rate is less than the limiting value, a climbing lane is not warranted by this second element of criterion 3.

The remaining issue to examine if neither of the other elements of criterion 3 are satisfied is whether there is a two-level reduction in the level of service between the approach and the upgrade. To evaluate this criterion, the level of service for the grade and the approach segment should both be determined. Since this criterion needs consideration in only a very limited number of cases, it is not discussed in detail here.

The HCM (**14**) provides additional details and worksheets to perform the computations needed for analysis in the preceding criteria. This procedure is also available in computer software, reducing the need for manual calculations.

Because there are so many variables involved that hardly any given set of conditions can be properly described as typical, a detailed analysis such as the one described is recommended wherever climbing lanes are being considered.

The location where an added lane should begin depends on the speeds at which trucks approach the grade and on the extent of sight distance restrictions on the approach. Where there are no sight distance restrictions or other conditions that limit speeds on the approach, the added lane may be introduced on the upgrade beyond its beginning because the speed of trucks will not be reduced beyond the level tolerable to following drivers until they have traveled some distance up the grade. This optimum point for capacity would occur for a reduction in truck speed to 60 km/h [40 mph], but a 15 km/h [10 mph] decrease in truck speed below the average running speed, as discussed in the preceding section on "Critical Lengths of Grade for Design," is the most practical reduction obtainable from the standpoint of level of service and safety. This 15-km/h [10-mph] reduction is the accepted basis for determining the location at which to begin climbing lanes. The distance from the bottom of the grade to the point where truck speeds fall to 15 km/h [10 mph] below the average running speed may be determined from Exhibit 3-59 or Exhibit 3-63. Different curves would apply for trucks with other than a weight/power ratio of 120 kg/kW [200-lb/hp]. For example, assuming an approach condition on which trucks with a 120-kg/kW [200-lb/hp] weight/power ratio are traveling within a flow having an average running speed of 110 km/h [70 mph], the resulting 15-km/h [10-mph] speed reduction occurs at distances of approximately 175 to 350 m [600 to 1,200 ft] for grades varying from 7 to 4 percent. With a downgrade approach, these distances would be longer and, with an upgrade approach, they would be shorter. Distances thus determined may be used to establish the point at which a climbing lane should begin. Where restrictions, upgrade approaches, or other conditions indicate the likelihood of low speeds for approaching trucks, the added lane should be introduced near the foot of the grade. The beginning of the added lane should be preceded by a tapered section with a desirable taper ratio of 25:1 that should be at least 90 m [300 ft] long.

The ideal design is to extend a climbing lane to a point beyond the crest, where a typical truck could attain a speed that is within 15 km/h [10 mph] of the speed of the other vehicles with a desirable speed of at least 60 km/h [40 mph]. This may not be practical in many instances because of the unduly long distance needed for trucks to accelerate to the desired speed. In such situations, a practical point to end the added lane is where trucks can return to the normal lane without undue interference with other traffic—in particular, where the sight distance becomes sufficient to permit passing when there is no oncoming traffic or, preferably, at least 60 m [200 ft] beyond that point. An appropriate taper length should be provided to permit trucks to return smoothly to the normal lane. For example, on a highway where the passing sight distance becomes available 30 m [100 ft] beyond the crest of the grade, the climbing lane should extend

90 m [300 ft] beyond the crest (i.e., 30 m [100 ft] plus 60 m [200 ft]), and an additional tapered section with a desirable taper ratio of 50:1 that should be at least 180 m [600 ft] long.

A climbing lane should desirably be as wide as the through lanes. It should be so constructed that it can immediately be recognized as an added lane for one direction of travel. The centerline of the normal two-lane highway should be clearly marked, including yellow barrier lines for no-passing zones. Signs at the beginning of the upgrade such as "Slower Traffic Keep Right" or "Trucks Use Right Lane" may be used to direct slow-moving vehicles into the climbing lane. These and other appropriate signs and markings for climbing lanes are presented in the MUTCD (**6**).

The cross slope of a climbing lane is usually handled in the same manner as the addition of a lane to a multilane highway. Depending on agency practice, this design results in either a continuation of the cross slope or a lane with slightly more cross slope than the adjacent through lane. On a superelevated section, the cross slope is generally a continuation of the slope used on the through lane.

Desirably, the shoulder on the outer edge of a climbing lane should be as wide as the shoulder on the normal two-lane cross section, particularly where there is bicycle traffic. However, this may be impractical, particularly when the climbing lane is added to an existing highway. A usable shoulder of 1.2 m [4 ft] in width or greater is acceptable. Although not wide enough for a stalled vehicle to completely clear the climbing lane, a 1.2-m [4-ft] shoulder in combination with the climbing lane generally provides sufficient width for both the stalled vehicle and a slow-speed passing vehicle without need for the latter to encroach on the through lane.

In summary, climbing lanes offer a comparatively inexpensive means of overcoming reductions in capacity and providing improved operation where congestion on grades is caused by slow trucks in combination with high traffic volumes. As discussed earlier in this section, climbing lanes also improve safety. On some existing two-lane highways, the addition of climbing lanes could defer reconstruction for many years or indefinitely. In a new design, climbing lanes could make a two-lane highway operate efficiently, whereas a much more costly multilane highway would be needed without them.

Climbing Lanes on Freeways and Multilane Highways

General. Climbing lanes, although they are becoming more prevalent, have not been used as extensively on freeways and multilane highways as on two-lane highways, perhaps for the reason that multilane facilities more frequently have sufficient capacity to handle their traffic demands, including the typical percentage of slow-moving vehicles with high weight/power ratios, without being congested. Climbing lanes are generally not as easily justified on multilane facilities as on two-lane highways because, on two-lane facilities, vehicles following other slower moving vehicles on upgrades are frequently prevented by opposing traffic from using the adjacent traffic lane for passing, whereas there is no such impediment to passing on multilane facilities. A slow-moving vehicle in the normal right lane does not impede the following vehicles that can readily move left to the adjacent lane and proceed without difficulty, although there is evidence that safety is enhanced when vehicles in the traffic stream move at the same speed.

Because highways are normally designed for 20 years or more in the future, there is less likelihood that climbing lanes will be justified on multilane facilities than on two-lane roads for several years after construction even though they are deemed desirable for the peak hours of the design year. Where this is the case, there is economic advantage in designing for, but deferring construction of, climbing lanes on multilane facilities. In this situation, grading for the future climbing lane should be provided initially. The additional grading needed for a climbing lane is small when compared to that needed for the overall cross section. If, however, even this additional grading is impractical, it is acceptable, although not desirable, to use a narrower shoulder adjacent to the climbing lane rather than the full shoulder provided on a normal section.

Although primarily applicable in rural areas, there are instances where climbing lanes are need in urban areas. Climbing lanes are particularly important for freedom of operation on urban freeways where traffic volumes are high in relation to capacity. On older urban freeways and arterial streets with appreciable grades and no climbing lanes, it is a common occurrence for heavy traffic, which may otherwise operate well, to platoon on grades.

Trucks. The principal determinants of the need for climbing lanes on multilane highways are critical lengths of grade, effects of trucks on grades in terms of equivalent passenger-car flow rates, and service volumes for the desired level of service and the next poorer level of service.

Critical length of grade has been discussed previously in this chapter. It is the length of a particular upgrade that reduces the speed of low-performance trucks 15 km/h [10 mph] below the average running speed of the remaining traffic. The critical length of grade that results in a 15-km/h [10-mph] truck speed reduction is found using Exhibit 3-63 and is then compared to the length of the particular grade being examined. If the critical length of grade is less than the length of grade being evaluated, consideration of a climbing lane is warranted.

In determining service volume, the passenger-car equivalent for trucks is a significant factor. It is generally agreed that trucks on multilane facilities have less effect in deterring following vehicles than on two-lane roads. Comparison of passenger-car equivalents in the HCM (**14**) for the same percent of grade, length of grade, and percent of trucks clearly illustrates the difference in passenger-car equivalents of trucks for two-lane and multilane facilities.

To justify the cost of providing a climbing lane, the existence of a low level of service on the grade should be the criterion, as in the case of justifying climbing lanes for two-lane roads, because highway users will accept a higher degree of congestion (i.e., a lower level of service) on individual grades than over long sections of highway. As a matter of practice, the service volume on an individual grade should not exceed that for the next poorer level of service from that used for the basic design. The one exception is that the service volume for level-of-service D should not be exceeded.

Generally, climbing lanes should not be considered unless the directional traffic volume for the upgrade is equal to or greater than the service volume for level-of-service D. In most cases when the service volume, including trucks, is greater than 1700 vehicles per hour per lane and the length of the grade and the percentage of trucks are sufficient to consider climbing lanes, the volume in terms of equivalent passenger cars is likely to approach or even exceed the capacity. In this situation, an increase in the number of lanes throughout the highway section would represent a better investment than the provision of climbing lanes.

Climbing lanes are also not generally warranted on four-lane highways with directional volumes below 1000 vehicles per hour per lane regardless of the percentage of trucks. Although a truck driver will occasionally pass another truck under such conditions, the inconvenience with this low volume is not sufficient to justify the cost of a climbing lane in the absence of appropriate criteria.

The procedures in the HCM (**14**) should be used to consider the traffic operational characteristics on the grade being examined. The maximum service flow rate for the desired level of service, together with the flow rate for the next poorer level of service, should be determined. If the flow rate on the grade exceeds the service flow rate of the next poorer level of service, consideration of a climbing lane is warranted. In order to use the HCM procedures, the free-flow speed must be determined or estimated. The free-flow speed can be determined by measuring the mean speed of passenger cars under low to moderate flow conditions (up to 1300 passenger cars per hour per lane) on the facility or similar facility.

Recent data (**14, 41**) indicates that the mean free-flow speed under ideal conditions for multilane highways ranges from 0.6 km/h [1 mph] lower than the 85th percentile speed of 65 km/h [40 mph] to 5 km/h [3 mph] lower than the 85th percentile speed of 100 km/h [60 mph]. Speed limit is one factor that affects free-flow speed. Recent research (**14, 41**) suggests that the free-flow speed is approximately 11 km/h [7 mph] higher than the speed limit on facilities with 65- and 70-km/h [40- and 45-mph] speed limits and 8 km/h [5 mph] higher than the speed limit on facilities with 80- and 90-km/h [50- and 55-mph] speed limits. Analysis based on these rules of thumb should be used with caution. Field measurement is the recommended method of determining the free-flow speed, with estimation using the above procedures employed only when field data are not available.

Where the grade being investigated is located on a multilane highway, other factors should sometimes be considered; such factors include median type, lane widths, lateral clearance, and access point density. These factors are accounted for in the capacity analysis procedures by making adjustments in the free-flow speed and are not normally a separate consideration in determining whether a climbing lane would be advantageous.

For freeways, adjustments are made in traffic operational analyses using factors for restricted lane widths, lateral clearances, recreational vehicles, and unfamiliar driver populations. The HCM (**14**) should be used for information on considering these factors in analysis.

Under certain circumstances there should be consideration of additional lanes to accommodate trucks in the downgrade direction. This is accomplished using the same procedure as described above and using the passenger-car equivalents for trucks on downgrades in place of the values for trucks and recreational vehicles on upgrades.

Climbing lanes on multilane roads are usually placed on the outer or right-hand side of the roadway as shown in Exhibit 3-66. The principles for cross slopes, for locating terminal points, and for designing terminal areas or tapers for climbing lanes are discussed earlier in this chapter in conjunction with two-lane highways; these principles are equally applicable to climbing lanes on multilane facilities. A primary consideration is that the location of the uphill terminus of the climbing lane should be at the point where a satisfactory speed is attained by trucks, preferably about 15 km/h [10 mph] below the average running speed of the highway. Passing sight distance need not be considered on multilane highways.

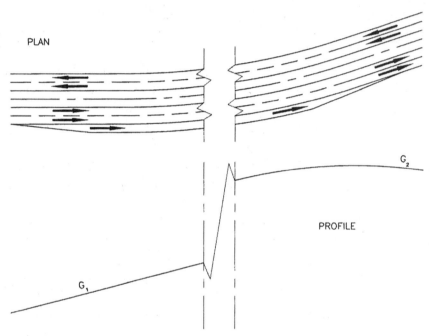

Exhibit 3-66. Climbing Lane on Freeways and Multilane Highways

Methods for Increasing Passing Opportunities on Two-Lane Roads

Several highway agencies have pioneered successful methods for providing more passing opportunities along two-lane roads. Some of the more recognized of these methods, including passing lanes, turnouts, shoulder driving, and shoulder use sections are described in the FHWA informational guide *Low Cost Methods for Improving Traffic Operations on Two-Lane Roads* (**42**). A synopsis of portions of material found in this guide pertaining to these designs is presented in the succeeding sections. More detailed criteria for these methods are found in the guide.

Passing Lanes

An added lane can be provided in one or both directions of travel to improve traffic operations in sections of lower capacity to at least the same quality of service as adjacent road sections. Passing lanes can also be provided to improve overall traffic operations on two-lane highways by reducing delays caused by inadequate passing opportunities over significant lengths of highways, typically 10 to 100 km [6 to 60 miles]. Where passing lanes are used to improve traffic operations over a length of road, they frequently are provided systematically at regular intervals.

The location of the added lane should appear logical to the driver. The value of a passing lane is more obvious at locations where passing sight distance is restricted than on long tangents that may provide passing opportunities even without passing lanes. On the other hand, the location of a passing lane should recognize the need for adequate sight distance at both the lane addition and lane drop tapers. A minimum sight distance of 300 m [1,000 ft] on the approach to each taper is recommended. The selection of an appropriate location also needs to consider the location of intersections and high-volume driveways in order to minimize the volume of turning movements on a road section where passing is encouraged. Furthermore, other physical constraints such as bridges and culverts should be avoided if they restrict provision of a continuous shoulder.

The following is a summary of the design procedure to be followed in providing passing sections on two-lane highways:

1. Horizontal and vertical alignment should be designed to provide as much of the highway as practical with passing sight distance (see Exhibit 3-7).
2. Where the design volume approaches capacity, the effect of lack of passing opportunities in reducing the level of service should be recognized.
3. Where the critical length of grade is less than the physical length of an upgrade, consideration should be given to constructing added climbing lanes. The critical length of grade is determined as shown in Exhibits 3-63 and 3-64.
4. Where the extent and frequency of passing opportunities made available by application of Criteria 1 and 3 are still too few, consideration should be given to the construction of passing lane sections.

Passing-lane sections, which may be either three or four lanes in width, are constructed on two-lane roads to provide the desired frequency of passing zones or to eliminate interference from low-speed heavy vehicles, or both. Where a sufficient number and length of passing sections cannot be obtained in the design of horizontal and vertical alignment alone, an occasional added lane in one or both directions of travel may be introduced as shown in Exhibit 3-67 to provide more passing opportunities. Such sections are particularly advantageous in rolling terrain, especially where alignment is winding or the profile includes critical lengths of grade.

In rolling terrain a highway on tangent alignment may have restricted passing conditions even though the grades are below critical length. Use of passing lanes over some of the crests provides added passing sections in both directions where they are most needed. Passing-lane sections should be sufficiently long to permit several vehicles in line behind a slow-moving vehicle to pass before returning to the normal cross section of two-lane highway

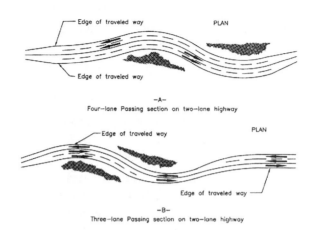

Exhibit 3-67. Passing Lanes Section on Two-Lane Roads

A minimum length of 300 m [1,000 ft], excluding tapers, is needed to assure that delayed vehicles have an opportunity to complete at least one pass in the added lane. Where such a lane is provided to reduce delays at a specific bottleneck, the needed length is controlled by the extent of the bottleneck. A lane added to improve overall traffic operations should be long enough, over 0.5 km [0.3 mi], to provide a substantial reduction in traffic platooning. The optimal length is usually 0.8 to 3.2 km [0.5 to 2.0 mi], with longer lengths of added lane appropriate where traffic volumes are higher. The HCM (**14**) provides guidance in the selection of a passing lane of optimal length. Operational benefits typically result in reduced platooning for 5 to 15 km [3 to 10 miles] downstream depending on volumes and passing opportunities. After that, normal levels of platooning will occur until the next added lane is encountered.

The introduction of a passing-lane section on a two-lane highway does not necessarily involve much additional grading. The width of an added lane should normally be the same as the lane widths of the two-lane highway. It is also desirable for the adjoining shoulder to be at least 1.2 m [4 ft] wide and, whenever practical, the shoulder width in the added section should match that of the adjoining two-lane highway. However, a full shoulder width is not as needed on a passing lane section as on a conventional two-lane highway because the vehicles likely to stop are few and there is little difficulty in passing a vehicle with only two wheels on the shoulder. Thus, if the normal shoulder width on the two-lane highway is 3.0 m [10 ft], a 1.8- to 2.4-m [6- to 8-ft] widening of the roadbed on each side is all that may be needed.

Four-lane sections introduced explicitly to improve passing opportunities need not be divided because there is no separation of opposing traffic on the two-lane portions of the highway. The use of a median, however, is advantageous and should be considered on highways carrying a total of 500 vehicles per hour or more, particularly on highways to be ultimately converted to a four-lane divided cross section.

The transition tapers at each end of the added-lane section should be designed to encourage safe and efficient operation. The lane drop taper length should be computed from the MUTCD (**6**) formula L=0.6WS (L=Length in meters, W=Width in meters, S=Speed in km/h) or L=WS

[L=Length in ft, W=Width in ft, S=Speed in mph] while the recommended length for the lane addition taper is half to two-thirds of the lane drop length.

The signing and marking of an added lane is partially addressed in the MUTCD (**6**), which indicates the appropriate centerline markings for such lanes as well as the signing and marking of lane drop transitions. However, the MUTCD (**6**) does not address signing in advance of and at the lane addition. A sign with the legend "Passing Lane 1 Kilometer" ["Passing Lane 1/2 Mile"] should be placed in advance of each added lane in order that drivers of both slow-moving vehicles and following vehicles can prepare to make effective use of the added lane. Additional signs 3 to 10 km [2 to 5 mi] in advance are also desirable because they may reduce the frustration and impatience of drivers following a slow-moving vehicle by assuring them that they will soon have an opportunity to pass. In addition, a sign should be installed at the beginning of the lane addition taper to encourage slower-moving vehicles to keep right.

The transitions between the two- and three- or four-lane pavements should be located where the change in width is in full view of the driver. Sections of four-lane highway, particularly divided sections, longer than about 3 km [2 mi] may cause the driver to lose his sense of awareness that the highway is basically a two-lane facility. It is essential, therefore, that transitions from a three- or four-lane cross section back to two lanes be properly marked and identified with pavement markings and signs to alert the driver of the upcoming section of two-lane highway. An advance sign before the end of the passing lane is particularly important to inform drivers of the narrower roadway ahead; for more information, see the MUTCD (**6**).

Turnouts

A turnout is a widened, unobstructed shoulder area that allows slow-moving vehicles to pull out of the through lane to give passing opportunities to following vehicles (**42, 43**). The driver of the slow-moving vehicle, if there are following vehicles, is expected to pull out of the through lane and remain in the turnout only long enough for the following vehicles to pass before returning to the through lane. When there are only one or two following vehicles, this maneuver can be accomplished without it being necessary for the driver of the vehicle in the turnout to stop. However, when this number is exceeded, the driver may need to stop in the turnout in order for all the following vehicles to pass. Turnouts are most frequently used on lower volume roads where long platoons are rare and in difficult terrain with steep grades where construction of an additional lane may not be cost effective. Such conditions are often found in mountain, coastal, and scenic areas where more than 10 percent of the vehicle volumes are large trucks and recreational vehicles.

The recommended length of turnouts including taper is shown in Exhibit 3-68. Turnouts shorter than 60 m [200 ft] are not recommended even for very low approach speeds. Turnouts longer than 185 m [600 ft] are not recommended for high-speed roads to avoid use of the turnout as a passing lane. The recommended lengths are based on the assumption that slow-moving vehicles enter the turnout at 8 km/h [5 mph] slower than the mean speed of the through traffic. This length allows the entering vehicle to coast to the midpoint of the turnout without braking, and then, if necessary, to brake to a stop using a deceleration rate not exceeding 3 m/s^2 [10 ft/s^2].

The recommended lengths for turnouts include entry and exit tapers. Typical entry and exit taper lengths range from 15 to 30 m [50 to 100 ft] (**42, 43**).

Metric		US Customary	
Approach speed (km/h)	Minimum length (m)[a]	Approach speed (mph)	Minimum length (ft)[a]
30	60	20	200
40	60	30	200
50	65	40	300
60	85	45	350
70	105	50	450
80	135	55	550
90	170	60	600
100	185		
[a] Maximum length should be 185 m (600 ft) to avoid use of the turnout as a passing lane.			

Exhibit 3-68. Recommended Lengths of Turnouts Including Taper

The minimum width of the turnout is 3.6 m [12 ft] with widths of 5 m [16 ft] considered desirable. Turnouts wider than 5 m [16 ft] are not recommended.

A turnout should not be located on or adjacent to a horizontal or vertical curve that limits sight distance in either direction. The available sight distance should be at least 300 m [1,000 ft] on the approach to the turnout.

Proper signing and pavement marking are also needed both to maximize turnout usage and assure safe operation. An edge line marking on the right side of the turnout is desirable to guide drivers, especially in wider turnouts.

Shoulder Driving

In parts of the United States, a long-standing custom has been established for slow-moving vehicles to move to the shoulder when another vehicle approaches from the rear and return to the traveled way after that following vehicle has passed. The practice generally occurs where adequate paved shoulders exist and, in effect, these shoulders function as continuous turnouts. This custom is regarded as a courtesy to other drivers requiring little or no sacrifice in speed by either driver. While highway agencies may want to permit such use as a means of improving passing opportunities without a major capital investment, they should recognize that in many States shoulder driving is currently prohibited by law. Thus, a highway agency considering shoulder driving as a passing aid may need to propose legislation to authorize such use as well as develop a public education campaign to familiarize drivers with the new law.

Highway agencies should evaluate the mileage of two-lane highways with paved shoulders as well as their structural quality before deciding whether to allow their use as a passing aid. It should be recognized that, where shoulder driving becomes common, it will not be limited to selected sites but rather will occur anywhere on the system where paved shoulders are provided.

Another consideration is that shoulder widths of at least 3.0 m [10 ft], and preferably 3.6 m [12 ft], are needed. The effect that shoulder driving may have on the use of the highway by bicyclists should also considered. Because the practice of shoulder driving has grown up through local custom, no special signing to promote such use has been created.

Shoulder Use Sections

Another approach to providing additional passing opportunities is to permit slow-moving vehicles to use paved shoulders at selected sites designated by specific signing. This is a more limited application of shoulder use by slow-moving vehicles than shoulder driving described in the previous section. Typically, drivers move to the shoulder only long enough for following vehicles to pass and then return to the through lane. Thus, the shoulder-use section functions as an extended turnout. This approach enables a highway agency to promote shoulder use only where the shoulder is adequate to handle anticipated traffic loads and the need for more frequent passing opportunities has been established by the large amount of vehicle platooning.

Shoulder use sections generally range in length from 0.3 to 5 km [0.2 to 3 mi]. Shoulder use should be allowed only where shoulders are at least 3.0 m [10 ft] and preferably 3.6 m [12 ft] wide. Adequate structural strength to support the anticipated loads along with good surface conditions are needed. Particular attention needs to be placed on the condition of the shoulder because drivers are unlikely to use a shoulder if it is rough, broken, or covered with debris. Signs should be erected at both the beginning and end of the section where shoulder use is allowed. However, since signing of shoulder-use sections is not addressed in the MUTCD (**6**), special signing should be used.

Emergency Escape Ramps

General

Where long, descending grades exist or where topographic and location controls require such grades on new alignment, the design and construction of an emergency escape ramp at an appropriate location is desirable to provide a location for out-of-control vehicles, particularly trucks, to slow and stop away from the main traffic stream. Out-of-control vehicles are generally the result of a driver losing braking ability either through overheating of the brakes due to mechanical failure or failure to downshift at the appropriate time. Considerable experience with ramps constructed on existing highways has led to the design and installation of effective ramps that save lives and reduce property damage. Reports and evaluations of existing ramps indicate that they provide acceptable deceleration rates and afford good driver control of the vehicle on the ramp (**44**).

Forces that act on every vehicle to affect the vehicle's speed include engine, braking, and tractive resistance forces. Engine and braking resistance forces can be ignored in the design of escape ramps because the ramp should be designed for the worst case, in which the vehicle is out of gear and the brake system has failed. The tractive resistance force contains four subclasses: inertial, aerodynamic, rolling, and gradient. Inertial and negative gradient forces act to maintain motion of the vehicle, while rolling, positive gradient, and air resistance forces act to retard its motion. Exhibit 3-69 illustrates the action of the various resistance forces on a vehicle.

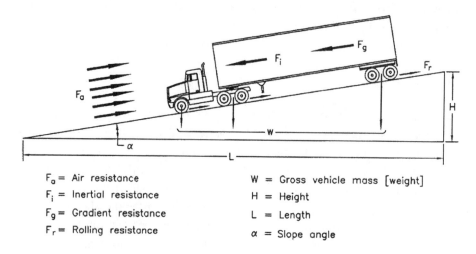

F_a = Air resistance   W = Gross vehicle mass [weight]
F_i = Inertial resistance   H = Height
F_g = Gradient resistance   L = Length
F_r = Rolling resistance   α = Slope angle

Exhibit 3-69. Forces Acting on a Vehicle in Motion

Inertial resistance can be described as a force that resists movement of a vehicle at rest or maintains a vehicle in motion, unless the vehicle is acted on by some external force. Inertial resistance must be overcome to either increase or decrease the speed of a vehicle. Rolling and positive gradient resistance forces are available to overcome the inertial resistance. Rolling resistance is a general term used to describe the resistance to motion at the area of contact between a vehicle's tires and the roadway surface and is only applicable when a vehicle is in motion. It is influenced by the type and displacement characteristics of the surfacing material of the roadway. Each surfacing material has a coefficient, expressed in kg/1,000 kg [lb/1,000 lb] of gross vehicle weight, which determines the amount of rolling resistance of a vehicle. The values shown in Exhibit 3-70 for rolling resistance have been obtained from various sources throughout the country and are a best available estimate.

Surfacing material	Metric		US Customary	
	Rolling resistance (kg/1000 kg GVM)	Equivalent grade (%)[a]	Rolling resistance (lb/1000 lb GVW)	Equivalent grade (%)[a]
Portland cement concrete	10	1.0	10	1.0
Asphalt concrete	12	1.2	12	1.2
Gravel, compacted	15	1.5	15	1.5
Earth, sandy, loose	37	3.7	37	3.7
Crushed aggregate, loose	50	5.0	50	5.0
Gravel, loose	100	10.0	100	10.0
Sand	150	15.0	150	15.0
Pea gravel	250	25.0	250	25.0
[a] Rolling resistance expressed as equivalent gradient.				

Exhibit 3-70. Rolling Resistance of Roadway Surfacing Materials

Gradient resistance is due to the effect of gravity and is expressed as the force needed to move the vehicle through a given vertical distance. For gradient resistance to provide a beneficial force on an escape ramp, the vehicle must be moving upgrade, against gravity. In the case where the vehicle is descending a grade, gradient resistance is negative, thereby reducing the forces available to slow and stop the vehicle. The amount of gradient resistance is influenced by the total weight of the vehicle and the magnitude of the grade. For each percent of grade, the gradient resistance is 10 kg/1,000 kg [10 lb/1,000 lb] whether the grade is positive or negative.

The remaining component of tractive resistance is aerodynamic resistance, the force resulting from the retarding effect of air on the various surfaces of the vehicle. Air causes a significant resistance at speeds above 80 km/h [50 mph], but is negligible under 30 km/h [20 mph]. The effect of aerodynamic resistance has been neglected in determining the length of the arrester bed, thus introducing a small safety factor.

Need and Location for Emergency Escape Ramps

Each grade has its own unique characteristics. Highway alignment, gradient, length, and descent speed contribute to the potential for out-of-control vehicles. For existing highways, operational problems on a downgrade will often be reported by law enforcement officials, truck drivers, or the general public. A field review of a specific grade may reveal damaged guardrail, gouged pavement surfaces, or spilled oil indicating locations where drivers of heavy vehicles had difficulty negotiating a downgrade. For existing facilities an escape ramp should be provided as soon as a need is established. Crash experience (or, for new facilities, crash experience on similar facilities) and truck operations on the grade combined with engineering judgment are frequently used to determine the need for a truck escape ramp. Often the impact of a potential runaway truck on adjacent activities or population centers will provide sufficient reason to construct an escape ramp.

Unnecessary escape ramps should be avoided. For example, a second escape ramp should not be needed just beyond the curve that created the need for the initial ramp.

While there are no universal guidelines available for new and existing facilities, a variety of factors should be considered in selecting the specific site for an escape ramp. Each location presents a different array of design needs; factors that should be considered include topography, length and percent of grade, potential speed, economics, environmental impact, and crash experience. Ramps should be located to intercept the greatest number of runaway vehicles, such as at the bottom of the grade and at intermediate points along the grade where an out-of-control vehicle could cause a catastrophic crash.

A technique for new and existing facilities available for use in analyzing operations on a grade, in addition to crash analysis, is the *Grade Severity Rating System* (**45**). The system uses a predetermined brake temperature limit (260°C [500°F]) to establish a safe descent speed for the grade. It also can be used to determine expected brake temperatures at 0.8 km [0.5 mi] intervals along the downgrade. The location where brake temperatures exceed the limit indicates the point that brake failures can occur, leading to potential runaways.

Escape ramps generally may be built at any practical location where the main road alignment is tangent. They should be built in advance of horizontal curves that cannot be negotiated safely by an out-of-control vehicle and in advance of populated areas. Escape ramps should exit to the right of the roadway. On divided multilane highways, where a left exit may appear to be the only practical location, difficulties may be expected by the refusal of vehicles in the left lane to yield to out-of-control vehicles attempting to change lanes.

Although crashes involving runaway trucks can occur at various sites along a grade, locations having multiple crashes should be analyzed in detail. Analysis of crash data pertinent to a prospective escape ramp site should include evaluation of the section of highway immediately uphill, including the amount of curvature traversed and distance to and radius of the adjacent curve.

An integral part of the evaluation should be the determination of the maximum speed that an out-of-control vehicle could attain at the proposed site. This highest obtainable speed can then be used as the minimum design speed for the ramp. The 130- to 140-km/h [80- to 90-mph] entering speed, recommended for design, is intended to represent an extreme condition and therefore should not be used as the basis for selecting locations of escape ramps. Although the variables involved make it impractical to establish a maximum truck speed warrant for location of escape ramps, it is evident that anticipated speeds should be below the range used for design. The principal factor in determining the need for an emergency escape ramp should be the safety of the other traffic on the roadway, the driver of the out-of-control vehicle, and the residents along and at the bottom of the grade. An escape ramp, or ramps if the conditions indicate the need for more than one, should be located wherever grades are of a steepness and length that present a substantial risk of runaway trucks and topographic conditions will permit construction.

Types of Emergency Escape Ramps

Emergency escape ramps have been classified in a variety of ways. Three broad categories used to classify ramps are gravity, sandpile, and arrester bed. Within these broad categories, four basic emergency escape ramp designs predominate. These designs are the sandpile and three types of arrester beds, classified by grade of the arrester bed: descending grade, horizontal grade, and ascending grade. These four types are illustrated in Exhibit 3-71.

The gravity ramp has a paved or densely compacted aggregate surface, relying primarily on gravitational forces to slow and stop the runaway. Rolling resistance forces contribute little to assist in stopping the vehicle. Gravity ramps are usually long, steep, and are constrained by topographic controls and costs. While a gravity ramp stops forward motion, the paved surface cannot prevent the vehicle from rolling back down the ramp grade and jackknifing without a positive capture mechanism. Therefore, the gravity ramp is the least desirable of the escape ramp types.

Sandpiles, composed of loose, dry sand dumped at the ramp site, are usually no more than 120 m [400 ft] in length. The influence of gravity is dependent on the slope of the surface. The

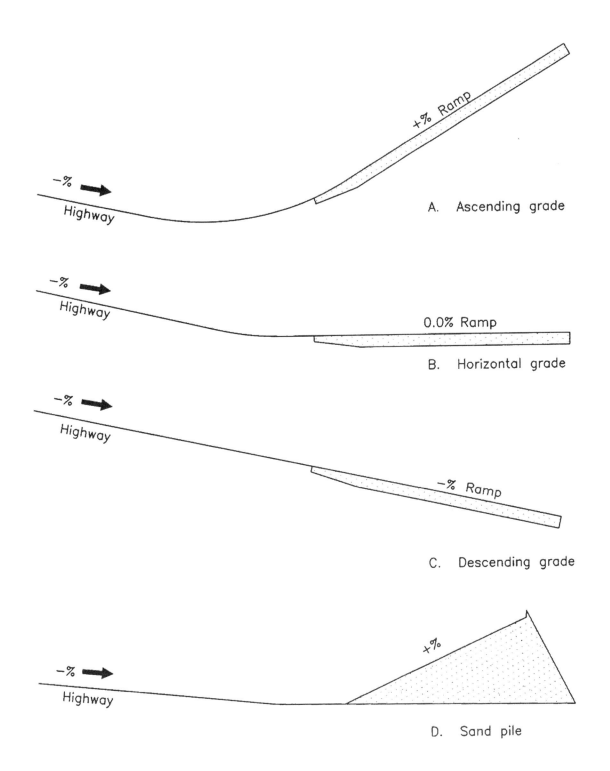

Note: Profile is along the baseline of the ramp.

Exhibit 3-71. Basic Types of Emergency Escape Ramps

increase in rolling resistance is supplied by loose sand. Deceleration characteristics of sandpiles are usually severe and the sand can be affected by weather. Because of the deceleration characteristics, the sandpile is less desirable than the arrester bed. However, at locations where inadequate space exists for another type of ramp, the sandpile may be appropriate because of its compact dimensions.

Descending-grade arrester-bed escape ramps are constructed parallel and adjacent to the through lanes of the highway. These ramps use loose aggregate in an arrester bed to increase rolling resistance to slow the vehicle. The gradient resistance acts in the direction of vehicle movement. As a result, the descending-grade ramps can be rather lengthy because the gravitational effect is not acting to help reduce the speed of the vehicle. The ramp should have a clear, obvious return path to the highway so drivers who doubt the effectiveness of the ramp will feel they will be able to return to the highway at a reduced speed.

Where the topography can accommodate, a horizontal-grade arrester-bed escape ramp is another option. Constructed on an essentially flat gradient, the horizontal-grade ramp relies on the increased rolling resistance from the loose aggregate in an arrester bed to slow and stop the out-of-control vehicle, since the effect of gravity is minimal. This type of ramp is longer than the ascending-grade arrester bed.

The most commonly used escape ramp is the ascending-grade arrester bed. Ramp installations of this type use gradient resistance to advantage, supplementing the effects of the aggregate in the arrester bed, and generally, reducing the length of ramp needed to stop the vehicle. The loose material in the arresting bed increases the rolling resistance, as in the other types of ramps, while the gradient resistance acts downgrade, opposite to the vehicle movement. The loose bedding material also serves to hold the vehicle in place on the ramp grade after it has come to a safe stop.

Each of the ramp types is applicable to a particular situation where an emergency escape ramp is desirable and should be compatible with established location and topographic controls at possible sites. The procedures used for analysis of truck escape ramps are essentially the same for each of the categories or types identified. The rolling resistance factor for the surfacing material used in determining the length needed to slow and stop the runaway safely is the difference in the procedures.

Design Considerations

The combination of the above external resistance and numerous internal resistance forces not discussed acts to limit the maximum speed of an out-of-control vehicle. Speeds in excess of 130 to 140 km/h [80 to 90 mph] will rarely, if ever, be attained. Therefore, an escape ramp should be designed for a minimum entering speed of 130 km/h [80 mph], with a 140-km/h [90-mph] design speed being preferred. Several formulas and software programs have been developed to determine the runaway speed at any point on the grade. These methods can be used to establish a design speed for specific grades and horizontal alignments (**44, 45, 46**).

The design and construction of effective escape ramps involve a number of considerations as follows:

- To safely stop an out-of-control vehicle, the length of the ramp should be sufficient to dissipate the kinetic energy of the moving vehicle.
- The alignment of the escape ramp should be tangent or on very flat curvature to minimize the driver's difficulty in controlling the vehicle.
- The width of the ramp should be adequate to accommodate more than one vehicle because it is not uncommon for two or more vehicles to have need of the escape ramp within a short time. A minimum width of 8 m [26 ft] may be all that is practical in some areas, though greater widths are preferred. Desirably, a width of 9 to 12 m [30 to 40 ft] would more adequately accommodate two or more out-of-control vehicles. Ramp widths less than indicated above have been used successfully in some locations where it was determined that a wider width was unreasonably costly or not needed. Widths of ramps in use range from 3.6 to 12 m [12 to 40 ft].
- The surfacing material used in the arrester bed should be clean, not easily compacted, and have a high coefficient of rolling resistance. When aggregate is used, it should be rounded, uncrushed, predominantly a single size, and as free from fine-size material as practical. Such material will maximize the percentage of voids, thereby providing optimum drainage and minimizing interlocking and compaction. A material with a low shear strength is desirable to permit penetration of the tires. The durability of the aggregate should be evaluated using an appropriate crush test. Pea gravel is representative of the material used most frequently, although loose gravel and sand are also used. A gradation with a top size of 40 mm [1.5 in] has been used with success in several States. Material conforming to the AASHTO gradation No. 57 is effective if the fine-sized materials is removed.
- Arrester beds should be constructed with a minimum aggregate depth of 1 m [3 ft]. Contamination of the bed material can reduce the effectiveness of the arrester bed by creating a hard surface layer up to 300 mm [12 in] thick at the bottom of the bed. Therefore, an aggregate depth up to 1,100 mm [42 in] is recommended. As the vehicle enters the arrester bed, the wheels of the vehicle displace the surface, sinking into the bed material, thus increasing the rolling resistance. To assist in decelerating the vehicle smoothly, the depth of the bed should be tapered from a minimum of 75 mm [3 in] at the entry point to the full depth of aggregate in the initial 30 to 60 m [100 to 200 ft] of the bed.
- A positive means of draining the arrester bed should be provided to help protect the bed from freezing and avoid contamination of the arrester bed material. This can be accomplished by grading the base to drain, intercepting water prior to entering the bed, underdrain systems with transverse outlets or edge drains. Geotextiles or paving can be used between the subbase and the bed materials to prevent infiltration of fine materials that may trap water. Where toxic contamination from diesel fuel or other material spillage is a concern, the base of the arrester bed may be paved with concrete and holding tanks to retain the spilled contaminants may be provided.
- The entrance to the ramp should be designed so that a vehicle traveling at a high rate of speed can enter safely. As much sight distance as practical should be provided preceding the ramp so that a driver can enter safely. The full length of the ramp should

be visible to the driver. The angle of departure for the ramp should be small, usually 5 degrees or less. An auxiliary lane may be appropriate to assist the driver to prepare to enter the escape ramp. The main roadway surface should be extended to a point at or beyond the exit gore so that both front wheels of the out-of-control vehicle will enter the arrester bed simultaneously; this also provides preparation time for the driver before actual deceleration begins. The arrester bed should be offset laterally from the through lanes by an amount sufficient to preclude loose material being thrown onto the through lanes.

- Access to the ramp should be made obvious by exit signing to allow the driver of an out-of-control vehicle time to react, so as to minimize the possibility of missing the ramp. Advance signing is needed to inform drivers of the existence of an escape ramp and to prepare drivers well in advance of the decision point so that they will have enough time to decide whether or not to use the escape ramp. Regulatory signs near the entrance should be used to discourage other motorists from entering, stopping, or parking at or on the ramp. The path of the ramp should be delineated to define ramp edges and provide nighttime direction; for more information, see the MUTCD (**6**). Illumination of the approach and ramp is desirable.

- The characteristic that makes a truck escape ramp an effective safety device also makes it difficult to retrieve a vehicle captured by the ramp. A service road located adjacent to the arrester bed is needed so tow trucks and maintenance vehicles can use it without becoming trapped in the bedding material. The width of this service road should be at least 3 m [10 ft]. Preferably this service road should be paved but may be surfaced with gravel. The road should be designed such that the driver of an out-of-control vehicle will not mistake the service road for the arrester bed.

- Anchors, usually located adjacent to the arrester bed at 50 to 100 m [150 to 300 ft] intervals, are needed to secure a tow truck when removing a vehicle from the arrester bed. One anchor should be located about 30 m [100 ft] in advance of the bed to assist the wrecker in returning a captured vehicle to a surfaced roadway. The local tow-truck operators can be very helpful in properly locating the anchors.

As a vehicle rolls upgrade, it loses momentum and will eventually stop because of the effect of gravity. To determine the distance needed to bring the vehicle to a stop with consideration of the rolling resistance and gradient resistance, the following simplified equation may be used (**33**):

Metric	**US Customary**	
$$L=\frac{V^2}{254(R\pm G)}$$	$$L=\frac{V^2}{30(R\pm G)}$$	(3-41)
where: L = length of arrester bed, m; V = entering velocity, km/h; G = percent grade divided by 100; R = rolling resistance, expressed as equivalent percent gradient divided by 100 (see Exhibit 3-70)	where: L = length of arrester bed, ft; V = entering velocity, mph; G = percent grade divided by 100; R = rolling resistance, expressed as equivalent percent gradient divided by 100 (see Exhibit 3-70)	

For example, assume that topographic conditions at a site selected for an emergency escape ramp limit the ramp to an upgrade of 10 percent (G =+ 0.10). The arrester bed is to be constructed with loose gravel for an entering speed of 140 km/h [90 mph]. Using Exhibit 3-70, R is determined to be 0.10. The length of the arrester bed should be determined using the Equation (3-41). For this example, the length of the arrester bed is about 400 m [1,350 ft].

When an arrester bed is constructed using more than one grade along its length, such as shown in Exhibit 3-72, the speed loss occurring on each of the grades as the vehicle traverses the bed should be determined using the following equation:

Metric	US Customary	
$$V_f^2 = V_i^2 - 254\,L(R \pm G)$$	$$V_f^2 = V_i^2 - 30\,L(R \pm G)$$	(3-42)
where: Vf = speed at end of grade, km/h; Vi = entering speed at beginning of grade, km/h; L = length of grade, m; R = rolling resistance, expressed as equivalent percent gradient divided by 100 (see Exhibit 3-70); G = percent grade divided by 100	where: Vf = speed at end of grade, mph; Vi = entering speed at beginning of grade, mph; L = length of grade, ft; R = rolling resistance, expressed as equivalent percent gradient divided by 100 (see Exhibit 3-70); G = percent grade divided by 100	

The final speed for one section of the ramp is subtracted from the entering speed to determine a new entering speed for the next section of the ramp and the calculation repeated at each change in grade on the ramp until sufficient length is provided to reduce the speed of the out-of-control vehicle to zero.

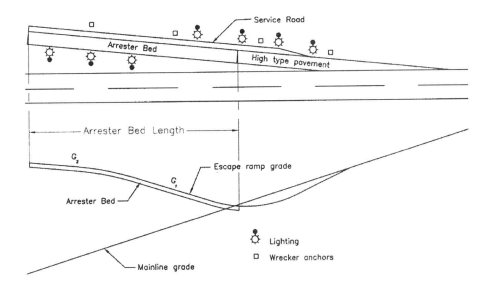

Exhibit 3-72. Typical Emergency Escape Ramp

Exhibit 3-72 shows a plan and profile of an emergency escape ramp with typical appurtenances.

Where the only practical location for an escape ramp will not provide sufficient length and grade to completely stop an out-of-control vehicle, it should be supplemented with an acceptable positive attenuation device.

Where a full-length ramp is to be provided with full deceleration capability for the design speed, a "last-chance" device should be considered when the consequences of leaving the end of the ramp are serious.

Any ramp-end treatment should be designed with care to ensure that its advantages outweigh the disadvantages. The risk to others as the result of an out-of-control truck overrunning the end of an escape ramp may be more important than the harm to the driver or cargo of the truck. The abrupt deceleration of an out-of-control truck may cause shifting of the load, shearing of the fifth wheel, or jackknifing, all with potentially harmful occurrences to the driver and cargo.

Mounds of bedding material between 0.6 and 1.5 m [2 and 5 ft] high with 1V:1.5H slopes (i.e., slopes that change in elevation by one unit of length for each 1 to 5 units of horizontal distance) have been used at the end of ramps in several instances as the "last-chance" device. At least one escape ramp has been constructed with an array of crash cushions installed to prevent an out-of-control vehicle from leaving the end of the ramp. Furthermore, at the end of a hard-surfaced gravity ramp, a gravel bed or attenuator array may sufficiently immobilize a brakeless runaway vehicle to keep it from rolling backward and jackknifing. Where barrels are used, the barrels should be filled with the same material as used in the arrester bed, so that any finer material does not result in contamination of the bed and reduction of the expected rolling resistance.

Brake Check Areas

Turnouts or pulloff areas at the summit of a grade can be used for brake-check areas or mandatory-stop areas to provide an opportunity for a driver to inspect equipment on the vehicle and to ensure the brakes are not overheated at the beginning of the descent. In addition, information about the grade ahead and the location of escape ramps can be provided by diagrammatic signing or self-service pamphlets. An elaborate design is not needed for these areas. A brake-check area can be a paved lane behind and separated from the shoulder or a widened shoulder where a truck can stop. Appropriate signing should be used to discourage casual stopping by the public.

Maintenance

After each use, aggregate arrester beds should be reshaped using power equipment to the extent practical and the aggregate scarified as appropriate. Since aggregate tends to compact over time, the bedding material should be cleaned of contaminants and scarified periodically to retain the retarding characteristics of the bedding material and maintain free drainage. Using power equipment for work in the arrester bed reduces the exposure time for the maintenance workers to

the potential that a runaway truck may need to use the facility. Maintenance of the appurte
should be accomplished as appropriate.

Vertical Curves

General Considerations

Vertical curves to effect gradual changes between tangent grades may be any one of the crest
or sag types depicted in Exhibit 3-73. Vertical curves should be simple in application and should
result in a design that is safe and comfortable in operation, pleasing in appearance, and adequate
for drainage. The major control for safe operation on crest vertical curves is the provision of
ample sight distances for the design speed; while research (4) has shown that vertical curves with
limited sight distance do not necessarily experience safety problems, it is recommended that all
vertical curves should be designed to provide at least the stopping sight distances shown in
Exhibit 3-1. Wherever practical, more liberal stopping sight distances should be used.
Furthermore, additional sight distance should be provided at decision points.

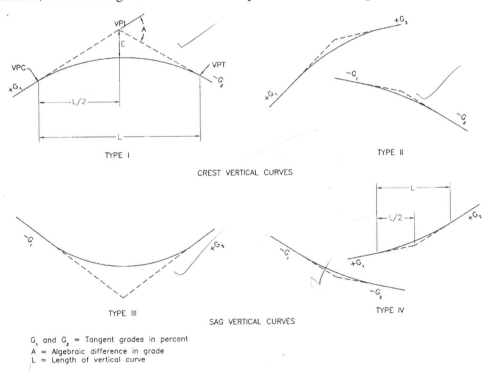

Exhibit 3-73. Types of Vertical Curves

For driver comfort, the rate of change of grade should be kept within tolerable limits. This
consideration is most important in sag vertical curves where gravitational and vertical centripetal
forces act in opposite directions. Appearance also should be considered in designing vertical

curves. A long curve has a more pleasing appearance than a short one; short vertical curves may give the appearance of a sudden break in the profile due to the effect of foreshortening.

Drainage of curbed roadways on sag vertical curves (Type III in Exhibit 3-73) needs careful profile design to retain a grade of not less than 0.5 percent or, in some cases, 0.30 percent for the outer edges of the roadway. Although not desirable, flatter grades may be appropriate in some situations.

For simplicity, a parabolic curve with an equivalent vertical axis centered on the vertical point of intersection (VPI) is usually used in roadway profile design. The vertical offsets from the tangent vary as the square of the horizontal distance from the curve end (point of tangency). The vertical offset from the tangent grade at any point along the curve is calculated as a proportion of the vertical offset at the VPI, which is AL/800, where the symbols are as shown in Exhibit 3-73. The rate of change of grade at successive points on the curve is a constant amount for equal increments of horizontal distance, and is equal to the algebraic difference between intersecting tangent grades divided by the length of curve in meters [feet], or A/L in percent per meter [percent per foot]. The reciprocal L/A is the horizontal distance in meters [feet] needed to make a 1-percent change in gradient and is, therefore, a measure of curvature. The quantity L/A, termed "K," is useful in determining the horizontal distance from the vertical point of curvature (VPC) to the high point of Type I curves or to the low point of Type III curves. This point where the slope is zero occurs at a distance from the VPC equal to K times the approach gradient. The value of K is also useful in determining minimum lengths of vertical curves for various design speeds. Other details on parabolic vertical curves are found in textbooks on highway engineering.

On certain occasions, because of critical clearance or other controls, the use of asymmetrical vertical curves may be appropriate. Because the conditions under which such curves are appropriate are infrequent, the derivation and use of the relevant equations have not been included herein. For use in such limited instances, refer to asymmetrical curve data found in a number of highway engineering texts.

Crest Vertical Curves

Minimum lengths of crest vertical curves based on sight distance criteria generally are satisfactory from the standpoint of safety, comfort, and appearance. An exception may be at decision areas, such as sight distance to ramp exit gores, where longer lengths are needed; for further information, refer to the section of this chapter concerning decision sight distance.

Exhibit 3-74 illustrates the parameters used in determining the length of a parabolic crest vertical curve needed to provide any specified value of sight distance. The basic equations for length of a crest vertical curve in terms of algebraic difference in grade and sight distance follow:

Metric	US Customary	
When S is less than L,	When S is less than L,	
$$L = \dfrac{AS^2}{100\left(\sqrt{2h_1} + \sqrt{2h_2}\right)^2}$$	$$L = \dfrac{AS^2}{100\left(\sqrt{2h_1} + \sqrt{2h_2}\right)^2}$$	(3-43)
When S is greater than L,	When S is greater than L,	
$$L = 2S - \dfrac{200\left(\sqrt{h_1} + \sqrt{h_2}\right)^2}{A}$$	$$L = 2S - \dfrac{200\left(\sqrt{h_1} + \sqrt{h_2}\right)^2}{A}$$	(3-44)
where: L = length of vertical curve, m; S = sight distance, m; A = algebraic difference in grades, percent; h_1 = height of eye above roadway surface, m; h_2 = height of object above roadway surface, m	where: L = length of vertical curve, ft; S = sight distance, ft; A = algebraic difference in grades, percent; h_1 = height of eye above roadway surface, ft; h_2 = height of object above roadway surface, ft	

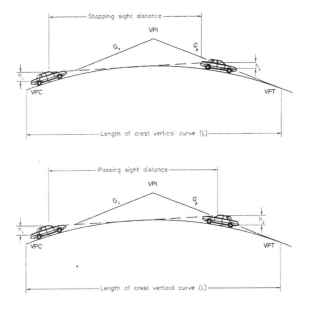

Exhibit 3-74. Parameters Considered in Determining the Length of a Crest Vertical Curve to Provide Sight Distance

When the height of eye and the height of object are 1,080 mm and 600 mm [3.5 ft and 2.0 ft], respectively, as used for stopping sight distance, the equations become:

Metric	US Customary	
When S is less than L, $$L = \frac{AS^2}{658}$$	When S is less than L, $$L = \frac{AS^2}{2158}$$	(3-45)
When S is greater than L, $$L = 2S - \frac{658}{A}$$	When S is greater than L, $$L = 2S - \frac{2158}{A}$$	(3-46)

Design controls—stopping sight distance. The minimum lengths of vertical curves for different values of A to provide the minimum stopping sight distances for each design speed are shown in Exhibit 3-75. The solid lines give the minimum vertical curve lengths, on the basis of rounded values of K as determined from Equations (3-45) and (3-46).

The short dashed curve at the lower left, crossing these lines, indicates where S = L. Note that to the right of the S = L line, the value of K, or length of vertical curve per percent change in A, is a simple and convenient expression of the design control. For each design speed this single value is a positive whole number that is indicative of the rate of vertical curvature. The design control in terms of K covers all combinations of A and L for any one design speed; thus, A and L need not be indicated separately in a tabulation of design value. The selection of design curves is facilitated because the minimum length of curve in meters [feet] is equal to K times the algebraic difference in grades in percent, L = KA. Conversely, the checking of plans is simplified by comparing all curves with the design value for K.

Exhibit 3-76 shows the computed K values for lengths of vertical curves corresponding to the stopping sight distances shown in Exhibit 3-1 for each design speed. For direct use in design, values of K are rounded as shown in the right column. The rounded values of K are plotted as the solid lines in Exhibit 3-75. These rounded values of K are higher than computed values, but the differences are not significant.

Where S is greater than L (lower left in Exhibit 3-75), the computed values plot as a curve (as shown by the dashed line for 70 km/h [45 mph]) that bends to the left, and for small values of A the vertical curve lengths are zero because the sight line passes over the high point. This relationship does not represent desirable design practice. Most States use a minimum length of vertical curve, expressed as either a single value, a range for different design speeds, or a function of A. Values now in use range from about 30 to 100 m [100 to 325 ft]. To recognize the distinction in design speed and to approximate the range of current practice, minimum lengths of vertical curves are expressed as about 0.6 times the design speed in km/h, $L_{min} = 0.6V$, where V is in kilometers per hour and L is in meters, or about three times the design speed in mph, [$L_{min} = 3V$], where V is in miles per hour and L is in feet. These terminal adjustments show as the vertical lines at the lower left of Exhibit 3-75.

METRIC

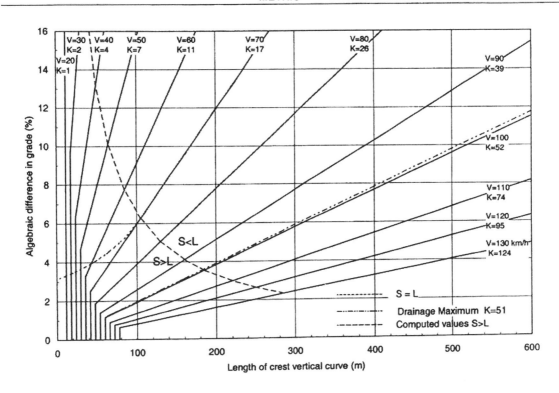

Exhibit 3-75. continued

US CUSTOMARY

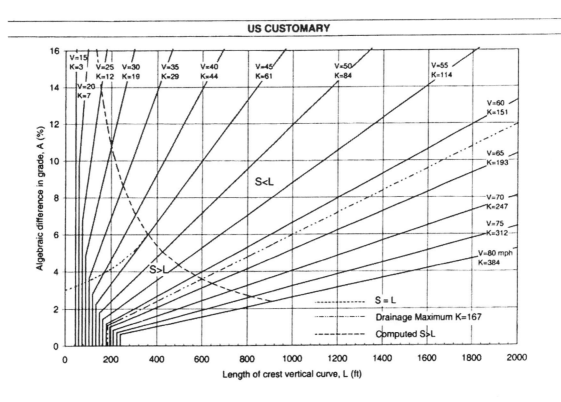

Exhibit 3-75. Design Controls for Crest Vertical Curves—Open Road Conditions

273

Metric				US Customary			
Design speed (km/h)	Stopping sight distance (m)	Rate of vertical curvature, K[a]		Design speed (mph)	Stopping sight distance (ft)	Rate of vertical curvature, K[a]	
		Calculated	Design			Calculated	Design
20	20	0.6	1	15	80	3.0	3
30	35	1.9	2	20	115	6.1	7
40	50	3.8	4	25	155	11.1	12
50	65	6.4	7	30	200	18.5	19
60	85	11.0	11	35	250	29.0	29
70	105	16.8	17	40	305	43.1	44
80	130	25.7	26	45	360	60.1	61
90	160	38.9	39	50	425	83.7	84
100	185	52.0	52	55	495	113.5	114
110	220	73.6	74	60	570	150.6	151
120	250	95.0	95	65	645	192.8	193
130	285	123.4	124	70	730	246.9	247
				75	820	311.6	312
				80	910	383.7	384

[a] Rate of vertical curvature, K, is the length of curve per percent algebraic difference in intersecting grades (A). $K = L/A$

Exhibit 3-76. Design Controls for Stopping Sight Distance and for Crest Vertical Curves

The values of K derived above when S is less than L also can be used without significant error where S is greater than L. As shown in Exhibit 3-75, extension of the diagonal lines to meet the vertical lines for minimum lengths of vertical curves results in appreciable differences from the theoretical only where A is small and little or no additional cost is involved in obtaining longer vertical curves.

For night driving on highways without lighting, the length of visible roadway is that roadway that is directly illuminated by the headlights of the vehicle. For certain conditions, the minimum stopping sight distance values used for design exceed the length of visible roadway. First, vehicle headlights have limitations on the distance over which they can project the light intensity levels that are needed for visibility. When headlights are operated on low beams, the reduced candlepower at the source plus the downward projection angle significantly restrict the length of visible roadway surface. Thus, particularly for high-speed conditions, stopping sight distance values exceed road-surface visibility distances afforded by the low-beam headlights regardless of whether the roadway profile is level or curving vertically. Second, for crest vertical curves, the area forward of the headlight beam's point of tangency with the roadway surface is shadowed and receives only indirect illumination.

Since the headlight mounting height (typically about 600 mm [2 ft]) is lower than the driver eye height used for design (1,080 mm [3.5 ft]), the sight distance to an illuminated object is controlled by the height of the vehicle headlights rather than by the direct line of sight. Any object within the shadow zone must be high enough to extend into the headlight beam to be directly illuminated. On the basis of Equation (3-43), the bottom of the headlight beam is about

400 mm [1.3 ft] above the roadway at a distance ahead of the vehicle equal to the stopping sight distance. Although the vehicle headlight system does limit roadway visibility length as mentioned above, there is some mitigating effect in that other vehicles, whose taillight height typically varies from 450 to 600 mm [1.5 to 2.0 ft], and other sizable objects receive direct lighting from headlights at stopping sight distance values used for design. Furthermore, drivers are aware that visibility at night is less than during the day, regardless of road and street design features, and they may therefore be more attentive and alert.

There is a level point on a crest vertical curve of Type I (see Exhibit 3-73), but no difficulty with drainage on highways with curbs is typically experienced if the curve is sharp enough so that a minimum grade of 0.30 percent is reached at a point about 15 m [50 ft] from the crest. This corresponds to K of 51 m [167 ft] per percent change in grade, which is plotted in Exhibit 3-75 as the drainage maximum. All combinations above or to the left of this line satisfy the drainage criterion. The combinations below and to the right of this line involve flatter vertical curves. Special attention is needed in these cases to ensure proper pavement drainage near the high point of crest vertical curves. It is not intended that K of 51 m [167 ft] per percent grade be considered a design maximum, but merely a value beyond which drainage should be more carefully designed.

Design controls—passing sight distance. Design values of crest vertical curves for passing sight distance differ from those for stopping sight distance because of the different sight distance and object height criteria. The general Equations (3-43) and (3-44) apply, but the 1,080 mm [3.5 ft] height of object results in the following specific formulas with the same terms as shown above:

Metric	US Customary	
When S is less than L,	When S is less than L,	
$$L = \frac{AS^2}{864}$$	$$L = \frac{AS^2}{2800}$$	(3-47)
When S is greater than L,	When S is greater than L,	
$$L = 2S - \frac{864}{A}$$	$$L = 2S - \frac{2800}{A}$$	(3-48)

For the minimum passing sight distances shown in Exhibit 3-7, the minimum lengths of crest vertical curves are substantially longer than those for stopping sight distances. The extent of difference is evident by the values of K, or length of vertical curve per percent change in A, for passing sight distances shown in Exhibit 3-77. These lengths are 7 to 10 times the corresponding lengths for stopping sight distance.

Metric			US Customary		
Design speed (km/h)	Passing sight distance (m)	Rate of vertical curvature, K* design	Design speed (mph)	Passing sight distance (ft)	Rate of vertical curvature, K* design
30	200	46	20	710	180
40	270	84	25	900	289
50	345	138	30	1090	424
60	410	195	35	1280	585
70	485	272	40	1470	772
80	540	338	45	1625	943
90	615	438	50	1835	1203
100	670	520	55	1985	1407
110	730	617	60	2135	1628
120	775	695	65	2285	1865
130	815	769	70	2480	2197
			75	2580	2377
			80	2680	2565
Note: *Rate of vertical curvature, K, is the length of curve per percent algebraic difference in intersecting grades (A). K=L/A					

Exhibit 3-77. Design Controls for Crest Vertical Curves Based on Passing Sight Distance

Generally, it is impractical to design crest vertical curves to provide for passing sight distance because of high cost where crest cuts are involved and the difficulty of fitting the resulting long vertical curves to the terrain, particularly for high-speed roads. Passing sight distance on crest vertical curves may be practical on roads with unusual combinations of low design speeds and gentle grades or higher design speeds with very small algebraic differences in grades. Ordinarily, passing sight distance is provided only at locations where combinations of alignment and profile do not need the use of crest vertical curves.

Sag Vertical Curves

At least four different criteria for establishing lengths of sag vertical curves are recognized to some extent. These are (1) headlight sight distance, (2) passenger comfort, (3) drainage control, and (4) general appearance.

Headlight sight distance has been used directly by some agencies and for the most part is the basis for determining the length of sag vertical curves recommended here. When a vehicle traverses a sag vertical curve at night, the portion of highway lighted ahead is dependent on the position of the headlights and the direction of the light beam. A headlight height of 600 mm [2 ft] and a 1-degree upward divergence of the light beam from the longitudinal axis of the vehicle is commonly assumed. The upward spread of the light beam above the 1-degree divergence angle provides some additional visible length of roadway, but is not generally considered in design. The following equations show the relationships between S, L, and A, using S as the distance between the vehicle and point where the 1-degree upward angle of the light beam intersects the surface of the roadway:

Metric	US Customary	
When S is less than L, $$L=\frac{AS^2}{200\left[0.6+S\left(\tan 1°\right)\right]}$$	When S is less than L, $$L=\frac{AS^2}{200\left[2.0+S\left(\tan 1°\right)\right]}$$	(3-49)
or, $$L=\frac{AS^2}{120+3.5\,S}$$	or, $$L=\frac{AS^2}{400+3.5\,S}$$	(3-50)
When S is greater than L, $$L=2\,S-\frac{200\left[0.6+S\left(\tan 1°\right)\right]}{A}$$	When S is greater than L, $$L=2S-\frac{200\left[2.0+S\left(\tan 1°\right)\right]}{A}$$	(3-51)
or, $$L=2\,S-\left(\frac{120+3.5\,S}{A}\right)$$	or, $$L=2\,S-\left(\frac{400+3.5\,S}{A}\right)$$	(3-52)
where: L = length of sag vertical curve, m; S = light beam distance, m; A = algebraic difference in grades, percent	where: L = length of sag vertical curve, ft; S = light beam distance, ft; A = algebraic difference in grades, percent	

For overall safety on highways, a sag vertical curve should be long enough that the light beam distance is nearly the same as the stopping sight distance. Accordingly, it is appropriate to use stopping sight distances for different design speeds as the value of S in the above equations. The resulting lengths of sag vertical curves for the recommended stopping sight distances for each design speed are shown in Exhibit 3-78 with solid lines using rounded values of K as was done for crest vertical curves.

The effect on passenger comfort of the change in vertical direction is greater on sag than on crest vertical curves because gravitational and centripetal forces are in opposite directions, rather than in the same direction. Comfort due to change in vertical direction is not readily measured because it is affected appreciably by vehicle body suspension, vehicle body weight, tire

METRIC

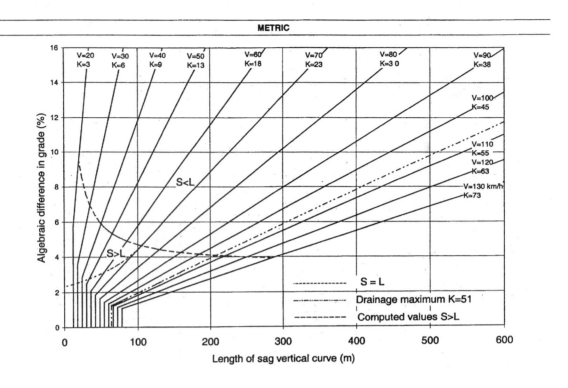

US CUSTOMARY

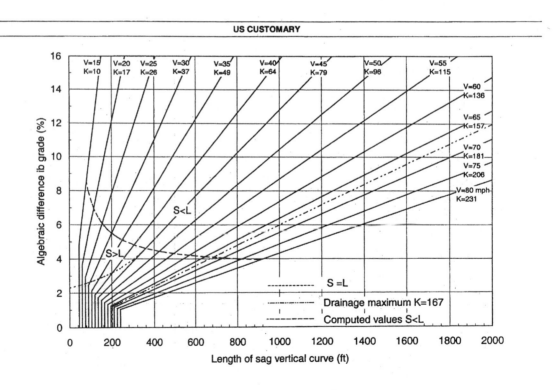

Exhibit 3-78. Design Controls for Sag Vertical Curves—Open Road Conditions

flexibility, and other factors. Limited attempts at such measurements have led to the broad conclusion that riding is comfortable on sag vertical curves when the centripetal acceleration does not exceed 0.3m/s² [1 ft/s²]. The general expression for such a criterion is:

Metric	US Customary	
$$L = \frac{AV^2}{395}$$	$$L = \frac{AV^2}{46.5}$$	(3-53)
where: L = length of sag vertical curve, m; A = algebraic difference in grades, percent; V = design speed, km/h	where: L = length of sag vertical curve, ft; A = algebraic difference in grades, percent; V = design speed, mph	

The length of vertical curve needed to satisfy this comfort factor at the various design speeds is only about 50 percent of that needed to satisfy the headlight sight distance criterion for the normal range of design conditions.

Drainage affects design of vertical curves of Type III (see Exhibit 3-73) where curbed sections are used. An approximate criterion for sag vertical curves is the same as that expressed for the crest conditions (i.e., a minimum grade of 0.30 percent should be provided within 15 m [50 ft] of the level point). This criterion corresponds to K of 51 m [167 ft] per percent change in grade, which is plotted in Exhibit 3-78 as the drainage maximum. The drainage criterion differs from other criteria in that the length of sag vertical curve determined for it is a maximum, whereas, the length for any other criterion is a minimum. The maximum length of the drainage criterion is greater than the minimum length for other criteria up to 100 km/h [65 mph].

For general appearance of sag vertical curves, some use was formerly made of a rule-of-thumb for minimum curve length of 30A [100A] or, in Exhibit 3-78, K = 30 [K = 100]. This approximation is a generalized control for small or intermediate values of A. Compared with headlight sight distance, it corresponds to a design speed of approximately 80 km/h [50 mph]. On high-type highways, longer curves are appropriate to improve appearance.

From the preceding discussion, it is evident that design controls for sag vertical curves differ from those for crests, and separate design values are needed. The headlight sight distance appears to be the most logical criterion for general use, and the values determined for stopping sight distances are within the limits recognized in current practice. The use of this criterion to establish design values for a range of lengths of sag vertical curves is recommended. As in the case of crest vertical curves, it is convenient to express the design control in terms of the K rate for all values of A. This entails some deviation from the computed values of K for small values of A, but the differences are not significant. Exhibit 3-79 shows the range of computed values and the rounded values of K selected as design controls. The lengths of sag vertical curves on the basis of the design speed values of K are shown by the solid lines in Exhibit 3-78. It is to be emphasized that these lengths are minimum values based on design speed; longer curves are desired wherever practical, but special attention to drainage should be exercised where values of K in excess of 51 [167] are used.

Minimum lengths of vertical curves for flat gradients also are recognized for sag conditions. The values determined for crest conditions appear to be generally suitable for sags. Lengths of sag vertical curves, shown as vertical lines in Exhibit 3-78, are equal to 0.6 times the design speed in km/h [three times the design speed in mph].

Sag vertical curves shorter than the lengths computed from Exhibit 3-79 may be justified for economic reasons in cases where an existing feature, such as a structure not ready for replacement, controls the vertical profile. In certain cases, ramps may also be designed with shorter sag vertical curves. Fixed-source lighting is desirable in such cases. For street design, some engineers accept design of a sag or crest where A is about 1 percent or less without a length of calculated vertical curve. However, field modifications during construction usually result in constructing the equivalent to a vertical curve, even if short.

Metric				US Customary			
Design speed (km/h)	Stopping sight distance (m)	Rate of vertical curvature, K[a]		Design speed (mph)	Stopping sight distance (ft)	Rate of vertical curvature, K[a]	
		Calculated	Design			Calculated	Design
20	20	2.1	3	15	80	9.4	10
30	35	5.1	6	20	115	16.5	17
40	50	8.5	9	25	155	25.5	26
50	65	12.2	13	30	200	36.4	37
60	85	17.3	18	35	250	49.0	49
70	105	22.6	23	40	305	63.4	64
80	130	29.4	30	45	360	78.1	79
90	160	37.6	38	50	425	95.7	96
100	185	44.6	45	55	495	114.9	115
110	220	54.4	55	60	570	135.7	136
120	250	62.8	63	65	645	156.5	157
130	285	72.7	73	70	730	180.3	181
				75	820	205.6	206
				80	910	231.0	231

[a] Rate of vertical curvature, K, is the length of curve (m) per percent algebraic difference intersecting grades (A). K = L/A

Exhibit 3-79. Design Controls for Sag Vertical Curves

Sight Distance at Undercrossings

Sight distance on the highway through a grade separation should be at least as long as the minimum stopping sight distance and preferably longer. Design of the vertical alignment is the same as at any other point on the highway except in some cases of sag vertical curves underpassing a structure illustrated in Exhibit 3-80. While not a frequent problem, the structure fascia may cut the line of sight and limit the sight distance to less that otherwise is attainable. It is generally practical to provide the minimum length of sag vertical curve discussed above at grade

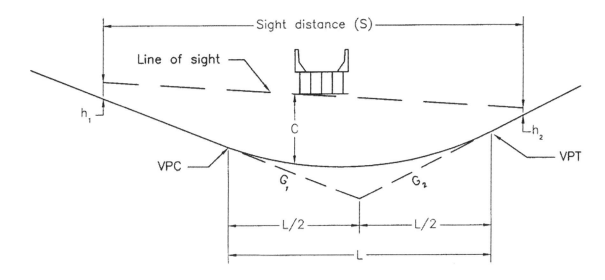

Exhibit 3-80. Sight Distance at Undercrossings

separation structures, and even where the recommended grades are exceeded, the sight distance should not need to be reduced below the minimum recommended values for stopping sight distance.

For some conditions, the designer may wish to check the available sight distance at an undercrossing, such as at a two-lane undercrossing without ramps where it would be desirable to provide passing sight distance. Such checks are best made graphically on the profile, but may be performed through computations.

The general equations for sag vertical curve length at undercrossings are:

Case 1—Sight distance greater than length of vertical curve (S>L):

Metric	US Customary	
$$L = 2S - \dfrac{800\left(C - \left(\dfrac{h_1 + h_2}{2}\right)\right)}{A}$$	$$L = 2S - \dfrac{800\left(C - \left(\dfrac{h_1 + h_2}{2}\right)\right)}{A}$$	(3-54)
where:	where:	
L = length of vertical curve, m; S = sight distance, m; A = algebraic difference in grades, percent; C = vertical clearance, m; h_1 = height of eye, m; h_2 = height of object, m	L = length of vertical curve, ft; S = sight distance, ft; A = algebraic difference in grades, percent; C = vertical clearance, ft; h_1 = height of eye, ft; h_2 = height of object, ft	

Case 2—Sight distance less than length of vertical curve (S<L):

Metric	US Customary	
$$L = \dfrac{AS^2}{800\left(C - \left(\dfrac{h_1 + h_2}{2}\right)\right)}$$	$$L = \dfrac{AS^2}{800\left(C - \left(\dfrac{h_1 + h_2}{2}\right)\right)}$$	(3-55)
where: L $\;=\;$ length of vertical curve, m; S $\;=\;$ sight distance, m; A $\;=\;$ algebraic difference in grades, percent; C $\;=\;$ vertical clearance, m; h_1 $\;=\;$ height of eye, m; h_2 $\;=\;$ height of object, m	where: L $\;=\;$ length of vertical curve, ft; S $\;=\;$ sight distance, ft; A $\;=\;$ algebraic difference in grades, percent; C $\;=\;$ vertical clearance, ft; h1 $\;=\;$ height of eye, ft; h2 $\;=\;$ height of object, ft	

Using an eye height of 2.4 m [8.0 ft] for a truck driver and an object height of 0.6 m [2.0 ft] for the taillights of a vehicle, the following equations can be derived:

Case 1—Sight distance greater than length of vertical curve (S>L):

Metric	US Customary	
$$L = 2S - \left(\dfrac{800(C - 1.5)}{A}\right)$$	$$L = 2S - \left(\dfrac{800(C - 5)}{A}\right)$$	(3-56)

Case 2—Sight distance less than length of vertical curve (S<L):

Metric	US Customary	
$$L = \dfrac{AS^2}{800(C - 1.5)}$$	$$L = \dfrac{AS^2}{800(C - 5)}$$	(3-57)

General Controls for Vertical Alignment

In addition to the above specific controls for vertical alignment discussed above, there are several general controls that should be considered in design.

- A smooth gradeline with gradual changes, as consistent with the type of highways, roads, or streets and the character of terrain, should be sought for in preference to a line with numerous breaks and short lengths of grades. Specific design criteria are the maximum grade and the critical length of grade, but the manner in which they are applied and fitted to the terrain on a continuous line determines the suitability and appearance of the finished product.

- The "roller-coaster" or the "hidden-dip" type of profile should be avoided. Such profiles generally occur on relatively straight horizontal alignment where the roadway profile closely follows a rolling natural ground line. Examples of such undesirable profiles are evident on many older roads and streets; they are unpleasant aesthetically and difficult to drive. Hidden dips may create difficulties for drivers who wish to pass because the passing driver may be deceived if the view of the road or street beyond the dip is free of opposing vehicles. Even with shallow dips, this type of profile may be disconcerting because the driver cannot be sure whether or not there is an oncoming vehicle hidden beyond the rise. This type of profile is avoided by use of horizontal curves or by more gradual grades.

- Undulating gradelines, involving substantial lengths of momentum grades, should be evaluated for their effect on traffic operation. Such profiles permit heavy trucks to operate at higher overall speeds than is possible when an upgrade is not preceded by a downgrade, but may encourage excessive speeds of trucks with attendant conflicts with other traffic.

- A "broken-back" gradeline (two vertical curves in the same direction separated by a short section of tangent grade) generally should be avoided, particularly in sags where the full view of both vertical curves is not pleasing. This effect is particularly noticeable on divided roadways with open median sections.

- On long grades, it may be preferable to place the steepest grades at the bottom and flatten the grades near the top of the ascent or to break the sustained grade by short intervals of flatter grade instead of providing a uniform sustained grade that is only slightly below the recommended maximum. This is particularly applicable to roads and streets with low design speeds.

- Where at-grade intersections occur on roadway sections with moderate to steep grades, it is desirable to reduce the grade through the intersection. Such profile changes are beneficial for vehicles making turns and serve to reduce the potential for crashes.

- Sag vertical curves should be avoided in cuts unless adequate drainage can be provided.

COMBINATIONS OF HORIZONTAL AND VERTICAL ALIGNMENT

General Considerations

Horizontal and vertical alignment are permanent design elements for which thorough study is warranted. It is extremely difficult and costly to correct alignment deficiencies after a highway is constructed. On freeways, there are numerous controls such as multilevel structures and costly right-of-way. On most arterial streets, heavy development takes place along the property lines, which makes it impractical to change the alignment in the future. Thus, compromises in the alignment designs should be weighed carefully, because any initial savings may be more than offset by the economic loss to the public in the form of crashes and delays.

Horizontal and vertical alignment should not be designed independently. They complement each other, and poorly designed combinations can spoil the good points and aggravate the deficiencies of each. Horizontal alignment and profile are among the more important of the permanent design elements of the highway. Excellence in the design of each and of their combination increases usefulness and safety, encourages uniform speed, and improves appearance, nearly always without additional cost (**23, 47, 48, 49, 50, 51, 52, 53**).

General Design Controls

It is difficult to discuss combinations of horizontal alignment and profile without reference to the broader issue of highway location. These subjects are interrelated and what is said about one is generally applicable to the other. It is assumed in this discussion that the general location of a facility has been fixed and that the remaining task is the development of a specific design harmonizing of the vertical and horizontal lines, such that the finished highway, road, or street will be an economical, pleasant, and safe facility on which to travel. The physical constraints or influences that act singly or in combination to determine the alignment are the character of roadway based on the traffic, topography, and subsurface conditions, the existing cultural development, likely future developments, and the location of the roadway's terminals. Design speed is considered in determining the general roadway location, but as design proceeds to the development of more detailed alignment and profile it assumes greater importance. The selected design speed serves to keep all elements of design in balance. Design speed determines limiting values for many elements such as curvature and sight distance and influences many other elements such as width, clearance, and maximum gradient, which are all discussed in the preceding parts of this chapter.

Appropriate combinations of horizontal alignment and profile are obtained through engineering studies and consideration of the following general guidelines:

* Curvature and grades should be in proper balance. Tangent alignment or flat curvature at the expense of steep or long grades and excessive curvature with flat grades both represent poor design. A logical design that offers the best combination of safety, capacity, ease and uniformity of operation, and pleasing appearance within the practical limits of terrain and area traversed is a compromise between these two extremes.

* Vertical curvature superimposed on horizontal curvature, or vice versa, generally results in a more pleasing facility, but such combinations should be analyzed for their effect on traffic. Successive changes in profile not in combination with horizontal curvature may result in a series of humps visible to the driver for some distance which, as previously discussed, represents an undesirable condition. The use of horizontal and vertical alignments in combination, however, may also result in certain undesirable arrangements, as discussed below.

* Sharp horizontal curvature should not be introduced at or near the top of a pronounced crest vertical curve. This condition is undesirable because the driver may not perceive the horizontal change in alignment, especially at night. The disadvantages of this arrangement are avoided if the horizontal curvature leads the vertical curvature (i.e., the horizontal curve is made longer than the vertical curve). Suitable designs can also be developed by using design values well above the appropriate minimum values for the design speed.

- Somewhat related to the preceding guideline, sharp horizontal curvature should not be introduced near the bottom of a steep grade approaching or near the low point of a pronounced sag vertical curve. Because the view of the road ahead is foreshortened, any horizontal curvature other than a very flat curve assumes an undesirable distorted appearance. Further, vehicle speeds, particularly for trucks, are often high at the bottom of grades, and erratic operations may result, especially at night.

- On two-lane roads and streets, the need for passing sections at frequent intervals and including an appreciable percentage of the length of the roadway often supersedes the general guidelines for combinations of horizontal and vertical alignment. In such cases, it is appropriate to work toward long tangent sections to assure sufficient passing sight distance in design.

- Both horizontal curvature and profile should be made as flat as practical at intersections where sight distance along both roads or streets is important and vehicles may have to slow or stop.

- On divided highways and streets, variation in width of median and the use of independent profiles and horizontal alignments for the separate one-way roadways are sometimes desirable. Where traffic justifies provision of four lanes, a superior design without additional cost generally results from such practices.

- In residential areas, the alignment should be designed to minimize nuisance to the neighborhood. Generally, a depressed facility makes a highway less visible and less noisy to adjacent residents. Minor horizontal adjustments can sometimes be made to increase the buffer zone between the highway and clusters of homes.

- The alignment should be designed to enhance attractive scenic views of the natural and manmade environment, such as rivers, rock formations, parks, and outstanding structures. The highway should head into, rather than away from, those views that are outstanding; it should fall toward those features of interest at a low elevation, and it should rise toward those features best seen from below or in silhouette against the sky.

Alignment Coordination in Design

Coordination of horizontal alignment and profile should not be left to chance but should begin with preliminary design, at which time adjustments can be readily made. Although a specific order of study cannot be stated for all highways, a general procedure applicable to most facilities is described below.

The designer should use working drawings of a size, scale, and arrangement so that he or she can study long, continuous stretches of highway in both plan and profile and visualize the whole in three dimensions. Working drawings should be of a small scale, with the profile plotted jointly with the plan. A continuous roll of plan-profile paper usually is suitable for this purpose. To assist in this visualization, there also are programs available for personal computers (PCs) that allow designers to view proposed vertical and horizontal alignments in three dimensions.

After study of the horizontal alignment and profile in preliminary form, adjustments in either, or both, can be made jointly to obtain the desired coordination. At this stage, the designer should not be concerned with line calculations other than known major controls. The study should be made largely on the basis of a graphical or computer analysis. The criteria and elements of design covered in this and the preceding chapter should be kept in mind. For the selected design speed, the values for controlling curvature, gradient, sight distance, and superelevation runoff length should be obtained and checked graphically or with a PC or CADD system. Design speed may have to be adjusted during the process along some sections to conform to likely variations in speeds of operation. This need may occur where noticeable changes in alignment characteristics are needed to accommodate unusual terrain or right-of-way controls. In addition, the general design controls, as enumerated separately for horizontal alignment, vertical alignment, and their combination, should be considered. All aspects of terrain, traffic operation, and appearance should be considered and the horizontal and vertical lines should be adjusted and coordinated before the costly and time-consuming calculations and the preparation of construction plans to large scale are started.

The coordination of horizontal alignment and profile from the standpoint of appearance usually can be accomplished visually on the preliminary working drawings or with the assistance of PC programs that have been developed for this purpose. Generally, such methods result in a satisfactory product when applied by an experienced designer. This means of analysis may be supplemented by models, sketches, or images projected by a PC at locations where the appearance of certain combinations of line and grade is unclear. For highways with gutters, the effects of superelevation transitions on gutter-line profiles should be examined. This can be particularly significant when flat grades are involved and can result in local depressions. Slight shifts in profile in relation to horizontal curves can sometimes eliminate the problem.

The procedures described above should obviously be modified for the design of typical local roads or streets, as compared to higher type highways. The alignment of any local road or street, whether for a new roadway or for reconstruction of an existing roadway, is governed by the existing or likely future development along it. The crossroad or street intersections and the location of driveways are dominant controls. Although they should be fully considered, they should not override the broader desirable features described above. Even for street design, it is desirable to work out long, flowing alignment and profile sections rather than a connected series of block-by-block sections. Some examples of poor and good practice are illustrated in Exhibit 3-81.

OTHER ELEMENTS AFFECTING GEOMETRIC DESIGN

In addition to the design elements discussed previously, several other elements affect or are affected by the geometric design of a roadway. Each of these elements is discussed only to the extent needed to show its relation to geometric design and how it, in turn, is affected thereby. Detailed design of these elements is not covered here.

Drainage

Highway drainage facilities carry water across the right-of-way and remove storm water from the roadway itself. Drainage facilities include bridges, culverts, channels, curbs, gutters, and

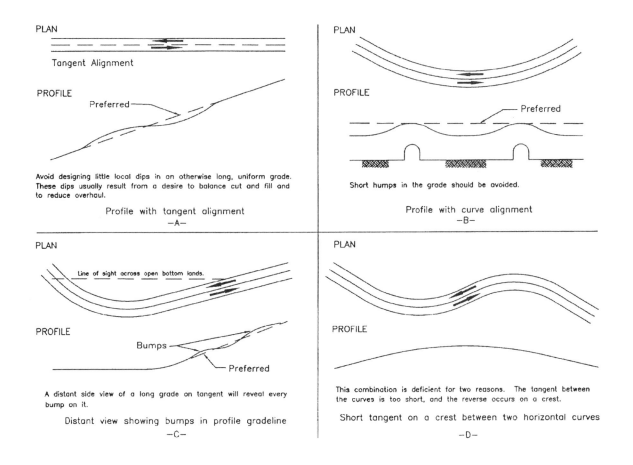

PLAN

Tangent Alignment

PROFILE

Preferred

Avoid designing little local dips in an otherwise long, uniform grade. These dips usually result from a desire to balance cut and fill and to reduce overhaul.

Profile with tangent alignment
—A—

PLAN

PROFILE

Preferred

Short humps in the grade should be avoided.

Profile with curve alignment
—B—

PLAN

Line of sight across open bottom lands.

PROFILE

Bumps

Preferred

A distant side view of a long grade on tangent will reveal every bump on it.

Distant view showing bumps in profile gradeline
—C—

PLAN

PROFILE

This combination is deficient for two reasons. The tangent between the curves is too short, and the reverse occurs on a crest.

Short tangent on a crest between two horizontal curves
—D—

Exhibit 3-81. Alignment and Profile Relationships in Roadway Design (48)

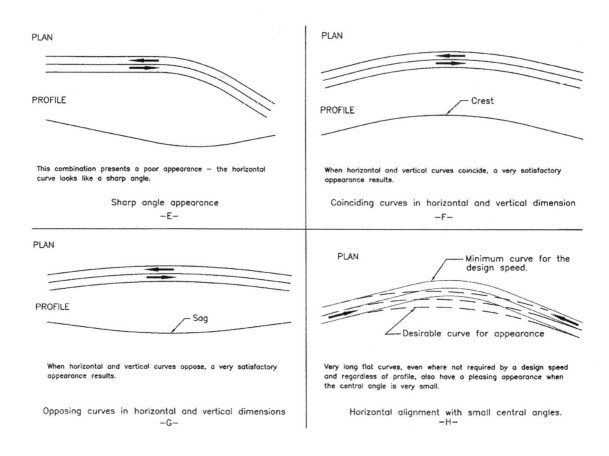

PLAN

PROFILE

This combination presents a poor appearance — the horizontal curve looks like a sharp angle.

Sharp angle appearance
—E—

PLAN

PROFILE   ⌐ Crest

When horizontal and vertical curves coincide, a very satisfactory appearance results.

Coinciding curves in horizontal and vertical dimension
—F—

PLAN

PROFILE   ⌐ Sag

When horizontal and vertical curves oppose, a very satisfactory appearance results.

Opposing curves in horizontal and vertical dimensions
—G—

PLAN   ⌐ Minimum curve for the
                             design speed.

                            Desirable curve for appearance

Very long flat curves, even where not required by a design speed and regardless of profile, also have a pleasing appearance when the central angle is very small.

Horizontal alignment with small central angles.
—H—

Exhibit 3-81. Alignment and Profile Relationships in Roadway Design (Continued)

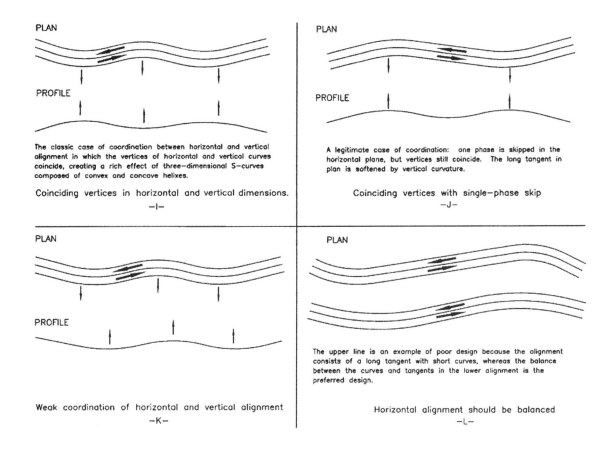

Exhibit 3-81. Alignment and Profile Relationships in Roadway Design (Continued)

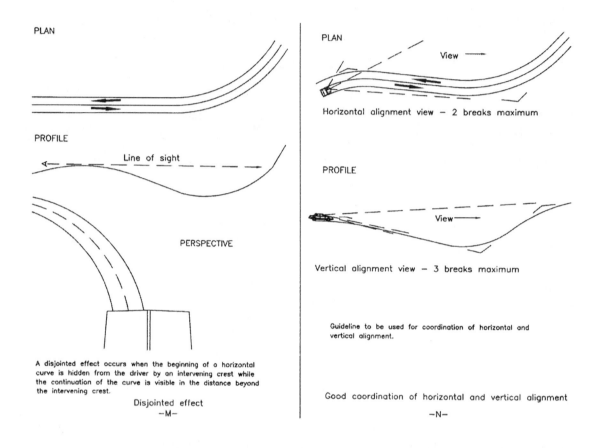

PLAN

PROFILE

Line of sight

PERSPECTIVE

A disjointed effect occurs when the beginning of a horizontal
curve is hidden from the driver by an intervening crest while
the continuation of the curve is visible in the distance beyond
the intervening crest.

Disjointed effect
—M—

PLAN

View ⟶

Horizontal alignment view — 2 breaks maximum

PROFILE

View ⟶

Vertical alignment view — 3 breaks maximum

Guideline to be used for coordination of horizontal and
vertical alignment.

Good coordination of horizontal and vertical alignment
—N—

Exhibit 3-81. Alignment and Profile Relationships in Roadway Design (Continued)

various types of drains. Hydraulic capacities and locations of such structures should be designed to take into consideration damage to upstream and downstream property and to secure as low a degree of risk of traffic interruption by flooding as is consistent with the importance of the road, the design traffic service needs, Federal and State regulations, and available funds. While drainage design considerations are an integral part of highway geometric design, specific drainage design criteria are not included in this policy. The AASHTO *Highway Drainage Guidelines* (**54**) should be referred to for a general discussion of drainage, and the AASHTO *Model Drainage Manual* (**55**) should be referred to for guidelines on major areas of highway hydraulic design.

Many State highway agencies have excellent highway drainage manuals that may be used for reference for hydraulic design procedures. Alternatively, the AASHTO *Model Drainage Manual* (**55**) and computer software (**56**) may be referenced. In addition, other publications on drainage are widely used and are available to highway agencies from FHWA or the National Technical Information Service (**56**).

Hydraulic requirements for stream crossings and flood plain encroachments frequently affect highway alignment and profile (**57**). The probable effects of a highway encroachment on the risk of flood damage to other property and the risk of flood damage to the highway should be evaluated when a flood plain location is under consideration. Water surface elevations for floods of various return periods will influence decisions regarding the highway profile where an encroachment on the flood plain is considered. Highway profiles at stream crossings will often be determined by hydraulic considerations. To the extent practical, stream crossings and other highway encroachments on flood plains should be located and aligned to preserve the natural flood flow distribution and direction. Stream stability and the stream environment are also important and complex considerations in highway location and design.

Surface channels are used to intercept and remove surface runoff from roadways, wherever practical. They should have adequate capacity for the design runoff and should be properly located and shaped. Channels are usually lined with vegetation, and rock or paved channel linings are used where vegetation will not control erosion. Runoff from roadway surfaces normally drains down grass slopes to roadside or median channels. Curbs or dikes, inlets, and chutes or flumes are used where runoff from the roadway would erode fill slopes. Where storm drains are needed, curbs are usually provided. Care should be exercised to ensure that these curbs do not encroach on the clear zone of the highway. For further guidance, refer to the discussion on horizontal clearance to obstructions in Chapter 4.

Drainage inlets should be designed and located to limit the spread of water on the traveled way to tolerable widths. Because grates may become blocked by trash accumulation, curb openings or combination inlets with both grate and curb openings are advantageous for urban conditions. Grate inlets and depressions or curb-opening inlets should be located outside the through-traffic lanes to minimize the shifting of vehicles attempting to avoid riding over them. Inlet grates should also be designed to accommodate bicycle and pedestrian traffic where appropriate. Discontinuous sections of curbing, as at the gore of ramps, and variable curb offsets should not be used as expedients to handle pavement drainage where these features would detract from highway safety. Inlets should be designed and located to prevent silt and debris carried in suspension from being deposited on the traveled way where the longitudinal gradient is decreased. Extra inlets should be installed near low points of sag vertical curves to take any overflow from blocked inlets. Inlets should be located so that concentrated flow and heavy sheet flow will not cross traffic lanes. Where roadway surfaces are warped, as at cross streets or ramps, surface water should be intercepted just before the change in cross slope. Also, inlets should be located just upgrade of pedestrian crossings. Storm drains should have adequate capacity to avoid ponding of water on the roadway and bridges, especially in sag vertical curves. The general effect of drainage on the geometry of roadways, shoulder ditches, or gutters and side slopes is discussed further in Chapter 4.

Drainage is usually more difficult and costly for urban than for rural highways because of more rapid rates and larger volumes of runoff, costlier potential damage to adjacent property by flooding, higher overall costs because of more inlets and underground systems, greater restrictions because of urban development, lack of natural areas of water bodies to receive flood water, and higher volumes of traffic, including pedestrians. There is greater need to intercept concentrated storm water before it reaches the highway and to remove over-the-curb flow and surface water without interrupting traffic flow or causing a problem for vehicle occupants or pedestrians. To accommodate such runoff, underground systems and numerous inlets, curbs, and gutters are usually needed. Often new outfall drains of considerable length must be constructed because existing storm water systems often lack capacity for highway surface drainage volumes. A joint use storm water system, shared by the highway agency with others, can have economic advantages to both parties, because it is normally more economical to build a common system rather than two independent systems. Urban drainage design is discussed in the FHWA *Urban Drainage Design Manual* (**58**).

Reduction of peak flows can be achieved by the storage of water that falls on the site in detention basins, storm drainage pipes, swales and channels, parking lots, and rooftops. Storm water is released to the downstream conveyance facility or stream at a reduced flow rate. This concept should be considered for use in highway drainage design where existing downstream conveyance facilities are inadequate to handle peak flow rates from highway storm drainage facilities, where the highway would contribute to increased peak flow rates and aggravate downstream flooding problems, and as a technique to reduce the construction costs of outfalls from highway storm drainage facilities. Storm water detention may also be needed in order to conform with Federal and State water quality regulations. Some States have environmental regulations that require specific pollution/erosion measures.

The cost of drainage is neither incidental nor minor on most roads. Careful attention to needs for adequate drainage and protection of the highway from floods in all phases of location and design will prove to be effective in reducing costs in both construction and maintenance.

Erosion Control and Landscape Development

Erosion prevention is one of the major factors in design, construction, and maintenance of highways. It should be considered early in the location and design stages. Some degree of erosion control can be incorporated into the geometric design, particularly in the cross section elements.

Of course, the most direct application of erosion control occurs in drainage design and in the writing of specifications for landscaping and slope planting.

Erosion and maintenance are minimized largely by the use of flat side slopes, rounded and blended with natural terrain; serrated cut slopes; drainage channels designed with due regard to width, depth, slopes, alignment, and protective treatment; inlets located and spaced with erosion control in mind; prevention of erosion at culvert outlets; proper facilities for groundwater interception; dikes, berms, and other protective devices to trap sediment at strategic locations; and protective ground covers and planting.

Landscape development should be in keeping with the character of the highway and its environment. Programs include the following general areas of improvement: (1) preservation of existing vegetation, (2) transplanting of existing vegetation where practical, (3) planting of new vegetation, (4) selective clearing and thinning, and (5) regeneration of natural plant species and material.

The objectives in planting or the retention and preservation of natural growth on roadsides are closely related. In essence, they are to provide (1) vegetation that will be an aid to aesthetics and safety, (2) vegetation that will aid in lowering construction and maintenance costs, and (3) vegetation that creates interest, usefulness, and beauty for the pleasure and satisfaction of the traveling public.

Landscaping of urban highways and streets assumes additional importance in mitigating the many nuisances associated with urban traffic. Landscaping can reduce this contribution to urban blight and make the urban highways and streets better neighbors.

Further information concerning landscape development and erosion control is presented in the AASHTO *Guide for Transportation Landscape and Environmental Design* (**49**).

Rest Areas, Information Centers, and Scenic Overlooks

Rest areas, information centers, and scenic overlooks are functional and desirable elements of the complete highway facility and are provided for the safety and convenience of highway users. A safety rest area is a roadside area, with parking facilities separated from the roadway, provided for the travelers to stop and rest for short periods. The area may provide drinking water, restrooms, tables and benches, telephones, information displays, and other facilities for travelers. A rest area is not intended to be used for social or civic gatherings or for such active forms of recreation as boating, swimming, or organized games. An information center is a staffed or unstaffed facility at a rest area for the purpose of furnishing travel and other information or services to travelers. A scenic overlook is a roadside area provided for motorists to park their vehicles, beyond the shoulder, primarily for viewing the scenery or for taking photographs in safety. Scenic overlooks need not provide comfort and convenience facilities.

Site selection for rest areas, information centers, and scenic overlooks should consider the scenic quality of the area, accessibility, and adaptability to development. Other essential considerations include an adequate source of water and a means to treat and/or properly dispose of sewage. Site plans should be developed through the use of a comprehensive site planning process that should include the location of ramps, parking areas for cars and trucks, buildings, picnic areas, water supply, sewage treatment facilities, and maintenance areas. The objective is to give maximum weight to the appropriateness of the site rather than adherence to uniform distance or driving time between sites.

Facilities should be designed to accommodate the needs of older persons and persons with disabilities. Further information concerning rest area design is presented in the AASHTO *Guide for Development of Rest Areas on Major Arterials and Freeways* (**59**).

Lighting

Lighting may improve the safety of a highway or street and the ease and comfort of operation thereon. Statistics indicate that nighttime crash rates are higher than daytime crash rates. To a large extent, this may be attributed to reduced visibility at night. There is evidence that in urban and suburban areas, where there are concentrations of pedestrians and roadside intersectional interferences, fixed-source lighting tends to reduce crashes. Lighting of rural highways may be desirable, but the need for it is much less than on streets and highways in urban areas. The general consensus is that lighting of rural highways is seldom justified except in certain critical areas, such as interchanges, intersections, railroad grade crossings, long or narrow bridges, tunnels, sharp curves, and areas where roadside interferences are present. Most modern rural highways should be designed with an open cross section and horizontal and vertical alignment of a fairly high type. Accordingly, they offer an opportunity for near maximum use of vehicle headlights, resulting in reduced justification for fixed highway lighting.

On freeways where there are no pedestrians, roadside entrances, or other intersections at grade, and where rights-of-way are relatively wide, the justification for lighting differs from that of noncontrolled streets and highways. The AASHTO *Informational Guide for Roadway Lighting* (**60**) was prepared to aid in the selection of sections of freeways, highways, and streets for which fixed-source lighting may be warranted, and to present design guide values for their illumination. This guide also contains a section on the lighting of tunnels and underpasses.

Whether or not rural at-grade intersections should be lighted depends on the layout and the traffic volumes involved. Intersections that do not have channelization are frequently left unlighted. On the other hand, intersections with substantial channelization, particularly multi-road layouts and those designed on a broad scale, are often lighted. It is especially desirable to illuminate large-scale channelized intersections. Because of the sharp curvatures, little of such intersections is within the lateral range of headlights, and the headlights of other vehicles are a hindrance rather than an aid because of the variety of directions and turning movements. There is need to obtain a reduction in the speed of vehicles approaching some intersections. The indication of this need should be definite and visible at a distance from the intersection that is beyond the range of headlights. Illumination of the intersection with fixed-source lighting accomplishes this.

At interchanges it also is desirable, and sometimes essential, to provide fixed-source lighting. Drivers should be able to see not only the road ahead, but also the entire turning roadway area to properly discern the paths to be followed. They should also see all other vehicles that may influence their own behavior. Without lighting, there may be a noticeable decrease in the usefulness of the interchange at night; there would be more cars slowing down and moving with uncertainty at night than during daylight hours. Consideration should be given to improving visibility at night by roadway lighting (or reflectorizing devices) the parts of grade separation structures that particularly should be avoided by motorists, such as curbs, piers, and abutments. The greater the volume of traffic, particularly turning traffic, the more important the fixed-source lighting at interchanges becomes. Illumination should also be considered on those sections of major highways where there are turning movements to and from roadside development.

Floodlighting or highway lighting may be desirable at railroad-highway grade crossings when there are nighttime movements of trains. In some cases, such treatments may apply also to crossings operated with flashing signals, or gates, or both.

Tunnels, toll plazas, and movable bridges are nearly always lighted, as are bridges of substantial length in urban and suburban areas. It is questionable whether the cost of lighting long bridges in rural areas is justified or desirable.

To minimize the effect of glare and to provide the most economical lighting installation, luminaires are mounted at heights of at least 9 m [30 ft]. Lighting uniformity is improved with higher mounting heights, and in most cases, mounting heights of 10 to 15 m [35 to 50 ft] are usually preferable. High mast lighting, special luminaires on masts of 30 m [100 ft], is used to light large highway areas such as interchanges and rest areas. This lighting furnishes a uniform light distribution over the whole area and may provide alignment guidance. However, it also has a disadvantage in that the visual impact on the surrounding community from scattered light is increased.

Luminaire supports (poles) should be placed outside the roadside clear zones whenever practical. The appropriate clear zone dimensions for the various functional classifications will be found in the discussion of horizontal clearance to obstructions in Chapter 4. Where poles are located within the clear zone, regardless of distances from the traveled way, they should be designed to have a suitable impact attenuation feature; normally, a breakaway design is used. Breakaway poles should not be used on streets in densely developed areas, particularly with sidewalks. When struck, these poles could interfere with pedestrians and cause damage to adjacent buildings. Because of lower speeds and parked vehicles, there is much less chance of injuries to vehicle occupants from striking fixed poles on a street as compared to a highway. Poles should not be erected along the outside of curves on ramps where they are more susceptible to being struck. Poles located behind longitudinal barriers (installed for other purposes) should be offset sufficiently to allow for deflection of the longitudinal barriers under impact.

On a divided highway or street, luminaire supports may be located either in the median or on the right side of the roadway. Where luminaire supports are located on the right side of the roadway, the light source is usually closer to the more heavily used traffic lanes. However, with median installation, the cost is generally lower and illumination is greater on the high-speed lanes. For median installations, dual-mast arms should be used, for which 12 to 15 m [40 to 50 ft] mounting heights are favored. These should be protected with a suitable longitudinal barrier. On narrow medians, it is usually preferable to place the luminaire supports so they are integral with the median barrier.

Where highway lighting is being considered for future installation, considerable savings can be effected through design and installation of necessary conduits under roadways and curbs as part of initial construction.

Highway lighting for freeways is intimately associated with the type and location of highway signs. For full effectiveness, the two should be designed jointly.

Utilities

Highway and street improvements, whether upgraded within the existing right-of-way or entirely on new right-of-way, generally entail adjustment of utility facilities. Although utilities generally have little effect on the geometric design of the highway or street, full consideration should be given to measures, reflecting sound engineering principles and economic factors, needed to preserve and protect the integrity and visual quality of the highway or street, its maintenance efficiency, and the safety of traffic. The costs of utility adjustments vary considerably because of the large number of companies, type and complexity of the facility, and the degree of involvement with the improvement. Depending on the location of a project, the utilities involved could include (1) sanitary sewers; (2) water supply lines; (3) oil, gas, and petroleum product pipelines; (4) overhead and underground power and communications lines including fiber optic cable; (5) cable television; (6) wireless communication towers; (7) drainage and irrigation lines; (8) heating mains; and (9) special tunnels for building connections.

General

Utility lines should be located to minimize need for later adjustment, to accommodate future highway or street improvements, and to permit servicing such lines with minimum interference to traffic.

Longitudinal installation should be located on uniform alignment as near as practical to the right-of-way line so as to provide a safe environment for traffic operation and preserve space for future highway or street improvements or other utility installations. To the extent practical, utilities along freeways should be constructed so they can be serviced from outside the controlled access lines.

To the extent practical, utility line crossings of the highway should cross on a line generally normal to the highway alignment. Those utility crossings that are more likely to need future servicing should be encased or installed in tunnels to permit servicing without disrupting the traffic flow.

The horizontal and vertical location of utility lines within the highway right-of-way limits should conform to the clear roadside policies applicable for the system, type of highway or street, and specific conditions for the particular section involved. Safety of the traveling public should be a prime consideration in the location and design of utility facilities on highway and street rights-of-way. The clear roadside dimension to be maintained for a specific functional classification is discussed in the section on "Horizontal Clearance to Obstructions" in Chapter 4.

Sometimes attachment of utility facilities to highway structures, such as bridges, is a practical arrangement and may be authorized. Where it is practical to locate utility lines elsewhere, attachment to bridge structures should be avoided.

On new installations or adjustments to existing utility lines, provision should be made for known or planned expansion of the utility facilities, particularly those located underground or attached to bridges.

All utility installations on, over, or under highway or street right-of-way and attached structures should be of durable materials designed for long service-life expectancy, relatively free from routine servicing and maintenance, and meet or exceed the applicable industry codes or specifications.

Utilities that are to cross or otherwise occupy the right-of-way of rural or urban freeways should conform to the AASHTO *Policy on the Accommodation of Utilities Within Freeway Right-of-Way* (**61**). Those on non-controlled access highways and streets should conform to the AASHTO *Guide for Accommodating Utilities Within Highway Right-of-Way* (**62**).

Urban

Because of lack of space in most metropolitan areas, special consideration should be given in the initial highway design to the potential for joint usage of the right-of-way that is consistent with the primary function of the highway or street.

Appurtenances to underground installations, such as vents, drains, markers, manholes, and shutoffs, should be located so as not to interfere with the safety or maintenance of the highway or street, and so as not to be concealed by vegetation. Preferably they should be located near the right-of-way line.

Where there are curbed sections, utilities should be located in the border areas between the curb and sidewalk, at least 0.5 m [1.5 ft] behind the face of the curb, and where practical, behind the sidewalk. Where shoulders are provided rather than curbs, a clear zone commensurate with rural conditions should be provided.

Existing development and limited right-of-way widths may preclude location of some or all utility facilities outside the roadway of the street or highway. Under some conditions, it may be appropriate to reserve the area outside the roadway exclusively for the use of overhead lines with all other utilities located under the roadway, and in some instances the location of all the facilities under the roadway may be appropriate. Location under the roadway is an exception to the stated policy and as such requires special consideration and treatment. Accommodation of these facilities under the roadway should be accomplished in a manner that will ensure a minimum adverse effect on traffic as a result of future utility service and maintenance activities.

Rural

On new construction no utility should be situated under any part of the roadway, except where it crosses the highway.

Normally, no poles should be located in the median of divided highways. Utility poles, vent standpipes, and other above-ground utility appurtenances that may be struck by errant vehicles should not be permitted within the highway clear zone. The only exceptions permitted would be where the appurtenance is breakaway or could be installed behind a traffic barrier erected to protect errant vehicles from some other potential risk. The AASHTO *Roadside Design Guide* (**63**) discusses clear zone widths and may be used as a reference to determine appropriate widths for freeways, rural arterials, and high-speed rural collectors. For low-speed rural collectors and rural local roads, except for very low-volume local roads with ADTs less than or equal to 400 vehicles per day, a minimum clear zone of 3 m [10 ft] should be provided.

Traffic Control Devices

Signing and Marking

Signing and marking are directly related to the design of the highway or street and are features of traffic control and operation that the designer should consider in the geometric layout of such a facility. The signing and marking should be designed concurrently with the geometrics. The potential for future operational problems can be significantly reduced if signing and marking are treated as an integral part of design. The extent to which signs and markings are used depends on the traffic volume, the type of facility, and the extent of traffic control appropriate for safe and efficient operation. Arterial highways are usually numbered routes of fairly high type and have relatively high traffic volumes. On such highways, signs and markings are employed extensively. Local roads are low-type highways and usually have low volumes and speeds. On these facilities the use of complex traffic control devices is limited.

Although safety and efficiency of operation depend to a considerable degree on the geometric design of the facility, the physical layout should also be supplemented by effective signing as a means of informing, warning, and controlling drivers. Signing plans coordinated with horizontal and vertical alignment, sight distance obstructions, operational speeds and maneuvers, and other applicable items should be worked out before completion of design.

Highway signs are of three general types: regulatory signs, used to indicate the rules for traffic movement; warning signs, used to indicate conditions that may involve risk to highway users; and guide signs, used to direct traffic along a route or toward a destination. Uniformity in the use of signs and other traffic control devices is the main objective of the policies contained in the MUTCD (**6**).

Location, reflectorization, and lighting of signs are important considerations in signing. For details regarding design, location, and application of signs, reference should be made to the MUTCD (**6**).

Because supports for highway signs have the potential of being struck by motorists, signs should be placed on structures, outside the clear zone, or behind traffic barriers placed for other reasons. If these measures are not practical, the sign supports should be breakaway or, for overhead sign supports, shielded by appropriate traffic barriers. The AASHTO *Standard Specifications for Structural Supports for Highway Signs, Luminaires, and Traffic Signals* (**64**) provides the criteria for breakaway sign supports. Likewise, sign supports should not be placed to block sidewalks. Sign supports in sidewalks can severely impact pedestrians with vision impairments and are obstacles to all pedestrians.

Markings and markers, like signs, have the function of controlling traffic to encourage safe and efficient operation. Markings or markers either supplement regulatory or warning signs or serve independently to indicate certain regulations or warn of certain conditions present on the highway. For highways and streets there are three general types of markings in use—pavement markings, object markings, and delineators.

Pavement markings include centerline stripes, lane lines, and edge striping. These may be supplemented by other pavement markings, such as approach to obstructions, stop and crosswalk lines, and various word and symbol markings.

Physical obstructions in or near the roadway should be removed in order to provide the appropriate clear zone. Where removal is impractical, such objects should be adequately marked by painting or by use of other high-visibility material. Where the object is in the direct line of traffic, the obstruction and marking thereon preferably should be illuminated at night by floodlighting; where this is not practical, the object markings should be effectively reflectorized.

Post-mounted delineators are another type of marking device used to guide traffic, particularly at night. Reflector units are installed at certain heights and spacings to delineate the roadway where alignment changes may be confusing and not clearly defined. Refer to the MUTCD (**6**) for marking criteria, methods, and policies.

Traffic Signals

Traffic-control signals are devices that control vehicular and pedestrian traffic by assigning the right-of-way to various movements for certain pre-timed or traffic-actuated intervals of time. They are one of the key elements in the function of many urban streets and of some rural intersections. For this reason the planned signal operation for each intersection of a facility should be integrated with the design so as to achieve optimum operational efficiency. Careful consideration should be given in plan development to intersection and access locations, horizontal and vertical curvature with respect to signal visibility, pedestrian needs, and geometric schematics to ensure effective signal operation (individual signal phasing and traffic coordination between signals). In addition to initial installation, potential future signal needs should also be evaluated. The design of traffic signal devices and warrants for their use are covered in the MUTCD (**6**).

Lane arrangement is the key to successful operation of signalized intersections. The crossing distances for both vehicles and pedestrians should be kept as short as practical to reduce exposure to conflicting movements. Therefore, the first step in the development of intersection geometrics should be a complete analysis of current and future traffic demand. The need to provide right- and left-turn lanes to minimize the interference of turning traffic with the movement of through traffic should be evaluated concurrently with the potential for obtaining any additional right-of-way needed. Along a highway or street with a number of signalized intersections, the locations where turns will, or will not, be accommodated should also be examined to ensure a good fit with two-way signal coordination. Because of the large volume of traffic turning into and out of large parking areas, parking area entrances and exits should be designed in a manner that will simplify the operation of the affected traffic signals.

Noise Barriers

In recognition of the adverse effect that noise can have on people living on, working on, or otherwise using land adjacent to highways, noise barriers are being used to an increasing extent. Such noise barriers have been constructed on both new and existing highways.

Careful consideration should be exercised to ensure that the construction of these noise barriers will not compromise the safety of the highway. Every effort should be made to locate noise barriers to allow for sign placement and to provide the horizontal clearances to obstructions outside the edge-of-traveled way established in Chapter 4. It is recognized, however, that such a setback may sometimes be impractical. In such situations, the largest practical width commensurate with cost-effectiveness consideration should be provided. Stopping sight distance is another important design consideration. Therefore, horizontal clearances should be checked for adequate sight distances. Construction of a noise barrier should be avoided at a given location if it would limit stopping sight distance below the minimum values shown in Exhibit 3-1. This situation could be particularly critical where the location of the noise barrier is along the inside of a curve. Some designs use a concrete safety shape either as an integral part of the noise barrier or as a separate roadside barrier between the edge of roadway and the noise barrier. On non-tangent alignments a separate concrete roadside barrier may obstruct sight distance even though the noise barrier does not. In such instances it may be appropriate to install metal rather than concrete roadside barriers in order to retain adequate sight distance. Care should be exercised in the location of noise barriers near gore areas. Barriers at these locations should begin or terminate, as the case may be, at least 60 m [200 ft] from the theoretical nose.

For further discussion on noise barriers, see the section on noise control in Chapter 4.

Fencing

Highway agencies use fencing extensively to delineate the acquired control of access for a highway. While provision of fencing is not a duty, fencing may also serve to reduce the likelihood of encroachment onto the highway right-of-way.

Any portion of a highway with full control of access may be fenced except in areas of precipitous slopes, natural barriers, or where it can be established that fencing is not needed to preserve access control. Fencing is usually located at or near the right-of-way line or, where frontage roads are used, in the area between the through highway and the frontage road (outer separation).

Fencing for access control is usually owned by the highway agency so that the agency has control of the type and location of fence. The lowest cost type of fence best suited to the specific adjacent land use is generally provided. If fencing is not needed for access control, the fence should be the property of the adjacent landowner.

Maintenance of Traffic Through Construction Areas

Maintenance of a safe flow of traffic during construction should be carefully planned and executed. Although it is often better to provide detours, this is frequently impractical and flow of traffic is maintained through the construction area. Sometimes traffic lanes are closed, shifted, or encroached upon in order that the construction can be undertaken. When this occurs, designs for traffic control should minimize the effect on traffic operations by minimizing the frequency or duration of interference with normal traffic flow. The development of traffic control plans is an essential part of the overall project design and may affect the design of the facility itself. The traffic control plan depends on the nature and scope of the improvement, volumes of traffic, highway or street pattern, and capacities of available highways or streets. A well-thought-out and carefully developed plan for the movement of traffic through a work zone will contribute significantly to the safe and efficient flow of traffic as well as the safety of the construction forces. It is desirable that such plans have some built-in flexibility to accommodate unforeseen changes in work schedule, delays, or traffic patterns.

The goal of any traffic control plan should be to safely route vehicle, bicycle, worker access, and pedestrian traffic, including persons with disabilities, through or around construction areas with geometrics and traffic control devices as nearly comparable to those for normal operating situations as practical, while providing room for the contractor to work effectively. Policies for the use and application of signs and other traffic control devices when highway construction occurs are set forth in the MUTCD (**6**). It cannot be emphasized too strongly that the MUTCD (**6**) principles should be applied and a plan developed for the particular type of work performed.

Adequate advance warning and sufficient follow-up information should be provided to drivers to prepare them for the changed operating conditions in construction areas. The distance that such signing should be located in advance of the work zone varies with the speed on the affected facility. Size of signs may vary depending on the need for greater legibility and emphasis or the type of highway. Construction operations frequently create the need for adjustments in traffic patterns including the shifting of lanes. The minimum taper length for lane transitions in construction areas can be computed by a formula found in the MUTCD (**6**). Various configurations are illustrated in the MUTCD (**6**) and should be used in developing traffic control plans.

The stopping of traffic by a flagger or any other means should be avoided wherever practical. Designs that provide for constant movement around an obstruction in the roadway, even if it is slow, are more acceptable and are less irritating to drivers than designs that require them to stop.

When construction operations are scheduled to take place adjacent to passing traffic, a clear zone should be included in the traffic control plans, wherever practical, between the work space and the passing traffic. Under certain conditions, a positive barrier is justified.

Traffic operational considerations for the design of a detour are speed, capacity, travel distance, and safety. The speed for a detour may be less than that on the facility being improved but should be high enough so as not to affect the capacity. When an existing highway or street is used as a detour, higher volumes result and it may be appropriate to increase the capacity of such a route in advance. The capacity is generally increased by eliminating troublesome turning movements, rerouting transit vehicles and trucks, banning parking, adopting and enforcing a loading/unloading ban during peak hours, eliminating or adjusting certain transit stops, coordinating signal timing, and sometimes physically widening the traveled way. An effective means of increasing capacity is by instituting a one-way detour system, coupled with parking restrictions. A detour plan is tested by comparing the traffic volumes expected to use the rearranged plan to the calculated capacity of the detour system.

The roadway near construction access points should be well lighted and delineated. Channelization of traffic should be accomplished by the use of signing on yielding supports, pavement markings, and barricades.

Construction areas, detours, and temporary connections often include geometric features and roadway environments that may need more caution and alertness than is normally expected of drivers. Care in the layout of these areas, in the use of delineation and warning devices, and in the establishment of areas for contractor operations is appropriate to minimize the impact on the safety of both motorists and workers. Items that should be considered in developing traffic control plans include the following:

- Diversion and detour alignments to allow traffic to pass smoothly around the work zones. The surface of the traveled way, whether located within the construction area or on a detour, should be maintained in a condition that will permit the safe movement of traffic at a reasonable speed.
- Adequate tapers for lane drops or where traffic is shifted laterally. Appropriate values for taper lengths can be found in the MUTCD (**6**).

- In urban areas, diversion provisions for all existing pedestrian flows. The selected diversion paths should include safe roadway crossings, a smooth surface, and adequate width to accommodate persons with disabilities.
- Adequate traffic control devices and pavement markings for both daytime and nighttime effectiveness, including specifying temporary marking materials that can be removed when traffic-lane patterns change.
- Roadway illumination and warning lights where justified. Steady burning lights are used to delineate a continuous travel path through or around a work zone. The very short "on" time of flashing lights does not enable motorists to focus on the light and make a depth perception estimate. The use of flashers should be limited to marking a single object or condition, marking the start of a section using steady burn lights, and for use with traffic control signs.
- The location of cones, delineators, drums, barriers, or barricades, to channelized traffic, when special conditions exist or if not shown in the standard plans.
- Policies concerning the removal of signs and markings from the job site, when they are no longer needed, if not provided for in the specifications.
- Except in extenuating circumstances, the removal of contractor equipment completely off the roadways, medians, and shoulders at night, on weekends, and whenever equipment is not in operation. In those instances where such removal is not practical, appropriate signing, lighting, barricades, barriers, and similar devices to protect the motorist from collision with the equipment should be specified. The storage of hazardous materials, however, should not be permitted on roadways, medians, or shoulders near the flow of traffic.
- A requirement in the plans or specifications for controlling or prohibiting the parking of employee's private vehicles in those areas on the project that may compromise the safety of workers and through traffic.

REFERENCES

1. Johansson, G., and K. Rumar. "Drivers' Brake Reaction Times," *Human Factors*, Vol. 13, No. 1, February 1971: 23-27.

2. Massachusetts Institute of Technology. *Report of the Massachusetts Highway Accident Survey*, CWA and ERA project, Cambridge, Mass.: Massachusetts Institute of Technology, 1935.

3. Normann, O. K. "Braking Distances of Vehicles from High Speeds," *Proceedings HRB*, Vol. 22, Highway Research Board, 1953: 421-436.

4. Fambro, D. B., K. Fitzpatrick, and R. J. Koppa. *Determination of Stopping Sight Distances*, NCHRP Report 400, Washington, D.C.: Transportation Research Board, 1997.

5. AASHTO. *Guidelines for Skid Resistant Pavement Design*, Washington, D.C.: AASHTO, 1976.

6. U.S. Department of Transportation, Federal Highway Administration. *Manual on Uniform Traffic Control Devices for Streets and Highways*, Washington, D.C.: 1988 or most current edition.

7. Alexander, G. J., and H. Lunenfeld. *Positive Guidance in Traffic Control*, Washington, D.C.: U.S. Department of Transportation, Federal Highway Administration, 1975.

8. King, G. F., and H. Lunenfeld. *Development of Information Requirements and Transmission Techniques for Highway Users*, NCHRP Report 123, Washington, D.C.: Transportation Research Board, 1971.

9. McGee, H. W., W. Moore, B. G. Knapp, and J. H. Sanders. *Decision Sight Distance for Highway Design and Traffic Control Requirements*, Report No. FHWA-RD-78-78, McLean, Virginia: U.S. Department of Transportation, Federal Highway Administration, February 1978.

10. Robinson, G. H., D. J. Erickson, G. L. Thurston, and R. L. Clark. "Visual Search by Automobile Drivers," *Human Factors*, Vol. 14, No. 4, August 1972: 315-323.

11. Prisk, C. W. "Passing Practices on Rural Highways," *Proceedings HRB*, Vol. 21, Highway Research Board, 1941: 366-378.

12. Weaver, G. D., and J. C. Glennon. *Passing Performance Measurements Related to Sight Distance Design, Report 134-6*, College Station, Texas: Texas Transportation Institute, Texas A&M University, July 1971.

13. Weaver, G. D., and D. L. Woods. *Passing and No-Passing Signs, Markings, and Warrants*, Report No. FHWA-RD-79-5, Washington, D.C.: U.S. Department of Transportation, Federal Highway Administration, September 1978.

14. Transportation Research Board. *Highway Capacity Manual*, Special Report 209, Washington, D.C.: Transportation Research Board, 2000 or most current edition.

15. Harwood, D. W., J. M. Mason, R. E. Brydia, M. T. Pietrucha, and G. L. Gittings. *Intersection Sight Distance*, NCHRP Report 383, Washington, D.C.: Transportation Research Board, 1996.

16. Moyer, R. A. "Skidding Characteristics of Automobile Tires on Roadway Surfaces and Their Relation to Highway Safety," *Bulletin No. 120*, Ames, Iowa: Iowa Engineering Experiment Station, 1934.

17. Stonex, K. A., and C. M. Noble. "Curve Design and Tests on the Pennsylvania Turnpike," *Proceedings HRB*, Vol. 20, Highway Research Board, 1940: 429-451.

18. Moyer, R. A., and D. S. Berry. "Marking Highway Curves with Safe Speed Indications." *Proceedings HRB*, Vol. 20, Highway Research Board, 1940: 399-428.

19. Barnett, J. "Safe Side Friction Factors and Superelevation Design," *Proceedings HRB*, Vol. 16, Highway Research Board, 1936: 69-80.

20. Bonneson, J. A. *Superelevation Distribution Methods and Transition Designs*, NCHRP Project 439, Washington, D.C.: Transportation Research Board, 2000.

21. Hajela, G. P. *Compiler, Resume of Tests on Passenger Cars on Winter Driving Surfaces, 1939-1966*, Chicago: National Safety Council, Committee on Winter Driving Hazards, 1968.

22. MacAdam, C. C., P. S. Fancher, and L. Segal. *Side Friction for Superelevation on Horizontal Curves*, Report No. FHWA-RD-86-024, McLean, Virginia: U.S. Department of Transportation, Federal Highway Administration, August 1985.

23. Tunnard, C., and B. Pushkarev. *Man Made America: Chaos or Control?* New Haven: Yale University Press, 1963.

24. Barnett, J. *Transition Curves for Highways*, Washington, D.C.: Federal Works Agency, Public Roads Administration, 1940.

25. Shortt, W. H. "A Practical Method for Improvement of Existing Railroad Curves," *Proceedings Institution of Civil Engineering*, Vol. 76, London: Institution of Civil Engineering, 1909: 97-208.

26. Bureau of Public Roads. *Study of Speed Curvature Relations of Pentagon Road Network Ramps*, unpublished data, Washington, D.C.: Federal Works Agency, Public Roads Administration, 1954.

27. Cysewski, G. R. "Urban Intersectional Right Turning Movements," *Traffic Engineering*, Vol. 20, No. 1, October 1949: 22-37.

28. George, L. E. "Characteristics of Left-Turning Passenger Vehicles," *Proceedings HRB*, Vol. 31, Highway Research Board, 1952: 374-385.

29. Harwood, D. W., J. M. Mason, W. D. Glauz, B. T. Kulakowski, and K. Fitzpatrick. *Truck Characteristics for Use in Highway Design and Operation*, Report No. FHWA-RD-89-226, McLean, Virginia: U.S. Department of Transportation, Federal Highway Administration, August 1990.

30. *Offtracking Characteristics of Trucks and Truck Combinations*, Research Committee Report No. 3, San Francisco, California: Western Highway Institute, February 1970.

31. AASHTO. *Standard Specifications for Highway Bridges*, Washington, D.C.: AASHTO, 1996.

32. Raymond, Jr., W. L. "Offsets to Sight Obstructions Near the Ends of Horizontal Curves," *Civil Engineering*, ASCE, Vol. 42, No. 1, January 1972: 71-72.

33. Taragin, A. "Effect of Length of Grade on Speed of Motor Vehicles," *Proceedings HRB*, Vol. 25, Highway Research Board, 1945: 342-353.

34. Willey, W. E. "Survey of Uphill Speeds of Trucks on Mountain Grades," *Proceedings HRB*, Vol. 29, Highway Research Board, 1949: 304-310.

35. Huff, T. S., and F. H. Scrivner. "Simplified Climbing-Lane Design Theory and Road-Test Results," *Bulletin 104*, Highway Research Board, 1955: 1-11.

36. Schwender, H. C., O. K. Normann, and J. O. Granum. "New Method of Capacity Determination for Rural Roads in Mountainous Terrain," *Bulletin 167*, Highway Research Board, 1957: 10-37.

37. Hayhoe, G. F., and J. G. Grundmann. *Review of Vehicle Weight/Horsepower Ratio as Related to Passing Lane Design Criteria*, Final Report of NCHRP Project 20-7(10), University Park, Pennsylvania: Pennsylvania State University, October 1978.

38. Gillespie, T. *Methods for Predicting Truck Speed Loss on Grades*, Report No. FHWA/RD-86/059, McLean, Virginia: U.S. Department of Transportation, Federal Highway Administration, October 1986.

39. Fancher, Jr., P. S., and T. D. Gillespie. *Truck Operating Characteristics*, NCHRP Synthesis of Highway Practice 241, Washington, D.C.: Transportation Research Board, 1997.

40. Walton, C. M., and C. E. Lee. *Speed of Vehicles on Grades*, Research Report 20-1F, Austin, Texas: Center for Highway Research, University of Texas at Austin, August 1975.

41. Glennon, J. C. "An Evaluation of Design Criteria for Operating Trucks Safely on Grades," *Highway Research Record 312*, Highway Research Board, 1970: 93-112.

42. Harwood, D. W., and C. J. Hoban. *Low Cost Methods for Improving Traffic Operations on Two-Lane Roads*, Report No. FHWA-IP-87-2, McLean, Virginia: U.S. Department of Transportation, Federal Highway Administration, 1987.

43. Harwood, D. W., and A. D. St. John. *Passing Lanes and Other Operational Improvements on Two-Lane Highways*, Report No. FHWA/RD-85/028, McLean, Virginia: Federal Highway Administration, December 1985.

44. Witheford, D. K. *Truck Escape Ramps*, NCHRP Synthesis of Highway Practice 178, Washington, D.C.: Transportation Research Board, May 1992.

45. *Grade Severity Rating System Users Manual*, Report No. FHWA-IP-88-015, McLean, Virginia: Federal Highway Administration, August 1989.

46. Institute of Transportation Engineers. *Truck Escape Ramps, Recommended Practice*, Washington, D.C.: Institute of Transportation Engineers, 1989.

47. Cron, F. W. "The Art of Fitting the Highway to the Landscape," in W. B. Snow, ed., *The Highway and the Landscape*, New Brunswick, New Jersey: Rutgers University Press, 1959.

48. Leisch, J. E. *Application of Human Factors in Highway Design*, unpublished paper presented at AASHTO Region 2 meeting, June 1975.

49. AASHTO. *A Guide for Transportation Landscape and Environmental Design*, Washington, D.C.: AASHTO, 1991.

50. Smith, B. L., and Lamm, Ruediger. "Coordination of Horizontal and Vertical Alignment with Regard to Highway Aesthetics," *Transportation Research Record 1445*, Transportation Research Board, 1994.

51. "Roads," Chapter Four in *National Forest Landscape Management*, Vol. 2, Forest Service, U.S. Department of Agriculture, March 1977.

52. *Practical Highway Aesthetics*. New York, New York: ASCE, 1977.

53. *Design Guidelines for the Control of Blowing and Drifting Snow*, Strategic Highway Research Program, National Research Council, 1994.

54. AASHTO. *Highway Drainage Guidelines*, Vols. 1-11, Washington, D.C.: AASHTO, 1993.

55. AASHTO. *Model Drainage Manual*, Washington, D.C.: AASHTO, 1991.

56. Federal Highway Administration computer software and related publications are available from McTRANS, 512 Weil Hall, University of Florida, Gainesville, Florida 32611-2083. Phone (904) 392-0378 or PC-TRANS, 2011 Learned Hall, University of Kansas, Lawrence, Kansas, 66045. Telephone (913) 864-3199:

    HY 7. *Bridge Waterways Analysis Model*, (WSPRO), 1998. WSPRO Research Report. FHWA-RD-86-108, NTIS PB87-216107, WSPRO Users Manual (Version P60188), 1990. FHWA-IP-89-27, NTIS PB218420.

    HY 8. *FHWA Culvert Analysis* (Version 6.1), 1999. Research Report (Version 1.0), 1987. HY 8 Applications Guide, 1987, FHWA-ED-87-101.

    HY 22, *Urban Drainage Design Programs*, Version 2.1, 1998.

    HYDRAIN. *Drainage Design System* (Version 6.1), 1999.

    HYDRAIN Users Manual, 1999.

57. Richardson, E. V., et al. *Highways in the River Environment: Hydraulic and Environmental Design Considerations*, prepared by the Civil Engineering Department, Engineering Research Center, Colorado State University for the U.S. Department of Transportation, Federal Highway Administration. Washington, D.C.: February 1990.

58. Federal Highway Administration Publications, Hydraulic Design Series (HDS) and Hydraulic Engineering Circulars (HEC). Washington, D.C.: U.S. Department of Transportation. Available from National Technical Information Service (NTIS), 5285 Port Royal Road, Springfield, VA 22161. Telephone (703) 487-4650:

    HDS 1. *Hydraulics of Bridge Waterways*, 1978. FHWA-EPD-86-101. NTIS PB86-181708.

    HDS 2. *Highway Hydrology* (SI), 1996. FHWA-SA-96-067, NTIS PB97-134290.

    HDS 3. *Design Charts for Open-Channel Flow*, 1961. FHWA-EPD-86-102. NTIS PB86-179249.

    HDS 4. *Introduction to Hydraulics* (SI), 1997. FHWA-HI-97-028. NTIS PB97-186761.

HDS 5. *Hydraulic Design of Highway Culverts*, 1985. FHWA-IP-65-15. NTIS PB86-196961.

HEC 9. *Debris-Control Structures*, 1971. FHWA-EPD-86-106. NTIS PB86-179801.

HEC 11. *Design of Riprap Revetments*, 1989, FHWA-IP-89-0106. NTIS PB89-218424.

HEC 12. *Drainage of Highway Pavements*, 1984. FHWA-TS-84-202. NTIS PB84-215003.

HEC 14. *Hydraulic Design of Energy Dissipaters for Culverts and Channels*, 1983. FHWA-EPD-86-110. NTIS PB86-180205.

HEC 15. *Design of Roadside Channels with Flexible Linings*, 1988. FHWA-IP-87-7. NTIS PB89-122584.

HEC 17. *Design of Encroachments on Flood Plains Using Risk Analysis*, 1981. FHWA EPD86-112. NTIS PB86-182110.

HEC 18. *Evaluating Scour at Bridges*, 1995. FHWA-HI-96-301. NTIS PB96-163498.

HEC 20. *Stream Stability at Highway Structures*, Edition 2, (SI), 1995. FHWA-HI-96-032. NTIS PB96-163480.

HEC 21. *Bridge Deck Drainage Systems*, 1993. FHWA-SA-92-010. HTIS PB94-109584.

HEC 22. *Urban Drainage Design Manual* (SI), 1996. FHWA-SA-96-078. NTIS PB97-199491.

HEC 23. *Bridge Scour and Stream Instability Countermeasures* (SI), 1997. FHWA-HI-97-030. NTIS PB97-199491.

59. AASHTO. *A Guide for Development of Rest Areas on Major Arterials and Freeways*, Washington, D.C.: AASHTO, 2001.

60. AASHTO. *An Informational Guide for Roadway Lighting*, Washington, D.C.: AASHTO, 1984.

61. AASHTO. *A Policy on the Accommodation of Utilities Within Freeway Right-of-Way*, Washington, D.C.: AASHTO, 1989.

62. AASHTO. *A Guide for Accommodating Utilities Within Highway Right-of-Way*, Washington, D.C.: AASHTO, 1994.

63. AASHTO. *Roadside Design Guide*, Washington, D.C.: AASHTO, 1996.

64. AASHTO. *Standard Specifications for Structural Supports for Highway Signs, Luminaires, and Traffic Signals*, Washington, D.C.: AASHTO, 1994.

65. Brudis and Associates, Inc. *Advisory Speeds on Maryland Roads,* Hanover, Maryland: Maryland Department of Transportation, Office of Traffic and Safety, August 1999.

CHAPTER 4
CROSS SECTION ELEMENTS

GENERAL

To assure consistency in this policy, the terms "roadway" and "traveled way" are defined by AASHTO as follows:

Roadway: The portion of a highway, including shoulders, for vehicular use. A divided highway has two or more roadways (see Exhibits 4-1 and 4-2).

Traveled way: The portion of the roadway for the movement of vehicles, exclusive of shoulders (see Exhibits 4-1 and 4-2).

PAVEMENT

Surface Type

The selection of pavement type is determined based on the traffic volume and composition, soil characteristics, weather, performance of pavements in the area, availability of materials, energy conservation, initial cost, and the overall annual maintenance and service-life cost. The structural design of pavements is not included in this policy, but is addressed in the AASHTO *Guide for Design of Pavement Structures* (**1**).

Important pavement characteristics that are related to geometric design are the effect on driver behavior and the ability of a surface to retain its shape and dimensions, to drain, and to retain adequate skid resistance. High-type pavements retain their shape and do not ravel at the edges if placed on a stable subgrade. Their smoothness and proper cross-slope design enable drivers to steer easily and keep their vehicles moving in the proper path. At the other extreme, low-type surfaces have a tendency toward raveling, which reduces their effective width and requires greater steering effort to maintain a correct path. Accordingly, low-type surfaces are used where traffic volume is light.

While the selection of design speed is dependent on many factors other than pavement surface type, high-type surfaces provide for higher operating speeds than do low-type surfaces. Therefore, the surface type provided should be consistent with the selected design speed for the highway.

Cross Slope

Undivided traveled ways on tangents, or on flat curves, have a crown or high point in the middle and a cross slope downward toward both edges. Unidirectional cross slopes across the

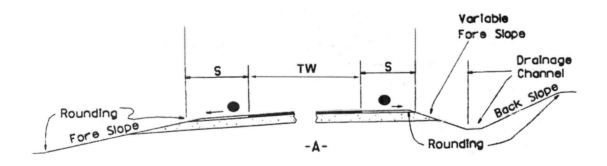

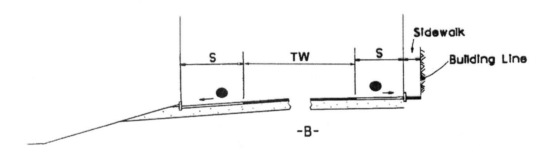

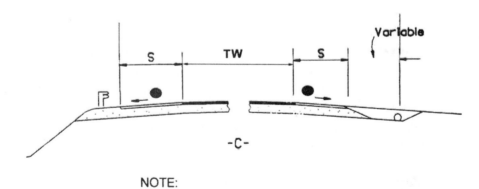

NOTE:
TW = Traveled Way
S = Usable Shoulder
● = Rate of Slope 2 to 6 Percent

Exhibit 4-1. Typical Cross Section, Normal Crown

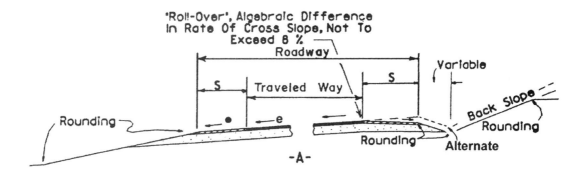

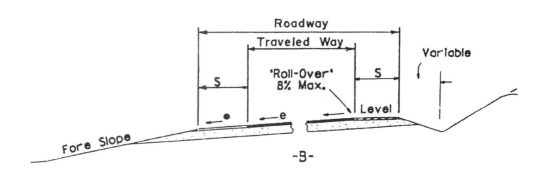

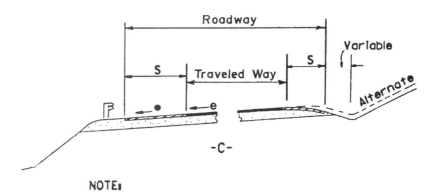

NOTE:

S = Usable Shoulder

● = Superelevation Rate (e) Where Greater
Than Normal Shoulder Slope

Exhibit 4-2. Typical Cross Section, Superelevated

entire width of the traveled way may be utilized. The downward cross slope may be a plane or rounded section or a combination. With plane cross slopes, there is a cross slope break at the crown line and a uniform slope on each side. Rounded cross sections usually are parabolic, with a slightly rounded surface at the crown line and increasing cross slope toward the edge of the traveled way. Because the rate of cross slope is variable, the parabolic section is described by the crown height (i.e., the vertical drop from the center crown line to the edge of the traveled way). The rounded section is advantageous in that the cross slope steepens toward the edge of the traveled way, thereby facilitating drainage. Disadvantages are that rounded sections are more difficult to construct, the cross slope of the outer lanes may be excessive, and warping of pavement areas at intersections may be awkward or difficult to construct.

On divided highways each one-way traveled way may be crowned separately as on two-lane highways, or it may have a unidirectional cross slope across the entire width of the traveled way, which is almost always downward to the outer edge. A cross section with each roadway crowned separately, as shown in Exhibit 4-3A through Exhibit 4-3C, has an advantage in rapidly draining the pavement during rainstorms. In addition, the difference between high and low points in the cross section is minimal. Disadvantages are that more inlets and underground drainage lines are needed, and treatment of intersections is more difficult because of the number of high and low points on the cross section. Use of such sections should preferably be limited to regions of high rainfall or where snow and ice are major factors. Sections having no curbs and a wide depressed median are particularly well-suited for these conditions.

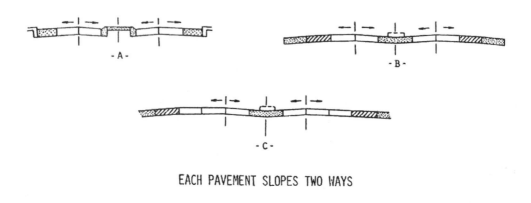

EACH PAVEMENT SLOPES TWO WAYS

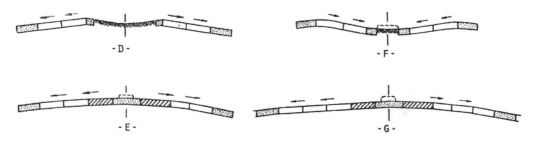

EACH PAVEMENT SLOPES ONE WAY

Exhibit 4-3. Roadway Sections for Divided Highway (Basic Cross Slope Arrangements)

Roadways with unidirectional cross slopes, as shown in Exhibit 4-3D through Exhibit 4-3G, tend to provide more comfort to drivers when they change lanes and may either drain away from or toward the median. Drainage away from the median may effect a savings in drainage structures, minimize drainage across the inner, higher-speed lanes, and simplify treatment of intersecting streets. Drainage toward the median is advantageous in that the outer lanes, which are used by most traffic, are more free of surface water. This surface runoff, however, should then be collected into a single conduit under the median. Where curbed medians exist, drainage is concentrated next to or on higher-speed lanes. When the median is narrow, this concentration results in splashing on the windshields of opposing traffic.

The rate of cross-slope is an important element in cross-section design. Superelevation on curves is determined by the speed-curvature relationships given in Chapter 3, but cross slope or crown on tangents or on long-radius curves are complicated by two contradictory controls. On one hand, a reasonably steep lateral slope is desirable to minimize ponding of water on pavements with flat profile grades as a result of pavement imperfections or unequal settlement. A steep cross slope is also desirable on curbed pavements to confine water flow to a narrow width of pavement adjacent to the curb. On the other hand, steep cross slopes are undesirable on tangents because of the tendency of vehicles to drift toward the low edge of the traveled way. This drifting becomes a major concern in areas where snow and ice are common. Cross slopes up to and including 2 percent are barely perceptible in terms of vehicle steering. However, cross slopes steeper than 2 percent are noticeable and require a conscious effort in steering. Furthermore, steep cross slopes increase the susceptibility to lateral skidding when vehicles brake on icy or wet pavements or when stops are made on dry pavements under emergency conditions.

The prevalence of high winds may significantly alter the effect of cross slope on steering. In rolling or mountainous terrain with alternate cut-and-fill sections or in areas alternately forested and cleared, any substantial cross wind produces an intermittent impact on a vehicle moving along the highway and affects its steering. In areas where such conditions are likely, it is desirable to avoid high rates of cross slope.

On high-type two-lane roadways, crowned at the center, the accepted rate of cross slope ranges from 1.5 to 2 percent. When three or more lanes are inclined in the same direction on multilane highways, each successive pair of lanes or portion thereof outward from the first two lanes from the crown line may have an increased slope. The two lanes adjacent to the crown line should be pitched at the normal minimum slope, and on each successive pair of lanes or portion thereof outward, the rate may be increased by about 0.5 to 1 percent. As shown in Exhibit 4-3G, the left side has a continuous sloped pavement while the right has an increased slope on the outer lane.

Use of cross slopes steeper than 2 percent on high-type, high-speed highways with a central crown line is not desirable. In passing maneuvers, drivers cross and recross the crown line and negotiate a total rollover or cross-slope change of over 4 percent. The reverse curve path of travel of the passing vehicle causes a reversal in the direction of centrifugal force, which is further exaggerated by the effect of the reversing cross slopes. Trucks with high centers of gravity crossing over the crown line are caused to sway from side to side when traveling at high speed, at which time control may be difficult to maintain. Exhibits 4-3A through 4-3C are examples of roadway conditions where this situation would be encountered.

In areas of intense rainfall, a somewhat steeper cross slope may be needed to facilitate roadway drainage. In such cases, the cross slope on high-type pavements may be increased to 2.5 percent, with a corresponding crown line crossover of 5 percent. Where three or more lanes are provided in each direction, the maximum cross slope should be limited to 4 percent. Use of this increased cross slope should be limited to the condition described in the preceding discussion. For all other conditions, a maximum cross slope of 2 percent should be used for high-type pavements. In locations of intense rainfall and where the maximum cross slope is used, consideration should be given to the use of grooving or open-graded mixes.

The cross slope rates discussed above pertain largely to high-type surfaces. A greater cross slope should be utilized for low-type surfaces. Exhibit 4-4 shows a range of values applicable to each type of surface.

Surface type	Range in cross-slope rate (%)
High	1.5–2
Low	2–6

Exhibit 4-4. Normal Traveled-Way Cross Slope

Because of the nature of the surfacing materials used and surface irregularities, low-type surfaces such as earth, gravel, or crushed stone need an even greater cross slope on tangents to prevent the absorption of water into the surface. Therefore, cross slopes greater than 2 percent may be used on these types of surfaces.

Where roadways are designed with outer curbs, the lower values in the ranges of cross slopes in Exhibit 4-4 are not recommended because of the increased likelihood of there being a sheet of water over a substantial part of the traveled way adjacent to the curb. For any rate of rainfall, the width of traveled way that is inundated with water varies with the rate of cross slope, roughness of gutter, frequency of discharge points, and longitudinal grade. A cross slope greater than 1 percent is desirable, and in some cases, a cross slope of more than 1.5 percent is needed to limit inundation to about half of the outer traffic lane. A cross slope of 1.5 percent is suggested as a practical minimum for curbed high-type pavement. Curbs with steeper adjacent gutter sections may permit the use of lesser rates of cross slope. A preferred cross-section treatment is the use of a straight shoulder slope and the avoidance of curbs, whenever practical.

Skid Resistance

Skidding crashes are a major concern in highway safety. It is not sufficient to attribute skidding crashes merely to "driver error" or "driving too fast for existing conditions." The roadway should provide a level of skid resistance that will accommodate the braking and steering maneuvers that can reasonably be expected for the particular site.

Research has demonstrated that highway geometrics affect skidding (**2**). Therefore, skid resistance should be a consideration in the design of all new construction and major reconstruction projects. Vertical and horizontal alignments can be designed in such a way that the potential for skidding is reduced. Also, improvements to the vertical and horizontal alignments should be considered as a part of any reconstruction project.

Pavement types and textures also affect a roadway's skid resistance. The four main causes of poor skid resistance on wet pavements are rutting, polishing, bleeding, and dirty pavements. Rutting causes water accumulation in the wheel tracks. Polishing reduces the pavement surface microtexture and bleeding can cover it. In both cases, the harsh surface features needed for penetrating the thin water film are diminished. Pavement surfaces will lose their skid resistance when contaminated by oil drippings, layers of dust, or organic matter. Measures taken to correct or improve skid resistance should result in the following characteristics: high initial skid resistance durability, the ability to retain skid resistance with time and traffic, and minimum decrease in skid resistance with increasing speed.

Tining during placement leaves indentations in the pavement surface and has proved to be effective in reducing the potential for hydroplaning on roadways with portland cement concrete surfaces. The use of surface courses or overlays constructed with polish-resistant coarse aggregate is the most widespread method for improving the surface texture of bituminous pavements. Overlays of open-graded asphalt friction courses are quite effective because of their frictional and hydraulic properties. For further discussion, refer to the AASHTO *Guidelines for Skid Resistant Pavement Design* (**3**).

LANE WIDTHS

The lane width of a roadway greatly influences the safety and comfort of driving. Lane widths of 2.7 to 3.6 m [9 to 12 ft] are generally used, with a 3.6-m [12-ft] lane predominant on most high-type highways. The extra cost of providing a 3.6-m [12-ft] lane width, over the cost of providing a 3.0-m [10-ft] lane width is offset to some extent by a reduction in cost of shoulder maintenance and a reduction in surface maintenance due to lessened wheel concentrations at the pavement edges. The wider 3.6-m [12-ft] lane provides desirable clearances between large commercial vehicles traveling in opposite directions on two-lane, two-way rural highways when high traffic volumes and particularly high percentages of commercial vehicles are expected.

Lane widths also affect highway level of service. Narrow lanes force drivers to operate their vehicles closer to each other laterally than they would normally desire. Restricted clearances have much the same effect. In a capacity sense the effective width of traveled way is reduced when adjacent obstructions such as retaining walls, bridge trusses or headwalls, and parked cars restrict the lateral clearance. Further information on the effect of lane width on capacity and level of service is presented in the *Highway Capacity Manual* (HCM) (**4**). In addition to the capacity effect, the resultant erratic operation has an undesirable effect on driver comfort and crash rates.

Where unequal-width lanes are used, locating the wider lane on the outside (right) provides more space for large vehicles that usually occupy that lane, provides more space for bicycles, and allows drivers to keep their vehicles at a greater distance from the right edge. Where a curb is used adjacent to only one edge, the wider lane should be placed adjacent to that curb. The basic design decision is the total roadway width, while the placement of stripes actually determines the lane widths.

Although lane widths of 3.6 m [12 ft] are desirable on both rural and urban facilities, there are circumstances where lanes less than 3.6 m [12 ft] wide should be used. In urban areas where pedestrian crossings, right-of-way, or existing development become stringent controls, the use of 3.3-m [11-ft] lanes is acceptable. Lanes 3.0 m [10 ft] wide are acceptable on low-speed facilities, and lanes 2.7 m [9 ft] wide are appropriate on low-volume roads in rural and residential areas. For further information, see NCHRP Report 362, *Roadway Widths for Low-Traffic Volume Roads* (5). In some instances, on multilane facilities in urban areas, narrower inside lanes may be utilized to permit wider outside lanes for bicycle use. In this situation, 3.0-m to 3.3-m [10- to 11-ft] lanes are common on inside lanes with 3.6-m to 3.9-m [12- to 13-ft] lanes utilized on outside lanes.

Auxiliary lanes at intersections and interchanges often help to facilitate traffic movements. Such added lanes should be as wide as the through-traffic lanes but not less than 3.0 m [10 ft]. Where continuous two-way left-turn lanes are provided, a lane width of 3.0 m to 4.8 m [10 to 16 ft] provides the optimum design.

It may not be cost-effective to design the lane and shoulder widths of local and collector roads and streets that carry less than 400 vehicles per day using the same criteria applicable to higher volume roads or to make extensive operational and safety improvements to such very low-volume roads. AASHTO is currently evaluating alternative design criteria for local and collector roads and streets that carry less than 400 vehicles per day based on a safety risk assessment.

SHOULDERS

General Characteristics

A shoulder is the portion of the roadway contiguous with the traveled way that accommodates stopped vehicles, emergency use, and lateral support of subbase, base, and surface courses. In some cases, the shoulder can accommodate bicyclists. It varies in width from only 0.6 m [2 ft] on minor rural roads where there is no surfacing, or the surfacing is applied over the entire roadbed, to approximately 3.6 m [12 ft] on major roads where the entire shoulder may be stabilized or paved.

The term "shoulder" is variously used with a modifying adjective to describe certain functional or physical characteristics. The following meanings apply to the terms used here:

- The "graded" width of shoulder is that measured from the edge of the traveled way to the intersection of the shoulder slope and the foreslope planes, as shown in Exhibit 4-5A.

- The "usable" width of shoulder is the actual width that can be used when a driver makes an emergency or parking stop. Where the sideslope is 1V:4H or flatter, the "usable" width is the same as the "graded" width since the usual rounding 1.2 to 1.8 m [4 to 6 ft] wide at the shoulder break will not lessen its useful width appreciably. Exhibits 4-5B and 4-5C illustrate the usable shoulder width.

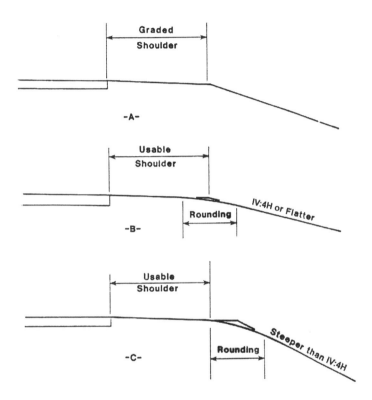

Exhibit 4-5. Graded and Usable Shoulders

Shoulders may be surfaced either full or partial width to provide a better all-weather load support than that afforded by native soils. Materials used to surface shoulders include gravel, shell, crushed rock, mineral or chemical additives, bituminous surface treatments, and various forms of asphaltic or concrete pavements.

The shoulder on minor rural roads with low traffic volume serves essentially as structural lateral support for the surfacing and as an additional width for the traveled way. This permits drivers meeting or passing other vehicles to drive on the edge of the roadway without leaving the surfacing, thus making use of the shoulder itself. Roads with a narrow traveled way, narrow shoulders, and an appreciable traffic volume tend to provide poor service, have a relatively higher crash rate, and need frequent and costly maintenance.

Well-designed and properly maintained shoulders are needed on rural highways with an appreciable volume of traffic, on freeways, and on some types of urban highways. Their advantages include:

- Space is provided away from the traveled way for vehicles to stop because of mechanical difficulties, flat tires, or other emergencies.
- Space is provided for motorists to stop occasionally to consult road maps or for other reasons.
- Space is provided for evasive maneuvers to avoid potential crashes or reduce their severity.
- The sense of openness created by shoulders of adequate width contributes to driving ease and reduced stress.
- Sight distance is improved in cut sections, thereby potentially improving safety.
- Some types of shoulders enhance highway aesthetics.
- Highway capacity is improved because uniform speed is encouraged.
- Space is provided for maintenance operations such as snow removal and storage.
- Lateral clearance is provided for signs and guardrails.
- Storm water can be discharged farther from the traveled way, and seepage adjacent to the traveled way can be minimized. This may directly reduce pavement breakup.
- Structural support is given to the pavement.
- Space is provided for pedestrian and bicycle use, for bus stops, for occasional encroachment of vehicles, for mail delivery vehicles, and for the detouring of traffic during construction.

For further information on other uses of shoulders, refer to NCHRP Report 254, *Shoulder Geometrics and Use Guidelines* (**6**).

Urban highways generally have curbs along the outer lanes. A stalled vehicle, during peak hours, disturbs traffic flow in all lanes in that direction when the outer lane serves through-traffic. Where on-street parking is permitted, the parking lane provides some of the same services listed above for shoulders. Parking lanes are discussed later in this chapter in the section on "On-Street Parking."

Width of Shoulders

Desirably, a vehicle stopped on the shoulder should clear the edge of the traveled way by at least 0.3 m [1 ft], and preferably by 0.6 m [2 ft]. This preference has led to the adoption of 3.0 m [10 ft] as the normal shoulder width that should be provided along high-type facilities. In difficult terrain and on low-volume highways, shoulders of this width may not be practical. A minimum shoulder width of 0.6 m [2 ft] should be considered for the lowest-type highway, and a 1.8- to 2.4-m [6- to 8-ft] shoulder width is preferable. Heavily traveled, high-speed highways and highways carrying large numbers of trucks should have usable shoulders at least 3.0 m [10 ft] wide and preferably 3.6 m [12 ft] wide; however, widths greater than 3.0 m [10 ft] may encourage unauthorized use of the shoulder as a travel lane. Where bicyclists and pedestrians are to be accommodated on the shoulders, a minimum usable shoulder width (i.e., clear of rumble strips) of

1.2 m [4 ft] should be used. For additional information on shoulder widths to accommodate bicycles, see the AASHTO *Guide for the Development of Bicycle Facilities* (**7**). Shoulder widths for specific classes of highways are discussed in Chapters 5 through 8.

Where roadside barriers, walls, or other vertical elements are present, it is desirable to provide a wide enough graded shoulder that the vertical elements will be offset a minimum of 0.6 m [2 ft] from the outer edge of the usable shoulder. To provide lateral support for guardrail posts and/or clear space for lateral dynamic deflection of the particular barrier in use, it may be appropriate to provide a graded shoulder that is wider than the shoulder where no vertical elements are present. On low-volume roads, roadside barriers may be placed at the outer edge of the shoulder; however, a minimum clearance of 1.2 m [4 ft] should be provided from the traveled way to the barrier.

Although it is desirable that a shoulder be wide enough for a vehicle to be driven completely off the traveled way, narrower shoulders are better than none at all. For example, when a vehicle making an emergency stop can pull over onto a narrow shoulder such that it occupies only 0.3 to 1.2 m [1 to 4 ft] of the traveled way, the remaining traveled way width can be used by passing vehicles. Partial shoulders are sometimes used where full shoulders are unduly costly, such as on long (over 60 m [200 ft]) bridges or in mountainous terrain.

Regardless of the width, a shoulder should be continuous. The full benefits of a shoulder are not realized unless it provides a driver with refuge at any point along the traveled way. A continuous shoulder provides a sense of security such that almost all drivers making emergency stops will leave the traveled way. With intermittent sections of shoulder, however, some drivers will find it necessary to stop on the traveled way, creating an undesirable situation. A continuous paved shoulder provides an area for bicyclists to operate without obstructing faster moving motor vehicle traffic. Although continuous shoulders are preferable, narrow shoulders and intermittent shoulders are superior to no shoulders. Intermittent shoulders are briefly discussed below in the section on "Turnouts."

Shoulders on structures should normally have the same width as usable shoulders on the approach roadways. As previously discussed, the narrowing or loss of shoulders, especially on structures, may cause serious operational and safety problems. Long, high-cost structures usually warrant detailed special studies to determine practical dimensions. Reduced shoulder widths may be considered in rare cases. A discussion of these conditions is provided in Chapters 7 and 10.

Shoulder Cross Sections

Important elements in the lateral drainage systems, shoulders should be flush with the roadway surface and abut the edge of the traveled way. All shoulders should be sloped to drain away from the traveled way on divided highways with a depressed median. With a raised narrow median, the median shoulders may slope in the same direction as the traveled way. However, in regions with snowfall, median shoulders should be sloped to drain away from the traveled way to avoid melting snow draining across travel lanes and refreezing. All shoulders should be sloped sufficiently to rapidly drain surface water, but not to the extent that vehicular use would be restricted. Because the type of shoulder construction has a bearing on the cross slope, the two should be determined jointly. Bituminous and concrete-surfaced shoulders should be sloped from 2 to 6 percent, gravel or crushed-rock shoulders from 4 to 6 percent, and turf shoulders from 6 to 8 percent. Where curbs are used on the outside of shoulders, the cross slope should be appropriately designed with the drainage system to prevent ponding on the traveled way.

It should be noted that rigid adherence to the slope rates outlined in this chapter may present minor traffic operational problems if they are applied without regard to the cross section of the paved surface. On tangent or long-radius curved alignment with normal crown and turf shoulders, the maximum algebraic difference in the traveled way and shoulder grades should be from 6 to 7 percent. Although this maximum algebraic difference in slopes is not desirable, it is tolerable due to the benefits gained in pavement stability by avoiding storm water detention at the pavement edge.

Shoulder slopes that drain away from the paved surface on the outside of well-superelevated sections should be designed to avoid too great a cross-slope break. For example, use of a 4 percent shoulder cross slope in a section with a traveled way superelevation of 8 percent results in a 12 percent algebraic difference in the traveled way and shoulder grades at the high edge-of-traveled way. Grade breaks of this order are not desirable and should not be permitted (Exhibit 4-2A). It is desirable that all or part of the shoulder should be sloped upward at about the same rate or at a lesser rate than the superelevated traveled way (see the dashed line labeled Alternate in Exhibit 4-2A). Where this is not desirable because of storm water or melting snow and ice draining over the paved surface, a compromise might be used in which the grade break at the edge of the paved surface is limited to approximately 8 percent by flattening the shoulder on the outside of the curve (Exhibit 4-2B).

One means of avoiding too severe of a grade break is the use of a continuously rounded shoulder cross section on the outside of the superelevated traveled way (Exhibit 4-2C). The shoulder in this case is a convex section continuing from the superelevation slope instead of a sharp grade break at the intersection of the shoulder and traveled way slopes. In this method, some surface water will drain upon the traveled way; however, this disadvantage is offset by the benefit of a smoother transition for vehicles that may accidentally or purposely drive upon the shoulder. It should also be noted that convex shoulders present more difficulties in construction than do planar sections. An alternate method to the convex shoulder consists of a planar shoulder section with multiple breaks in the cross slope. Shoulder cross slopes on the high side of a superelevated section that are substantially less than those discussed above are generally not detrimental to shoulder stability. There is no discharge of storm water from the traveled way to the shoulder and, therefore, little likelihood of shoulder erosion damage.

In some areas, shoulders are designed with a curb or gutter at the outer edge to confine runoff to the paved shoulder area. Drainage for the entire roadway is handled by these curbs, with the runoff directed to selected outlets. The outer portion of the paved shoulder serves as the longitudinal gutter. Cross slopes should be the same as for shoulders without a curb or gutter, except that the slope may be increased somewhat on the outer portion of the shoulder. This type of shoulder is advantageous in that the curb on the outside of the shoulder does not deter motorists from driving off the traveled way, and the shoulder serves as a gutter in keeping storm water off the traveled lanes. Proper delineation should adequately distinguish the shoulder from the traveled way.

Shoulder Stability

If shoulders are to function effectively, they should be sufficiently stable to support occasional vehicle loads in all kinds of weather without rutting. Evidence of rutting, skidding, or vehicles being mired down, even for a brief seasonal period, may discourage and prevent the shoulder from being used as intended.

All types of shoulders should be constructed and maintained flush with the traveled way pavement if they are to fulfill their intended function. Regular maintenance is needed to provide a flush shoulder. Unstabilized shoulders generally undergo consolidation with time, and the elevation of the shoulder at the traveled way edge tends to become lower than the traveled way. The drop-off can adversely affect driver control when driving onto the shoulder at any appreciable speed. In addition, when there is no visible assurance of a flush stable shoulder, the operational advantage of drivers staying close to the pavement edge is reduced.

Paved or stabilized shoulders offer numerous advantages, including: (1) provision of refuge for vehicles during emergency situations, (2) elimination of rutting and drop-off adjacent to the edge of the traveled way, (3) provision of adequate cross slope for drainage of roadway, (4) reduction of maintenance, and (5) provision of lateral support for roadway base and surface course.

Shoulders with turf growth may be appropriate, under favorable climatic and soil conditions, for local roads and some collectors. Turf shoulders are subject to a buildup that may inhibit proper drainage of the traveled way unless adequate cross slope is provided. When wet, the turf may be slippery unless closely mowed and on granular soil. Turf shoulders offer good traveled-way delineation and do not invite use as a traffic lane. Stabilized turf shoulders need little maintenance other than mowing.

Based on experience, drivers are wary of unstabilized shoulders, especially on high-volume highways, such as suburban expressways. Such experience has led to the replacement of unstabilized shoulders with some form of stabilized or surfaced shoulders.

In some areas, rural highways are built with surfacing over the entire width, including shoulders. Depending upon the conditions, this surfacing may be from about 8.4 to 13.2 m [28 to 44 ft] wide for two-lane roads. This type of treatment protects shoulders from erosion and also protects the subgrade from moisture penetration, thereby enhancing the strength and durability of the pavement. Also, edge stripes are generally used to delineate the edge of the traveled way, but in some cases there is no indication of the edge of traveled way. This design is desirable because a continuous shoulder is provided, even if its separate width is not apparent.

Experience on heavy-volume facilities shows that, on occasion, traffic will use smooth-surfaced shoulders as through-traffic lanes. On moderate-to-steep grades, trucks may pull to the right and encroach upon the shoulder. While such shoulder encroachments are undesirable, this does not warrant the elimination of the surfaced shoulder because of factors such as high-volume traffic and truck usage.

Shoulder Contrast

It is desirable that the color and texture of shoulders be different from those of the traveled way. This contrast serves to clearly define the traveled way at all times, particularly at night and during inclement weather, while discouraging the use of shoulders as additional through lanes. Bituminous, crushed stone, gravel, and turf shoulders all offer excellent contrast with concrete pavements. Satisfactory contrast with bituminous pavements is more difficult to achieve. Various types of stone aggregates and turf offer good contrast. Several states have attempted to achieve contrast by seal-coating shoulders with lighter color stone chips. Unfortunately, the color distinction may diminish in a few years. The use of edge lines as described in the *Manual on Uniform Traffic Control Devices* (MUTCD) (**8**) reduces the need for shoulder contrast. Edge lines should be applied where shoulder use by bicycles is expected. Some states have provided depressed rumble strips in the shoulder to provide an audible alert to the motorists that they have crossed over onto the shoulder. This is particularly effective at night and during inclement weather. However, care should be used if the shoulders are to be used by bicyclists.

Turnouts

It is not always economically practical to provide wide shoulders continuously along the highway, especially where the alignment passes through deep rock cuts or where other conditions limit the cross-section width. In such cases, consideration should be given to the use of intermittent sections of shoulder or turnouts along the highway. Such turnouts provide an area for emergency stops and also allow slower moving vehicles to pull out of the through lane to permit following vehicles to pass.

Proper design of turnouts should consider turnout length, including entry and exit tapers, turnout width, and the location of the turnout with respect to horizontal and vertical curves where sight distance is limited. Turnouts should be located so that approaching drivers have a clear view of the entire turnout in order to determine whether the turnout is available for use (**9**). Where bicycle traffic is expected, turnouts should be paved so bicyclists may move aside to allow faster traffic to pass.

HORIZONTAL CLEARANCE TO OBSTRUCTIONS

The term "clear zone" is used to designate the unobstructed, relatively flat area provided beyond the edge of the traveled way for the recovery of errant vehicles. The clear zone includes any shoulders or auxiliary lanes.

The AASHTO *Roadside Design Guide* (**10**) discusses clear zone widths as related to speed, volume, and embankment slope. The Guide may be used as a reference for determination of clear-zone widths for freeways, rural arterials, and high-speed rural collectors. For low-speed rural collectors and rural local roads, a minimum clear-zone width of 3.0 m [10 ft] should be provided.

For urban arterials, collectors, and local streets where curbs are utilized, space for clear zones is generally restricted. A minimum offset distance of 500 mm [18 in] should be provided beyond the face of the curb, with wider offsets provided where practical. This "operational" offset will generally permit curbside parking and will not have a negative impact on traffic flow. However, since most curbs do not have a significant capability to redirect vehicles, a minimum clear zone distance commensurate with prevailing traffic volumes and vehicle speeds should be provided where practical.

CURBS

General Considerations

The type and location of curbs affects driver behavior and, in turn, the safety and utility of a highway. Curbs serve any or all of the following purposes: drainage control, roadway edge delineation, right-of-way reduction, aesthetics, delineation of pedestrian walkways, reduction of maintenance operations, and assistance in orderly roadside development. A curb, by definition, incorporates some raised or vertical element.

Curbs are used extensively on all types of low-speed urban highways, as defined in the Design Speed section in Chapter 2. In the interest of safety, caution should be exercised in the use of curbs on high-speed rural highways. Where curbs are needed along high-speed rural highways due to drainage considerations, the need for access control, restricted right-of-way, or other reasons, they should always be located at the outside edge of the shoulder.

While cement concrete curbs are installed by some highway agencies, granite curbs are used where the local supply makes them economically competitive. Because of its durability, granite is preferred over cement concrete where deicing chemicals are used for snow and ice removal.

Conventional concrete or bituminous curbs offer little visible contrast to normal pavements, particularly during fog or at night when surfaces are wet. The visibility of channelizing islands with curbs and of continuous curbs along the edges of the traveled way may be improved through the use of reflectorized markers that are attached to the top of the curb.

In another form of high-visibility treatment, reflectorized paints or other reflectorized surfaces, such as applied thermoplastic, can make curbs more conspicuous. However, to be kept fully effective, reflectorized curbs need periodic cleaning or repainting, which usually involves substantial maintenance costs. Curb markings should be placed in accordance with the MUTCD (**8**).

Curb Configurations

Curb configurations include both vertical and sloping curbs. Exhibit 4-6 illustrates several curb configurations that are commonly used. A curb may be designed as a separate unit or integrally with the pavement. Vertical and sloping curb designs may include a gutter, forming a combination curb and gutter section.

Vertical curbs may be either vertical or nearly vertical and are intended to discourage vehicles from leaving the roadway. As shown in Exhibit 4-6A, they range from 150 to 200 mm [6 to 8 in] in height. Vertical curbs should not be used along freeways or other high-speed roadways because an out-of-control vehicle may overturn or become airborne as a result of an impact with such a curb. Since curbs are not adequate to prevent a vehicle from leaving the roadway, a suitable traffic barrier should be provided where redirection of vehicles is needed.

Vertical curbs and safety walks may be desirable along the faces of long walls and tunnels, particularly if full shoulders are not provided. These curbs tend to discourage vehicles from driving close to the wall, and thus the safety walk, reducing the risk to persons walking from disabled vehicles.

Sloping curbs are designed so vehicles can cross them readily when the need arises. As shown in Exhibit 4-6B through 4-6G, sloping curbs are low with flat sloping faces. The curbs shown in Exhibits 4-6B, 4-6C, and 4-6D are considered to be mountable under emergency conditions although such curbs will scrape the undersides of some vehicles. For ease in crossing, sloping curbs should be well rounded as in Exhibits 4-6B through 4-6G.

Extruded curbs of either cement or bituminous concrete are used in many states. Extruded curbs usually have sloping faces because they provide better initial stability, are easier to construct, and are more economical than steep faces. Typical extruded curb designs are shown in Exhibits 4-6C, 4-6E, and 4-6G.

When the slope of the curb face is steeper than 1V:1H, vehicles can mount the curb more readily when the height of the curb is limited to at most 100 mm [4 in] and preferably less. However, when the face slope is between 1V:1H and 1V:2H, the height should be limited to about 150 mm [6 in]. Some highway agencies construct a vertical section on the lower face of the curb (Exhibits 4-6C, 4-6D, and 4-6F) as an allowance for future resurfacing. This vertical portion should not exceed approximately 50 mm [2 in], and where the total curb height exceeds 150 mm [6 in], it may be considered a vertical curb rather than a sloping curb.

Sloping curbs can be used at median edges, to outline channelizing islands in intersection areas, or at the outer edge of the shoulder. For example, any of the sloping configurations in Exhibit 4-6 might be used for a median curb. When curbs are used to outline channelizing islands, an offset should be provided. Offsets to curbed islands are discussed in Chapter 9.

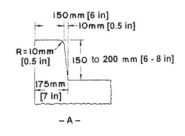

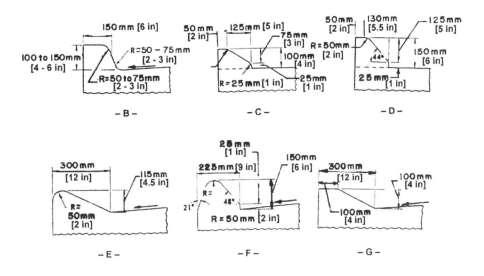

Exhibit 4-6. Typical Highway Curbs

Shoulder curbs are placed at the outer edge of the shoulder to control drainage, improve delineation, control access, and reduce erosion. These curbs, combined with a gutter section, may be part of the longitudinal drainage system. If the surfaced shoulders are not wide enough for a vehicle to park, the shoulder curb should appear to be easily mountable to encourage motorists to park clear of the traveled way. Where it is expected that bicyclists will use the roadway, sufficient width from the face of the curb should be provided so bicyclists can avoid conflict with motorists while not having to travel too close to the curb. For further information, see the AASHTO *Guide for the Development of Bicycle Facilities* (**7**).

Gutter sections may be provided on the traveled-way side of a vertical or sloping curb to form the principal drainage system for the roadway. Inlets are provided in the gutter or curb, or both. Gutters are generally 0.3 to 1.8 m [1 to 6 ft] wide, with a cross slope of 5 to 8 percent to increase the hydraulic capacity of the gutter section. In general, the 5 to 8 percent slope should be confined to the 0.6 to 0.9 m [2 to 3 ft] adjacent to the curb. Shallow gutters without a curb have small flow capacity and thus limited value for drainage. Generally, it is not practical to design gutter sections to contain all of the runoff; some overflow onto the surface can be expected. The spread of water on the traveled way is kept within tolerable limits by the proper size and spacing of inlets. Grate inlets and depressions for curb-opening inlets should not be placed in the lane because of their adverse effect on drivers who veer away from them. Bicycle-safe grates should be used everywhere bicyclists are permitted. Warping of the gutter for curb-opening inlets should be limited to the portion within 0.6 to 0.9 m [2 to 3 ft] of the curb to minimize adverse driving effects.

The width of a vertical or sloping curb is considered a cross-section element entirely outside the traveled way. Also, a gutter of contrasting color and texture should not be considered part of the traveled way. When a gutter has the same surface color and texture as the traveled way, and is not much steeper in cross slope than the adjoining traveled way, it may be considered as part of the traveled way. This arrangement is used frequently in urban areas where restricted right-of-way width does not allow for the provision of a gutter. However, with any form of curb there is some effect on the lateral position of drivers; drivers tend to move away from a curb, which reduces effective through-lane width. A gutter with an evident longitudinal joint and somewhat steeper cross slope than the adjacent lane is a greater deterrent to driving near the gutter than the situation in which the traveled way and gutter are integral.

Curb Placement

Vertical or sloping curbs located at the edge of the traveled way may have some effect on lateral placement of moving vehicles, depending on the curb configuration and appearance. Curbs with low, sloping faces may encourage drivers to operate relatively close to them. Curbs with less sloping faces may encourage drivers to shy away from them and, therefore, should incorporate some additional roadway width. Sloping curbs placed at the edge of the traveled way, although considered mountable in emergencies, can be mounted satisfactorily only at reduced speeds. For low-speed urban street conditions, curbs may be placed at the edge of the traveled way, although it is preferable that the curbs be offset 0.3 to 0.6 m [1 to 2 ft].

Data on the lateral placement of vehicles with respect to high vertical curbs show that drivers will shy away from curbs that are high enough to damage the underbody and fenders of vehicles (4). The exact relationship is not known precisely, but it has been established that the lateral placement varies with the curb height and steepness and the location of other obstructions outside the curb. The lateral placement with respect to the curb is somewhat greater where the curb is first introduced than where the curb is continuous for some distance. The shying away at the beginning of the curb will be lessened if the curb is introduced with the end flared away from the pavement edge.

Vertical curbs should not be used along freeways or other high-speed arterials, but if a curb is needed, it should be of the sloping type and should not be located closer to the traveled way than the outer edge of the shoulder. In addition, sloping-end treatments should be provided. Vertical curbs introduced intermittently along streets should be offset 0.6 m [2 ft] from the edge of the traveled way. Where a continuous curb is used along a median or channelizing island through an intersection or interchange, curbs should be offset at least 0.3 m [1 ft], and preferably 0.6 m [2 ft], from the traveled way.

When using curbs in conjunction with traffic barriers, such as on bridges, consideration should be given to the type and height of barrier. Curbs placed in front of traffic barriers can result in unpredictable impact trajectories. If a curb is used in conjunction with a traffic barrier, the height of a vertical curb should be limited to 100 mm [4 in] or it should be of the sloping type, ideally, located flush with or behind the face of the barrier. Curbs should not be used with concrete median barriers. Improperly placed curbs may cause errant vehicles to vault the concrete median barrier or to strike it, causing the vehicle to overturn. For a more detailed discussion on curb usage and location in relation to railings, refer to the AASHTO *Roadside Design Guide* (**10**).

DRAINAGE CHANNELS AND SIDESLOPES

General Considerations

Modern highway drainage design should incorporate safety, good appearance, control of pollutants, and economical maintenance. This may be accomplished with flat sideslopes, broad drainage channels, and liberal warping and rounding.

An important part of highway design is consistency, which prevents discontinuities in the highway environment and considers the interrelationship of all highway elements. The interrelationship between the drainage channel and sideslopes is important for safety because good roadside design can reduce the potential severity of crashes that occur when a vehicle leaves the roadway.

Drainage Channels

Drainage channels perform the vital function of collecting and conveying surface water from the highway right-of-way. Drainage channels, therefore, should have adequate capacity for the design runoff, provide for unusual storm water with minimum damage to the highway, and be located and shaped to provide a safe transition from the roadway to the backslope. Channels should be protected from erosion with the least expensive protective lining that will withstand the expected flow velocities. Channels should be kept clean and free of material that would lower the channel's capacity. Channel deterioration can reduce the capacity of the channel, which may result in overflow, often with erosion or deposition in the area adjacent to the channel.

Where the construction of a highway would have an adverse effect on drainage conditions downstream, drainage channels can be an effective means of flood storage within the highway right-of-way. Drainage channels include (1) roadside channels in cut sections to remove water from the highway cross section, (2) toe-of-slope channels to convey the water from any cut section and from adjacent slopes to the natural watercourse, (3) intercepting channels placed back of the top of cut slopes to intercept surface water, and (4) flumes to carry collected water down steep cut or fill slopes.

The primary purpose for construction of roadside channels is to control surface drainage. The most economical method of constructing a roadside channel usually entails the formation of open-channel ditches by cutting into the natural roadside terrain to produce a drainage channel. From a standpoint of hydraulic efficiency, the most desirable channel contains steep sides. However, limitations on slope stability usually require somewhat flatter slopes. Construction and maintenance factors also impose restrictions on the degree of slope steepness that is practical alongside a highway. The offsetting factor of right-of-way costs should also be considered when selecting combinations of slopes to be used.

The effect of slope combinations and safety during traversal by an errant vehicle is also an important consideration in designing the roadside. In general, the severity of traversal of roadside channels less than about 1.2 to 2.4 m [4 to 8 ft] wide is essentially the same for comparable slope combinations regardless of channel shape. Slope combinations forming these narrow channels can be selected to produce cross sections that can be safely traversed by an unrestrained vehicle occupant.

The use of foreslopes steeper than 1V:4H severely limits the range of backslopes. Flatter foreslopes permit greater flexibility in the selection of backslopes to permit safe traversal. The flatter foreslope also provides greater recovery distance for an errant vehicle. For additional information, refer to the AASHTO *Roadside Design Guide* (**10**).

The depth of channel should be sufficient to remove surface water without saturation of the subgrade. The depth of water that can be tolerated, particularly on flat channel slopes, depends upon the soil characteristics. In regions with severe winter climates, channel sideslopes of 1V:5H or 1V:6H are preferable to reduce snow drifts.

A broad, flat, rounded drainage channel also provides a sense of openness that reduces driver tension. With a channel sideslope of 1V:4H or flatter and a 3.0 m [10 ft] shoulder, the entire roadside channel is visible to the driver. This lessens the driver's feeling of restriction and adds measurably to the driver's willingness to use the shoulder in an emergency.

The minimum desirable grade for channels should be based upon the drainage velocities needed to avoid sedimentation. The maximum desirable grade for unpaved channels should be based upon a tolerable velocity for vegetation and shear on soil types. Refer to the AASHTO *Highway Drainage Guidelines* (**11**) for further guidance in this area. The channel grade does not have to follow that of the roadbed, particularly if the roadbed is flat. Although desirable, it is unnecessary to standardize the design of roadside drainage channels for any length of highway. Not only can the depth and width of the channel be varied to meet different amounts of runoff, slopes of channel, types of lining, and distances between discharge points, but the lateral distance between the channel and the edge of the traveled way can also be varied. Usually, liberal offsets can be obtained where cuts are slight and where cuts end and fills begin. Care should be taken, however, to avoid abrupt major changes in the roadway section that would result in such a discontinuity of the highway environment as to violate driver expectancy. Care should also be taken to avoid major breaks in channel grade that would cause unnecessary scour or silt deposition.

Intercepting channels generally have a flat cross section, preferably formed by a dike made with borrow material to avoid disturbing the natural ground surface. Intercepting channels should have ample capacity and should follow the contour as much as practical, except when located on top of a slope that is subject to sliding. In slide areas, storm water should be intercepted and removed as rapidly as practical. Sections of channels that cross highly permeable soil might need lining with impermeable material.

Median drainage channels are generally shallow depressed areas, or swales, located at or near the center of the median, and formed by the flat sideslopes of the divided roadways. The swale is sloped longitudinally for drainage and water is intercepted at intervals by inlets or transverse channels and discharged from the roadway in storm drains or culverts. Flat, traversable drainage dikes are sometimes used to increase the efficiency of the inlets. Refer to the section on medians in this chapter for further discussion. Safety grates on median drains and cross drains, while enhancing safety for errant vehicles, can reduce the hydraulic efficiency of the drainage structures if not properly designed. The reduced inlet capacity is compounded by the accumulation of debris on the grates, occasionally resulting in roadway flooding. If the use of grates significantly reduces the hydraulic capacity or causes clogging problems to occur, other methods of drainage, or shielding of the structure, should be considered.

Flumes are used to carry the water collected by intercepting channels down cut slopes and to discharge the water collected by shoulder curbs. Flumes can either be open channels or pipes. High velocities preclude sharp turns in open flumes and generally need some means of dissipating the energy of flow at the outlet of the flume. Closed flumes or pipes are preferred to avoid failure due to settlement and erosion. Generally in highly erodible soil, watertight joints should be provided to prevent failure of the facility. Caution should be exercised to avoid splash, which causes erosion.

Channel erosion may be prevented with the use of linings that withstand the velocity of storm runoff. The type of linings used in roadside channels depends upon the velocity of flow, type of soil, and grade and geometry of the channel. Grass is usually the most economical channel lining except on steep slopes where the velocity of flow exceeds the permissible velocities for grass protection. Other materials that can be used for channel lining where grass will not provide adequate protection include concrete, asphalt, stone, and nylon. Smooth linings generate higher velocities than rough linings such as stone and grass. Provision should be made to dissipate the energy of the high-velocity flow before it is released to avoid scour at the outlet and damage to the channel lining. If erosive velocities are developed, a special channel design or energy dissipater may be needed.

Refer to the AASHTO *Highway Drainage Guidelines* (11) and drainage design manuals, as well as handbooks and publications from the Soil Conservation Service, U. S. Army Corps of Engineers, and Bureau of Reclamation, for details on design and protective treatments, including filter requirements. In addition, FHWA publications, such as *Design of Stable Channels With Flexible Linings* (12), provide excellent references. For further information on drainage design, see Chapter 3.

Sideslopes

Sideslopes should be designed to ensure roadway stability and to provide a reasonable opportunity for recovery for an out-of-control vehicle.

Three regions of the roadside are important to safety: the top of the slope (hinge point), the foreslope, and the toe of the slope (intersection of the foreslope with level ground or with a backslope, forming a ditch). Exhibit 4-7 illustrates these three regions.

The hinge point contributes to loss of steering control because vehicles tend to become airborne in crossing this point. The foreslope region is important in the design of high slopes where a driver could attempt a recovery maneuver or reduce speed before impacting the ditch area. The toe of the slope is often within the roadside clear zone and therefore, the probability that an out-of-control vehicle will reach the ditch is high. In this case, a safe transition between fore- and backslopes should be provided.

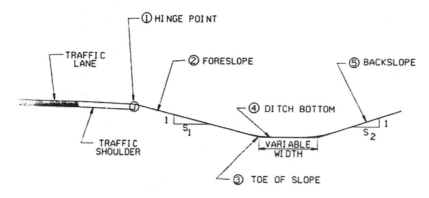

Exhibit 4-7. Designation of Roadside Regions

Research on these three regions of the roadside has found that rounding at the hinge point, though not essential to reduce vehicle rollovers, can increase the general safety of the roadside (**13**). Rounded slopes reduce the chances of an errant vehicle becoming airborne, thereby reducing the likelihood of encroachment and affording the driver more control over the vehicle. Foreslopes steeper than 1V:4H are not desirable because their use severely limits the choice of backslopes. Slopes 1V:3H or steeper are recommended only where site conditions do not permit use of flatter slopes. When slopes steeper than 1V:3H are used, consideration should be given to the use of a roadside barrier.

Another important safety factor for intersecting roadways is the angle of break between a sideslope and a transverse slope. Field observations indicate that more consideration should be given in roadway design to carrying the desirable flat sideslopes through intersections, driveway approaches, median openings, and cut sections. Providing a flatter slope between the shoulder edge and the ditch bottom, locating the ditch a little farther from the roadway, or even enclosing short sections of drainage facilities will enhance the safety of the roadside, often at a small increase in cost.

Earth cut and fill slopes should be flattened and liberally rounded as fitting with the topography and consistent with the overall type of highway. Effective erosion control, low-cost maintenance, and adequate drainage of the subgrade are largely dependent upon proper shaping of the sideslopes. Slope and soil data are used in combination to approximate the stability of the slopes and the erosion potential. Overall economy depends not only on the initial construction cost but also on the cost of maintenance, which is dependent on slope stability. Furthermore, flat or rounded natural slopes with good overall appearance are appropriate for any roadside located near developed and populated areas.

Normally, backslopes should be 1V:3H or flatter, to accommodate maintenance equipment. In developed areas, sufficient space may not be available to permit the use of desirable slopes. Backslopes steeper than 1V:3H should be evaluated with regard to soil stability and traffic safety. Retaining walls should be considered where space restrictions would otherwise result in slopes steeper than 1V:2H. On the other hand, soil characteristics may necessitate the use of slopes flatter than 1V:2H or even 1V:3H. If adequate width is not available in such cases, retaining walls may be needed. The type of retaining structure should be compatible with the area traversed and the grade separation structures. To minimize the feeling of constriction, walls should be set back as far as practical from the traveled way. Where retaining walls are used in combination with earth slopes, the walls may be located either at the roadway level adjacent to the shoulder or on the outer portion of the separation width above the depressed roadway.

On freeways and other arterials with relatively wide roadsides, sideslopes should be designed to provide a reasonable opportunity for recovery of an out-of-control vehicle. Where the roadside at the point of departure is reasonably flat, smooth, and clear of fixed objects, many potential crashes can be averted. A rate of slope of 1V:6H or flatter on embankments can be negotiated by a vehicle with a good chance of recovery and should, therefore, be provided where practical. For moderate heights with good roundings, steeper slopes up to about 1V:3H can also be traversable (though not recoverable). On intermediate-height fills, the cost of a continuous flat slope may be prohibitive, but it may be practical to provide a recovery area that is reasonably flat and rounded adjacent to the roadway. The recovery area should extend well beyond the edge of the shoulder as specific conditions may permit.

Consistent with traffic demand, roads and streets with wide borders should also be designed with a similar clear roadside. However, because of generally lower speeds and narrower side clearances along streets, the clear roadside area concept, at best, can only be partially used. This is also true for widening and other reconstruction within limited right-of-way.

Desirably, slope combinations would be selected so that unrestrained occupants could be expected to sustain no injury, or only minor injuries, and the vehicle would not incur major damage during traversal. However, site conditions such as restricted right-of-way or the cost effectiveness of such design may dictate the use of slope combinations steeper than desirable. If constraints make it impractical to provide the appropriate roadside recovery distance, the need for a roadside barrier should be considered. Where the height and slope of roadway embankments are such that the severity of potential crashes will be reduced by the placement of a roadside barrier, the cross section should be designed to allow adequate slope rounding and to support the barrier.

Flat and well-rounded sideslopes simplify the establishment of turf and its subsequent maintenance. Grasses usually can be readily established on sideslopes as steep as 1V:2H in favorable climates and 1V:3H in semiarid climates. With slopes of 2V:3H and steeper, it is difficult to establish turf, even in areas of abundant rainfall. Because of the greater velocity of runoff, sufficient water for the maintenance of grass does not seep into the soil. Deep-rooted plants that do not depend upon surface water alone may be appropriate where slopes are excessively steep. Slopes of the order of 1V:3H and flatter can be mechanically mowed. Although steeper slopes reduce the mowing area considerably, the slow, time-consuming manual methods required to mow the area add substantially to maintenance costs.

With some types of soils, it is essential for stability that slopes be reasonably flat. Soils that are predominantly clay or gumbo are particularly susceptible to erosion, and slopes of 1V:3H or flatter should be used. The intersections of slope planes in the highway cross section should complement the earth forms of the terrain being traversed. Some earth forms are well-rounded and others are steeply sloped. The designer should strive to create a natural look that is aesthetically pleasing. Since rounded landforms are the natural result of erosion, such rounded forms are stable; therefore, use of well-rounded forms in the design of the highway cross section is likely to result in stability.

To attain a natural appearance along the roadside, flat, well-rounded sideslopes should be provided. A uniform slope through a cut or fill section often results in a formal or stilted appearance. This appearance can be softened and made more natural by flattening the slopes on the ends where the cut or fill is minimal and by gradually steepening it toward the controlling maximum slope of the cut or fill. This design may be readily accomplished by liberal rounding of the hinge point in the transition area. On short cut or fill sections the result may be one of continuous longitudinal rounding whereas, on sections of substantial length the effect will be one of funneling. The transitioning of sideslopes is especially effective at the ends of cuts when combined with an increased lateral offset of the drainage channel and a widened shoulder.

The combination of flat slopes and rounding is frequently referred to as a streamlined cross section. With this shape, the cross winds sweep along the surface without forming eddies that contribute to the wind erosion and drifting of snow. The streamlined cross section usually results in a minimum expenditure for snow removal because the winds blow the snow off the traveled way instead of drifting it, as happens in cross sections with steep slopes and no rounding. When combined with the design of an elevated roadway on earth embankment to ensure drainage of the subgrade, the streamlined cross section results in a roadway that needs minimal maintenance and operating costs and operates safely.

In some cases, an irregular slope stake line results from the strict adherence to specified cut or fill slopes. It may be more aesthetically pleasing to vary the slope to yield a neat stake line.

Design slopes for rock vary widely, depending upon the materials. A commonly used slope for rock cuts is 2V:1H. With modern construction methods, such as pre-splitting, slopes ranging as steep as 6V:1H may be used in good-quality rock. Deep cuts in rock often require the construction of benches in the slopes.

Slope stability as well as appearance may be enhanced in poor-quality rock by the establishment of vegetative cover. In some parts of the country, serrated cut slopes aid in the establishment of vegetative cover on decomposed rock or shale slopes. Serration may be constructed in any material that can be ripped or that will hold a vertical face long enough to establish vegetation (**14**).

Desirably, the toe of the rock-cut slope should be located beyond the minimum lateral distance from the edge of the traveled way needed by the driver of an errant vehicle to either regain control and begin a return to the roadway or to slow the vehicle. Wide shelves at the bottom of rock cuts have advantages in that a safe landing area is provided for falling boulders and space is available for snow storage in colder climates. This width can also be shaped to provide a clear roadside recovery area.

Rock outcroppings are frequently left in place during construction of new highways for economic or aesthetic purposes. These should be eliminated within the clear roadside recovery area where removal is practical. Alternatively, if they cannot be removed, they should be shielded by the installation of a roadside barrier.

For additional guidance on sideslope design, refer to the AASHTO *Roadside Design Guide* (**10**).

ILLUSTRATIVE OUTER CROSS SECTIONS

Exhibits 4-1 and 4-2 illustrate typical combinations of outer cross-section elements— shoulders, side-drainage channels, sidewalks, curbs, and sideslopes—for normal crowned and superelevated sections, respectively. Only a few of the desirable arrangements are illustrated, but other practical arrangements are discussed.

Normal Crown Sections

Exhibit 4-1A shows the most widely used cross section in modern highway practice. The combination of elements is simple and forms a streamlined cross section. Usable shoulder widths are included on both the fill and cut sections. The controlling shoulder slopes range from 2 percent, for a paved or impervious surface, to 8 percent, the maximum slope applicable to a turf surface.

In Exhibit 4-1A the drainage channel at the right is formed by the foreslope on the roadway side and the cut slope, or backslope, on the outer side. The foreslope and backslope combination should be designed such that it can be safely traversed by an errant vehicle. The channel should be wide enough to provide sufficient drainage capacity and deep enough to ensure roadbed stability. A depth of 0.3 to 1.2 m [1 to 4 ft] below the shoulder break is recommended.

In areas where an errant vehicle may tend to encroach the roadside, it is desirable to provide rounding at the intersection of slope planes. Rounding of all slope intersections also improves appearance and simplifies maintenance. In general, 1.2 to 1.8 m [4 to 6 ft] of rounding is the minimum desirable at the edge of the shoulder. The rounding needed at the top of cut slopes is dependent upon a number of factors, including the type of soil, slope ratio and height, and the natural ground slopes. The rounding may vary from 1.2 m [4 ft] to 4.5 m [15 ft]. Toe-of-slope rounding minimizes slope change and offers an increase in fill stability. Toe-of-slope rounding also varies with slopes and fill heights, and has the same general dimensions as on cut slopes.

Exhibit 4-1B illustrates a type of curb treatment that can be used for drainage control or to separate roadways and sidewalks. The left side of the exhibit shows a curb or dike that is used for fill slope protection. The shoulder slope of this section should be designed in conjunction with the drainage system to prevent ponding upon the roadway. Frequent outlets are needed for drainage. To the extent possible, sidewalks should be separated from the roadway. In areas fully developed with retail stores and offices, it may not be practical to offset the sidewalk from the roadway because of the right-of-way considerations. In such cases, curbs are used to separate the sidewalk from the edge of the roadway. This section is shown on the right side of the exhibit.

Exhibit 4-1C shows a steep fill section with guardrail at the edge of the shoulder on the left side of the roadway. Where a sidewalk is needed, it should be located behind the guardrail. For shallow fill sections, roadside safety may be enhanced by enclosing sections of drainage facilities, as shown on the right side of the roadway.

Superelevated Sections

The low sides of the three superelevated cross sections of Exhibit 4-2 are similar to those of Exhibit 4-1 except for the shoulder slope in those cases where the superelevation rate is greater than the normal shoulder slope. It is desirable from an operational standpoint that the shoulder slope on the low side be the same as the traveled way superelevation slope.

In Exhibit 4-2A the direction of shoulder slope on the high side of the cross section is the same as that for normal crowned traveled ways except that its rate of slope should be limited. To avoid an undesirable rollover effect, the algebraic difference in cross slopes at the edge of the traveled way should not exceed 8 percent. Accordingly, use of this cross section should be reserved for low rates of superelevation and shoulder slope. The shoulder slope on the alternate section of Exhibit 4-2A is a projection of the superelevated traveled way.

In Exhibit 4-2B the level shoulder on the high edge of this cross section represents a compromise that prevents the shoulder from draining to the traveled way while complying with the 8 percent rollover control. The use of this cross section should be reserved for stable soils where the percolation, caused by the water falling directly upon the shoulder, is not very great. Where snowfall is prevalent, this cross section would tend to allow snow melt from a windrow on the shoulder to flow across the traveled way, creating a potential icing situation when refreezing occurs.

Exhibit 4-2C shows the high-side shoulder rolled over in a well-rounded transverse vertical curve so that the water falling upon the shoulder is divided between the traveled way and the side channel or fill slope. On this rounded shoulder, any vehicle would stand nearly level as needed to facilitate tire changes and other repairs. The vertical curve should not be less than 1.2 m [4 ft] long, and at least the inner 0.6 m [2 ft] of the shoulder should be held at the superelevated slope. The shoulder slope on the alternate section of Exhibit 4-2C is a planar section with multiple breaks.

Superelevation is advantageous for traffic operations on less developed arterials, as well as for rural highways and urban freeways; however, in built-up areas, the combination of wide pavements, proximity of adjacent development, control of cross slope and profile for drainage, frequency of cross streets, and other urban features combine to make superelevation impractical or undesirable. Usually, superelevation is not provided on local streets in residential, commercial, or industrial areas. For further information on superelevation, refer to Chapter 3.

TRAFFIC BARRIERS

General Considerations

Traffic barriers are used to prevent vehicles that leave the traveled way from hitting an object that has greater crash severity potential than the barrier itself. Because barriers are a source of crash potential themselves, their use should be carefully considered. For more detailed information regarding traffic barriers, refer to the AASHTO *Roadside Design Guide* (**10**).

Research continues to develop improved and more cost-effective barriers. The criteria discussed herein will undoubtedly be refined and amended in the future. Therefore, the designer should remain current on new barrier concepts and criteria.

Traffic barriers include both longitudinal barriers and crash cushions. The primary function of longitudinal barriers is to redirect errant vehicles. The primary function of crash cushions is to decelerate errant vehicles to a stop.

Longitudinal barriers are located along the roadside and in medians. Bridge parapets or rails are covered in AASHTO design criteria and specifications for highway bridges. Longitudinal barriers are generally denoted as one of three types: flexible, semirigid, or rigid. The major difference between these types is the amount of barrier deflection that takes place when the barrier is struck.

Flexible barrier systems undergo considerable dynamic deflection upon impact and generally impose lower impact forces on the vehicle than semirigid and rigid systems. The resistance of this system is derived from tensile force in the longitudinal member. Within the impact zone, the cable or beams tear away from the support post upon impact; thus, the post offers negligible resistance. However, the posts outside the impact zone provide sufficient resistance to keep the deflection of the longitudinal member within an acceptable limit. This system is designed primarily to contain rather than redirect the vehicle and needs more lateral clearance from fixed objects due to the deflection during impact.

In the semirigid system, resistance is achieved through the combined flexure and tensile strength of the rail. The posts near the point of impact are designed to break or tear away, thereby distributing the impact force by beam action to adjacent posts. However, posts outside the impact zone provide sufficient resistance to control the deflection of the longitudinal member to an acceptable limit and redirect the errant vehicle along the path of traffic flow.

A rigid system does not deflect substantially upon impact. During collisions, energy is dissipated by the raising and lowering of the vehicle and by deformation of the vehicle sheet metal. As the angle of impact increases, barrier deceleration forces increase because of the absence of barrier deflection. Therefore, installation of a rigid system is most appropriate where shallow impact angles are expected, such as along narrow medians or shoulders. The rigid system has proved to be very effective as a protective shield where deflection cannot be tolerated, such as at a work zone. Because this system suffers little or no damage on impact, hence needing little maintenance, it should be considered where heavy traffic volumes hamper replacement of damaged rail.

Important factors to consider in the selection of a longitudinal system include barrier performance, lateral deflection characteristics, and the space available to accommodate barrier deflection. Consideration should also be given to the adaptability of the system to operational transitions and end treatments and to the initial and future maintenance cost.

Six options are available for the treatment of roadside obstacles: (1) remove or redesign the obstacle so it can be safely traversed, (2) relocate the obstacle to a point where it is less likely to be struck, (3) reduce impact severity by using an appropriate breakaway device, (4) redirect a vehicle by shielding the obstacle with a longitudinal traffic barrier and/or crash cushion, (5) delineate the obstacle if the above alternatives are not appropriate, or (6) take no action.

Roadway cross section significantly affects traffic barrier performance. Curbs, dikes, sloped shoulders, and stepped medians can cause errant vehicles to vault or submarine a barrier or to strike a barrier so that the vehicle overturns. Optimum barrier system performance is provided by a relatively level surface in front of the barrier and, for semirigid and flexible barriers, beneath and behind the barrier. Where curbs and dikes are used to control drainage, they should be located flush with the face of the barrier or slightly behind it.

In new construction, all curbs and dikes that are not an integral part of the barrier system should be avoided; drainage should be controlled by gentle swales or other means that will not adversely affect barrier performance. Where a barrier is to be installed in the vicinity of an existing curb and the cost of removing the curb cannot be justified, the designer should select a barrier and locate it so that the adverse effect of the curb on barrier performance is minimized.

Longitudinal Barriers

Roadside Barriers

A roadside barrier is a longitudinal system used to shield motorists from obstacles or slopes located along either side of a roadway. It may occasionally be used to protect pedestrians, bystanders, and cyclists from vehicular traffic. Elements which may warrant shielding by a roadside barrier include embankment obstacles, roadside obstacles, and sensitive areas such as playgrounds.

Recent studies indicate that rounding at the shoulder and at the toe of an embankment slope can reduce its crash severity potential. Rounded slopes reduce the chances that an errant vehicle will become airborne, thereby reducing the potential consequences of an encroachment and affording the driver more vehicle control.

The height and slope of an embankment are the key factors in determining barrier need through a fill section. The designer should refer to current warrants and criteria for determination of barrier needs (**10**).

A clear, unobstructed, flat roadside is desirable. When these conditions do not exist, criteria to determine the need for a barrier should be consulted. Roadside obstacles include non-traversable areas and fixed objects. If it is not practical to remove, modify, or relocate an obstacle, then a barrier may be needed. The purpose of a barrier is to enhance safety. Therefore, a barrier should be installed only if it is clear that the barrier will have lower crash severity potential than the roadside obstacle.

Short lengths of roadside barriers are discouraged. Where a barrier is needed in two or more closely spaced locations, continuous barrier should be provided.

Barriers should be located beyond the edge of the shoulder to ensure that the full shoulder width may be used. The fill supporting the barrier should be sufficiently wide to provide lateral support. At bridge locations, roadside barriers should be aligned with the bridge rail and properly secured to the bridge to minimize the possibility of a vehicle striking the barrier and snagging or colliding with a bridge rail or curb. Proper treatment of the exposed end of the barrier is also important. An untreated or square approach end of a barrier presents a formidable roadside obstacle. To provide safe barriers, ends may be buried, covered with a mound of earth, flared back, or protected with a crash cushion or an approved crash tested terminal. Buried barrier ends should be designed to minimize ramping of impacting vehicles. The AASHTO *Roadside Design Guide* (**10**) provides more information on crashworthy end treatments.

The need for a barrier in rock cuts and near large boulders is a matter of judgment by the highway designer and depends on the potential severity of a crash and the lateral clearance available.

For additional material on roadside barriers, refer to the AASHTO *Roadside Design Guide* (**10**).

Median Barriers

A median barrier is a longitudinal system used to minimize the possibility of an errant vehicle crossing into the path of traffic traveling in opposite directions. When traffic volumes are low, the probability of a vehicle crossing a median and colliding with a vehicle in the opposing direction is relatively low. Likewise, for relatively wide medians the probability of a vehicle crossing the median and colliding with a vehicle in the opposing roadway is also relatively low. In these instances, median barriers are generally recommended only when there has been a history of cross-median collisions or, for new roadways, where an incidence of high crash rates of this type would be expected. Although cross-median collisions may be reduced by median barriers, total crash frequency will generally increase because the space available for return-to-the-road maneuvers is decreased.

Special consideration should be given to barrier needs for medians separating traveled ways at different elevations. The ability of an errant driver leaving the higher elevated roadway to return to the road or to stop diminishes as the difference in elevations increases. Thus, the potential for cross-median head-on collisions increases.

An important safety consideration in the design of median barriers is shielding motorists from the exposed end of the barrier. As discussed previously, exposed ends may be buried, covered with a mound of earth, flared back, or protected with an end terminal end or a crash cushion. For more information on crashworthy end treatments, refer to the AASHTO *Roadside Design Guide* (**10**).

For all divided highways, regardless of median width and traffic volume, the median roadside should also be examined for other factors, such as obstacles and lateral drop-off, as discussed earlier.

Careful consideration should be given to the installation of median barriers on multilane expressways or other highways with partial control of access. Even medians that are narrow permit inadvertent encroachments with a chance for motorist recovery and can also include geometric features to accommodate crossing or left-turn traffic. With the addition of a barrier, barrier ends at median openings present formidable obstacles. Crash cushions, although needing maintenance and imposing a high initial cost, may be needed to shield an errant motorist from barrier ends. Consequently, an evaluation of the number of median openings, crash history, alignment, sight distance, design speed, traffic volume, and median width should be conducted prior to installation of median barriers on non-freeway facilities.

Barriers should also be considered on outer separations of 15 m [50 ft] or less where the frontage roads carry two-way traffic.

Common types of median barrier include double-faced steel W-beam (blocked-out) installed on strong posts, box beam installed on weak posts, and concrete barrier. Less common types of median barrier include two- or three-cable barrier installed on light steel posts, double-faced steel W-beam installed on weak posts, double-faced steel three-beam (blocked-out) installed on strong posts, and a cable-chainlink fence combination. For additional data on median barrier types, refer to the AASHTO *Roadside Design Guide* (**10**).

In selecting the type of median barrier, it is important to match the dynamic lateral deflection characteristics to the site. The maximum deflection should be less than one-half the median width to prevent penetration into the opposing lanes of traffic. The median barrier should be designed to redirect the colliding vehicle in the same direction as the traffic flow. In addition, the design should be aesthetically pleasing.

On heavily traveled facilities, a concrete barrier with a sloping face has many advantages. For example, this type of barrier deflects a vehicle striking it at a slight impact angle. It is aesthetically pleasing and needs little maintenance. The latter is an important consideration on highways with narrow medians since maintenance operations encroach on the high-speed traveled way and may require closure of one of the traffic lanes during repair time. The designer should also bear in mind that even though a concrete barrier does not deflect, there may be significant intrusion into the air space above and beyond the barrier by high-center-of-gravity vehicles striking the barrier at high speeds or large angles. A bus or tractor-trailer may lean enough to strike objects mounted on top of the barrier or within a distance of up to 3.0 m [10 ft] of the barrier face. While piers and abutments may be able to withstand such impacts, other structures such as sign trusses and luminaire supports may become involved in secondary collisions.

The appropriate types of median barriers are different for stepped median sections (i.e., where the median is between roadways of different elevations). Cable, W-beam on weak posts, and box-beam systems are generally limited to relatively flat medians and may not be appropriate for some stepped median sections. The AASHTO *Roadside Design Guide* (**10**) provides further guidance in this area.

It is important that, during the selection and design of a median barrier, consideration is given to the potential effect of the barrier on sight distance on horizontal curves.

Due to ongoing research and development, the design of median barriers and terminals is continually improving. Reference should be made to the latest developments in median barrier and terminal design.

Precast concrete median barrier can be used for temporary protection of work areas and for guiding traffic during construction. It can also be incorporated permanently as part of the completed facility.

Bridge Railings

Bridge railings prevent vehicles, pedestrians, or cyclists from falling off the structure. AASHTO's *Standard Specifications for Highway Bridges* (**15**) specifies geometric, design load, and maximum allowable material stress requirements for the design of traffic railings for pedestrians, bicycles, and combination types. Bridge railings are longitudinal traffic barriers that differ from other traffic barriers primarily in their foundations. These railings are a structural extension of a bridge while other traffic barriers are usually set in or on soil.

The need for a traffic barrier rarely ends at the end of a bridge. Therefore, the bridge railing should be extended with a roadside barrier, which in turn should have a crash-worthy terminal. At the juncture between a bridge railing and roadside barrier, an incompatibility usually exists in the stiffness of the two barrier types. This stiffness should be carefully transitioned over a length to prevent the barrier system from pocketing or snagging an impacting vehicle.

Where a roadside barrier is provided between the edge of the traveled way and the bridge railing so that a sidewalk can be included, special attention should be given to the barrier end treatment. End treatments that are both functional and safe are difficult to design. The end treatments should safely accommodate vehicles, yet not impede pedestrian usage of the walkway.

The recommended lateral clearances between the traveled way and bridge railings usually exceed curb offset distances. This may create a problem in a bridge railing where a curbed cross section is used on a bridge approach and a flush cross section is used on the bridge. Such problems may result if the length of the bridge and its approaches make the bridge resemble a controlled-access facility where traffic will operate at speeds in excess of 80 km/h [50 mph], even though the approach speeds are less than 80 km/h [50 mph]. Such high speeds may render curb usage acceptable away from but not on the bridge. In such cases, it may be reasonable to drop the curb at the first intersection away from the end of the bridge. Another option is to reduce the curb to a low, sloping curb with a gently sloped traffic face, well in advance of the introduction of the traffic barrier. This would be reasonably compatible with the traffic barrier even if continued into the high-speed region of the bridge.

Crash Cushions

Crash cushions are protective systems that prevent errant vehicles from impacting roadside obstacles by decelerating the vehicle to a safe stop when hit head-on or redirecting it away from the obstacle (**10**). A common application of a crash cushion is at the end of a bridge rail located in a gore area. Where site conditions permit, a crash cushion should also be considered as an alternative to a roadside barrier for shielding rigid objects such as bridge piers, overhead sign supports, abutments, and retaining-wall ends. Crash cushions may also be used to shield roadside and median barrier terminals.

Site preparation is important in using crash cushion design. Inappropriate site conditions may compromise cushion effectiveness. Crash cushions should be located on a level area free from curbs or other physical obstacles. The design of new highway facilities should consider alternatives to use of crash cushions where appropriate.

MEDIANS

A median is the portion of a highway separating opposing directions of the traveled way. Medians are highly desirable on arterials carrying four or more lanes. Median width is expressed as the dimension between the edges of traveled way and includes the left shoulders, if any. The principal functions of a median are to separate opposing traffic, provide a recovery area for out-of-control vehicles, provide a stopping area in case of emergencies, allow space for speed changes and storage of left-turning and U-turning vehicles, minimize headlight glare, and provide width for future lanes. Additional benefits of a median in an urban area are that it may offer an open green space, may provide a refuge area for pedestrians crossing the street, and may control the location of intersection traffic conflicts. For maximum efficiency, a median should be highly visible both night and day and should contrast with the traveled way. Medians may be depressed, raised, or flush with the traveled way surface.

In determining median width, consideration should be given to the potential need for median barrier. Where practical, median widths should be such that a median barrier is not needed. The general range of median widths is from 1.2 to 24 m [4 to 80 ft] or more. Economic factors often limit the median width that can be provided. Cost of construction and maintenance increases as median width increases, but the additional cost may not be appreciable compared with the total cost of the highway and may be justified in view of the benefits gained.

At unsignalized intersections on rural divided highways, the median should generally be as wide as practical. In urban and suburban areas, however, narrower medians appear to operate better at unsignalized intersections; therefore, wider medians should only be used in urban and suburban areas where needed to accommodate turning and crossing maneuvers by larger vehicles (**16**). Medians at unsignalized intersections should be wide enough to allow selected design vehicles to safely make a selected maneuver. The appropriate design vehicle for determining the median width should be chosen based on the actual or anticipated vehicle mix of crossroad and U-turn traffic. A consideration in the use of wider medians on roadways other than freeways is the provision of adequate storage area for vehicles crossing the highway at unsignalized intersections and at median openings serving commercial and private driveways. Such median openings may need to be controlled as intersections (see Chapter 9). Wide medians may be a disadvantage when signalization is needed. The increased time for vehicles to cross the median can lead to inefficient signal operation.

If right-of-way is restricted, a wide median may not be justified if provided at the expense of narrowed border areas. A reasonable border width is needed to adequately serve as a buffer between the private development along the road and the traveled way, particularly where zoning is limited or non-existent. Space should be provided on the borders for sidewalks, highway signs, utility lines, parking, drainage channels, structures, proper slopes, clear recovery zones, and any retained native growth. Narrowing the border areas may create obstacles and hindrances similar to those that the median is designed to avoid.

A depressed median is generally preferred on freeways for more efficient drainage and snow removal. Median side slopes should preferably be 1V:6H, but slopes of 1V:4H may be adequate.

Drainage inlets in the median should be designed either with the top of the inlet flush with the ground or with culvert ends provided with traversable safety grates.

Raised medians have application on arterial streets where it is desirable to regulate left-turn movements. They are also frequently used where the median is to be planted, particularly where the width is relatively narrow. Careful consideration should be given to the location and type of plantings. Plantings, particularly in narrow medians, may create problems for maintenance activities. Also, plantings such as trees in the median can also cause visual obstructions for turning motorists if not carefully located. Plantings and other landscaping features in median areas may constitute roadside obstacles and should be consistent with the AASHTO *Roadside Design Guide* (**10**).

Flush medians are commonly used on urban arterials. Where used on freeways, a median barrier may be needed. The crowned type is frequently used because it eliminates the need for collecting drainage water in the median. In general, however, the slightly depressed median is preferred either with a cross slope of about 4 percent or with a minor steepening of the roadway cross slope.

The concept of converting flush medians to two-way left-turn lanes on urban streets has become widely accepted. This concept offers several advantages when compared to no median. Among these advantages are reduced travel time; improved capacity; reduced crash frequency, particularly of the rear-end type; more flexibility (because the median lane can be used as a travel lane during closure of a through lane); and public preference both from drivers and owners of abutting properties (**33**). Median widths of 3.0 to 4.8 m [10 to 16 ft] provide the optimum design for two-way left-turn lanes. Refer to the MUTCD (**8**) for appropriate lane markings and to Chapter 2 for additional discussion and details.

Two-way left-turn lanes may be inappropriate at many locations and conversion of existing two-way left-turn lanes to nontraversable medians should be considered. Two-way left-turn lanes have been widely used to provide access to closely spaced, low-volume commercial driveways along arterial roads. From an access management perspective, they increase rather than control access opportunities. Highway agencies have installed raised-curb or concrete median barriers on existing highways in place of flush medians to better manage highway access as traffic and safety concerns increase. In addition, some median openings for minor streets have been closed, permitting only right turns in and out of these streets. This median treatment can reduce the number and location of conflicts along a section of roadway. It should be recognized that diverted left-turn volumes may increase congestion and collisions at downstream intersections; provisions to accommodate U-turn traffic should also be considered at downstream locations.

Where there is no fixed-source lighting, headlight glare across medians or outer separations can be a nuisance, particularly where the highway has relatively sharp curves or if the profiles of the opposing roadways are uneven. Under these conditions, some form of antiglare treatment should be considered as part of the median barrier installation, provided it does not act as a snow fence and does not create drifting problems.

When medians are about 12 m [40 ft] or wider, drivers have a sense of separation from opposing traffic; thus, a desirable ease and freedom of operation is obtained, the noise and air pressure of opposing traffic is not noticeable, and the glare of headlights at night is greatly reduced. With widths of 18 m [60 ft] or more, the median can be pleasingly landscaped in a park-like manner. Plantings used to achieve this park-like appearance need not compromise the roadside recovery area.

There is demonstrated benefit in any separation, raised or flush. Wider medians are desirable at rural unsignalized intersections, but medians as wide as 18 m [60 ft] may not be desirable at urban and suburban intersections or at intersections that are signalized or may need signalization in the foreseeable future. For further guidance in the selection of median widths for divided highways with at-grade intersections, refer to NCHRP Report 375, *Median Intersection Design* (**16**).

FRONTAGE ROADS

Frontage roads serve numerous functions, depending on the type of arterial they serve and the character of the surrounding area. They may be used to control access to the arterial, function as a street facility serving adjoining properties, and maintain circulation of traffic on each side of the arterial. Frontage roads segregate local traffic from the higher speed through-traffic and intercept driveways of residences and commercial establishments along the highway. Cross connections provide access between the traveled way and frontage roads and are usually located in the vicinity of the crossroads. Thus, the through character of the highway is preserved and unaffected by subsequent development of the roadsides.

Frontage roads are used on all types of highways. Each chapter pertaining to a particular type of highway includes a discussion on the use of frontage roads with that highway type. Frontage roads are used most frequently on freeways where their primary function is to distribute and collect traffic between local streets and freeway interchanges. In some circumstances, frontage roads are desirable on arterial streets both in downtown and suburban areas. Frontage roads not only provide more favorable access for commercial and residential development than the faster moving arterial street but also help to preserve the safety and capacity of the latter. In rural areas, development of expressways may need separated frontage roads that are somewhat removed from the right-of-way and serve as access connections between crossroads and adjacent farms or other development.

Despite the advantages of using frontage roads on arterial streets, the use of continuous frontage roads on relatively high-speed arterial streets with intersections may be undesirable. Along cross streets, the various through and turning movements at several closely spaced intersections may greatly increase crash potential. Multiple intersections are also vulnerable to wrong-way entrances. Traffic operations are improved if the frontage roads are located a considerable distance from the main line at the intersecting cross roads in order to lengthen the spacing between successive intersections along the crossroads. In urban areas, a minimum spacing of about 50 m [150 ft] between the arterial and the frontage roads is desirable. For further discussion on frontage roads at intersections, refer to the section in Chapter 9 on "Intersection Design Elements With Frontage Roads."

In general, frontage roads are parallel to the traveled way, may be provided on one or both sides of the arterial, and may or may not be continuous. Where the highway crosses a grid street system on a diagonal course or where the street pattern is irregular, the frontage roads may be a variable distance from the traveled way. Arrangements and patterns of frontage roads are shown in Exhibits 4-8 and 4-9. Exhibit 4-8A illustrates the most common arrangement, two frontage roads running parallel and approximately equidistant from a freeway. In urban areas, continuous frontage roads that are parallel to the freeway permit the use of the frontage roads as a backup system in case of an accident on the freeway or other freeway disruption. Exhibit 4-8B shows a freeway with one frontage road. On the side without the frontage road, the local streets serve to collect and distribute the traffic. Exhibit 4-9 shows an irregular pattern of frontage roads.

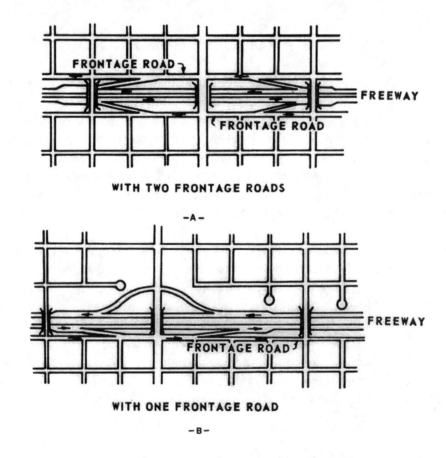

Exhibit 4-8. Typical Frontage Road Arrangements

From an operational and safety standpoint, one-way frontage roads are much preferred to two-way frontage roads. While one-way operation inconveniences local traffic to some degree, the reduction in vehicular and pedestrian conflicts at intersecting streets generally compensate for this inconvenience. In addition, there is some reduction in the roadway and right-of-way width required. Two-way frontage roads at busy intersections complicate crossing and turning movements. Where offramps join a two-way frontage road, the potential for wrong-way entry is increased. This problem is greatest where the ramp joins the frontage road at an acute angle, thus giving the appearance of an onramp to the wrong-way driver.

Two-way frontage roads may be considered for partially developed urban areas where the adjoining street system is so irregular or so disconnected that one-way operation would introduce considerable added travel distance and cause undue inconvenience. Two-way frontage roads may also be appropriate for suburban or rural areas where points of access to the through facility are infrequent, where only one frontage road is provided, or where roads or streets connecting with the frontage roads are widely spaced. In urban areas that are developed or likely to be developed, two-way frontage roads should be considered where there is no parallel street within reasonable distance of the frontage roads.

Connections between the arterial and frontage road are an important element of design. On arterials with slow-moving traffic and one-way frontage roads, slip ramps or simple openings in a narrow outer separation may work reasonably well. Slip ramps from a freeway to two-way frontage roads are generally unsatisfactory because they may induce wrong-way entry to the freeway traveled way and create an increased crash potential at the intersection of the ramp and frontage road. On freeways and other arterials with high operating speeds, the ramps and their terminals should be liberally designed to provide for speed changes and storage. Details of ramp design are covered in later chapters.

Exhibits 4-10 and 4-11 each illustrate an arrangement of frontage roads with entrance and exit ramps that are applicable to freeways and other higher speed arterials. The one-way frontage roads illustrated in Exhibit 4-10 are designed to ensure good operation on both freeways and frontage roads. Exhibit 4-11 shows an arrangement of entrance and exit ramps at two-way frontage roads. This design incorporates a wide outer separation that is not always practical in

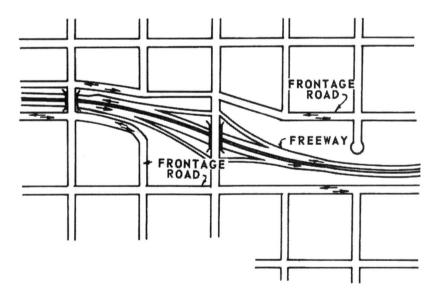

Exhibit 4-9. Frontage Roads, Irregular Pattern

urban areas. The actual width would depend on the design of the ramps and their terminals. In most cases, the width of outer separation would be greater than 60 m [200 ft] in the area of the ramp terminals. The offramp is connected to the frontage road at a right angle to discourage wrong-way entry. Careful attention needs to be given to the placement of signs and the use of traffic markings to prohibit wrong-way movements. Because of the potential for wrong-way movements, the offramp should not intersect the frontage road opposite a two-way side street access.

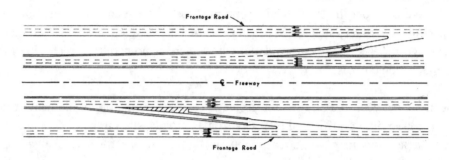

Exhibit 4-10. One-way Frontage Roads, Entrance and Exit Ramps

The design of a frontage road is influenced by the type of service it is intended to provide. Where a frontage road is continuous and passes through highly developed areas, it assumes the character of an important street, serving both local traffic as well as overflow from the traveled way. Where the frontage roads are not continuous or are only a few blocks in length, follow an irregular pattern, border the rear and sides of buildings, or serve only scattered development, traffic will be light and operation will be local in character. Refer to Chapter 6 for guidelines on the widths of two-lane frontage roads for rural and urban collectors.

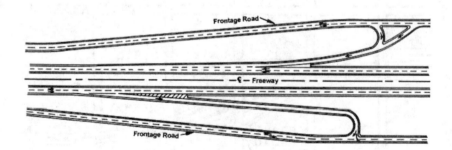

Exhibit 4-11. Two-way Frontage Roads, Entrance and Exit Ramps

OUTER SEPARATIONS

The area between the traveled way of a through-traffic roadway and a frontage road or street is referred to as the "outer separation." Such separations function as buffers between the through-traffic on the arterial and the local traffic on the frontage road and provide space for a shoulder for the through roadway and ramp connections to or from the through facility.

The wider the outer separation, the less influence local traffic will have on through-traffic. Wide separations lend themselves to landscape treatment and enhance the appearance of both the highway and the adjoining property. A substantial width of outer separation is particularly advantageous at intersections with cross streets because it minimizes vehicle and pedestrian conflicts.

Where ramp connections are provided between the through roadway and the frontage road, the outer separation should be substantially wider than typical. The needed width will depend mostly upon the design of the ramp termini.

Where two-way frontage roads are provided, a driver on the through facility faces approaching traffic on the right (opposing frontage road traffic) as well as opposing arterial traffic on the left. Desirably, the outer separation should be sufficiently wide to minimize the effects of approaching traffic, particularly the potentially confusing and distracting nuisance of headlight glare at night. With one-way frontage roads the outer separation need not be as wide as with two-way frontage roads.

The one-lane, one-way frontage road with parking illustrated in Exhibit 4-12 serves businesses along a major undivided arterial street in a densely developed area of a large city. The raised and curbed outer separation creates a buffer between through-traffic and local traffic and provides a refuge for pedestrians.

Exhibit 4-12. Frontage Road in Business Area With Narrow Outer Separation

The cross section and treatment of an outer separation depend largely upon its width and the type of arterial and frontage road. Preferably, the strip should drain away from the through roadway either to a curb and gutter at the frontage road or to a swale within the strip. Typical cross sections of outer separations for various types of arterials are illustrated in Exhibit 4-13.

The cross section in Exhibit 4-13A is applicable to low-speed arterial streets in densely developed areas. Exhibit 4-13B shows a minimal outer separation that may be applicable to ground-level freeways and high-speed arterial streets. This outer separation consists simply of the shoulders of the through roadway and the frontage road, as well as a physical barrier. Exhibit 4-13C shows a depressed arterial with a cantilevered frontage road. In this example, the inside edge of the frontage road is located directly over the outside edge of the through roadway. Exhibit 4-13D illustrates a common type of outer separation along a section of depressed freeway, Exhibit 4-13E shows a walled section at a depressed arterial with a ramp, and Exhibit 4-13F shows a typical freeway outer separation with a ramp.

NOISE CONTROL

General Considerations

Noise may be defined as unwanted sound. Motor vehicles generate traffic noise from the motor, aerodynamics, exhaust, and interaction of the tires with the roadway. Efforts should be made to minimize the radiation of noise into noise-sensitive areas along the highway. The designer should evaluate existing or potential noise levels and estimate the effectiveness of reducing highway traffic noise through location and design considerations.

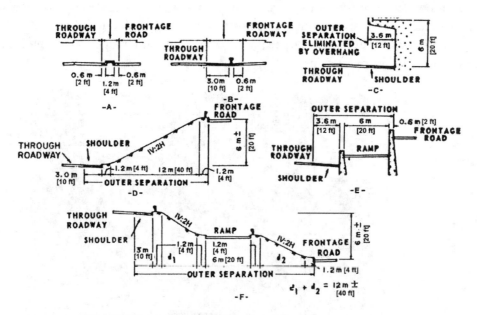

Exhibit 4-13. Typical Outer Separations

The physical measurement of human reaction to sound is difficult because there is no instrument that will measure this directly. A close correlation can be obtained by using the A-scale on a standard sound level meter. The meter yields a direct reading in effective decibels (dBA).

A few general relationships may be helpful in understanding some of the principles of sound generation and transmission. Because noise is measured on a logarithmic scale, a decrease of 10 dBA will appear to an observer as only half of the original noise level. For example, a noise of 70 dBA sounds only one-half as loud as 80 dBA, assuming the same frequency composition and other things being equal. A doubling of the noise source produces a 3 dBA increase in the noise level. For example, if a single vehicle produces a noise level of 60 dBA at a certain distance from the receiver, two of these vehicles at a common point of origin will produce 63 dBA, four vehicles will produce 66 dBA, eight vehicles will produce 69 dBA, and so forth.

Noise decreases with distance, but not as quickly as one might expect. For example, the sound level will decrease approximately 3 to 4.5 dBA for each doubling of distance from a highway.

The same traffic noise level will produce different human reactions depending on the environment in which the noise is heard. The actual noise level is not, in itself, a good predictor of public annoyance. For example, the reaction is usually less if the noise source is hidden from view. The type of development in an area is another factor that affects the annoyance level. High traffic noise levels are usually more tolerable in industrial than in residential areas. Other factors that influence human reactions to noise are pitch and intermittency. The higher the pitch or more pronounced the intermittency of the noise, the greater the degree of annoyance. For further information, see the AASHTO *Guide on Evaluation and Abatement of Traffic Noise* (**32**).

General Design Procedures

The first step in analyzing the effects of noise from a proposed highway facility is to define the criteria for noise impacts. With these criteria defined, the location of noise-sensitive areas can be identified. These may include residential areas, schools, churches, motels, parks, hospitals, nursing homes, libraries, etc. The existing noise levels are determined by measurement of identified noise-sensitive land uses or activities.

The highway-generated noise level is then predicted by one of the noise prediction methods presently available. Pertinent factors are traffic characteristics (speed, volume, and composition), topography (vegetation, barriers, and distance), and roadway characteristics (configuration, pavement type, grades, and type of facility). The prediction is normally based on the highway traffic that will yield the worst hourly traffic noise on a regular basis for the design year. More detailed information on noise prediction is available (**17, 18, 19, 20, 21**).

Exhibit 4-14 provides FHWA noise-abatement criteria for various land uses. These sound levels are used to determine the noise impact on each land use. Traffic noise impacts occur under two criteria: 1) when the predicted levels approach or exceed the noise-abatement criteria and, 2) when predicted noise levels substantially exceed the existing noise level, even though the predicted levels may not exceed the noise-abatement criteria. To adequately assess the traffic noise impact of a proposed project, both criteria should be analyzed.

Activity category	Category description	Design noise levels (dBA)[a]	
		$L_{eq}(h)$[b]	$L_{10}(h)$
A	Tracts of land in which serenity and quiet are of extraordinary significance and serve an important public need and where the preservation of those qualities is essential if the area is to continue to serve its intended purpose. Such areas could include amphitheaters, particular parks or portions of parks, open spaces or historic districts which are dedicated or recognized by appropriate local officials for activities requiring special qualities of serenity and quiet.	57	60 (Exterior)
B	Picnic areas, recreation areas, playgrounds active sports areas and parks not included in Category A and residences, motels, hotels, public meeting rooms, schools, churches, libraries, and hospitals.	67	70 (Exterior)
C	Developed lands, properties or activities not included in Categories A or B above.	72	75 (Exterior)
D	Undeveloped lands which do not contain improvements or activities devoted to frequent human habitation or use and for which such improvements or activities are unplanned and not programmed.		_[c]
E	Residences, motels, hotels, public meeting rooms, schools, churches, libraries, hospitals, and auditoriums.	52	55[d] (Interior)

[a] Source: *Federal Aid Highway Program Manual*, Vol. 7, Ch. 7, Sec. 3 Transmittal 348, August 9, 1982.

[b] Either $L_{10}(h)$ or $L_{eq}(h)$ (but not both) may be used for a specific project.

[c] Noise-abatement criteria have not been established for these lands. They may be treated as developed lands if the probability for development is high. Provisions for noise abatement would be based on the need, expected benefits, and costs of such measures.

[d] Interior noise abatement criteria in this category apply to (1) indoor activities where no extreme noise-sensitive land use or activity is identified, and (2) exterior activities that are either remote from the highway or shielded so that they will not be significantly affected by the noise, but the interior activities will.

Exhibit 4-14. Noise-Abatement Criteria for Various Land Uses

Noise Reduction Designs

Potential noise problems should be identified early in the design process. Line, grade, earthwork balance, and right-of-way should all be worked out with noise in mind. Noise attenuation may be inexpensive and practical if built in the design and expensive if not considered until the end of the design process. An effective method of reducing traffic noise from adjacent areas is to design the highway so that some form of solid material blocks the line of sight between the noise source and the receptors. Advantage should be taken of the terrain in forming a natural barrier so that the appearance remains aesthetically pleasing.

In terms of noise considerations, a depressed highway section is the most desirable. Depressing the roadway below ground level has the same general effect as erecting barriers (i.e., a shadow zone is created where noise levels are reduced [see Exhibit 4-15]). Where a highway is constructed on an embankment, the embankment beyond the shoulders will sometimes block the line of sight to receptors near the highway, thus reducing the potential noise impacts (see Exhibit 4-16).

Special sound barriers may be justified at certain locations, particularly along ground-level or elevated highways through noise-sensitive areas. Concrete, wood, metal, or masonry walls are very effective. One of the more aesthetically pleasing barriers is the earth berm that has been graded to achieve a natural form blending with the surrounding topography. The practicality of berm construction should be considered as part of the overall grading plan for the highway. There will be instances where an effective earth berm can be constructed within normal right-of-way or with a minimal additional right-of-way purchase. If right-of-way is insufficient to accommodate a full-height earth berm, a lower earth berm can be constructed in combination with a wall or screen to achieve the desired height.

Shrubs, trees, or ground covers are not very efficient in shielding sound because of their permeability to air flow. However, almost all buffer plantings offer some noise reduction and exceptionally wide and dense plantings may result in substantial reductions in noise levels. Even where the noise reduction is not considered significant, the aesthetic effects of the plantings will produce a positive effect.

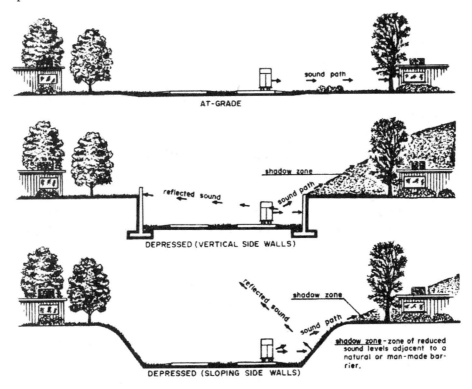

Exhibit 4-15. Effects of Depressing the Highway

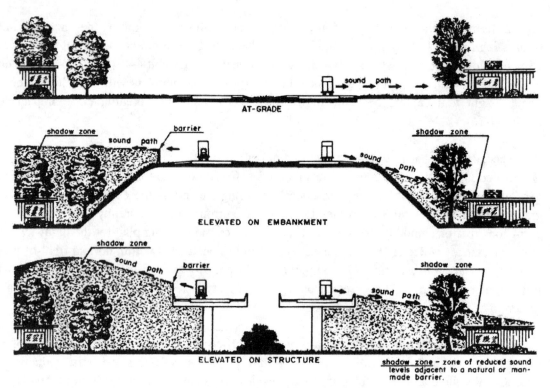

Exhibit 4-16. Effects of Elevating the Highway

ROADSIDE CONTROL

General Considerations

The efficiency and safety of a highway without control of access depend greatly upon the amount and character of roadside interference, characterized by vehicle movements to and from businesses, residences, or other development along the highway. Abutting property owners have rights of access, but it is desirable that the highway authority be empowered to regulate and control the location, design, and operation of access driveways and other roadside elements such as mailboxes. Such access control minimizes interference to through traffic on the highway. Interference resulting from indiscriminate roadside development and uncontrolled driveway connections results in lowered capacity, increased conflict, and early obsolescence of the highway.

Driveways

Driveway terminals are, in effect, low-volume intersections; thus, their design and location merit special consideration. The operational effects of driveways are directly related to the functional classification of the roadway to which they provide access. For example, whereas driveways might adversely affect the operation of arterials, they become important links on local streets that provide access to local establishments.

Driveways used for right turns only are desirable where the cross section includes a curbed median or a flush median and median barrier. Driveways used for both right and left turns offer considerably more interference to through traffic and are undesirable on arterial streets. However, on major streets with numerous motorist-oriented businesses, the elimination of left turns at driveways may worsen traffic operations by forcing large volumes of traffic to make U-turns or travel around the block in order to reach their destination.

The regulation and design of driveways is intimately linked with the available right-of-way and the land use and zoning control of the adjacent property. On new facilities, the needed right-of-way can be obtained to provide the desired degree of driveway regulation and control. To prohibit undesirable access conditions on existing facilities, either additional right-of-way can be acquired or agreements can be made with property owners to improve existing conditions. Often the desired degree of driveway control must be effected through the use of police powers by requiring permits for all new driveways and adjustment of existing that do not conform to established regulations. The objective of driveway regulations is to preserve efficiency and promote operational efficiency by prescribing desirable spacing and proper layout of driveways. The attainment of these objectives is dependent upon the type and extent of legislative authority granted to the highway agency. Many states and local municipalities have developed design policies for driveways and formed separate units to issue permits for new, or for changes in existing, driveway connections to main highways. For further information on the regulation and design of driveways, refer to *Guidelines for Driveway Design and Location* (**22**).

Driveway regulations generally control right-of-way encroachment, driveway location, driveway design, sight distance, drainage, use of curbs, parking, setback, lighting, and signing. Some of the principles of intersection design can also be applied directly to driveways. An important feature of driveway design is the elimination of large graded or paved areas adjacent to the traveled way upon which drivers can enter and leave the facility at will. Another feature is the provision of adequate driveway widths, throat dimensions, and proper layout to accommodate the types of vehicles patronizing the roadside establishment.

Sight distance, another important design control, can be limited by the presence of unnecessary roadside structures. Therefore, no advertising signs should be permitted in the right-of-way. Billboards or other elements outside the right-of-way that obstruct sight distance should be controlled by statutory authority or by purchase of easements.

For roadways without access control but with concentrated business development along the roadside, consideration should be given to the use of a frontage road. This type of control and design is particularly pertinent to a main highway or street on a new location for which sufficient right-of-way can be acquired. In the first stage, intermittent sections of frontage roads are constructed to connect the few driveways initially needed. Then, in succeeding stages, extensions or additional sections of frontage roads are provided to intercept driveways resulting from further development of the roadsides. Thus, serious roadside interference is prevented at all stages, and the through character of the highway or street is preserved by gradual and judicious provision of frontage roads.

Mailboxes

Mailboxes and appurtenant newspaper tubes served by carriers in vehicles may very well constitute a risk to motorists either directly or indirectly, depending upon the placement of the mailbox, the cross-section dimensions of the highway or street, sight distance conditions in the vicinity of the mailbox, traffic volume, and impact resistance of the mailbox support. The safety of both the carrier and the motoring public is affected whenever the carrier slows for a stop and then resumes travel along the highway. The risk is greatly increased if the cross section of the highway and the lateral placement of mailboxes are such that the vehicle occupies a portion of the traveled way while the mailbox is being serviced.

The mounting height of the box places the box in a direct line with the windshield on many vehicles. This situation is more critical where multiple box installations are encountered. In many areas, the typical multiple mailbox installation consists of two or more posts supporting a horizontal member, usually a timber plank, which carries the group of mailboxes. The horizontal support element tends to penetrate the windshield and enter the passenger compartment when struck by a vehicle. Such installations are to be avoided where exposed to traffic. In fact, the mailbox and support should be, where practical, located in an area not exposed to through traffic.

Mailboxes should be placed for maximum convenience to the patron, consistent with safety considerations for highway traffic, the carrier, and the patron. Consideration should be given to minimum walking distance within the roadway for the patron, available stopping sight distance in advance of the mailbox site (especially on older roads), and potential restriction to corner sight distance at driveway entrances. The placing of mailboxes along high-speed, high-volume highways should be avoided if other practical locations are available. New installations should, where practical, be located on the far right side of an intersection with a public road or private driveway entrance. Boxes should be placed only on the right-hand side of the highway in the direction of travel of the carrier except on one-way streets where they may also be placed on the left-hand side.

Preferably, a mailbox should be placed so that it is not susceptible to being struck by an out-of-control vehicle. Where this placement is not practical, the supports should be of a type that will yield or break away safely if struck. The mailbox should be firmly attached to the support to prevent it from breaking loose and flying through the windshield. The same safety criteria also apply to multiple box installations.

One of the primary considerations is the location of the mailbox in relation to the traveled way. Basically, a vehicle stopped at a mailbox should be clear of the traveled way. The higher the traffic volume or the speed, the greater the clearance should be. An exception to this may be considered on low-volume, low-speed roads and streets.

Most vehicles stopped at a mailbox will be clear of the traveled way when the mailbox is placed outside a 2.4 m [8 ft] wide usable shoulder or turnout. This position is recommended for most rural highways. For high-volume, high-speed highways, it is recommended that the width of shoulder in front of the mailbox or turnout be increased to 3.0 m [10 ft] or even 3.6 m [12 ft] for some conditions. However, it may not be practical to consider even a 2.4-m [8-ft] shoulder or turnout on low-volume, low-speed roads or streets. To provide space for opening the mailbox door, it is recommended that the roadside face of a mailbox be set 200 to 300 mm [8 to 12 in] outside the shoulder or turnout. Current postal regulations should be consulted for specific set-back criteria.

In areas of heavy or frequent snowfall, mailboxes may be placed at about the customary line of the plowed windrow, but no closer than about 3.0 m [10 ft] to the edge-of-traveled way if the shoulder is wider than 3.0 m [10 ft]. Cantilever mailbox supports may prove advantageous for snow-plowing operations. Wherever practical, mailboxes should be located behind existing guardrail.

In some urban and suburban areas, mailboxes are located along selected streets and highways where the local post office has established delivery routes. In these areas when the roadway has a curb and gutter section, mailboxes should be located with the front of the box 150 to 300 mm [6 to 12 in] back of the face of curb. On residential streets without curbs or shoulder and which carry low-traffic volumes operating at low speeds, the roadside face of a mailbox should be offset between 200 to 300 mm [8 to 12 in] behind the edge of the traveled way.

For guidance on mailbox installations, refer to the latest editions of AASHTO's *A Guide for Erecting Mailboxes on Highways* (**23**) and *Roadside Design Guide* (**10**).

TUNNELS

General Considerations

Development of streets or highways may include sections constructed in tunnels either to carry the streets or highways under or through a natural obstacle or to minimize the impact of the freeway on the community. General conditions under which tunnel construction may be warranted include:

- Long, narrow terrain ridges where a cut section may either be costly or carry environmental consequences
- Narrow rights-of-way where all of the surface area is needed for street purposes
- Large intersection areas or a series of adjoining intersections on an irregular or diagonal street pattern
- Railroad yards, airport runways, or similar facilities
- Parks or similar land uses, existing or planned
- Where right-of-way acquisition costs exceed cost of tunnel construction and operation.

Although the costs of operation and maintenance of tunnels are beyond the scope of this policy, these costs should nevertheless be considered.

Additional construction and design features of tunnel sections are discussed below. It is not intended that this section be considered complete on the subject of highway tunnels. Instead, the material that follows provides highway planners and designers with general background information. To accomplish this basic objective, some simplification of subject matter is appropriate. As with any highly specialized branch of engineering, such simplified information should be used with caution. In addition, the ventilating, lighting, pumping, and other mechanical or electrical considerations in tunnel design are regarded as outside the scope of this policy.

Types of Tunnels

Tunnels can be classified into two major categories: (1) tunnels constructed by mining methods, and (2) tunnels constructed by cut-and-cover methods.

The first category refers to those tunnels that are constructed without removing the overlying rock or soil. Usually this category is subdivided into two very broad groups according to the appropriate construction method. The two groups are named to reflect the overall character of the material to be excavated: hard rock and soft ground.

Of particular interest to the highway designer are the structural requirements of these construction methods and their relative costs. As a general rule, hard-rock tunneling is less expensive than soft-ground tunneling. A tunnel constructed through solid, intact, and homogeneous rock will normally represent the lower end of the scale with respect to structural demands and construction costs. A tunnel located below water in material needing immediate and heavy support will require extremely expensive soft-ground tunneling techniques such as shield and compressed air methods.

The shape of the structural cross section of the tunnel varies with the type and magnitude of loadings. In those cases where the structure will be subjected to roof loads with little or no side pressures, a horseshoe-shaped cross section is used. As side pressures increase, curvature is introduced into the sidewalls and invert struts added. When the loadings approach a distribution similar to hydrostatic pressures, a full circular section is usually more efficient and economical. All cross sections are dimensioned to provide adequate space for ventilation ducts.

The second category of tunnel classification deals with the two types of tunnels that are constructed from the surface: trench and cut-and-cover tunnels. The latter are used exclusively for subaqueous work. In the trench method, prefabricated tunnel sections are constructed in shipyards or dry docks, floated to the site, sunk into a dredged trench, and joined together underwater. The trench is then backfilled. When conditions are favorable with respect to subsurface soil, amount of river current, volume and character of river traffic, availability of construction facilities, and type of existing waterfront structures, the trench method may prove more economical than alternative methods.

The cut-and-cover method is by far the most common type of tunnel construction for shallow tunnels, which often occurs in urban areas. As the name implies, the method consists of excavating an open cut, building the tunnel within the cut, and backfilling over the completed structure. Under ideal conditions, this method is the most economical for constructing tunnels located at a shallow depth. However, it should be noted that surface disruption and problems with utilities generally make this method very expensive and difficult.

General Design Considerations

Tunnels should be made as short as practical because the feeling of confinement and magnification of traffic noise can be unpleasant to motorists, and tunnels are the most expensive highway structures to construct. The horizontal alignment through the tunnel is an important design consideration as well. Keeping as much of the tunnel length as practical on tangent will not only minimize the length but also improve operating efficiency. Tunnels designed with extreme curvature may result in limited stopping sight distance. Therefore, sight distance across the face of the tunnel wall should be carefully examined.

The vertical alignment through the tunnel is another important design consideration. Grades in tunnels should be determined primarily on the basis of driver comfort while striving to reach a point of economic balance between construction costs and operating and maintenance expenses. Many factors have to be considered in tunnel lengths and grades and their effects on tunnel lighting and ventilation. For example, lighting expenses are highest near portals and depend heavily on availability of natural light and the need to make a good light transition. Ventilation costs depend on length, grades, natural and vehicle-induced ventilation, type of system, and air quality constraints.

The overall roadway design should avoid the need for guide signs within tunnels, because normal vertical and lateral clearances are usually insufficient for such signing and additional clearance can be provided only at very great expense. Exit ramps should be located a sufficient distance downstream from the tunnel portal to permit needed guide signs between the tunnel and the point of exit. This distance should be a minimum of 300 m [1,000 ft]. It is also highly undesirable that traffic be expected to merge, diverge, or weave within a tunnel, as might be the case if the tunnel is located between two closely spaced interchanges. Therefore, forks and exit or entrance ramps should be avoided within tunnels.

Tunnel Sections

From the standpoint of service to traffic, the design criteria used for tunnels should not differ materially from those used for grade separation structures. The same design criteria for alignment and profile and for vertical and horizontal clearances generally apply to tunnels except that minimum values are typically used because of high cost and restricted right-of-way.

Full left- and right-shoulder widths of the approach freeway desirably should be carried through the tunnel. Actually, the need for added lateral space is greater in tunnels than under separation structures because of the greater likelihood of vehicles becoming disabled in the longer lengths. If shoulders are not provided, intolerable delays may result when vehicles become disabled during periods of heavy traffic. However, the cost of providing shoulders in tunnels may be prohibitive, particularly on long tunnels that are constructed by the boring or shield-drive methods. Thus, the determination of the width of shoulders to be provided in a tunnel should be based on an in-depth analysis of all factors involved. Where it is not practical to provide shoulders in a tunnel, arrangements should be made for around-the-clock emergency service vehicles that can promptly remove any stalled vehicles.

Exhibit 4-17 illustrates the minimum and desirable cross sections for two-lane tunnels. The minimum roadway width between curbs, as shown in Exhibit 4-17A, should be at least 0.6 m [2 ft] greater than the approach traveled way, but not less than 7.2 m [24 ft]. The curb or sidewalk on either side should be a minimum of 0.5 m [1.5 ft]. The total clearance between walls of a two-lane tunnel should be a minimum of 9 m [30 ft]. The roadway width and the curb or sidewalk width can be varied as needed within the 9-m [30-ft] minimum wall clearance; however, each width should not be less than the minimum value stated above.

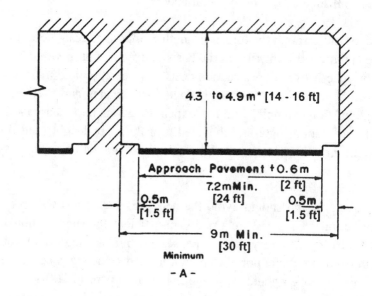

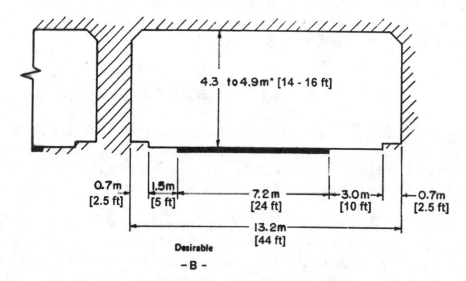

*An allowance should be added for future repaving.

Exhibit 4-17. Typical Two-lane Tunnel Sections

The minimum vertical clearance is 4.9 m [16 ft] for freeways and 4.3 m [14 ft] for other highways. However, the minimum clear height should not be less than the maximum height of load that is legal in a particular state, and it is desirable to provide an allowance for future repaving of the roadways.

Exhibit 4-17B illustrates the desirable section with two 3.6-m [12-ft] lanes, a 3.0-m [10-ft] right shoulder, a 1.5-m [5-ft] left shoulder, and a 0.7-m [2.5-ft] curb or sidewalk on each side. The roadway width may be distributed to either side in a different manner if needed to better fit the dimensions of the tunnel approaches. The vertical clearance for the desirable section is 4.9 m [16 ft] for freeways and 4.3 m [14 ft] for other highways.

Normally, pedestrians are not permitted in freeway tunnels; however, space should be provided for emergency walking and for access by maintenance personnel. Raised sidewalks, 0.7 m [2.5 ft] wide, are desirable beyond the shoulder areas to serve the dual purpose of a safety walk and a buffer to prevent the overhang of vehicles from damaging the wall finish or the tunnel lighting fixtures. Separate tunnels may be warranted for pedestrians or other special uses, such as bicycle routes.

Exhibit 4-18 shows several tunnel sections as well as a partially covered highway. Directional traffic should be separated for safety reasons and to relieve the dizzying effect of two-way traffic in a confined space. This separation can be achieved by providing a twin opening as shown in Exhibit 4-18A, by multilevel sections as shown in Exhibits 4-18B and 4-18C, or by terraced structures as shown in Exhibit 4-18D. The terraced roadways are open on the outside for light, view, and ventilation. Exhibit 4-18E illustrates roadways that are tunneled under hillside buildings. A partially covered section, as shown in Exhibit 4-18F, provides light and ventilation to the motorist while minimizing freeway intrusion on the community traversed. This type of cross section is covered in the section "Depressed Freeways" in Chapter 8.

Examples of Tunnels

Exhibit 4-19 shows a freeway tunneling through a hillside. The portals are staggered and attractively designed. The interchange is located a sufficient distance from the tunnel to allow space for effective signing and the necessary traffic maneuvers.

Exhibit 4-20 illustrates the interior of a three-lane directional tunnel. Note the two rows of lighting fixtures on each wall in the foreground. The upper row of lights provides supplemental daytime lighting at the entrance portal to reduce the optical shock of traveling from natural to artificial lighting. The ceramic-tile finish on the walls and ceiling provides reflective surfaces that increase the brightness level and uniformity of lighting. A curb-to-curb width of 12.3 m [41 ft] is provided with 0.7 m [2.5 ft] wide safety walks along each wall.

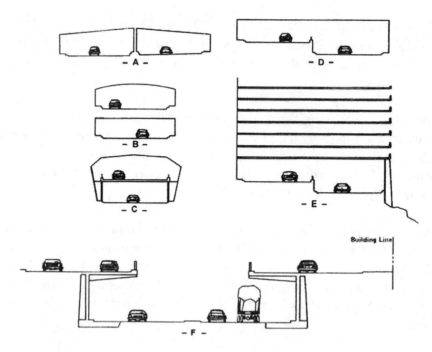

Exhibit 4-18. Diagrammatic Tunnel Sections

Exhibit 4-19. Entrance to a Freeway Tunnel

Exhibit 4-20. Interior of a 3-lane One-way Tunnel

PEDESTRIAN FACILITIES

Sidewalks

Sidewalks are an integral part of city streets but are rarely provided in rural areas. However, the potential for collisions with pedestrians is higher in many rural areas due to the higher speeds and general absence of lighting. The limited data available suggest that sidewalks in rural areas do reduce pedestrian collisions.

Sidewalks in rural and suburban areas are more often justified at points of community development, such as residential areas, schools, local businesses, and industrial plants, that result in pedestrian concentrations near or along the highways. When suburban residential areas are developed, initial roadway facilities are needed for the development to function, but the construction of sidewalks is sometimes deferred. However, if pedestrian activity is anticipated, sidewalks should be included as part of the construction. Shoulders may obviate the need for sidewalks if they are of a type that encourages pedestrian use in all weather conditions. If sidewalks are utilized, they should be separated from the shoulder. If the sidewalk is raised above the level of the shoulder, the cross section typically approaches that of an urban highway.

In suburban and urban locations, a border area generally separates the roadway from a community's homes and businesses. The main function of the border is to provide space for sidewalks. Other functions are to provide space for streetlights, fire hydrants, street hardware, and aesthetic vegetation and to serve as a buffer strip. Border width varies considerably, but 2.4 m

[8 ft] is considered an appropriate minimum width. Swale ditches may be located in these borders to provide an economical alternative to curb and gutter sections.

Sidewalk widths in residential areas may vary from 1.2 to 2.4 m [4 to 8 ft]. The width of a planted strip between the sidewalk and traveled-way curb, if provided, should be a minimum of 0.6 m [2 ft] to allow for maintenance activities. Sidewalks covering the full border width are generally justified and often appropriate in commercial areas, through adjoining multiple-residential complexes, near schools and other pedestrian generators, and where border width is restricted.

Where sidewalks are placed adjacent to the curb, the widths should be approximately 0.6 m [2 ft] wider than those widths used when a planted strip separates the sidewalk from the curb. This additional width provides space for roadside hardware and snow storage outside the width needed by pedestrians. It also allows for the proximity of moving traffic, the opening of doors of parked cars, and bumper overhang on angled parking.

Justification for the construction of sidewalks depends upon the potential for vehicle-pedestrian conflicts. Traffic volume-pedestrian warrants for sidewalks along highways have not been established. In general, wherever roadside and land development conditions affect regular pedestrian movement along a highway, a sidewalk or path area, as suitable to the conditions, should be furnished.

As a general practice, sidewalks should be constructed along any street or highway not provided with shoulders, even though pedestrian traffic may be light. Where sidewalks are built along a high-speed highway, buffer areas should be established so as to separate them from the traveled way.

Sidewalks should have all-weather surfaces to ensure their intended use. Without them, pedestrians often choose to use the traveled way. Pedestrian crosswalks are regularly marked in urban areas but are rarely marked on rural highways. However, where there are pedestrian concentrations, appropriate traffic-control devices should be used, together with appropriate walkways constructed within the right-of-way.

When two urban communities are in proximity to one another, consideration should be given to connecting the two communities with sidewalks, even though pedestrian traffic may be light. This may avoid driver-pedestrian conflicts on these sections of a through route.

Pedestrian facilities such as sidewalks must be designed to accommodate persons with disabilities. See the sections on "Grade-separated Pedestrian Crossings" and "Sidewalk Curb Ramps" later in this chapter for further discussion on this point.

Generally, the guidelines set forth in this section for the accommodation of pedestrians are also applicable to bridges. However, because of the high cost of bridges and the operational features that may be unique to bridge sites, pedestrian-way details on a bridge will often differ from those on its approaches. For example, where a planted strip between a sidewalk and the traveled way approaches a bridge, continuation of the offset, affected by the planted strip, will seldom be justified.

Where flush shoulders approach a bridge and light pedestrian traffic is anticipated on the shoulders, the shoulder width should be continued across the bridge, and possibly increased, to account for the restriction to pedestrian escape imposed by the bridge rail. A flush roadway shoulder should not be interrupted by a raised walkway on a bridge. Where such installations already exist, and removal is not economically justified, the ends of the walkway should be ramped into the shoulder at a rate of approximately 1:20 with the shoulder grade.

Provisions for pedestrians are often appropriate on street overcrossings and on longer bridge crossings. On lower-speed streets, a vertical curb at the edge of the sidewalk is usually sufficient to separate pedestrians from vehicular traffic. Continuity of curb height should be maintained on the approaches to and over structures. For higher-speed roadways on structures, a barrier-type rail of adequate height may be used to separate the walkway and the traveled way. A pedestrian-type rail or screen should be used at the outer edge of the walkway. On long bridges (greater than 60 m [200 ft]), a single walkway may be provided. However, care should be taken to ensure that approach walkways provide safe and relatively direct access to the bridge walkway. Fences may need to be erected to channelize pedestrians and prevent or control conflicts between pedestrians and vehicular traffic.

For a discussion of the potential problems associated with the introduction of a traffic barrier between a roadway and a walkway, see the section on "Bridge Railings," earlier in this chapter. For a discussion on designing sidewalks to accommodate persons with disabilities, see the section on "Sidewalk Curb Ramps," later in this chapter. Further guidance on sidewalk and pedestrian crossing design can be found in current *Americans with Disabilities Act Accessibility Guidelines* (ADAAG) (**24**) and in the AASHTO *Guide for the Planning, Design, and Operation of Pedestrian Facilities* (**25**).

Grade-Separated Pedestrian Crossings

A grade-separated pedestrian facility allows pedestrians and motor vehicles to cross at different levels, either over or under a roadway. It provides pedestrians with a safe refuge for crossing the roadway without vehicle interference. Pedestrian separations should be provided where pedestrian volume, traffic volume, intersection capacity, and other conditions favor their use, although their specific location and design require individual study. They may be warranted where there are heavy peak pedestrian movements, such as at central business districts, factories, schools, or athletic fields, in combination with moderate to heavy vehicular traffic or where unusual risk or inconvenience to pedestrians would otherwise result. Pedestrian separations, usually overpasses, may be needed at freeways or expressways where cross streets are terminated. On many freeways, highway overpasses for cross streets may be limited to three- to five-block intervals. As this situation imposes an extreme inconvenience on pedestrians desiring to cross the freeway at the terminated streets, pedestrian separations may be provided. Local, State, and Federal laws and codes should be consulted for possible additional criteria concerning need, as well as additional design guidance.

Where there are frontage roads adjacent to the arterial highway, the pedestrian crossing may be designed to span the entire or only the through roadway. Separations of both through roadways and frontage roads may not be justified if the frontage roads carry light and relatively slow-moving traffic; however, in some cases the separation should span the frontage roads as well. Fences may be needed to prevent pedestrians from crossing the arterial at locations where a separation is not provided.

Pedestrian crossings or overcrossing structures at arterial streets are not likely to be used unless it is obvious to the pedestrian that it is easier to use such a facility than to cross the traveled way. Generally, pedestrians are more reluctant to use undercrossings than overcrossings. This reluctance may be minimized by locating the undercrossing on line with the approach sidewalk and ramping the sidewalk gently to permit continuous vision through the undercrossing from the sidewalk. Good sight lines and lighting are needed to enhance a sense of security. Ventilation may be needed for very long undercrossings.

Pedestrian ramps should be provided at all pedestrian separation structures. Where warranted and practical, a stairway can be provided in addition to the ramp. Elevators should be considered where the length of ramp would result in a difficult path of travel for a person with or without a disability.

Walkways for pedestrian separations should have a minimum width of 2.4 m [8 ft]. Greater widths may be needed where there are exceptionally high volumes of pedestrian traffic, such as in the downtown areas of large cities and around sports stadiums.

A serious problem associated with pedestrian overcrossings and highway overpasses with sidewalks is vandals dropping objects into the path of traffic moving under the structure. The consequences of objects being thrown from bridges can be very serious. In fact, there are frequent reports of fatalities and major injuries caused by this type of vandalism. There is no practical device or method yet devised that can be universally applied to prevent a determined individual from dropping an object from an overpass. For example, small objects can be dropped through mesh screens. A more effective deterrent is a solid plastic enclosure. However, these are expensive and may be insufferably hot in the summer. They also obscure and darken the pedestrian traveled way, which may be conducive to other forms of criminal activity. Any completely enclosed pedestrian overpass has an added problem that children may walk or play on top of the enclosure. In areas subject to snow and icing conditions, the possibility that melting snow and ice may drop from the roof of a covered overpass and fall onto the roadway below should be considered.

At present it is not practical to establish absolute warrants as to when or where barriers should be installed to discourage the throwing of objects from structures. The general need for economy in design and the desire to preserve the clear lines of a structure unencumbered by screens should be carefully balanced against the need to provide safe operations for both motorists and pedestrians.

Locations where screens definitely should be considered at the time the overpass is constructed include:

364

- On an overpass near a school, a playground, or elsewhere where it would be expected that the overpass would be frequently used by children unaccompanied by adults.
- On all overpasses in large urban areas used exclusively by pedestrians and not easily kept under surveillance by police.
- On an overpass where the history of incidents on nearby structures indicates a need for screens.

Screens should also be installed on existing structures where there have been incidents of objects being dropped from the overpass and where it seems evident that increased surveillance, warning signs, or apprehension of a few individuals involved will not effectively alleviate the problem.

More complete information on the use of protective screens on pedestrian overpasses is available in the AASHTO *Roadside Design Guide* (**10**).

Exhibit 4-21 illustrates two typical pedestrian overcrossings of major highways.

Sidewalk Curb Ramps

When designing a project that includes curbs and adjacent sidewalks, proper attention should be given to the needs of persons with disabilities whose means of mobility are dependent upon wheelchairs and other devices (**26**). The street intersection with steep-faced curbs need not be an obstacle to persons with disabilities. In fact, adequate and reasonable access can be provided for sidewalk curb ramps.

Design details of sidewalk and wheelchair curb ramps will vary in relation to the following factors:

- Sidewalk width
- Sidewalk location with respect to the back face of curb
- Height and width of curb cross section
- Design turning radius and length of curve along the curb face
- Angle of street intersections
- Planned or existing location of sign and signal control devices
- Storm water inlets and public service utilities
- Potential sight obstructions
- Street width
- Border width
- Roadway grade in combination with the grades of the sidewalk, curb, ramps and gutter

Exhibit 4-21. Typical Pedestrian Overpasses on Major Highways

As a result, basic curb ramp types have been established and used in accordance with the geometric characteristics of each intersection. Currently, ADAAG (**24**) requires a 0.9-m [3-ft] minimum curb ramp width and an 8.33 percent maximum grade. Cross slopes on adjacent sidewalks should be no greater than 2 percent. A level landing area is required at the top of each curb ramp. In addition, 0.6-m [2-ft] detectable warning strips that comply with ADAAG are recommended at the bottom of curb ramps to improve detectability by people with visual impairments.

Exhibit 4-22 illustrates various sidewalk curb ramp designs. Exhibit 4-22A shows the condition where the entire grade differential is totally achieved outside the sidewalk. This condition is desirable since it does not require anyone to walk across the ramped area. In this case, a steep face curb can be used along the curb ramp if the presence of landscaping or other fixed obstructions constrain pedestrians from walking across the curb ramp.

In most areas where sidewalks are needed, the curb ramp should be incorporated in the sidewalk, as shown in Exhibits 4-22B and 4-22C. Exhibit 4-22B reflects a normal design with adequate room for curb ramp slope development. Exhibit 4-22C shows an example where a width restriction results in the curb ramp being constructed totally within the sidewalk area.

When other options are not practical, a built-up curb ramp, such as the one illustrated in Exhibit 4-22D, may be desirable. However, the curb ramp should not project into the traveled way. Also, drainage may be adversely affected if not properly considered. The curb ramp area should be protected and should only be used at locations that include a parking lane.

The location of the sidewalk curb ramp should be carefully coordinated with respect to the pedestrian crosswalk lines. This planning should ensure that the bottom of the curb ramp is situated within the parallel boundaries of the crosswalk markings. The bottom of the curb ramp should be perpendicular to the face of the curb with the least amount of warping in the sidewalk, curb ramp, and street transitions.

Exhibit 4-23 shows a typical sidewalk ramp at the middle of the curb radius. In areas where pedestrian and/or vehicular traffic volumes are moderate to high, use of this configuration should be discouraged. Such placement forces the curb ramp users to enter diagonally into the intersection, perhaps misdirecting them and exposing them to conflicts with traffic from two directions. This situation is of special concern to people who are visually impaired.

Exhibit 4-24 illustrates sidewalk curb ramps at the beginning and end points of the curb radius.

Curb ramps for persons with disabilities are not limited to intersections and marked crosswalks. Curb ramps should also be provided at other appropriate or designated points of pedestrian concentration, such as loading islands and midblock pedestrian crossings. Because non-intersection pedestrian crossings are generally unexpected by the motorist, warning signs

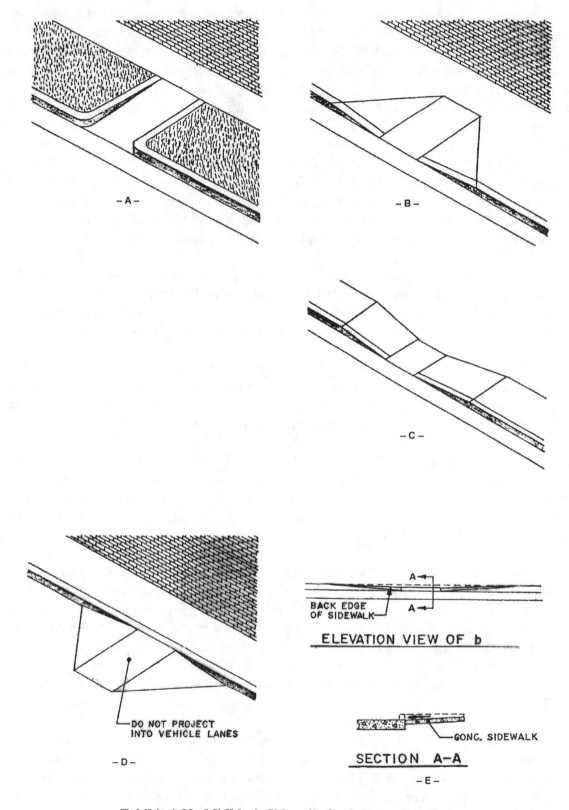

Exhibit 4-22. Midblock Sidewalk Curb Ramp Details

Exhibit 4-23. Sidewalk Curb Ramp at Middle of Radius—Discouraged Where Pedestrian and/or Vehicular Volumes are Moderate to High

should be installed and parking should be prohibited to ensure adequate visibility. For additional design guidance and recommendations with respect to pedestrian crosswalk markings, refer to the MUTCD (**8**), ADAAG (**24**), and AASHTO *Guide for the Planning, Design, and Operation of Pedestrian Facilities* (**25**).

Exhibit 4-25 shows a midblock sidewalk curb ramp. Curb ramps should have a non-skid surface.

As shown in Exhibit 4-26, when a major highway or secondary intersecting road involves pedestrian traffic and the roadway geometrics involve convex islands or median dividers, the plan should include curb ramps for persons with disabilities. People in wheelchairs cannot safely take refuge on islands that are less than 1.2 m [4 ft] wide because of their vulnerability to moving traffic. A 1.8-m [6-ft] island width is desirable.

Each intersection will differ with respect to the intersection angles, turning roadway widths, size of islands, drainage inlets, traffic-control devices, and other variables previously described. An appropriate plan should be prepared that indicates all of the desired geometrics, including vertical profiles at the curb flow line. The plan should then be evaluated to determine convenient

Exhibit 4-24. Sidewalk Curb Ramp at End of Curb Radius

Exhibit 4-25. Sidewalk Curb Ramp at Midblock

and safe locations of the ramps to accommodate usage by persons with disabilities. Drainage inlets should be located on the upstream side of all crosswalks and sidewalk ramps. This design operation will govern the pedestrian crosswalk patterns, stop bar locations, regulatory signs, and, in the case of new construction, establish the most desirable location of signal supports.

Curb ramps should be provided at all intersections where curb and sidewalk are provided even though the highway grade may exceed the allowable sidewalk grade. This provision allows for wheelchairs to be easily maneuvered. For further information on sidewalk curb ramps for persons with disabilities, see the current *ADAAG* (**24**), the AASHTO *Guide for the Planning, Design, and Operation of Pedestrian Facilities* (**25**), and *Designing Sidewalks*

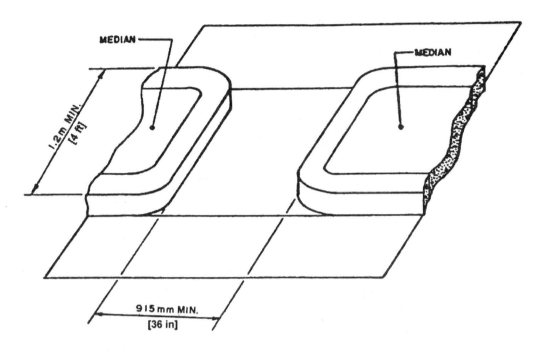

Exhibit 4-26. Median and Island Openings

and Trails for Access, Part I: Review of Existing Guidelines and Practices (**27**) and *Part II: Best Practices Design Guide* (**28**).

BICYCLE FACILITIES

Most of the facilities needed for bicycle travel can consist of the street and highway system generally as it presently exists. However, at certain locations, or in certain corridors, it is appropriate to supplement the existing highway system by providing specifically designated bikeways.

Provisions for bicycle facilities should be in accordance with the AASHTO *Guide for the Development of Bicycle Facilities* (**7**). Even if specific bicycle facilities are not provided, consideration should be given to other practical measures for enhancing bicycle travel on the highway.

Chapter 2 provides further discussion on the subject of bicycle facilities.

BUS TURNOUTS

Bus travel is an increasingly important mode of mass transportation. Bus turnouts serve to remove the bus from the traveled way. The location and design of turnouts should provide ready access in the safest and most efficient manner practical.

Freeways

The basic design objective for a freeway bus turnout is that the deceleration, standing, and acceleration of buses take place on pavement areas clear of and separated from the traveled way. Other elements in the design of bus turnouts include passenger platforms, ramps, stairs, railings, signs, and markings. Speed-change lanes should be long enough to enable the bus to leave and enter the traveled way at approximately the average running speed of the highway without undue discomfort to passengers. The length of acceleration lanes from bus turnouts should be well above the normal minimum values, as the buses start from a standing position and the loaded bus has a lower acceleration capability than passenger cars. Normal-length deceleration lanes are suitable. The width of the bus standing area and speed-change lanes, including the shoulders, should be 6.0 m [20 ft] to permit the passing of a stalled bus. The pavement areas of turnouts should contrast in color and texture with the traveled way to discourage through-traffic from encroaching on or entering the bus stop.

The dividing area between the outer edge of freeway shoulder and the edge of bus turnout lane should be as wide as practical, preferably 6.0 m [20 ft] or more. However, in extreme cases, this width could be reduced to a minimum of 1.2 m [4 ft]. A barrier is usually needed in the dividing area, and fencing is desirable to keep pedestrians from entering the freeway. Pedestrian loading platforms should not be less than 1.5 m [5 ft] wide and preferably 1.8 m to 3.0 m [6 to 10 ft] wide. Some climates may warrant the covering of platforms. Exhibit 4-27 illustrates typical cross sections of turnouts including a normal section, a section through an underpass, and a section on an elevated structure.

Arterials

The interference between buses and other traffic can be considerably reduced by providing turnouts on arterials. It is somewhat rare that sufficient right-of-way is available on the lower type arterial streets to permit turnouts in the border area, but advantage should be taken of every opportunity to do so.

To be fully effective, bus turnouts should incorporate (1) a deceleration lane or taper to permit easy entrance to the loading area, (2) a standing space sufficiently long to accommodate the maximum number of vehicles expected to occupy the space at one time, and (3) a merging lane to enable easy reentry into the traveled way.

The deceleration lane should be tapered at an angle flat enough to encourage the bus operator to pull completely clear of the through lane before stopping. Usually it is not practical to

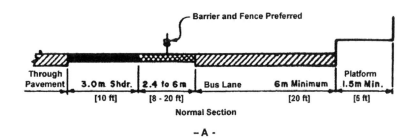

Normal Section

- A -

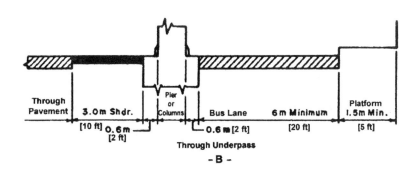

Through Underpass

- B -

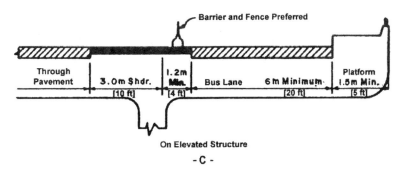

On Elevated Structure

- C -

Exhibit 4-27. Bus Turnouts

provide a length sufficient to permit deceleration from highway speeds clear of the traveled way. A taper of about 5:1, longitudinal to transverse, is a desirable minimum. When the bus stop is on the far side of an intersection, the intersection area may be used as the entry area to the stop.

The loading area should provide about 15 m [50 ft] of length for each bus. The width should be at least 3.0 m [10 ft] and preferably 3.6 m [12 ft]. The merging or reentry taper may be somewhat more abrupt than the deceleration taper but, preferably, should not be sharper than 3:1. Where the turnout is on the near side of an intersection, the width of cross street is usually great enough to provide the needed merging space.

The minimum total length of turnout for a two-bus loading area should be about 55 m [180 ft] for a midblock location, 45 m [150 ft] for a near-side location, and 40 m [130 ft] for a far-side location. These dimensions are based on a loading area width of 3.0 m [10 ft]. They should be increased by 4 to 5 m [13 to 16 ft] for a width of 3.6 m [12 ft]. Greater lengths of bus turnouts expedite bus maneuvers, encourage full compliance on the part of bus drivers, and lessen interference with through traffic.

Exhibit 4-28 shows a bus turnout at a midblock location. The width of the turnout is 3.0 m [10 ft], and the length of the turnout, including tapers, is 63 m [210 ft]. The deceleration and acceleration tapers are 4:1.

For more information on bus turnouts, see the AASHTO *Guide for Design of High-Occupancy Vehicle and Public Transportation Facilities* (**29**) and *Guidelines for the Location and Design of Bus Stops* (**30**).

Exhibit 4-28. Midblock Bus Turnout

Park-and-Ride Facilities

Location

Park-and-ride facilities should be located adjacent to the street or highway and be visible to the commuters whom they are intended to attract. Preferably, the parking areas should be located at points that precede the bottlenecks or points where there is significant traffic congestion. They should be located as close to residential areas as practical in order to minimize travel by vehicles with only one occupant and should be located far enough out that land costs are not prohibitive. In addition, bicycle and pedestrian access to park-and-ride facilities should be considered.

Other considerations that affect parking lot location are impacts on surrounding land uses, available capacity of the highway system between the roadway and proposed sites, terrain, and the costs to acquire the land.

Design

The size of the park-and-ride parking lot is dependent upon the design volume, the available land area, and the size and number of other parking lots in the area. Twenty to sixty spaces represent a reasonable range.

Each parking area should provide a drop-off facility close to the station entrance, plus a holding or short-term parking area for passenger pickup. This area should be clearly separated from the park-and-ride areas.

Consideration should be given to the location for bus loading and unloading, taxi service, bicycle parking, and special parking for persons with disabilities. Conflicts between pedestrians and vehicles should be minimized. Parking aisles should be located perpendicular to the bus roadway so that pedestrians do not need to cross the driveways between parking aisles. All bus roadways should have a minimum width of 6.0 m [20 ft] to permit the passing of standing buses. Facilities should be designed for self-parking. Parking spaces should be 2.7 m by 6.0 m [9 ft by 20 ft] for full-sized cars. Where a special section is provided for subcompact cars, 2.4 m by 4.5 m [8 ft by 15 ft] spaces are sufficient. Parking requirements for persons with disabilities should be in accordance with the ADAAG (**24**).

Sidewalks should be a minimum of 1.5 m [5 ft] wide and loading areas should be 3.6 m [12 ft] wide. Principal loading areas should be provided with sidewalk curb ramps. Preferably, pedestrians should not have to walk more than 120 m [400 ft], although slightly longer distances may be permitted under some circumstances. Pedestrian paths from parking spaces to loading areas should be as direct as practical. Facilities for locking bicycles should be provided where needed.

Grades of parking areas should be set so that drainage can be effective. Recommended grades along vehicle paths within the parking area are 1 percent minimum and 2 percent desirable with a maximum of 5 percent. Grades of over 8 percent parallel to the length of the parked vehicles should be avoided. Climatic conditions should be considered in establishing the maximum acceptable grade. Curvature, radius of planned vehicular paths within the parking area, and access roads should be sufficiently large to accommodate the vehicles that they are intended to serve.

Access to the lots should be at points where they will disrupt through traffic as little as practical. Access points should be at least 90 m [300 ft] from other intersections, and there should be sufficient sight distance for vehicles to exit and enter the lot. This means that exits and entrances should not be located on crest vertical curves. There should be at least 90 m [300 ft] corner sight distance.

There should be at least one exit and entrance for each 500 spaces in a lot. Exits and entrances should be provided at separate locations and should access different streets, if practical. It is also desirable to provide separate access for public transit vehicles.

Curb returns should be at least 9.0 m [30 ft] in radius, although 4.5 m [15 ft] radii are suitable for access points used exclusively by passenger vehicles.

Principal passenger-loading areas should be provided with shelters to protect public transit patrons. Such shelters should, as a minimum, be sufficiently large to accommodate off-peak passenger volumes but should be larger wherever practical. To determine the size of the shelter, the number of passengers that the shelter is to service is multiplied by a factor of 0.3 to 0.5 m^2 [3 to 5 ft^2]. Because the shelter can be expanded relatively easily at a later date, provided sufficient platform space is installed initially, it is not critical to provide a shelter that accommodates all passengers at the time of original construction. Accessories that should be provided with the shelter include lighting, benches, route information, trash receptacles, and often telephones.

The design of the bus-loading area can be of a parallel or a sawtooth design; the best arrangement depends on the number of buses expected to use the facility. Where more than two buses are expected to be using a facility at one time, the sawtooth arrangement is generally preferable, as it is easier for buses to bypass a standing bus. A recommended design of a sawtooth arrangement is shown in Exhibit 4-29. The length of space that should be provided for a parallel design is 29 m [95 ft]. This length will permit loading of two buses. For each additional space, 14 m [45 ft] should be allowed. The loading area should be at least 7.2 m [24 ft] wide in order to permit the passing of a standing bus. The area delineating the passenger refuge area should be curbed in order to reduce the height between the ground and the first bus step and to reduce encroachment by buses on the passenger areas. Parallel-type loading areas should not be located on curves as it makes it very difficult for drivers to park with both the front and rear doors close to the curb.

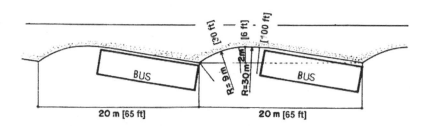

Exhibit 4-29. Sawtooth Bus Loading Area

Special designs may be needed to accommodate articulated buses, particularly where a sawtooth arrangement is used. A well-designed parking lot includes a buffer area around the lot with appropriate landscaping, often with a fence to separate land areas. The buffer should be at least 3.0 m [10 ft] wide.

Lighting should be provided on all but the smaller lots. A level of 2.2 to 5.4 lux (lx) [0.2 to 0.5 foot candles (fc)] average maintained intensity will generally suffice.

Drainage systems should be designed so that parked cars will not be damaged by storm water. Under some circumstances, minimal ponding of water may be permitted or may even be desirable when the drainage is designed as part of a storm water management system. The storm intensity that the drainage system should accommodate may depend on the practice of the municipality. Permissible depths of ponding should generally not exceed 75 to 100 mm [3 to 4 in] in areas where cars are parked, and there should be no ponding on pedestrian and bicycle routes or where persons wait for transit vehicles.

Exhibit 4-30 shows a typical park-and-ride facility. For additional information, refer to the AASHTO *Guide for Design of High-Occupancy Vehicle and Public Transportation Facilities* (**29**), TCRP Report 19, *Guidelines for the Location and Design of Bus Stops* (**30**), and the AASHTO *Guide for the Design of Park-and-Ride Facilities* (**31**).

ON-STREET PARKING

A roadway network should be designed and developed to provide for the safe and efficient movement of vehicles operating on the system. Although the movement of vehicles is the primary function of a roadway network, segments of the network may, as a result of land use, also provide on-street parking.

In the design of freeways and other control of access-type facilities, as well as on most rural arterials, collectors, and local streets, stopping or parking should be permitted only in emergencies. On-street parking generally decreases through-traffic capacity, impedes traffic flow, and increases crash potential. Since the primary service of an arterial is the movement of vehicles, it is desirable to prohibit parking on urban arterial streets and rural arterial highway sections. However, within urban areas and in rural communities located on arterial highway routes, existing and developing land uses may necessitate the consideration of on-street parking. Usually, adequate off-street parking facilities are not available. Therefore, the designer should consider on-street parking so that the proposed street or highway improvement will be compatible with the land use.

When a proposed roadway improvement is to include on-street parking, parallel parking should be considered. Under certain circumstances, angle parking is an allowable form of street parking. The type of on-street parking selected should depend on the specific function and width of the street, the adjacent land use, traffic volume, as well as existing and anticipated traffic operations. Angle parking presents special problems because of the varying length of vehicles and the sight distance problems associated with vans and recreational vehicles. The extra length of such vehicles may interfere with the traveled way.

Exhibit 4-30. Typical Park-and-Ride Facility

An important part of the urban parking problem is the uneven distribution of off-street parking facilities within urban central business districts and the lack of off-street facilities in urban neighborhood commercial areas. As a consequence, there is a demand for on-street parking to provide for the delivery and pick-up of goods. Frequently, alleys and other off-street loading areas are not provided in many communities. Short-duration parking for business or shopping should therefore be accommodated.

Curb parking on urban arterial streets is acceptable when the available through-traffic lanes can accommodate traffic demand. On rural arterials, provisions should be made for emergency stopping only. On urban arterial street reconstruction projects or on projects where additional right-of-way is being acquired to upgrade an existing route to arterial status, parking should be eliminated whenever practical to increase capacity and safety. The impacts on abutting land uses should, however, be carefully considered, as the loss of existing on-street parking can cause significant loss in the economic well-being of the abutting property.

It has been found that most vehicles will parallel park within 150 to 300 mm [6 to 12 in] of the curb face and on the average will occupy approximately 2.1 m [7 ft] of actual street space. Therefore, the desirable minimum width of a parking lane is 2.4 m [8 ft]. However, to provide better clearance from the traveled way and to accommodate use of the parking lane during peak periods as a through-travel lane, a parking lane width of 3.0 to 3.6 m [10 to 12 ft] is desirable. This width is also sufficient to accommodate delivery vehicles and serve as a bicycle route, allowing a bicyclist to maneuver around an open door on a motor vehicle.

On urban collector streets, the demand for land access and mobility is equal. The desirable parking lane width on urban collectors is 2.4 m [8 ft] to accommodate a wide variety of traffic operations and land uses. To provide better clearance and the potential to use the parking lane during peak periods as a through-travel lane, a parking lane width of 3.0- to 3.6-m [10- to 12-ft] is desirable. A 3.0 to 3.6 m [10 to 12 ft] parking lane will also accommodate urban transit operations. On urban collector streets within residential neighborhoods where only passenger vehicles need to be accommodated in the parking lane, 2.1-m [7-ft] parking lanes have been successfully used. In fact, a total width of 10.8 m [36 ft], consisting of two travel lanes of 3.3 m [11 ft] and parking lanes of 2.1 m [7 ft] on each side, are frequently used.

On-street parking is generally permitted on local streets. A 7.8 m [26 ft] wide roadway is the typical cross section used in many urban residential areas. This width assures one through lane even where parking occurs on both sides. Specific parking lanes are not usually designated on such local streets. The lack of two moving lanes may be inconvenient to the user in some cases; however, the frequency of such concerns has been found to be remarkably low. Random intermittent parking on both sides of the street usually results in areas where two-way movement can be accommodated.

Construction procedures on new roadways should be carefully considered so as to provide a longitudinal joint at the boundary of the proposed parking lane. It has been found that such joints aid in ensuring that the parked vehicle clears the parallel travel lane. On asphalt-surfaced streets, traffic markings are recommended to identify the parking lane. The marking of parking spaces encourages more orderly and efficient use of parking spaces where parking turnover occurs and tends to prevent encroachment on fire hydrant zones, bus stops, loading zones, and approaches to corners.

In urban areas, central business districts, and commercial areas where significant pedestrian crossings are likely to occur, the design of the parking lane/intersection relationship should be given consideration. When the parking lane is carried up to the intersection, motorists may utilize the parking lane as an additional lane for right-turn movements. Such movements may cause operational problems and often result in turning vehicles mounting the curb and possibly striking such intersection elements as traffic signals, utility poles, or luminaire supports. The transitioning out of the parking lane of a minimum of 6.0 m [20 ft] in advance of the intersection is one method of eliminating this problem. An example of such treatment is shown in Exhibit 4-31. A second method is to prohibit parking for such a distance as to create a short turn lane.

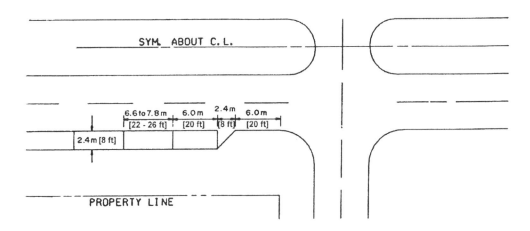

Exhibit 4-31. Parking Lane Transition at Intersection

REFERENCES

1. AASHTO. *Guide for the Design of Pavement Structures*, Washington, D.C.: AASHTO, 1986.

2. Dunlap, D. F., P. S. Fancher, R. E. Scott, C. C. MacAdam, and L. Segal. *Influence of Combined Highway Grade and Horizontal Alignment on Skidding*, NCHRP Report 184, Washington, D.C.: Transportation Research Board, 1978.

3. AASHTO. *Guidelines for Skid Resistant Pavement Design*, Washington, D.C.: AASHTO, 1976.

4. Transportation Research Board. *Highway Capacity Manual*, Special Report 209, Washington, D.C.: Transportation Research Board, 1985.

5. Zegeer, C. V., R. Stewart, F. M. Council, and T. R. Neuman. *Roadway Widths for Low-Traffic Volume Roads,* NCHRP Report 362, Transportation Research Board, 1994.

6. Downs, Jr., H. G., and D. W. Wallace. *Shoulder Geometrics and Use Guidelines,* NCHRP Report 254, Washington, D.C.: Transportation Research Board, 1982.

7. AASHTO. *Guide for the Development of Bicycle Facilities*, Washington, D.C.: AASHTO, 1999.

8. U.S. Department of Transportation, Federal Highway Administration. *Manual on Uniform Traffic Control Devices for Streets and Highways*, Washington, D.C.: 1988.

9. Harwood, D. W., and C. J. Hoban. *Low Cost Methods for Improving Traffic Operation on Two-Lane Roads*, Report FHWA IP-87-2, Washington, D.C.: U.S. Department of Transportation Federal Highway Administration, 1987.

10. AASHTO. *Roadside Design Guide*, Washington, D.C.: AASHTO, 1996.

11. AASHTO. *Highway Drainage Guidelines*, Washington, D.C.: AASHTO, 1988.

12. Federal Highway Administration. "Design of Stable Channels with Flexible Linings," *HEC-15*. Washington, D.C.: FHWA, Office of Engineering, Bridge Division, 1988.

13. Marquis, E. L., R. M. Olson, and G. D. Weaver. *Selection of Safe Roadside Cross Sections,* NCHRP Report 158, Washington, D.C.: Transportation Research Board, 1975.

14. Richards, D., and D. Ham. "Serrated Soft-Rock Cut Slopes," *FHWA Report FHWA-RDDP-5-1*. Washington, D.C.: Federal Highway Administration, 1973.

15. AASHTO. *Standard Specifications for Highway Bridges*, Washington, D.C.: AASHTO, 1996.

16. Harwood, D. W., M. T. Pietrucha, M. D. Wooldridge, R. E. Brydia, and K. Fitzpatrick. *Median Intersection Design,* NCHRP Report 375, Washington, D.C.: Transportation Research Board, 1995.

17. Simpson, M. A. *Noise Barrier Design Handbook*. Report No. FHWA-RD-76-58, McLean, Virginia: U.S. Department of Transportation, Federal Highway Administration, 1976.

18. Barry, T. M., and J. A. Reagan. *FHWA Highway Traffic Noise Prediction Model,* Report No. FHWA-RD-77-108, McLean, Virginia: U.S. Department of Transportation, Federal Highway Administration, 1978.

19. Bowlby, W. *SNAP 1.1—Revised Program and User's Manual for the FHWA Level I Highway Traffic Noise Prediction Computer Program,* Report No. FHWA-DP-45-4, Washington, D.C.: U.S. Department of Transportation, Federal Highway Administration, 1980.

20. Bowlby, W., J. Higgins, and J. Reagan. *Noise Barrier Cost Reduction Procedure STAMINA 2.0/OPTIMS: Manual,* Report No. FHWA-DP-58-1, Washington, D.C.: Federal Highway Administration, 1982, Revised March 1983.

21. Bowlby, W. *Sound Procedures for Measuring Highway Noise: Final Report,* Report No. FHWA-DP-45-1R, Washington, D.C.: U.S. Department of Transportation, Federal Highway Administration, 1981.

22. Institute of Transportation Engineers. *Guidelines for Driveway Design and Location,* Washington, D.C.: 1985.

23. AASHTO. *A Guide for Erecting Mailboxes on Highways*, Washington, D.C.: AASHTO, 1994.

24. Architectural and Transportation Barriers Compliance Board (Access Board). *Americans with Disabilities Act Accessibility Guidelines* (ADAAG), Washington, D.C.: July 1994 or most current edition.

25. AASHTO. *Guide for the Planning, Design, and Operation of Pedestrian Facilities,* Washington, D.C.: AASHTO, forthcoming.

26. Templer, J. *Development of Priority Accessible Networks: An Implementation Manual. Provisions for the Elderly and Handicapped Pedestrians,* Report No. FHWA-IP-80-8, Washington, D.C.: U.S. Department of Transportation, Federal Highway Administration, 1980.

27. Federal Highway Administration. *Designing Sidewalks and Trails for Access, Part I: Review of Existing Guidelines and Practices*, Report No. FHWA-HEP-99-006, Washington, D.C., 1999.

28. Federal Highway Administration. *Designing Sidewalks and Trails for Access, Part II: Best Practices Guide*, Washington, D.C., forthcoming.

29. AASHTO. *Guide for Design of High-Occupancy Vehicle and Public Transportation Facilities*, Washington, D.C.: AASHTO, 1983.

30. *Guidelines for the Location and Design of Bus Stops,* TCRP Report 19, Washington, D.C.: Transportation Research Board, 1996.

31. AASHTO. *Guide for the Design of Park-and-Ride Facilities*, Washington, D.C.: AASHTO, 1992.

32. AASHTO. *Guide on Evaluation and Abatement of Traffic Noise*, Washington, D.C.: AASHTO, 1993.

33. Bonneson, J. A., and P. T. McCoy. *Capacity and Operational Effects of Midblock Left-turn Lanes.* NCHRP Report 395, Washington, D.C.: Transportation Research Board, 1997.

CHAPTER 5
LOCAL ROADS AND STREETS

INTRODUCTION

This chapter presents guidance on the application of geometric design criteria to facilities functionally classified as local roads and streets. The chapter is subdivided into sections on rural, urban, and special-purpose local roads.

A local road or street serves primarily to provide access to farms, residences, businesses, or other abutting properties. Although local roads and streets may be planned, constructed, and operated with the predominant function of providing access to adjacent property, some local roads and streets serve a limited amount of through traffic. Such roads properly include geometric design and traffic control features more typical of collectors and arterials to encourage the safe movement of through traffic. On these roads the through traffic is local in nature and extent rather than regional, intrastate, or interstate.

Local roads and streets constitute a high proportion of the roadway mileage in the United States. The traffic volume generated by the abutting land uses are largely short trips or a relatively small part of longer trips where the local road connects with major streets or highways of higher classifications. Because of the relatively low traffic volumes and the extensive mileage, design criteria for local roads and streets are of a comparatively low order as a matter of practicality. However, to provide the requisite traffic mobility and safety—together with the essential economy in construction, maintenance, and operation—they must be planned, located, and designed to be suitable for predictable traffic operations and must be consistent with the development and culture abutting the right-of-way.

In restricted or unusual conditions, it may not be practical to meet the design criteria presented in this chapter. In such cases, every effort should be made to obtain the best possible alignment, grade, sight distance, and drainage that are consistent with terrain, development (present and anticipated), safety, and available funds.

Drainage, both on the pavement itself and from the sides and subsurface, is an important design consideration. Inadequate drainage can lead to high maintenance costs and adverse operational conditions. In snow regions, roadways should be designed so that there is sufficient storage space for plowed snow and proper drainage for melting conditions.

Safety is an important factor in all roadway improvements. However, it may not be practical to provide an obstacle-free roadside on local roads and streets. Every effort should be made to provide as much clear roadside as is practical. This becomes more important as speeds increase. Flatter slopes, guardrail, and warning signs all help to achieve roadside safety.

It may not be cost-effective to design local roads and streets that carry less than 400 vehicles per day using the same criteria applicable to higher volume roads or to make extensive traffic operational or safety improvements to such very low-volume roads. AASHTO is currently evaluating alternative design criteria for local roads that carry less than 400 vehicles per day based on a safety risk assessment.

LOCAL RURAL ROADS

General Design Considerations

A major part of the rural highway system consists of two-lane local roads. These roadways should be designed to accommodate the highest practical criteria compatible with traffic and topography.

Design Traffic Volume

Roads should be designed for a specific traffic volume and a specified acceptable level of service. The average daily traffic (ADT) volume, either current or projected to some future design year, should be the basis for design. Usually, the design year is about 20 years from the date of completion of construction but may range from the current year to 20 years depending on the nature of the improvement.

Design Speed

Design speed is a selected speed used to determine the various design features of the roadway. Geometric design features should be consistent with a specific design speed selected as appropriate for environmental and terrain conditions. Designers are encouraged to select design speeds equal to or greater than the minimum values shown in Exhibit 5-1. Low design speeds are generally applicable to roads with winding alignment in rolling or mountainous terrain or where environmental conditions dictate. High design speeds are generally applicable to roads in level terrain or where other environmental conditions are favorable. Intermediate design speeds would be appropriate where terrain and other environmental conditions are a combination of those described for low and high speed. Exhibit 5-1 lists values for minimum design speeds as appropriate for traffic needs and types of terrain; terrain types are discussed further in Chapters 2 and 3.

Sight Distance

Minimum stopping sight distance and passing sight distance should be as shown in Exhibits 5-2 and 5-3. Criteria for measuring sight distance, both vertical and horizontal, are as follows: For stopping sight distance, the height of eye is 1,080 mm [3.5 ft] and the height of object is 600 mm [2 ft]; for passing sight distance, the height of object is 1,080 mm [3.5 ft]. Chapter 3 provides a general discussion of sight distance.

	Metric						US Customary					
	Design speed (km/h) for specified design volume (veh/day)						Design speed (mph) for specified design volume (veh/day)					
Type of terrain	under 50	50 to 250	250 to 400	400 to 1500	1500 to 2000	2000 and over	under 50	50 to 250	250 to 400	400 to 1500	1500 to 2000	2000 and over
Level	50	50	60	80	80	80	30	30	40	50	50	50
Rolling	30	50	50	60	60	60	20	30	30	40	40	40
Mountainous	30	30	30	50	50	50	20	20	20	30	30	30

Exhibit 5-1. Minimum Design Speeds for Local Rural Roads

Metric				US Customary			
Initial speed (km/h)	Design stopping sight distance (m)	Rate of vertical curvature, K^a (m/%)		Initial speed (mph)	Design stopping sight distance (ft)	Rate of vertical curvature, K^a (ft/%)	
		Crest	Sag			Crest	Sag
20	20	1	3	15	80	3	10
30	35	2	6	20	115	7	17
40	50	4	9	25	155	12	26
50	65	7	13	30	200	19	37
60	85	11	18	35	250	29	49
70	105	17	23	40	305	44	64
80	130	26	30	45	360	61	79
90	160	39	38	50	425	84	96
100	185	52	45	55	495	114	115
				60	570	151	136

[a] Rate of vertical curvature, K, is the length of curve per percent algebraic difference in the intersecting grades (i.e., K = L/A). (See Chapter 3 for details.)

Exhibit 5-2. Design Controls for Stopping Sight Distance and for Crest and Sag Vertical Curves

Metric			US Customary		
Design speed (km/h)	Design passing sight distance (m)	Rate of vertical curvature, K[a] (m/%)	Design speed (mph)	Design passing sight distance (ft)	Rate of vertical curvature, K[a] (ft/%)
30	200	46	20	710	180
40	270	84	25	900	289
50	345	138	30	1090	424
60	410	195	35	1280	585
70	485	272	40	1470	772
80	540	338	45	1625	943
90	615	438	50	1835	1203
100	670	520	55	1985	1407
			60	2135	1628

[a] Rate of vertical curvature, K, is the length of curve per percent algebraic difference in the intersecting grades (i.e., K = L/A). (See Chapter 3 for details.)

Exhibit 5-3. Design Controls for Crest Vertical Curves Based on Passing Sight Distance

Grades

Suggested maximum grades for local rural roads are shown in Exhibit 5-4.

	Metric									US Customary								
	Maximum grade (%) for specified design speed (km/h)									Maximum grade (%) for specified design speed (mph)								
Type of terrain	20	30	40	50	60	70	80	90	100	15	20	25	30	40	45	50	55	60
Level	9	8	7	7	7	7	6	6	5	9	8	7	7	7	7	6	6	5
Rolling	12	11	11	10	10	9	8	7	6	12	11	11	10	10	9	8	7	6
Mountainous	17	16	15	14	13	12	10	10	–	17	16	15	14	13	12	10	10	–

Exhibit 5-4. Maximum Grades for Local Rural Roads

Alignment

Alignment between control points should be designed to be as favorable as possible consistent with the environmental impact, topography, terrain, design traffic volume, and the amount of reasonably obtainable right-of-way. Sudden changes between curves of widely different radii or between long tangents and sharp curves should be avoided. Where practical, the design should include passing opportunities. Where crest vertical curves and horizontal curves occur together, there should be greater than minimum sight distance to ensure that the horizontal curves are visible to approaching drivers.

Cross Slope

Pavement cross slope should be adequate to provide proper drainage. Normally, cross slopes range from 1.5 to 2 percent for high-type pavements and 2 to 6 percent for low-type pavements.

High-type pavements are those that retain smooth riding qualities and good non-skid properties in all weather with little maintenance.

Low-type pavements are those with treated earth surfaces and those with loose aggregate surfaces. A 3 percent cross slope is desirable for low-type pavements. For further information on pavement cross slope, see Chapter 4.

Superelevation

For rural roads with paved surfaces, superelevation should be not more than 12 percent except where snow and ice conditions prevail, in which case the superelevation should be not more than 8 percent. For aggregate roads, superelevation should be not more than 12 percent.

Superelevation runoff is the length of highway needed to accomplish the change in cross slope from a section with the adverse crown removed to a fully superelevated section. Minimum lengths of runoff are given in Chapter 3. Adjustments in design runoff lengths may be desirable for smooth riding, surface drainage, and good appearance. For a general discussion on this topic, see Chapter 3.

Number of Lanes

Two travel lanes usually can accommodate the normal traffic volume on rural local roads. If exceptional traffic volumes occur in specific areas, additional lanes may be provided in accordance with design criteria in Chapter 2. Provisions for climbing and passing lanes are covered in Chapter 3.

Width of Traveled Way, Shoulder, and Roadway

Graded shoulder width is measured from the edge of the traveled way to the point of intersection of shoulder slope and foreslope. The minimum roadway width is the sum of the traveled way and graded shoulder widths given in Exhibit 5-5. Where roadside barriers are proposed, it is desirable to provide a minimum offset of 1.2 m [4.0 ft] from the traveled way to the barrier whenever practical. For further information, see the sections on "Shoulders" and

Metric				US Customary				
	Minimum width of traveled way (m) for specified design volume (veh/day)				Minimum width of traveled way (ft) for specified design volume (veh/day)			
Design speed (km/h)	under 400	400 to 1500	1500 to 2000	over 2000	Design speed (mph)	under 400	400 to 1500	1500 to 2000	over 2000
20	5.4	6.0[a]	6.0	6.6	15	18	20[a]	20	22
30	5.4	6.0[a]	6.6	7.2[c]	20	18	20[a]	22	24[c]
40	5.4	6.0[a]	6.6	7.2[c]	25	18	20[a]	22	24[c]
50	5.4	6.0[a]	6.6	7.2[c]	30	18	20[a]	22	24[c]
60	5.4	6.0[a]	6.6	7.2[c]	40	18	20[a]	22	24[c]
70	6.0	6.6	6.6	7.2[c]	45	20	22	22	24[c]
80	6.0	6.6	6.6	7.2[c]	50	20	22	22	24[c]
90	6.6	6.6	7.2[c]	7.2[c]	55	22	22	24[c]	24[c]
100	6.6	6.6	7.2[c]	7.2[c]	60	22	22	24[c]	24[c]
	Width of graded shoulder on each side of the road (m)				Width of graded shoulder on each side of the road (ft)			
All speeds	0.6	1.5[a,b]	1.8	2.4	All speeds	2	5[a,b]	6	8

[a] For roads in mountainous terrain with design volume of 400 to 600 veh/day, use 5.4-m [18-ft] traveled way width and 0.6-m [2-ft] shoulder width.

[b] May be adjusted to achieve a minimum roadway width of 9 m [30 ft] for design speeds greater than 60 km/h [40 mph].

[c] Where the width of the traveled way is shown as 7.2 m [24 ft], the width may remain at 6.6 m [22 ft] on reconstructed highways where alignment and safety records are satisfactory.

See text for roadside barrier and offtracking considerations.

Exhibit 5-5. Minimum Width of Traveled Way and Shoulders

"Longitudinal Barriers" in Chapter 4 and for information on vehicle offtracking, see the section on "Derivation of Design Values for Widening on Horizontal Curves" in Chapter 3.

When bicycle facilities are included as part of the design, refer to AASHTO's *Guide for the Development of Bicycle Facilities* (**1**).

Structures

The design of bridges, culverts, walls, tunnels, and other structures should be in accordance with the current *Standard Specifications for Highway Bridges* (**2**). Except as otherwise indicated in this chapter and in Chapter 4, the dimensional design of structures should also be in accordance with Reference (**2**).

The minimum design loading for new bridges on local rural roads should be MS-18 [HS-20].

The minimum clear roadway widths for new and reconstructed bridges should be as given in Exhibit 5-6. For general discussion of structure widths, see Chapter 10.

Bridges to Remain in Place

Existing substandard structures should be improved, but because of their high replacement cost, reasonably adequate bridges and culverts that meet tolerable criteria may be retained. Some of the non-technical factors that should be considered are the aesthetic value and the historical significance attached to famous structures, covered bridges, and stone arches.

Where an existing road is to be reconstructed, an existing bridge that fits the proposed alignment and profile may remain in place when its structural capacity in terms of design loading and clear roadway width are at least equal to the values given in Exhibit 5-7 for the applicable traffic volume.

The values in Exhibit 5-7 do not apply to structures with total lengths greater than 30 m [100 ft]. These structures should be analyzed individually, taking into consideration the clear width provided, traffic volume, remaining life of the structure, pedestrian volume, snow storage, design speed, crash history, and other pertinent factors.

Vertical Clearance

Vertical clearance at underpasses should be at least 4.3 m [14 ft] over the entire roadway width, with an allowance for future resurfacing.

Metric			US Customary		
Design volume (veh/day)	Minimum clear roadway width for bridges[a]	Design loading structural capacity	Design volume (veh/day)	Minimum clear roadway width for bridges[a]	Design loading structural capacity
400 and under	Traveled way + 0.6 m (each side)	MS-18	400 and under	Traveled way + 2 ft (each side)	HS-20
400 to 2000	Traveled way + 1.0 m (each side)	MS-18	400 to 2000	Traveled way + 3 ft (each side)	HS-20
over 2000	Approach roadway width[b]	MS-18	over 2000	Approach roadway width[b]	HS-20

[a] Where the approach roadway width (traveled way plus shoulders) is surfaced, that surface width should be carried across the structures.

[b] For bridges in excess of 30 m [100 ft] in length, the minimum width of traveled way plus 1 m [3 ft] on each side is acceptable.

Exhibit 5-6. Minimum Clear Roadway Widths and Design Loadings for New and Reconstructed Bridges

Metric			US Customary		
Design volume (veh/day)	Design loading structural capacity	Minimum clear roadway width (m)[a,b,c]	Design volume (veh/day)	Design loading structural capacity	Minimum clear roadway width (ft)[a,b,c]
0 to 50	M 9	6.0[d]	0 to 50	H 10	20[d]
50 to 250	M 13.5	6.0	50 to 250	H 15	20
250 to 1500	M 13.5	6.6	250 to 1500	H 15	22
1500 to 2000	M 13.5	7.2	1500 to 2000	H 15	24
over 2000	M 13.5	8.4	over 2000	H 15	28

[a] Clear width between curbs or rails, whichever is the lesser.

[b] Minimum clear widths that are 0.6 m [2 ft] narrower may be used on roads with few trucks. In no case shall the minimum clear width be less than the approach traveled way width.

[c] Does not apply to structures with total length greater than 30 m [100 ft].

[d] For single-lane bridges, use 5.4 m [18 ft].

Exhibit 5-7. Minimum Structural Capacities and Minimum Roadway Widths for Bridges to Remain in Place

Right-of-Way Width

The provision of right-of-way widths that accommodate construction, adequate drainage, and proper maintenance of a highway is a very important part of the overall design. Wide rights-of-way permit the construction of gentle slopes, resulting in greater safety for the motorist and providing for easier and more economical maintenance. The procurement of sufficient right-of-way at the time of the initial improvement permits the widening of the roadway and the widening and strengthening of the pavement at a reasonable cost as traffic volumes increase.

In developed areas, it may be desirable to limit the right-of-way width. However, the right-of-way width should not be less than that required for all the elements of the design cross sections, utility accommodation, and appropriate border areas.

Foreslopes

The maximum rate of foreslope depends on the stability of local soils as determined by soil investigation and local experience. Slopes should be as flat as practical, and other factors should be considered to determine the design slope. Flat foreslopes increase safety by providing a maneuver area in emergencies, are more stable than steep slopes, aid in the establishment of plant growth, and simplify maintenance work. Vehicles that leave the traveled way can often be kept under control if slopes are gentle and drainage ditches are well-rounded. Such recovery areas should be provided where terrain and right-of-way controls permit.

Combinations of rate and height of slope should provide for vehicle recovery. Where controlling conditions (such as high fills, right-of-way restrictions, or the presence of rocks, watercourses, or other roadside features) make this impractical, consideration should be given to the provision of guardrail, in which case the maximum rate of foreslope could be used.

Cut sections should be designed with adequate ditches. Preferably, the foreslope should not be steeper than 1V:2H, and the ditch bottom and slopes should be well-rounded. The backslope should not exceed the maximum required for stability.

Horizontal Clearance to Obstructions

A clear zone of 2 to 3 m [7 to 10 ft] or more from the edge of the traveled way, appropriately graded with relatively flat slopes and rounded cross-sectional design, is desirable. An exception may be made where guardrail protection is provided. The recovery area should be clear of all unyielding objects such as trees, sign supports, utility poles, light poles, and any other fixed objects that might severely damage an out-of-control vehicle.

To the extent practical, where another highway or railroad passes over, the structure should be designed so that the pier or abutment supports have lateral clearance as great as the clear roadside area on the approach roadway. For further information on providing roadside lateral clearance, see the AASHTO *Roadside Design Guide* (3).

Where it is not practical to carry the full-width approach roadway across an overpass or other bridge, an appropriately transitioned roadside barrier should be provided. At selected locations, such as the outside of a sharp curve, a broader recovery area with greater horizontal clearances should be provided to any roadside obstruction.

Curbs

The use of curbs in conjunction with intermediate or high design speeds should be limited, as discussed in Chapter 4. Where curbs are to be used, refer to the discussion on curbs in the section "Local Urban Streets" in this chapter.

Intersection Design

Intersections should be carefully located to avoid steep profile grades and to ensure adequate approach sight distance. An intersection should not be situated just beyond a short-crest vertical curve or on a sharp horizontal curve. When there is no practical alternate to such a location, the approach sight distance on each leg should be carefully checked, and where practical, backslopes should be flattened and horizontal or vertical curves lengthened to provide additional sight distance. The driver of a vehicle approaching an intersection should have an unobstructed view of the entire intersection and sufficient lengths of the intersecting roadways to permit the driver to anticipate and avoid potential collisions. Sight distances at intersections with six different types of traffic control are presented in Chapter 9.

Intersections should be designed with a corner radius of the pavement or surfacing that is adequate for a selected design vehicle, representing a larger vehicle that is anticipated to use the intersection with some frequency. For minimum edge radius, see Chapter 9. Where turning volumes are significant, consideration should be given to speed change lanes and channelization.

Intersection legs that operate under stop control should intersect at right angles wherever practical, and should not intersect at an angle less than 60 degrees. For further details, see Chapter 9.

Railroad-Highway Grade Crossings

Appropriate grade-crossing warning devices shall be installed at all railroad-highway grade crossings on local roads and streets. Details of the devices to be used are given in the *Manual on Uniform Traffic Control Devices* (MUTCD) (**4**). In some States, the final approval of the devices to be used may be vested in an agency having oversight over railroads.

Sight distance is an important consideration at railroad-highway grade crossings. There should be sufficient sight distance along the road and along the railroad tracks for an approaching driver to recognize the crossing, perceive the warning device, determine whether a train is approaching, and stop if necessary. For further information on railroad-highway grade crossings, see Chapter 9.

The roadway width at all railroad crossings should be the same as the width of the approach roadway. Crossings that are located on bicycle routes that are not perpendicular to the railroad may need additional paved shoulder for bicycles to maneuver over the crossing. For further information, see the AASHTO *Guide for the Development of Bicycle Facilities* (**1**).

Traffic Control Devices

Signs, pavement and other markings, and, where appropriate, traffic signal controls are essential elements for all local roads and streets. Refer to the MUTCD (**4**) for details of the devices to be used and, for some conditions, warrants for their use.

Bicycle Facilities

The local roadway may be sufficient to accommodate bicycle traffic. Where special facilities for bicycles are desired, they should be in accordance with the AASHTO *Guide for the Development of Bicycle Facilities* (**1**).

Erosion Control

All slopes and drainage areas should be designed with proper regard for the desired natural ground cover and growth regeneration on areas opened during construction. Various acceptable methods of erosion control, including seeding and mulching of slopes, sodding, or other protection of swales and other erodible areas, should be included in the local road design. Consideration should also be given to maintenance requirements and overall economics.

In roadside design, the preservation of natural ground covers and desirable growth of shrubs and trees should be considered, provided that such growth does not constitute an obstruction in the recovery area.

LOCAL URBAN STREETS

General Design Considerations

A local urban street is a public roadway for vehicular travel including public transit and refers to and includes the entire area within the right-of-way. The street also serves pedestrian and bicycle traffic and usually accommodates public utility facilities within the right-of-way. The development or improvement of streets should be based on a functional street classification that is part of a comprehensive community development plan. The design criteria should be appropriate for the ultimately planned development.

Most urban functional classifications include three classes of streets: arterials, collectors, and local access routes, which are discussed in Chapter 1. Geometric design guidance for collector streets is provided in Chapter 6 and is provided for arterial streets in Chapter 7. It is not practical to present separate design criteria for local streets for each design feature discussed below. However, where there are substantial differences from the criteria used in design of other functional classes, specific design guidance is given below.

The design features of local urban streets are governed by practical limitations to a greater extent than those of similar roads in rural areas. The two major design controls are (1) the type and extent of urban development with its limitations on rights-of-way, and (2) zoning or regulatory restrictions. Some streets serve primarily to provide access to adjacent residential development areas. In such cases, the overriding consideration is to foster a safe and pleasant environment whereas the convenience of the motorist is secondary. Other local streets not only provide access to adjacent development but also serve limited through traffic. Traffic service features may be an important concern on such streets.

On streets serving industrial or commercial areas, the vehicle dimensions, traffic volumes, and vehicle loads differ greatly from those on residential streets, and different dimensional and structural design values are appropriate. Here, safety and traffic service are usually the major design controls. Where a particular design feature varies depending on the area served, such as residential, commercial, or industrial, different design guidelines are presented for each condition. The designer should be apprised of local ordinances and resolutions that affect certain design features.

Design Traffic Volume

Traffic volume is not usually a major factor in determining the geometric criteria to be used in designing residential streets. Traditionally, such streets are designed with a standard two-lane cross section, but a four-lane cross section may be appropriate in certain urban areas, as governed by traffic volume, administrative policy, or other community considerations.

For streets serving industrial or commercial areas, however, traffic volume is a major factor. The ADT projected to some future design year should be the design basis. It usually is difficult and costly to modify the geometric design of an existing street unless provision is made at the time of initial construction. Design traffic volumes in such areas should be that estimated for at least 10 years, and preferably 20 years, from the date of construction completion.

Design Speed

Design speed is not a major factor for local streets. For consistency in design elements, design speeds ranging from 30 to 50 km/h [20 to 30 mph] may be used, depending on available right-of-way, terrain, likely pedestrian presence, adjacent development, and other area controls. In the typical street grid, the closely spaced intersections usually limit vehicular speeds, making the effect of design speed less important. Since the function of local streets is to provide access to adjacent property, all design elements should be consistent with the character of activity on and adjacent to the street, and should encourage speeds generally not exceeding 50 km/h [30 mph].

Sight Distance

Minimum stopping sight distance for local streets should range from 30 to 60 m [100 to 200 ft] depending on the design speed (see Exhibit 3-1). Design for passing sight distance seldom is applicable on local streets.

Grades

Grades for local residential streets should be as level as practical, consistent with the surrounding terrain. The gradient for local streets should be less than 15 percent. Where grades of 4 percent or steeper are necessary, the drainage design may become critical. On such grades special care should be taken to prevent erosion on slopes and open drainage facilities.

For streets in commercial and industrial areas, gradient design desirably should be less than 8 percent, grades should desirably be less than 5 percent, and flatter grades should be encouraged.

To provide for proper drainage, the desirable minimum grade for streets with outer curbs should be 0.30 percent, but a minimum grade of 0.20 percent may be used.

Alignment

Alignment in residential areas should closely fit with the existing topography to minimize the need for cuts or fills without sacrificing safety. The alignment of local streets in residential areas should be arranged to discourage through traffic. Street alignment in commercial and industrial areas should be commensurate with the topography but should be as direct as possible.

Street curves should be designed with as large a radius curve as practical, with a minimum radius of 30 m [100 ft]. Where curves are superelevated, lower values may apply, but the radius should not be less than approximately 25 m [75 ft] for a 30-km/h [20-mph] design speed.

Cross Slope

Pavement cross slope should be adequate to provide proper drainage. Normally, cross slopes where there are flush shoulders range from 1.5 to 2 percent for high-type pavements and 2 to 6 percent for low-type pavements.

High-type pavements are those that retain smooth riding qualities and good non-skid properties in all weather with little maintenance. Low type pavements or with treated earth surfaces and those with either loose or stabilized aggregate surfaces. A 3 percent cross slope is desirable for low-type pavements. For further information on pavement cross slope, see Chapter 4.

Where there are outer curbs, cross slopes steeper by about 0.5 to 1 percent are desirable for the lane adjacent to the curb.

Superelevation

Superelevation may be advantageous for local street traffic operations in specific locations, but in built-up areas the combination of wide pavement areas, proximity of adjacent development, control of cross slope, profile for drainage, frequency of cross streets, and other urban features often combine to make the use of superelevation impractical or undesirable. Usually, superelevation is not provided on local streets in residential and commercial areas; it may be considered on local streets in industrial areas to facilitate operation.

If superelevation is used, street curves should be designed for a maximum superelevation rate of 4 percent. If terrain dictates sharp curvature, a maximum superelevation of 6 percent is justified if the curve is long enough to provide an adequate superelevation transition. Minimum lengths of superelevation runoff and a detailed discussion of superelevation are found in Chapter 3.

Number of Lanes

On residential streets where the primary function of the street is to provide access to adjacent development and foster a safe and pleasant environment, at least one unobstructed moving lane must be ensured even where parking occurs on both sides. The level of user inconvenience occasioned by the lack of two moving lanes is remarkably low in areas where single-family units prevail. Local residential street patterns are such that travel distances are less than 1 km [0.5 mi] from the trip origin to a collector street. In multifamily-unit residential areas, a minimum of two moving traffic lanes to accommodate opposing traffic may be desirable. In many residential areas an 8 m [26 ft] wide roadway is typical. This curb-face-to-curb-face width provides for a 3.6-m [12-ft] center travel lane and two 2.2-m [7-ft] parking lanes. Opposing conflicting traffic will yield and pause on the parking lane area until there is sufficient width to pass.

In commercial areas where there are midblock left turns, it may be advantageous to provide an additional continuous two-way, left-turn lane in the center of the roadway.

Width of Traveled Way

Street lanes for moving traffic preferably should be at least 3.0 m [10 ft] wide. Where practical, they should be 3.3 m [11 ft] wide, and in industrial areas they should be 3.6 m [12 ft] wide. Where the available or attainable width of right-of-way imposes severe limitations, 2.7-m [9-ft] lanes can be used in residential areas, as can 3.3-m [11-ft] lanes in industrial areas. Added turning lanes where used at intersections should be at least 2.7 m [9 ft] wide, and desirably 3.0 to 3.6 m [10 to 12 ft] wide, depending on the percentage of trucks.

Where bicycle facilities are included as part of the design, refer to the AASHTO *Guide for the Development of Bicycle Facilities* (**1**).

Parking Lanes

Where used in residential areas, a parallel parking lane a minimum of 2.1 m [7 ft] wide should be provided on one or both sides, as appropriate to the conditions of lot size and intensity of development. In commercial and industrial areas, parking lanes should be a minimum of 2.4 m [8 ft] wide and are usually provided on both sides.

Parking lane width determination in commercial and industrial areas should include consideration for use of the parking lane for moving traffic during peak periods that may occur where industries have high employment concentrations. Where curb and gutter sections are used, the gutter pan width should be considered as part of the parking lane width.

Median

Medians provided on local urban streets primarily to enhance the environment and to act as buffer strips should be designed to minimize interference with access to the land abutting the roadway. A discussion of the various median types appears in Chapter 4.

Median openings should be situated only where there is adequate sight distance. The shape and length of the median openings depend on the width of median and the vehicle types to be accommodated. The desirable length of median openings, measured between the inner edge of the lane adjacent to the median and the centerline of the intersection roadway, should be great enough to provide for a 12-m [40-ft] turning control radius for left-turning P vehicles. The minimum length of median openings should be that of the width of the projected roadway of the intersecting cross street or driveway.

Curbs

Streets normally are designed with curbs to allow greater use of available width and for control of drainage, protection of pedestrians, and delineation. The curb should be 100 to 150 mm [4 to 6 in] high, depending on drainage considerations, traffic control, and safety.

On divided streets the type of median curbs should be compatible with the width of the median and the type of turning movement control to be effected.

Vertical curbs with heights of 150 mm [6 in] or more adjacent to the traveled way should be offset at least 0.3 m [1 ft]. Where a curb-and-gutter section is provided, the gutter pan width, normally 0.6 m [2 ft], should be used as the offset distance. For additional information regarding curbs, see Chapter 4.

Drainage

Drainage is an important consideration in urban areas because of high runoff and flood potential. Surface flow from adjacent tributary areas may be intercepted by the street system, where it is collected within the roadway by curbs, gutters, and ditches, and conveyed to an appropriate drainage system. Where drains are available under or near the roadway, the flow is transferred at frequent intervals from the street cross section by gratings or curb-opening inlets to basins and from there by connectors to drainage channels or underground drains.

Economic considerations usually dictate that maximum practical use be made of the street sections for surface drainage. To avoid undesirable flowline conditions, the minimum gutter grade should be 0.30 percent. However, in very flat terrain and where no drainage outlet is available, gutter grades as low as 0.20 percent may be used. Where a drainage system is available, the inlets should be spaced to provide a high level of drainage protection in areas of high pedestrian use or where adjacent property has an unusually important public or community purpose (e.g., schools and churches). For further details, see the drainage section in Chapter 3.

Cul-De-Sacs and Turnarounds

A local street open at one end only should have a special turning area at the closed end. This turning area desirably should be circular and have a radius appropriate to the vehicle types expected. Minimum outside radii of 10 m [30 ft] in residential areas and 15 m [50 ft] in commercial and industrial areas are commonly used.

A dead-end street narrower than 12 m [40 ft] usually should be widened to enable passenger vehicles, and preferably delivery trucks, to make U-turns or at least turn around by backing only once. The design commonly used is a circular pavement symmetrical about the centerline of the street sometimes with a central island, as shown in Exhibit 5-8C, which also shows minimum dimensions for the design vehicles. Although this type of cul-de-sac operates satisfactorily and

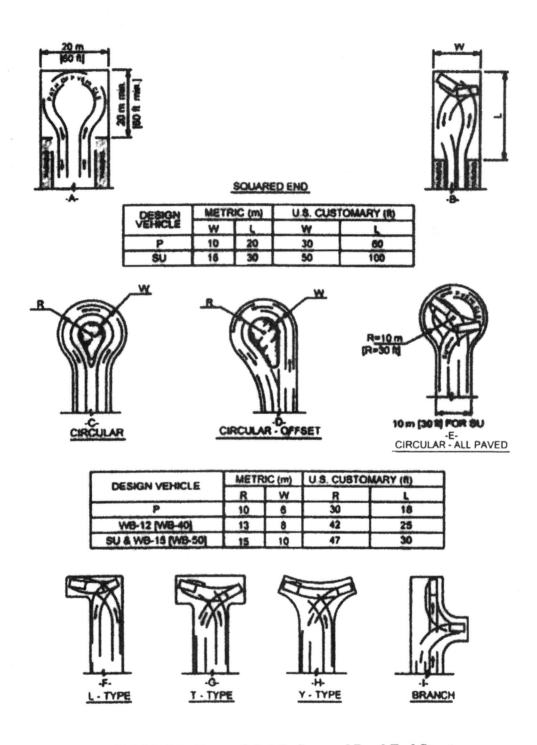

Exhibit 5-8. Types of Cul-de-Sacs and Dead-End Streets

looks well, better operation is obtained if the design is offset so that the entrance-half of the pavement is in line with the approach-half of the street, as shown in Exhibit 5-8D. One steering reversal is avoided on this design. Where a radius of less than 15 m [50 ft] is used, the island should be bordered by sloping curbs to permit the maneuvering of an occasional oversized vehicle.

An all-paved plan, as opposed to an island configuration, with a 10-m [30-ft] outer radius, shown in Exhibit 5-8E, requires little additional paving. If the approach pavement is at least 10 m [30 ft] wide, the result is a cul-de-sac on which passenger vehicles can make the customary U-turn and SU design trucks can turn by backing only once. A radius of about 12 m [40 ft] will enable a WB-15 [WB-50] vehicle to turn around by maneuvering back and forth.

Other variations or shapes of cul-de-sacs that include right-of-way and site controls may be provided to permit vehicles to turn around by backing only once. Several types (Exhibit 5-8F, 5-8G, 5-8H, and 5-8I) may also be suitable for alleys. The geometry of a cul-de-sac should be altered if adjoining residences also use the area for parking.

Alleys

Alleys provide access to the side or rear of individual land parcels. They are characterized by a narrow right-of-way and range in width from 5 to 6 m [16 to 20 ft] in residential areas and up to 10 m [30 ft] in industrial areas.

Alleys should be aligned parallel to, or concentric with, the street property lines. It is desirable to situate alleys in such a manner that both ends of the alley are connected either to streets or to other alleys. Where two alleys intersect, a triangular corner cutoff of not less than 3 m [10 ft] along each alley property line should be provided. Dead-end alleys should include a turning area in accordance with Exhibit 5-9. This dead-end turning area design may be suitable for application on some very low-volume roads.

Curb return radii at street intersections may range from 1.5 m [5 ft] in residentially zoned areas to 3 m [10 ft] in industrial and commercial areas where large numbers of trucks are expected. Alleys should have grades established to meet as closely as possible the existing grades of the abutting land parcels. The longitudinal grade should not be less than 2 percent.

Alley cross sections may be V-shaped with transverse slopes of 2.5 percent toward a center V gutter. Runoff is thereby directed to a catch basin in the alley or to connecting street gutters.

Sidewalks

Sidewalks used for pedestrian access to schools, parks, shopping areas, and transit stops and placed along all streets in commercial areas should be provided along both sides of the street.

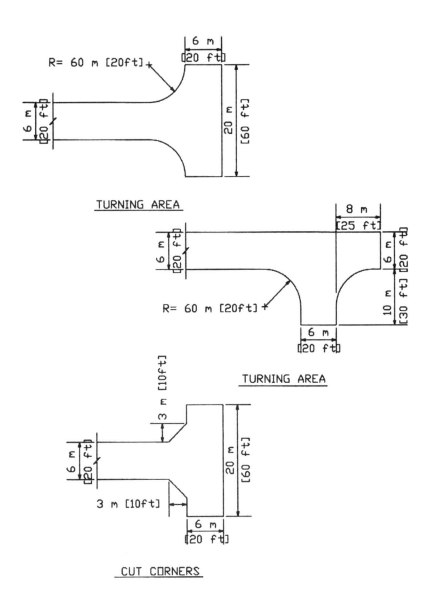

Exhibit 5-9. Alley Turnarounds

In residential areas, sidewalks should be provided on at least one side of all local streets and are desirable on both sides of the street. The sidewalks should be located as far as practical from the traveled way and usually close to the right-of-way lines.

The minimum sidewalk width should be 1.2 m [4 ft]; sidewalk widths of 2.4 m [8 ft] or greater may be needed in commercial areas. If roadside appurtenances are situated on the sidewalk adjacent to the curb, additional width may be needed to secure the clear width. Greater sidewalk widths should be considered for higher volume sidewalks and where the sidewalk is against the curb or wall. Further guidance on designing sidewalks can be found in the AASHTO Guide for the *Planning, Design, and Operation of Pedestrian Facilities* (**5**).

Sidewalk Curb Ramps

Sidewalk curb ramps should be provided at crosswalks to accommodate persons with disabilities. Such ramps may be the same width as the approach sidewalks; the suggested minimum width should be 1.0 m [3 ft] exclusive of sideslopes. Further discussion of this topic appears in Chapter 4. Further guidance on designing sidewalk-driveway interfaces can be found in the AASHTO *Guide for the Planning, Design, and Operation of Pedestrian Facilities* (**5**).

Driveways

A driveway is an access constructed within a public right-of-way, connecting a public roadway with adjacent property and intended to provide vehicular access into that property in a manner that will not cause the blocking of any sidewalk, border area, or street roadway.

Some of the principles of intersection design apply directly to driveways. In particular, driveways should have well-defined locations. Large graded or paved areas adjacent to the traveled way, which allow drivers to enter or leave the street randomly, should be discouraged.

Sight distance is an important design control for driveways. Driveway locations where sight distance is not sufficient should be avoided. Vertical obstructions to essential sight distances should be controlled by regulations. Driveways should be regulated as to width of entrance, spacing, and placement with respect to property lines and intersecting streets, angle of entry, vertical alignment, and number of entrances to a single property to provide for traffic safety and maximum use of curb space for parking where permitted. Driveways should be situated as far away from intersections as practical, particularly if the driveway is located near an arterial street.

Driveway returns should not be less than 1 m [3 ft] in radius. Flared driveways are preferred because they are distinct from intersection delineations, can properly handle turning movements, and can minimize problems for persons with disabilities. Further guidance on the design of sidewalk-driveway interfaces can be found in the AASHTO *Guide for the Planning, Design, and Operation of Pedestrian Facilities* (**5**).

Roadway Widths for Bridges

The clear width for all new bridges on streets with curbed approaches should be the same as the curb-to-curb width of the approaches. For streets with shoulders and no curbs, the clear roadway width preferably should be the same as the approach roadway width and in no case less than the width shown in Exhibit 5-6. Sidewalks on the approaches should be carried across all new structures. There should be at least one sidewalk on all street bridges.

Horizontal Clearance to Obstructions

On all streets a minimum clearance of 0.5 m [1.5 ft] should be provided between the curb face and obstructions such as utility poles, lighting poles, and fire hydrants. In areas of dense pedestrian traffic, the construction of vertical curbing (typically 150 to 225 mm [6 to 9 in] high) aids in protecting the high-volume pedestrian traffic. For facilities without curbs, a clear zone commensurate with a rural cross section and design speed should be provided, as described in the AASHTO *Roadside Design Guide* (3).

Trees are acceptable along local streets where speeds are low (60 km/h [40 mph] or below), where curbs are present, and where adequate sight distance is available from intersecting streets and driveways.

Guardrail is not used extensively on local streets except where there is a significant risk to motorists and pedestrians, such as along sections with steep foreslopes and at approaches to overcrossing structures.

Vertical Clearance

Vertical clearance at underpasses should be at least 4.3 m [14 ft] over the entire roadway width, with an allowance for future resurfacing.

Border Area

A border area should be provided along streets for the safety of motorists and pedestrians as well as for aesthetic reasons. The street alignment should be selected to minimize roadside slopes. However, the preservation and enhancement of the environment is of major importance in the design and construction of local streets.

The border area between the roadway and the right-of-way line should be wide enough to serve several purposes, including provision of a buffer space between pedestrians and vehicular traffic, sidewalk space, snow storage, an area for placement of underground and aboveground utilities, and an area for maintainable aesthetic features such as grass or other landscaping. The border width may be a minimum of 1.5 m [5 ft], but desirably should be 3.0 m [10 ft] or wider.

Where the available right-of-way is limited and in areas of high right-of-way costs, as in some industrial and commercial areas, a buffer width of 0.6 m [2 ft] may be tolerated.

Wherever practical, an additional obstacle-free buffer width of 3.6 m [12 ft] or more should be provided between the curb and the sidewalk for safety and environmental enhancement. In residential areas, wider building setback controls can be used to attain these features.

Right-of-Way Width

The right-of-way width should be sufficient to accommodate the ultimate planned roadway including median (if used), shoulder (if used), landscaping strip, sidewalks, utility strips in the border areas, and necessary outer slopes.

Provision for Utilities

In addition to the primary purpose of serving vehicular traffic and in accordance with State law or municipal ordinance, streets also often accommodate public utility facilities within the street right-of-way. Use of the rights-of-way by utilities should be planned to minimize interference with traffic using the street. References (**6**) and (**7**) provide general principles for location and construction of utilities to minimize conflict between the use of the street right-of-way for vehicular movement and for its secondary purpose of providing space for location of utilities.

Intersection Design

Intersections, including median openings, should be designed with adequate corner sight distance, as described in Chapter 9, and the intersection area should be kept free of obstacles. To maintain the minimum sight distance, restrictions on height of embankment, locations of buildings, on street parking, and screening fences may be appropriate. Any landscaping in the clear sight triangle should be low growing and should not be higher than 1.0 m [3 ft] above the level of the intersecting street pavements.

Intersecting streets should meet at approximately a 90° angle. The alignment design should be adjusted to avoid an angle of intersection of less than 60°. Closely spaced offset intersections are undesirable.

The intersection and approach areas where vehicles are stored while waiting to enter the intersection should be designed with a relatively flat grade; the maximum grade on the approach leg should not exceed 5 percent where practical. Where ice and snow may create poor driving conditions, the desirable grade on the approach leg should be 0.5 percent with no more than 2 percent wherever practical.

At street intersections, there are two distinct radii that need to be considered—the effective turning radius of the turning vehicle and the radius of the curb return (see Exhibit 5-10.) The effective turning radius is the minimum radius appropriate for turning from the right-hand travel lane on the approach street to the appropriate lane of the receiving street. This radius is determined by the selection of a design vehicle appropriate for the streets being designed and the lane on the receiving street into which that design vehicle will turn. Desirably this should be at least 7.5 m [25 ft].

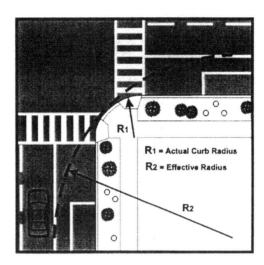

Exhibit 5-10. Actual Curb Radius and Effective Radius for Right-Turn Movements at Intersections

The radius of the curb return should be no greater than that needed to accommodate the design turning radius. However, the curb return radius should be at least 1.5 m [5 ft] to enable effective use of street-sweeping equipment.

In industrial areas the radius of the curb return should not be less than 10 m [30 ft], and desirably, use should be made of a three-centered curve of sufficiently large radius to accommodate the largest vehicles expected.

Further information pertaining to intersection design appears in Chapter 9.

Railroad-Highway Grade Crossings

Appropriate grade-crossing warning devices shall be installed at all railroad-highway grade crossings on local roads and streets. Details of the devices to be used are given in the MUTCD (**4**). In some States, the final approval of the devices to be used may be vested in an agency having oversight over railroads.

Sight distance is an important consideration at railroad-highway grade crossings. There must be sufficient sight distance along the road and railroad tracks for an approaching driver to recognize the crossing, perceive the warning device, determine whether a train is approaching, and stop if necessary. (For further information on railroad-highway grade crossings, see Chapter 9.) Signalized intersections adjacent to signalized railroad grade crossings should be designed with railroad preemption.

The roadway width at all railroad crossings should be the same as the width of the approach roadway. Sidewalks should be provided at railroad grade crossings to connect existing or future walkways that approach these crossings. Crossings that are located on bicycle routes that are not perpendicular to the railroad may need additional paved shoulder for bicycles to maneuver over the crossing. For further information, see the AASHTO *Guide for the Development of Bicycle Facilities* (**1**).

Street and Roadway Lighting

Good visibility under day or night conditions is one of the fundamental needs for motorists to travel on roadways in a safe and coordinated manner. Properly designed and maintained street lighting will produce comfortable and accurate visibility at night, which will facilitate and encourage both vehicular and pedestrian traffic. Thus, where adequate illumination is provided, efficient night use can be made of existing streets. Determinations of need for lighting should be coordinated with crime prevention programs and other community needs.

Warrants for the justification of street lighting involve more than just identifying the functional classification of the roadway. Pedestrian and vehicular volume, night-to-day crash ratios, roadway geometry, merging lanes, curves, and intersections all need careful consideration in establishing illumination levels.

At locations where illumination levels provided for the various roadway, walkway, and area classifications are not influenced by these considerations, the suggested minimum levels of illumination (expressed in average maintained horizontal lux) are presented in Exhibit 5-11. Illumination levels at intersections should be the sum of illumination levels on intersecting streets at the intersection.

Classification	Metric		US Customary	
	Industrial/ commercial (lux)	Residential (lux)	Industrial/ commercial (ft-c)	Residential (ft-c)
Local	9.7	4.3	0.9	0.4
Alleys	6.5	2.2	0.6	0.2
Sidewalks	9.7	2.2	0.9	0.2

Exhibit 5-11. Minimum Illumination Levels

Uniformity of lighting is an indication of the quality of illumination and should be considered along with illumination levels. Uniformity of illumination can be represented by a uniformity ratio of the average-to-minimum lux values on the roadway or walkway surface. Recommended uniformity ratios are as follows: residential roadways, 6:1; commercial roadways, 3:1; residential walkways, 10:1; and commercial walkways, 4:1.

Because glare is also an indication of the quality of lighting, the type of fixtures and the height at which the light sources are mounted are also factors in designing street lighting systems. The objectives of the designer should be to minimize visual discomfort and impairment of driver and pedestrian vision due to glare. Where only intersections are lighted, a gradual lighting transition from dark to light to dark should be provided so that drivers may have time to adapt their vision. More detailed discussion of this topic is contained in AASHTO's *An Informational Guide for Roadway Lighting* (**8**).

Traffic Control Devices

Consistent and uniform application of traffic control devices is important. Details of the standard devices and warrants for many conditions are found in the MUTCD (**4**).

Geometric design of streets should include full consideration of the types of traffic control to be used, especially at intersections where multiphase or actuated traffic signals are likely to be needed.

Erosion Control

Design of streets should include considerations for preservation of natural ground cover and desirable growth of shrubs and trees within the right-of-way. Seeding, mulching, sodding, or other acceptable measures of covering slopes, swales, and other erodible areas should be incorporated in urban local street design. For further information, see the section on "Erosion Control and Landscape Development" in Chapter 3.

Landscaping

Landscaping in keeping with the character of the street and its environment should be provided for aesthetic and erosion-control purposes. Landscape designs should be arranged to permit a sufficiently wide, clear, and safe pedestrian walkway. Individuals with disabilities, bicyclists, and pedestrians should all be considered. Combinations of turf, shrubs, and trees should be considered in continuous border areas along the roadway. However, care should be exercised to ensure that sight distances and clearance to obstruction guidelines are observed, especially at intersections. The roadside should be developed to serve both the community and the traveling motorist. Landscaping should also consider maintenance problems and costs, future sidewalks, utilities, additional lanes, and possible bicycle facilities. For further information on landscaping, see the AASHTO *Guide for Transportation Landscape and Environmental Design* (**9**).

Bicycle Facilities

Local roadways and streets are generally sufficient to accommodate bicycle traffic. However, where special facilities are desired, they should be planned and designed in accordance with the AASHTO *Guide for the Development of Bicycle Facilities* (**1**).

SPECIAL-PURPOSE ROADS

Introduction

For the purpose of design, highways have been classified in this book by function with specific design criteria for each functional class. Subsequent chapters discuss the design of collectors, arterials, and freeways. The first two sections of this chapter discuss the design of local roads and streets. Another type of local road, however, is different because of its purpose and does not fit into any of the classifications identified above. This type of local road is referred to as a special-purpose road and because of its unique character, separate design criteria are provided. Special-purpose roads include recreational roads, resource recovery roads, and local service roads. Roads in the special-purpose category are generally lightly traveled and operate with low traffic speeds; for these reasons, the design criteria for special-purpose roads differ from those for other roadway types.

Recreational Roads

General Considerations

Roads serving recreational sites and areas are unique in that they are also part of the recreational experience. Design criteria described below meet the unusual demands on roads for access to, through, and within recreational sites, areas, and facilities for the complete enjoyment of the recreationist. The criteria are intended to protect and enhance the existing aesthetic, ecological, environmental, and cultural amenities that form the basis for distinguishing each particular recreational site or area.

Visitors to a recreational site need access to the general area, usually by a statewide or principal arterial highway. Secondly, they need access to the specific recreational site. This is the most important link from the statewide road system. For continuity beyond this point, design criteria assume that the visitor is aware of the recreational nature of the area. The design should be accomplished by a multidisciplinary team of varied backgrounds and experience, to ultimately provide a road system that is an integral part of the recreational site. Depending on the conditions, internal tributaries will have a variety of lower design features.

The criteria discussed in this chapter are applicable for public roads within all types of recreational sites and areas. Design criteria for recreational roads are discussed for primary access roads, circulation roads, and area roads. Primary access roads are defined as roads that allow through movement into and between access areas. Circulation roads allow movement between activity sites within an access area, whereas area roads allow direct access to individual activity areas, such as campgrounds, park areas, boat launching ramps, picnic groves, and scenic and historic sites.

Exhibit 5-12 depicts a potential road system serving a recreational area. Road links are labeled in accordance with the classification system noted.

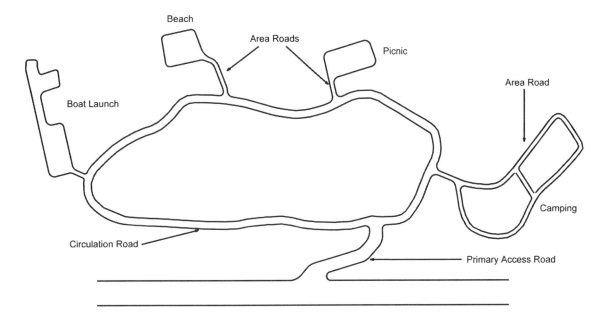

Exhibit 5-12. Potential Road Network

Design Speed

The effect of design speed on various roadway features is considered in its selection; however, the speed is selected primarily on the basis of the character of the terrain and the functional classification of the road. The design speeds should be approximately 60 km/h [40 mph] for primary access roads, 50 km/h [30 mph] for circulation roads, and 30 km/h [20 mph] for area roads. There may be instances where design speeds less than these may be appropriate because of severe terrain conditions or major environmental concerns. Design speeds on one-lane roads would usually be less than 50 km/h [30 mph]. If a design speed of greater than 60 km/h [40 mph] is used, the first section of this chapter should be consulted.

Design speed is the principal factor that should be correlated with the physical features of design to achieve a roadway that will accommodate the traffic safely for the planned use. Once a design speed is selected, all geometric features should be related to this speed to obtain a balanced design. Changes in terrain and other physical controls may dictate a change in design speed in certain sections. A decrease in design speed along the road should not be introduced abruptly, but be extended over a sufficient distance to allow the driver to adjust and make the transition to the slower speed.

Design Vehicle

The physical dimensions and operating characteristics of vehicles and the percentage of vehicles of various sizes using recreational roads are primary geometric design controls. Existing and anticipated vehicle types should be reviewed to establish representative vehicles for each functional roadway class. Each design vehicle considered should represent a substantial percentage of the vehicles expected to use the facility during its design life.

Three categories of vehicles are common to recreational areas: motor homes, vehicles with trailers, and standard passenger vehicles. Critical physical dimensions for geometric design are the overall length, width, and height of these units. Minimum turning paths of the design vehicles are influenced by the vehicle steering mechanism, track width, and wheelbase arrangement. Figures in Chapter 2 show minimum turn paths for motor homes (MH), passenger cars with 9-m [30-ft] travel trailers (P/T), passenger cars with 6.1-m [20-ft] boats (P/B), and motor homes with 6.1-m [20-ft] boats (MH/B). Turning path dimensions for other vehicle types such as buses and passenger cars are also presented in Chapter 2.

Sight Distance

Minimum stopping sight distance and passing sight distance are a direct function of the design speed. The subject of sight distance for two-lane roads is addressed in Chapter 3; however, sight distance design criteria are not included in Chapter 3 for roads with very low design speeds and for two-way single-lane roads. On two-way single-lane roads, sufficient sight distance should be available whenever two vehicles might approach one another for one vehicle to reach a turnout or for both vehicles to stop before colliding. Stopping sight distance should be measured using an eye height of 1,080 mm [3.5 ft] and a height of opposing vehicle of 1,300 mm [4.25 ft]. The stopping sight distance for a two-way, single-lane road should be approximately twice the stopping sight distance that would be used in design of a comparable two-lane road. Suggested stopping sight distances for two-way, single-lane roads are given in Exhibit 5-13.

Passing Sight Distance

Because of low operating speeds and the nature of travel on recreational roads, frequent passing maneuvers are not anticipated. Nevertheless, minimum passing sight distance should be

Metric				US Customary			
Initial speed (km/h)	Design stopping sight distance (m)	Rate of vertical curvature, K[a] (m/%)		Initial speed (mph)	Design stopping sight distance (ft)	Rate of vertical curvature, K[a] (ft/%)	
		Crest	Sag			Crest	Sag
Two-lane roads and one-way, single-lane roads				Two-lane roads and one-way, single-lane roads			
20	20	1	3	15	80	3	10
30	35	2	6	20	115	7	17
40	50	4	9	25	155	12	26
50	65	7	13	30	200	19	37
60	85	11	18	35	250	29	49
70	105	17	23	40	305	44	64
Two-way, single-lane roads				Two-way, single-lane roads			
20	40	2	6	15	160	12	27
30	70	7	13	20	230	25	44
40	100	15	21	25	310	45	65
50	130	26	29	30	400	74	89
60	170	44	40	35	500	116	117
70	210	67	52	40	610	172	147

[a] Rate of vertical curvature, K, is the length of curve per percent algebraic difference in the intersecting grades (i.e., K = L/A). (See Chapter 3 for details.)

Exhibit 5-13. Design Controls for Stopping Sight Distance and for Crest and Sag Vertical Curves—Recreational Roads

provided as frequently as possible, particularly on primary access roads where users travel considerable distances to reach activity sites. Suggested minimum passing sight distances for two-lane recreational roads are given in Exhibit 5-14. Passing sight distance is not a factor on single-lane roads. Where a faster vehicle approaches a slower vehicle from behind, it is assumed that, where appropriate, the slower vehicle will pull into a turnout and allow the faster vehicle to pass.

Grades

Grade design for recreational roads differs substantially from that for rural highways in that the weight/power ratio of recreational vehicles (RVs) seldom exceeds 30 kg/kW [50 lb/hp], and this fact indicates that gradeability of RVs approaches that for passenger cars. Furthermore, because vehicle operating speeds on recreational roads are relatively low, large speed reductions on grades are not anticipated.

Metric			US Customary		
Design speed (km/h)	Design passing sight distance (m)	Rate of vertical curvature, K[a] (m/%)	Design speed (mph)	Design passing sight distance (ft)	Rate of vertical curvature, K[a] (ft/%)
30	200	46	20	710	180
40	270	84	25	900	289
50	345	138	30	1090	424
60	410	195	35	1280	585
70	485	272	40	1470	772
80	540	338	45	1625	943
90	615	438	50	1835	1203
100	670	520	55	1985	1407
110	730	617	60	2135	1628
120	775	695	65	2285	1865
130	815	769	70	2480	2197
			75	2580	2377
			80	2680	2565

[a] Rate of vertical curvature, K, is the length of curve per percent algebraic difference in the intersecting grades; i.e., K = L/A. (See Chapter 3 for details.)

Exhibit 5-14. Design Controls for Passing Sight Distance for Crest Vertical Curves— Recreational Roads

When grades are kept within the suggested limits, critical length of grade is not a major concern for most recreational roads. Critical length of grade may be a factor on primary access roads to recreational areas, and appropriate consideration should be given to this element in the design for these roads.

Exhibit 5-15 identifies suggested maximum grades for given terrain and design speed based primarily on the operational performance of vehicles that use recreational roads. Chapter 3 contains a more detailed discussion on the selection of an appropriate maximum grade. A major item to be considered in selection of a maximum grade is the capability of the soil for erosion resistance. In many instances, grades considerably less than those shown in Exhibit 5-15 should be chosen to satisfy this concern. In addition, the surface type should also be a factor in grade selection. Steep grades with dirt or gravel surfaces may cause driving problems in the absence of continued maintenance, whereas a bituminous surface generally will offer better vehicle performance.

Type of terrain	Metric					US Customary				
	Maximum grade (%) for a specified design speed (km/h)					Maximum grade (%) for a specified design speed (mph)				
	20	30	40	50	60	10	20	25	30	40
Level	8	8	7	7	7	8	8	7	7	7
Rolling	12	11	10	10	9	12	11	10	10	9
Mountainous	18	16	15	14	12	18	16	15	14	12

Exhibit 5-15. Maximum Grades for Recreational Roads

Vertical Alignment

Vertical curves should be safe, comfortable in operation, pleasing in appearance, and adequate for drainage. Minimum or greater-than-minimum stopping sight distance should be provided in all cases. The designer should exercise considerable judgment in designing vertical curves because lengths in excess of the minimum may be needed at driver decision points, where drainage or aesthetic problems exist, or simply to provide an additional margin of safety.

Vertical curve design for two-lane roads is discussed in Chapter 3, which also presents specific design values. Exhibit 5-13 also includes additional information for very low design speeds not tabulated elsewhere. For two-way, single-lane roads, crest vertical curves should be significantly longer than those for two-lane roads. As discussed above, the stopping sight distance for a two-way, single-lane road should be approximately twice the stopping sight distance for a comparable two-lane road. Exhibit 5-13 includes K values for single-lane roads, from which vertical curve lengths can be determined.

Horizontal Alignment

Because the use of straight sections of roadway would be physically impractical and aesthetically undesirable, horizontal curves are essential elements in the design of recreational roads. The proper relationship between design speed and horizontal curvature and the relationship of both to superelevation are discussed in detail in Chapter 3. The guidance provided in Chapter 3 is generally applicable to paved recreational-roads; however, in certain instances variations are appropriate. At locations where there is a tendency to drive slowly, as with local and some circulation roads, a maximum superelevation rate of 6 percent is suggested. On roads with design speeds of 30 km/h [20 mph] or less, superelevation may not be warranted.

The design values for maximum curvature and superelevation discussed in Chapter 3 are based on friction data for paved surfaces. Some lower volume recreational facilities may not be paved, and because friction values for gravel surfaces are less than those for paved surfaces, friction values should be considered in curvature selection. Exhibit 5-16 shows the relationship

Metric

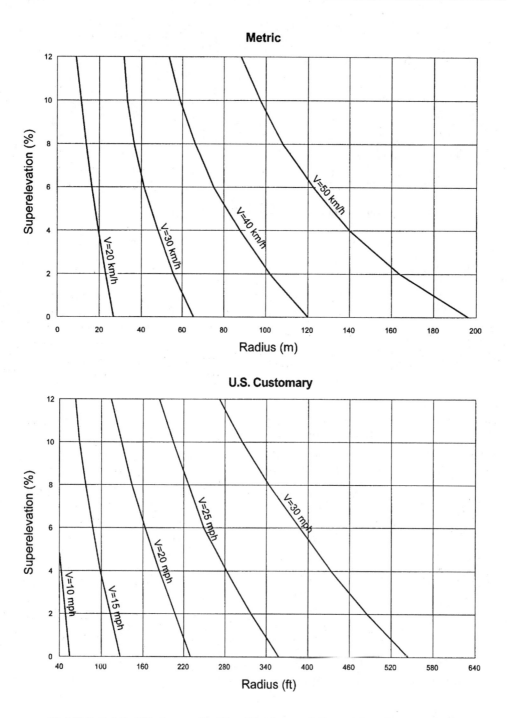

U.S. Customary

Exhibit 5-16. Minimum-Radius Horizontal Curve for Gravel Surface

between minimum radius and superelevation for gravel-surfaced roads. This figure was developed by using f values from 0.12 at 15 km/h [10 mph] to 0.10 at 50 km/h [30 mph].

Number of Lanes

The number of lanes should be sufficient to accommodate the design traffic volume. For low-volume recreational roads, capacity conditions do not normally govern design and provision of two travel lanes is appropriate. In some cases where traffic volume is less than 100 vehicles per day, it may be practical to use a two-way, single-lane roadway. This type of road is often desirable from an economic and environmental standpoint. When single-lane roadways with two-way traffic are used, turnouts for passing should be provided at intervals. Such turnouts should be intervisible, provided on all sight-restricted curves, and located so that the maximum distance between turnouts is no more than 300 m [1,000 ft]. The turnouts should be a minimum of 3 m [10 ft] wide for a length of 15 m [50 ft] and should have an 8-m [25-ft] taper on each end. For roads that serve substantial proportions of over-wide and extra-long vehicles, the turnout design criteria should be adjusted to accommodate these larger vehicles. Exhibit 5-17 shows a typical design that may be used for turnouts on tangent and curve sections for two-way, single-lane roads.

Widths of Traveled Way, Shoulder, and Roadway

A roadway is defined as that portion of the highway including shoulders for vehicular use. Proper roadway width is selected on the basis of numerous factors including existing and anticipated vehicular and bicycle traffic, safety, terrain, and design speed. Exhibit 5-18 gives recommended traveled way widths and shoulder widths for the various types of roadways. The sum of the traveled way and shoulder widths given in Exhibit 5-18 constitutes the roadway width.

The low operating speeds and relatively low traffic volume on recreational roads do not warrant wide shoulders. In addition, wide shoulders may be aesthetically objectionable. These considerations are reflected in the shoulder width values given in Exhibit 5-18. Under adverse terrain conditions, intermittent shoulder sections or turnouts may be suitable alternatives to continuous shoulders, particularly on lower functional roadway classes. Where guardrail is used, the graded width of the shoulder should be increased by about 0.6 m [2 ft].

Cross Slope

Cross slope is provided on roadways to ensure adequate drainage. However, excessive surface sloping can cause steering difficulties. Cross slope rates given in the section of this chapter on local rural roads are generally applicable to recreational roads.

W₁	L	W₂
3.6 m [12 ft]	30 m [100 ft]	6.6 m [22 ft]
4.2 m [14 ft]	30 m [100 ft]	7.2 m [24 ft]

Exhibit 5-17. Turnout Design

Type of road	Metric		US Customary	
	Traveled way width (m)[a]	Shoulder width (m)	Traveled way width (ft)[a]	Shoulder width (ft)
Primary access roads (two lanes)	6.6–7.2	0.6–1.2	22–24	2–4
Circulation roads (two lanes)	6.0–6.6	0.6–1.2	20–22	2–4
Area roads (two lanes)	5.4–6.0	0.0–0.6	18–20	0–2
Area roads (one lane)[b]	3.6	0.0–0.3	12	0–1
[a] Widening on the inside of sharp curves should be provided; additional width equal to 35 [400] divided by the curve radius in meters [feet] is recommended. [b] Roadway widths greater than 4.2 m [14 ft] should not be used because drivers will tend to use the facility as a two-lane road.				

Exhibit 5-18. Widths of Traveled Way and Shoulders—Recreational Roads

On single-lane roads with low-type surfaces, a crown would not usually be provided. Roads of this type would be inslope graded (toward the cut ditch) or outslope graded (toward the embankment fill), depending on the resistance of the soil to erosion.

Clear Recovery Area

Providing a clear zone adjacent to a road involves a trade-off between safety and aesthetics. A driver who leaves the road should be provided a reasonable chance to regain control and avoid serious injury. On the other hand, the philosophy of recreational roads dictates that natural roadside features should be preserved where practical. Because of the character of the traffic and the relatively low operating speeds on recreational roads, wide clear zones are not as important as on high-speed, high-volume facilities. For these reasons, dimensions smaller than those used on these higher order roads are appropriate. Desirably, 3 m [10 ft] or more of recovery area, measured from the edge of the traveled way, should be provided on the higher order recreational roads, (i.e., the primary access roads). These values are recommended for the general case; however, where economic and environmental concerns are great, even smaller values are appropriate. Clear zone widths on the lower order recreational roads, i.e., circulation roads and areas, are even less critical than on primary access roads. In areas where the crash potential is greater than normal, such as on the outside of sharp horizontal curves at the end of long, steep downgrades, liberal clear zone widths should be provided.

Roadside Slopes

Where terrain conditions permit, backslopes, foreslopes, and roadside drainage channels should have gentle well-rounded transitions. Foreslopes of 1V:4H or flatter are safer and more stable than steeper slopes and permit establishment and maintenance of turf. The maximum rate of foreslope depends on terrain conditions and the stability of local soils as determined by local experience. Cut sections should be designed with adequate ditches.

The ditch should be deep enough to accommodate the design flow and provide for satisfactory drainage of the pavement base and subbase. While foreslopes of 1V:4H or flatter are preferable, there are other important considerations in ditch design for recreational roads. Surrounding terrain and physical feature preservation may dictate narrow-width ditches. The lower speeds prevailing on recreational roads reduce the chance of personal injury for passengers in vehicles that drive into shallow-sided ditches.

Roadside Barrier

Roadside barriers should be installed at points of unusual risk, particularly those points that are unusual compared with the overall characteristics of the road. The criteria used in freeway design do not fit the low-volume recreational road situation. The AASHTO *Roadside Design Guide* (3) provides some insight into the application of roadside barriers on low-speed, low-volume facilities.

Signing and Marking

Although safety and efficiency of operation depend to a major extent on the geometric design of a road, they should be supplemented by standard signing and marking to provide information and warning to drivers. The extent to which signs and markings are used depends on the traffic volume, the type of highway, and the frequency and use by drivers unfamiliar with the area. The MUTCD (**4**) contains details regarding design, location, and application of highway signs and markings.

Structures

The design of bridges, culverts, walls, tunnels, and other structures should be in accordance with the AASHTO *Standard Specifications for Highway Bridges* (**2**). The minimum design loading for new bridges should be M 13.5 [H15]. Higher design loadings are appropriate for highways carrying other than just recreational traffic. The vertical clearance at underpasses should be at least 4.3 m [14 ft] over the entire roadway width. The clear roadway widths for new and reconstructed bridges should be a minimum of the surface width plus 1 m [3 ft]. Where the approach roadway is surfaced for the full crown width, that surfaced width should be carried across structures.

Recreational roads should be reviewed to determine if they are sufficient to accommodate bicycle traffic. When special facilities are desired, they should be in accordance with the AASHTO *Guide for the Development of Bicycle Facilities* (**1**).

Resource Recovery Roads

Resource recovery roads include mining and logging roads. Design criteria appropriate for this type of road in many areas are not significantly different from those for recreational roads. For this reason the criteria developed for recreational roads should be followed to the extent they are applicable. Several items are unique to this category of road and deserve special attention.

Traffic on resource recovery roads is primarily composed of large, slow-moving, heavily loaded vehicles. For this reason, particular attention should be paid to superelevation of horizontal curves. The center of gravity of trucks is much higher than that of passenger cars, and this fact increases the tendency of trucks to overturn. When semitrailers are used, only part of the payload is on the drive axles. This situation increases the tendency of the drive wheels to spin and sideslip on slippery surfaces. For these reasons the maximum superelevation should be limited to 6 percent. On long sustained grades adverse to the direction of haul, the superelevation should be reduced to accommodate slow-moving trucks.

Gradients on this type of facility have an effect on the road maintenance costs and costs to users. An economic analysis is usually appropriate to determine the most economical grade for the specific conditions encountered. Such an analysis should consider the increase in culvert installations to prevent ditch erosion on steeper grades and the more frequent surface replacement needs. Adverse grades are a special problem on roads planned for heavy hauling. Sections of adverse grades should not be so long that they slow a loaded truck to crawl speed. Except for short sections that can be overcome largely by momentum, adverse grades merit special analysis. In many instances, failure to use flatter grades may result in additional expenses for transportation during the life of the road far in excess of any savings in construction costs.

Geometric design features for resource recovery roads are similar to those for recreational roads in that they should be consistent with the design speed selected. Low design speeds 60 km/h [40 mph] or below are generally applicable to roads with winding alignment in rolling mountainous terrain. Exhibit 5-19 lists those minimum design speeds for both single-lane and two-lane roads for varying terrain conditions.

Type of terrain	Metric		US Customary	
	Design speed (km/h) for roads with specified number of lanes		Design speed (mph) for roads with specified number of lanes	
	Single lane	Two lanes	Single lane	Two lanes
Level	50	60	30	40
Rolling	30	50	20	30
Mountainous	15	30	10	20

Exhibit 5-19. Design Speeds for Resource Recovery and Local Service Roads

Because of the mechanical limitations of many of the vehicles using these roads, special attention should be given to the need for warning signs and markings. On long descending grades, consideration should be given to providing escape lanes for use by heavy vehicles that lose their brakes and run out of control. Deceleration may be artificially induced by the use of loose material or by providing a combination of sufficient length and upgrade for freewheeling deceleration. Further information is provided in the section on "Emergency Escape Ramps" in Chapter 3.

Many design considerations for resource development roads are based on the economics of the equipment operating on the facility. The effects of grades and curvature on operational cost are discussed in considerable detail in the *Logging Road Handbook* (**10**).

In many instances, resource development roads are ultimately used for other (e.g., recreational) purposes. In instances such as these, the original design should take into account all the possible ultimate usages.

Local Service Roads

Local service roads are those serving isolated areas that have little or no potential for further development (or that would need a higher type facility if further developed) and those serving a minimal number of parcels of land. Most of these roads will not be through roads (connected to public roads on both ends) but will dead end at the service to the last parcel on the road.

Traffic on this type of road is very low (less than 100 vehicles per day) and generally consists of drivers who are familiar with the road. The design criteria for local service roads, therefore, should be basically the same as those developed for recreational roads.

REFERENCES

1. AASHTO. *Guide for the Development of Bicycle Facilities*, Washington, D.C.: AASHTO, 1999.

2. AASHTO. *Standard Specifications for Highway Bridges*, Washington, D.C.: AASHTO, 1996.

3. AASHTO. *Roadside Design Guide*, Washington, D.C.: AASHTO, 1996, or more recent edition.

4. U.S. Department of Transportation, Federal Highway Administration. *Manual on Uniform Traffic Control Devices for Streets and Highways*, Washington, D.C.: 1988 or most current edition.

5. AASHTO. *Guide for the Planning, Design, and Operation of Pedestrian Facilities*, Washington, D.C.: AASHTO, forthcoming.

6. Bert, K. E., et al. *Accommodation of Utility Plants Within the Rights-of-Way of Urban Streets and Highways, Manual of Improved Practice,* ASCE Manual No. 14. Chicago: American Public Works Association, and New York: American Society of Civil Engineers, July 1974.

7. AASHTO. *Guide for Accommodating Utilities on Highway Rights-of-Way*, Washington, D.C.: AASHTO, 1994.

8. AASHTO. *An Informational Guide for Roadway Lighting*, Washington, D.C.: AASHTO, 1984.

9. AASHTO. *A Guide for Transportation Landscape and Environmental Design,* Washington, D.C., AASHTO, 1991.

10. U.S. Department of Agriculture, Forest Service. *Logging Road Handbook—The Effect of Road Design on Hauling Costs,* Handbook No. 183, Washington, D.C.: U.S. Department of Agriculture, Forest Service, 1960.

11. U.S. Army Corps of Engineers, Construction Engineering Laboratory. *Design Guidelines for Recreational Roads*, Washington, D.C.: 1975.

12. U.S. Department of Agriculture, Forest Service. *Manual Network Analysis Method for the Transportation of Resources,* No. 7710-6, Washington, D.C.: U.S. Department of Agriculture, Forest Service, Engineering Management Services, 1991.

13. Glennon, J. C. *Design and Traffic Control Guidelines for Low-Volume Rural Roads,* NCHRP Report 214. Transportation Research Board, 1979.

14. Zegeer, C. V., R. Stewart, F. M. Council, and T. R. Neuman. *Roadway Widths for Low Traffic Volume Roads*, NCHRP Report 362, Transportation Research Board, 1994.

15. Neuman, T. R. "Design Guidelines for Very Low-Volume Local Roads (<400 ADT)," Final Report of NCHRP Project 20-7(75), Chicago: CH2M Hill, unpublished, 1999.

16. American Society of Civil Engineers, National Association of Home Builders, and the Urban Land Institute. *Residential Streets*, Second Edition. New York, 1990.

CHAPTER 6
COLLECTOR ROADS AND STREETS
INTRODUCTION

This chapter presents guidance on the application of geometric design criteria to facilities functionally classified as collector roads and streets. The chapter is subdivided into sections on rural and urban collectors.

The function of a collector may be understood by referring to those functional classes above and below it—the arterial and the local road or street. The collector has aspects of both arterials and local roads and often serves as a connection between them. Since the function of a collector combines aspects of both arterials and local streets, collectors serve a dual function: collecting traffic for movement between arterial streets and local roads and providing access to abutting properties.

Collector streets link neighborhoods or areas of homogeneous land use with the arterial street system. These streets not only serve traffic movements between arterials and local streets, but also serve through traffic within local areas. Collector streets should be planned so as not to disrupt the activities within the areas they serve.

The collector street is a public highway, usually serving moderate traffic volumes. There may be few discernible differences between collectors and local streets within a neighborhood, since collectors provide access function to adjacent residential development and to some neighborhood facilities. However, the design of a collector street should reflect its function as a collector and should not be conceived or developed simply as a continuous access street. The collector should allow access to abutting properties consistent with the level of service desired.

The use of design criteria exceeding those described in this chapter is encouraged, where practical. Every effort should be made to obtain the best possible alignment, grade, sight distance, and drainage that are consistent with terrain, present and anticipated development, safety, and available funds.

Drainage, both on the pavement itself and from the sides and subsurface, is an important design consideration. Inadequate drainage can lead to high maintenance costs and adverse operational conditions. In areas of significant snowfall, roadways should be designed so that there is sufficient storage space, outside the traveled way, for plowed snow and proper drainage for melting conditions.

Safety is an important factor in all roadway improvements. On low-volume roads or streets or in urban areas, it may not be practical to provide an obstacle-free roadside. However, every effort should be made to provide as much clear roadside as is practical. The judicious use of flatter slopes, roadside barriers, and warning signs helps to improve roadside safety. Proper placement of utility features also assists in achieving safer roadsides.

423

It may not be cost-effective to design collector roads and streets that carry less than 400 vehicles per day using the same criteria applicable to higher volume roads or to make extensive traffic operational or safety improvements to such very low-volume roads. AASHTO is currently evaluating alternative design criteria for collector roads that carry less than 400 vehicles per day based on a safety risk assessment.

Noise abatement may need to be considered on collector roads and streets; for further information, see the section on "Noise Control" in Chapter 4.

The *Highway Capacity Manual* (HCM) (**1**) provides the designer with a tool to evaluate level of service for the highway facility under consideration. Collector streets should generally be designed for level-of-service C to D. In rural areas, level-of-service C is desirable for collector roads. In heavily developed portions of metropolitan areas, conditions may necessitate the use of level-of-service D. Level-of-service D is also a practical choice where unusually high traffic volumes exist or where terrain is rolling or mountainous. For further information, see the section on "Levels of Service" in Chapter 2. Collector roads and streets cannot be designed entirely on the basis of functional classification. Many other facets of design must also be considered, as described below.

RURAL COLLECTORS

General Design Considerations

Two-lane collector highways constitute an important part of the rural highway system. Rural collectors should be designed with the most favorable alignment and cross section practical, consistent with traffic and topography. Basic information needed for design of rural collectors includes crash history, traffic volumes, terrain, and alignment.

Design Traffic Volumes

Rural collector highways should be designed for specific traffic volumes and specified acceptable levels of service. Usually, the design year is 20 years from the date of construction completion but may be any number of years within a range from the present (for restoration projects on existing roads) to 20 years in the future (for new construction projects).

The average daily traffic (ADT) volume for the design year should serve as the basis for the project design.

Design Speed

Geometric design features should be consistent with a design speed appropriate for the conditions. Low design speeds of 70 km/h [45 mph] and below are generally applicable to highways with curvilinear alignment in rolling or mountainous terrain, or where environmental conditions dictate. High design speeds of 80 km/h [50 mph] and above are generally applicable to highways in level terrain or where other environmental conditions are favorable. Exhibit 6-1 identifies minimum design speeds for rural collector roads as a function of the type of terrain and design traffic volumes. The designer should strive for higher values than those shown where specific safety concerns are present and costs are not prohibitive.

Sight Distance

Stopping sight distance and passing sight distance are a direct function of the design speed. An eye height of 1,080 mm [3.5 ft] and an object height of 600 mm [2.0 ft] are used to determine stopping sight distance. An eye height of 1,080 mm [3.5 ft] and an object height of 1,080 mm [3.5 ft] are used to determine passing sight distance. For further information on sight distance, see Exhibits 6-2 and 6-3 and the section on "Sight Distance" in Chapter 3.

Grades

Exhibit 6-4 identifies suggested maximum grades for rural collectors in specific terrain and design conditions.

Alignment

The designer should provide the most favorable alignment as practical for rural collectors. Horizontal and vertical alignment should complement each other and should be considered in combination to achieve appropriate safety, capacity, and appearance for the type of improvement proposed. Topography, traffic volume and composition, and right-of-way conditions are controlling features. Abrupt changes in horizontal alignment should be avoided. Vertical curves should meet the sight distance criteria for the design speed. In addition, frequent opportunities for passing should be provided, where practical. For further information, see the sections on "Horizontal Alignment" and "Vertical Alignment" in Chapter 3.

Cross Slope

Pavement cross slope should be adequate to provide proper drainage. Normally, cross slopes range from 1.5 to 2 percent for high-type pavements. High-type pavements are those that retain smooth riding qualities and good non-skid properties in all weather under heavy traffic volumes and loadings with little maintenance required.

Low-type pavements are those with treated earth surfaces and those with loose aggregate surfaces. A cross slope of 3 to 6 percent is desirable for low-type pavements. For further information, see the section on "Cross Slope" in Chapter 4.

	Metric			US Customary		
	Design speed (km/h) for specified design volume (veh/day)			Design speed (mph) for specified design volume (veh/day)		
Type of terrain	0 to 400	400 to 2000	over 2000	0 to 400	400 to 2000	over 2000
Level	60	80	100	40	50	60
Rolling	50	60	80	30	40	50
Mountainous	30	50	60	20	30	40
Note: Where practical, design speeds higher than those shown should be considered.						

Exhibit 6-1. Minimum Design Speeds for Rural Collectors

Metric				US Customary			
Design speed	Design stopping sight distance	Rate of vertical curvature, K[a] (m/%)		Design speed	Design stopping sight distance	Rate of vertical curvature, K[a] (ft/%)	
(km/h)	(m)	Crest	Sag	(mph)	(ft)	Crest	Sag
20	20	1	3	15	80	3	10
30	35	2	6	20	115	7	17
40	50	4	9	25	155	12	26
50	65	7	13	30	200	19	37
60	85	11	18	35	250	29	49
70	105	17	23	40	305	44	64
80	130	26	30	45	360	61	79
90	160	39	38	50	425	84	96
100	185	52	45	55	495	114	115
				60	570	151	136

[a] Rate of vertical curvature, K, is the length of curve per percent algebraic difference in the intersecting grades; i.e., K = L/A (see Chapter 3 for details).

Exhibit 6-2. Design Controls for Stopping Sight Distance and for Crest and Sag Vertical Curves

Metric

Design speed (km/h)	Design passing sight distance (m)	Rate of vertical curvature, K[a] (m/%)
30	200	46
40	270	84
50	345	138
60	410	195
70	485	272
80	540	338
90	615	438
100	670	520

US Customary

Design speed (mph)	Design passing sight distance (ft)	Rate of vertical curvature, K[a] (ft/%)
20	710	180
25	900	289
30	1090	424
35	1280	585
40	1470	772
45	1625	943
50	1835	1203
55	1985	1407
60	2135	1628

[a] Rate of vertical curvature, K, is the length of curve per percent algebraic difference in the intersecting grades; i.e., K = L/A (See Chapter 3 for details).

Exhibit 6-3. Design Controls for Crest Vertical Curves Based on Passing Sight Distance

Metric

Type of terrain	Maximum grade (%) for specified design speed (km/h)							
	30	40	50	60	70	80	90	100
Level	7	7	7	7	7	6	6	5
Rolling	10	10	9	8	8	7	7	6
Mountainous	12	11	10	10	10	9	9	8

US Customary

Type of terrain	Maximum grade (%) for specified design speed (mph)								
	20	25	30	35	40	45	50	55	60
Level	7	7	7	7	7	7	6	6	5
Rolling	10	10	9	9	8	8	7	7	6
Mountainous	12	11	10	10	10	10	9	9	8

Note: Short lengths of grade in rural areas, such as grades less than 150 m [500 ft] in length, one-way downgrades, and grades on low-volume rural collectors may be up to 2 percent steeper than the grades shown above.

Exhibit 6-4. Maximum Grades for Rural Collectors

Superelevation

Many rural collector highways have curvilinear alignments. A superelevation rate compatible with the design speed should be used. For rural collectors, superelevation should not exceed 12 percent. Where snow and ice conditions may be a factor, the superelevation rate should not exceed 8 percent. Superelevation runoff denotes the length of highway needed to accomplish the change in cross slope from a section with the adverse crown removed to a fully superelevated section and vice versa. Adjustments in design runoff lengths may be needed to provide a smooth ride, surface drainage, and good appearance. The section on "Horizontal Alignment" in Chapter 3 provides a detailed discussion on superelevation for appropriate design speeds.

Number of Lanes

The number of lanes should be sufficient to accommodate the design volumes for the desired level of service. Normally, capacity conditions do not govern rural collector roads, and two lanes are appropriate. For further information, see the section on "Highway Capacity" in Chapter 2.

Width of Roadway

For high-type surfaces, the minimum roadway width is the sum of the traveled way and shoulder widths shown in Exhibit 6-5. Shoulder width is measured from the edge of the traveled way to the point of intersection of shoulder slope and foreslope. Where roadside barriers are included, a minimum offset of 1.2 m [4 ft] from the traveled way to the barrier should be provided, wherever practical. For further information, see the sections on "Shoulders" and "Longitudinal Barriers" in Chapter 4 and the section in Chapter 3 on "Traveled Way Widening on Horizontal Curves" for vehicle offtracking information.

Where bicycle facilities are included as part of the design, refer to the AASHTO *Guide for the Development of Bicycle Facilities* (**2**).

Foreslopes

The maximum rate of foreslope should depend on the stability of local soils as determined by soil investigation and local experience. Slopes should be as flat as practical, taking into consideration other design constraints. Flat foreslopes improve safety by providing a maneuvering area in emergencies, are more stable than steep slopes, aid in the establishment of plant growth, and simplify maintenance work. Roadside barriers may be used where topography and right-of-way are restrictive and a need is justified.

Metric					US Customary				
	Minimum width of traveled way (m) for specified design volume (veh/day)[a]					Minimum width of traveled way (ft) for specified design volume (veh/day)[a]			
Design speed (km/h)	under 400	400 to 1500	1500 to 2000	over 2000	Design speed (mph)	under 400	400 to 1500	1500 to 2000	over 2000
30	6.0[b]	6.0	6.6	7.2	20	20[b]	20	22	24
40	6.0[b]	6.0	6.6	7.2	25	20[b]	20	22	24
50	6.0[b]	6.0	6.6	7.2	30	20[b]	20	22	24
60	6.0[b]	6.6	6.6	7.2	35	20[b]	22	22	24
70	6.0	6.6	6.6	7.2	40	20[b]	22	22	24
80	6.0	6.6	6.6	7.2	45	20	22	22	24
90	6.6	6.6	7.2	7.2	50	20	22	22	24
100	6.6	6.6	7.2	7.2	55	22	22	24	24
					60	22	22	24	24
	Width of shoulder on each side of road (m)					Width of shoulder on each side of road (ft)			
All speeds	0.6	1.5[c]	1.8	2.4	All speeds	2.0	5.0[c]	6.0	8.0

[a] On roadways to be reconstructed, a 6.6-m [22 ft] traveled way may be retained where the alignment and safety records are satisfactory.

[b] A 5.4-m [18-ft] minimum width may be used for roadways with design volumes under 250 veh/day.

[c] Shoulder width may be reduced for design speeds greater than 50 km/h [30 mph] as long as a minimum roadway width of 9 m [30 ft] is maintained.

See text for roadside barrier and offtracking considerations.

Exhibit 6-5. Minimum Width of Traveled Way and Shoulders

Drivers who inadvertently leave the traveled way can often recover control of their vehicles if foreslopes are 1V:4H or flatter and shoulders and ditches are well rounded or otherwise made traversable. Such recoverable slopes should be provided where terrain and right-of-way conditions allow.

Where provision of recoverable slopes is not practical, the combinations of rate and height of slope provided should be such that occupants of an out-of-control vehicle have a good chance of survival. Where high fills, right-of-way restrictions, watercourses, or other problems render such designs impractical, roadside barriers should be considered, in which case the maximum rate of fill slope may be used. Reference should be made to the current edition of the AASHTO *Roadside Design Guide* (3). For further information, see the section on "Traffic Barriers" in Chapter 4.

Cut sections should be designed with adequate ditches. Preferably, the foreslope should not be steeper than 1V:3H and, where practical, should be 1V:4H or flatter. The ditch bottom and slopes should be well rounded, and the backslope should not exceed the maximum needed for stability.

Structures

The design of bridges, culverts, walls, tunnels, and other structures should be in accordance with the current AASHTO *Standard Specifications for Highway Bridges* (**4**), or with the AASHTO *LRFD Bridge Design Specification* (**5**). Except as otherwise indicated herein, the dimensional design of structures should also be in accordance with these standard specifications.

The minimum design loading for bridges on collector roads should be MS-18 [HS 20]. The minimum roadway widths for new and reconstructed bridges should be as shown in Exhibit 6-6.

Metric			US Customary		
Design volume (veh/day)	Minimum clear roadway width for bridges[a]	Design loading structural capacity	Design volume (veh/day)	Minimum clear roadway width for bridges[a]	Design loading structural capacity
400 and under	Traveled way + 0.6 m (each side)	MS-18	400 and under	Traveled way + 2 ft (each side)	HS-20
400 to 1500	Traveled way + 1 m (each side)	MS-18	400 to 1500	Traveled way + 3 ft (each side)	HS-20
1500 to 2000	Traveled way + 1.2 m (each side)[b]	MS-18	1500 to 2000	Traveled way + 4 ft (each side)[b]	HS-20
over 2000	Approach roadway (width)[b]	MS-18	over 2000	Approach roadway (width)[b]	HS-20

[a] Where the approach roadway width (traveled way plus shoulders) is surfaced, that surface width should be carried across the structures.

[b] For bridges in excess of 30 m [100 ft] in length, the minimum width of traveled way plus 1 m [3 ft] on each side is acceptable.

Exhibit 6-6. Minimum Roadway Widths and Design Loadings for New and Reconstructed Bridges

Bridges to Remain in Place

Because highway geometric and roadway improvements may encourage higher speeds and attract larger vehicles, existing structures also should be improved correspondingly. Because of their high cost, reasonably adequate bridges and culverts that meet tolerable criteria may be retained.

Where an existing highway is to be reconstructed, an existing bridge that fits the proposed alignment and profile may remain in place when its structural capacity in terms of design loading and roadway width are at least equal to the values shown for the applicable traffic volume in Exhibit 6-7.

The values in Exhibit 6-7 do not apply to structures with a total length greater than 30 m [100 ft]. Such structures should be analyzed individually by taking into consideration the clear width provided, crash history, traffic volumes, remaining life of the structure, design speed, and other pertinent factors.

Metric			US Customary		
Design volume (veh/day)	Design loading structural capacity	Minimum clear roadway width (m)[a]	Design volume (veh/day)	Design loading structural capacity	Minimum clear roadway width (ft)[a]
under 400	MS 13.5	6.6	under 400	H 15	22
400 to 1500	MS 13.5	6.6	400 to 1500	H 15	22
1500 to 2000	MS 13.5	7.2	1500 to 2000	H 15	24
over 2000	MS 13.5	8.4	over 2000	H 15	28
[a] Clear width between curbs or railings, whichever is less, should be equal to or greater than the approach traveled way width, wherever practical.					

Exhibit 6-7. Structural Capacities and Minimum Roadway Widths for Bridges to Remain in Place

Vertical Clearance

Vertical clearance at underpasses should be at least 4.3 m [14 ft] over the entire roadway width, with an additional allowance for future resurfacing.

Horizontal Clearance to Obstructions

For rural collector roads with a design speed of 70 km/h [45 mph] or less, a minimum clear zone of 3 m [10 ft] measured from the edge of the traveled way should be provided. This recovery area should be clear of all unyielding objects such as trees, sign supports, utility poles, light poles, and other fixed objects. The benefits of removing these obstructions should be weighed against any environmental and aesthetic effects.

For rural collector roads with a design speed of 80 km/h [50 mph] or more, the AASHTO *Roadside Design Guide* (3) should be used for guidance in selecting an appropriate clear zone width.

The approach roadway width (traveled way plus shoulders) should be carried across an overpass or bridge, where practical. Approach roadside barriers, anchored to the bridge rails or parapets, should be provided. Sidewalks should extend across a bridge if the approach roadway has sidewalks or sidewalk areas. To the extent practical, where another highway or railroad passes over the roadway, the overpass structure should be designed so that the pier or abutment supports have lateral clearance as great as the clear zone on the approach roadway. Where a setback beyond the clear zone is not practical, roadside barrier protection should be provided at the piers.

Right-of-Way Width

The provision of right-of-way widths that accommodate construction, adequate drainage, and proper maintenance of a highway is an important part of the overall design. Wide rights-of-way permit the construction of gentle slopes, resulting in greater safety for the motorist and provide for easier and more economical maintenance. The acquisition of sufficient right-of-way, at the time of initial construction, permits subsequent widening of the roadway and the widening and strengthening of the pavement at a reasonable cost as traffic volumes increase.

In developed areas it may be desirable to limit the right-of-way width. However, the right-of-way width should not be less than that needed for all elements of the design cross section, utility accommodation, and appropriate border areas.

Intersection Design

Intersections should be carefully located to avoid steep profile grades and to ensure adequate approach sight distance. An intersection should not be situated just beyond a sharp crest vertical curve or on a sharp horizontal curve. Where there is no practical alternative to such a location, the approach sight distance on each leg should be checked and, where practical, backslopes should be flattened and horizontal or vertical curves lengthened, to provide additional sight distance. The driver of a vehicle approaching an intersection should have an unobstructed view of the entire intersection and sufficient lengths of the intersecting roadway to permit the driver to anticipate and avoid potential collisions. Sight distances at intersections with six different types of traffic control cases are presented in Chapter 9.

Intersections should be designed with a corner radius for pavement or surfacing adequate for the larger vehicles anticipated; for information on minimum edge radius, see Chapter 9. Where turning volumes are substantial speed-change lanes and channelization should be considered.

Intersection legs that operate under stop sign control should intersect at right angles, wherever practical, and should not intersect at an angle less than 60 degrees. For more information on intersection angle, see Chapter 9.

A stopping area that is as level as practical should be provided for approaches on which vehicles may be required to stop.

Railroad-Highway Grade Crossings

Appropriate grade crossing warning devices should be installed at all railroad-highway grade crossings on collector roads and streets. Details of the devices to be used are given in the *Manual on Uniform Traffic Control Devices* (MUTCD) (**6**). In some states, the final approval of these devices may be vested in an agency having oversight over railroads.

Sight distance is an important consideration at railroad-highway grade crossings. There should be sufficient sight distance along the road for an approaching driver to recognize the railroad crossing, perceive the warning device, determine whether a train is approaching, and stop if necessary. Adequate sight distance along the track is needed for drivers of stopped vehicles to decide when it is safe to proceed across the tracks. For further information on railroad-highway grade crossings, see Chapter 9.

The roadway width at railroad crossings should be the same as the width of the approach roadway. Crossings that are located on bicycle routes that are not perpendicular to the railroad may need additional paved shoulder width for bicycles to maneuver over the crossing. For further information, see the AASHTO *Guide for the Development of Bicycle Facilities* (**2**).

Traffic Control Devices

Traffic control devices should be applied consistently and uniformly. Details of the standard traffic control devices and warrants for various conditions are found in the MUTCD (**6**). Geometric design of rural collectors should include full consideration of the types of traffic control to be used, especially at intersections where multi-phase or actuated traffic signals are likely to be needed. For further information, see the section on "Traffic Control Devices" in Chapter 3.

Erosion Control

Design of rural collectors should consider preservation of the natural ground cover and desirable growth of shrubs and trees within the right-of-way. Shrubs, trees, and other vegetation should be considered in assessing the driver's sight line and the clear recovery distance. Seeding, mulching, sodding, or other acceptable measures for covering slopes, swales, and other erodible areas should also be considered in the rural collector design. For further information, see the section on "Erosion Control and Landscape Development" in Chapter 3.

URBAN COLLECTORS

General Design Considerations

A collector street is a public facility for vehicular travel and includes the entire area within the right-of-way. The urban collector street also serves bicycle and pedestrian traffic and often accommodates public utility facilities within the right-of-way. The development or improvement of streets should be based on a functional street classification established as part of a comprehensive community development plan. The design criteria should be those for the ultimate planned development. Design criteria for collector streets should exceed those shown below, where practical.

The function of urban collectors is equally divided between mobility and access. Few cities have effective access control restrictions along collector streets; almost all such streets permit access to abutting properties, except where access rights have been acquired. Many new collectors are planned and constructed with little or no access restriction. However, uncontrolled access may eventually result in the obsolescence of a collector facility. Therefore, management of driveway access to urban collectors is desirable.

When a major objective of the design is to expedite traffic mobility, there are many additional criteria for which guidelines are appropriate. Such criteria include minimizing conflict points, providing adequate storage length for all turning movements, minimizing conflicts with pedestrians and bicyclists, coordinating driveway locations on opposite sides of the roadway, locating signals to meet progression needs, and maintaining efficient circulation while providing adequate ingress and egress capacity. By using these design criteria, an optimum system of access can be developed.

Access control on urban collector streets should be used primarily to ensure that access points conform to the adopted criteria for safety, location, design, construction, and maintenance.

Design Traffic Volumes

Traffic volumes are a major factor in determining the geometric criteria to be used in designing urban collector streets. Specifically, the design traffic volumes projected to some future design year should be the basis of design. It usually is difficult and costly to modify the geometric design of an existing collector street unless provisions are made at the time of initial construction. The design traffic should be estimated for at least 10 and preferably 20 years from the anticipated completion of construction.

Design Speed

Design speed is a factor in the design of collector streets. For consistency in design, a design speed of 50 km/h [30 mph] or higher should be used for urban collector streets, depending on available right-of-way, terrain, adjacent development, likely pedestrian presence, and other site controls. See Exhibit 6-1 and the section on "Design Speed" in Chapter 2 for additional information.

In the typical urban street grid, closely spaced intersections often limit vehicular speeds and thus make the consideration of design speed of lesser significance. Nevertheless, the longer sight distances and curve radii commensurate with higher design speeds result in safer highways and should be used to the extent practical.

Sight Distance

Stopping sight distance for urban collector streets varies with design speed. Design for passing sight distance seldom is applicable on urban collector streets. For further information, see Exhibits 6-2 and 6-3 and the section on "Sight Distance" in Chapter 3.

Grades

Grades for urban collector streets should be as level as practical, consistent with the surrounding terrain. A minimum grade of 0.30 percent is acceptable to facilitate drainage. However, it is recommended that a grade of 0.50 percent grade or more be used, where practical, for drainage purposes. Where adjacent sidewalks are present, a maximum grade of 5 percent is recommended to meet the *Americans with Disabilities Act Accessibility Guidelines* (ADAAG) and other applicable criteria, where terrain conditions permit (**7, 8**). The grade of an urban street is generally depressed below the surrounding terrain to direct drainage from adjacent property to the curb area so that it can reach the storm drain system. Applicable gradients, vertical curve lengths, and other pertinent features are discussed in the section on "Vertical Alignment" in Chapter 3. Maximum grades for urban collector streets should be as presented in Exhibit 6-8.

Alignment

Alignment in residential areas should fit closely the existing topography to minimize the need for cuts or fills without sacrificing safety.

Cross Slope

Pavement cross slope should be adequate to provide proper drainage. Cross slope should normally be from 1.5 to 3 percent where there are flush shoulders adjacent to the traveled way or where there are outer curbs.

Superelevation

Superelevation, in specific locations, may be advantageous for urban collector street traffic operation. However, in built-up areas, the combination of wide pavement areas, proximity of adjacent development, control of cross slope, profile for drainage, frequency of cross streets, and other urban features often combine to make its use impractical or undesirable. Where used, superelevation on urban collector streets should be 6 percent or less. The absence of superelevation on urban collectors for low speeds of 70 km/h [45 mph] and below generally is not detrimental to the motorist. Often, some warping or partial removal or reversal of the tangent pavement crown may facilitate operations. When warping or removing the pavement crown, drainage should be considered. For further information, see the sections on "Horizontal Alignment" and "Design for Low-Speed Urban Streets" in Chapter 3.

	Metric								**US Customary**								
	Maximum grade (%) for specified design speed (km/h)								Maximum grade (%) for specified design speed (mph)								
Type of terrain	30	40	50	60	70	80	90	100	20	25	30	35	40	45	50	55	60
Level	9	9	9	9	8	7	7	6	9	9	9	9	9	8	7	7	6
Rolling	12	12	11	10	9	8	8	7	12	12	11	10	10	9	8	8	7
Mountainous	14	13	12	12	11	10	10	9	14	13	12	12	12	11	10	10	9

Note: Short lengths of grade in urban areas, such as grades less than 150 m [500 ft] in length, one-way downgrades, and grades on low-volume urban collectors may be up to 2 percent steeper than the grades shown above.

Exhibit 6-8. Maximum Grades for Urban Collectors

Number of Lanes

Two moving traffic lanes plus additional width for shoulders and parking are sufficient for most urban collector streets. Where the street is developed in stages, initially a rural cross section with shoulders may be used. The street should be planned for later conversion of the shoulder width to a parking lane or a through lane, usually with outer curbs. Where the initial development utilizes a rural cross section, a clear zone consistent with rural conditions and commensurate with the design speed should be provided. When the conversion of the shoulder occurs, the clear zone can be modified to that appropriate for urban conditions. If practical and economically feasible, the initial construction should be four lanes with curbs, allowing parking on the two outer lanes until later development necessitates the use of all four lanes for traffic movement.

In some cases, in commercial areas where there are mid-block left turns, it may be advantageous to provide an additional continuous two-way left-turn lane in the center of the roadway.

The number of lanes to be provided on urban collector streets with high traffic volumes should be determined from a capacity analysis. This analysis should consider both intersections and mid-block locations, when appropriate, in assessing the ability of a proposed design to provide the desired level of service. Such analyses should be made for the future design year traffic volume utilizing the procedures in the most recent edition of the *Highway Capacity Manual* (**1**). For further information, see the section on "Highway Capacity" in Chapter 2.

Width of Roadway

The width of an urban collector street should be planned as the sum of the widths of the ultimate lanes for moving traffic, parking, and bicycles, including median width where appropriate.

Lanes within the traveled way should range in width from 3.0 to 3.6 m [10 to 12 ft]. In industrial areas, lanes should be 3.6 m [12 ft] wide except where lack of space for right-of-way imposes severe limitations; in such cases, lane widths of 3.3 m [11 ft] may be used. Added turning lanes at intersections, where used, should range in width from 3.0 to 3.6 m [10 to 12 ft], depending on the percentage of trucks. Where shoulders are used, roadway widths should be determined by referring to Exhibit 6-5.

Where bicycle facilities are included as part of the design, refer to the AASHTO *Guide for the Development of Bicycle Facilities* (**2**).

Parking Lanes

Although on-street parking may constitute a safety problem and may impede traffic flow, provision of parking lanes parallel to the curb is conventional on many collector streets. Parallel parking is normally acceptable on urban collectors where sufficient street width is available to provide a parking lane. In residential areas, a parallel parking lane from 2.1 to 2.4 m [7 to 8 ft] in width should be provided on one or both sides of the street, as appropriate for the lot size and density of development. In commercial and industrial areas, parking lane widths should range from 2.4 to 3.3 m [8 to 11 ft] and are usually provided on both sides of the street.

The principal problem of diagonal or angle parking, in comparison to parallel parking, is the lack of adequate visibility for the driver during the back-out maneuver. Collector street designs with diagonal or angle parking should only be considered in special cases. ADA guidelines concerning parking should be taken into consideration (**7, 8**). For further information, see the section concerning "On-Street Parking" in Chapter 4.

The determination of parking lane width should consider the appropriate width for any likely future use as a lane for moving traffic either continuously or during peak hours. Where curb-and-gutter sections are used, the gutter pan width may be considered as part of the parking lane width, but, where practical, the parking lane widths discussed above should be in addition to the gutter pan width.

Medians

Urban collector streets designed for four or more lanes should include width for an appropriate median treatment, where practical. For general types of median treatments for collector streets, the following widths may be considered: (1) paint-striped separation, 0.6 to 1.2 m [2 to 4 ft] wide; (2) narrow raised-curbed sections, 0.6 to 1.8 m [2 to 6 ft] wide; (3) raised curbed sections, 3.0 to 4.8 m [10 to 16 ft] wide, providing space for left-turn lanes; (4) paint-striped sections, 3.0 to 4.8 m [10 to 16 ft] wide, providing space for two-way left-turn lanes; and (5) raised-curb sections, 5.4 to 7.6 m [18 to 25 ft] wide to provide more space for left-turn lanes and for passenger cars to stop in median crossovers. Wider medians from 8 to 12 m [27 to 40 ft] may be used for a parkway design where space is available for landscaping. Thus, each increment in additional median width provides specific operational advantages. Median should be as wide as practical within the constraints of each particular site.

On urban collector streets with raised-curb medians, openings should be provided only at intersections with other streets and at reasonably spaced driveways serving major traffic generators such as industrial plants and shopping centers. Where practical, median openings should be designed to include left-turn lanes.

Median openings should be situated only where there is adequate sight distance. The shape and length of the median openings depend on the width of the median and the vehicle types that are to be accommodated. The minimum length of median openings should be that of the projected roadway width of the intersecting cross street or driveway. Desirably, the length of median openings should be great enough to provide a 15-m [50-ft] turning radius or the turning radius for the design vehicle for left-turning vehicles between the inner edge of the lane adjacent to the median and the centerline of the intersection roadway.

On many urban collectors it may be impractical to use a raised-curb median. A continuous center two-way left-turn lane, flush with the adjacent traveled way, is an alternative design that may also be considered. A further discussion on medians is found in the section on "Medians" in Chapter 4 and the section on "Median Openings" in Chapter 9.

Curbs

Collector streets normally are designed with curbs to allow greater use of available width and for control of drainage, protection of pedestrians, and delineation. The curb on the right side of the traveled way should be a vertical curb, 150 mm [6 in] high, usually with an appropriate batter. On lightly traveled residential streets with grades less than 2 percent, a sloping curb that is lower and does not require modification at driveway entrances may be used. The curb slope should be 1V:6H or flatter.

On divided streets, the type of median curbs should be determined in conjunction with the median width and the type of turning movement control to be provided. Where mid-block left-turn movements are permitted and the median width is less than 3 m [10 ft], a well-delineated flush or rounded raised median separator 50 to 100 mm [2 to 4 in] high is effective in channelizing traffic and in avoiding excessive travel distances and concentrations of turns at intersections. Where wider traversable medians are appropriate, they may be either flush or bordered with low curbs 25 to 50 mm [1 to 2 in] high. On narrow and intermediate-width medians, and on some wide medians, where cross-median movements are undesirable or create problems, a vertical curb should be used on the median side of the traveled way, usually 150 mm [6 in] high and with an appropriate batter. A median barrier should be used where positive separation of opposing traffic is essential, where there is no need for pedestrian crossings, and where local regulations permit. For further information, see the section on "Curbs" in Chapter 4.

Vertical curbs with heights of 150 mm [6 in] or more, adjacent to the traveled way, should be offset by 0.3 to 0.6 m [1 to 2 ft] from the edge of the traveled way. Where there is combination curb-and-gutter construction, the gutter pan width, which is normally 0.3 to 0.6 m [1 to 2 ft], may provide the offset distance.

Drainage

Surface runoff is gathered by a system of gutters, inlets, catch basins, and storm sewers. The gutter grade should be 0.3 percent or more. However, a gutter grade of 0.5 percent or more should be used, where practical, for better drainage. Inlets or catch basins with an open grate should be located in the gutter line and should be so spaced that ponding of water on the pavement does not exceed tolerable limits. In addition, grates should be designed to accommodate bicycle and pedestrian traffic. For additional details, see the drainage portions of Chapters 3 and 4.

Sidewalks

Sidewalks should be provided along both sides of urban collector streets that are used for pedestrian access to schools, parks, shopping areas, and transit stops and along all collectors in commercial areas. In residential areas, sidewalks are desirable on both sides of collector streets, but should be provided on at least one side. The sidewalk should be situated as far as practical from the traveled way, usually close to the right-of-way line. For further information, see the section on "Sidewalks" in Chapter 4. Additional design guidance on sidewalks can also be found in the AASHTO *Guide for the Planning, Design, and Operation of Pedestrian Facilities* (**9**).

The minimum sidewalk width should be at least 1.2 m [4 ft] in residential areas and should range from 1.2 to 2.4 m [4 to 8 ft] in commercial areas. Sidewalk widths of at least 1.5 m [5 ft] are recommended by the ADAAG (**7, 8**).

Sidewalk curb ramps should be provided at crosswalks to accommodate persons with disabilities. The section on "Pedestrian Facilities" in Chapter 4 discusses various design applications at such ramps.

Driveways

Driveways should be regulated as to width of entrance, placement with respect to property lines and intersecting streets, angle of entrance, vertical alignment, and number of entrances to a single property. ADA guidelines should be considered in the design of driveways (**6, 7**). Further guidance on the design of sidewalk-driveway interfaces can be found in the AASHTO *Guide for the Planning, Design, and Operation of Pedestrian Facilities* (**9**).

Roadway Widths for Bridges

The clear width for all new bridges on urban collector streets with curbed approaches should be the same as the curb-to-curb width of the approaches. The bridge rail should be placed flush with the front face of the curb if no sidewalk is present to minimize the likelihood that vehicles will vault the rail. For urban collector streets with shoulders and no curbs, the full width of approach roadways should preferably be extended across bridges. Sidewalks on the approaches should be extended across all new structures. In addition, a sidewalk should be included on at least one side on all bridges on collector streets. Further discussion of roadway widths for bridges is presented in the section on "Traffic Barriers" in Chapter 4. Exhibits 6-6 and 6-7 apply to bridge widths on urban collector streets.

Vertical Clearance

Vertical clearance at underpasses should be at least 4.3 m [14 ft] over the entire roadway width, with an additional allowance for future resurfacing.

Horizontal Clearance to Obstructions

Roadside obstructions on urban collector streets should preferably be located at or near the right-of-way line and outside of the sidewalks. On urban collector streets that have curbs but no shoulders, a clearance of 0.5 m [1.5 ft] or more beyond the face of the curb should be provided to roadside obstructions, where practical. Where a continuous parking lane is provided, no clearance is needed, but a setback of 0.5 m [1.5 ft] to obstructions is desirable to avoid interference with opening car doors. In areas of dense pedestrian traffic, the provision of vertical curbing between the traveled way and adjacent street fixtures will discourage drivers from encroaching on the sidewalk. Urban collector streets with shoulders and without curbs should have clear zones, as described previously for rural collectors.

Roadside obstacles, such as trees, that might seriously damage out-of-control vehicles should be removed wherever practical. However, the potential benefits of removing such obstacles should be weighed against the adverse environmental and aesthetic effects of their removal. Therefore, trees should be removed only when considered essential for safety. However, it may only be practical to remove those fixed objects in very vulnerable locations. For further information, see the section on "Horizontal Clearance to Obstructions" in Chapter 4.

A wide and level border area should be provided along collector streets for the safety of the motorist and pedestrian, as well as for aesthetic reasons. However, the preservation and enhancement of the environment are of major importance in the design and construction of collector streets and may preclude provision of a border area. The street alignment should be selected to minimize cut and fill slopes.

Roadside barriers are not used extensively on urban collector streets except where there are safety concerns or environmental considerations such as along sections with steep foreslopes and at approaches to structures. Roadside barriers may also be needed to shield vehicles from over-crossing structures.

Right-of-Way Width

The right-of-way width should be sufficient to accommodate the ultimate planned roadway, including median, shoulder, grass border, sidewalks, bicycle facilities, public utilities, and outer slopes. The width of right-of-way for a two-lane urban collector street should generally range from 12 to 18 m [40 to 60 ft], depending on the conditions listed above.

Provision for Utilities

In addition to the primary purpose of serving vehicular traffic, urban collector streets may accommodate public utility facilities within the street right-of-way in accordance with state law or municipal ordinance. Use of the right-of-way by utilities should be planned to minimize interference with traffic using the street. The AASHTO *Guide for Accommodating Utilities Within Highway Right-of-Way* (**10**) presents general principles for utility location and construction to minimize conflicts between the use of the street right-of-way for vehicular movements and the secondary objective of providing space for locating utilities.

Border Area

The border area between the roadway and the right-of-way line should be wide enough to serve several purposes, including the provision of a buffer space between pedestrians, bicyclists, and vehicular traffic; a sidewalk; and an area for underground and above-ground utilities such as traffic signals, parking meters, and fire hydrants. A portion of the border area should accommodate snow storage and may include aesthetic features such as grass or landscaping. The border width should range from 2.4 to 3.3 m [8 to 11 ft], including the sidewalk width. For safety reasons, traffic signals, utility poles, fire hydrants, and other utilities should be placed as far back from the curb as practical. Breakaway features may be built into such obstacles, where practical, as an aid to safety.

Intersection Design

The pattern of traffic movements at intersections and the volume of traffic on each approach during one or more peak periods of the day, including pedestrian and bicycle traffic, are indicative of the appropriate type of traffic control devices, the widths of lanes (including auxiliary lanes), and where applicable, the type and extent of channelization needed to expedite the movement of traffic. The arrangement of islands and the shape and length of auxiliary lanes may differ depending on whether or not signal control is used. The composition and character of traffic is a design control; movements involving large trucks need larger intersection areas and flatter approach grades than those used at intersections where traffic consists predominantly of passenger cars. Bus stops located near an intersection may create a need for additional modification to the intersection design. Approach speeds of traffic also have a bearing on the geometric design as well as on the appropriate traffic control devices and pavement markings. For further information, see the section on "Traffic Control Devices" in Chapter 3.

The number and location of approach roadways and their angles of intersection are major controls for the intersection geometric design, the location of islands, and the types of control devices. Intersections at grade preferably should be limited to no more than four approach legs. When two crossroads intersect the collector highway in close proximity, they should be combined into a single intersection.

Important design considerations for at-grade intersections fall into two major categories: the geometric design of the intersection (including a capacity analysis) and the location and type of traffic control devices. For the most part, these considerations are applicable to both new and existing intersections although, for existing intersections in built-up areas, heavy development may make extensive design changes impractical.

Chapter 9 presents a discussion of all major aspects of intersection design.

Railroad-Highway Grade Crossings

Appropriate grade crossing warning devices should be installed at all railroad-highway grade crossings on collector streets. Details of these devices are given in the MUTCD (**5**). In some states, the final approval of these devices may be vested in an agency having oversight over railroads.

Sight distance is an important consideration at railroad-highway grade crossings on collector streets. There should be sufficient sight distance along the street for the approaching driver to recognize the railroad crossing, perceive the warning device, determine whether a train is approaching, and stop if necessary. Adequate sight distance along the tracks is also needed for drivers of stopped vehicles to decide when it is safe to proceed across the tracks.

The roadway width at all crossings should be the same as the curb-to-curb width of the approaches. Where street sections are not curbed, the crossing width should be consistent with the approach street and shoulder widths. Sidewalks should be provided at railroad crossings where approach sidewalks exist or are planned within the near future. Provisions for future sidewalks should be incorporated into design, if they can be anticipated, to avoid future crossing work on the railroad facility.

Crossings that are located on bicycle routes that are not perpendicular to the railroad may need additional paved shoulder width for bicycles to maneuver over the crossing. For further information, see the AASHTO *Guide for the Development of Bicycle Facilities* (**2**).

The design of railroad-highway grade crossings is discussed more fully in Chapter 9.

Street and Roadway Lighting

Good visibility under both day and night conditions is fundamental to enabling motorists, pedestrians, and bicyclists to travel on roadways in a safe and coordinated manner. Properly designed and maintained street lighting should provide comfortable and accurate night visibility, which should facilitate vehicular, bicycle, and pedestrian traffic.

Decisions concerning appropriate street lighting should be coordinated with safety management, crime prevention, and other community concerns. The AASHTO publication *An Informational Guide for Roadway Lighting* (**11**) provides discussion on street and roadway lighting. Further information is also provided in the section on "Lighting" in Chapter 3.

Traffic Control Devices

Traffic control devices should be applied consistently and uniformly. Details of the standard devices and warrants for many conditions are found in the MUTCD (**6**).

Geometric design of streets should include full consideration of the types of traffic control to be used, especially at intersections where multi-phase or actuated traffic signals are likely to be needed. Signal progression, signal phasing (including pedestrian and bicycle phases), and traffic flow rates are important considerations in major signalized intersection design. For further information, see the section on "Traffic Control Devices" in Chapter 3.

Erosion Control

Design of streets should consider preservation of natural ground cover and desirable growth of shrubs and trees within the right-of-way. Seeding, mulching, sodding, or other acceptable measures for covering slopes, swales, and other erodible areas should also be considered in urban collector street design. For further information, see the section on "Erosion Control and Landscape Development" in Chapter 3.

Landscaping

Landscaping should be provided in keeping with the character of the street and its environment for both aesthetic and erosion control purposes. Landscape designs should be arranged to permit a sufficiently wide, clear, and safe pedestrian walkway. The needs of individuals with disabilities, bicyclists, and pedestrians should be considered. Combinations of turf, shrubs, and trees should be considered in continuous border areas along the roadway. However, care should be exercised to ensure that sight distances and guidelines on clearance to obstructions are observed, especially at intersections. The roadside should be developed to serve both the community and the traveling motorist. Landscaping should also consider maintenance problems and costs, future sidewalks, utilities, additional lanes, and possible bicycle facilities. For further information on landscaping, see the AASHTO *Guide for Transportation Landscape and Environmental Design* (**12**).

REFERENCES

1. Transportation Research Board. *Highway Capacity Manual*, Fourth Edition, Washington, D.C.: Transportation Research Board, 2000 or most current edition.
2. AASHTO. *Guide for the Development of Bicycle Facilities*, Washington, D.C.: AASHTO, 1999.
3. AASHTO. *Roadside Design Guide*, Washington, D.C.: AASHTO, 1996.
4. AASHTO. *Standard Specifications for Highway Bridges*, Washington, D.C.: AASHTO, 1996.
5. AASHTO. *LRFD Bridge Design Specification,* second edition, Washington, D.C.: AASHTO, 1998.
6. U.S. Department of Transportation, Federal Highway Administration. *Manual on Uniform Traffic Control Devices for Streets and Highways*, Washington, D.C.: 1988 or most current edition.

7. Architectural and Transportation Barriers Compliance Board (Access Board). *Americans with Disabilities Act Accessibility Guidelines* (ADAAG), Washington, D.C.: July 1994 or most current edition.

8. UFAS. Uniform Federal Accessibility Standards, most current edition.

9. AASHTO. *Guide for the Planning, Design, and Operation of Pedestrian Facilities*, Washington, D.C.: AASHTO, forthcoming.

10. AASHTO. *Guide for Accommodating Utilities Within Highway Right-of-Way*, Washington, D.C.: AASHTO, 1994.

11. AASHTO. *An Informational Guide for Roadway Lighting*, Washington, D.C.: AASHTO, 1985.

12. AASHTO. *A Guide for Transportation Landscape and Environmental Design*, Washington, D.C.: AASHTO, 1991.

13. U.S. Department of Transportation, Federal Highway Administration. HEC 12. *Drainage of Highway Pavements*, FHWA-15-84-202. Washington, D.C.: Office of Engineering, Bridge Division, 1984.

14. Schoppert, D. W., and D. W. Hoyt. *Factors Influencing Safety at Highway-Rail Grade Crossings*, NCHRP Report 50, Washington, D.C.: Highway Research Board, 1968.

15. American Society of Civil Engineers, National Association of Home Builders, and the Urban Land Institute. *Residential Streets*, Washington, D.C.: American Society of Civil Engineers, 1974.

16. JHK and Associates. *Design of Urban Streets*, prepared for the U.S. Department of Transportation, Federal Highway Administration, Washington, D.C.: 1980.

17. Zegeer, C. V., R. Stewart, F. M. Council, and T. R. Neuman. *Roadway Widths for Low-Traffic Volume Roads*, NCHRP Report 362, Transportation Research Board, 1994.

CHAPTER 7
RURAL AND URBAN ARTERIALS
INTRODUCTION

The principal and minor arterial road systems provide a high-speed, high-volume network for travel between major points in both rural and urban areas. Chapter 1 discusses extensively the functional purposes of this class of facility, both rural and urban. This chapter provides the general information needed to establish the basis of design for these roadways.

The design of these arterials covers a broad range of roadways, from two-lane to multilane, and is the most difficult class of roadway design because of the need to provide safe and efficient operations under sometimes unusual or constrained conditions. The designer should be thoroughly familiar with the material in all chapters of this publication in order to skillfully blend the various types of arterials into the functional network. Although freeways are included in the functional description of an arterial, they have distinctive design requirements and are therefore treated separately in Chapter 8.

This chapter considers rural and urban arterials separately because each has distinctive features. However, the designer should be prepared to use design features from both arterial types to provide for suitable transitions as an arterial moves between rural and urban settings.

RURAL ARTERIALS
General Characteristics

Rural arterials constitute an important part of the rural highway system, including cross sections that range from two-lane roadways to multilane, divided controlled-access highways. The first portion of this chapter relates to the design of new rural arterials and the reconstruction of existing ones. Such roadways are designed on the basis of traffic volume needs and should be constructed to the most favorable design criteria practical.

Rural principal arterials comprise the Interstate system and most rural freeways. They also include other multilane roadways and some two-lane highways that connect urban centers. Minor rural arterials link urban centers to larger towns and are spaced to provide a relatively high level of service to developed areas of a state.

The appropriate design geometrics for an arterial may be readily determined from the selected design speed and the design traffic volumes, with consideration of the type of terrain, the general character of the alignment, and the composition of traffic. Operational characteristics, design features, cross sections, and rights-of-way are also discussed in this chapter.

Two-lane highways constitute the majority of the rural arterial system. Such roadways are adequate where traffic volumes are light and long sight distances are generally available.

Two-lane arterials generally have all-weather surfaces and are marked and signed in accordance with the current edition of the *Manual on Uniform Traffic Control Devices* (MUTCD) (**1**).

General Design Considerations

Design Speed

Rural arterials, excepting freeways, should be designed for speeds of 60 to 120 km/h [40 to 75 mph] depending on terrain, driver expectancy and, in the case of reconstruction projects, the alignment of the existing facility. Design speeds in the higher range—100 to 120 km/h [60 to 75 mph]—are normally used in level terrain, design speeds in the midrange—80 to 100 km/h [50 to 60 mph]—are normally used in rolling terrain, and design speeds in the lower range—60 to 80 km/h [40 to 50 mph]—are used in mountainous terrain. Where a lower design speed is used, refer to Chapters 2, 3, and 4 to select appropriate design features.

Design Traffic Volume

Before an existing rural arterial is improved or a new rural arterial is constructed, the design traffic volume should be determined. The first step in determining the design traffic volume is to determine the current average daily traffic (ADT) volume for the roadway; in the case of new construction, the ADT can be estimated. These ADT values should then be projected to the design year, normally 20 years into the future. The design of low-volume rural arterials is normally based on ADT values alone because neither capacity nor intersection operations typically govern the overall operation. Such roadways normally provide free flow under all conditions. By contrast, it is usually appropriate to design high-volume rural arterials using an hourly volume as the design traffic volume. The design hourly volume (DHV) that should generally be used in design is the 30th highest hourly volume of the year, abbreviated as 30 HV, which is typically about 15 percent of the ADT on rural roads. For further information on the determination of design traffic volumes, see the section on "Traffic Characteristics" in Chapter 2.

Levels of Service

Procedures for estimating the traffic operational performance of particular highway designs are presented in the *Highway Capacity Manual* (HCM) (**2**), which also presents a thorough discussion of the level-of-service concept. Although the choice of an appropriate design level of service is left to the highway agency, designers should strive to provide the highest level of service practical and consistent with anticipated conditions. Level-of-service characteristics are discussed in Chapter 2 and summarized in Exhibit 2-31. For acceptable degrees of congestion, rural arterials and their auxiliary facilities (i.e., turning lanes, passing sections, weaving sections, intersections, and interchanges) should generally be designed for level-of-service B, except in mountainous areas where level-of-service C is acceptable.

Sight Distance

Sight distance is directly related to and varies appreciably with design speed. Stopping sight distance, a key safety-related design element, should be provided through the length of the roadway. Passing and decision sight distances influence roadway operations and should be provided wherever practical. Providing decision sight distance at locations where complex decisions are made greatly enhances the chances that drivers will be able to safely accomplish maneuvers. Examples of locations where complex decisions are required include high-volume intersections, transitions in roadway width, and transitions in the number of lanes. Provision for adequate sight distance on rural arterials, which may combine both high speeds and high traffic volumes, can be complex. Exhibit 7-1 presents the recommended minimum values of stopping and passing sight distance. Refer to Chapter 3 for a comprehensive discussion of sight distance and for tabulated values for decision sight distance.

Metric			US Customary		
Design speed (km/h)	Minimum stopping sight distance (m)	Minimum passing sight distance (m)	Design speed (mph)	Minimum stopping sight distance (ft)	Minimum passing sight distance (ft)
50	65	345	30	200	1090
60	85	410	35	250	1280
70	105	485	40	305	1470
80	130	540	45	360	1625
90	160	615	50	425	1835
100	185	670	55	495	1985
110	220	730	60	570	2135
120	250	775	65	645	2285
130	285	815	70	730	2480
			75	820	2580
			80	910	2680

Exhibit 7-1. Minimum Sight Distances for Arterials

Ideally, intersections and railroad crossings should be grade separated or provided with adequate sight distance. Intersections should be placed in sag and/or tangent locations, where practical, to allow maximum visibility of the roadway and pavement markings.

Alignment

A smooth flowing alignment is desirable on a rural arterial. Changes in alignment, both horizontal and vertical, should be so gradual that they will not surprise the driver. Roads with adequate alignment usually operate more efficiently and safely than roads with poor alignment, even where improved signing and pavement marking are provided; therefore, adequate alignment should be provided wherever practical.

Grades

The length and steepness of grades directly affect the operational characteristics of an arterial. Exhibit 7-2 presents recommended maximum grades for rural arterials. When vertical curves for stopping sight distance are considered, there are seldom advantages to using the maximum grade values except when grades are long.

Type of terrain	Metric								US Customary								
	Maximum grade (%) for specified design speed (km/h)								Maximum grade (%) for specified design speed (mph)								
	60	70	80	90	100	110	120	130	40	45	50	55	60	65	70	75	80
Level	5	5	4	4	3	3	3	3	5	5	4	4	3	3	3	3	3
Rolling	6	6	5	5	4	4	4	4	6	6	5	5	4	4	4	4	4
Mountainous	8	7	7	6	6	5	5	5	8	7	7	6	6	5	5	5	5

Exhibit 7-2. Maximum Grades for Rural Arterials

Number of Lanes

The number of lanes on an arterial roadway should be determined based on consideration of volume, level of service, and capacity conditions. A multilane arterial, as discussed in this chapter, refers to four or more lanes.

Superelevation

Where curves are used on a rural arterial alignment, a superelevation rate compatible with the design speed should be used. Superelevation rates should not exceed 12 percent; however, where ice and snow conditions are a factor, the maximum superelevation rate should not exceed 8 percent. Superelevation runoff denotes the length of roadway needed to accomplish the change in cross slope from a section with adverse crown removed to a fully superelevated section and vice versa. Adjustments in design runoff lengths may be needed for smooth riding, drainage, and appearance. Chapter 3 provides a detailed discussion of superelevation and tables of appropriate superelevation rates and runoff lengths for various design speeds.

Cross Slope

Cross slope is provided to enhance roadway drainage. Two-lane rural roadways are normally designed with a centerline crown and cross slopes ranging from 1.5 to 2 percent with the higher values being most prevalent. Multilane roadways are either crowned at the centerline or sloped in one direction.

When three or more lanes are inclined in the same direction on multilane highways, each successive pair of lanes, or portion thereof, outward from the first two lanes from the crown line, may have an increased slope. For a more complete discussion, see the section on "Cross Slope" in Chapter 4.

Vertical Clearances

New or reconstructed structures should provide 4.9-m [16-ft] clearance over the entire roadway width. Existing structures that provide clearance of 4.3 m [14 ft], if allowed by local statute, may be retained. In highly urbanized areas, a minimum clearance of 4.3 m [14 ft] may be provided if there is an alternate route with 4.9-m [16-ft] clearance. Structures should provide additional clearance for future resurfacing of the underpassing road.

Structures

The full width for the approach roadways should normally be provided across all new bridges. Long bridges, defined as bridges having an overall length in excess of 60 m [200 ft], may have a lesser width. On long bridges, offsets to parapet, rail, or barrier should be at least 1.2 m [4 ft] measured from the edge of the traveled way on both sides of the roadway. See Chapter 10 for further information on bridge widths.

For an existing bridge to remain in place, it should have adequate structural strength and a width at least equal to the width of the traveled way plus 0.6 m [2 ft] clearance on each side. Bridges should be considered for ultimate widening or replacement if they do not provide at least MS-18 [HS-20] loadings. As an interim measure, narrow bridges should be considered for special narrow bridge treatments such as signing and pavement marking.

Traffic Control Devices

Signs, pavement delineation, and pavement marking play an important role in the optimum operation of rural arterials. Placement of these items should be considered early in the design stage while adjustments to the alignment and intersection design can be easily considered. Refer to the current MUTCD (1) for guidance in signing and marking.

Erosion Control

Consideration of erosion control features is important to the proper design of a rural arterial. By controlling erosion, the safety of the roadside is maintained and the environment downstream is protected from siltation and other possible harmful effects. Providing adequate ground treatment and cover has the additional benefit of assuring a pleasing roadside appearance.

Widths

The logical approach to determining appropriate lane and shoulder widths is to provide a width related to the traffic demands. Exhibit 7-3 provides values for the width of traveled way and usable shoulder that should be considered for the volumes indicated. Regardless of weather conditions, shoulders should be usable at all times. On high-volume highways, shoulders should generally be paved, but because of economic constraints, paved shoulders may not always be practical. As a minimum, 0.6 m [2 ft] of the shoulder width should be paved to provide for pavement support, wide vehicles, collision avoidance, and additional pavement width for bicyclists. The shoulder should be constructed to a uniform width for relatively long stretches of roadway. For additional information concerning shoulders, refer to Chapter 4.

Metric					US Customary				
	Minimum width of traveled way (m)[a] for specified design volume (veh/day)					Minimum width of traveled way (ft)[a] for specified design volume (veh/day)			
Design speed (km/h)	under 400	400 to 1500	1500 to 2000	over 2000	Design speed (mph)	under 400	400 to 1500	1500 to 2000	over 2000
60	6.6	6.6	6.6	7.2	40	22	22	22	24
70	6.6	6.6	6.6	7.2	45	22	22	22	24
80	6.6	6.6	7.2	7.2	50	22	22	24	24
90	6.6	6.6	7.2	7.2	55	22	22	24	24
100	7.2	7.2	7.2	7.2	60	24	24	24	24
110	7.2	7.2	7.2	7.2	65	24	24	24	24
120	7.2	7.2	7.2	7.2	70	24	24	24	24
130	7.2	7.2	7.2	7.2	75	24	24	24	24
All speeds	Width of usable shoulder (m)[b]				All speeds	Width of usable shoulder (ft)[b]			
	1.2	1.8	1.8	2.4		4	6	6	8

[a] On roadways to be reconstructed, an existing 6.6-m [22-ft] traveled way may be retained where alignment and safety records are satisfactory.

[b] Usable shoulders on arterials should be paved; however, where volumes are low or a narrow section is needed to reduce construction impacts, the paved shoulder may be reduced to 0.6 m [2 ft].

Exhibit 7-3. Minimum Width of Traveled Way and Usable Shoulder for Rural Arterials

Horizontal Clearance to Obstructions

A clear unobstructed roadside is highly desirable. Where fixed objects or nontraversable slopes fall within the clear roadside zones discussed in the section on "Horizontal Clearance to Obstructions" in Chapter 4, refer to AASHTO *Roadside Design Guide* (**3**) for guidance in selecting the appropriate treatment. Utilities and trees that will grow to 100 mm [4 in] or more in diameter should be located near the right-of-way line and should be outside the selected clear zone.

Cross Section and Right-of-Way

The type of surfacing and shoulder treatment should fit the volume and composition of traffic. Two-lane rural arterials are normally crowned to drain away from the centerline except where superelevation is provided. The treatment of cross slopes, drainage channels, and sideslopes is discussed in Chapter 4. The right-of-way should be wide enough to accommodate all of the cross-sectional elements throughout the project. This usually precludes a uniform right-of-way width since there are typically many situations where additional width is very desirable. Such situations occur where the sideslopes extend beyond the normal right-of-way, for clear areas at the bottom of traversable slopes, for wide clear areas on the outside of curves, where greater sight distance is desirable, at intersections and junctions with highways, at railroad-highway grade crossings, for environmental considerations, and for maintenance access.

Local conditions such as drainage and snow storage should be considered in determining right-of-way widths. Where additional lanes may be needed in the future, the initial right-of-way width should be adequate to provide the wider roadway section. It may be desirable to construct the initial two lanes off center within the right-of-way, so the future construction will cause less interference with traffic and the investment in initial grading and surfacing can be salvaged.

Provision for Passing

In designing two-lane, two-way arterials, the alignment and profile should provide sections for safe passing at frequent intervals. Design of the horizontal and vertical alignment should provide as great a proportion of the highway length as practical with adequate passing sight distance. Exhibit 7-1 presents the minimum passing sight distances for design speeds greater than 50 km/h [30 mph]. Restrictive cases may exist where passing sight distance is economically difficult to justify. Even in those instances, passing sections should be provided with at least the frequency to attain the desired level of service. Where achievement of sufficient passing sight distance is not practical, auxiliary lanes such as truck climbing lanes or passing lanes should be considered as a means to obtain the desired level of service.

Although truck climbing lanes are normally provided to prevent unreasonable reductions in operating speeds on upgrades, they also provide opportunities for passing in areas that normally would not permit passing. Adequately designed and well-marked climbing lanes will usually be used by slow-moving vehicles and allow passing by drivers who prefer to move at normal speeds. Climbing lanes are usually provided to the right of the normal traffic lane and should be the same width as the through lanes with a somewhat reduced shoulder width. A usable shoulder width of 1.2 m [4 ft] or greater is acceptable. The design elements and warrants for the use of climbing lanes are discussed in Chapter 3. An example of a climbing lane on a two-lane rural arterial is shown in Exhibit 7-4.

Passing lanes should be considered where climbing lanes are not warranted and where the extent and frequency of passing sections are too few. The use of passing lanes to increase passing opportunities on two-lane highways is addressed in Chapter 3.

Exhibit 7-4. Climbing Lane on Two-Lane Rural Arterial

In conclusion, a summary of the design procedures to be followed in providing passing sections on two-lane highways includes:

1. Design of the horizontal and vertical alignment should provide as great a proportion of the highway length as practical with adequate passing sight distance.

2. For design volumes approaching capacity, the effect of passing sections on increasing capacity should be considered.

3. For further information for climbing lane warrants, refer to Chapter 3.

4. Where the extent and frequency of passing opportunities made available by application of items 1 and 3 are insufficient, consideration should be given to the construction of passing lanes utilizing a three-lane cross section.

Ultimate Development of Four-Lane Divided Arterials

Although many two-lane arterials will adequately serve the traffic demands in the future, there are numerous instances, particularly near urban areas, where two-lane arterials will require ultimate development to a higher type arterial to handle the expected traffic.

Where it is anticipated that the DHV for the design year will be in excess of the service volume of the two-lane arterial for its desired level of service, the initial improvement should be patterned to the ultimate development of a four-lane divided arterial and provision made for acquisition of the needed right-of-way. Refer to the *HCM* (**2**) for traffic operational analysis procedures to determine whether a two-lane arterial can provide the desired level of service or whether a four-lane arterial should be considered. The eventual need for additional lanes should be considered during the design of a two-lane arterial. Even where right-of-way is restricted, some form of separator should be used in the ultimate facility, with a median at least 1.2 m [4 ft] wide and preferably much wider. (Four-lane undivided arterials are discussed in a subsequent section of this chapter).

In the ultimate development of a four-lane divided arterial, the initial two-lane roadway should be constructed so that it can eventually form one of the two-lane, one-way roadways. The advantages of this approach over building the initial two lanes in the center of the right-of-way are as follows:

1. There is no loss of investment in existing surfacing and in highway and railroad overcrossings when the second roadway is constructed.

2. This approach allows grading of the entire roadway and/or the construction of undercrossings and overcrossings to accommodate the ultimate improvement when a decision to do so is warranted. The economics of such a decision needs to be carefully considered, as do the benefits associated with minimization of future impacts.

   If the entire roadway is graded initially, traffic will be subjected to little restriction or delay when the additional two-lane surfacing is constructed. The two-lane surfacing originally constructed continues in use as a two-way highway, no detours are needed, and contact with construction operations is restricted to intersections and turnouts on one side.

   If the decision is to construct undercrossings and overcrossings, similar benefits also occur.

3. It is often desirable to initially acquire sufficient right-of-way for the ultimate development, including that required for future intersection improvements and grade separations. The economics of such a decision are important to consider, but the preservation of the right-of-way for the ultimate improvement is typically the compelling factor. Increase in land value, particularly after the construction or improvement of the arterial may more than offset the investment in additional right-of-way.

4. Later adjustment of minor road structures and plant growth are reduced to a minimum. When the entire grade for the ultimate four-lane divided arterial is constructed initially, all structures such as drains and culverts usually are completed and remain undisturbed when the final two lanes are added. If grading for only one of the two-lane roadways is economically advisable, road structures may be completed on one side, and temporary headwalls and open drains may be provided on the side where additional lanes will be placed later.

5. By grading the entire roadway for four lanes, future impacts to wetlands created by roadside ditches and recharge basins are avoided.

Care should be exercised, however, to ensure that an appropriate clear zone is provided in the initial stage. A similar procedure may be adopted for topsoiling, seeding, planting, and any other work that is done to prevent soil erosion, the value of which increases with time.

Two-lane arterials planned for ultimate conversion to a divided arterial usually have sufficient volume initially to warrant a traveled way of 7.2 m [24 ft] wide and usable shoulders, 2.4 m [8 ft] wide, as shown in Exhibit 7-5A. These traveled way and shoulder dimensions are commensurate with those recommended for four-lane divided arterials, as discussed later in this chapter. Where an arterial will ultimately be developed to a four-lane divided arterial with a wide median and the initial roadway is offset to one side of the right-of-way centerline, the roadway generally is crowned to drain both ways. Ultimately, the wide median is depressed to be self-draining and may receive surface runoff from one-half of each roadway (Exhibit 7-5B). Grading for the future development generally is deferred when the median is wide.

Where the right-of-way for the future four-lane arterial is restricted, a narrow median, which should be not less than 1.2 m [4 ft] wide, may need to be used. If provision of a median barrier is anticipated for the ultimate improvement, space for a wider median should be provided to accommodate the width of the barrier plus the appropriate clearance between the edge of the traveled way and the face of the barrier. As in the case of a wide median, the initial two-lane construction should be offset so that the ultimate development is centered on the right-of-way. To economize on the cost of drainage structures and to simplify construction, the initial and future two-lane roadways may be positioned to drain to the outside (Exhibit 7-5C). Future grading may or may not be deferred, depending on local conditions and on the probable length of time to the full development.

On most two-lane arterials constructed many years ago, no provision was made for future improvement to a higher roadway type. In such instances, where practical, a new two-lane, one-way roadway should be provided approximately parallel to the first, which is then converted to one-way operation to form a divided arterial. Where there is adjacent development, it may be more practical to construct another one-way, two-lane roadway nearby without disturbing the existing development. This method also may be advantageous where topography is not favorable to direct widening of the existing roadway section. If this construction cannot be accomplished, it may be practical to obtain a divided section by widening 4.2 m [14 ft] on each side of the existing roadway (Exhibit 7-5D). When none of these methods is practical, it may be necessary to find a new location. The old road then becomes a local facility and may also serve as an alternate route. From the standpoint of adequacy and service provided to through traffic, the last method is preferred because the arterial on a new location will not be influenced by the old facility and can be built to modern design criteria, preferably with some control of access.

For roadways that will ultimately be developed with narrow medians (Exhibits 7-5C and 7-5D), all of the cross sections shown have minimum combined widths of roadways and median of 20 m [70 ft]. About 3.6 m [12 ft] or more of additional width should be obtained so that median lanes for left turns may be provided at intersections.

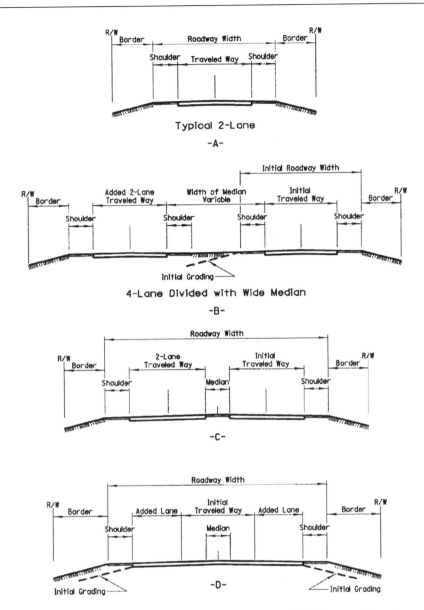

**Exhibit 7-5. Two-Lane Arterial Cross Section With Ultimate Development
to a Four-Lane Arterial**

Multilane Undivided Arterials

A multilane undivided arterial is the narrowest arterial on which each traffic lane is intended to be used by traffic in one direction of travel, and all passing is accomplished on lanes not subject to use by opposing traffic. The ability to pass without traveling in the lane of opposing traffic results in a freer and smoother operation and a large increase in arterial capacity over that of two-lane arterials. Because of the generally higher volumes, drivers on multilane arterials are confronted with additional traffic friction—from opposing traffic, roadsides, and traffic in the same direction. The crash rate on multilane undivided arterials is often higher than on two-lane arterials because multilane arterials generally carry heavier traffic volumes, have a higher frequency of intersections and other access points, and have greater development of adjacent land. Frequency of at-grade crossings has appreciable bearing on crash experience and capacity. Turn lanes and adequate intersection sight distance greatly improve the safety of intersection operations.

The elements of design discussed in preceding chapters are applicable generally to multilane undivided arterials, except that passing sight distance is not essential. The sight distance that should be provided at all points is the stopping sight distance because passing can be accomplished without the necessity of using an opposing traffic lane. In addition, intersection sight distance, as described in Chapter 9, should be provided at intersections.

Undivided arterials with four or more lanes are most applicable in urban and suburban areas where there is concentrated development of adjacent land. If traffic volumes justify the construction of multilane arterials in rural areas where speeds are apt to be high, it is generally considered that opposing traffic should be separated by a median. All arterials on new locations that need four or more lanes should be designed with a median. Preferably a median should be provided for improved safety in the improvement of an existing two-lane arterial to a multilane facility.

Divided Arterials

General Features

A divided arterial is one with separated lanes for traffic in opposite directions. It may be situated on a single roadbed or may consist of two widely separated roadways. The width of the median may vary and is governed largely by the type of area, character of terrain, intersection treatment, and economics. An arterial is not normally considered to be divided unless two full lanes are provided in each direction of travel and the median has a width of 1.2 m [4 ft] or more and is constructed or marked in a manner to preclude its use by moving vehicles except in emergencies or for left turns. A four-lane rural facility should have adequate median width to provide for protected left turns, which is a very important safety consideration. For example, vehicles making left turns should not be required to stop in the passing lane of a roadway designed for high volumes and speeds.

The principal advantages of dividing multilane arterials are increased safety, comfort, and ease of operation. A key reason for providing a median is the reduction in head-on collisions, which are usually serious, and the virtual elimination of such crashes on sections with wide medians or with a median barrier. Where median lanes for left turns are provided, rear-end collisions and other inconveniences to through traffic resulting from left-turn movements are greatly reduced. Pedestrians crossing the divided arterial are required to watch traffic in only one direction at a time and are provided a refuge at the median, particularly if a raised island is provided. Where the median is wide enough, crossing and left-turning vehicles can slow down or stop between the one-way roadways to take advantage of breaks in traffic and cross when it is safe to do so. Research has shown that four-lane undivided facilities have significantly more collisions than four-lane facilities with medians. Therefore, four-lane undivided facilities should be proposed only as a last resort when a median or turn lanes cannot be provided. Divided multilane arterials provide for more relaxed and pleasant operation, particularly in inclement weather and at night when headlight glare is bothersome. Headlight glare is reduced somewhat by narrow medians but can almost be eliminated by wide medians or glare screens on a median barrier.

Lane Widths

Roadways on divided arterials should be designed with lanes 3.6 m [12 ft] wide. The high speeds and large volumes associated with divided arterials will justify the construction of 3.6-m [12-ft] lanes. On reconstructed arterials, it may be acceptable to retain 3.3-m [11-ft] lanes if the alignment and safety record are satisfactory.

Cross Slope

Each roadway of a divided arterial may be sloped to drain to both edges, or each roadway may be sloped to drain to its outer edge, depending on climatic conditions and the width of median. Roadways on divided arterials should have a normal cross slope of 1.5 to 2 percent. Traveled ways with unidirectional slope may have the outer lane on a steeper slope than the inner lane. On an auxiliary lane, the cross slope should not normally exceed 3 percent on tangent alignment. In no case should the cross slope of an outer and/or auxiliary lane be less than the adjacent lane.

Shoulders

Arterials with sufficient traffic volume to justify the construction of four lanes also justify the provision of full-width shoulders. The width of usable outside shoulders should be at least 2.4 m [8 ft] and be usable during all seasons. Paving of the usable width of shoulder is preferred.

The normal roadway section, including usable shoulders, should be extended across all structures except for long bridges (over 60 m [200 ft] in length, which may have 1.2-m [4-ft] shoulders).

Shoulder space on the left side of the individual roadways of a four-lane divided arterial (i.e., within the median) is not intended to serve the same purpose as the right shoulder. The shoulder on the right, through customary use on undivided arterials, is accepted by all drivers as a suitable refuge space for stops. Where the median is flush with the roadway or has sloping curbs, vehicles may encroach or drive on it momentarily when forced to do so to avoid a crash. Only on rare occasions should drivers need to use the median for deliberate stops.

On divided arterials with two lanes in each direction, a paved shoulder strip 1.2 m [4 ft] wide should satisfy the needs for a shoulder within the median. Such a shoulder strip will preclude rutting at the edge-of-traveled way and will alleviate possible loss of driver control of vehicles that inadvertently encroach on the median.

On divided arterials with three or more lanes in each direction, a driver in distress in the lane nearest the median may have difficulty maneuvering to the right-hand shoulder. Consequently, a full-width shoulder within the median is desirable on divided arterials having six or more lanes.

Guardrail and median barrier should be considered in accordance with the AASHTO *Roadside Design Guide* (3).

Median Barrier Clearance

In cases where a wall or median barrier is used in the median, the AASHTO *Roadside Design Guide* (3) should be consulted for guidance in selecting an appropriate lateral clearance from the normal edge of the traveled way to the base of the wall or barrier and the type of barrier to be used.

Medians

On highways without at-grade intersections, the median may be as narrow as 1.2 to 1.8 m [4 to 6 ft] under very restricted conditions, but wider medians should be provided, wherever practical. A wide median allows the use of independent profiles. In addition, reduced frequency of cross-median crashes and reduction of headlight glare are safety features associated with a wide median.

Where intersections are to be provided, special concern should be given to median width. NCHRP Report 375 (4) has found that most types of undesirable driving behavior in the median areas of divided highway intersections are associated with competition for space by vehicles traveling through the median in the same direction. The potential for such problems is limited where crossroad and U-turn volumes are low, but may increase at higher volumes. Types of undesirable driving behavior observed include side-by-side queuing, angle stopping, and encroaching on the through lanes of a divided highway. At rural unsignalized intersections, the frequency of undesirable driving behavior and crashes was observed to decrease as the median width increased; this implies that medians should be as wide as practical. It was also found that the frequency of undesirable driving behavior increased as the median opening length increased.

While medians as narrow as 1.2 to 1.8 m [4 to 6 ft] may be used under very restricted conditions, medians 3.6 to 9 m [12 to 30 ft] wide provide protection for left-turning vehicles at intersections. Medians of 1.2 to 2.4 m [4 to 8 ft] wide should be avoided, if practical, where left turns are common. Such widths do not provide sufficient space for turning vehicles and may encourage other motorists to encroach into the adjacent lane when attempting to pass a turning vehicle that is only partially in the median.

In many cases, the median width at rural unsignalized intersections is a function of the design vehicle selected for turning and crossing maneuvers. Where a median width of 7.5 m [25 ft] or more is provided, a passenger car making a turning or crossing maneuver will have space to stop safely in the median area. Medians less than 7.5 m [25 ft] wide should be avoided at rural intersections because drivers may be tempted to stop in the median with part of their vehicles unprotected from through traffic. The school bus is often the largest vehicle to use the median roadway frequently. The selection of a school bus as the design vehicle results in a median width of 15 m [50 ft]. Larger design vehicles, including trucks, may be used at intersections where enough turning or crossing trucks are present; median widths of 25 m [80 ft] or more may be needed to accommodate large tractor-trailer trucks without encroaching on the through lanes of a major road.

There was concern that median widths in the range of 15 to 25 m [50 to 80 ft] at divided highway intersections could cause some drivers to become confused. No evidence of such confusion at rural intersections has been found (**4**). However, an intersection with a wider median may become confusing to some drivers if the median is so wide that a driver on the crossroad approach cannot see the far roadway of the divided highway. Such designs should be avoided and, where they are used, signing should be provided to discourage wrong-way movements.

Median widths of more than 18 m [60 ft] are undesirable at intersections that are signalized or may need signalization in the foreseeable future. The efficiency of signal operations decreases as the median width increases, because drivers need more time to traverse the median and special detectors may be needed to avoid trapping drivers in the median at the end of the green phase for traffic movements that pass through the median. Furthermore, if the median is so wide that separate signals are needed on the two roadways of the divided highway, delays to motorists will increase substantially and careful attention should be given to vehicle storage needs on the median roadway between the two signals.

The subsequent section in this chapter concerning median widths at intersections on urban arterials indicates that wider medians may increase crashes and lead to undesirable driving behavior. Therefore, consideration should be given to limiting use of wider medians at rural intersections that are likely to undergo urban or suburban development in the foreseeable future.

Undesirable driving behavior at rural unsignalized intersections increases as the median opening length increases (**4**). The median opening length should be equal to a least that described in Chapter 9, but median openings at rural unsignalized intersections should not be unnecessarily long.

Medians should be designed to provide a forgiving roadside and guardrail or median barrier should be considered in accordance with the AASHTO *Roadside Design Guide* (**3**). Further information on median design is presented in Chapter 4.

Alignment and Profile

A divided arterial generally serves high-volume and high-speed traffic for which a smooth flowing alignment should be provided. Because a divided arterial consists of two separated roadways, there may be instances where median widths and roadway elevations can be varied. Special topographic or intersection considerations may necessitate such treatments for economic or operational reasons. Precaution should be taken that such variations do not adversely affect operations. Potential problems associated with sharp reverse curves, headlight glare, roadside design, and grades of intersection crossings should be considered.

Profile design is less difficult for multilane highways than for two-lane highways. With two or more lanes for travel in each direction, the profile grade is generally governed by stopping sight distance, except at intersections. For volumes well below capacity, grades may be steeper and longer on multilane highways than on two-lane arterials, because there is a continuous lane for passing of heavy, slow vehicles on upgrades.

Although vertical design controls may be less restrictive for divided arterials than for two-lane arterials because passing sight distance need not be considered, the design of appropriate profiles for divided arterials involves design judgment and careful study. Even though a profile may satisfy all of the design controls, the finished product can appear forced and angular. A smooth-flowing roadway with gradual changes in horizontal and vertical alignment should be designed to the extent practical. Such design is of primary importance where a median of constant width is used in rolling terrain. The lack of a need to provide passing sight distance may tempt designers to use a roller coaster profile, which appears more displeasing on a divided arterial than on a two-lane arterial. With a wide divided arterial of uniform cross section, the driver's longitudinal perspective of distance is compressed and can make the combination of horizontal and vertical alignment appear abrupt and disjointed. The relation of horizontal and vertical alignment should be studied to obtain a suitable combination. To avoid an undesirable appearance, profile designs should be checked in long continuous plots, wherein the foreshortened aspect can be simulated. The section on "Combinations of Horizontal and Vertical Alignment" in Chapter 3 provides additional guidance on this topic.

Climbing Lanes on Multilane Arterials

Multilane rural roads usually have sufficient capacity to handle their traffic load, including the normal percentage of heavy trucks, without becoming severely congested. Climbing lanes generally are not as easily justified on multilane arterials as on two-lane arterials, because on two-lane arterials vehicles following slow-moving trucks on upgrades may be unable or psychologically discouraged from using an adjacent traffic lane for passing. By contrast, on multilane arterials, drivers have an adjacent lane available to them in which to pass slow-moving vehicles.

Furthermore, a climbing lane on a two-lane, two-way road is useful during both peak and non-peak hours, whereas on a multilane arterial, a climbing lane is likely to have only limited use during non-peak hours. During periods of lower traffic volumes, a vehicle following a slow-moving truck in the right lane can readily move to the adjacent lane and proceed without difficulty, although there is evidence that slow vehicles on through-traffic lanes may cause crashes.

Because new or reconstructed arterials are designed for 20 years or more in the future, there is little likelihood of climbing lanes being justified on multilane arterials for several years after their initial construction, even though they are deemed desirable for the peak hours of the design year. Thus, there may be an economic advantage in designing for, but deferring construction of, climbing lanes on multilane arterials. In this situation, grading for the future climbing lane should be provided initially. Very little additional grading is needed because a full shoulder is likely to be provided where there is no climbing lane; however, only a narrow shoulder is typically used outside of a climbing lane, because the climbing lane itself can serve as an emergency stopping area when needed. A full discussion on the need for climbing lanes is found in Chapter 3.

Superelevated Cross Sections

A divided arterial on a curve should be superelevated to ensure safe traffic operation, pleasing appearance, and economy. Care should be taken in the superelevation transition to fit site conditions and to meet controls of intersection design.

General methods of attaining superelevated cross sections for divided arterials are discussed in Chapter 3. In the design of arterials, the inclusion of a median in the cross section alters the manner in which superelevation is attained. Depending on the width of median and its cross section, there are three general cases for attaining superelevation.

Case I—*The whole of the traveled way, including the median, is superelevated as a plane section.* Case I should necessarily be limited to narrow medians and moderate superelevation rates to avoid substantial differences in elevation of the extreme edges of the traveled way arising from the median tilt. Specifically, Case I should be applied only to median with widths of 4.5 m [15 ft] or less.

Case II—*The median is held in a horizontal plane and the two traveled ways are rotated separately around their median edges.* Case II can apply to any width of median but is most appropriate for medians with widths between 4 and 18 m [15 and 60 ft]. By holding the median edges level, the difference in elevation between the extreme traveled way edges can be limited to that needed to superelevate the roadway. Superelevation transition design for Case II usually has the median-edge profiles as the control. One traveled way is rotated about its lower edge and the other about its higher edge.

Case III—*The two traveled ways are treated separately for superelevation with a resulting variable difference in elevation at the median edges.* Case III design can be used on wide medians (i.e., those with widths of 18 m [60 ft] or more). For this case, the difference in elevation of the extreme edges of the traveled way is minimized by a compensating slope across the median. With a wide median, it is possible to design the profiles and superelevation transition separately for the two roadways.

Chapter 3 and, specifically, Exhibit 3-37, contains additional guidance concerning methods for attaining superelevation for Cases I, II, and III.

Exhibit 7-6 demonstrates treatment of cross sections for narrow and wide medians with superelevated roadway in relation to the width of median for the three cases noted. In Cross Sections A and D both roadways lie in the same plane. The roadways are superelevated by rotating them about a profile control on the centerline of the median. The same effect can be obtained by rotation about the edge of the traveled way or any other convenient control line.

Where Cross Section A of Exhibit 7-6 is used, the median should be graded in accordance with the AASHTO *Roadside Design Guide* (**3**) and designed so that surface water from the higher roadway does not drain across the lower roadway. On tangent alignment, a shallow drainage swale can be provided in a median about 4.5 m [15 ft] wide and a well-rounded drainage channel with a width of about 18 m [60 ft] as shown in Exhibit 7-7F. On a superelevated section rotated about the median centerline, as in Cross Section A of Exhibit 7-6, approximately 9 m [30 ft] of median width is needed for a rounded drainage channel and adequate left shoulders. In a median less than 9 m [30 ft] wide, a channel with flat sideslopes can be provided if the superelevation rate is small, or a paved channel can be used in conjunction with higher rates of superelevation.

The projection of superelevation across wide medians may be fitting in some instances as in Cross Section A of Exhibit 7-6, but its general use in conjunction with large rates of superelevation is not satisfactory in appearance and generally not economical. It may fit at highway intersections where the profile of the intersecting road approximates the superelevated slope. Occasionally, it may fit the natural slope of the terrain. However, unless these conditions prevail, the large difference in elevation between the outer shoulder edges is likely to be objectionable. For example, the difference in elevation between the outer shoulder edges of a four-lane divided arterial with a median of 12 m [40 ft] and a superelevation rate of 8 percent is about 2.4 m [8 ft].

In level terrain and in terrain where the natural slope of the land is adverse to the cross-sectional slope, substantial improvement in appearance and economy in earthwork results if the wide median is made level as in Cross Section B of Exhibit 7-6, or sloped opposite to the superelevation plane as in Cross Section C.

Superelevation runoff lengths may vary for each of the three cases (refer to Exhibit 3-29). For Case I designs, the length of runoff should be based on the total rotated width (including the median width). Runoff lengths for Case II designs should be the same as those for undivided highways with a similar number of lanes. Finally, runoff lengths for Case III designs are based on the needs of the separate one-way roadways, as defined by their superelevation rates and rotated widths.

In Cross Sections B and E of Exhibit 7-6 the edges of roadways on the median sides are at the same elevation. Designs on this basis are pleasing in appearance and generally are desirable for safe operation. With a wide separation between the one-way roadways, Cross Section B has considerable advantage over Cross Section A in the reduction in difference in elevation across the entire roadbed. On roadways having a superelevation rate near 10 percent, the treatment in Cross-Section B requires a minimum median width of about 9 m [30 ft] to provide fully effective shoulder areas and a well-rounded traversable swale.

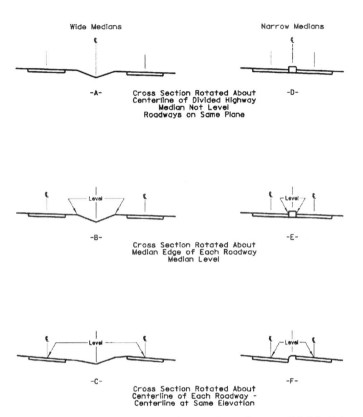

Exhibit 7-6. Methods of Attaining Superelevation on Divided Arterials

In Cross Sections C and F of Exhibit 7-6, the two one-way roadways have a common centerline grade. The difference in elevation of the outer extremities of the superelevated roadways is minimal, being the product of the superelevation rate and the width of one of the one-way roadways. The method of attaining superelevation runoff is directly applicable to each one-way roadway.

With a wide median, the treatment of Cross-Section C of Exhibit 7-6 allows the desired appearance to be maintained and permits economy in the wide-graded cross section. The roadway as a whole will appear fairly level to the motorist, who will not readily perceive the difference in elevation of the inside edges of roadway. This cross section generally is not suitable for important at-grade intersections unless the median is very wide. The median should be sufficiently wide in relation to superelevation to afford a smooth S-shaped profile across its width. The width for this shape is somewhat more than that needed for the previous sections. About 12 m [40 ft] is needed, with a superelevation rate of 10 percent and adequate shoulder areas. This width can be reduced to about 9 m [30 ft] when a paved channel is provided.

On a divided arterial with variable width of median and difference in elevation between the two roadways, each roadway is designed with a separate profile. With a reasonably wide median, each roadway can be superelevated in any manner suitable for a single roadway with little effect on the median slope. With a narrow median, an appreciable difference in elevation might require a retaining wall in the median. The manner of superelevating the roadways has some effect on the height of wall, but this amount is minimal and should have little bearing on design. Exhibit 7-7 shows various median configurations that may be used on rural arterials, excepting Configurations C, D and E, which are more appropriate for urban situations as described later in this Chapter. Refer to the AASHTO *Roadside Design Guide* (**3**) for guidance on designing a forgiving roadside.

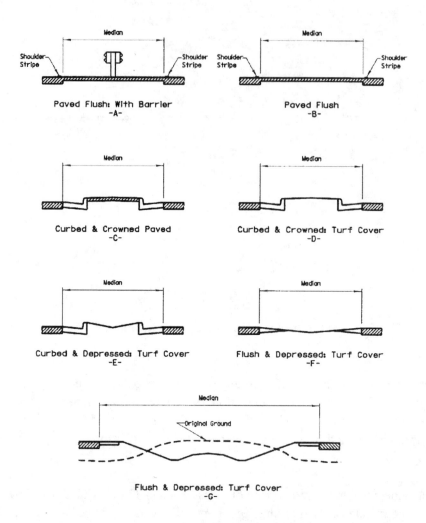

Exhibit 7-7. Typical Medians on Divided Arterials

Cross Section and Right-of-Way Widths

Cross-sectional elements of divided arterials—the widths and details of traveled ways, shoulders, medians, sideslopes, clear zones and drainage channels—have been discussed separately in this and other chapters. The appropriate right-of-way widths, including all elements in a composite arterial cross section, are presented in Exhibit 7-8. In an ideal situation, the topography, other physical constraints, and economic feasibility permit the design of a well-balanced cross section of desirable dimensions, for which an adequate width of right-of-way is established and procured. On the other hand, the constraints may be so tight that if a divided arterial is to be provided at all, it should be designed within a limited width of right-of-way, using minimum or near-minimum dimensions for each element of the arterial cross section. In the first instance, the right-of-way is based on the most favorable design criteria for the cross-sectional elements; in the latter case the cross section is determined on the basis of the available width of right-of-way.

The widths of cross-sectional elements should be proportioned to provide a well-balanced arterial section. Recommended traveled way and shoulder widths are shown in Exhibit 7-3. The border width is affected directly by the depth of cut or fill. If the right-of-way is restricted, the border area or median width, rather than the lane or shoulder width, should be reduced. The extent to which the border area and/or median width is reduced respectively should be decided carefully. Provision of a median width greater than that which eliminates the need for a median barrier is generally not warranted if doing so would subsequently require provision of substantial amounts of roadside guardrail that would otherwise not be needed, or if adjacent roadside development is present or anticipated. Consideration should be given to achieving approximately the same clear zone width for both the median and roadside.

Exhibit 7-8C shows a desirable divided arterial cross section warranted for a high-type facility where liberal width of right-of-way is attainable. Where these wider widths cannot be obtained, attempts should be made to provide a right-of-way width that permits the use of a median 9 m [30 ft] or more and sufficient borders to provide for the needed clear zone. For additional information on clear zones, refer to the AASHTO *Roadside Design Guide* (**3**).

Sometimes the right-of-way may be so restricted that minimum or near-minimum widths of cross-sectional elements must be used. If at all practical, the right-of-way should be wide enough to permit the use of median and borders of not less than 4.5 m [15 ft] (see Exhibit 7-8A). A 4.5-m [15-ft] median is near the minimum median width within which a median lane can be provided at intersections. Exhibit 7-7 shows some sections with curbs, which are generally not recommended along rural roadways. Sloping curbs may be used in restricted areas where needed to control drainage, or where special treatment is needed at locations such as intersections.

The cross sections and right-of-way widths shown in Exhibit 7-8 pertain to four-lane facilities. Where provision is to be made for ultimate conversion to a six- or eight-lane facility, the right-of-way widths should be increased by the width of lanes to be added. It is preferable to include this additional width in the median.

The cross-sectional arrangements shown in Exhibit 7-8 indicate generally balanced sections for what are termed "desirable," "minimum," and "restricted" rights-of-way. Some variation in these arrangements may be appropriate in individual cases. The right-of-way width need not be uniform and may be varied along the course of the arterial as needed for grading, for safe roadside design, and other conditions. Where controls become rigid, the two roadways may have to be brought closer together. Where physical conditions are favorable and land is readily available, the roadways of a divided highway may be spread farther apart. Where future grade separations and ramps are envisioned, provision for the initial acquisition of additional rights-of-way should be considered.

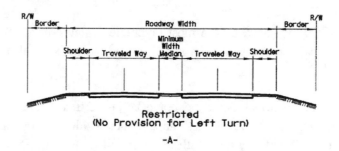

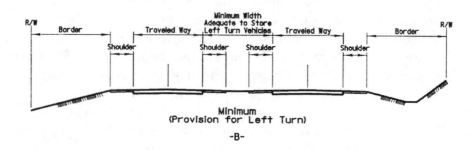

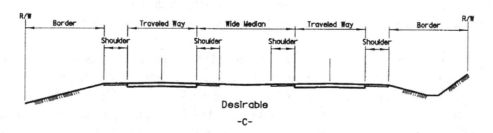

Exhibit 7-8. Cross Sectional Arrangements on Divided Arterials

The cross sections depicted in Exhibit 7-8 represent normally divided facilities in rural areas. Sometimes in rural areas, and particularly in and near urban districts, it is appropriate to separate through traffic from local traffic. Where such is the case, frontage roads may be provided along the outer limits of the highway cross section (Exhibit 7-9). Frontage roads serve to collect and distribute local traffic to and from adjacent development and provide parking and service thereto removed from the main traveled way, thus freeing through traffic from the disturbance introduced by local operation. The component parts of a typical cross section with frontage roads in generally flat terrain are shown in Exhibit 7-9A. The frontage roads are shown within the right-of-way limits, which is the typical arrangement. Frontage roads sometimes are provided outside the right-of-way limits, in which case the right-of-way can be narrower than shown. Where the profile of the through traveled way passes over or cuts through the natural ground, frontage roads are generally held at the level of the existing development, and the difference in elevation between the main traveled ways and the frontage roads is affected within the outer separations by earth slopes or retaining walls.

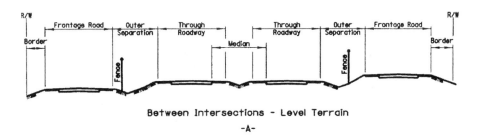

Between Intersections - Level Terrain
-A-

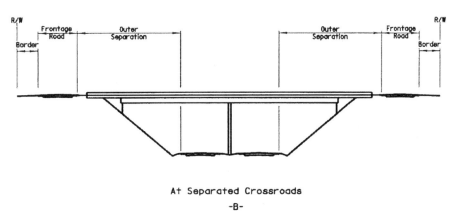

At Separated Crossroads
-B-

Exhibit 7-9. Cross Sectional Arrangements on Divided Arterials

Some crossroads in divided arterials may be grade separated from the through traveled way with local service provided by frontage or other roads. If all crossroads were grade separated in this manner, the facility would be a freeway. However, grade separation on divided arterials may be appropriate at some crossroads but not at others. A typical cross section at a separated crossroad with a depressed arterial is depicted in Exhibit 7-9B. Where frontage roads are provided, the outer separations should be wider on arterials with two-way frontage roads and on arterials with grade separations than on arterials crossing at grade to allow for roadside slopes and ramps. Further discussion on interchanges is presented in Chapter 10.

Sections With Widely Separated Roadways

Occasionally, it is practical to widely separate the one-way roadways of a divided arterial. This design may be appropriate where an existing two-lane arterial proves inadequate and is improved to a four-lane section but for which direct widening is not practical because of topography or adjacent development. In such a case, the old roadway is not disturbed but is converted to one-way operation and another, completely separate, one-way roadway is constructed. This action sometimes results in acquisition of two separate rights-of-way to contain the individual roadways of the divided arterial.

Widely separated one-way roadways may be particularly appropriate for certain topographic conditions. In valleys where drainage makes the location difficult, individual roadways may be situated on each side of the valley. Drainage of roadways is then simplified, with both sides draining directly to the natural channel. Along ridges or where there is a continual change in ground cross slopes, the separate roadways may be better fitted to the terrain than an arterial on a single roadbed. Such arrangements simplify location problems because only one roadway is considered at a time. With reduced roadway prisms, construction scars are kept to a minimum and more of the natural growth is retained, particularly between the separate roadways. In areas where right-of-way is not restricted, designs involving widely separated roadways often result in lower construction costs.

Intersections between a crossroad and a one-way roadway are greatly simplified in design and operation. Crash potential is generally reduced and the capacity of intersections is increased. Moreover, operation on widely separated roadways provides the maximum in driver comfort. Strain is lessened by largely eliminating the view and influence of opposing traffic. Substantial reduction or elimination of headlight glare at night is especially helpful in easing driver tension.

Operational problems of intersections on roadways with very wide medians should be considered. Desirably, a wide median is adequate to store the longest legal vehicles. To determine the number of intersection lanes needed, all movements and their volumes should be considered. The need for turn-arounds, connecting roadways, and frontage roads should be considered along with the effect on adjacent property owners. Signing to prevent wrong-way operation should be provided in accordance with the MUTCD (1), particularly when both roadways of the divided highway are not visible to drivers stopped at the crossroad. Additional discussion on wide medians is presented in the earlier section of this chapter on "Medians."

If arterials of appreciable length have roadways separated so widely that each roadway cannot be seen from the other, drivers may assume that they are on a two-way instead of a one-way roadway and hesitate to pass slow-moving vehicles. This situation can be alleviated by an occasional open view between the two roadways.

Intersections

The liberal use of high-type intersections and interchanges is highly desirable on arterials that do not have full control of access. Adequate turning widths with acceleration and deceleration tapers will provide a minimum design for minor intersections on a minor arterial. Where practical, principal arterials that intersect should be served by interchanges, possibly of the free-flow type. A comprehensive study of all intersections is needed for new and reconstruction projects, and a suitable design, consistent with the desired level of service, should be selected. Rural intersection control by traffic signals is not desirable. Drivers generally do not anticipate signals in rural areas that have high operating speeds, especially when traffic volumes are relatively low. Curbed islands present an obstacle to drivers and may become snow traps in regions that receive frequent snowfalls. Therefore, curbs at intersections should be avoided in high-speed areas.

If interchanges are intermixed with intersections, adequate merging distances should be provided to allow ramp traffic to operate freely. The merging driver should not have to be concerned with cross traffic at a downstream intersection while making a merging maneuver. Design of intersections and interchanges should be in accordance with Chapters 9 and 10, respectively.

Access Management

Arterials are designed and built with the intention of providing better traffic service than is available on local roads and streets. Although an arterial may not have more traffic lanes, its ability to carry greater volumes is usually related to the amount of crossroad interference or side friction to which it is subjected. One of the most important considerations in arterial development is the amount of access control, full or partial, that can be acquired. The ability to control access on an arterial will often relate directly to the project's safety.

Controlling access is vital to maintaining operations and safety and to preserving the level of service for which the arterial was initially designed. Access control is usually not too difficult to obtain in a rural area where development is light. Adequate access can normally be provided without great interference to traffic operations. However, rural areas do pose distinct access-related problems. The movement of large, slow-moving farm machinery is not uncommon and numerous field entrances are also requested by landowners. Because of these unique problems, access points should be situated to minimize their detrimental effects to through traffic. If access points are needed on opposite sides of the roadway, they should be situated directly opposite each other to allow vehicles to cross the arterial in the shortest possible time. Where access is needed for two adjacent properties or where different land uses adjoin one another, providing one driveway to serve both properties will reduce the number of access locations needed. Adequate and uniform spacing between access points will also help eliminate many conditions where a large vehicle at an intersection hides another vehicle on a nearby approach. Consideration should also be given to the location of access points in relationship to intersection sight distance restrictions and other intersections. High-volume access points can lead to particular operational problems if not properly situated. Short sections of rural frontage roads may be used to combine access points and minimize their operational effect to the arterial.

The appropriate degree of access control or access management depends on the type and importance of an arterial. Anticipation of future land use is a critical factor in determining the degree of access control. Provision of access management is vital to the concept of an arterial route if it is to provide the service life for which it is designed. For additional guidance on access management techniques for arterials, refer to NCHRP Report 420, *Impacts of Access-Management Techniques* (**5**).

Bikeways and Pedestrian Facilities

Rural arterials often provide the only direct connection between populated areas and locations to which the public wishes to travel. Schools, parks, and rural housing developments are usually located to be readily accessible by automobile. However, pedestrians and bicyclists may also wish to travel to the same destination points. Where demands for pedestrian and bicycle travel exist, the designer should consider their effects on the safety and operation of the arterial.

Where pedestrian use of the travel lanes or paved shoulders is deemed undesirable, pedestrians may be accommodated by sidewalks on one or both sides of the roadway. If off-roadway bicycle facilities are desired, they should be designed as shared-use paths in accordance with the AASHTO *Guide for the Development of Bicycle Facilities* (**6**).

Bus Turnouts

Where bus routes are located on a rural arterial, provision should be made for loading and unloading of passengers. Because of its size, a bus cannot easily leave the roadway unless special provisions are made. A well-marked, widened shoulder or an independent turnout is highly desirable and should be provided, if practical. Although it may be impossible or impractical to provide, for example, school bus turnouts for every dwelling, they should be provided at locations where there are known concentrations of passengers. Adequate provisions for buses will provide greater capacity and safety for a rural arterial. For additional guidance concerning bus turnouts, refer to Chapter 4.

Railroad-Highway Crossings

Desirably, all railroad crossings on the rural arterial system should be grade separated. However, practical considerations make it likely that many crossings will be at grade. Various treatments can be applied at railroad-highway crossings to encourage safe operation including adequate signing, signals, signals with gates, and grade separations. Judgment should be used in the selection of appropriate design and traffic control treatments for railroad-highway crossings; factors to be considered in such judgments include the volume and speed of traffic on both roadways and railroads, the available sight distance, and the anticipated safety benefits of specific treatments. Given the high traffic volumes and speeds on many arterials, the designer should attempt to provide the most protection practical. For further guidance on traffic control systems for railroad-highway grade crossings, refer to the MUTCD (**1**). For further information on design criteria for railroad-highway grade crossings, see Chapter 9.

Rest Areas

The provision of rest areas on the rural arterial system is a desirable feature particularly on principal arterials. Rest areas provide the high-speed, long-distance traveler with the opportunity for short periods of relaxation, which relieves driver fatigue. These facilities serve as a desirable safety feature, as evidenced by public recognition of the fatigue problem and their use of rest areas.

The location of rest areas should be considered early in development of a rural multilane arterial. Sites of special interest or visual quality provide additional reasons for the motorist to stop and usually extend the length of their stay. The spacing of rest areas depends on many considerations. For example, construction and operating costs for rest areas are significant, but benefits to drivers should also be considered. Additional information on rest areas may be found in the AASHTO *Guide for Development of Rest Areas on Major Arterials and Freeways* (**7**).

URBAN ARTERIALS

General Characteristics

Urban arterials carry large traffic volumes within and through urban areas. Their design varies from freeways with fully controlled access to two-lane streets. The type of arterial selected is closely related to the level of service desired. The principal objective for an urban arterial should be mobility with limited or restricted service to local development. If restriction of local access is not practical, special designs that incorporate access management are desirable. Such designs can vary from roadways that provide separate turn lanes to one-way streets.

Urban arterials are capable of providing some access to abutting property. Such access service should, however, be only incidental to the arterial's primary function of serving major traffic movements.

Before designing an urban arterial, it is important to establish the extent and need for such a facility. Once the need is established, steps should then be taken to protect the ability of the arterial to serve traffic at the desired level of service from future changes, such as strip development or the unplanned location of a major traffic generator. Development along an arterial should be anticipated regardless of a city's size. However, with proper planning and design, such development need not seriously affect the arterials' major function of safely serving through travel. Rather, it can complement such development and continue to provide the desired level of service.

Urban arterials are functionally divided into two classes, principal and minor. These classes are discussed in detail in Chapter 1. The urban arterial system, which includes arterial streets and freeways, serves the major activity centers of a metropolitan area, the highest traffic volume corridors, and the longest trips. The portion of the arterial system, either planned or existing, on which access is not fully controlled constitutes the arterial street system for the urban area. From the standpoint of design characteristics, all such urban streets are treated as a single class and are addressed in this chapter. Design of freeways is addressed in Chapter 8.

General Design Considerations

In the development of a transportation improvement program, routes selected for improvement as arterials may comprise portions of an existing street system, or they may be projected locations on new alignments through relatively undeveloped areas. Usually, they will be existing streets because, historically, the need for improving existing streets has surpassed the availability of resources. As a consequence, street improvements tend to lag, rather than lead, land-use development.

Major improvement of existing arterials can be extremely costly, particularly where additional rights-of-way need to be acquired through highly developed areas. Accordingly, it is often necessary to use design values that are less than desirable and below the design values that are used where sufficient right-of-way is available or can be acquired economically. When restricted conditions are encountered, consideration should be given to providing above-desirable values for other design or traffic control elements that tend to offset those created by the restriction (e.g., eliminating left-turn movements through an area having less-than-normal lane widths).

Design Speed

Design speeds for urban arterials generally range from 50 to 100 km/h [30 to 60 mph]. Lower speeds apply in central business districts and in more developed areas, while higher speeds are more applicable to outlying suburban and developing areas. Design speed should be selected as described in Chapter 2.

Design Traffic Volume

The design of urban arterials should be based on traffic data developed for the design year, normally 20 years from the date of construction completion. The DHV is the most reliable traffic volume measure representing the traffic demand for use in design of urban arterials. Sometimes, capacity analysis, which is used to determine whether a particular design can provide a desired level of service for those conditions represented by the design traffic volume, is also used as a design tool. Refer to Chapter 2 for further information on design traffic volumes and capacity analysis.

Levels of Service

For acceptable degrees of congestion, rural and suburban arterials and their auxiliary facilities (i.e., turning lanes, weaving sections, intersections, interchanges, and traffic control systems [traffic signals], etc.), should generally be designed for level-of-service C. In heavily developed sections of metropolitan areas, the use of level-of-service D may be appropriate. When level-of-service D is selected, it may be desirable to consider the use of one-way streets or alternative bypass routes to improve the level of service. For additional guidance on determining the level of service for a specific facility, refer to the *HCM* (**2**).

Sight Distance

The provision of adequate sight distance is important in urban arterial design. Sight distance affects normal operational characteristics, particularly where roadways carry high traffic volumes. The sight distance values given in Exhibit 7-1 are also applicable to urban arterial design. Design values for intersection sight distance are presented in Chapter 9.

Alignment

The alignment of an urban arterial should be developed in strict accordance with its design speed, particularly where a principal arterial is to be constructed on a new location and is not restricted by normal right-of-way constraints. There are many situations, however, where this is not practical. An example of this is the necessity to shift (deflect) the alignment of through lanes to accommodate the inclusion of left turn lanes in an intersection area. Under such circumstances, the intersection alignment should be consistent with the guidance in Chapter 9. It is desirable to use the best alignment design practical since curves on urban arterials are often not superelevated in the low-speed range (see discussion in this chapter in the section below on "Superelevation" for further explanation).

Grades

The grades selected for an urban arterial may have a significant effect on its operational performance. For example, steep grades affect truck speeds and the overall capacity on the facility. On arterials having large numbers of trucks and operating near capacity, flat grades should be considered to avoid undesirable speed reductions. Steep grades may also result in operational problems at intersections, particularly during adverse weather conditions. For these reasons, it is desirable to provide the flattest grades practical while providing at least minimum gradients to ensure adequate longitudinal drainage in curbed sections. The recommended maximum grades for urban arterials are presented in Exhibit 7-10. Where steep grades cannot be flattened, climbing lanes should be considered, based on the warrants presented in Chapter 3.

Superelevation

Curves on low-speed, curbed arterial streets are often not superelevated. Difficulties associated with drainage, ice formation, driveways, pedestrian crossings, and the effect on adjacent developed property should be evaluated when superelevation is considered. The section on "Horizontal Alignment" in Chapter 3 provides a more detailed discussion of superelevation. When little or no superelevation is to be provided on curves for low-speed arterial streets, the Case II distribution of superelevation discussed in Chapter 3 usually is used. Supplemental guidance applicable to both rural and urban arterials is presented in the section on "Superelevated Cross Sections" in the earlier discussion of rural arterials in this chapter.

Type of terrain	Metric						US Customary						
	Maximum grade (%) for specified design speed (km/h)						Maximum grade (%) for specified design speed (mph)						
	50	60	70	80	90	100	30	35	40	45	50	55	60
Level	8	7	6	6	5	5	8	7	7	6	6	5	5
Rolling	9	8	7	7	6	6	9	8	8	7	7	6	6
Mountainous	11	10	9	9	8	8	11	10	10	9	9	8	8

Exhibit 7-10. Maximum Grades for Urban Arterials

Cross Slope

Sufficient cross slope for adequate pavement drainage is important on urban arterials. The typical problems related to splashing and hydroplaning are compounded by heavy traffic volumes and curbed sections, especially for high speeds. Cross slopes should range from 1.5 to 3 percent; the lower portion of this range is appropriate where drainage flow is across a single lane and higher values are appropriate where flow is across several lanes. Even higher cross-slope rates may be used for parking lanes. The overall cross section should provide a smooth appearance without sharp breaks. Because urban arterials are often curbed, it is necessary to provide for longitudinal as well as cross-slope drainage. The use of higher cross-slope rates also reduces flow on the roadway and ponding of water due to pavement irregularities and rutting. The section on "Cross Slopes" in Chapter 4 provides additional guidance.

Vertical Clearances

New or reconstructed structures should provide 4.9-m [16-ft] vertical clearance over the entire roadway width. Existing structures that provide clearance of 4.3 m [14 ft], if allowed by local statute, may be retained. In highly urbanized areas, a minimum clearance of 4.3 m [14 ft] may be provided if there is an alternate route with 4.9-m [16-ft] clearance. Structures should provide additional clearance for future resurfacing of the underpassing road.

Lane Widths

Lane widths may vary from 3.0 to 3.6 m [10 to 12 ft]. Lane widths of 3.0 m [10 ft] may be used in highly restricted areas having little or no truck traffic. Lane widths of 3.3 m [11 ft] are used quite extensively for urban arterial street designs. The 3.6-m [12-ft] lane widths are most desirable and should be used, where practical, on higher speed, free-flowing, principal arterials.

Under interrupted-flow operating conditions at low speeds (70 km/h [45 mph] or less), narrower lane widths are normally adequate and have some advantages. For example, reduced lane widths allow more lanes to be provided in areas with restricted right-of-way and allow shorter pedestrian crossing times because of reduced crossing distances. Arterials with reduced lane widths are also more economical to construct. A 3.3-m [11-ft] lane width is adequate for through lanes, continuous two-way left-turn lanes, and lanes adjacent to a painted median. Left-turn and combination lanes used for parking during off-peak hours and for traffic during peak hours may be 3.0 m [10 ft] in width. If provision for bicyclists is to be made, see the AASHTO *Guide for the Development of Bicycle Facilities* (**6**).

If substantial truck traffic is anticipated, additional lane width may be desirable. The widths needed for all lanes and intersection design controls should be evaluated collectively. For instance, a wider right-hand lane that provides for right turns without encroachment on adjacent lanes may be attained by providing a narrower left-turn lane. Local practice and experience regarding lane widths should also be evaluated.

Curbs and Shoulders

Shoulders are desirable on any highway, and urban arterials are no exception. Where four lanes are warranted, shoulders are desirable. They contribute to enhanced safety by affording maneuver room and providing space for immobilized vehicles. They offer a measure of safety to the occasional pedestrian in sparsely developed areas where sidewalks are not appropriate and provide space for bicyclists where allowed by law. They also serve as speed-change lanes for vehicles turning into driveways and provide storage space for plowed snow.

Despite the many advantages of shoulders on arterial streets, their use is generally limited due to restricted right-of-way and the necessity of using the available right-of-way for traffic lanes. Where the abutting property is used for commercial purposes or consists of high-density residential development, a shoulder, if provided, is subject to such heavy use in serving local traffic that the pavement strength of the shoulder must be about the same as that for the travel lanes. In urban and suburban areas, the outside edges of shoulders are often curbed and a closed drainage system provided to minimize the amount of right-of-way needed. In addition, curbs are often appropriate in heavily developed areas as a means of controlling access.

In those instances where sufficient right-of-way exists to consider shoulders, refer to the discussion on shoulders in the rural arterial section of this chapter for guidance. Where provision of shoulders is not practical, and curbs are to be used, refer to the section on "Curb Placement" in Chapter 4.

Number of Lanes

The number of lanes varies, depending on traffic demand and availability of the right-of-way, but the normal range for urban arterial streets is four to eight lanes in both directions of travel combined. A capacity analysis should be performed to determine the proper number of lanes. In addition, roadways are sometimes widened through intersections by the addition of one or two lanes to accommodate turning vehicles. Chapter 2 presents additional information on capacity analysis.

Width of Roadway

The roadway width should be adequate to accommodate through and turning traffic lanes, medians, curbs, and appropriate clearances from curb or barrier faces. Parking on arterial streets should only be considered when needed because of existing conditions. When parking lanes are provided, consideration should be given to providing a width adequate to allow ultimate operation as a traffic lane. In many instances at intersections, the parking lane is used to provide a right-turn lane or used as a through lane in order to provide additional width for a left-turn lane.

Medians

Medians are a desirable feature of arterial streets and should be provided where space permits. Medians and median barriers are discussed in Chapter 4. Where right-of-way is limited, it is frequently necessary to determine how best to allocate the available space between border areas, traveled way, and medians. On the lower volume arterials, the decision is often resolved in favor of no median at all. However, a median 1.2 m [4 ft] wide is better than none, and it should be noted that any additional median width provides an added increment of safety and improved operation between intersections.

At intersections in urban and suburban areas, median widths should be limited, whenever practical, to those widths needed to accommodate appropriate left-turn treatments for current and future traffic volumes. At intersections where left turns are made, a left-turn lane is always desirable to increase capacity and safety. To accommodate left-turn movements, the median should be at least 3.6 m [12 ft] wide. Desirably, the median should be at least 5.4 m [18 ft] wide to provide for a 3.6-m [12-ft] turning lane and a 1.8-m [6-ft] medial separator between the turning lane and the opposing traffic lane. At restricted locations, a 3.0-m [10-ft] lane with a 0.6-m [2-ft] medial separator may be used. Refer to Chapter 9 for additional guidance concerning provision of dual left-turn lanes and other special intersection treatments.

Exhibit 7-8 presents various configurations for medians that may be used on urban arterials. The type of median treatment used is usually dependent on local practice and available right-of-way widths. The median type selected should be compatible with the needs of drainage and street hardware.

Median openings on divided highways with depressed or raised curbed medians should be carefully considered. Such openings should only be provided for street intersections or for major developments. Spacing between median openings should be adequate to allow for introduction of left-turn lanes.

Where intersections are relatively infrequent (e.g., 1.0 km [0.5 mi] or more apart), the median width may be varied by using a narrow width between intersections for economy and then gradually widening the median on the intersection approaches to accommodate left-turn lanes. This solution is rarely practical, however, and should generally not be used where intersections are closely spaced because the curved alignment of the lane lines may result in excessive maneuvering by drivers to stay within the through lanes. It is far more desirable that the median be of uniform width. Where a narrow median is provided on a high-speed facility, consideration should be given to inclusion of a median barrier. Refer to the AASHTO *Roadside Design Guide* (**3**) for guidance on use and placement of median barriers.

For a street with an odd number of lanes, typically three or five, the center lane is often used to provide a deceleration and storage lane for left-turning vehicles. Left-turn bays are marked in advance of intersections. The center lane between left-turn bays is typically used for vehicles making midblock left turns. In some cases, the center lane is designated for "Left-Turn Only" from either direction, commonly referred to as two-way left-turn lane (TWLTL) design, without specially marked bays at minor intersections. This type of operation works well where the speed on the arterial highway is relatively low (40 to 70 km/h [25 to 45 mph]) and there are no heavy concentrations of left-turning traffic.

Where an arterial passes through a developed area having numerous cross streets and driveways, and where it is impractical to limit left turns, the two-way left-turn lane is often the only practical solution. Because left-turning vehicles are provided a separate space to slow and wait for gaps in traffic, the interference to traffic in through lanes is minimized. Continuous two-way left-turn lanes should be identified by lane and arrow markings placed in accordance with the MUTCD (**1**). Exhibit 7-11 shows an example of a two-way left-turn lane. For further information, see the section on "Medians" in Chapter 4 and the section on "Continuous Left-Turn Lanes" in Chapter 9.

A raised curbed median may be used on low-speed urban arterial streets. This median type is used where it is desirable to manage access along an arterial street by preventing midblock left turns. Raised curbed medians provide a refuge for pedestrians and a good location for signs and other appurtenances. In snow-belt areas, raised curbed medians provide positive delineation, in contrast to flush medians that may become indiscernible under the lightest of snowfall conditions.

However, raised curbed medians also present disadvantages that should be considered. On streets serving high-speed traffic, a raised curbed median does not prevent pedestrian or cross-median crashes unless a median barrier is provided. If accidentally struck, the raised curb may cause drivers to lose control of their vehicles. Also, such medians can be difficult to see at night without appropriate fixed-source lighting or proper delineation. Additionally, raised curb medians cast a shadow from oncoming headlights that not only makes the curb difficult to see but also places some of the adjacent lane in shadow. In some cases, the prevention of midblock left turns may cause operational problems at intersections because of increased concentrations of left-turning traffic or motorists making U-turns.

Exhibit 7-11. Continuous Two-Way Left-Turn Lane

The foregoing disadvantages of raised curbed medians can be largely eliminated by use of flush medians or low-profile sloped curbs. However, flush medians are difficult to see under wet nighttime conditions and may become indiscernible under the lightest of snowfall conditions. Visibility of flush medians can be improved by use of a contrasting pavement texture and by improved delineation, such as the use of reflectorized pavement markers. The use of raised bars or blocks has proven to be an ineffective median treatment and should not be considered.

When a two-lane suburban arterial is proposed for improvement to a multilane facility with a median, access management principles suggest that a raised curbed median is more desirable than a flush median. The limiting of left-turns except at intersections discourages uncontrolled development and access to the highway and promotes improved traffic operations.

Special consideration should be given to the median width where intersections are provided. Research in NCHRP Report 375 (**4**) found that most types of undesirable driving behavior in the median area of divided highway intersections are associated with competition for space by vehicles traveling through the median in the same direction. The potential for such problems is generally greater at urban and suburban rather than at rural intersections, where volumes of turning and crossing traffic are lower. Types of undesirable driving behavior observed include side-by-side queuing, angle stopping, and encroaching on through lanes of a divided highway. At urban and suburban unsignalized intersections, the frequency of crashes and undesirable driving behavior were observed to increase as the median width increased. Thus, medians at urban and suburban unsignalized intersections should not be wider than necessary. This trend is opposite to that observed at rural unsignalized intersections.

The median at an urban or suburban unsignalized intersection should be wide enough to accommodate the left-turn treatment selected by the designer. As at rural intersections, the appropriate design vehicle for turning and crossing maneuvers, based on the vehicle mix for crossroad and U-turn traffic, should be considered in determining the median width. At urban and suburban unsignalized intersections, medians less than 7.5 [25 ft] wide can be used effectively to discourage drivers from stopping in the median and multiple vehicles traveling in the same direction from using the median at the same time. At locations with substantial crossing and turning volumes of larger vehicles, such as school buses or trucks, it may be appropriate to provide enough width to store such vehicles in the median without encroaching on the through lanes of the major road.

Urban and suburban unsignalized intersections with median widths from 9 to 15 m [30 to 50 ft] generally experience slightly higher crash rates than intersections with narrower medians but appear to operate quite well (**4**). Such intersections apparently are within the realm of normal operational expectations of the driver. However, urban and suburban intersections with medians wider than 15 m [50 ft] have more crashes, and intersections with medians wider than 18 m [60 ft] are difficult to signalize properly.

Experience indicates that at unsignalized intersections drivers on divided arterials in urban areas prefer medians that are obviously narrow or those that provide an adequate refuge area to allow independent crossing of the two roadways of a divided highway.

At urban and suburban signalized intersections, crash frequency increases as the median width increases (**4**). Therefore, median widths at urban and suburban signalized intersections should not be wider than necessary and should be determined primarily by the space needed in the median for current or future left-turn treatments. Median widths of more than 18 m [60 ft] are undesirable at intersections that are signalized or that may need signalization in the foreseeable future. The efficiency of signal operations decreases as the median width increases, because drivers need more time to traverse the median and special detectors may be needed to avoid trapping drivers in the median at the end of the green phase for traffic movements passing through the median. Furthermore, if the median becomes so wide that separate signals are needed on the two roadways of the divided highway, delays to motorists will increase substantially. Careful attention should be given to vehicle storage needs in the median area between the two signals.

Uncurbed, unpaved narrow medians often present problems for turning movements at intersections in that vehicles tend to run off the roadway edges. To minimize this problem, the provision of edge lines and sufficient paved area beyond the edge lines will provide positive guidance and will accommodate the turning paths of passenger cars and occasional large vehicles.

A median barrier may be desirable on some arterial streets with fast-moving traffic. A barrier provides a positive separation of traffic and discourages indiscriminate pedestrian crossings. Where the median barrier is terminated at cross streets and other median openings, it should have a crashworthy terminal or terminal end appropriate for the speed of traffic. Further discussion on treatment of the ends of barriers is presented in the *Roadside Design Guide* (**3**). Additional information on median barriers and median treatments at intersection areas is found in Chapters 4 and 9, respectively. The information on medians and median barriers in Chapter 4 is especially pertinent to urban arterials since they need the most varied application of these features.

Drainage

An adequate drainage system to accommodate design runoff should be included in the design of every arterial street. Inlets that are safe for bicycles should be located adjacent to and upstream of intersections and at intermediate locations where necessary. Where a shoulder or parking lane is provided, the full width of the shoulder or parking lane may be utilized to conduct surface water to the drainage inlets. Where no shoulder or parking lane is provided, one-half of the outside traffic lane and curb offset may be utilized to conduct surface drainage, provided two or more traffic lanes exist in each direction. Ponding of water at low points in the traveled way on arterial streets is undesirable. The width of water spread on the roadway should not be substantially greater than the width of spread encountered on continuous grades. Highways with design speeds greater than 70 km/h [45 mph] will have a higher potential for hydroplaning if the traveled way is covered with water. Additional inlets should be provided in sag locations to avoid ponding of water where the grade flattens to zero percent and to mitigate flooding should an inlet become clogged. Chapters 3 and 4 have comprehensive discussions concerning drainage.

Parking Lanes

Where parking is needed and adequate off-street parking facilities are not available or practical, parallel parking may be considered as long as adequate capacity is provided by the through lanes. However, parking is highly undesirable on high-speed roadways.

Passenger vehicles parked adjacent to a curb will occupy, on the average, approximately 2.1 m [7 ft] of street width. Therefore, the total parking lane width for passenger cars should be 3.0 to 3.6 m [10 to 12 ft]. This width is also adequate for an occasional parked commercial vehicle. For desirable widths to accommodate usage by bicyclists, refer to the AASHTO *Guide for the Development of Bicycle Facilities* (**6**). Where it is unlikely that there will be a future need to use the parking lane as a through lane, a parking lane width as narrow as 2.4 m [8 ft] may be acceptable.

A parking lane less than 3.3 m [11 ft] in width is considered undesirable if future use of the parking lane for through traffic is anticipated. Such a lane can be used as an additional through-traffic lane during peak hours by prohibiting parking during these hours. A parking lane 3.0 m [10 ft] in width is acceptable for use as a storage lane for turning vehicles at signalized intersections by prohibiting parking for some distance upstream from the intersection. A parking lane of 2.7 m [9 ft] may be acceptable as a storage lane for turning vehicles where the design speed on the arterial is 60 km/h [40 mph] or less.

The marking of parking spaces on arterial streets encourages more orderly and efficient use where parking turnover is substantial and also tends to prevent encroachment on fire hydrant zones, bus stops, loading zones, approaches to corners, clearance space for islands, and other zones where parking is prohibited. Typical parking-space markings are shown in the MUTCD (**1**).

In downtown districts and in areas with large office or industrial buildings, it may be possible to provide parking turnouts as shown in Exhibit 7-12. The paved area of the turnout is 7.2 m [24 ft] wide, a curbed island with entrance and exit tapers separates the through lanes from the turnout.

Exhibit 7-12. Parking Turnouts in Downtown District

Borders and Sidewalks

The border is the area between the roadway and the right-of-way line that separates traffic from adjacent homes and businesses. For a minimum section in a residential area, the border area should include a sidewalk and a buffer strip between the sidewalk and curb. Exhibit 7-13 illustrates an arterial street in a residential area. The exhibit shows curbs, a parking lane, curb cuts for driveways, and sidewalks. In blocks that are fully developed with retail stores and offices, the entire border area usually is devoted to sidewalk.

Some factors to be considered in determining border widths are width of sidewalk for pedestrian needs, snow storage, storm drainage, traffic control devices, roadside appurtenances, and utilities. The minimum border should be 2.4 m [8 ft] wide and preferably 3.6 m [12 ft] or more. Every effort should be made to provide wide borders not only to serve functional needs but also as a matter of aesthetics, safety, and reducing the nuisance of traffic to adjacent development. Where sidewalks are not included as a part of the initial construction, the border should be sufficiently wide to provide for their future installation. For further information, see the section on "Pedestrian Facilities" presented later in this chapter.

Exhibit 7-13. Arterial Street in Residential Area

Where bicycle traffic is anticipated or is to be served on arterial streets, provisions to accommodate bicycles should be in accordance with the AASHTO *Guide for the Development of Bicycle Facilities* (**6**).

Exhibit 7-14 illustrates a divided arterial street in a residential area. This type of arterial features a turf buffer strip that is provided between the sidewalk and the curb. In addition, vertical-curb and gutter sections are employed on the outside of parking lanes that may also serve as shoulders.

Railroad-Highway Crossings

Railroad-highway crossings on an urban arterial can often be the most disruptive feature affecting its operation. Crossings that are frequently occupied or occupied during high-volume traffic periods should be treated by providing a grade separation. Crossings that are occupied only infrequently or during off-peak traffic periods may be operated as an at grade crossing with high-type traffic control, such as gate-equipped automatic flashing signals.

At-grade crossings that involve bicycle routes that are not perpendicular to the railroad may need additional paved shoulder width to allow bicyclists to maneuver over the crossing. For further information, see the AASHTO *Guide for the Development of Bicycle Facilities* (**6**).

Exhibit 7-14. Divided Arterial Street With Parking Lanes

Roadway Width for Bridges

The minimum clear width for new bridges on arterial streets should be the same as the curb-to-curb width of the street. On long bridges, defined as bridges with overall lengths in excess of 60 m [200 ft], the offsets to parapets, rails, or barriers may be reduced to 1.2 m [4 ft] where shoulders or parking lanes are provided on the arterial. For further relevant discussion, see the sections on "Curbs," "Sidewalks," "Traffic Barriers," and "Bridge Railings" in Chapter 4.

Bridges to Remain in Place

Reasonable attempts should be made to improve existing structures that do not meet current design policies or guidelines, but are otherwise suitable for retention. When making this decision, an important consideration is the extent to which such features that do not meet current policies and guidelines are likely to contribute to crash frequency and operational deficiencies. Other factors to be considered include the remaining life, the cost of improvements and/or rehabilitation compared to replacement, and the historical significance, aesthetic value, and notoriety of the structure.

Horizontal Clearance to Obstructions

Clear roadside design is recommended for urban arterials whenever practical. On curbed street sections, clear roadsides are often impractical, particularly in restricted areas. In such areas, a clearance between curb face and object of 0.5 m [1.5 ft] (or wider where practical) should be provided. A 1.0-m [3-ft] clearance to roadside objects should be provided particularly near turning radii at intersections and driveways. This offset provides sufficient clearance to keep the overhang of a truck from striking an object. Where pedestrians are not a factor, obstructions should be set well back, protected, or provided with breakaway features. For further guidance, refer to the AASHTO *Roadside Design Guide* (**3**).

Right-of-Way Width

The width of right-of-way for the complete development of an arterial street is influenced by traffic demands, topography, land use, cost, intersection design, and the extent of ultimate expansion. The width of right-of-way should be the summation of the various cross-sectional elements: through traveled ways, medians, auxiliary lanes, shoulders, borders, and, where appropriate, frontage roads, roadside clear zones, sideslopes, drainage facilities, utility appurtenances, and retaining walls. The width of right-of-way should be based on the preferable dimensions of each element to the extent practical in developed areas. The designer is confronted with the problem of providing an overall cross section that will give maximum service within a limited width of right-of-way. Right-of-way widths in urban areas are governed primarily by economic considerations, physical obstructions, or environmental concerns. Along any arterial route, conditions of development and terrain vary, and accordingly, the availability of right-of-way varies. For this reason, the right-of-way on a given facility should not be a fixed width predetermined on the basis of the most critical point along the facility. Instead, every opportunity should be taken to provide a desirable right-of-way width along most, if not all, of the facility.

Traffic Barriers

Traffic barriers are sometimes used on urban arterials in restricted areas, at separations, and in medians. The barrier should be compatible with the desired visual quality and should be installed in accordance with accepted practice. Exposed ends should be treated with crashworthy designs or other appropriate means. For further information, refer to the AASHTO *Roadside Design Guide* (**3**).

Access Management

General Features

Partial control of access and the application of access management techniques are highly desirable on an urban or suburban arterial. Effective access management will not only enhance the initial level of service of a facility but may also preserve that original level of service as further development occurs. While access to abutting property may be required, it should be carefully regulated to limit the number of access points and their locations. Access management is especially important on intersection approaches on both the arterial and cross streets where auxiliary and storage lanes may be needed.

Access control and access management may be exercised by statute or through application of zoning ordinances, driveway regulations, turning and parking regulations, and effective geometric highway design. Implementation of any of these options should involve coordination with the community and adjacent property owners. For additional discussion on access control and access management, refer to Chapter 2.

Access Control by Statute

Where a high degree of access control is desired, it is usually accomplished by statute. When statutory control is applied to an arterial street, access is usually limited to the cross streets or to other major traffic generators.

Access Control by Zoning

Zoning can be used effectively to control the type of property development along an arterial and thereby influence the type and volume of traffic generated. Property uses can be limited to those that attract very few people, excluding those uses that generate significant volumes of traffic during peak hours of through-traffic movements. In certain cases, it may be desirable to exclude land uses that generate heavy volumes of commercial traffic if, for various reasons, this class of vehicle cannot be accommodated readily due to limitations in the highway geometrics.

Zoning regulations should require ample off-street parking as a condition for approval of a building permit. Also, the internal arrangement of the land-use development should be such that the parking spaces most distant from the arterial street are the most attractive to the user. This type of internal design minimizes congestion in the vicinity of the entrance at the street. Vehicles exiting from the parking facility to the arterial (or preferably to a cross street) should not impede traffic entering the parking facility from the arterial.

Subdivision or zoning ordinances should require that the developer of a major traffic generator provide a suitable connection to the arterial street (or preferably to a cross street) comparable to that for a well-designed street intersection serving a similar volume of traffic. If direct access to the arterial is provided, it should be understood that the intersection is subject to the same traffic control measures, including restrictions to turning movements, as are applicable elsewhere on the arterial. The developer may be required to provide a frontage road along the property that connects the major entrances to the arterial in order to maintain a high level of operation and safety on the arterial.

Access Control Through Driveway Regulations

Driveway controls can be effective in preserving the functional character of arterial streets. In heavily built-up areas and areas with potential for intensive development, permits for driveways and entrances can be controlled to minimize interference with the free flow of traffic on the arterial. In more sparsely developed areas with little potential for dense development, driveway controls are also desirable to ensure that future driveways are located so that there will continue to be minimum interference with the free movement of traffic.

Access Control Through Geometric Design

Left turns in and out of adjacent properties can have a great effect on the safety and operation of an arterial. Such movements can be prohibited by constructing a raised curbed median or by installing a median barrier. Left turns can be accommodated by U-turns at intersections, jug-handle configurations, or around-the-block movements. The operational and safety effects of relocating mid-block turns to these alternative locations should be carefully considered when evaluating this option. Additional information concerning the effects of mid-block left-turn lanes can be found in NCHRP Report 395, *Capacity and Operational Effects of Midblock Left-Turn Lanes* (**8**). Right-turn-in and right-turn-out arrangements are another important geometric design feature to control access to an arterial.

Frontage roads and grade separations provide the ultimate in access control. Fully developed frontage roads effectively control access to through lanes on an arterial street while providing access to adjoining property, separating local from through traffic, and permitting circulation of traffic along each side of the arterial. When used in conjunction with grade-separation structures at major cross streets, an arterial takes on many of the operating characteristics of a freeway.

Due to right-of-way restrictions, frontage roads are usually located immediately adjacent to the arterial. For this reason, careful attention should be given to proper signing to minimize the potential for wrong-way entry into the through lanes of the arterial. Efforts should be made to provide adequate storage distance for turning vehicles on the crossroad between the frontage road and the arterial, although this is often difficult because of restricted right-of-way width. If signalization is needed at the intersection of the crossroad and the frontage road, the operation of this signal should be coordinated with the signal at the intersection of the crossroad and the arterial.

General features of frontage roads and their design are discussed in Chapter 4. The effect of frontage roads on the design of intersections is addressed in the section on "Intersection Design Elements With Frontage Roads" in Chapter 9. Additional information concerning access management can be found in NCHRP Report 420, *Impacts of Access-Management Techniques* (**5**).

Pedestrian Facilities

Arterial streets may accommodate both vehicles and pedestrians; therefore, the design should include sidewalks, crosswalks, and sometimes grade separations for pedestrians. Pedestrian facilities and control measures will vary, depending largely on the volume of pedestrian traffic, volume of vehicular traffic to be crossed, number of lanes to be crossed, and number of vehicles turning at intersections.

On some sections of arterial streets that traverse relatively undeveloped areas, no initial pedestrian demand may be present. Therefore, sidewalks may not be needed initially. Because these areas will usually be developed in the future, the design should allow for the ultimate installation of sidewalks. However, as a general practice, sidewalks should be constructed initially along all arterial streets that are not provided with shoulders, even through pedestrian traffic may be light.

The major pedestrian-vehicular conflict usually occurs at intersections. On the lower classes of arterials, especially at intersections with minor cross streets where turning movements are light, pedestrian facilities are usually limited to crosswalk markings. Features that help the pedestrian include fixed-source lighting, refuge islands, barriers, and signals. Such features are discussed in Chapter 4.

On the higher-volume arterials (i.e., six or eight lanes wide with heavy traffic volumes), the interference between pedestrians and vehicles at intersections sometimes presents a serious problem. The problem is especially acute where the arterial traverses a business district and there are intersections with higher-volume cross streets. Although grade separations for pedestrians are justified in some instances, crosswalks are the predominant form of crossing. Conflicts between pedestrians and vehicular traffic can be reduced by shortening pedestrian crossing distances by various means such as curb bulbs or narrower lanes, restricting left and/or right turns and separate pedestrian signal phases. The accommodation of pedestrians can have an effect on the capacity of intersections and should be evaluated during design.

The number of pedestrian crossings on heavily traveled arterials should be kept to a minimum, but in and near developed areas it is usually appropriate to provide crosswalks at every intersecting street. Enforcement of a ban on pedestrian crossings at an intersection is very difficult. A crossing should not be closed to pedestrians unless the benefits in improved safety and traffic operation are sufficient to offset the inconvenience to pedestrians. In addition, indiscriminate closing of pedestrian crossings will lead to illegal crossing maneuvers. Therefore, proper and reasonable design for pedestrians is important.

The pedestrian walk signal is especially desirable on wide arterials having frontage roads because of the great distances to be traversed. On exceptionally wide arterial streets, pedestrian signals may be mounted in the median as well as on the far side of the intersection and, where appropriate, in the outer separators. Refer to the current MUTCD (1) for additional information concerning installation of signals.

Where intersections are channelized or a median is provided, consideration should be given to the use of curbing for those areas likely to be used by pedestrians for refuge when crossing the roadway. For design speeds less than 80 km/h [50 mph], a 150-mm [6-in] sloping curb is the preferred type. For design speeds of 80 km/h [50 mph] or greater, a 100-mm [4-in] sloping curb is preferred. In all cases, the curb offset should be consistent with the design criteria in Chapter 4.

The use of crosswalks at typical curbed-street intersections may be difficult for persons with disabilities. Curb ramps of appropriate width and slope must be provided in curbed areas that have sidewalks. Sidewalk curb ramps are addressed in Chapter 4.

For further guidance on the accommodation of pedestrians, refer to the AASHTO *Guide for the Planning, Design, and Operation of Pedestrian Facilities* (**9**).

Provision for Utilities

The urban arterial system often serves as a utility corridor. Desirably utilities are located underground or at the outer edge of the right-of-way. In addition, poles should be located as near the right-of-way lines as practical. Whenever practical, service access openings and covers should not be located in the traveled way but should desirably be placed outside the entire roadway. However, locations in the medians or parking lanes may be acceptable under special conditions. Utilities should seldom be added to an arterial by the open-cut method. Additional installations should be bored or jacked to avoid interference with normal traffic movements.

Intersection Design

The design and operation of intersections have a significant effect on the operational quality of an arterial. Intersection and stopping sight distance, pedestrian movements, capacity, grades, and provision for turning movements all affect intersection operation. Turning movements onto arterials and crossroads should not encroach on adjacent lanes; where avoiding encroachments on adjacent lanes is not practical, the effects of such encroachments should be considered. It is recommended that each individual intersection be carefully evaluated in the early design phases. Chapter 9 discusses intersection development in detail.

Operational Control and Regulations

The efficiency of an arterial street system is strongly influenced by the adequacy of traffic control devices and the degree of enforcement of traffic regulations. If the demand for arterial route service exists within the traffic corridor, traffic control devices and traffic regulations are of such importance that it may be possible to convert a street with only moderately good design into a major traffic artery.

The potential of traffic control measures to improve capacity and level of service should be exploited to the maximum degree on properly designed arterial streets. Improvements to the arterial system may help to relieve congestion on the local street system by diverting traffic to the higher speed arterial. Traffic control measures may be divided into the following categories: (1) traffic control devices, (2) regulatory measures, and (3) directional lane usage.

Traffic Control Devices

Where traffic signals are anticipated during the initial planning of an arterial street, intersection design should integrate the ultimate signal operation. The design should consider the reduction of signal phases by providing for concurrent opposing left-turn phases and by constructing left-turn lanes in a manner that will allow their free operation. Channelization, which provides for single or double left turns and free-flow right turns, often results in better signal control and may assist pedestrian crossings. However, multiple lane shifts to accommodate the installation of turn lanes should be designed in accordance with Chapter 9.

Signal spacing to allow free-flow timing in both directions of travel is highly desirable and may be achieved by controlling intersection locations during early development stages. If this cannot be achieved, suitable time-space diagrams based on traffic forecasts may be used to determine signal timing and spacing for major access points. Such efforts will allow optimum signal progression to provide maximum vehicle capacity and minimum vehicle delay time. Driveway location also affects signal operation significantly. In addition, driveway locations that allow interference with major through movements should be avoided. Locations at which entering or exiting vehicles will generate false calls to the signal are also undesirable. The physical location of signal supports can often be improved for safety and signal visibility reasons when an intersection is considered in the roadway design stage.

The ultimate goal of any intersection design should be to serve the traffic demands safely at a level of service consistent with the overall arterial design. This goal requires that all intersection elements, including traffic signals, be integrated into all aspects of the design process. Traffic control devices such as signs, markings, signals, and islands are placed on or adjacent to an arterial to regulate, warn, or guide traffic. Each device is designed to fulfill a specific need with respect to traffic operation, control, or safety. The need for traffic control devices should be determined by an engineering study conducted in conjunction with the geometric design of the street or highway. To assure uniform design and installation application of the various traffic control devices, refer to the current MUTCD (1).

The importance of a signal system that is responsive to traffic demands cannot be overemphasized. Traffic demands fluctuate from hour to hour and from day to day. The signal system should likewise be flexible. The system should include detection and data processing components as well as a control component. The installation of a sophisticated signal system is generally the most economical method to improve traffic service in a densely developed area.

Among other considerations, signal systems should allocate green time to each movement in accordance with the demand for that movement. Signals should be coordinated for progressive movement in the direction of the predominant flow of traffic on the arterial street, although the optimum traffic service for the arterial system as a whole should also be considered. Where multiple phasing is needed to control various traffic movements at intersections, the detection and programming equipment should provide the capability to skip specific signal phases when there is no traffic demand. Idle and lost green time due to lack of demand or too many phases should be kept to a minimum. Signal equipment to accomplish this goal is available. Although it is expensive to improve traffic service through use of signal systems, it is usually less costly than purchasing additional right-of-way and constructing additional lanes, which may be the only other alternative to improve the arterial's capacity. Sophisticated traffic signal systems require considerable effort to maintain and operate; therefore, the operating agency should be prepared to assign adequate resources, including capable maintenance staff, to the system if optimum benefits are to be realized.

Progression of through movements on an arterial is one of the most effective means of improving operations on both arterials and cross streets. By moving queues of traffic through the corridor, the corridor remains open, allowing more time for cross-street movements. The selection and operation of the optimum traffic control equipment is related to the design of the arterial.

Successful operation of an urban arterial depends largely on proper pavement marking, especially on arterials having multiple lanes and particularly when special provision is made for left turns. Recent developments in pavement marking materials show considerable promise toward providing effective long-life markings, even for areas where snow removal often obliterates ordinary markings in very short time periods. Overhead lane signing is usually very helpful. Signs enable drivers to make maneuvers well in advance and are especially helpful under adverse weather conditions, such as rain or snow. Adequate crosswalk markings enhance pedestrian safety on urban arterials.

Regulatory Measures

Regulatory measures take many forms, as discussed in the *Uniform Vehicle Code* (**10**) and the *Model Traffic Ordinance* (**11**). The two measures that contribute most to the operational efficiency of arterial streets are restrictions on turning movements and prohibition of curbside parking, stopping, or standing.

Restrictions on turning movements at both midblock and intersection locations promote efficient operation if applied wisely; however, such restrictions should not be applied indiscriminately. Some alternative route must be provided for roadway users to reach their destinations, and this can only be accomplished by providing them with turning opportunities at another location.

Operational and Control Measures for Right-Turn Maneuvers

In most cases, vehicles making right turns from an arterial interfere little with through traffic. However, at intersections where there are heavy pedestrian movements, right-turning vehicles may delay through traffic as they wait for a pedestrian to cross. Where corner curb radii are short, vehicles slowing to turn right may also interfere with through traffic to an extent that is out of proportion to their numbers. In such cases, it may be desirable to prohibit turns during certain hours if there are alternative routes for motorists to reach their destinations.

Operational and Control Measures for Left-Turn Maneuvers

Vehicles turning left into cross streets or at midblock locations may cause substantial delays to through traffic and may contribute to crashes, thus diminishing arterial effectiveness. There is a popular belief that the effects of such left-turn movements can be eliminated simply by placing "No Left Turn" signs. In fact, motorists that desire to turn left do not just disappear, but reach their destinations by alternative routes. Thus, prohibition of left turns at some locations may create or increase operational or safety problems at other locations.

Effective control of turning movements lies in discovering or anticipating the extent of the problem and in providing for the movements through a combination of measures including selective prohibition of turns, geometric design, and traffic control. It is difficult to discuss these factors independently, and no firm rules are applicable to all situations. Several principles and methods that, if properly considered and applied, will lead to appropriate designs are outlined as follows:

1. The capability for motorists to reach their desired destinations must be provided. Left turns should not be prohibited unless alternative routings are available.

2. As a general rule, the fewer the number of left turns at any location, the less the interference with other traffic. Thus, for a given total number of left turns within a given length of highway, it may be better to encourage a few left turns at each of several locations than to concentrate the turns at a single location.

3. Separate signal phases for left-turn movements reduce the amount of green time available for other movements at the intersection. Multiphase signals are therefore advantageous only if traffic operation and safety are improved sufficiently to offset the loss in green time. This determination should be made on a case-by-case basis.

4. Where selective prohibition of left turns is necessary, there are operational advantages in concentrating left turns at intersections where the volume of cross traffic is low so that a large fraction of the signal time is available for the green phase on the arterial. Where two arterial streets intersect, there may be advantage in requiring left-turning vehicles to bypass the main intersection. For instance, one manner in which the left-turn maneuver from an arterial to another can be accomplished is to require the motorist to turn left from the first arterial a block in advance of the main intersection, then proceed one block, turn right, proceed another block, and turn left. Where such techniques are used, clear guide signing is essential.

5. It is sometimes advantageous to route left-turning traffic around a block, through a series of right turns after passing through the main intersection, rather than permitting a direct left-turn maneuver. However, this approach has disadvantages as well. Traffic volumes are increased because the left-turning vehicle now must pass through the intersection twice. In addition, the distance of travel by the vehicle that desires to turn left is increased, and the increased right-turn volumes may have an impact on the operation of three other intersections. This approach to left-turn maneuvers should generally be limited to locations where the left-turn volumes are small and the provision of a separate left-turn lane is not practical.

6. The desirability of exclusive left-turn lanes cannot be overemphasized. Such lanes may consist of separate left-turn lanes in the median or continuous center lanes used exclusively for left turns from both directions. Multiphase signal control is very inefficient if turning traffic and through traffic both use the same lane. Where turning traffic is light, a left-turn lane may eliminate the need for left-turn signal phasing because the storage of left-turning vehicles will not affect through traffic. Traffic safety

is greatly enhanced if turning vehicles can be stored separately from lanes used by through vehicles.

7. With a separate left-turn phase, dual left-turn lanes can accommodate up to about 180 percent of the volume that can be served by a single left-turn lane with the same available green time, depending on the width of the cross street and the radius of turn. Desirably, the turning radius for a dual left-turn lane is 27 m [90 ft]. Thus, where sufficient right-of-way, space for a long-radius turn, and a wide cross street are available, the installation of dual left-turn lanes may be a practical design to serve a heavy left-turn movement. Exhibit 7-15 shows an example of dual left-turn lanes at an intersection on an urban arterial. Further guidance concerning the design of dual left-turn lanes is presented in Chapter 9 and in the HCM (**2**).

Exhibit 7-15. Urban Arterial With Dual Left-Turn Lanes

8. Grade separations or other special treatments for left-turn movements are sometimes appropriate, as discussed in Chapter 10.

In summary, left-turn demands should be accommodated as near as practical to the point at which the motorist desires to turn left. Shifting the left-turn maneuvers away from this point of desire may lead to secondary problems. Nevertheless, if the point at which motorists desire to turn left is highly objectionable from the standpoint of design, traffic control, or safety, regulatory measures may be employed to move those left turns to a location that is more suitable. Only in exceptional cases should such maneuvers be shifted more than two blocks from the point of desire. Where left turns are permitted from an arterial street, the intersection design should incorporate left-turn storage lanes unless it is impractical to provide them.

Regulation of Curb Parking

Curb parking is permissible on arterials when speeds are low and traffic demand is well below capacity. At higher speeds and during periods of heavy traffic movement, curb parking is incompatible with arterial street service and should not be permitted.

Curb parking reduces capacity and interferes with free flow of adjacent traffic. Eliminating curb parking can increase the capacity of arterials with four- or six-lane curb-to-curb widths by 50 to 80 percent.

Where parking provisions are included in the design, cross-sectional dimensions should be such that the entire width can be used by moving traffic when parking is removed. At intersections, there should be a liberal distance from the corner of the intersection to the nearest parking stall. This distance should be at least 6.0 m [20 ft] from a crosswalk. This provides extra maneuvering space for turning traffic, reduces the conflict with through traffic, and increases sight distance.

No other single operational control can have as dramatic an effect on traffic flow on arterial streets as the proper regulation of parking, including the enforcement of "No Parking" regulations in loading zones and at bus stops.

Directional Lane Usage

Typically, the conventional arterial street is a multilane two-way facility with an equal number of lanes for traffic in each direction of travel. Often, however, one-way operation is employed where conditions are suitable. Somewhat less frequently, reversible lane operation is used to improve operational efficiency. The conditions under which each form of operation is most suitable depend largely on traffic flow characteristics, the street pattern, and the geometric features of the particular street. Where a street system is undergoing expansion or improvement, the ultimate form of directional usage should be anticipated, and the design should be prepared accordingly. Once an arterial street is completed, conversion from one form of directional usage to another may involve considerable expense and disruption to traffic. For existing streets of conventional design, this conversion may be a practical alternative for increasing traffic capacity. For information concerning signing for directional lane usage, refer to the current MUTCD (**1**).

One-way operation. An arterial facility consisting of one or more pairs of one-way streets can generally be employed where the following conditions exist: (1) a single two-way street does not have adequate capacity and does not lend itself readily to improvement to accommodate anticipated traffic demand, particularly where left-turning movements at numerous intersections are difficult to handle, (2) there are two parallel arterial streets a block or two apart, and (3) there are a sufficient number of cross streets and appropriate spacing to permit circulation of traffic.

One-way streets have the following advantages:

- Traffic capacity may be increased as a result of reduced mid-block and intersection conflicts and more efficient operation of traffic control devices.

- Travel speed is increased as a result of reducing mid-block conflicts and delays caused by slowing or stopped left-turning vehicles. The increase in the number of lanes in one direction permits ready passing of slow-moving vehicles. One-way operation permits good progressive timing of traffic signals.

- The number and severity of crashes is reduced by eliminating head-on crashes and reducing some types of intersection conflicts.

- Traffic capacity may be increased by providing an additional lane for through traffic. Although a two-way street with only one lane in each direction may not have sufficient width to accommodate two lanes in each direction, it may have sufficient width to accommodate three lanes in one direction when converted to one-way operation.

- The available street width is used fully through the elimination of need for a median.

- On-street parking that would otherwise have to be eliminated or curtailed may be retained.

Disadvantages to one-way operation are:

- Travel distances are increased because certain destinations can be reached only by around-the-block maneuvers. This disadvantage is more acute if the street grid is composed entirely of one-way streets.

- One-way streets may be confusing to strangers.

- Emergency vehicles may be blocked by cars in all lanes at intersections waiting for signals to change.

In summary, there are several advantages and disadvantages to one-way operation. The choice of one- or two-way operation depends largely on which type of operation can serve the traffic demands most economically and with greatest benefit to the adjacent property. Both types of operation should be considered. In many cases, the proper choice is immediately obvious. In other instances, a thorough study involving all relevant considerations may be needed.

Reverse-flow operation. The familiar imbalance in directional distribution of traffic during peak hours on principal radial streets in large and medium-sized cities often results in congestion in the direction of heavier flow and excess capacity for opposing traffic. Capacity during peak hours can be increased by using more than half of the lanes for the peak direction of travel.

Reverse-flow operation on undivided streets generally is justified where 65 percent or more of the traffic moves in one direction during peak periods, where the remaining lanes for the lighter flow are adequate for that traffic, where there is continuity in the route and width of street, where there is no median, and where left turns and parking can be restricted. Refer to the AASHTO *Guide for the Design of High Occupancy Vehicle Facilities* (**12**) for additional guidance concerning the appropriateness of reverse-flow operation.

The conventional undivided street need not be changed appreciably for conversion to reverse-flow operation, and the cost of additional control measures is not great. On a five-lane street, three lanes can be operated in the direction of heavier flow. On a six-lane street with directional distribution of approximately 65 to 35 percent, four lanes can be operated inbound and two lanes outbound during the morning peak. The assignment of the center lanes can be reversed during the evening peak so that two lanes are generated inbound and four lanes outbound. During off-peak periods, traffic is accommodated on three lanes in each direction or on two lanes in each direction with curb parking.

Streets with three or four lanes can also be operated with a reverse flow. However, with only one lane in the direction of lighter flow, a slow vehicle or one picking up or discharging a passenger delays all traffic in that direction of travel, and a vehicle breakdown blocks traffic in that direction completely. Occasionally, circumstances may be such that such streets can be adapted to complete reverse flow (i.e., one-way inbound in the morning and one-way outbound in the evening). At other times the street may be operated as a two-way street, with or without parking.

Direct left turns by traffic in the off-peak direction on a two-way reversible street should be carefully controlled. Left turns from the predominant flow are subject to the same considerations and regulations as they are for conventional operation with two-way traffic. By contrast, on a one-way reversible street, left turns at all intersections can be made readily.

Reverse-flow operation requires special signing and additional control devices. More policing and staffing are also needed to operate the control devices. Stanchions or traffic cones are usually desirable to separate opposing traffic, and "No Left Turn" and "Keep Right" signs on pedestals or stanchions are often used.

Assigning traffic to proper lanes can be accomplished by placing overhead signs indicating lane usage for specific times of day. These signs should be supplemented with traffic control signals located directly over each lane indicating when reversible lanes are open or closed to traffic in the specified direction. This is usually accomplished with a signal head displaying a red "X" for closed or a green directional arrow for open. Refer to the MUTCD (**1**) for further guidance. This combination of signs and signals will decrease the potential for motorists to pull out for left turns into a lane that is signed for traffic in the opposite direction. It is better to place separate lane-use control signals at intervals over each lane. This method is particularly adaptable to long bridges and sections of streets without side connections.

Further efficiency can be gained for the predominant direction of travel by progressive timing of signals. With a three-dial interconnected signal system, signals are set for proper progression of the major movement in the peak periods. A third setting is used for the traffic flow during off-peak periods. In some cases, the signals for the center lane or lanes are set red in both directions during off-peak hours, thus converting the unused traveled way into a median area for safety.

Reverse-flow operation on a divided facility is termed "contra-flow operation." While the principle of reverse-flow operation is applicable to divided arterials, the arrangement is more difficult than on an undivided roadway. The difficulty of handling cross and turning traffic, the potential confusion for pedestrians, and the potential for conflicts between opposing vehicles at high volumes may make other arrangements preferable to contra-flow operation. For example, the capacity of an undivided arterial with a reverse-flow lane allocation of three-two-three or three-three-three lanes (equivalent to the peak directional capacity of 10- or 12-lane conventional sections, respectively) may be comparable to the capacity of a six-lane freeway. For these widths, likely volumes would be 3,500 to 4,000 vehicles per hour in one direction, or two-way ADT volumes of 50,000 to 60,000 vehicles per day, for which a freeway is warranted. Furthermore, traffic flows that are currently unbalanced may not remain unbalanced in future years. Reverse-flow operation for at-grade facilities is applicable chiefly as a means of increasing capacity on existing highways.

Frontage Roads and Outer Separations

Frontage roads are sometimes used on arterial streets to control access, as discussed in the preceding section on "Access Control Through Geometric Design" in this chapter. Other important functions of frontage roads are minimizing interference with operations on the through-traffic lanes while still providing access to abutting properties. For data on widths and other design features of frontage roads and outer separations, refer to Chapters 4, 9, and 10.

Exhibit 7-16 is an example of a two-way frontage road along a divided arterial. Notice the distance from the edge of the arterial to the location of the intersection of the cross street and frontage road. Providing sufficient distance for turn-lane storage on the cross street is an important design feature in frontage road design.

Grade Separations and Interchanges

Grade separations and interchanges are addressed in Chapter 10 and many of the principles discussed are applicable to arterial streets. Although grade separations and interchanges are not often used on arterial streets, due to high cost and limited right-of-way, they may be the only means available for providing sufficient capacity at some critical intersections.

In some cases, grade separations can be constructed within the existing right-of-way. Locations where grade separations should be considered on urban arterial streets are:

- at very high-volume intersections between principal arterials
- at high-volume intersections having more than four approach legs
- at high-type arterial street intersections where all other principal intersections are grade separated
- at all railroad crossings
- at sites where terrain conditions favor separation of grades

Exhibit 7-16. Divided Arterial Street With Two-Way Frontage Road

Normally, where a grade separation is provided on an urban arterial street, it is included as part of a diamond interchange. A single point urban interchange (SPUI) can provide the benefits of a grade separation while reducing cross-street delays and right-of-way needs. Other types of interchanges have application where more than four legs are involved. These interchange types are discussed in Chapter 10.

Where a grade separation is proposed, it is desirable to carry the entire approach roadway width, including parking lanes or shoulders, across or under the grade separation. However, in cases with restricted right-of-way it may be appropriate to reduce the width. Such a reduction is not as objectionable on arterial streets as on freeways because of lower speeds. The reduction in parking-lane or shoulder width should be accomplished with a taper. See Chapter 10 for a discussion on taper design elements.

Interchange elements for arterial streets may be designed with lower dimensional values than with freeways. Desirably, loop ramps should have radii no less than 45 m [150 ft]. Diamond ramps may have lengths as short as the minimum distance necessary to overcome the difference in elevation between the two roadways at suitable gradients. The length of speed-change lanes should be consistent with design speed. Chapter 10 provides criteria for design of interchanges and grade separations.

Erosion Control

When an urban arterial is designed with an open-ditch cross section, rural erosion control measures should be applied and water quality impacts should be considered. Curbed cross sections usually need more intensive treatment to prevent damage to adjacent property and siltation in sewers and drainage systems. Seeding, mulching, and sodding are usually employed to protect disturbed areas from erosion. Landscaping features, such as ground cover plantings, bushes, and trees, are also employed to control erosion, enhance beauty, and provide a visual buffer for adjacent properties.

Lighting

Adequate lighting is very important for safe operation of an urban arterial and also aids older drivers. The higher volumes and speeds, that are typically found on an arterial, make it especially challenging for the driver to make correct decisions with adequate time to execute the proper maneuvers without creating undue conflict in the traveled way. Where lighting is adequate, sudden braking and swerving are minimized. The visibility of signing and pavement marking also helps to smooth traffic flow. A safely designed, adequate lighting system is more important to optimum operation of an urban arterial than for any other type of city street. The lighting should be continuous and of an energy-saving type. Lighting in an urban area is often a matter of civic pride and is a deterrent to crime. In the event that it is impractical to provide continuous lighting, consideration should be given to providing intermittent lighting at such locations as intersections and ramp termini. The AASHTO *Informational Guide for Roadway Lighting* (**13**) is recommended as a source of lighting information.

Bikeways

Appropriate consideration should be given to practical measures for enhancing bicycle travel. If special bicycle facilities are desired, roadway widths should be adequate and the design of such facilities should be in accordance with the AASHTO *Guide for the Development of Bicycle Facilities* (**6**). For additional information on bicycle facilities, Chapters 2 and 4 also should be consulted.

Public Transit Facilities

Wherever there is a demand for highways to serve passenger car traffic, there is likewise a potential demand for public transportation. With almost total elimination of fixed-rail transit vehicles from surface streets and the increased use of free-wheeled buses, public transit has become increasingly compatible with other highway traffic. Other high-volume passenger vehicles such as the minibus, taxicab, and limousine may merit serious consideration in the overall planning of a high-volume arterial. The transit vehicle is more efficient than the private automobile with respect to street space occupied per passenger carried. With proper recognition of bus needs and provisions for them in the design and operation of highways, buses can become even more compatible with highway traffic in the future. The detailed discussion of bus facilities presented below is not intended to indicate that other types of mass transit facilities should not be considered. The more sophisticated public transit modes, including fixed rail, present unique and varied problems that are outside the scope of roadway design. Therefore, this discussion concentrates on the transit arrangements that most directly affect roadway design. Other transit modes need studies appropriate for that specific mode.

The vehicle-carrying capacity of through traffic lanes is decreased when a transit vehicle and other traffic use the same lanes. A bus stopping for passenger loading, for example, not only blocks traffic in that particular lane but affects traffic operations in all lanes. It is desirable that such interferences be minimized through careful planning, design, and traffic control measures.

The needs of public transit should be considered in the development of an urban highway improvement program. The routings of transit vehicles (including turns and transfer points) and the volumes of buses (i.e., average or minimum headways) should be considered in highway design. Design and operational features of the highway that are affected by these considerations include: (1) locations of bus stops (spacing and location with respect to intersections), (2) design of bus stops and turnouts, (3) reservation of bus lanes, and (4) special traffic control measures. Because some of the design and control measures that are beneficial to bus operation have an adverse effect on other traffic, and vice versa, a compromise that is most favorable to all users is appropriate.

Location of Bus Stops

The demand for bus service is largely a function of land-use patterns. The general location of bus stops is largely dictated by patronage and by the locations of intersection bus routes and transfer points. Bus stops should be located for the convenience of patrons.

The specific location of a bus stop within the general area where a bus stop is needed is influenced not only by convenience to patrons but also by the design and operational characteristics of the highway. Except where cross streets are widely spaced, but stops are usually located in the immediate vicinity of intersections. This facilitates crossing of streets by patrons without the need for midblock crosswalks. Midblock locations for bus stops may be appropriate where blocks are exceptionally long, or where bus patrons are concentrated at places of employment or residences that are well removed from intersections.

Bus stops at intersections may be located on the near (approach) or far (departure) side of the intersection. There are advantages and disadvantages to both near- and far-side locations, and the specific location for each bus stop should examined separately to determine the most suitable arrangement. Factors that should be considered include service to bus patrons, efficiency of transit operations, and efficiency of traffic operations. Far-side bus stops are advantageous at intersections where (1) other buses may turn left or right from the arterial, (2) turning movements from the arterial by other vehicle types, particularly right turns, are heavy, and (3) approach volumes are heavy, creating a large demand for vehicle storage on the near-side approach. Far-side bus stops have also proven to be effective in reducing collisions involving pedestrians. Sight distance conditions generally favor far-side bus stops, especially at unsignalized intersections; a driver approaching a cross street on the through lanes of an arterial can better see any vehicles approaching from the right if no bus is present. At near-side bus stops, the view of through drivers to their right may be blocked by a stopped bus. If the intersection is signalized, the bus may block the view of one of the signal heads.

Another disadvantage of near-side bus stops is the difficulty encountered by other vehicles in making turns while a bus is loading. Drivers frequently proceed around the bus to turn right, which interferes first with other traffic on the arterial and then with the bus as it leaves the stop. This disadvantage is eliminated if the cross street is one way from right to left. Thus, where the street pattern consists of a one-way grid, there is some advantage in having stops at alternate cross streets in advance of the streets crossing from right to left.

Where buses turn left, the bus stop should be located at least one block before the turn. Even with this arrangement the bus must cross all traffic lanes in the direction of travel to reach the left lane for the turn. Midblock bus stops are used occasionally under these circumstances, but the bus maneuver may be more difficult.

On highly developed arterials with ample rights-of-way, bus turnouts, and speed-change lanes, there is a definite traffic advantage to the far-side bus stop, as shown in Exhibit 7-17A. Such stops can be combined with speed-change lanes for turning vehicles entering the arterial. Where the stop is located on the near side of an intersection, vehicles turning right from the through lanes of the arterial cannot use the deceleration lane when it is occupied by a transit vehicle and instead may maneuver around it on the through lanes. Where the bus stop is located on the far side of the intersection, traffic turning right from the arterial does so freely. A far-side bus stop should be situated a short distance beyond the intersection so that vehicles making right turns onto the arterial can use a part of the acceleration lane as a maneuvering area and so that succeeding buses have space to stop without blocking the cross street.

Bus stops on cross streets should be located and arranged so that transferring riders do not need to cross the arterial, regardless of the direction they wish to travel. Exhibit 7-17A shows such an arrangement.

In another bus-stop arrangement for an arterial with frontage roads (Exhibit 7-17B), buses leave and return to the arterial by special openings in the outer separation in advance of and beyond the intersection. This arrangement has the advantage that buses stop in a position well removed from the through lanes. Right-turning traffic to and from the arterial street may also use these special openings, thereby reducing conflicts at the intersection proper. In an alternate arrangement, no slot in advance of the intersection is provided, and buses cross to the frontage road at the intersection proper. Both slots may be eliminated where the frontage road is continuous between successive cross streets because buses can leave the through lanes at one intersection and use the frontage road to reenter the arterial at the next intersecting street. This type of operation is fitting where bus stops are widely spaced.

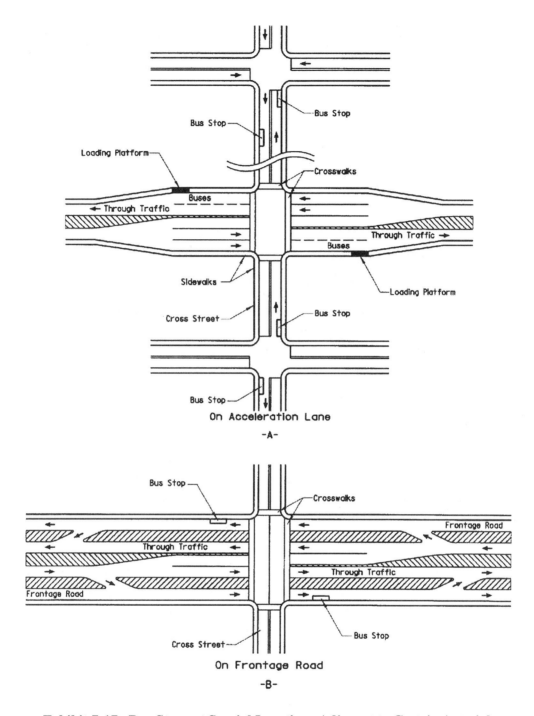

Exhibit 7-17. Bus Stops at Special Locations Adjacent to Certain Arterials

Midblock bus stops, like far-side stops, have an advantage over near-side stops in that the full roadway width on the intersection approach is made available for vehicle storage and turning maneuvers to maintain as high a capacity as practical. However, midblock bus stops are not generally suitable for streets where parking is permitted, as is the case on some arterials during off-peak hours. Usually, a crosswalk is needed at midblock bus stops to provide access to the stops from either side of the arterial and to serve as an intermediate crosswalk for other pedestrian traffic. In such cases, signal control may be needed to create safe crossing opportunities for pedestrians. Midblock signals violate driver expectations and merit careful consideration of their safety and operational implications. At a major transit stop and heavy pedestrian movements, a pedestrian grade separation may be warranted.

Additional information concerning the location and design of bus stops is presented in TCRP Report 19, *Guidelines for the Location and Design of Bus Stops* (**14**).

Bus Turnouts

The interference between buses and other traffic can be considerably reduced by providing stops clear of the lanes for through traffic. However, since bus operators may not use the turnout if they have difficulty maneuvering back into traffic, the bus turnout should be designed so that a bus can enter and leave easily. The preceding discussion and Exhibit 7-17 illustrate methods for reducing interference between buses and through traffic on high-type arterials. It is somewhat rare that sufficient right-of-way is available on lower-type arterial streets to permit turnouts in the border area, but advantage should be taken of every opportunity to do so. For geometric details, see the section on "Bus Turnouts" in Chapter 4.

Reserved Bus Lanes

Some improvement in transit service can be realized by excluding other traffic from the curb lane of arterial streets. The success of this regulatory measure is rather limited in most instances, however, for the reason that vehicles making right turns must occupy this same lane, it is not practical to exclude them, for distances up to a block or two in advance of the turn. Vehicles preparing to turn right cannot be distinguished from through traffic, so compliance with the exclusive bus lane regulation is largely on a voluntary basis. Nevertheless, there are certain combinations of conditions under which at least a modest improvement in transit service can be achieved. These conditions are not always apparent or definable, and the only way to determine conclusively that there will be overall benefit is to test the regulation in practice at locations where a preliminary investigation indicates likelihood of success.

There is one arrangement by which exclusive use of the reserved lane by buses can be ensured with little enforcement. This can be accomplished by offsetting the division line on a multilane street so that there is only one lane in a given direction of travel and that lane is reserved for the exclusive use of buses; the remaining lanes are available for normal traffic in the opposite direction of travel. The current MUTCD (**1**) specifies markings for the division line separating traffic in the two directions of travel. Permanently installed flexible stanchions may be used to further accentuate the division between the bus lane and lanes for ordinary traffic. Exhibits 7-18 illustrates a street operated in this manner. The lane on the left in the exhibit is reserved for buses traveling toward the viewer. This arrangement is sometimes referred to as a "contra-flow" system. This system may facilitate bus movements under high-density traffic conditions.

Exhibit 7-18. Exclusive Bus Lane

Reserved bus lanes of the type described above may be justified even for relatively few buses. Because the carrying capacity of a bus is many times that of a passenger car, a small number of buses can serve the same number of people as a large number of passenger cars.

Traffic Control Measures

Traffic control devices on arterial streets are usually installed with the intent of favoring automobile traffic, with only secondary consideration to transit vehicles. For express-bus operation, the control measures that are most favorable for one mode will generally be equally well suited for the other. However, where local service is provided by buses with frequent stops to pick up and discharge passengers, a signal system that provides for good progressive movement of privately operated vehicles may actually result in reverse progression for buses. The resulting slow travel speed for buses tends to discourage patronage, further adding to the already heavy volume of automobile traffic.

Limited effort has been devoted to development of traffic control systems that are more favorable for bus service without serious adverse effects on other traffic. This approach holds some promise of improving travel speeds for buses and making public transit more attractive. This result can be achieved only at the expense of reduced travel speed for passenger cars, but in some instances this solution may be the most beneficial from the overall point of view. Development of a suitable signal system requires careful investigation by those possessing the proper technical skills and should be a part of the arterial improvement program, involving the joint efforts of traffic specialists, the transit industry, and the design team.

Although the major emphasis in the application of traffic control measures is in minimizing delay, the control measures can be used to facilitate bus operation in other respects, particularly where buses turn from the arterial onto a cross street.

Buses making right turns may create a problem where the cross street is narrow and the adjoining property is developed so intensively that it is not practical to provide a sufficiently long curb return radius. Buses turning right from the curb lane may encroach beyond the centerline of the cross street. At signalized intersections, the space beyond the centerline is normally occupied by vehicles stopped for the red signal. Under such conditions, the stop line on the cross street should be relocated to provide sufficient space for a bus to turn, and where needed, an auxiliary signal head should be placed at the relocated stop line to obtain compliance.

REFERENCES

1. U.S. Department of Transportation, Federal Highway Administration. *Manual on Uniform Traffic Control Devices for Streets and Highways*, Washington, D.C.: 1988 or most current edition.
2. Transportation Research Board. *Highway Capacity Manual,* Special Report No. 209. Washington, D.C.: Transportation Research Board, 1998 or most current edition.
3. AASHTO. *Roadside Design Guide,* Washington, D.C.: AASHTO, 1996.
4. Harwood, D. W., M. T. Pietrucha, M. E. Wooldridge, R. E. Brydia, and K. Fitzpatrick. *Median Intersection Design.* NCHRP Report 375, Washington, D.C.: Transportation Research Board, 1995.
5. Gluck, J., H. S. Levinson, and V. Stover. *Impacts of Access-Management Techniques,* NCHRP Report 420, Washington, D.C.: Transportation Research Board, 1999.
6. AASHTO. *Guide for the Development of Bicycle Facilities,* Washington, D.C.: AASHTO, 1999.
7. AASHTO. *A Guide for Development of Rest Areas on Major Arterials and Freeways,* Washington, D.C.: AASHTO, 2001.
8. Bonneson, J. A., and P. T. McCoy. *Capacity and Operational Effects of Midblock Left-Turn Lanes,* NCHRP Report 395, Washington, D.C.: Transportation Research Board, 1997.
9. AASHTO. *Guide for the Planning, Design, and Operation of Pedestrian Facilities,* Washington, D.C.: AASHTO, forthcoming.
10. National Committee on Uniform Traffic Laws and Ordinances. *Uniform Vehicle Code.*
11. National Committee on Uniform Traffic Laws and Ordinances. *Model Traffic Ordinance.*
12. AASHTO. *Guide for the Design of High Occupancy Vehicle Facilities,* Washington, D.C.: AASHTO, 1992.
13. AASHTO. *Informational Guide for Roadway Lighting.* Washington, D.C.: AASHTO, 1984.
14. *Guidelines for the Location and Design of Bus Stops*, TCRP Report 19, Washington, D.C.: Transportation Research Board, 1996.

CHAPTER 8
FREEWAYS

INTRODUCTION

Freeways are arterial highways with full control of access. They are intended to provide for high levels of safety and efficiency in the movement of large volumes of traffic at high speeds. Control of access refers to the regulation of public access rights to and from properties abutting the highway. With full control of access, preference is given to through traffic by providing access connections with selected public roads only and by prohibiting crossings at grade and direct private driveway connections.

The principal advantages of access control include preservation of highway capacity, higher speeds, and improved safety for highway users. Highways with fully controlled access have grade separations at all railroads and grade separations or interchanges at selected public crossroads. The remaining crossroads are interconnected or terminated.

Essential freeway elements include the roadway, medians, grade separations at crossroads, ramps to and from the traveled way at selected locations, and in some cases, frontage roads. Chapters 2, 3, and 4 describe roadway design elements, controls, and criteria applicable to all highway classes. This chapter identifies the various types of freeways, emphasizes selected features, and discusses other design details unique to freeways. The design of freeway interchanges is discussed in Chapter 10.

This chapter is organized with an introductory section on the general design considerations for freeways, followed by separate discussions of the design of rural and urban freeways.

GENERAL DESIGN CONSIDERATIONS

Design Speed

As a general consideration, the design speed of urban freeways should not be so high as to exceed the limits of prudent construction, right-of-way, and socioeconomic costs. However, this design speed should not be less than 80 km/h [50 mph]. Wherever this minimum design speed is used, it is important to have a properly posted speed limit, which is enforced during off-peak hours. On many urban freeways, particularly in developing areas, a design speed of 100 km/h [60 mph] or higher can be provided with little additional cost. In addition, the corridor of the main line may be relatively straight with the character of the roadway and location of interchanges permitting an even higher design speed. Under these conditions, a design speed of 110 km/h [70 mph] is desirable because higher design speeds are closely related to the overall quality and safety of a facility. For rural freeways, a design speed of 110 km/h [70 mph] should be used. In mountainous terrain, a design speed of 80 to 100 km/h [50 to 60 mph], which is consistent with driver expectancy, may be used.

Design Traffic Volumes

Both urban and rural freeways should normally be designed to accommodate traffic projections for a 20-year period, particularly in the case of new construction. However, some elements of freeway reconstruction may be based on a lesser design period. For further guidance on the selection of appropriate periods for forecasting design traffic volumes, refer to Chapter 2. Specific capacity needs should be determined from directional design hourly volumes (DDHV) for the appropriate design period. In large metropolitan areas, the selection of appropriate design traffic volumes and design periods may be influenced by system planning. Segments of freeways may be constructed or reconstructed to be commensurate with either intermediate traffic demands or traffic based on the completed systems, whichever may be most appropriate.

Levels of Service

Procedures for traffic operational analyses for freeways, including appropriate adjustments for operational and highway factors, are found in the *Highway Capacity Manual* (HCM) (**1**), which also presents a thorough discussion of the level-of-service concept. Designers should strive to provide the highest level of service practical and consistent with anticipated conditions. The levels of service concept is discussed in Chapter 2 and the levels of service are summarized in Exhibit 2-31. For acceptable degrees of congestion, freeways and their auxiliary facilities (i.e., ramps, mainline weaving sections, and C-D roads in urban and developing areas) should generally be designed for level-of-service C. In heavily developed sections of metropolitan areas, achievement of level-of-service C may not be practical and the use of level-of-service D may be appropriate. In rural areas, level-of-service B is desirable for through and auxiliary lanes, although level-of-service C may be acceptable on auxiliary facilities carrying unusually high volumes.

Pavement and Shoulders

Freeways should have a minimum of two through-traffic lanes for each direction of travel. Through-traffic lanes should be 3.6 m [12 ft] wide. Pavements should have a high-type surface with adequate skid resistance and provide a high degree of structural adequacy. Cross slopes should range between 1.5 and 2 percent on tangent sections, with the higher value recommended for areas with moderate rainfall. For areas of heavy rainfall, a cross slope of 2.5 percent may be needed to provide adequate pavement drainage. Appropriate cross-slope rates are discussed in Chapter 4. For elevated freeways on viaducts, two-lane pavements usually are sloped to drain the full width of the roadway. On wider facilities, particularly in areas with heavy rainfall, the crown may be located on the lane line at one-third or one-half the total width from one edge, thus providing two directions for surface drainage. In areas that experience snow, the median and cross slopes of the traveled way should be designed to prevent snow stored in the median from melting and draining across the roadway. This may result in icing conditions during freezing temperatures.

Paved shoulders should be continuous on both the right and left sides of all freeway facilities. The usable paved width of the right shoulder should be at least 3.0 m [10 ft]; where the DDHV for truck traffic exceeds 250 veh/h, the right shoulder width should be 3.6 m [12 ft]. On four-lane freeways, the median (or left) shoulder is normally 1.2 to 2.4 m [4 to 8 ft] wide, at least 1.2 m [4 ft] of which should be paved and the remainder stabilized. On freeways of six or more lanes, the usable paved width of the median shoulder should also be 3.0 m [10 ft] and preferably 3.6 m [12 ft] where the DDHV for truck traffic exceeds 250 veh/h. Ramp shoulder widths are usually constructed adjacent to acceleration and deceleration lanes with transitions to the freeway shoulder width at the taper ends. Shoulder cross slope should range between 2 and 6 percent and can be at least 1 percent greater than the pavement cross slope on tangent sections to facilitate drainage. To provide visual contrast, the color or texture of the shoulders should be different from that of the traveled way. On viaducts, differentiation between traveled way and shoulders is sometimes accomplished by striping and pavement marking, or by corrugated depressions.

Curbs

In the interest of safety, caution should be exercised in the use of curbs on freeways; where curbs are provided in special cases, they should not be closer to the traveled way than the outer edge of shoulder and should be easily traversable. An example of a special case in which shoulder curbs are used on freeways is at locations where curbs are provided to control drainage and reduce erosion. For more information, refer to the discussion on curb types and their placement in Chapter 4.

Superelevation

The full superelevation rates used on freeways that are depressed, built at ground-level, or elevated on embankments are not generally applicable to elevated freeways on viaducts. Appearance and adjacent development somewhat limit the difference in elevation between the edges of multilane pavements. Superelevation rates of 6 to 8 percent are generally the maximum that should be used on viaducts. The lower value may be used where freezing and thawing conditions are likely, because bridge decks generally freeze more rapidly than other roadway sections. Combinations of design speed and curvature that result in superelevation rates greater than these values should be avoided. Where freeways are intermittently elevated on viaducts, the lower superelevation rates should be used throughout to promote consistently safe operation. Maximum superelevation rates of 8 to 12 percent are applicable for freeways if snow and ice conditions are not a factor. In lower speed situations, a maximum superelevation rate of 6 percent may be applicable.

Grades

Maximum grades for freeways are presented in Exhibit 8-1 as a function of design speed and terrain type. Grades on urban freeways should be comparable to those on rural freeways of the same design speed. Steeper grades may be tolerated in urban areas, but the closer spacing of interchange facilities and the need for frequent changes in speed make it desirable to use flat grades wherever practical. On sustained upgrades, the need for climbing lanes should be investigated, as discussed in Chapter 3.

Type of Terrain	Metric						US Customary						
	Design Speeds (km/h)						Design Speeds (mph)						
	80	90	100	110	120	130	50	55	60	65	70	75	80
	Grades (%)[a]						Grades (%)[a]						
Level	4	4	3	3	3	3	4	4	3	3	3	3	3
Rolling	5	5	4	4	4	4	5	5	4	4	4	4	4
Mountainous	6	6	6	5	–		6	6	6	5	5	–	–

[a] Grades 1% steeper than the value shown may be provided in mountainous terrain or in urban areas with crucial right-of-way controls.

Exhibit 8-1. Maximum Grades for Rural and Urban Freeways

Structures

The design of bridges, culverts, walls, tunnels, and other structures should be in accordance with the principles of the current *Standard Specifications for Highway Bridges* (**2**) or with the AASHTO *LRFD Bridge Design Specification* (**3**). Structures carrying freeway traffic should provide an MS-18 [HS20-44] design loading.

The clear width on bridges carrying freeway traffic should be as wide as the approach roadway, as discussed in Chapter 10. On bridges longer than 60 m [200 ft], some economy in substructure costs may be gained by building a single structure rather than twin parallel structures. In such cases, the approach shoulder widths are provided and a median barrier is extended across the bridge.

Structures carrying ramps should provide a clear width equal to the ramp width and paved shoulders. Clear widths for structures carrying auxiliary lanes are discussed in Chapter 10.

The structure width and lateral clearance of highways and streets overpassing or underpassing the freeway are dependent on the functional classification of the highway or street as discussed in Chapters 5, 6, and 7.

Vertical Clearance

The vertical clearance to structures passing over freeways should be at least 4.9 m [16 ft] over the entire roadway width, including auxiliary lanes and the usable width of shoulders (with an allowance for future resurfacing). In highly developed urban areas, where attainment of the

4.9-m [16-ft] clearance would be unreasonably costly, a minimum clearance of 4.3 m [14 ft] may be used if there is an alternate freeway facility with the minimum 4.9-m [16-ft] clearance.

Because of their lesser resistance to impacts, the vertical clearance to sign trusses and pedestrian overpasses should be 5.1 m [17 ft]. On urban routes with less than the 4.9-m [16-ft] clearance, the vertical clearance to sign trusses should be 0.3 m [1 ft] greater than the minimum clearance for other structures. Similarly, the vertical clearance from the deck to the cross bracing of through-truss structures should also be a minimum of 5.1 m [17 ft], with an allowance for future resurfacing.

Horizontal Clearance to Obstructions

Urban freeways at ground level and rural freeways should have clear zone widths consistent with their operating speed and side slopes, as discussed in the section on "Horizontal Clearance to Obstructions," the section in Chapter 4. Detailed discussions of clear zone are also included in the AASHTO *Roadside Design Guide* (**4**). Non-traversable obstacles within the clear zones should be shielded by an appropriate barrier as long as the barrier represents a lower potential for severe crashes. Fixed objects that cannot be relocated beyond the clear zone should be of a "breakaway" design or shielded by barriers or attenuators. Bridge piers and abutments should be located as near to the clear-zone edge as practical. Where the width of the median is less than twice the width of the clear zone, shielding of median piers may be appropriate depending on median width and traffic volumes.

In rural cut sections, drainage is carried in side ditches. The foreslope and backslope should provide an acceptable recovery area in case drivers lose control and leave the traveled way. Where the right-of-way on a depressed freeway is insufficient to provide for a swale, the drainage is usually carried in a gutter section along the outside edge of the shoulder. Details of gutter sections are covered in the section on curbs in Chapter 4.

Depressed freeways in urban areas have more restrictive rights-of-way and retaining walls or bridge piers may need to be placed within the clear zone. Such walls and piers should not be located on the shoulder and preferably should be at least 0.6 m [2 ft] beyond the outer edge of shoulder. Retaining walls and pier crash walls should incorporate an integral concrete barrier shape, or they should be offset from the shoulder to permit shielding with a barrier, as discussed in the section on "Lateral Clearances" in Chapter 10. Where walls are located beyond the clear zone or are not needed, back slopes should be traversable and fixed objects within the clear zone should be of a "breakaway" design or shielded.

Elevated freeways on embankments generally warrant roadside barriers where slopes are steeper than 1V:3H or where the area beyond the toe of slope remaining within the clear zone is not traversable. The tops of retaining walls used in conjunction with embankment sections should be located no closer to the roadway than the outer edge of shoulder, and the walls should incorporate the concrete barrier shape or be appropriately shielded.

Ramps and Terminals

The design of ramps and connections for all freeway types is covered in Chapter 10.

Outer Separations, Borders, and Frontage Roads

An outer separation is defined as the area between the traveled way of the main lanes and a frontage road or local street. A border is defined as the area between the frontage road or local street and the private development along the road. Where there are no frontage roads or local streets functioning as frontage roads, the area between the traveled way of the main lanes and the right-of-way limit should be referred to as the border. Because of the dense development along urban freeways, frontage roads are often needed to maintain local service and to collect and distribute ramp traffic entering and leaving the freeway. Where the freeway occupies a full block, the adjacent parallel streets are usually retained as frontage roads, which are discussed in detail in Chapter 4.

The outer separation or border provides for shoulders, sideslopes, drainage, access-control fencing, and in urban areas, retaining walls and ramps. In sensitive areas, the outer separation or border may also provide for noise abatement measures. Usually, the outer separation is the most flexible element of an urban freeway section. Adjustment in width of right-of-way, as may be needed through developed areas, ordinarily is made by varying the width of the outer separation.

The outer separation or border should be as wide as can be attained economically to provide a buffer zone between the freeway and its adjacent area. The border should extend beyond the construction limits, where practical, to facilitate maintenance operations and safety. Wide outer separations also permit well-designed ramps between the freeway and the frontage road. The typical range in widths of outer separations is 25 to 45 m [80 to 150 ft], but much narrower widths may be used in urban areas if retaining walls are employed.

RURAL FREEWAYS

Rural freeways are similar in concept to urban ground-level freeways, but the alignment and cross-sectional elements are more generous in design, which is commensurate with the higher design speed and generally with greater availability of right-of-way.

Where terrain permits, a design speed of 110 km/h [70 mph] should be used on rural freeways, which will result in a safer freeway than one designed for a lower speed. In some cases, fewer lanes will also be needed to accommodate a given volume of traffic at the desired level of service, which for rural freeways is usually level-of-service B. In addition to the increased safety and capacity, for a given level of service higher design speeds result in a more comfortable driving environment and reduce both fuel consumption and operating costs.

Freeways are initially designed to accommodate anticipated traffic for about 20 years and remain in service for a much longer time. Any cost savings realized by initially constructing for lower design speeds may be outweighed by the high costs, disruption to the environment, and inconvenience to traffic that accompany the reconstruction of major facilities.

Although level-of-service B is desirable for rural freeways, level-of-service C may be appropriate on auxiliary facilities where volumes are unusually high. Rural freeways generally have four through-traffic lanes except on approaches to metropolitan areas where six or more lanes may be provided. Where intersecting highways are classified as collectors and higher, interchanges are usually provided. Local roads may be terminated at the freeway, connected to frontage roads or other local roads for continuity of travel, or carried over or under the freeway by grade separation with or without interchange.

Alignment and Profile

Rural freeways are generally designed for high-volume and high-speed operation. They should, therefore, have smooth flowing horizontal and vertical alignments. Proper combinations of flat curvature, shorter tangents, gentle grades, variable median widths, and separate roadway elevations to enhance the safety and aesthetic aspects of freeways. Advantage should be taken of favorable topographic conditions to achieve the desired goals. Changing median widths on tangent alignments should be avoided where practical to avoid a distorted appearance.

Because there are usually fewer physical constraints in constructing the rural road network than its urban counterpart, rural freeways can usually be constructed near ground level with smooth and relatively flat profiles. The profile of a rural freeway is controlled more by drainage and earthwork considerations and less by the need for frequent grade separations and interchanges. If the need for elevated or depressed sections should occur, the guidelines for urban freeways are appropriate.

Even though the profile may satisfy all the design controls, the finished vertical alignment may appear forced and angular if minimum criteria are used. The designer should check profile designs in long continuous plots to help avoid an undesirable roller-coaster alignment in rolling terrain. The relation of horizontal and vertical alignment should be studied simultaneously to obtain a desirable combination.

Exhibit 8-2 illustrates a typical ground-level rural freeway with a curvilinear alignment.

Medians

Median widths of about 15 to 30 m [50 to 100 ft] are common on rural freeways. The 15-m [50-ft] dimension shown in Exhibit 8-3A provides for 1.8-m [6-ft] graded shoulders and 1V:6H foreslopes with a 1.0-m [3-ft] median ditch depth. Adequate space is provided for vehicle recovery; however, median piers may need shielding for heavier traffic volumes, in accordance with Chapter 4. The 30-m [100-ft] dimension shown in Exhibit 8-3B permits the designer to use independent profiles in rolling terrain to blend the freeway more appropriately with the environment while maintaining flat slopes for vehicle recovery. In flat terrain, the 30-m [100-ft] median is also suitable when stage construction will involve the future addition of two 3.6-m [12-ft] traffic lanes.

Exhibit 8-2. Typical Ground-Level Rural Freeway

Where the terrain is extremely rolling, or the land is not suitable for cultivation or grazing, a wide variable median with an average width of 45 m [150 ft] or more, as shown in Exhibit 8-3C, may be attainable. Such a width permits the use of independent roadway alignment, both horizontally and vertically, to its best advantage in blending the freeway into the natural topography. Foreslopes and backslopes used within the clear zone should provide for vehicle recovery. The remaining median width may be left in its natural state of vegetation, trees, and rock outcroppings to reduce maintenance costs and add scenic interest to passing motorists. The combination of independent alignment and a natural park-like median is pleasing to motorists. For driver reassurance, the opposing roadway should be in view at frequent intervals.

Median widths in the range of 3.0 to 9.0 m [10 to 30 ft], as shown in Exhibit 8-3D, may be needed where right-of-way restrictions dictate or in mountainous terrain. These medians are usually paved, and where roadways are crowned, underground drainage should be provided. Considering the usual developing-area traffic volumes as well as operational characteristics in mountainous areas, a median barrier is usually warranted as a safety measure.

To avoid extreme adverse travel for emergency and law-enforcement vehicles, emergency crossovers on rural freeways are normally provided where interchange spacing exceeds 8 km [5 mi]. Between interchanges, emergency crossovers are spaced at 5- to 6.5-km [3- to 4-mi] intervals. Maintenance crossovers may be needed at one or both ends of interchange facilities, depending on interchange type, for the purpose of snow removal and at other locations to facilitate maintenance operations. Maintenance or emergency crossovers generally should not be located closer than 450 m [1,500 ft] to the end of a speed-change taper of a ramp or to any structure. Crossovers should be located only where above-minimum stopping sight distance is provided and preferably should not be located on superelevated curves.

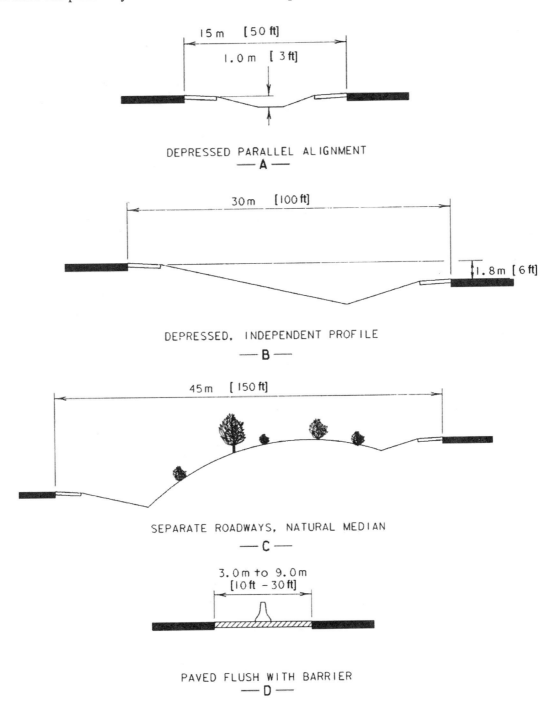

Exhibit 8-3. Typical Rural Medians

The width of the crossover should be sufficient to provide safe turning movements and should have a surface capable of supporting maintenance equipment used on it. The crossover should be depressed below shoulder level to be inconspicuous to traffic and should have 1V:10H or flatter sideslopes to minimize its effect as an obstacle to uncontrolled vehicles. Crossovers should not be placed in restricted-width medians unless the median width is sufficient to accommodate the vehicle length (i.e., 7.5 m [25 ft] or more). Where median barriers are employed, each end of the barrier at the median opening may need a crashworthy terminal. For further information, refer to the AASHTO *Roadside Design Guide* (4).

Sideslopes

Flat, rounded sideslopes, fitting with the topography and consistent with available right-of-way, should be provided on rural freeways. Foreslopes of 1V:6H or flatter are recommended in cut sections and for fills of moderate height, as discussed in Chapter 4. Where fill heights are intermediate, a combination of recoverable and non-recoverable slopes may be used to provide the acceptable vehicle recovery area, (see the AASHTO *Roadside Design Guide* [4] for further information). For high fills, steeper slopes protected by guardrail may be needed. In addition, backslopes of 1V:3H or flatter permit normal landscaping and erosion control practices and ease maintenance operations. In highly productive agricultural areas, steeper slopes may be used, but the combination of foreslope, backslope, and ditch configuration should permit vehicle recovery. Where rock or loess deposits are encountered, backslopes may be nearly vertical, but, where practical, should be located to provide an adequate recovery area for errant vehicles.

Frontage Roads

The need for local service across and along rural freeway corridors is usually considerably less than that along highly developed urban freeways. Therefore, along rural freeways, frontage roads are usually intermittent and relatively short. They either provide access to one or more severed properties or provide continuity of a local road by connecting it with a grade-separated crossroad.

Where a rural freeway is located parallel to and in close proximity to a major highway, the major highway is often converted to a continuous two-way frontage road and serves as a collector facility.

Because of the lack of continuity and the type of service being provided, newly constructed frontage roads are normally two-way facilities in rural areas. Since traffic operations at two-way frontage road intersections with grade-separated crossroads are more complex, such intersections are generally located as far as practical from grade-separation structures and interchange ramp terminals.

Rural frontage roads are generally outside the control-of-access line but within the right-of-way limits. Design details for rural frontage roads are similar to those used for local roads, as discussed in Chapters 3 and 5.

URBAN FREEWAYS

General Design Characteristics

Urban freeways are capable of carrying high traffic volumes. While freeways may have from 4 to 16 through-traffic lanes, typically there are no more than 4 through lanes in one direction. Urban freeways are classified as depressed, elevated, ground-level, or combination-type. These freeway types are used as is appropriate for conditions. Special freeway designs are appropriate for special conditions, including freeways with reverse-flow roadways, dual-divided freeways, and freeways with collector-distributor roads.

This section on urban freeways first discusses the design of freeway medians that is common to all freeway types. Then, separate discussions are presented on depressed, elevated, ground-level, and combination-type freeways, as well as special freeway designs that may be used in urban areas. The accommodation of transit and high-occupancy vehicle facilities within a freeway is also discussed.

Medians

A wide separation between traffic in opposing directions is safer and more comfortable for motorists, and therefore the median on urban freeways should be as wide and flat as practical. Additional median width may be used for mass transit or to provide additional lanes if more capacity is needed in the future. However, in densely developed areas with expensive right-of-way, the width available for a median is usually restricted. The minimum median width for a four-lane urban freeway should be 3.0 m [10 ft], which provides for two 1.2-m [4-ft] shoulders and a 0.6-m [2-ft] median barrier. For freeways with six or more lanes, the minimum width should be 6.6 m [22 ft], and preferably 7.8 m [26 ft] when the DDHV for truck traffic exceeds 250 veh/h to provide a wider median shoulder to accommodate a truck. The AASHTO *Roadside Design Guide* (4) should be utilized to determine median barrier warrants. When a median barrier is present, additional horizontal clearance may be needed to provide minimum stopping sight distance along the inside lane on sharper curves. Narrow medians should generally be paved for the full width with a material that is contrasting in color or texture.

Median crossovers for emergency or maintenance purposes are not generally warranted on urban freeways due to the close spacing of interchange facilities and the extensive development of the abutting street network.

Depressed Freeways

General Characteristics

A depressed freeway may occupy a full-block width and be parallel to the grid street system for most of its length. The roadways of a depressed freeway are typically located at an approximate depth of 4.9 m [16 ft] in addition to the clearance for structural depth below the surface of the adjacent streets. An allowance for future pavement overlays is frequently considered in setting the vertical clearance. In addition, depressed freeways are often flanked on one or both sides by frontage roads at the street level. All important streets pass over the through roadways while other streets are intercepted by frontage roads or terminated at the right-of-way line. Interchange with surface streets is usually accomplished by ramps that connect directly with frontage roads or by diamond interchanges where there are no frontage roads. Higher types of interchanges are usually provided at intersections with certain major arterials. For a more detailed discussion of interchange design, see Chapter 10.

Depressed freeways are desirable in that they reduce the impacts of the freeway on adjacent areas. They are less conspicuous than ground-level or elevated freeways, permit surface streets to cross at their normal grade, and reduce freeway noise. However, these advantages have to be balanced against the increased cost of providing for drainage. While gravity drainage facilities are sometimes feasible to accommodate the design storm without inundating the traveled way, pumping stations may be needed. For design guidance on pumping stations, refer to the AASHTO *Model Drainage Manual* (**5**).

Structures passing over the depressed freeway and retaining walls located in close proximity to the traveled way should be fenced to prevent objects being dropped or thrown onto vehicles below.

Slopes and Walls

Sideslopes of a depressed freeway are designed in the same manner as those for cut slopes, except that the slopes are more likely to be controlled by width restrictions. Foreslopes, if used beyond the shoulder, should be safely traversable.

Normally foreslopes are not used beyond the shoulder on depressed freeways and, in such cases, backslopes should not be steeper than 1V:3H. In developed areas, space may not be available for desirable slopes, particularly where ramps are present, and full- or partial-height retaining walls may be needed. Various forms of retaining walls are appropriate for depressed freeways, including those constructed of solid masonry, concrete, stone, precast panels, or metal. Wall types include cantilevered, crib or bin, mechanically stabilized, or sheet piling. Where retaining walls are used in combination with earth slopes, the walls may be located either at the roadway level adjacent to the shoulder or on the outer portion of the separation above the depressed roadway.

Retaining walls above the roadway are desirable from the driver's viewpoint, because they provide a more open feeling at the roadway level. This arrangement also provides space for storage of snow plowed from the freeway traveled way and shoulders. However, it may be more advantageous for the surrounding neighborhood if the wall is located at the roadway level and a slope is located on the upper portion of the cross section. This arrangement permits effective screening of the surrounding properties through planting. Slope maintenance may also be performed more safely, and noise abatement may be more effective. Both designs should be evaluated to determine which is best suited for the particular situation.

Retaining walls should be located no closer to the roadway than the outer edge of shoulder and preferably should be 0.6 m [2 ft] beyond the outer edge of shoulder. Where the wall is located at or near the shoulder edge, bridge columns, light fixtures, and sign supports should not protrude from the lower portion of the wall which should have the "safety shape" contour used on concrete median barriers, as shown in Exhibit 10-6. To minimize the angle of potential impacts, the base of the wall should not be placed farther than approximately 4.5 m [15 ft] from the outer edge of the traveled way.

Where the top of the retaining wall is at the level of a frontage road, the face of the parapet or rail should have a width equal to a normal shoulder width or be located at least 1.2 m [4 ft], and preferably 1.8 m [6 ft], from the edge of the traveled way. Where a retaining wall is located adjacent to an auxiliary lane or ramp, normal ramp shoulder widths should be provided.

Sight distance should be checked when designing slopes and retaining walls. On curved alignment, the slopes, walls, and other side obstructions should be sufficiently removed from the pavement edge to provide the design stopping sight distance for a vehicle in the traffic lane nearest the obstruction.

Typical Cross Section

Cross sections of depressed freeways vary considerably through urban and suburban areas. Whereas these cross sections are influenced primarily by the number of traffic lanes needed, another important factor is availability of right-of-way, which depends on the type and value of urban development, topography, soil and drainage conditions, and the frequency and type of interchanges to be used. The design of a cross section should be liberal, but sometimes it is appropriate to compromise on certain elements because of physical and economic limitations to fit the cross section within a relatively narrow right-of-way. Exhibits 8-4 through 8-6 illustrate depressed cross sections for various conditions.

Adjustment may be feasible in the widths of median and borders, but most adjustments must be accomplished within the width of the outer separation.

Where the freeway is bridged by closely spaced cross streets, a continuous full-depth section results. In outlying areas where separated crossroads are widely spaced, it usually is practical and economical to adjust the profile to decrease the depth of cut between structures, resulting in a combination of depressed and ground-level freeways. The benefits of this approach are that ramp design is simplified, excavation quantities are reduced, sideslopes can be flatter, and wider marginal areas at street level may be provided within the right-of-way. Generally, the result is a more pleasing freeway.

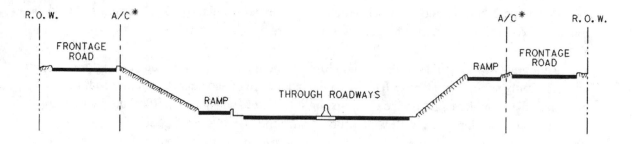

* ACCESS CONTROL LINE - PLACEMENT MAY VARY

Exhibit 8-4. Typical Cross Section for Depressed Freeways

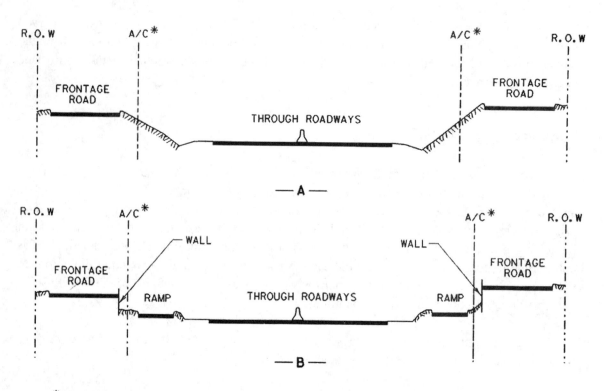

* ACCESS CONTROL LINE - PLACEMENT MAY VARY

Exhibit 8-5. Restricted Cross Sections for Depressed Freeways

Exhibit 8-4 shows a typical cross section for depressed freeways, providing for a 3.0- to 6.6-m [10- to 22-ft] median, 3.6-m [12-ft] traffic lanes, and 15 m [50 ft] for each frontage road plus border. The minimum median width of 3.0 to 6.6 m [10 to 22 ft] is based on the assumption that for depressed freeways the ultimate section is constructed initially. However, where additional width is needed in the median for staged construction, the median should be widened in multiples of 3.6 m [12 ft] (i.e., the width of a traffic lane). Where ramps are not needed, the uniform-width section should be graded to provide slopes as flat as practical within the available right-of-way.

Restricted Cross Section

Exhibit 8-5A presents a typical cross section that enables depressed sections to be constructed with earth slopes at locations without ramps but with retaining walls at ramps. The cross section in Exhibit 8-5A includes a 12-m [40-ft] frontage road plus border, 3.6-m [12-ft] traffic lanes, and a 3.0- to 6.6-m [10- to 22-ft] median.

Walls may be located at various points in the cross section, such as adjacent to the freeway shoulder, adjacent to the ramp shoulder, at the top of slopes, or at various combinations of these locations. Some variations in wall arrangements may be needed on the left and right sides, as shown in Exhibit 8-5B.

Walled Cross Section

Exhibit 8-6 shows walled cross sections that are appropriate for depressed freeways. In this example, the freeway is continuously walled and the ramps are omitted. Exhibit 8-6A shows a walled cross section with no overhang.

In special cases where even less right-of-way is obtainable, the design can be further consolidated by using an overhanging section where part of the frontage road is cantilevered over the freeway shoulder, as shown in Exhibit 8-6B. While the value of this alternative will vary depending on the restrictions and design selected, a typical overhang distance will range from 3.0 to 4.2 m [10 to 14 ft]. This design type may be applicable in special instances where large buildings or other obstructions cannot be avoided. A special feature of this design is its effectiveness in containing highway noise within the roadway and shielding abutting areas from such noise.

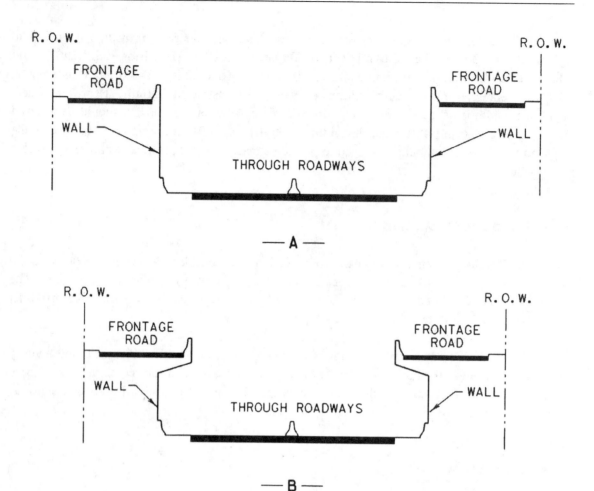

Exhibit 8-6. Cross Sections with Retaining Walls on Depressed Freeways Without Ramps

Although the restricted cross sections shown are acceptable, they should be used only where additional right-of-way would be extremely costly or where this type of cross section is needed to preserve the surrounding environment.

Examples of Depressed Freeways

Exhibit 8-7 shows an aerial view of a six-lane depressed freeway in the downtown district of a large city. This freeway fits in well with the existing street system. In this example, there is a four-level directional interchange in the upper center of the photograph. In the more densely developed portion (lower right), the freeway has full-height retaining walls and frontage roads. In the center of the photograph, sufficient right-of-way is available to use slopes in lieu of retaining walls.

Exhibit 8-8 shows a depressed freeway flanked by major surface streets on the upper level. The auxiliary lanes of the surface street on the right side are partially cantilevered over the

Exhibit 8-7. Depressed Freeway

Exhibit 8-8. Depressed Freeway

freeway shoulder. The cross streets overpass the freeway with level grades, thus facilitating traffic operations on the structures and at adjacent intersections.

Elevated Freeways

General Characteristics

An elevated freeway may be constructed on either a viaduct or an embankment. Continuous elevation of the freeway may be appropriate in level terrain where restricted right-of-way, high water table, extensive underground utilities, close pattern of streets to be retained, or other circumstances make construction of a depressed freeway undesirable and perhaps uneconomical.

Several structure types are used for viaducts carrying elevated freeways. Viaduct design is influenced by traffic demands, right-of-way, topography, foundation conditions, character of urban development, interchange needs, availability of materials, and economic considerations. Because of these multiple considerations, viaducts are perhaps the most difficult of all freeway types to fit harmoniously into the environment.

The supporting columns for viaducts are located in such a way as to provide reasonable clearance on each side and to leave much of the ground-level area free for other use. This design has advantages in that (1) practically all cross streets can be left open with little or no added expense, (2) existing utilities that cross the freeway right-of-way are minimally disturbed, and (3) surface traffic on cross streets usually can be maintained during construction with few, if any, detours. In addition, the space under the structure can be used for surface-street traffic, for parking, or for a transit line. If this space is not needed for these purposes, the area under the viaduct may have a high potential value to the community for some form of joint development or use. Such uses may include any of a wide variety of types, ranging from playgrounds to major buildings. Conversely, the disadvantages of the design are the high costs of maintaining the structure and its closed drainage system, susceptibility to icing, difficulty in obtaining a pleasant appearance, and possible need for police protection in the undeveloped space beneath the structure.

An elevated highway on an earth embankment should be of sufficient height to permit intersecting surface roads to pass under it. Freeways on embankments are feasible in suburban areas where crossing streets are widely spaced and where wide right-of-way and fill material are available. Usually, an embankment section occurs on a combination-type freeway in rolling terrain where excavation material from depressed portions is used for the embankment. Where appropriate, the fill may be confined by partial- or full-height walls on one or both sides. In addition, the sloped areas are available for planting to improve the appearance of the freeway.

Medians

Where a freeway is on a continuous viaduct, the median width should generally be the minimum width needed to accommodate the median shoulders and a barrier. When economically feasible, consideration should be given to decking over the opening between parallel structures. Where continuous decking is not feasible, median barriers or guardrails should be installed to stop or redirect an errant vehicle safely. Where a median barrier is used, decking also permits continuity of the barrier.

Ramps and Terminals

The design of ramps and connections for all types of freeways is covered in Chapter 10, but details and controls pertaining specifically to elevated sections are discussed below. Freeways on viaducts are generally located in densely developed areas where property values are high and space is limited. However, the various forms of ramp connections, such as loops, diagonal ramps, and semidirect connections, are as adaptable to elevated freeways as to depressed or other freeway types.

Despite the high cost of elevated freeways, the lengths of speed-change lanes should not be reduced. The length of acceleration and deceleration lanes should conform to the guidelines presented in Chapter 10. Long acceleration lanes are especially needed because a ramp leading to an elevated structure is usually on a relatively steep upgrade. Also, trucks need a considerable distance to accelerate to highway speed.

Gore areas at exits from an elevated structure have a higher than normal crash potential. The design should provide as much space in the gore area as practical, not only for recovery but also to permit the installation of an impact-attenuating device.

Frontage Roads

New frontage roads adjacent to viaduct freeways are not generally needed because the local street network is usually not disturbed. The existing parallel and cross streets are usually adequate to provide local circulation and access; however, frontage roads may be needed for use with embankment freeways to provide adequate local circulation and access. Frontage roads are discussed in Chapter 4, which presents their general features and develops their design values.

Clearance to Building Line

The minimum lateral clearance between a freeway viaduct and adjacent buildings may be a significant cross-sectional element. Major factors where buildings are close to the roadway are (1) working space for maintenance and repairs of structure or buildings, (2) space to prevent salt and water spray damage, (3) protective space against possible fire damage, and (4) space for ladders and other fire-fighting equipment to reach upper floors of buildings from the street below. Building offsets should be sufficient to ensure adequate sight distance to signs where the alignment is curvilinear. A clearance of 4.5 to 6.0 m [15 to 20 ft] is recommended to accommodate these space needs. Without such clearance, the use of some fire-fighting equipment, such as mechanically raised ladders, would be hampered. Some of these units might be operable from the elevated freeway.

Roadways directly under the structure are usually needed to accommodate surface traffic, but the cross section elements are not considered as controls where existing right-of-way determines the structure section.

Typical Cross Section

The total widths of elevated freeway sections, as well as the total right-of-way widths in which they are developed, can vary considerably. For elevated freeways on embankments, the total width needed is about the same as the total width needed for depressed freeways. Elevated freeways on structures may be cantilevered over parallel roadways or sidewalks.

The difference in elevation between the local street and the elevated freeway, except in the case of the multi-level viaduct shown in Exhibit 8-9B, should be approximately 6.0 m [20 ft]. The vertical clearance between the local street and the freeway bridge varies with the legal requirements of the various states, but usually ranges from 4.3 to 5.0 m [14 to 16.5 ft]. Where a railroad is overpassed, a vertical clearance of approximately 7.0 m [23 ft] is needed; thus, the gradeline of the elevated freeway should be about 8.4 m [28 ft] above the tracks. It is advantageous to have the viaduct as low as practical at ramp locations to allow for moderate ramp grades. This results in lower construction costs and greater ease of operation for vehicles using the ramps. These combined factors may justify a rolling freeway profile where it can be developed gracefully; however, a roller-coaster effect should be avoided. Where a viaduct would provide a clearance of less than about 3.0 m [10 ft] from the underside of the structure to the ground, retaining walls or fill are generally recommended unless the space underneath the structure could be used for other purposes such as off-street parking.

Viaduct Freeways Without Ramps

Exhibit 8-9 shows typical cross sections for elevated freeways without ramps on structures. The following dimensions are used for general illustration:

- Lane width is 3.6 m [12 ft].
- Parapet width is 0.6 m [2 ft].
- Shoulder width for four lanes is 3.0 m [10 ft] for the right shoulder and 1.2 m [4 ft] for the left shoulder; for six and eight lanes, shoulder width is 3.0 m [10 ft] for both right and left shoulders.
- Median width is 3.0 m [10 ft] for four lanes and 6.6 m [22 ft] for six and eight lanes.
- Minimum clearance between structure and building line is 4.5 m [15 ft]. For Exhibit 8-9B, the minimum clearance should be 6.0 m [20 ft].

Space under the structure may be available for surface streets or other use by the community.

In Exhibit 8-9A, the overhang enables surface roads to be provided outside the lines of columns, and the area between the columns can be used for vehicular traffic, public transit, or parking.

Where it is impractical to obtain the right-of-way widths needed for a conventional viaduct freeway, it may be practical to convert the normal two-way, one-level structure to a two-level structure. The double-deck design in Exhibit 8-9B is not a common type, but is adaptable to narrow rights-of-way, particularly where few ramps are needed. Double-deck structures may also be adaptable where it is not practical to continue the freeway as a single-deck structure because of large buildings or for other reasons. Conversion to double-deck construction through such confined areas may be the only practical solution. Double-deck structures have the disadvantage of long ramps on structures to allow vehicles to make the necessary change in elevation from the top roadway to the local city streets.

Sometimes an elevated freeway is constructed on two one-way structures, as shown in Exhibits 8-9C and 8-9D. These structures may be separated by one or more city blocks. In addition, the structure may be either a two-column section, as in Exhibit 8-9C, or a single-column, cantilevered section, as in Exhibit 8-9D, depending on the arrangements of understructure streets and other controls.

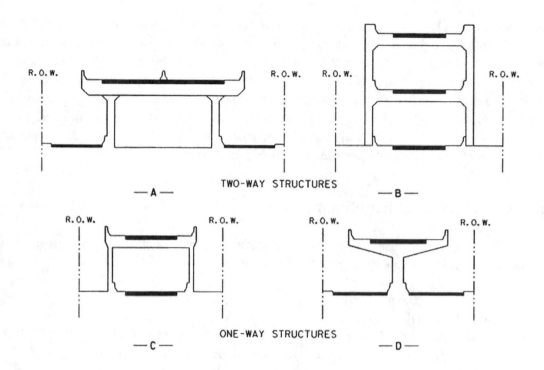

TWO-WAY STRUCTURES

—A—   —B—

ONE-WAY STRUCTURES

—C—   —D—

Exhibit 8-9. Typical Cross Sections for Elevated Freeways on Structures Without Ramps

Two-Way Viaduct Freeways With Ramps

Elevated freeways generally are developed as single-level, two-way structures, of which the basic structure section is shown in Exhibit 8-9A. Cross-sectional arrangements for elevated freeways on structure with ramps and frontage roads are illustrated in Exhibit 8-10. The following dimensions are used for general illustration:

Median width	3.0 to 6.6 m [10 to 22 ft]
Lane width	3.6 m [12 ft]
Right shoulder width:	
Four-lane	3.0 m [10 ft]
Six- and eight-lane	3.0 m [10 ft]
Left shoulder width:	
Four-lane	1.2 m [4 ft]
Six- and eight-lane	3.0 m [10 ft]
Parapet width	0.6 m [2 ft]
Clearance between structure and building line	4.5 m [15 ft]

An elevated section on structure has great flexibility in right-of-way arrangements. The most flexible element is the outer separation. In tight locations where ramps are not provided, the frontage roads can be located under a cantilevered section of the structure, as shown in

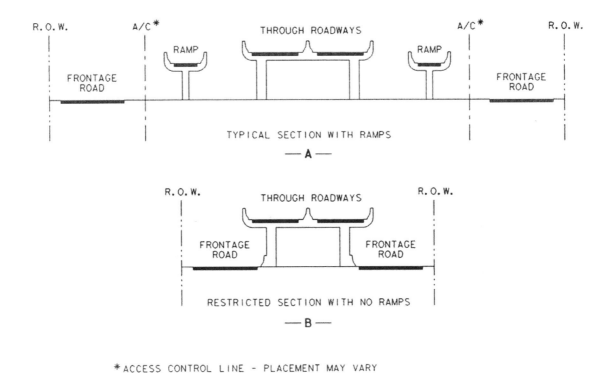

* ACCESS CONTROL LINE - PLACEMENT MAY VARY

Exhibit 8-10. Typical and Restricted Cross Sections for Elevated Freeways on Structure With Frontage Roads

Exhibit 8-10B. At these locations, the minimum building-line clearance may provide sufficient space for frontage roads.

Where there is no need to narrow the right-of-way width before and beyond ramps, it may be practical to obtain liberal clearance between the structure and building line. As a result, there may be space for a green belt strip, parking space off the frontage road, or a wider border and frontage road.

Freeways on Earth Embankment

Elevated freeways may be constructed on earth embankments provided that the embankment is high enough to permit cross streets to pass under the freeway. Such freeways are appropriate where the terrain is rolling and the right-of-way is sufficiently wide to allow gentle side slopes that can be pleasingly landscaped.

Exhibit 8-11 presents typical and restricted cross sections for elevated freeways on embankments. The left halves of these sections illustrate outer separations without ramps within the same right-of-way width. The difference in elevation between the frontage road and the through roadway is approximately 6.0 m [20 ft]. This section provides for median widths of 3.0 to 6.6 m [10 to 22 ft], lane widths of 3.6 m [12 ft], and right shoulder widths of 3.0 m [10 ft].

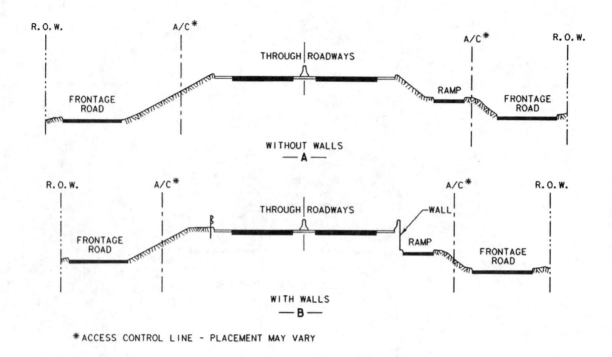

Exhibit 8-11. Typical and Restricted Cross Sections for Elevated Freeways on Embankment

The outer separation may permit the use of earth slopes at locations without ramps, but retaining walls are needed at ramps. In addition, embankment slopes greater than 1V:3H will generally warrant a roadside barrier. By omitting frontage roads and using walled sections, total widths may be reduced to widths typical of elevated structures on viaducts. Special wall treatment or planting of trees and shrubs may make the retaining walls aesthetically pleasing.

Examples of Elevated Freeways

Exhibit 8-12 shows a four-lane viaduct freeway on curved alignment located adjacent to an urban business district. The freeway is situated to minimize the amount of right-of-way needed. Existing cross streets have not been disturbed. In the upper part of the photograph, the freeway section changes from viaduct to embankment.

Exhibit 8-13 shows a two-level viaduct freeway in a densely developed area of a large city. Continuous frontage roads are not provided along this freeway segment. Also, the freeway is constructed on a minimum amount of right-of-way.

Exhibit 8-12. Viaduct Freeway

Ground-Level Freeways

General Characteristics

Many freeways have long segments that are constructed essentially at ground level. This design is often used in flat terrain and along railroads and water courses. Ground-level freeways are also suitable in suburban areas where cross streets are widely spaced. A major consideration in the design of ground-level freeways is the change in profile of each crossroad as it passes over or under the freeway. However, substantial lengths of ground-level freeways are generally not practical in heavily developed areas because the profiles of crossroads cannot be altered without severe impact on the community. The profile changes of cross streets are further discussed in the section on "Combination-Type Freeways" in this chapter.

Where a ground-level freeway follows the grid of a city, it is usually desirable to provide continuous one-way frontage roads that serve as a means of access to and from streets that are not carried across. However, there are situations where two-way frontage roads provide the only means to maintain local service, even though they are less desirable than one-way frontage roads.

Exhibit 8-13. Two-Level Viaduct Freeway

Ground-level freeways usually are employed in outlying sections of metropolitan areas where right-of-way is not as expensive as it is in downtown areas. As a result, the variable width elements of medians, outer separations, and borders are widened to increase the safety and appearance of the freeway.

Typical Cross Section

Exhibit 8-14A illustrates a typical cross section for a ground-level freeway with frontage roads, and Exhibit 8-14B shows a typical cross section without frontage roads. Where additional right-of-way is available, the outer separations and borders should be widened to provide aesthetically pleasing green belts and to insulate the freeway from the surrounding area. In areas where ramp connections are made to frontage roads, the width of outer separations should be increased to allow space for liberal design of ramp and ramp terminals.

Where only four or six lanes are provided initially, it may be desirable to provide the same right-of-way width as proposed for six- and eight-lane construction. In these situations, the median should be widened by multiples of 3.6 m [12 ft] in anticipation of a need for additional lanes. This step simplifies the construction of additional lanes, with nominal cost and minimal disruption to traffic.

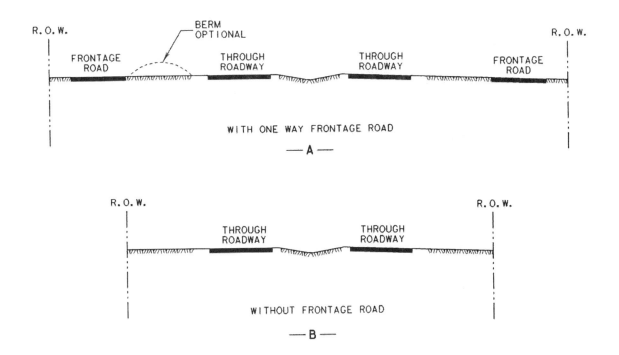

Exhibit 8-14. Typical Cross Sections for Ground-Level Freeways

Where fill material is available and the width of cross section is sufficient to construct safely traversable slopes, an earth berm may be desirable in the median, outer separation, or border. The earth berm shields the freeway from view, lessens highway noises in the adjacent areas, and minimizes headlight glare. Adequate provisions should be made for drainage to ensure that ponding of water does not occur on the shoulder area.

Restricted Cross Section

Exhibit 8-15 illustrates restricted cross sections for ground-level freeways. Specifically, Exhibit 8-15A shows a restricted cross section with a two-way frontage road while Exhibit 8-15B presents a restricted cross section without frontage roads. With restricted cross sections, both the median and outer separation should be paved. On these narrow medians, a median barrier is warranted. With two-way frontage roads, it is also desirable to provide a barrier in the outer separation in place of an access control fence. Preferably, this barrier should be located close to the frontage road to allow extra recovery space outside freeway shoulders. Where there is no fixed-source lighting, a glare screen may also be desirable in the outer separation.

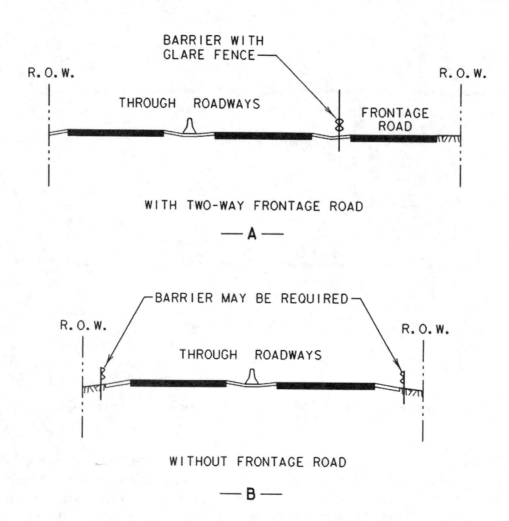

Exhibit 8-15. Restricted Cross Sections for Ground-Level Freeways

Example of a Ground-Level Freeway

Exhibit 8-2 in the previous section on rural freeways shows an example of a typical ground-level freeway. The curvilinear alignment creates an attractive driving environment for motorists.

Combination-Type Freeways

General Characteristics

In many cases, urban freeways incorporate some combination of depressed, elevated, or ground-level designs. Combination-type freeways result from variations in profile or cross section, and the following discussion is organized on the basis of these two controlling conditions.

Profile Control

Rolling terrain. The typical plan and profile of a combination-type freeway in rolling terrain are shown in Exhibit 8-16. The best profile is typically developed by underpassing some cross streets and overpassing others. The facility is neither generally depressed nor elevated, although for short lengths it embodies the design principles for fully depressed or fully elevated highways. For instance, in Exhibit 8-16, at A and C the facility is depressed, at B it is elevated on an earth embankment, and at each end of the illustration it approaches a ground-level section.

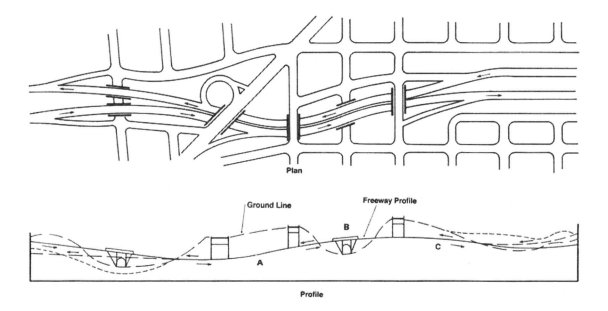

Exhibit 8-16. Profile Control—Rolling Terrain Combination-Type Freeway

Between A and C, the roadway has a fixed cross section with a narrow median because of lateral restrictions and cost of earthwork. Near each end of the illustration, the profile and cross section are varied to fit cross sloping terrain and less rigid controls, with an independently designed centerline and profile for each one-way roadway. This general type of design, which approaches the character of a rural freeway, should be considered in rolling terrain wherever sufficient right-of-way is available.

Flat terrain. A variation of a combination-type freeway in flat terrain is illustrated in Exhibit 8-17. Between grade separation structures, the highway profile closely follows the existing ground. The freeway also overpasses important cross streets by rolling the gradeline to the appropriate height above the surface streets; where practical, cross streets are carried over the freeway (as at A in Exhibit 8-17). This combination-type freeway design is suitable for flat terrain where (1) soil and groundwater conditions or underground utilities preclude depressing the freeway to any great extent below the existing ground, or (2) continuous viaduct construction is too costly or is otherwise objectionable. The freeway may be carried over a cross street on an earth embankment with a conventional grade separation structure (as at B in Exhibit 8-17) or on a relatively long structure (as at C in Exhibit 8-17). The factors that control the profile design are the availability of fill material and the soil conditions. In addition, this combination-type freeway design permits parallel or diagonal ramps to be provided between the grade separations.

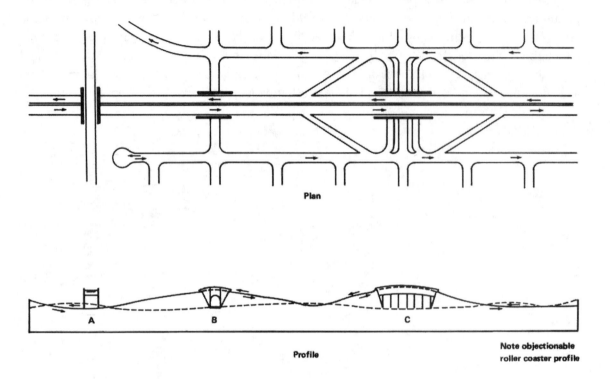

Exhibit 8-17. Profile Control—Flat Terrain Combination-Type Freeway

One of the disadvantages of this design is that a roller-coaster type of profile results where several successive cross streets are overpassed at close intervals. This situation is more pronounced on horizontal tangents where drivers can see two or more grade separations ahead. A moving vehicle that is ahead of a driver may disappear into dips and reappear again as the grade rises to a crest. Therefore, the profile should be designed to eliminate dips that would limit the recommended sight distance. Caution should also be exercised in designing the profile to ensure that adequate sight distance to exit ramps is provided. Where truck traffic is heavy, maximum grades of approximately 2 percent are desirable to prevent queuing at the base of the grade.

To minimize the overall rise and fall and make the rolling profile less pronounced, the cross streets may be depressed several feet below the ground surface and the freeway grade may be raised several feet above the ground level between grade separation structures. The profile may be further improved by raising some cross streets to overpass the freeway. Minimum governing distances for grade separation design are discussed in Chapter 10.

Cross-Section Control

The examples in Exhibit 8-18 are also considered combination-type freeways, but the primary influence on their design is the cross section. These special designs usually apply to relatively short lengths of roadway to meet specific conditions.

Exhibit 8-18A illustrates a design in which one roadway of the freeway is located above the other roadway, with one roadway above the existing ground level and the other below the existing ground. In effect, it is a one-way depressed and a one-way elevated facility separated in elevation to permit the cross streets to pass through on the intermediate (surface) level. This arrangement may be appropriate where the right-of-way is not sufficiently wide for either a two-way elevated or a two-way depressed facility, and where a two-level elevated structure would be objectionable. Wider rights-of-way are needed where ramps are provided.

A city area left underdeveloped because of extremely steep ground might serve as a practical location for a freeway section. A special design with partly elevated and walled sections at staggered levels can be employed, as shown in Exhibit 8-18B. A variety of other designs may also be used, including a one-level, two-way deck structure or a one-level, two-way cut-and-fill section retained by walls. The design selected would depend on the slope of the ground, soil conditions, and right-of-way width. Difficulty may be encountered at cross streets that are likely to have steep grades; however, areas of such topography usually have few cross streets.

Another variation of a combination-type freeway shown in Exhibit 8-18C consists of one through roadway at surface level and the other on an elevated structure; this design may be appropriate along a waterfront or a railroad where the right-of-way is relatively narrow and there are no cross streets. Access to and from the one-way through roadway at surface level is provided directly to streets that cross the frontage road; by contrast, access to and from the elevated roadway is provided by lateral ramps overpassing the frontage road.

Examples of Combination-Type Freeways

Exhibit 8-19 shows an eight-lane, combination-type freeway in a densely developed residential area of a large city. The freeway profile changes smoothly from fill to cut sections so that the profile blends with the local street system.

Exhibit 8-20 shows a six-lane freeway within a very narrow right-of-way in a densely developed area of a large city. The four-level cantilever structure provided is similar to Exhibit 8-18B. The second and third levels carry the directional roadways of the freeway. The top level is a promenade, which enhances the value of the adjacent apartment buildings and shields traffic noise, while the ground level serves local traffic.

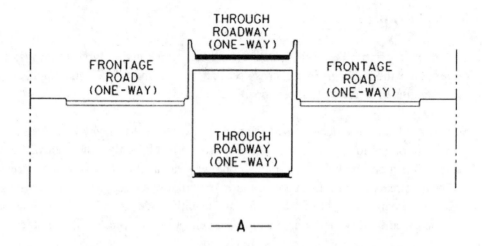

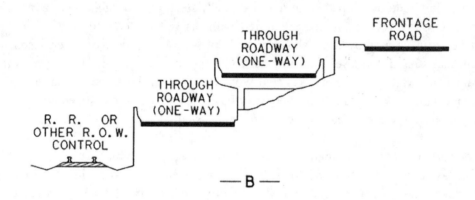

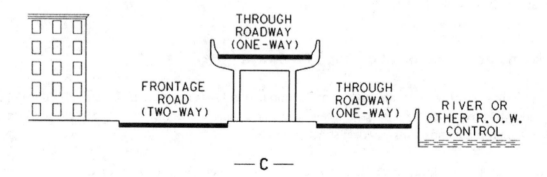

Exhibit 8-18. Cross-Section Control—Combination-Type Freeway

Exhibit 8-19. Combination-Type Freeway

Exhibit 8-20. Four-Level Cantilevered Freeway

Special Freeway Designs

The following section discusses three special freeway designs that may be appropriate in some urban locations: freeways with reverse-flow roadways, dual-divided freeways, and freeways with collector-distributor roadways.

Reverse-Flow Roadways

A reverse-flow roadway is a separate roadway, usually between the main roadways of a freeway, that serves traffic for opposite directions of travel at different times of day. Special conditions may warrant the use of a reverse-flow roadway as a part of the freeway design. This is usually accomplished by placing a separate reversible roadway within the normal median area, as shown in Exhibit 8-21A. Reverse-flow roadways are advantageous in that they provide an opportunity for better operations for motorists, but are disadvantageous in that they may have unused capacity much of the time because of the limited numbers of access points. The costs of construction, maintenance, and operation of a freeway with a reverse-flow roadway also may differ considerably from those of a conventional freeway.

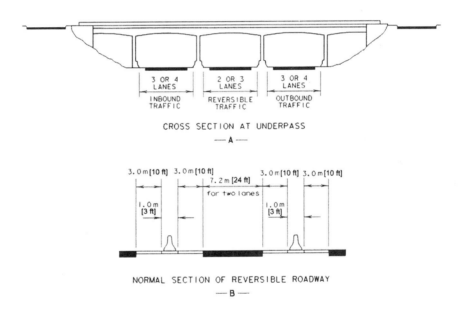

Exhibit 8-21. Typical Cross Sections for Reverse-Flow Operation

A separate reverse-flow roadway may be considered when (1) the directional distribution during peak hours is substantially unbalanced (e.g., a 65:35 percent split) and capacity analysis indicates a need for a conventional facility more than eight lanes wide, (2) design controls and right-of-way limitations are such that providing two or more parallel facilities on separate rights-of-way is not feasible, and (3) a sizable portion of traffic in the predominant direction during peak hours is destined for an area between the central portion of the city or another area of concentrated development and the outlying area (i.e., a large percentage of peak-hour traffic travels a long distance between principal points of origin and destination with little or no need for intermediate interchanges). In some large metropolitan areas, demand may be sufficiently great to justify the use of a reversible roadway exclusively for buses or other high-occupancy vehicles.

The right-of-way width needed for a reverse-flow freeway is not substantially different from that needed for a conventional freeway serving an equivalent traffic volume. In fact, with the dimensions shown in Exhibit 8-21B, the right-of-way needed for a three-two-three reversible freeway is the same as that needed for a conventional 10-lane freeway with a 7.2-m [24-ft] median. It should be noted that the cross section in Exhibit 8-21B uses full right and left shoulders on the reverse-flow roadway because it carries one-way traffic in different directions at different times.

In the central business district, it may be desirable to provide a separate collection and distribution system for the reversible roadway on radial freeways. The normal directional roadways would serve through traffic and freeway-freeway interchanges. Only the reversible roadway would connect directly to downtown streets. In particular, this arrangement enhances the usefulness of the reversible roadway for express bus operation.

Entrance and exit ramps on the reverse-flow roadway should be well spaced, and the entering traffic volume should be balanced with the capacity of the reverse-flow facility. Where there are major connectors, the design should provide for ramps going to and from the reverse-flow roadway and separated in grade from the outer freeway roadways. In cases to date, very few intermediate crossover connections (slip ramps) between the inner and outer roadways have been provided. Such connections need substantial width and length for proper design, usually in areas where the needed space is not available. Furthermore, the resulting weaving maneuvers and left-side exits or entrances are operationally undesirable on a freeway that warrants a reverse-flow roadway. In reverse-flow operations, there will normally be two intervals daily during which the central roadway is closed to change the direction of flow.

Adequate reverse-flow roadway terminals are needed to transfer traffic between the section of freeway with reverse-flow lanes and the conventional freeway section or the local street system. A reversible roadway section is usually terminated by transitioning the three roadways into two normal directional roadways, as shown in Exhibit 8-22A.

As illustrated in Exhibit 8-22A, the end of the reversible roadway is Y-shaped and has the entrance and exit connections on the median side of the normal roadways. The entrance connection leading into the reversing roadway is relatively easy to provide, and there are usually no operational problems at this point. However, the exit connection from the reversing roadway needs careful consideration during the design process to prevent undesirable merging situations and backups during peak flows. As a minimum, the connections should be designed as major forks that are 350 to 600 m [1,200 to 1,800 ft] long. Preferably, additional lanes should be provided on the normal roadway beyond the junction point to the next exit or for a distance of 750 to 1,000 m [2,500 to 3,000 ft]. Such lanes will provide for adequate merging.

Where there is a prominent exit from through lanes in the vicinity of the reversible roadway's terminal, the reversible roadway should be terminated beyond that exit. Conversely,

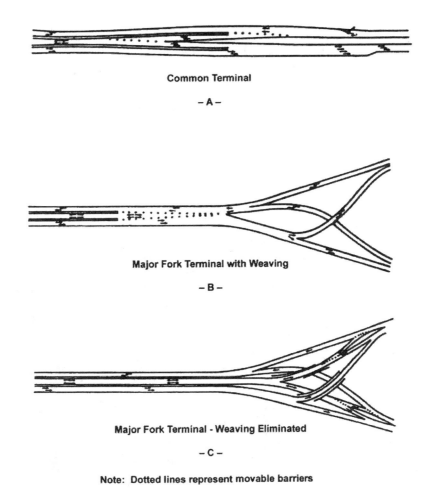

Common Terminal

– A –

Major Fork Terminal with Weaving

– B –

Major Fork Terminal - Weaving Eliminated

– C –

Note: Dotted lines represent movable barriers

Exhibit 8-22. Typical Reverse Roadway Terminals

where there is a prominent entrance near the terminal, the reversible terminal should be located in advance of the entrance. This arrangement minimizes congestion and weaving conflicts.

Where the reversible lanes terminate at a major fork on the freeway, the single-structure arrangement shown in Exhibit 8-22B would involve weaving for traffic entering or exiting from the reversible roadway. This design is not desirable if weaving movements are heavy. Such designs may result in considerable traffic operational problems. When the single-structure arrangement is used, the reversible lanes should be extended along one leg of the fork. This design eliminates weaving but denies access from the reversible lanes to the other leg. Where such an arrangement is not compatible with traffic desires, the weaving can be eliminated by providing another structure and designing the terminal as shown in Exhibit 8-22C.

The devices used for controlling traffic at terminals of a reverse-flow roadway include variable message signs, pavement markings, warning lights, lane-use traffic signals, and mechanically and electronically operated barricades at each terminal of the reverse-flow roadway and at intermediate ramp terminals. Such devices have been under development for years, and several types of installations now in operation are being evaluated.

Exhibit 8-23 illustrates a reverse-flow freeway in a suburban area. The facility illustrated has a three-two-four lane configuration, with the center two lanes operating in the peak flow direction in the morning and evening peak periods. However, the same concept can be used with a three-two-three or a four-two-four lane configuration. The reverse-flow roadway is 7.2 m [24 ft] wide and has 3.0-m [10-ft] shoulders on both sides. The normal directional roadway has 3.6-m [12-ft] lanes and has a 3.0-m [10-ft] shoulder on the right and a 1.8-m [6-ft] shoulder on the left. Each separator between the reverse-flow roadway and the normal roadway is 1.2 m [4 ft] wide. Within each separator is a barrier 0.6 m [2 ft] wide.

Exhibit 8-23. Reverse-Flow Freeway

Dual-Divided Freeways

Where more than eight through lanes are needed and the directional distribution of traffic is sufficiently balanced so that a reversible roadway is not applicable, a dual-divided freeway made up of two one-way roadways in each direction of travel may provide the optimum facility. All four roadways are within the control-of-access lines. This type of cross section is sometimes referred to as "dual-dual." The outer freeway roadways usually serve all of the interchange traffic, but they may also serve a substantial portion of the through traffic. For example, all trucks might be required to use the outer roadways only. Various arrangements are possible, depending on the character of traffic and crossroad conditions.

Dual-divided freeways usually function smoothly and carry extremely high volumes of traffic efficiently. Motorists using the inner roadways are removed from the weaving movements at frequently spaced interchanges, and disabled vehicles in either the inner or outer roadway can quickly be steered to a shoulder by traversing a minimum number of lanes.

Dual-divided construction may be the most practical solution to widening an existing freeway where the present traffic volumes are so great that the disruption in traffic during complete reconstruction cannot be tolerated. Where the future need can be anticipated and sufficient right-of-way can be reserved, it may be practical to develop a dual-divided facility in two stages.

Dual-divided facilities have great flexibility in their operation and maintenance. For example, during maintenance or reconstruction operations, one of the directional roadways may be temporarily closed during off-peak hours. Crash potential is greatly reduced by eliminating traffic conflicts with construction or maintenance work. Another advantage of dual-divided design is that the affected roadway can also be closed in case of a crash or other emergency, thus facilitating clean-up operations.

Disadvantages of dual-divided facilities are the wide expanse of pavement and heavy traffic volumes that may have a disruptive influence on an established community and tend to cut off the continuity of the area. The dual-roadway system reduces the flexibility of traffic distribution, resulting in uneven distribution among lanes. The costs for right-of-way, construction, and maintenance may be greater than those on a normal divided facility with an equal number of lanes. Snow removal from dual-divided facilities is also difficult in the inner lanes.

Roadway arrangements for a dual-divided freeway include four-four-four-four, three-three-three-three, three-two-two-three, two-three-three-two, or other suitable combinations of lanes. Typical cross sections are comparable to those previously described for various types of freeways with frontage roads, except that there are four, rather than two, main roadways. Each of the outer medians would have a median barrier and full shoulders on each side.

Exhibit 8-24 shows a typical layout for a dual-divided freeway. All interchange connections are made to the outer roadways, and intermediate transfer connections are provided between the inner and outer roadways so that traffic on the inner roadways can use the interchanges. The number of such transfer connections should be kept to a minimum, with one set serving several successive interchanges. There should be a spacing of 750 m [2,500 ft] or more between the terminal of a transfer connection and an exit ramp. The adequacy of all weaving lengths should be checked.

Exhibit 8-25 is an example of a dual-divided freeway with a four-three-three-four roadway arrangement and frontage roads. Rail rapid transit is carried in the median, and pedestrians have access to the transit station from the cross street. As explained above, all ramp connections are made to the outer roadways.

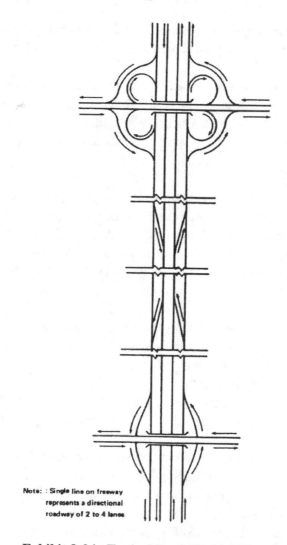

Note: : Single line on freeway
represents a directional
roadway of 2 to 4 lanes

Exhibit 8-24. Typical Dual-Divided Freeway

Exhibit 8-25. Dual-Divided Freeway With a 4-3-3-4 Roadway Arrangement

Freeways With Collector-Distributor Roads

An arrangement having cross-sectional elements similar to the dual-divided freeway is a freeway with a collector-distributor (C-D) road system. The purpose of a C-D road is to eliminate weaving on the mainline freeway lanes and reduce the number of entrance and exit points on the through roadways while satisfying the demand for access to and from the freeway. C-D roads may be provided within a single interchange (as discussed in Chapter 10), through two adjacent interchanges, or continuously through several interchanges of a freeway segment. Continuous C-D roads are similar to continuous frontage roads except that access to abutting property is not permitted.

The inside high-speed through roadways are identified as core roads, and the outside slower speed roadways are identified as C-D roads. Usually, the traffic volumes on the C-D system are less than those encountered on the dual-divided freeway, with fewer lanes. The minimum lane arrangement for a C-D system is two C-D, two core, two core, two C-D; however, other combinations may be developed as capacity needs warrant. Continuous C-D roads should be integrated into a basic lane design to develop an overall system. Capacity analysis and basic lane determination should be performed for the overall system rather than for the separate roadways.

Connections between the core roadways and C-D roads are called "transfer roads." Transfer roads may be either one or two lanes in width, and the principle of lane balance applies to the design of transfer roads on both the core and C-D roadways. Both transfer and C-D roads should have shoulders equal in width to those of the core roadways. The outer separation should be as wide as practical with an appropriate barrier. Terminals of transfer roads should be designed in accordance with guidelines for ramp terminals, as presented in Chapter 10.

The design speed of C-D roadways is usually less than that of the core roadways because most of the turbulence caused by weaving occurs on the C-D roadways. A maximum reduction in design speed of 20 km/h [10 mph] is preferable for continuous C-D road systems.

Accommodation of Transit and
High-Occupancy Vehicle Facilities

General Considerations

Combining mass transit or high-occupancy vehicle facilities with freeways is a means for providing optimum transportation services in larger cities. This type of improvement can be accomplished by the joint use of right-of-way to include rail transit or separate roadway facilities for buses and other high-occupancy vehicles, such as car and van pools. The total right-of-way cost not only is less than those for two separate land strips, but the combination also preserves taxable property, reduces the displacement of businesses and persons, and lessens impact on neighborhoods. In some cases, mass transit is incorporated into existing freeway systems. Reverse-flow roadways in the median and reserved lanes work well for exclusive bus and high-occupancy vehicle use during rush hours.

When transit, either bus or rail, is located within the freeway median, access to the transit vehicles is generally obtained from the crossroad at interchange locations. Such an arrangement does not lend itself to intermodal transfer. Transfer to and from buses or passenger cars adds congestion to the interchange area, and off-street parking is usually so remote from interchange areas that it discourages some transit ridership. Reverse-flow roadways, like the one in the median of the freeway shown in Exhibit 8-26, can also be operated as exclusive bus roadways. Bus roadways within the median essentially restrict operations to the line-haul or express type, because ramps that would permit collection and distribution from the median area are expensive or operationally undesirable. Furthermore, when freeways undergo major repair or reconstruction, it is often desirable to construct crossovers and temporarily shift all traffic onto one roadway. Where transit is located within the median, such temporary crossovers are not possible without a complete disruption of transit operations.

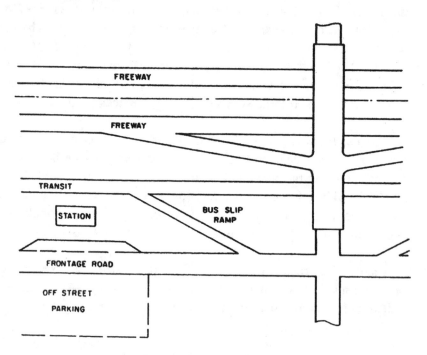

Exhibit 8-26. Bus Roadway Located Between a Freeway and a Parallel Frontage Road

Where the transit facility is parallel to the freeway but located to one side rather than in the median area, these objections are overcome. Exhibit 8-26 shows a bus roadway located between the freeway and a parallel frontage road. Access to the bus roadway is obtained from the frontage road. The station is removed from the congestion of the interchange area, adequate space is available for auto or bus turnouts, and space for off-street parking may be more readily available. All factors combine to enhance intermodal transfers. Slip ramps from the bus roadway to the frontage road permit collection and distribution, in addition to line-haul or express operation, without disruption of freeway operations. A similar arrangement can serve fixed-rail transit except that the slip ramp would be omitted.

Buses

True rapid transit service by bus has had only limited application because normal bus service usually combines collection and distribution with suburb-to-city transportation, and most street or highway facilities for such bus routes are not adaptable to high-speed operation. Many metropolitan areas have non-stop freeway express buses that operate on the freeway system from suburban pickup points near the freeway to locations within the central business district or to other heavy traffic generators. The number of buses operating during peak hours, the spacing of bus stops, and the design of bus turnouts determine the efficiency of bus operation and its effect on highway operations. Buses operating with short headways and frequent pickup and discharge points are likely to accumulate at stops and interfere with through traffic. On the other hand, express bus operation with few, if any, stops along the freeway provides superior transit service for outer urban areas and affects freeway operation the least.

Exclusive HOV lanes. In addition to express service, other operational means should be considered to reduce the travel time of the public transportation user when demand warrants. An exclusive HOV roadway is an entire highway facility reserved at all times solely for the use of buses or buses and other HOVs. This facility offers buses and HOVs a high level of service and decreases travel time for the users. HOV lanes and roadways are discussed in the AASHTO *Guide for the Design of High-Occupancy Vehicle Facilities* (**6**). A discussion of the park-and-ride facilities that are often provided with HOV lanes is contained in the AASHTO *Guide for the Design of Park-and-Ride Facilities* (**7**).

Bus stops. The spacing of bus stops largely determines the overall speed of buses. Bus stops on freeways should be spaced to permit buses to operate at or near the prevailing speed of traffic on the highway. To achieve this goal, a spacing of at least 3.5 km [2 mi] between bus stops is normally appropriate.

Bus stops along freeways are usually located at intersecting streets where passengers transfer to or from other lines or passenger cars. These stops may be provided at the freeway level, which requires stairs, ramps, or escalators, or at the street level, which requires bus access via interchange ramps. Bus turnouts should be located where site conditions are favorable and, if practical, where gradients on the acceleration lane are flat or downward. The design of bus turnouts is discussed in Chapter 4.

Bus-stop arrangements. The benefit of bus stops located at the freeway level is that buses consume little additional time other than that for stopping, loading, and starting. The disadvantage is that turnouts, stairways, and possibly extra spans at separations may be needed. With bus stops at street level, less special construction is needed and passengers do not need to use stairs or ramps. However, buses have to mix with traffic on the ramps and frontage roads and generally must cross the intersecting street at grade. Where traffic on the surface street is light, these disadvantages are lessened; however, where the streets are operating at or near capacity, buses crossing them will experience some delay. Generally, street-level stops are appropriate in and near downtown districts, and either street- or freeway-level stops are appropriate in suburban and outlying areas. Combinations of these two types may be used on any one freeway.

Bus stops at freeway level. Bus stops logically are located at street crossings where passengers can use the grade-separation structure for access from either side of the freeway. Exhibit 8-27A shows an arrangement at an overcrossing street without an interchange. The turnouts and loading platforms are under the structure, requiring greater span lengths or additional openings. Each stairway should be located on the side of the cross street used by most passengers. Two additional stairways can eliminate any crossings of surface streets by transferring riders.

Exhibit 8-27B shows an arrangement at an undercrossing street without an interchange. As indicated at the top left of this exhibit, platform exits and entrances may be connected directly to adjoining developments such as public buildings and department stores.

Sometimes transit stops are needed at locations other than at overcrossing streets, such as in outlying areas or in built-up districts where it is neither feasible nor desirable to provide stops at cross-street structures. Such stops preferably should be located opposite cross streets intercepted by frontage roads or major passenger walkways. A pedestrian overpass is needed to make bus stops usable from either side of the freeway. Exhibit 8-27C illustrates two likely layout plans. In the lower half of the exhibit, the turnout is located at the freeway level under the pedestrian structure. Pedestrians may reach this structure by stairs or ramps. An alternative layout, shown in the upper half of the exhibit, features a turnout located at the level of the frontage road, eliminating the need for passengers to climb stairs or ramps.

Exhibit 8-28 illustrates bus stops located at freeway level on a depressed section of freeway with diamond-type interchange ramps connecting to one-way frontage roads. The bus stops are located under the cross streets. In Exhibit 8-28A, the entrance to the turnout is located beyond the exit ramp nose, and the exit from the turnout is located in advance of the entrance ramp nose. In Exhibit 8-28B, buses use the freeway ramp exit to enter the turnout. In this case, the bus stop is usually located through a separate structure opening. Such consolidation of access points improves the efficiency of through and ramp traffic. Bus drivers readily adapt themselves to the appropriate route to enter and exit the bus turnout.

Exhibit 8-29 shows a bus stop between the outer connection and the loop of a cloverleaf interchange. A collector or distributor road is desirable so that the bus turnouts will not connect directly to the through roadway. The bus turnout should preferably be located beyond the structure to minimize conflicts. When the turnout is located in advance of the structure, buses must merge with traffic from the entrance loop and weave with traffic destined for the exit loop.

Bus stops at street level. Street-level bus stops can be provided at interchanges. For example, on diamond ramps, the bus stop may consist of a widened shoulder area adjacent to the ramp roadway or may be on a separate roadway. Generally, street-level bus stops adjacent to on-ramps are preferred.

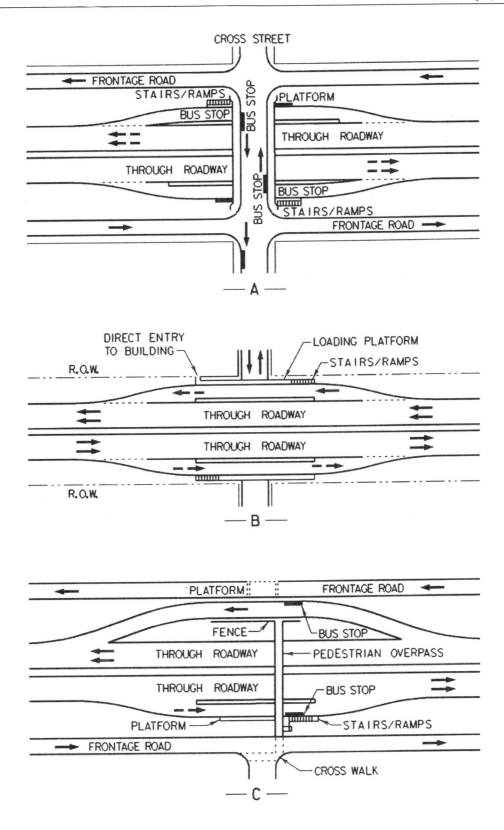

Exhibit 8-27. Bus Stops at Freeway Level

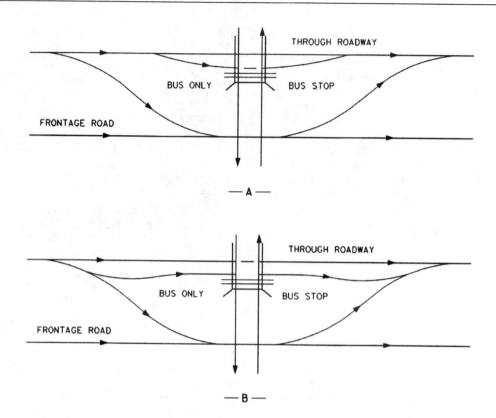

Exhibit 8-28. Bus Stops at Freeway-Level Diamond Interchange

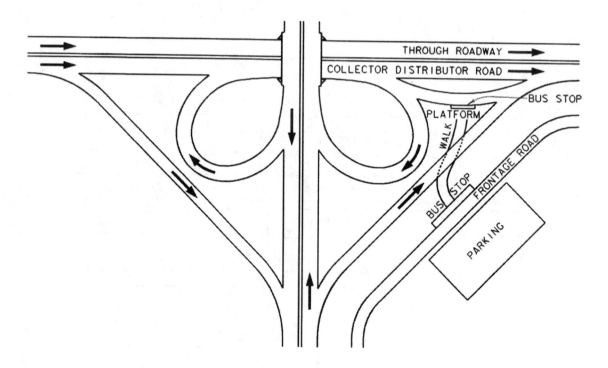

Exhibit 8-29. Freeway-Level Bus Stop at Cloverleaf Interchange

Exhibit 8-30 shows several examples of street-level bus stops on diamond interchanges. Exhibit 8-30A illustrates two possible locations for a bus stop at a simple diamond interchange without frontage roads. The bus stop can be provided on either the on-ramp or off-ramp by widening the ramp. An analysis of turning conflicts should be made to determine the feasibility and appropriateness of either option.

Exhibit 8-30B illustrates a street-level bus stop on a one-way frontage road at diamond interchanges. Buses use the off-ramp to reach the surface level, discharge and load their passengers at the cross street, and proceed via the on-ramp. Added travel distance is minimal, and where traffic on the cross street is light, little time is lost. However, where cross-street traffic is heavy and buses are numerous, operation may be difficult because buses must weave with the frontage road traffic to reach the sidewalk, cross the cross street, and then weave again on their way to the on-ramp.

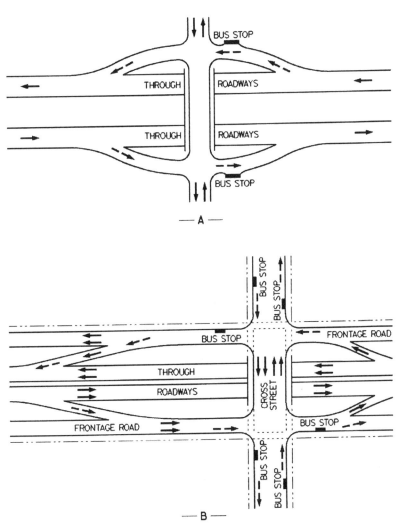

Exhibit 8-30. Bus Stops at Street Level on Diamond Interchange

Street-level bus stops are difficult to provide effectively within cloverleaf or directional interchanges. Consequently, bus stops should be omitted at such interchanges or be located on the cross street beyond the limits of the interchange.

Stairs, ramps, and escalators. With bus stops at the freeway level, stairs, ramps, escalators, or combinations of these are needed for passenger access between the freeway and local street levels. Transit facilities must be accessible to persons with disabilities. Therefore, stair-only access at transit stops is not permitted. Stairways and ramps at transit stops should be easy to climb and present an inviting appearance. This effect is partially accomplished by providing railings and ample lighting both day and night and by providing landings at every 1.8- to 2.4-m [6- to 8-ft] change in elevation. A covering over the stairways, ramps, and platforms may also be desirable. Stairways should be located where the climb will be minimal, preferably not more than 5.4 to 6.0 m [18 to 20 ft]. Where space is available, and only buses are to be served, the bus roadway under the structure might be raised 0.6 to 1.2 m [2 to 4 ft] by reducing vertical clearance to about 3.8 m [12.5 ft] (Most intra-city buses are less than 3.0 m [10 ft] high). When the stairs are located a little distance from the point of loading and unloading, the connecting walkway may be inclined at about a 4 percent grade, and another 0.3 to 0.6 m [1 to 2 ft] may be gained in elevation. Thus, it may be practical in some instances to reduce stairway height to 4.5 m [15 ft] or less.

Stairs and ramps are likely to be installed at bus stops in built-up districts. In addition, pedestrian ramps are well adapted to bus stops in suburban or park-like areas. Railings are desirable and usually necessary and combinations of ramps and stairs may be appropriate at some locations. If the bus line serves a large percentage of older passengers, is extremely busy, or the climb is extra long, the use of escalators should be considered. Provisions for persons with disabilities are to be included, such as the use of ramps and elevators, the widening of passageways and doors, and the elimination of other barriers. Chapter 4 and the *Americans with Disabilities Act Accessibility Guidelines* (ADAAG) (**8**) provide guidance on the design of facilities for persons with disabilities.

Rail Transit

Several metropolitan areas have incorporated, or plan to incorporate, rail transit into freeway rights-of-way. Exhibit 8-31 illustrates various arrangements of the joint freeway-transit use of a right-of-way.

Because rail transit installations are so unique and their design is so highly specialized, discussion of only a few general items is appropriate in this policy. Location and design of a rail transit facility are joint undertakings involving several specialized fields of interest. The location and design of stations, terminals, and parking facilities should be considered from the standpoint of serving these facilities by urban streets. Where rail is contiguous to the freeway traveled way, the entire highway design is affected. The design should ensure the safety of both highway and transit users.

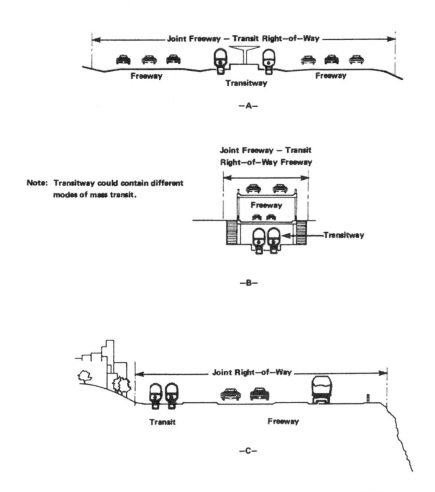

Exhibit 8-31. Joint Freeway-Transit Right-of-Way

The most common arrangement is to place the transit line within the median of a depressed or ground-level freeway, as shown in Exhibits 8-25, 8-31A, 8-32, and 8-34. When a rail transit line is placed in the middle of a freeway, it becomes an island separated from its passengers by lanes of rapidly moving vehicles. Access is generally provided by stairs or ramps connected to grade-separation structures. Where a rail transit line runs down one side of the freeway, accessibility to the transit is simplified, but the construction of interchange ramps becomes more costly. In some situations an alternate solution may be to stack freeway lanes above the transit line, as shown in Exhibit 8-31B, using a minimum of right-of-way. An additional level for cross traffic of vehicles and pedestrians may be needed to complete the movements to be handled.

Exhibit 8-31C illustrates an arrangement where a topographic feature, such as the river on the right, presents a natural deterrent to development on one side. The transit line is placed on the inside for easier access to the community. Where the area has scenic value, this arrangement presents motorists with an open view.

Typical sections. Exhibit 8-32 illustrates typical sections with rail transit provided in the freeway median. The dimensions given are illustrative and should not be considered as guidelines

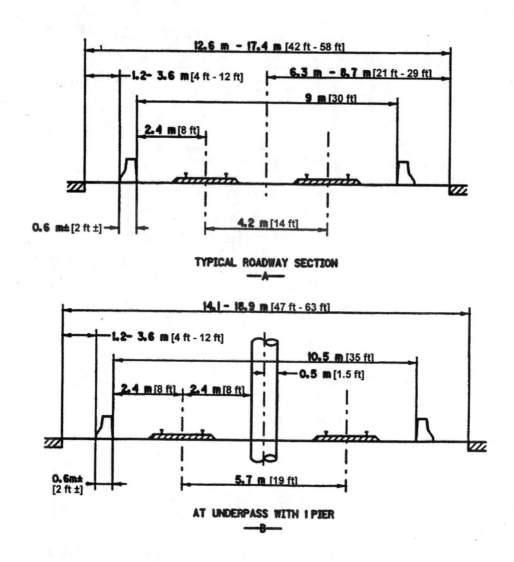

Exhibit 8-32. Typical Sections With Rail Transit in Freeway Median

or requirements. The rail transit dimensions and clearances are typical of guide dimensions that provide for general space needs. The American Railway Engineering and Maintenance-of-Way Association's *Manual for Railway Engineering* (**9**) is one source of current design criteria for railway dimensions and clearances. Exhibit 8-32A illustrates a minimum section without piers; by contrast, Exhibit 8-32B illustrates a minimum section with one median pier. Piers should be provided with crash walls to deflect transit cars in the event of a derailment. A fence should be mounted on or adjacent to the barrier to prevent pedestrians from entering the rail area. A screen or fence may also be needed to reduce the glare to motorists from the headlights of the mass transit vehicle. If a semirigid barrier is utilized at the shoulder edge, the dynamic deflection of the barrier should be added to the dimensions shown.

Where the rail line is placed along the outside of the freeway, reference can be made to the section on "Horizontal Clearance to Obstructions" in this chapter for additional information on clearances.

Stations. The transit station location and spacing should be in keeping with the environment and passenger flow. Frequent stations may be needed within the central business district and other heavy traffic generators, but few stations would be needed in the outlying or suburban areas. Downtown stations should be within easy walking distance of the business and working centers or a feeder bus system. Outlying stations should provide ample parking and storage for vehicles waiting to pick up passengers. Access to local buses and taxis also should be available. Two general layouts for a rail transit station at a local cross street or pedestrian overcrossing are shown in Exhibit 8-33 with typical control dimensions. The dimensions given are illustrative and should not be considered as guidelines or requirements. The station shown in Exhibit 8-33B makes better use of the available space by allowing more separation between the transit passengers and the freeway, while still maintaining ample distance between the train and the traveled way.

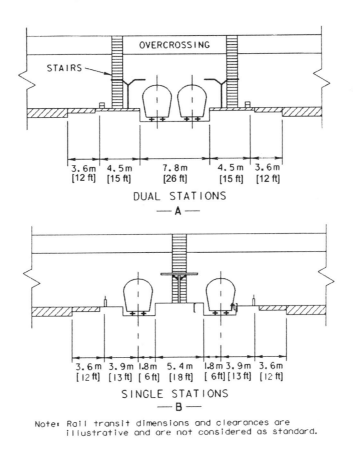

Exhibit 8-33. Example of Transit Station Layout

Example of rail transit combined with a freeway. Exhibit 8-34 presents an eight-lane depressed freeway with rail rapid transit in a median of minimum width. A single station is provided between two major cross streets. Pedestrian rampways connect to the cross streets, with the shelters being at the cross-street level. The traffic density on the near cross street illustrates why it is generally undesirable to have pedestrian access on a cross street with freeway ramps.

Exhibit 8-34. Depressed Freeway With Rail Rapid Transit in the Median

REFERENCES

1. Transportation Research Board. *Highway Capacity Manual*, Fourth Edition, Washington, D.C.: Transportation Research Board, 2000 or most current edition.
2. AASHTO. *Standard Specifications for Highway Bridges*, Washington, D.C.: AASHTO, 1996.
3. AASHTO. *LRFD Bridge Design Specification,* second edition, Washington, D.C.: AASHTO, 1998.
4. AASHTO. *Roadside Design Guide*, Washington, D.C.: AASHTO, 1996.
5. AASHTO. *Model Drainage Manual*, Washington D.C.: AASHTO, 1999.
6. AASHTO. *Guide for the Design of High-Occupancy Vehicle Facilities*, Washington, D.C.: AASHTO, 1992.
7. AASHTO. *Guide for the Design of Park-and-Ride Facilities*, Washington, D.C.: AASHTO, 1992.
8. Architectural and Transportation Barriers Compliance Board (Access Board). *Americans with Disabilities Act Accessibility Guidelines* (ADAAG), Washington, D.C.: July 1994 or most current edition.
9. American Railway Engineering and Maintenance-of-Way Association. *Manual for Railway Engineering*, Hyattsville, Maryland: AREMA, 1999.

CHAPTER 9
INTERSECTIONS

INTRODUCTION

An intersection is defined as the general area where two or more highways join or cross, including the roadway and roadside facilities for traffic movements within the area. Each highway radiating from an intersection and forming part of it is an intersection leg. The most common intersection at which two highways cross one another has four legs. It is not recommended that an intersection have more than four legs.

Intersections are an important part of a highway facility because, to a great extent, the efficiency, safety, speed, cost of operation, and capacity of the facility depend on their design. Each intersection involves through- or cross-traffic movements on one or more of the highways and may involve turning movements between these highways. Such movements may be facilitated by various geometric design and traffic control, depending on the type of intersection.

The three general types of highway crossings are at-grade intersections, grade separations without ramps, and interchanges. This chapter deals primarily with the design of intersections at grade; the latter two intersection types are discussed separately in Chapter 10. Certain intersection design elements, primarily those concerning the accommodation of turning movements, are common and applicable to intersections and to some parts of certain interchanges. The design elements in the following discussions apply to intersections and their appurtenant features.

GENERAL DESIGN CONSIDERATIONS AND OBJECTIVES

The main objective of intersection design is to facilitate the convenience, ease, and comfort of people traversing the intersection while enhancing the efficient movement of motor vehicles, buses, trucks, bicycles, and pedestrians. Intersection design should be fitted closely to the natural transitional paths and operating characteristics of its users.

Five basic elements should be considered in intersection design.

- Human Factors
  - Driving habits
  - Ability of drivers to make decisions
  - Driver expectancy
  - Decision and reaction time
  - Conformance to natural paths of movement
  - Pedestrian use and habits
  - Bicycle traffic use and habits

- Traffic Considerations
  - Design and actual capacities

- Design-hour turning movements
- Size and operating characteristics of vehicle
- Variety of movements (diverging, merging, weaving, and crossing)
- Vehicle speeds
- Transit involvement
- Crash experience
- Bicycle movements
- Pedestrian movements

- Physical Elements
  - Character and use of abutting property
  - Vertical alignments at the intersection
  - Sight distance
  - Angle of the intersection
  - Conflict area
  - Speed-change lanes
  - Geometric design features
  - Traffic control devices
  - Lighting equipment
  - Safety features
  - Bicycle traffic
  - Environmental factors
  - Cross walks

- Economic Factors
  - Cost of improvements
  - Effects of controlling or limiting rights-of-way on abutting residential or commercial properties where channelization restricts or prohibits vehicular movements
  - Energy consumption

- Functional Intersection Area

An intersection is defined by both its functional and physical areas (**1**), as illustrated in Exhibit 9-1. The functional area of an intersection extends both upstream and downstream from the physical intersection area and includes any auxiliary lanes and their associated channelization.

The functional area on the approach to an intersection or driveway consists of three basic elements: (1) perception-reaction distance, (2) maneuver distance, and (3) queue-storage distance. These elements are shown in Exhibit 9-2. The distance traveled during the perception-reaction time will depend upon vehicle speed, driver alertness, and driver familiarity with the location. Where there is a left-or right-turn lane, the maneuver distance includes the length needed for both

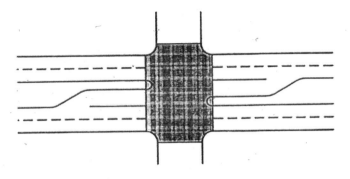

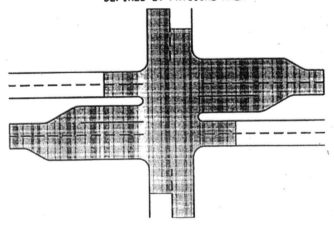

DEFINED BY FUNCTIONAL INTERSECTION AREA

Exhibit 9-1. Physical and Functional Intersection Area

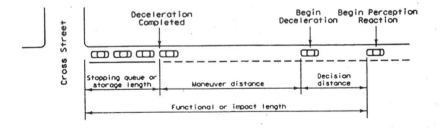

Exhibit 9-2. Elements of the Functional Area of an Intersection

braking and lane changing. In the absence of turn lanes, it involves braking to a comfortable stop. The storage length should be sufficient to accommodate the longest queue expected most of the time.

Ideally, driveways should not be located within the functional area of an intersection, as described above and shown in Exhibit 9-1, or in the influence area of an adjacent driveway.

TYPES AND EXAMPLES OF INTERSECTIONS

General Considerations

The basic types of intersections are the three-leg or T, the four-leg, and the multileg. At each particular location, the intersection type is determined primarily by the number of intersecting legs, the topography, the character of the intersecting highways, the traffic volumes, patterns, and speeds, and the desired type of operation.

Any of the basic intersection types can vary greatly in scope, shape, and degree of channelization. Once the intersection type is established, the design controls and criteria discussed in Chapter 2 and the elements of intersection design presented in Chapter 3, as well as in this chapter, should be applied to arrive at a suitable geometric plan. In this section each type of intersection is discussed separately, and likely variations of each are shown. It is not practical to show all possible variations, but those presented are sufficient to illustrate the general application of intersection design. Many other variations of types and treatment may be found in the NCHRP Report 279, *Intersection Channelization Design Guide* (**2**), which shows detailed examples that are not included in this policy.

Although many of the intersection design examples are located in urban areas, the principles involved apply equally to design in rural areas. Some minor design variations occur with different kinds of traffic control, but all of the intersection types shown lend themselves to cautionary or non-stop control, stop control for minor approaches, four-way stop control, and both fixed-time and traffic-actuated signal control. Right turns without stop or yield control are sometimes provided at channelized intersections. Such free-flow right turns should be used only where an adequate merge is provided. Where motor vehicle conflicts with pedestrians or bicyclists are anticipated, provisions for pedestrians and bicycle movements must be considered in the design. In built-up areas, the use of free-flow right-turn lanes should be considered only where significant traffic capacity or safety problems may occur without them and adequate pedestrian crossings can be provided.

Simple intersections are presented first, followed by more complex types, some of which are special adaptations. In addition, conditions for which each intersection type may be suited are discussed below.

Three-Leg Intersections

Basic Types of Intersections

Basic forms of three-leg or T intersections are illustrated in Exhibits 9-3 through 9-8.

The most common type of T intersection is shown in Exhibit 9-5A. The normal pavement widths of both highways should be maintained except for the paved returns or where widening is needed to accommodate the selected design vehicle. This type of unchannelized intersection is generally suitable for junctions of minor or local roads and junctions of minor roads with more important highways where the angle of intersection is not generally more than 30 degrees from

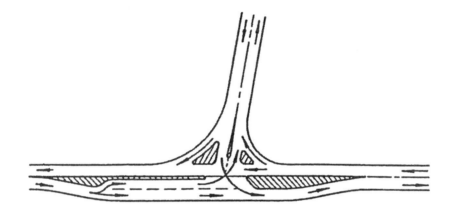

Exhibit 9-3. Channelized High-Type "T" Intersections

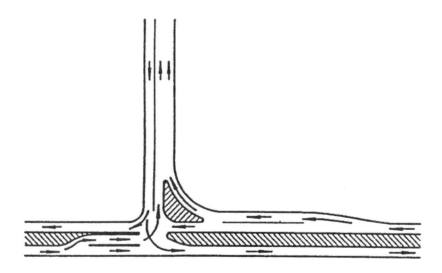

Exhibit 9-4. Three-Leg Rural Intersection, Channelized "T"

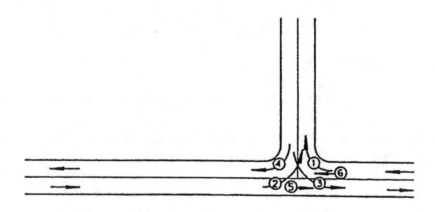

PLAIN 'T' INTERSECTION
-A-

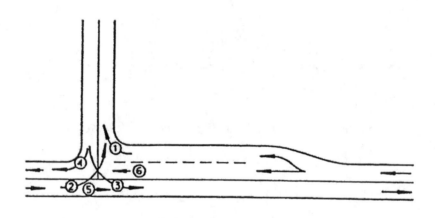

'T' INTERSECTION
(With Right Turn Lane)

- B -

Exhibit 9-5. "T" Intersections

perpendicular (i.e., from approximately 60 to 120 degrees). In rural areas, this intersection type is usually used in conjunction with two-lane highways carrying light traffic. In suburban or urban areas, it may be satisfactory for higher volumes and for multilane roads.

Where speeds and turning movements are high, an additional area of surfacing or flaring may be provided for maneuverability, as shown in Exhibit 9-5B and 9-6.

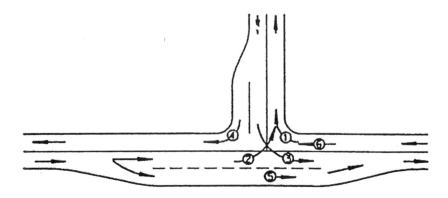

**'T' INTERSECTION (WITH RIGHT
HAND PASSING LANE)**

- A-

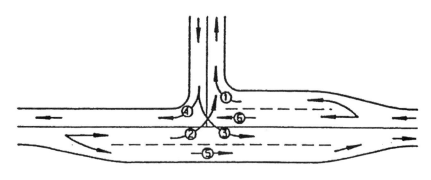

**'T' INTERSECTION (WITH RIGHT HAND
PASSING LANE AND RIGHT TURN LANE)**

- B -

Exhibit 9-6. Channelized "T" Intersections

The use of auxiliary lanes, such as left- and right-turn lanes, increases capacity and creates better operational conditions for turning vehicles. Left turns from the through highways are particularly difficult because a vehicle must slow down and perhaps stop before completing the turn. Intersections with separate left-turn lanes permit following-through vehicles to maneuver around these slower turning vehicles. Existing intersections can have an auxiliary lane added with minimal difficulties to provide the intersection types shown in Exhibits 9-5B and 9-6.

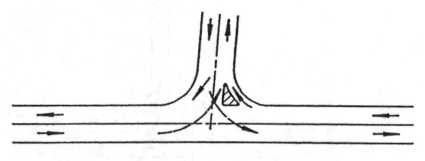

WITH SINGLE TURNING ROADWAY
– A –

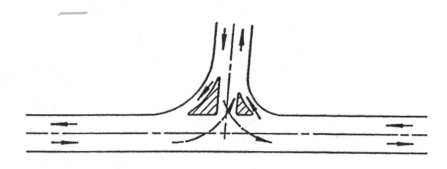

WITH A PAIR OF TURNING ROADWAYS
– B –

Exhibit 9-7. "T" Intersections

Exhibit 9-5B shows an added lane, on the side of the through highway adjacent to the intercepted road, acting as a right-turn lane for vehicles turning right off the major highway. This arrangement is suitable where the right-turning movement from the through highway (movement 1) is substantial and the left-turning movement from the through highway (movement 2) is minor.

Exhibit 9-6 shows an added lane on the side of the through highway opposite the intercepted road. This type of added lane is commonly referred to as a "left-turn lane" or a "right-hand passing lane." This arrangement is suitable where the left-turning movement from the through highway (movement 2) and the through movement (movement 5) are substantial, and the right-turning movement (movement 1) is minor. Here the added lane affords an opportunity for a through driver to pass to the right of a slower moving or stopped vehicle preparing to turn left. A driver turning left from the through highway naturally moves toward the center of the roadway,

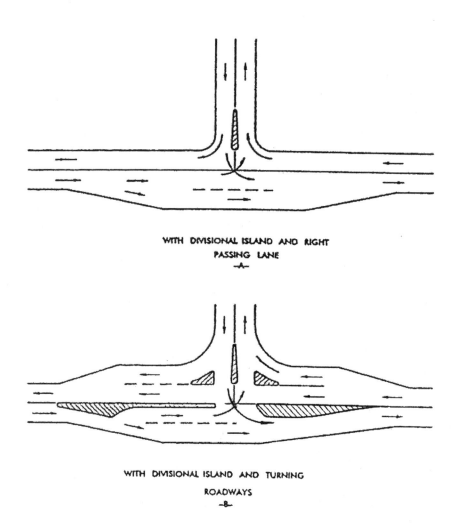

WITH DIVISIONAL ISLAND AND RIGHT
PASSING LANE
-A-

WITH DIVISIONAL ISLAND AND TURNING
ROADWAYS
-B-

Exhibit 9-8. Channelized "T" Intersections

and through traffic is encouraged to pass to the right of the vehicle slowing down or stopping to turn left.

Another flared arrangement, shown in Exhibit 9-6B, may be effected by adding auxiliary lanes on each side of the through highway approaching the intersection. Such an arrangement may be appropriate where the traffic demand at the intersections approaches or exceeds the capacity of a two-lane highway and where signal control may be needed, usually in developing areas. For such conditions in rural areas, the two-lane highway preferably would be converted to a divided section through the intersection, as shown in Exhibit 9-3. In addition to adding auxiliary lanes on the through highway, the road (i.e., the stem of the T intersection) may be widened on one or both sides, as shown in Exhibit 9-6A, for better maneuverability and increased capacity on the intercepted road.

Channelized Three-Leg Intersections

Exhibits 9-7 and 9-8 depict channelized intersections, supported with discussion of general application and functional characteristics. These exhibits, discussed in this section, illustrate various geometric designs that use channelization islands at three-leg intersections. Where an island with a convex section is appropriate, the island should have sufficient cross-sectional area to ensure its proper function. Specifically, to channelize and separate turning movements, the total cross-sectional area of islands should be at least 7 m^2 [75 ft^2]. The undesirable turning paths described below in conjunction with some types of intersections should be given particular attention. Unusual or sophisticated design geometrics should be avoided where practical.

Exhibit 9-7A presents an intersection with a right-turning roadway from the through roadway. This is accomplished by increasing the return radius between the two roadways sufficiently to permit a separate turning roadway that is separated from the normal traveled ways of the intersecting approaches by an island. The approach roadway may include a separate right-turn lane leading to the turning roadway for the accommodation of right-turn traffic. The need for such a right-turning roadway depends on the number of vehicles desiring to turn right as well as the approach speeds and the number of vehicles desiring to continue in the through direction.

Exhibit 9-7B shows an intersection with a pair of right-turning roadways, which is suitable where above-minimum speeds or turning paths are to be provided for these movements. However, this arrangement does not facilitate the left turn from the through highway. Usually, on two-lane highways where right-turning roadways are justified, the flaring of the through highway is also appropriate, as shown in Exhibit 9-8. The right-turning roadway for traffic entering the through highway should be made as narrow as practical to discourage drivers turning left from the through highway from entering this roadway improperly, while still providing sufficient width for anticipated turning trucks.

Exhibit 9-8A depicts a channelized intersection incorporating one divisional island on the intercepted road. Space for this island is made by flaring the pavement edges of the intercepted road and by using larger-than-minimum pavement edge radii for right-turning movements. To fit the paths of left-turning vehicles, the end of the island should generally be located about 2.4 to 3.6 m [8 to 12 ft] from the pavement edge of the through highway. This design is adaptable to two-lane highways over a wide range of volumes, particularly where space is not available for turning roadways and where simplicity is desired. For intermediate-to-heavy volumes (relative to the capacity of the highways), the through highway preferably should be flared, as shown in Exhibit 9-8B.

Exhibit 9-8B shows an intersection with a divisional island and right-turning roadways, a desirable configuration for intersections on important two-lane highways carrying intermediate-to-heavy traffic volumes (e.g., peak-hour volumes greater than 500 vehicles on the through highway with substantial turning movements.) All movements through the intersection are accommodated on separate lanes. The divisional islands shown in Exhibit 9-8 differ in location with respect to the centerline. Either one may be used on each configuration.

Exhibit 9-4 presents a three-leg rural intersection. The two-lane major highway has been converted to a divided highway through the intersection. This intersection is well designed, with liberal use of painted bars in the median. A right-turn lane in the upper right quadrant accommodates a non-restricted exit from the major route, and a separate left-turn storage lane serves higher turning volumes than can be accommodated by the intersection treatment shown in Exhibit 9-6B. A stop sign or signal normally controls the traffic intersecting the through highway.

Four-Leg Intersections

Basic Types

Basic types of four-leg intersections are shown in Exhibits 9-9 through 9-13. The overall design principles, island arrangements, use of auxiliary lanes, and many other aspects of the previous discussion of three-leg intersection design also apply to four-leg intersections.

Exhibit 9-9A illustrates the simplest form of an unchannelized four-leg intersection suitable for intersections of minor or local roads and often suitable for intersections of minor roads with major highways. The angle intersection should not be more than 30 degrees from perpendicular (i.e., from approximately 60 to 120 degrees). Approach pavements are continued through the intersection, and the corners are rounded to accommodate turning vehicles.

Exhibit 9-9B illustrates a flared intersection with additional capacity for through and turning movements at the intersection. Auxiliary lanes on each side of the normal pavement at the intersection enable through vehicles to pass slow-moving and standing vehicles preparing to turn. Depending on the relative volumes of traffic and the type of traffic control used, flaring of the intersecting roadways can be accomplished by parallel auxiliary lanes, as on the highway shown horizontally, or by pavement tapers, as shown on the crossroad. Flaring generally is similar on opposite legs. Parallel auxiliary lanes are essential where traffic volume on the major highway is near the uninterrupted-flow capacity of the highway or where through and cross traffic volumes are sufficiently high to warrant signal control. Auxiliary lanes are also desirable for lower volume conditions. The length of added pavement should be determined as it is for speed-change lanes, and the length of uniform lane width, exclusive of taper, should normally be greater than 45 m [150 ft] on the approach side of the intersection. The length of added pavement on the exit side of the intersection should be 60 m [200 ft] as shown in Exhibit 9-12B.

Exhibit 9-9C shows a flared intersection with a marked pavement area that divides traffic approaching the intersection. This configuration makes provision for a median lane suitable for two-lane highways where speeds are high, intersections are infrequent, and the left-turning movements from the highway could create a conflict. Pavement widening should be effected gradually, preferably with pavement-edge reverse curves with radii of 1,500 m [5,000 ft] or more or by the use of taper rates appropriate for the design speed. The marked pavement area should be

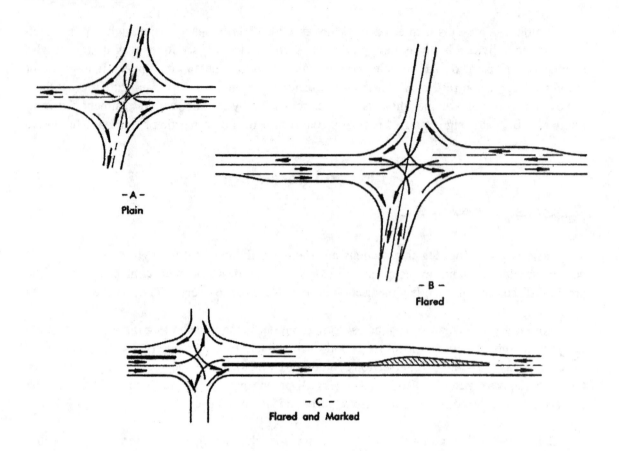

Exhibit 9-9. Unchannelized Four-Leg Intersections, Plain and Flared

at least 3.6 m [12 ft] wide at its widest point, and the through-traffic lane on each side of it should be 0.5 to 1 m [2 to 3 ft] wider than the normal lane width on the approaches. Near the crossroad, where the full widening is attained, the overall width of pavement is about 12 m [40 ft]. This configuration affords better protection for vehicles turning left from the major highway than does the arrangement in Exhibit 9-9B, which is better suited for intersections with signal control. An island marked on the pavement is not as positive a separator as a curbed divisional island, but it is appropriate where sand or snow may be a maintenance problem and where any curbed island may be an obstruction, as on high-speed rural highways.

Channelized Four-Leg Intersections

The usual configurations of four-leg intersections with simple channelization are shown in Exhibits 9-10 and 9-11. Except at minor intersections, right-turning roadways are often provided, as shown in Exhibit 9-10A, for the more important turning movements, where large vehicles are to be accommodated, and at minor intersections in quadrants where the angle of turn greatly exceeds 90 degrees.

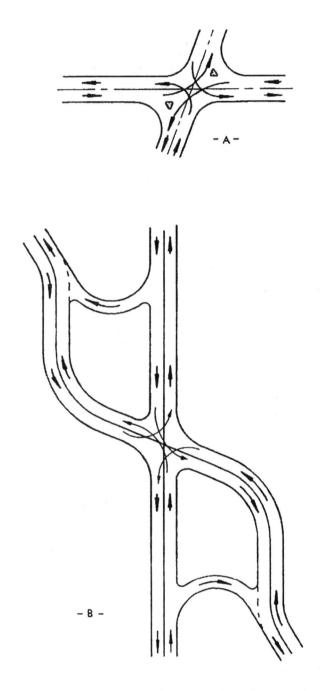

Exhibit 9-10. Channelized Four-Leg Intersections

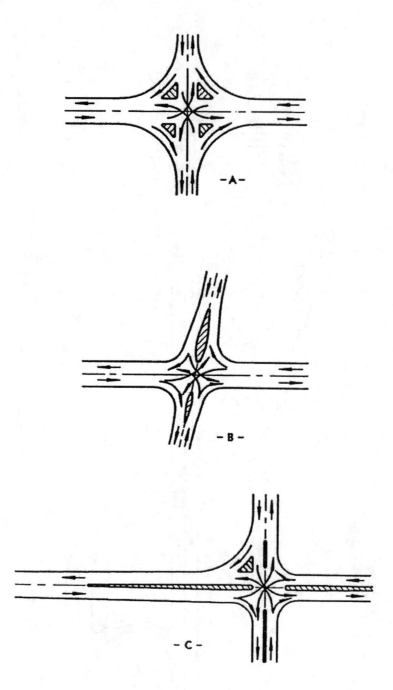

Exhibit 9-11. Channelized Four-Leg Intersections

Exhibit 9-10B shows an oblique-angle intersection with a skew angle of 45 degrees or more with separate turning roadways for two-way traffic in the acute-angle quadrants. Vehicles can turn readily to the right or left in this configuration, and awkward maneuvers and encroachments at the intersection proper are eliminated. The multiple points of intersection, the large skew angle, and driver expectancy may combine to make this type of intersection undesirable. Preferably, one or both highways should be realigned to reduce the skew angle. When realignment cannot be obtained, extensive application of appropriate signing and signal controls is recommended.

Exhibit 9-11A shows an intersection configuration with right-turning roadways in all four quadrants. This configuration is suitable where sufficient space is available and turning volumes are high, particularly in suburban areas where pedestrians are present. However, this arrangement is not common where the intersecting highways are only two lanes wide. Where one or more of the right-turning movements need separate turning roadways, additional lanes are generally needed for the complementary left-turning movements. In the latter case, the highway is normally widened, as shown in Exhibits 9-9B and 9-11C.

Exhibit 9-11B illustrates an intersection with divisional islands on the crossroad. This configuration fits a wide range of volumes, and its capacity is governed by the roadway widths provided through the intersection. The simplicity of this configuration, in many cases, makes it preferable to that shown in Exhibit 9-11A.

Exhibit 9-11C shows a configuration that is appropriate, except at a minor crossroad, for an intersection on a two-lane highway operating near capacity or carrying moderate volumes at high speeds. The two-lane approach on the major highway can be converted to a four-lane section with a divisional island. The additional areas are used for speed changes, maneuvering, and storage of turning vehicles. The form of channelization on the crossroad should be determined based on the cross and turning volumes and the sizes of vehicles to be accommodated.

The simplest form of intersection on a divided highway has paved areas for right turns and a median opening conforming to designs shown in later discussions in this chapter. Often the speeds and volumes of through and turning traffic justify a higher type of channelization suitable for the predominant traffic movements.

Exhibit 9-12A shows a high-type intersection on a divided highway. The approach on the right has a heavy left-turn volume that can utilize the auxiliary lane provided in the median. The lower leg of the intersection has a significant right-turn volume that is channelized with a triangular island and added auxiliary lane.

Exhibit 9-12B illustrates another configuration for the intersection of a high-speed divided highway and a major crossroad. Right-turning roadways with speed-change lanes and median lanes for left turns afford both a high degree of efficiency in operation and high capacity and permit through traffic on the highway to operate at reasonable speed. Traffic signal controls should be properly used.

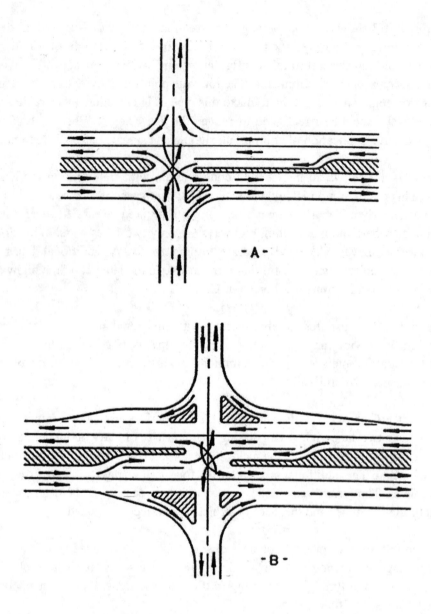

Exhibit 9-12. Four-Leg Intersections (Channelized High-type)

Exhibit 9-13A shows an intersection configuration with dual left-turn lanes for one of the left-turning movements. This configuration needs traffic signal control with a separate signal phase for the dual left-turn movement and is particularly suitable for locations in urban areas where there is a heavy turning movement in one quadrant of the intersection. The auxiliary lanes in the median should be separated from the through lanes by either an elongated island, as shown, or by pavement markings. Furthermore, pavement markings, contrasting pavements, and signs should be used to discourage through drivers from entering the median lane inadvertently. Left-turning vehicles typically leave the through lane to enter the median lane in single file but, once within it, are stored in two lanes. On receiving the green signal indication, left-turn maneuvers are accomplished simultaneously from both lanes. The median opening and the crossroad pavement should be sufficiently wide to receive the two side-by-side traffic streams.

Exhibit 9-13B shows a suitable configuration for an intersection with unusually heavy through volumes and a high left-turning volume in one quadrant. The high-volume left-turn movement is removed from the main intersection by providing a separate diagonal roadway and creating two additional intersections. A high degree of traffic operational efficiency can be attained by a system of progressively synchronized traffic signals and proper signal timing based on the distances and pavement widths between the three intersections. The three intersections should be at least 60 m [200 ft], and preferably 90 m [300 ft] or more, apart. A median lane for the left-turning movement onto the diagonal roadway should be two lanes wide. The right-turning movement using the diagonal roadway may flow continuously, and an auxiliary lane along each of the major roadways may be desirable. This design may be used where a grade separation is not practical, as in flat terrain with traffic having a high volume of heavy trucks, or where it is desired to defer the construction of a grade separation. Where movements in the other quadrants reach the proportions of through movements, additional diagonal roadways might be provided, but with major turning movements in more than one quadrant, a grade separation is generally preferred. Before using the configuration shown in Exhibit 9-13B, careful consideration should be given to its overall operational performance (i.e., delay to motorists) since this design, in effect, creates two additional intersections.

Multileg Intersections

Multileg intersections—those with five or more intersection legs—should be avoided wherever practical. At locations where multileg intersections are used, it may be satisfactory to have all intersection legs intersect at a common paved area, where volumes are light and stop control is used. At other than minor intersections, traffic operational efficiency can often be improved by reconfigurations that remove some conflicting movements from the major intersection. Such reconfigurations are accomplished by realigning one or more of the intersecting legs and combining some of the traffic movements at adjacent subsidiary intersections, as shown in Exhibit 9-14, or in some cases, by converting one or more legs to one-way operation away from the intersection.

Exhibit 9-14A shows the simplest application of this principle on an intersection with five approach legs. The diagonal leg is realigned to join the upper road at sufficient distance from the

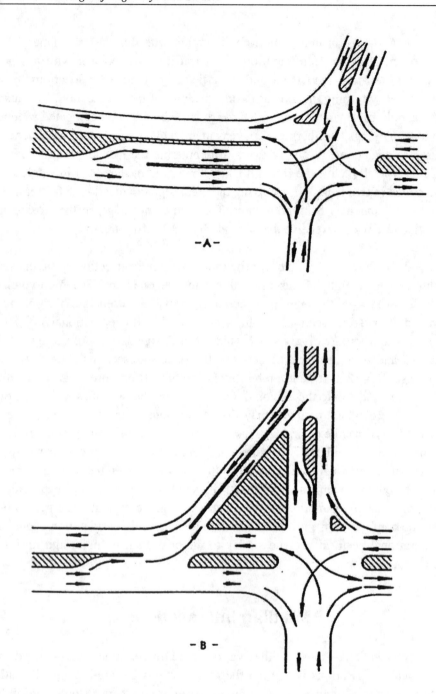

Exhibit 9-13. Four-Leg Intersections (Channelized High-type)

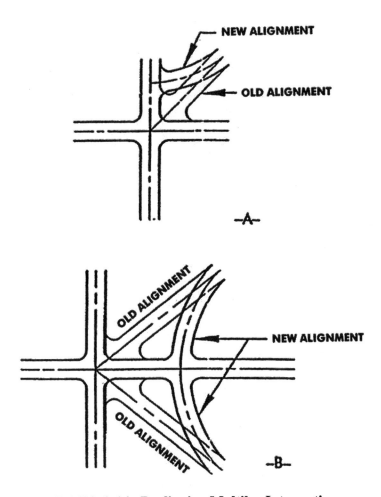

Exhibit 9-14. Realigning Multileg Intersections

main intersection to form two distinct intersections, each of which can be operated simply. The left-to-right highway is likely to be the more important route, and for this reason the diagonal leg is realigned to locate the new intersection on the less important road.

Exhibit 9-14B illustrates an intersection with six approach legs, two of which are realigned in adjacent quadrants to form a simple four-leg intersection at an appropriate distance to the right of the main intersection, which is itself converted to a simple four-leg intersection. This pattern applies where the top-to-bottom highway at the left is the more important route. If the left-to-right highway is more important, it may be preferable to realign the diagonal legs toward the other highway and thereby create three separate intersections along the minor highway. The intersection configurations in Exhibit 9-14 are shown in their simplest form. For example, separate turning lanes and divisional islands may be used, as appropriate, to fit the particular situation.

Modern Roundabouts

A recent synthesis of literature has summarized current practice and experience with modern roundabouts (**3**). Although the United States was home to the first one-way rotary intersection in the world (implemented at New York City's Columbus Circle in 1904), traffic circles fell out of favor in this country by the 1950s. Older traffic circles encountered serious operational and safety problems, including the tendency for traffic to lock up at higher volumes. The modern roundabout, although following different design principles from those of the old circles, has been used less in the United States than abroad, in part because of this country's experience with the traffic circles and rotaries built in the first half of the 20th century.

Since 1990, however, there has been an emergence of interest in modern roundabouts in some parts of the United States. This interest is due partially to the success of modern roundabouts in several countries in Europe and in Australia. France, which leads the world with an estimated 15,000 modern roundabouts, has been building roundabouts at a rate of about 1,000 per year. By comparison, the number of roundabouts in the United States, although growing, remains small. As of mid-1997, there were fewer than 50 modern roundabouts in the United States, in contrast with more than 35,000 in the rest of the world. Survey results from various states and municipalities documenting their experiences with roundabouts have been published (**3**).

The term "modern roundabout" is used in the United States to differentiate modern roundabouts from the nonconforming traffic circles or rotaries that have been in use for many years. Modern roundabouts are defined by two basic operational and design principles:

- Yield-at-Entry: Also known as off-side priority or the yield-to-left rule, yield-at-entry requires that vehicles on the circulatory roadway of the roundabout have the right-of-way and all entering vehicles on the approaches have to wait for a gap in the circulating flow. To maintain free flow and high capacity, yield signs are used as the entry control. As opposed to nonconforming traffic circles, modern roundabouts are not designed for weaving maneuvers, thus permitting smaller diameters. Even for multilane roundabouts, weaving maneuvers are not considered a design or capacity criterion.

- Deflection of Entering Traffic: Entrance roadways that intersect the roundabout along a tangent to the circulatory roadway are not permitted. Instead, entering traffic is deflected to the right by the central island of the roundabout and by channelization at the entrance into an appropriate curved path along the circulating roadway. Thus, no traffic is permitted to follow a straight path through the roundabout.

To provide for increased capacity, modern roundabouts often incorporate flares at the entry by adding lanes before the yield line and have wide circulatory roadways.

Modern roundabouts range in size from mini-roundabouts with inscribed circle diameters as small as 15 m [50 ft], to compact roundabouts with inscribed circle diameters between 30 and 35 m [98 to 115 ft], to large roundabouts, often with multilane circulating roadways and more than four entries up to 150 m [492 ft] in diameter. The greater speeds permitted by larger roundabouts, with inscribed circle diameters greater than 75 m [246 ft], may reduce their safety benefits to some degree.

Exhibit 9-15 shows the typical geometric elements of a single-lane modern roundabout and Exhibit 9-16 shows a pictorial roundabout example. Designing the geometry of a roundabout involves choosing the best operational and capacity performance while retaining the best safety enhancements. Roundabouts operate most safely when their geometry forces traffic to enter and circulate at slow speeds. Horizontal curvature and narrow pavement widths are used to produce this reduced-speed environment. However, the capacity of roundabouts is negatively affected by these low-speed design elements. As the widths and radii of the entry and circulatory roadways are reduced, the capacity of the roundabout is also reduced. Furthermore, many of the geometric criteria used in design of roundabouts are governed by the maneuvering needs of the largest

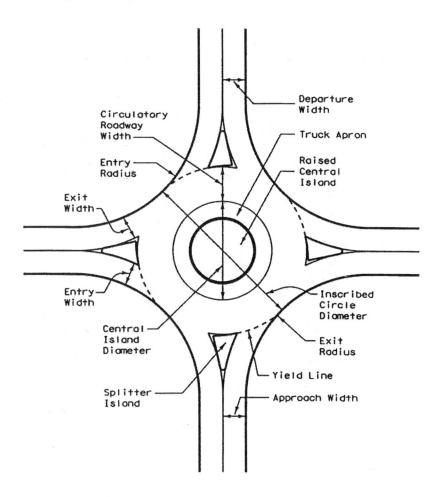

Exhibit 9-15. Geometric Elements of a Single-Lane Modern Roundabout

vehicles expected to travel through the intersection. Thus, designing a roundabout is a process of determining the optimal balance between safety provisions, operational performance, and accommodation of over-sized vehicles (**4**).

Achieving appropriate vehicular speeds through a roundabout is the most critical design objective. A well-designed roundabout reduces the relative speeds between conflicting traffic streams by requiring vehicles to negotiate the roundabout along a curved path. Increasing the curvature of the vehicle path decreases the relative speed between the entering and circulating vehicles. To determine the speed of a roundabout, the fastest path allowed by the geometry should be drawn. This is the smoothest, least-curved path that can be followed by a single vehicle, in the absence of other traffic and ignoring all lane markings, traversing the entry, around the central island, and out the exit. Usually the fastest path is the through movement, but in some cases it may be a right-turn movement (**4**).

Exhibit 9-16. Typical Modern Roundabout

Entry width is the largest determinant of a roundabout's capacity. The capacity of an approach is not only dependent on the number of entering lanes but also on the total width of the entry. In other words, the entry capacity increases with increasing entry width. Therefore, entries and circulatory roadways are generally described in terms of width, not number of lanes. Entries that are of sufficient width to accommodate multiple traffic streams are striped to designate separate lanes. However, the circulatory roadway is usually not striped, even when more than one lane of traffic is expected to circulate. The circulatory roadway should be at least as wide as the widest entry and should maintain a constant width throughout (**4**). For some single-lane roundabouts, the use of a mountable apron around the perimeter of the central island to provide the additional width needed to accommodate off tracking by combination trucks may be appropriate. At double-lane roundabouts, large vehicles may track across the whole width of the circulatory roadway to negotiate the roundabout. In some cases, roundabouts have been designed with aprons or gated roadways through the central island to accommodate over-sized trucks, emergency vehicles, or trains.

To maximize a roundabout's safe and efficient operation, entry widths should be kept to a minimum. Capacity needs and performance objectives should be considered in determining the width and number of lanes for each entry. In addition, the turning needs of the design vehicle may make an even wider entry appropriate. Therefore, determining the entry width and circulatory roadway width involves achieving an optimal capacity and operational balance. When the capacity needs can only be met by increasing the entry width, this can be accomplished in two ways: (1) by adding a full lane upstream of the roundabout and maintaining parallel lanes through the entire entry, or (2) by widening the approach gradually (flaring) through the entire entry. An example of entry flaring in two quadrants of a roundabout is shown in Exhibit 9-17 (**4**).

The installation of a roundabout for traffic-calming purposes should be supported by a demonstrated need for traffic calming along the intersecting roadways. Roundabouts installed for traffic-calming purposes are usually used on local roads. In such cases, capacity is not a consideration since traffic volumes are typically well below congestion levels.

Pedestrian crossing locations at roundabouts should achieve a balance among pedestrian convenience, pedestrian safety, and roundabout operations. Pedestrians generally want crossing locations as close as practical to the roundabout to minimize out-of-direction travel. The further the crossing is from the roundabout, the more likely it is that pedestrians will choose a shorter route that may present unintended conflicts. Both crossing location and crossing distance are important considerations. Crossing distance should be minimized to reduce exposure to pedestrian-vehicle conflicts. Pedestrian movements may be compromised at a yield-line crosswalk because driver attention is directed to the left looking for gaps in the circulating traffic stream. Therefore, the location of pedestrian crosswalks at the yield-line is discouraged. Crosswalks should be located to take advantage of the splitter island. The pedestrian refuge within the splitter island should be designed at street grade, rather than elevated to the height of the splitter island, provided that drainage can be accommodated. This arrangement eliminates the need for ramps within the refuge area. Crossings should also be located at a distance from the yield line that is approximately an even increment of a vehicle length to reduce the likelihood that vehicles will be queued across the crosswalk. Curb-cut ramps should be provided at each end of the crosswalk to connect the crosswalk to the sidewalk network and, thus, to other crosswalks around the roundabout (**4**).

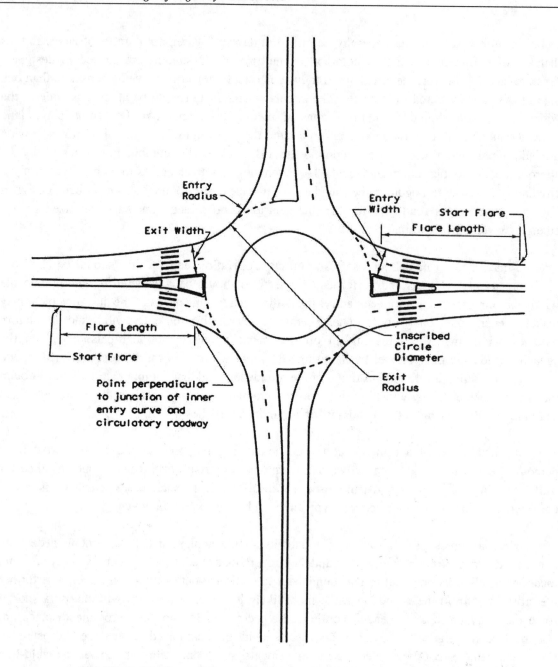

Exhibit 9-17. Roundabout with Entry Flaring in Two Quadrants

The designer should attempt, where practical, to provide bicyclists the choice of proceeding through the roundabout as either a vehicle or a pedestrian. In general, bicyclists are better served by operating as vehicles. Provisions for both options may allow bicyclists with varying degrees of skill to choose the method of navigating the roundabout with which they are most comfortable. To accommodate bicyclists traveling as vehicles, bicycle lanes should be terminated in advance of the roundabout to encourage bicyclists to mix with vehicle traffic. This method will generally be most successful at smaller roundabouts where bicycle speeds can most closely match vehicle speeds. It may be difficult for bicyclists to traverse double-lane roundabouts. In such cases, consideration of an alternative route along another street or bicycle path may be appropriate. To accommodate bicyclists traveling as pedestrians, a bicycle path or a shared bicycle/pedestrian path, physically separated from the circulatory roadway, should be provided. Refer to the AASHTO *Guide for the Development of Bicycle Facilities* (**5**) for more discussion of bicycle and shared-use path design.

Roundabouts offer the opportunity to provide attractive entries or centerpieces to communities. However, rigid objects such as monuments in the central island of a roundabout directly facing the entries may pose a safety concern. Aesthetic landscaping of the center island and, to a lesser degree, the splitter islands is frequently used. When placement of any landscaping treatment is provided, consideration should be given to motorist's sight distance needs. Pavement textures and aesthetic paving treatments are also frequently utilized at roundabouts.

CAPACITY ANALYSIS

Capacity and level-of-service analysis is one of the most important considerations in the design of intersections. This subject is discussed at length in Chapter 2 and is discussed throughout this chapter as it relates to the various elements of intersection design. Optimum capacities and levels of service can be obtained when intersections include auxiliary lanes, appropriate channelization, and traffic control devices. For more complete discussion of capacity and level-of-service analysis for intersections, including operational analysis procedures, refer to the *Highway Capacity Manual* (HCM) (**6**) and to Chapter 2 for guidance for its use.

ALIGNMENT AND PROFILE
General Considerations

Intersections are points of conflict between vehicles, pedestrians, and bicycles. The alignment and grade of the intersecting roads, therefore, should permit users to recognize the intersection and the other vehicles using it, and readily perform the maneuvers needed to pass through the intersection with minimum interference. To these ends, the alignment should be as straight and the gradients as flat as practical. The sight distance should be equal to or greater than the minimum values for specific intersection conditions, as derived and discussed later in this chapter. If design objectives are not met, users may have difficulty in discerning the actions of other users, in reading and discerning the messages of traffic control devices, and in controlling their operations.

Site conditions generally establish definite alignment and grade constraints on the intersecting roads. It may be practical to modify the alignment and grades, however, thereby improving traffic operations.

Alignment

Regardless of the type of intersection, for safety and economy, intersecting roads should generally meet at or nearly at right angles. Roads intersecting at acute angles need extensive turning roadway areas and tend to limit visibility, particularly for drivers of trucks. When a truck turns on an obtuse angle, the driver has blind areas on the right side of the vehicle. Acute-angle intersections increase the exposure time for the vehicles crossing the main traffic flow. The practice of realigning roads intersecting at acute angles in the manner shown in Exhibits 9-18A and 9-18B has proved to be beneficial. The greatest benefit is obtained when the curves used to realign the roads allow operating speeds nearly equivalent to the major-highway approach speeds.

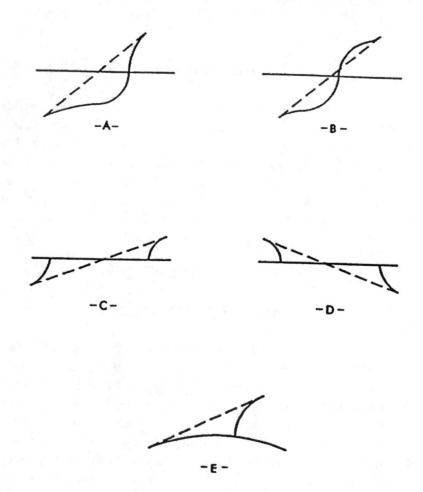

Exhibit 9-18. Realignment Variations at Intersections

The practice of constructing short-radius horizontal curves on side road approaches to achieve right-angle intersections should be avoided whenever practical. Such curves result in increased lane encroachments because drivers tend to reduce their path radius using a portion of the opposing lane. Also, the traffic control devices at the intersection may be located outside the driver's line of sight, resulting in the need to install advanced signing.

Another method of realigning a road that originally intersected another road at an acute angle is to make an offset intersection, as shown in Exhibits 9-18C and 9-18D. Only a single curve is introduced on each crossroad leg, but crossing vehicles must turn onto the major road and then reenter the minor road. (The terms "major road" and "minor road" are used here to indicate the relative importance of the roads that pass through the intersection rather than their functional classification.)

Realignment of the minor road, as shown in Exhibit 9-18C, provides poor access continuity because a crossing vehicle must reenter the minor road by making a left turn off the major highway. This design arrangement should only be used where traffic on the minor road is moderate, the anticipated minor road destinations are local, and the through traffic on the minor road is low.

Where the alignment of the minor road is as shown in Exhibit 9-18D, access continuity is better because a crossing vehicle first turns left onto the major road (e.g., a maneuver that can be done by waiting for an opening in the through-traffic stream) and then turns right to reenter the minor road, thus interfering little with through traffic on the major road.

Once a decision has been made to realign a minor road that intersects a major road at an acute angle, the angle of the realigned intersection should be as close to 90 degrees as practical. Although a right-angle crossing is normally desired, some deviation from a 90-degree angle is permissible. Reconstructing an intersection to provide an angle of at least 60 degrees provides most of the benefits of a 90-degree intersection angle while reducing the right-of-way takings and construction costs often associated with providing a right-angle intersection. The width of the roadway on the approach curves should be consistent with Exhibit 9-31 in order to reduce the potential for encroachment on adjacent lanes.

Where a large portion of the traffic from the minor road turns onto the major road, rather than continuing across the major road, the offset-intersection design may be advantageous regardless of the right or left entry. A road alignment that intersects two other roads at their junction to form an intersection with five or more legs should also be avoided.

Intersections on sharp curves should be avoided wherever practical because the superelevation and widening of pavements on curves complicate the intersection design and may reduce sight distance.

Where the major road curves and a minor road is located along a tangent to that curve, it is desirable to realign the minor road, as shown in Exhibit 9-18E, to guide traffic onto the main highway and improve the visibility at the point of intersection. This practice may have the disadvantage of adverse superelevation for turning vehicles and may need further study where curves have high superelevation rates and where the minor-road approach has adverse grades and a sight distance restriction due to the grade line.

Profile

Combinations of grade lines that make vehicle control difficult should be avoided at intersections. Substantial grade changes should be avoided at intersections, but it is not always practical to do so. Adequate sight distance should be provided along both intersecting roads and across their included corners, as discussed below, even where one or both intersecting roads are on vertical curves.

The gradients of intersecting roads should be as flat as practical on those sections that are to be used for storage of stopped vehicles, sometimes referred to as "storage platforms."

The calculated stopping and accelerating distances for passenger cars on grades of 3 percent or less differ little from the corresponding distances on the level. Grades steeper than 3 percent may need changes in several design elements to sustain operations equivalent to those on level roads. Most drivers are unable to judge the effect of steep grades on stopping or accelerating distances. Their normal deductions and reactions may thus be in error at a critical time. Accordingly, grades in excess of 3 percent should be avoided on the intersecting roads in the vicinity of the intersection. Where conditions make such designs too expensive, grades should not exceed about 6 percent, with a corresponding adjustment in specific geometric design elements.

The profile gradelines and cross sections on the legs of an intersection should be adjusted for a distance back from the intersection proper to provide a smooth junction and proper drainage. Normally, the gradeline of the major road should be carried through the intersection and that of the minor road should be adjusted to it. This design involves a transition in the crown of the minor road to an inclined cross section at its junction with the major road. For simple unchannelized intersections involving low design speeds and stop or signal control, it may be desirable to warp the crowns of both roads into a plane at the intersection; the appropriate plane depends on the direction of drainage and other conditions. Changes from one cross slope to another should be gradual. Intersections at which a minor road crosses a multilane divided highway with a narrow median on a superelevated curve should be avoided whenever practical because of the difficulty in adjusting grades to provide a suitable crossing. Gradelines of separate turning roadways should be designed to fit the cross slopes and longitudinal grades of the intersection legs.

The alignment and grades are subject to greater constraints at or near intersections than on the open road. At or near intersections, the combination of horizontal and vertical alignment should provide traffic lanes that are clearly visible to drivers at all times, clearly understandable for any desired direction of travel, free from the potential for conflicts to appear suddenly, and consistent in design with the portions of the highway just traveled.

The combination of vertical and horizontal curvature should allow adequate sight distance at an intersection. As discussed in Chapter 3, "Combinations of Horizontal and Vertical Alignment," a sharp horizontal curve following a crest vertical curve is undesirable, particularly on intersection approaches.

TYPES OF TURNING ROADWAYS

General

The widths of turning roadways for intersections are governed by the volumes of turning traffic and the types of vehicles to be accommodated. In almost all cases, turning roadways are designed for use by right-turning traffic. The widths for right-turning roadways may also be applied to other roadways within an intersection. There are three typical types of right-turning roadways at intersections: (1) a minimum edge-of-traveled-way design, (2) a design with a corner triangular island, and (3) a free-flow design using a simple radius or compound radii. The turning radii and the pavement cross slopes for free-flow right turns are functions of design speed and type of vehicles. For an in-depth discussion of the appropriate design criteria, see Chapter 3.

Minimum Edge-of-Traveled-Way Designs

Where it is appropriate to provide for turning vehicles within minimum space, as at unchannelized intersections, the corner radii should be based on minimum turning path of the selected design vehicles. The sharpest turn that can be made by each design vehicle is shown in Chapter 2, and the paths of the inner rear wheel and the front overhang are illustrated. The swept path widths indicated in Chapter 2, which are slightly greater than the minimum paths of nearly all vehicles in the class represented by each design vehicle, are the minimum paths attainable at speeds equal to or less than 15 km/h [10 mph] and consequently offer some leeway in driver behavior. These turning paths of the design vehicles shown in Exhibits 2-3 through 2-23 are considered satisfactory as minimum designs. Exhibits 9-19 and 9-20 summarize minimum-edge-of-traveled-way designs for various design vehicles.

The dimensions and turning radii of each design vehicle are identified in Chapter 2. In this chapter, the following design vehicles are presented: passenger car (P), single-unit truck (SU), city transit bus (CITY-BUS) intermediate semitrailer combination (WB-12 [WD-40]), semitrailer combination (WB-15 [WB-50]), interstate semitrailers (WB-19 and WB-20 [WB-62 and WB-65]), triple semitrailer/trailers combination (WB-30T [WB-100T]), turnpike-double combination (WB-33D [WB-109D]), and the conventional school bus (S-BUS11 [S-BUS36]).

The remaining design vehicles, including WB-20D [WB-67D] trucks, articulated buses, motor homes, motor coaches, and passenger cars pulling trailers or boats, are not addressed in this chapter. Should any of these vehicles be selected as the design vehicle for an intersection, refer to Chapter 2. Additional information on design characteristics of large trucks can be found in published sources (**7**, **8**).

Metric

Angle of turn (degrees)	Design vehicle	Simple curve radius (m)	Simple curve radius with taper		
			Radius (m)	Offset (m)	Taper H:V
30	P	18	–	–	–
	SU	30	–	–	–
	WB-12	45	–	–	–
	WB-15	60	–	–	–
	WB-19	110	67	1.0	15:1
	WB-20	116	67	1.0	15:1
	WB-30T	77	37	1.0	15:1
	WB-33D	145	77	1.1	20:1
45	P	15	–	–	–
	SU	23	–	–	–
	WB-12	36	–	–	–
	WB-15	53	36	0.6	15:1
	WB-19	70	43	1.2	15:1
	WB-20	76	43	1.3	15:1
	WB-30T	60	35	0.8	15:1
	WB-33D	–	60	1.3	20:1
60	P	12	–	–	–
	SU	18	–	–	–
	WB-12	28	–	–	–
	WB-15	45	29	1.0	15:1
	WB-19	50	43	1.2	15:1
	WB-20	60	43	1.3	15:1
	WB-30T	46	29	0.8	15:1
	WB-33D	–	54	1.3	20:1

US Customary

Angle of turn (degrees)	Design vehicle	Simple curve radius (ft)	Simple curve radius with taper		
			Radius (ft)	Offset (ft)	Taper H:V
30	P	60	–	–	–
	SU	100	–	–	–
	WB-40	150	–	–	–
	WB-50	200	–	–	–
	WB-62	360	220	3.0	15:1
	WB-67	380	220	3.0	15:1
	WB-100T	260	125	3.0	15:1
	WB-109D	475	260	3.5	20:1
45	P	50	–	–	–
	SU	75	–	–	–
	WB-40	120	–	–	–
	WB-50	175	120	2.0	15:1
	WB-62	230	145	4.0	15:1
	WB-67	250	145	4.5	15:1
	WB-100T	200	115	2.5	15:1
	WB-109D	–	200	4.5	20:1
60	P	40	–	–	–
	SU	60	–	–	–
	WB-40	90	–	–	–
	WB-50	150	120	3.0	15:1
	WB-62	170	140	4.0	15:1
	WB-67	200	140	4.5	15:1
	WB-100T	150	95	2.5	15:1
	WB-109D	–	180	4.5	20:1

Exhibit 9-19. Edge-of-Traveled-Way Designs for Turns at Intersections

Metric

Angle of turn (degrees)	Design vehicle	Simple curve radius (m)	Simple curve radius with taper		
			Radius (m)	Offset (m)	Taper H:V
75	P	11	8	0.6	10:1
	SU	17	14	0.6	10:1
	WB-12	–	18	0.6	15:1
	WB-15	–	20	1.0	15:1
	WB-19	–	43	1.2	20:1
	WB-20	–	43	1.3	20:1
	WB-30T	–	26	1.0	15:1
	WB-33D	–	42	1.7	20:1
90	P	9	6	0.8	10:1
	SU	15	12	0.6	10:1
	WB-12	–	14	1.2	10:1
	WB-15	–	18	1.2	15:1
	WB-19	–	36	1.3	30:1
	WB-20	–	37	1.3	30:1
	WB-30T	–	25	0.8	15:1
	WB-33D	–	35	0.9	15:1
105	P	–	6	0.8	–
	SU	–	11	1.0	–
	WB-12	–	12	1.2	–
	WB-15	–	17	1.2	15:1
	WB-19	–	35	1.0	15:1
	WB-20	–	35	1.0	15:1
	WB-30T	–	22	1.0	15:1
	WB-33D	–	28	2.8	20:1

US Customary

Angle of turn (degrees)	Design vehicle	Simple curve radius (ft)	Simple curve radius with taper		
			Radius (ft)	Offset (ft)	Taper H:V
75	P	35	25	2.0	10:1
	SU	55	45	2.0	10:1
	WB-40	–	60	2.0	15:1
	WB-50	–	65	3.0	15:1
	WB-62	–	145	4.0	20:1
	WB-67	–	145	4.5	20:1
	WB-100T	–	85	3.0	15:1
	WB-109D	–	140	5.5	20:1
90	P	30	20	2.5	10:1
	SU	50	40	2.0	10:1
	WB-40	–	45	4.0	10:1
	WB-50	–	60	4.0	15:1
	WB-62	–	120	4.5	30:1
	WB-67	–	125	4.5	30:1
	WB-100T	–	85	2.5	15:1
	WB-109D	–	115	2.9	15:1
105	P	–	20	2.5	–
	SU	–	35	3.0	–
	WB-40	–	40	4.0	–
	WB-50	–	55	4.0	15:1
	WB-62	–	115	3.0	15:1
	WB-67	–	115	3.0	15:1
	WB-100T	–	75	3.0	15:1
	WB-109D	–	90	9.2	20:1

Exhibit 9–19. Edge-of-Traveled-Way Designs for Turns at Intersections (Continued)

Metric

Angle of turn (degrees)	Design vehicle	Simple curve radius (m)	Simple curve radius with taper		
			Radius (m)	Offset (m)	Taper H:V
120	P	—	6	0.6	—
	SU	—	9	1.0	—
	WB-12	—	11	1.5	—
	WB-15	—	14	1.2	15:1
	WB-19	—	30	1.5	15:1
	WB-20	—	31	1.6	15:1
	WB-30T	—	20	1.1	15:1
	WB-33D	—	26	2.8	20:1
135	P	—	6	0.5	10:1
	SU	—	9	1.2	10:1
	WB-12	—	9	2.5	15:1
	WB-15	—	12	2.0	15:1
	WB-19	—	24	1.5	20:1
	WB-20	—	25	1.6	20:1
	WB-30T	—	19	1.7	15:1
	WB-33D	—	25	2.6	20:1
150	P	—	6	0.6	10:1
	SU	—	9	1.2	8:1
	WB-12	—	9	2.0	8:1
	WB-15	—	11	2.1	6:1
	WB-19	—	18	3.0	10:1
	WB-20	—	19	3.1	10:1
	WB-30T	—	19	2.2	10:1
	WB-33D	—	20	4.6	10:1

US Customary

Angle of turn (degrees)	Design vehicle	Simple curve radius (ft)	Simple curve radius with taper		
			Radius (ft)	Offset (ft)	Taper H:V
120	P	—	20	2.0	—
	SU	—	30	3.0	—
	WB-40	—	35	5.0	—
	WB-50	—	45	4.0	15:1
	WB-62	—	100	5.0	15:1
	WB-67	—	105	5.2	15:1
	WB-100T	—	65	3.5	15:1
	WB-109D	—	85	9.2	20:1
135	P	—	20	1.5	10:1
	SU	—	30	4.0	10:1
	WB-40	—	30	8.0	15:1
	WB-50	—	40	6.0	15:1
	WB-62	—	80	5.0	20:1
	WB-67	—	85	5.2	20:1
	WB-100T	—	65	5.5	15:1
	WB-109D	—	85	8.5	20:1
150	P	—	18	2.0	10:1
	SU	—	30	4.0	8:1
	WB-40	—	30	6.0	8:1
	WB-50	—	35	7.0	6:1
	WB-62	—	60	10.0	10:1
	WB-67	—	65	10.2	10:1
	WB-100T	—	65	7.3	10:1
	WB-109D	—	65	15.1	10:1

Exhibit 9–19. Edge-of-Traveled-Way Designs for Turns at Intersections (Continued)

Metric						US Customary					
Angle of turn (degrees)	Design vehicle	Simple curve radius (m)	Simple curve radius with taper			Angle of turn (degrees)	Design vehicle	Simple curve radius (ft)	Simple curve radius with taper		
			Radius (m)	Offset (m)	Taper H:V				Radius (ft)	Offset (ft)	Taper H:V
180	P	—	5	0.2	20:1	180	P	—	15	0.5	20:1
	SU	—	9	0.5	10:1		SU	—	30	1.5	10:1
	WB-12	—	6	3.0	5:1		WB-40	—	20	9.5	5:1
	WB-15	—	8	3.0	5:1		WB-50	—	25	9.5	5:1
	WB-19	—	17	3.0	15:1		WB-62	—	55	10.0	15:1
	WB-20	—	16	4.2	10:1		WB-67	—	55	13.8	10:1
	WB-30T	—	17	3.1	10:1		WB-100T	—	55	10.2	10:1
	WB-33D	—	17	6.1	10:1		WB-109D	—	55	20.0	10:1

Exhibit 9–19. Edge-of-Traveled-Way Designs for Turns at Intersections (Continued)

Metric						US Customary					
Angle of turn (degrees)	Design vehicle	3-centered compound Curve radii (m)	Symmetric offset (m)	3-centered compound Curve radii (m)	Asymmetric offset (m)	Angle of turn (degrees)	Design vehicle	3-centered compound Curve radii (ft)	Symmetric offset (ft)	3-centered compound Curve radii (ft)	Asymmetric offset (ft)
30	P	—	—	—	—	30	P	—	—	—	—
	SU	—	—	—	—		SU	—	—	—	—
	WB-12	—	—	—	—		WB-40	—	—	—	—
	WB-15	—	—	—	—		WB-50	—	—	—	—
	WB-19	—	—	—	—		WB-62	—	—	—	—
	WB-20	140-53-140	1.2	91-53-168	0.6-1.4		WB-67	460-175-460	4.0	300-175-550	2.0-4.5
	WB-30T	67-24-67	1.4	61-24-91	0.8-1.5		WB-100T	220-80-220	4.5	200-80-300	2.5-5.0
	WB-33D	168-76-168	1.5	76-61-198	0.5-2.1		WB-109D	550-250-550	5.0	250-200-650	1.5-7.0
45	P	—	—	—	—	45	P	—	—	—	—
	SU	—	—	—	—		SU	—	—	—	—
	WB-12	—	—	—	—		WB-40	—	—	—	—
	WB-15	60-30-60	1.0	—	—		WB-50	200-100-200	3.0	—	—
	WB-19	140-72-140	0.6	36-43-150	1.0-2.6		WB-62	460-240-460	2.0	120-140-500	3.0-8.5
	WB-20	140-53-140	1.2	76-38-183	0.3-1.8		WB-67	460-175-460	4.0	250-125-600	1.0-6.0
	WB-30T	76-24-76	1.4	61-24-91	0.8-1.7		WB-100T	250-80-250	4.5	200-80-300	2.5-5.5
	WB-33D	168-61-168	1.5	61-52-198	0.5-2.1		WB-109D	550-200-550	5.0	200-170-650	1.5-7.0
60	P	—	—	—	—	60	P	—	—	—	—
	SU	—	—	—	—		SU	—	—	—	—
	WB-12	—	—	—	—		WB-40	—	—	—	—
	WB-15	60-23-60	1.7	60-23-84	0.6-2.0		WB-50	200-75-200	5.5	200-75-275	2.0-7.0
	WB-19	120-30-120	4.5	34-30-67	3.0-3.7		WB-62	400-100-400	15.0	110-100-220	10.0-12.5
	WB-20	122-30-122	2.4	76-38-183	0.3-1.8		WB-67	400-100-400	8.0	250-125-600	1.0-6.0
	WB-30T	76-24-76	1.4	61-24-91	0.6-1.7		WB-100T	250-80-250	4.5	200-80-300	2.0-5.5
	WB-33D	198-46-198	1.7	61-43-183	0.5-2.4		WB-109D	650-150-650	5.5	200-140-600	1.5-8.0

Exhibit 9-20. Edge of Traveled Way for Turns at Intersections

Metric

Angle of turn (degrees)	Design vehicle	3-centered compound Curve radii (m)	Symmetric offset (m)	3-centered compound Curve radii Asymmetric (m)	offset (m)
75	P	30-8-30	0.6	–	–
	SU	36-14-36	0.6	–	–
	WB-12	36-14-36	1.5	36-14-60	0.6-2.0
	WB-15	45-15-45	2.0	45-15-69	0.6-3.0
	WB-19	134-23-134	4.5	43-30-165	1.5-3.6
	WB-20	128-23-128	3.0	61-24-183	0.3-3.0
	WB-30T	76-24-76	1.4	30-24-91	0.5-1.5
	WB-33D	213-38-213	2.0	46-34-168	0.5-3.5
90	P	30-6-30	0.8	–	–
	SU	36-12-36	0.6	–	–
	WB-12	36-12-36	1.5	36-12-60	0.6-2.0
	WB-15	55-18-55	2.0	36-12-60	0.6-3.0
	WB-19	120-21-120	3.0	48-21-110	2.0-3.0
	WB-20	134-20-134	3.0	61-21-183	0.3-3.4
	WB-30T	76-21-76	1.4	61-21-91	0.3-1.5
	WB-33D	213-34-213	2.0	30-29-168	0.6-3.5
105	P	30-6-30	0.8	–	–
	SU	30-11-30	1.0	–	–
	WB-12	30-11-30	1.5	30-17-60	0.6-2.5
	WB-15	55-14-55	2.5	45-12-64	0.6-3.0
	WB-19	160-15-160	4.5	110-23-180	1.2-3.2
	WB-20	152-15-152	4.0	61-20-183	0.3-3.4
	WB-30T	76-18-76	1.5	30-18-91	0.5-1.8
	WB-33D	213-29-213	2.4	46-24-152	0.9-4.6

US Customary

Angle of turn (degrees)	Design vehicle	3-centered compound Curve radii (ft)	Symmetric offset (ft)	3-centered compound Curve radii (ft)	Asymmetric offset (ft)
75	P	100-25-100	2.0	–	–
	SU	120-45-120	2.0	–	–
	WB-40	120-45-120	5.0	120-45-195	2.0-6.5
	WB-50	150-50-150	6.5	150-50-225	2.0-10.0
	WB-62	440-75-440	15.0	140-100-540	5.0-12.0
	WB-67	420-75-420	10.0	200-80-600	1.0-10.0
	WB-100T	250-80-250	4.5	100-80-300	1.5-5.0
	WB-109D	700-125-700	6.5	150-110-550	1.5-11.5
90	P	100-20-100	2.5	–	–
	SU	120-40-120	2.0	–	–
	WB-40	120-40-120	5.0	120-40-200	2.0-6.5
	WB-50	180-60-180	6.5	120-40-200	2.0-10.0
	WB-62	400-70-400	10.0	160-70-360	6.0-10.0
	WB-67	440-65-440	10.0	200-70-600	1.0-11.0
	WB-100T	250-70-250	4.5	200-70-300	1.0-5.0
	WB-109D	700-110-700	6.5	100-95-550	2.0-11.5
105	P	100-20-100	2.5	–	–
	SU	100-35-100	3.0	–	–
	WB-40	100-35-100	5.0	100-55-200	2.0-8.0
	WB-50	180-45-180	8.0	150-40-210	2.0-10.0
	WB-62	520-50-520	15.0	360-75-600	4.0-10.5
	WB-67	500-50-500	13.0	200-65-600	1.0-11.0
	WB-100T	250-60-250	5.0	100-60-300	1.5-6.0
	WB-109D	700-95-700	8.0	150-80-500	3.0-15.0

Exhibit 9-20. Edge of Traveled Way for Turns at Intersections (Continued)

Metric

Angle of turn (degrees)	Design vehicle	3-centered compound Curve radii (m)	Symmetric offset (m)	3-centered compound Curve radii (m)	Asymmetric offset (m)
120	P	30-6-30	0.6	–	–
	SU	30-9-30	1.0	–	–
	WB-12	36-9-36	2.0	30-9-55	0.6-2.7
	WB-15	55-12-55	2.6	45-11-67	0.6-3.6
	WB-19	160-21-160	3.0	24-17-160	5.2-7.3
	WB-20	168-14-168	4.6	61-18-183	0.6-3.8
	WB-30T	76-18-76	1.5	30-18-91	0.5-1.8
	WB-33D	213-26-213	2.7	46-21-152	2.0-5.3
135	P	30-6-30	0.5	–	–
	SU	30-9-30	1.2	–	–
	WB-12	36-9-36	2.0	30-8-55	1.0-4.0
	WB-15	48-11-48	2.7	40-9-56	1.0-4.3
	WB-19	180-18-180	3.6	30-18-195	2.1-4.3
	WB-20	168-14-168	5.0	61-18-183	0.6-3.8
	WB-30T	76-18-76	1.7	30-18-91	0.8-2.0
	WB-33D	213-21-213	3.8	46-20-152	2.1-5.6
150	P	23-6-23	0.6	–	–
	SU	30-9-30	1.2	–	–
	WB-12	30-9-30	2.0	28-8-48	0.3-3.6
	WB-15	48-11-48	2.1	36-9-55	1.0-4.3
	WB-19	145-17-145	4.5	43-18-170	2.4-3.0
	WB-20	168-14-168	5.8	61-17-183	2.0-5.0
	WB-30T	76-18-76	2.1	30-18-91	1.5-2.4
	WB-33D	213-20-213	4.6	61-20-152	2.7-5.6

US Customary

Angle of turn (degrees)	Design vehicle	3-centered compound Curve radii (ft)	Symmetric offset (ft)	3-centered compound Curve radii (ft)	Asymmetric offset (ft)
120	P	100-20-100	2.0	–	–
	SU	100-30-100	3.0	–	–
	WB-40	120-30-120	6.0	100-30-180	2.0-9.0
	WB-50	180-40-180	8.5	150-35-220	2.0-12.0
	WB-62	520-70-520	10.0	80-55-520	24.0-17.0
	WB-67	550-45-550	15.0	200-60-600	2.0-12.5
	WB-100T	250-60-250	5.0	100-60-300	1.5-6.0
	WB-109D	700-85-700	9.0	150-70-500	7.0-17.4
135	P	100-20-100	1.5	–	–
	SU	100-30-100	4.0	–	–
	WB-40	120-30-120	6.5	100-25-180	3.0-13.0
	WB-50	160-35-160	9.0	130-30-185	3.0-14.0
	WB-62	600-60-600	12.0	100-60-640	14.0-7.0
	WB-67	550-45-550	16.0	200-60-600	2.0-12.5
	WB-100T	250-60-250	5.5	100-60-300	2.5-7.0
	WB-109D	700-70-700	12.5	150-65-500	14.0-18.4
150	P	75-20-75	2.0	–	–
	SU	100-30-100	4.0	–	–
	WB-40	100-30-100	6.0	90-25-160	1.0-12.0
	WB-50	160-35-160	7.0	120-30-180	3.0-14.0
	WB-62	480-55-480	15.0	140-60-560	8.0-10.0
	WB-67	550-45-550	19.0	200-55-600	7.0-16.4
	WB-100T	250-60-250	7.0	100-60-300	5.0-8.0
	WB-109D	700-65-700	15.0	200-65-500	9.0-18.4

Exhibit 9-20. Edge of Traveled Way for Turns at Intersections (Continued)

Metric						US Customary					
Angle of turn (degrees)	Design vehicle	3-centered compound Curve radii (m)	Symmetric offset (m)	3-centered compound Curve radii (m)	Asymmetric offset (m)	Angle of turn (degrees)	Design vehicle	3-centered compound Curve radii (ft)	Symmetric offset (ft)	3-centered compound Curve radii (ft)	Asymmetric offset (ft)
180	P	15-5-15	0.2	–	–	180	P	50-15-50	0.5	–	–
	SU	30-9-30	0.5	–	–		SU	100-30-100	1.5	–	–
	WB-12	30-6-30	3.0	26-6-45	2.0-4.0		WB-40	100-20-100	9.5	85-20-150	6.0-13.0
	WB-15	40-8-40	3.0	30-8-55	2.0-4.0		WB-50	130-25-130	9.5	100-25-180	6.0-13.0
	WB-19	245-14-245	6.0	30-17-275	4.5-4.5		WB-62	800-45-800	20.0	100-55-900	15.0-15.0
	WB-20	183-14-183	6.2	30-17-122	1.8-4.6		WB-67	600-45-600	20.5	100-55-400	6.0-15.0
	WB-30T	76-17-76	2.9	30-17-91	2.6-3.2		WB-100T	250-55-250	9.5	100-55-300	8.5-10.5
	WB-33D	213-17-213	6.1	61-18-152	3.0-6.4		WB-109D	700-55-700	20.0	200-60-500	10.0-21.0

Exhibit 9-20. Edge of Traveled Way for Turns at Intersections (Continued)

In the design of the edge of the traveled way based on the path of a given design vehicle, it is assumed that the vehicle is properly positioned within the traffic lane at the beginning and end of the turn (i.e., 0.6 m [2 ft] from the edge of traveled way on the tangents approaching and leaving the intersection curve). Curve designs for edge of traveled way conforming to this assumption are shown in Exhibits 9-21 through 9-28. Such designs follow closely the inner wheel path of the selected design vehicle, with a clearance of 0.6 m [2 ft] or more throughout most of the turn, and with a clearance at no point less than 0.2 m [9 in]. Differences in the inner paths of vehicles turning left and right are not sufficient to be significant in design. Although not shown explicitly in the figures, the edge designs illustrated also apply to left-turn maneuvers, such as a left turn by a vehicle leaving a divided highway at a very low speed.

Where the alignment includes a horizontal curve at the beginning or end of a return radius, the design should be modified accordingly. The most expeditious way to customize a design for such special conditions is to use the appropriate design vehicle as an overlay on a plan of the intersection.

At an intersection with a low right-turn volume, the designer may determine that a deceleration and right-turn lane is not warranted. In this instance, the composition of the shoulder may be improved for greater load capacities to permit right-turning vehicles to utilize the shoulder. In turn, where right-turning volumes are high, consideration should be given to providing a right-turn lane along with appropriate provisions for vehicle deceleration. In rural areas, the appropriate shoulder width should be considered in conjunction with the design of right-turn lanes.

Design for Specific Conditions (Right-Angle Turns)

The designs illustrated in Exhibits 9-21 through 9-28 are those that accommodate the sharpest turns for specific design vehicles. Combinations of curves with radii other than those shown may also provide satisfactory operations. The choice of design for a specific intersection or turning movement where pedestrians are present is a particular concern, and it is desirable to keep the intersection area to a minimum. The selection of any specific design depends on the type and size of vehicles that will be turning and the extent to which they should be accommodated. In addition, the appropriate design may depend on other factors such as the type, character, and location of the intersecting roads, the vehicular and pedestrian traffic volumes, the number and frequency of the larger vehicles involved in turning movements, and the effect of these larger vehicles on other traffic. For example, if turning traffic is nearly all passenger vehicles, it may not be cost-effective or pedestrian friendly to design for large trucks. However, the design should allow for an occasional large truck to turn by swinging wide and encroaching on other traffic lanes without disrupting traffic significantly. Therefore, the designer should analyze the likely paths and encroachments that will result when a turn is made by a larger vehicle.

From the analysis of these maneuvers and corresponding paths, together with other pertinent data, the appropriate type of minimum design can be selected. Applications of minimum designs for turning movements are common, even in rural areas. Minimum designs are appropriate for locations with low turning speeds, low turning volumes, and high property values. The selection of a design vehicle for minimum edge-of-traveled-way designs, illustrated in Exhibits 9-21 through 9-28, depends on the designer's judgment upon consideration of the site conditions and analysis of the operational needs of larger vehicles.

As a summary, three minimum edge-of-traveled-way designs for turns may be considered at an intersection based on the turning paths of the design vehicles identified below:

- P design vehicle (Exhibit 9-21). This design vehicle is used at intersections in conjunction with parkways where minimum turns are appropriate, at local road intersections with major roads where turns are made only occasionally, and at intersections of two minor roads carrying low volumes. However, if conditions permit, the SU vehicle (Exhibit 9-22) is the preferred design vehicle.

- SU design vehicle (Exhibit 9-22). Generally, this design vehicle provides the recommended minimum edge-of-traveled-way design for rural highways other than those described above. Turning movements for urban conditions are discussed in a separate section of this chapter. Important turning movements on major highways, particularly those involving a large percentage of trucks, should be designed with larger radii, speed-change lanes, or both.

- Semitrailer combination design vehicles (Exhibits 9-23 through 9-28). These design vehicles should be used where truck combinations will turn repeatedly. Where designs for such vehicles are warranted, the simpler symmetrical arrangements of three-centered compound curves (shown in Exhibits 9-23 and 9-25) are generally preferred if these smaller truck combinations make up a sizable percentage of the turning volume. Because designs for semitrailer combination vehicles, particularly when used in two or more quadrants of an intersection, produce large paved areas, it may be desirable to provide somewhat larger radii and use a corner triangular island.

A more detailed discussion of the minimum edge-of-traveled-way design for each of these design vehicle types is presented below.

Passenger Vehicles

Three minimum designs for the inner edge of the traveled way for a 90-degree right turn to accommodate the P design vehicle are shown in Exhibit 9-21. A 7.5-m [25-ft] radius for the inner edge of the traveled way (the solid line in Exhibit 9-21A) is the sharpest simple arc that clears the inner wheel path by about 0.2 m [8 in] near the end of the arc. A simple circular curve with a radius of 9 m [30 ft], shown as a dotted line in the same exhibit, provides a 0.4-m [16-in] clearance at the ends of the curve, but has a clearance of about 1.6 m [5.4 ft] at the middle of the curve. With a radius of more than 9 m [30 ft], most passenger car drivers will naturally use a

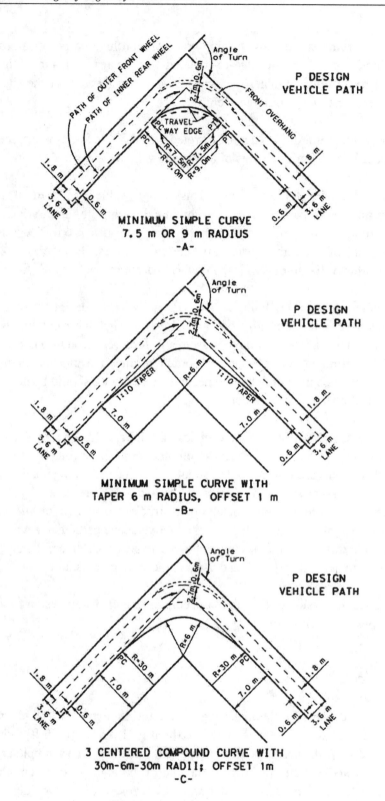

Metric

Exhibit 9-21. Minimum Traveled Way (Passenger Vehicles)

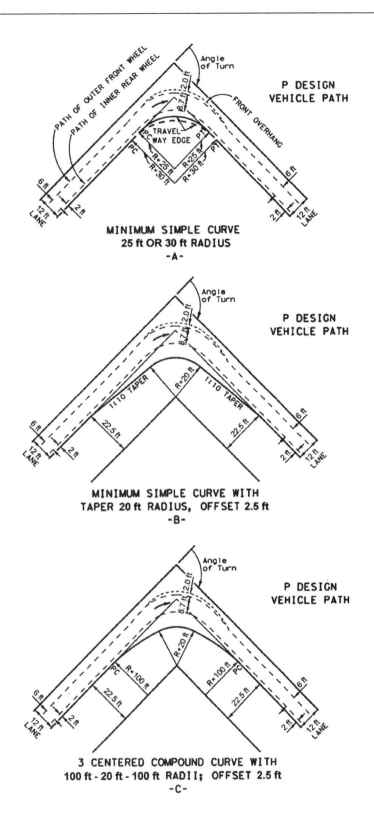

MINIMUM SIMPLE CURVE
25 ft OR 30 ft RADIUS
-A-

MINIMUM SIMPLE CURVE WITH
TAPER 20 ft RADIUS, OFFSET 2.5 ft
-B-

3 CENTERED COMPOUND CURVE WITH
100 ft - 20 ft - 100 ft RADII; OFFSET 2.5 ft
-C-

US Customary

Exhibit 9–21. Minimum Traveled Way (Passenger Vehicles) (Continued)

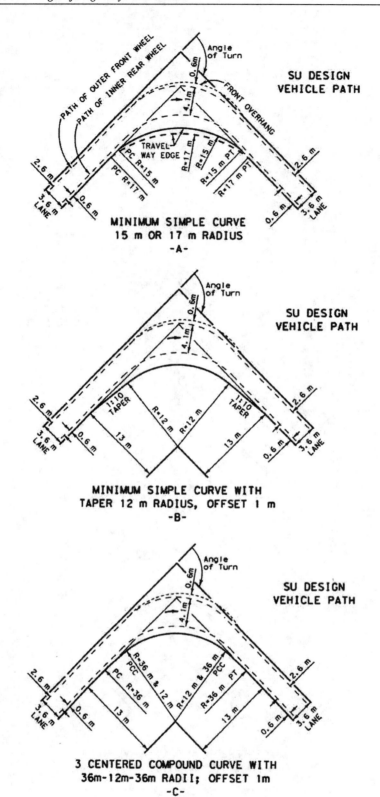

Metric

Exhibit 9-22. Minimum Traveled Way Designs (Single-Unit Trucks and City Transit Buses)

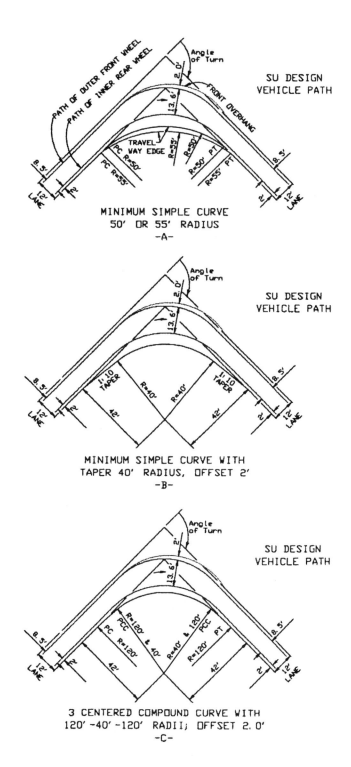

MINIMUM SIMPLE CURVE
50' OR 55' RADIUS
-A-

MINIMUM SIMPLE CURVE WITH
TAPER 40' RADIUS, OFFSET 2'
-B-

3 CENTERED COMPOUND CURVE WITH
120'-40'-120' RADII; OFFSET 2.0'
-C-

US Customary

Exhibit 9-22. Minimum Traveled Way Designs (Single-Unit Trucks and City Transit Buses) (Continued)

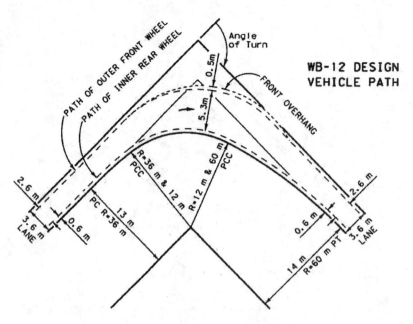

**3 CENTERED COMPOUND CURVE WITH
36m-12m-60m RADII; OFFSET 1m AND 2m
-A-**

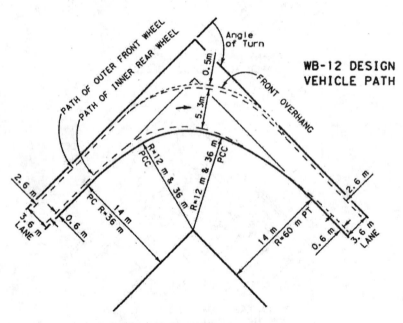

**3 CENTERED COMPOUND CURVE WITH
36m-12m-36m RADII; OFFSET 2m
-B-**

Metric

**Exhibit 9-23. Minimum Edge-of-Traveled-Way Designs (WB-12 [WB-40]
Design Vehicle Path)**

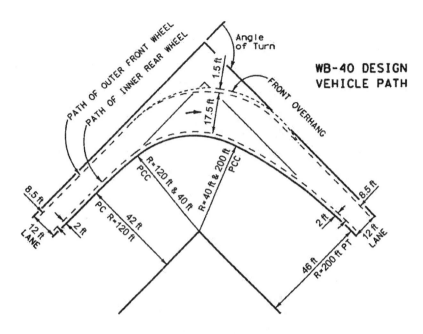

**3 CENTERED COMPOUND CURVE WITH
120 FT - 40 FT - 200 FT RADII; OFFSET 2 FT AND 6 FT
- A -**

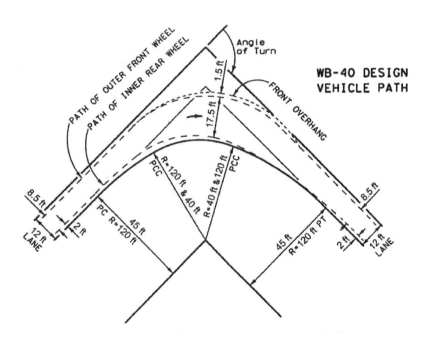

**3 CENTERED COMPOUND CURVE WITH
120 FT - 40 FT - 120 FT RADII; OFFSET 5 FT
- B -**

US Customary

**Exhibit 9–23. Minimum Edge-of-Traveled-Way Designs (WB-12 [WB-40]
Design Vehicle Path) (Continued)**

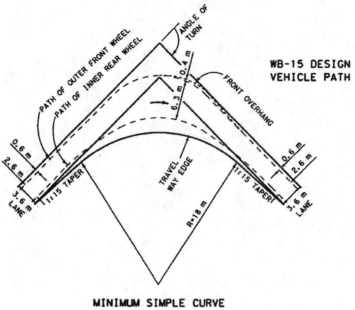

Metric

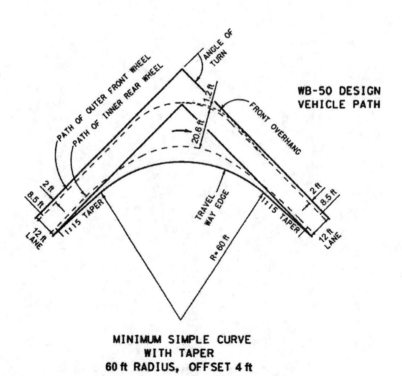

US Customary

Exhibit 9-24. Minimum Edge-of-Traveled-Way Designs (WB-15 [WB-50] Design Vehicle Path)

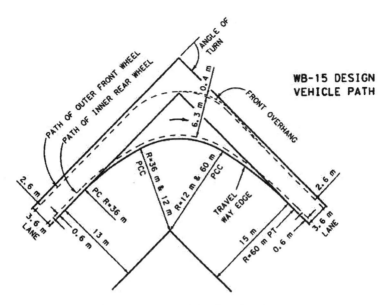

WB-15 DESIGN
VEHICLE PATH

WB-15 SEMITRAILER COMBINATION
3 CENTERED COMPOUND CURVE,
36m-12m-60m RADII, OFFSET 1m AND 3m

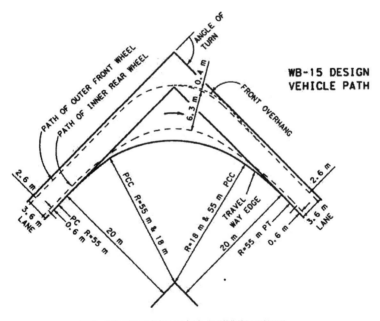

WB-15 DESIGN
VEHICLE PATH

WB-15 SEMITRAILER COMBINATION
3 CENTERED COMPOUND CURVE,
55m-18m-55m RADII, OFFSET 2m

Metric

**Exhibit 9-25. Minimum Edge-of-Traveled-Way Designs (WB-15 [WB-50]
Design Vehicle Path)**

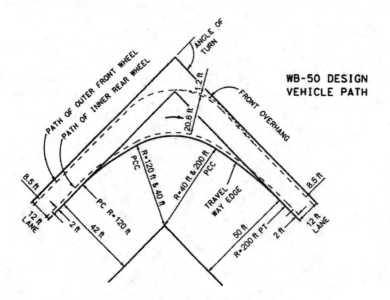

WB-50 SEMITRAILER COMBINATION
3 CENTERED COMPOUND CURVE,
120 ft - 40 ft - 200 ft RADII, OFFSET 2 ft AND 10 ft

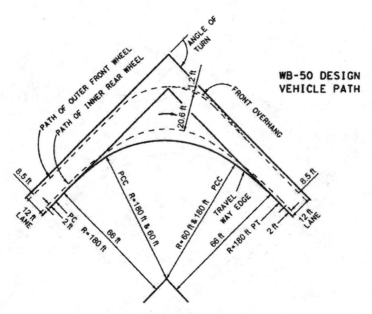

WB-50 SEMITRAILER COMBINATION
3 CENTERED COMPOUND CURVE,
180 ft - 60 ft - 180 ft RADII, OFFSET 6 ft

US Customary

**Exhibit 9–25. Minimum Edge-of-Traveled-Way Designs (WB-15 [WB-50])
Design Vehicle Path) (Continued)**

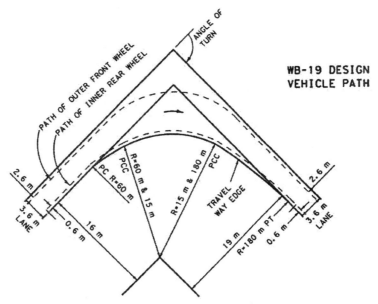

WB-19 INTERSTATE SEMITRAILER COMBINATION
3 CENTERED COMPOUND CURVE,
60m-15m-180m RADII, OFFSET 1m AND 4m

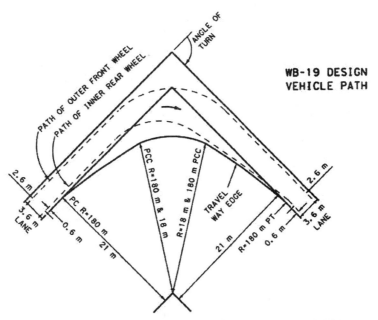

WB-19 INTERSTATE SEMITRAILER COMBINATION
3 CENTERED COMPOUND CURVE,
180m-18m-180m RADII, OFFSET 3m

Metric

**Exhibit 9-26. Minimum Edge-of-Traveled-Way Designs (WB-19 [WB-62]
Design Vehicle Path)**

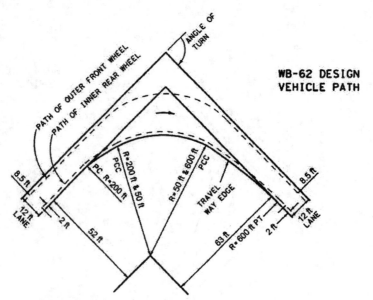

WB-62 INTERSTATE SEMITRAILER COMBINATION
3 CENTERED COMPOUND CURVE,
200 ft - 50 ft - 600 ft RADII, OFFSET 2 ft AND 13 ft

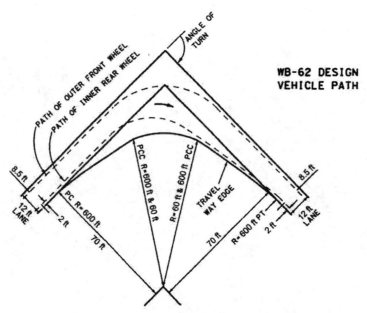

WB-62 INTERSTATE SEMITRAILER COMBINATION
3 CENTERED COMPOUND CURVE,
600 ft - 60 ft - 600 ft RADII, OFFSET 10 ft

US Customary

**Exhibit 9-26. Minimum Edge-of-Traveled-Way Designs (WB-19 [WB-62]
Design Vehicle Path) (Continued)**

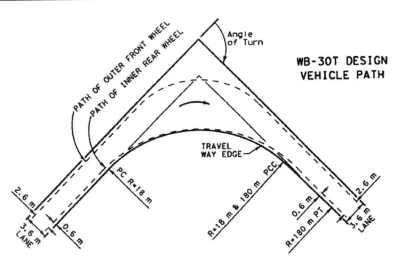

**WB-30T TRIPLE TRAILER COMBINATION
COMPOUND CURVE
R=18m AND 180m
-A-**

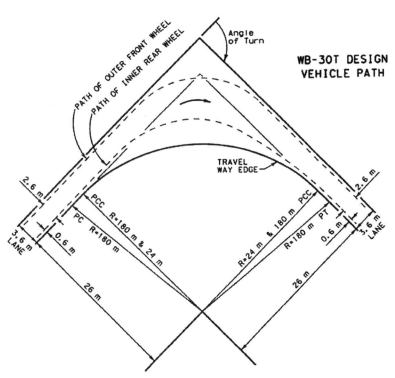

**WB-30T TRIPLE TRAILER COMBINATION
3 CENTERED COMPOUND CURVE
180m-24m-180m RADII; OFFSET 2m
-B-**

Metric

**Exhibit 9-27. Minimum Edge-of-Traveled-Way Designs (WB-30T [WB-100T] Design
Vehicle Path)**

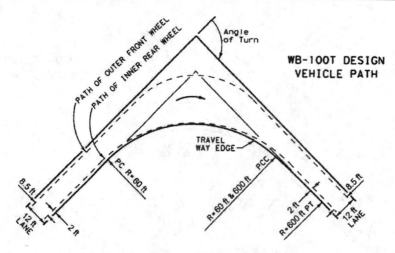

WB-100T TRIPLE TRAILER COMBINATION
COMPOUND CURVE
R = 60 ft AND 600 ft
-A-

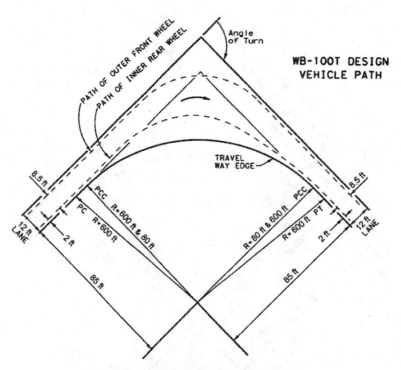

WB-100T TRIPLE TRAILER COMBINATION
3 CENTERED COMPOUND CURVE
600 ft - 80 ft - 600 ft RADII; OFFSET 5 ft
- B -

US Customary

Exhibit 9-27. Minimum Edge-of-Traveled-Way Designs (WB-30T [WB-100T] Design Vehicle Path) (Continued)

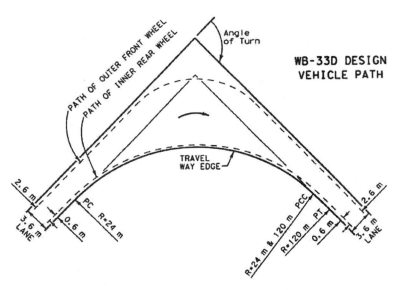

WB-33D DOUBLE TRAILER COMBINATION
COMPOUND CURVE
R=24m AND 120m
-A-

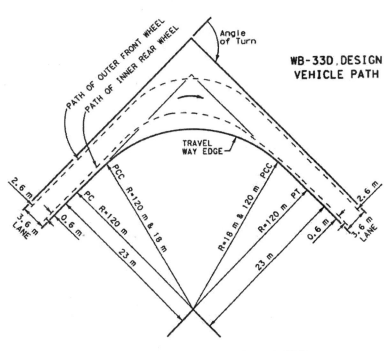

WB-33D DOUBLE TRAILER COMBINATION
3 CENTERED COMPOUND CURVE
120m-18m-120m RADII; OFFSET 5m
-B-

Metric

Exhibit 9-28. Minimum Edge-of-Traveled-Way Designs (WB-33D [WB-109D] Design Vehicle Path)

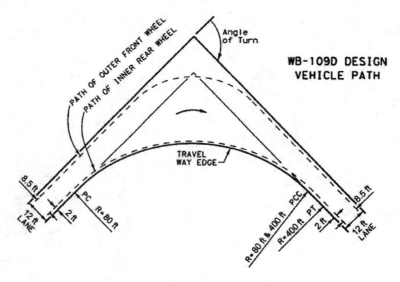

WB-109D DOUBLE TRAILER COMBINATION
COMPOUND CURVE
R=80 ft AND 400 ft
-A-

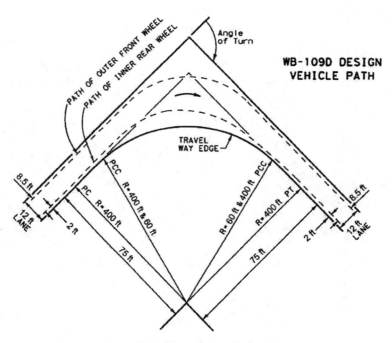

WB-109D DOUBLE TRAILER COMBINATION
3 CENTERED COMPOUND CURVE
400 ft - 60 ft - 400 ft RADII; OFFSET 15 ft
-B-

US Customary

Exhibit 9-28. Minimum Edge-of-Traveled-Way Designs (WB-33D [WB-109D] Design Vehicle Path) (Continued)

turning radius flatter than the minimum vehicles and will more or less follow the edge of the traveled way.

The edge design shown in Exhibit 9-21C is a practical equivalent to a circular curve. This design consists of a three-centered curve with radii of 30, 6, and 30 m [100, 20, and 100 ft], with the center of the middle curve being located 7 m [22.5 ft] from the extension of the tangent edges (measurement includes a 0.8-m [2.5-ft] offset). This design creates little extra pavement in contrast to the simple curve of 9 m [30 ft]; specifically, the paved area in a single quadrant between the extended tangent edges is only 20 m² [225 ft²] in comparison to 18 m² [200 ft²] for the simple circular curve. However, a minimum edge design is preferred because it more closely fits the design vehicle path. A layout consisting of a simple curve offset and connecting tapers that closely approximates the three-centered design is shown in Exhibit 9-21B.

Single-Unit Trucks and City Transit Buses

Minimum designs for the inner edge of the traveled way for a 90-degree right turn to accommodate the SU design vehicle are shown in Exhibit 9-22. The minimum travel way designs for the SU design vehicle will accommodate both the single-unit truck and the city transit bus. A 15-m [50-ft] radius for the inner edge of the traveled way (the solid line shown in Exhibit 9-22A) is the sharpest simple arc that accommodates the SU vehicle without encroachment on adjacent lanes. Toward the end of the turn, however, the inner wheel path closely approaches the edge of the traveled way. A simple circular curve of 17-m [55-ft] radius, shown as a dotted line in Exhibit 9-22A, allows for slightly more clearance at the far end of the intersection curve. Inner-edge radii of 18 m [60 ft] or more permit the SU vehicle to turn on a radius greater than the minimum.

The edge design shown in Exhibit 9-22C is a practical equivalent to a circular curve. It consists of a three-centered compound curve with radii of 36, 12, and 36 m [120, 40, and 120 ft], with the center of the middle curve located 13 m [42 ft] from the extension of the tangent edges (with a 0.6-m [2.0-ft] offset). In an operational sense, this design is much preferred over the simple circular curve because it better fits the minimum path of the inner rear wheel of the design vehicle. Because the resulting areas between the edges of the traveled way are 44 m² [475 ft²] for the compound-curve design and 50 m² [550 ft²] for the circular-curve with a 15-m [50-ft] radius, the former design needs less pavement. A simpler tapered layout that closely follows the three-centered design is shown in Exhibit 9-22B.

In any design that permits the SU design vehicle to turn on its minimum path without swinging wide, the front overhang will swing out 3.6 m [12 ft] from the edge of tangent traveled way at the far end of the turn, and the design vehicle thereby occupies the full width of a 3.6-m [12-ft] lane on the crossroad. With 3.0- or 3.3-m [10- or 11-ft] lanes, the turning vehicle would encroach on adjacent lanes. To preclude encroachment with lane widths less than 3.6 m [12 ft], edge-of-traveled-way radii larger than the minimum would need to be used.

Semitrailer Combination Trucks

It is not practical to fit simple circular arcs to the minimum design paths for semitrailer combination design vehicles. However, where traffic lanes are 3.6 m [12 ft] in width, such vehicles can turn without encroachment on adjacent lanes if the radius of a simple circular curve for the inner edge of traveled way is approximately 23 m [75 ft] for the WB-12 [WB-40] and 29 m [95 ft] for the WB-15 [WB-50] design vehicle. Such turns would be made with a turning radius of the outer front wheel greater than the minimums shown of these vehicles. To fit the edge of traveled way more closely to the minimum path of these design vehicles, an asymmetrical arrangement of three-centered compound curves should be used. For the WB-12 [WB-40] design vehicle, the curves should have radii of 36, 12, and 60 m [120, 40, and 200 ft] with the arc of the middle curve being offset 0.6 and 2.0 m [2.0 and 6.5 ft] from the extension of the tangent edges on the approach and exit sides, respectively. A simple curve with tapers is shown in Exhibit 9-24 for the WB-15 [WB-50] vehicle. Although not as efficient in the use of pavement area as the asymmetrical curve layout, it may be a preferred design because of its ease of construction.

Oblique-Angle Turns

For oblique-angle turns, minimum designs for the edge of the traveled way are developed in the same manner as those for right-angle intersections by plotting the paths of the design vehicles on the sharpest turns and fitting curves or combinations of curves to the paths of inner rear wheels. Suggested minimum designs in which three-centered compound curves are used for each design vehicle are given in Exhibit 9-20 for various angles of turn.

For convenience, the intersection angle condition is indicated by the angle of turn, which is the angle through which a vehicle travels in making a turn. It is measured from the extension of the tangent on which a vehicle approaches to the corresponding tangent on the intersecting road onto which the vehicle turns. This angle is the same as that commonly called the delta or central angle in surveying terminology. With angles of turn less than 90 degrees, the radii needed to fit the minimum paths of vehicles are longer than those suggested for right-angle turns. With angles of turn more than 90 degrees, the radii are decreased and larger offsets of the central arc should be provided.

The designs shown in Exhibit 9-20 are those suggested to fit the sharpest turns of the different design vehicles. Some other combinations of curves may also be used with satisfactory results. The use of tapers with simple curves is another method for design of the edge of the traveled way for turns at intersections, and dimensions for such combinations are shown in Exhibit 9-19. Tapers are needed to keep the intersection area to a minimum, and any of the designs shown in Exhibits 9-19 or 9-20 may be chosen, depending on the type and size of vehicles that will be turning and the extent to which those vehicles should be accommodated.

At 90-degree intersections with inner edges of traveled way designed for passenger vehicles, all trucks can turn by making a wide swing onto adjacent traffic lanes, which on two-lane roads are opposing traffic lanes. For angles of turn less than 90 degrees, trucks also can turn on an inner edge of traveled way designed for passenger vehicles with even less encroachment than for

90-degree turns. For turning angles more than 90 degrees, the minimum design for the P design vehicle should be adjusted to ensure that all turning trucks can remain within two lanes of traveled way on each of the intersecting roads. In this regard, for turning angles of 120 degrees or more, the same dimensions of three-centered curves, as needed for the P design vehicle (30, 6 and 30 m [90, 20, and 90 ft]) may be used, but the offset of the middle curve should be increased from 1 m [3 ft] to as much as 3 m [10 ft] for a 180-degree turn. Where space is available, even for minor roads, a design based on the SU design vehicle would be preferred. With edge design developed for the SU design vehicle, the WB-12 [WB-40] design vehicle will encroach only slightly, if at all, on adjacent traffic lanes, and the WB-15 [WB-50] design vehicle will encroach only partially on other lanes.

Design for angles of turn more than 90 degrees usually results in intersections with large paved areas, of which portions are often unused. This situation may lead to confusion among drivers and may create longer crossing paths for pedestrians. These conditions may be alleviated to a considerable extent by using three-centered asymmetric compound curves, two-centered curves, or larger radius circular curves together with corner triangular islands. On major highways intersecting at oblique angles, separate turning roadways with a corner island for right-turning traffic should be provided in quadrants where vehicles turn more than about 120 degrees.

Effect of Curb Radii on Turning Paths

The effect of curb radii on the right-turning paths of various design vehicles turning through an angle of 90 degrees (on streets without parking lanes) is shown in Exhibits 9-29 and 9-30.

Exhibit 9-29 shows the effects of a 4.5-m [15-ft] radius. With 3.6-m [12-ft] lanes, the design passenger vehicle can turn with no encroachment on an adjacent lane at the end of the turn, but the SU and BUS design vehicles will swing wide on both streets and will occupy two lanes at the end of the turn. To turn into two lanes on the cross street, the WB-15 [WB-50] design vehicle will occupy an area wider than those two lanes (i.e., the design vehicle would encroach on a shoulder or curb area, as well). The WB-33D [WB-109D] design vehicle would occupy an area as wide as four lanes on a cross street.

Exhibit 9-30 shows vehicle operation at a 12-m [40-ft] curb radius. The P vehicle can easily make the turn around this radius. The SU and BUS design vehicles can turn around the radius into one lane on the cross street by beginning its turn adjacent to the centerline of the major street. The WB-15 [WB-50] design vehicle needs the entire two-lane width of the cross street to complete the turn. This type of maneuver is practical for turns from an arterial street where the cross street is normally free of traffic because of signal or stop-sign control on the cross street. Turns from the cross street to the arterial about such radii can also be accommodated where signal control is used, but without signal control, drivers of vehicles turning from the cross street must wait for an appropriate gap in traffic to turn into the second lane of the arterial street. The WB-33D [WB-109D] design vehicle needs most of a third lane on the cross street to complete a turn.

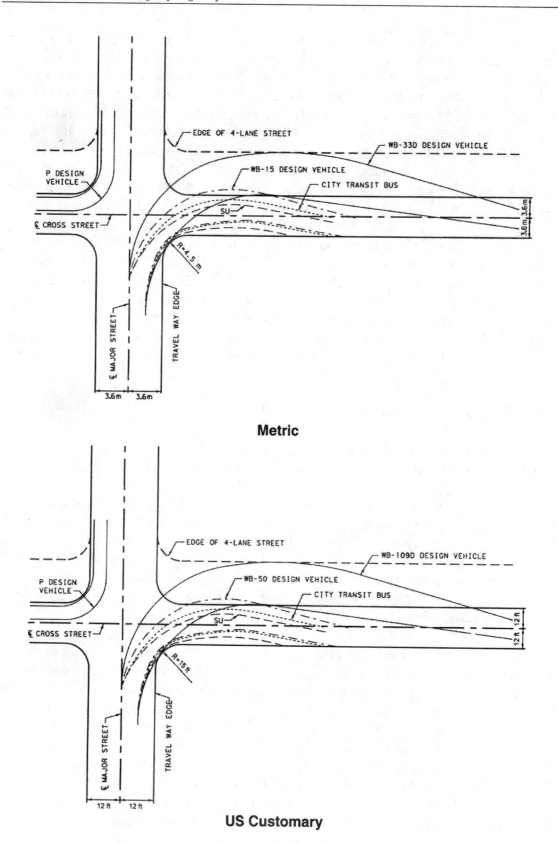

Metric

US Customary

Exhibit 9-29. Effect of Curb Radii on Right Turning Paths of Various Design Vehicles

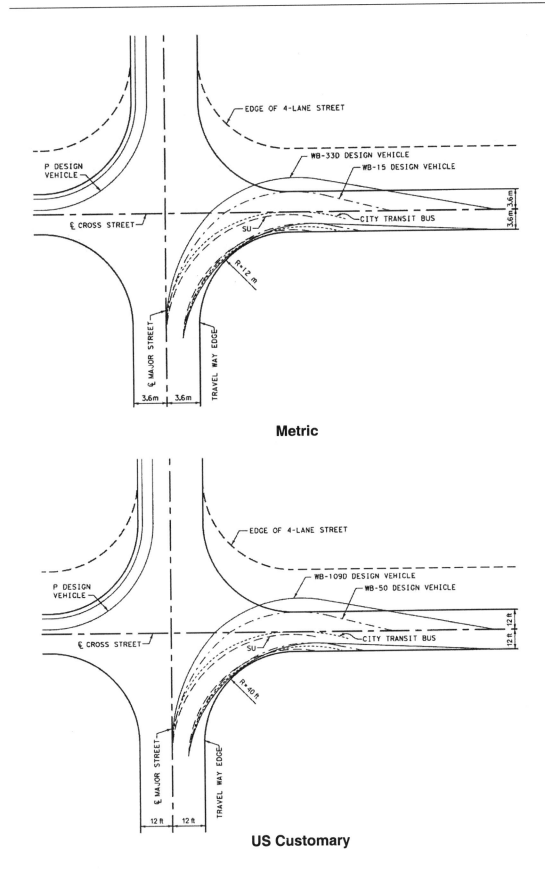

Metric

US Customary

Exhibit 9-30. Effect of Curb Radii on Right Turning Paths of Various Design Vehicles

Exhibit 9-31 shows the effect of the angle of intersection on turning paths of various design vehicles on streets without parking lanes. The dimensions d_1 and d_2 are the widths occupied by the turning vehicle on the major street and cross street, respectively, while negotiating turns through various angles. Both dimensions are measured from the right-hand curb to the point of maximum overhang. These widths, shown for various angles of turn and curb radii, and for two types of maneuvers, generally increase with the angle of turn.

Exhibit 9-31 also shows that a very large radius should be used or the streets should be very wide to accommodate the longer vehicles, particularly where the central angle is greater than 90 degrees. For this reason three-centered curves (or offset, simple curves in combination with tapers to fit the paths of vehicles properly) are much preferred. Exhibits 9-19 and 9-20 show curve radii suitable for accommodating the several classes of design vehicles for a wide range of angles of turn. Data are shown for simple curves and for two types of three-centered curves. The radii for simple curves have been omitted for angles of turn greater than 90 degrees for the reasons given above. However, they may be used for right-turn designs where sufficient right-of-way are available and where there is little pedestrian traffic.

With parking allowed along a curbed street, vehicles (except for WB-19 [WB-62] and larger vehicles) are able to turn without encroachment onto adjacent lanes, even where curb radii are relatively small. As shown in Exhibit 9-32, the SU and WB-12 [WB-40] design vehicles are able to turn at a 4.5-m [15-ft] curb radius with little, if any, encroachment on adjacent lanes. However, parking should be restricted for a distance of at least 4.5 m [15 ft] in advance of the right-turning radius and at least 9 m [30 ft] beyond the radius on the exit.

The BUS and WB-15 [WB-50] design vehicles will encroach onto the opposing lanes in making a turn unless the turning radius is at least 7.5 m [25 ft] and parking is restricted at the far end of the turn for at least 12 m [40 ft] beyond the radius. Because traffic volumes may increase to the point where all parking is prohibited either during rush hours or throughout the day, caution is advised in the use of radii of 4.5 or 7.5 m [15 or 25 ft] where parking is permitted. If parking is prohibited, the same turning conditions prevail as shown in Exhibits 9-29, 9-30, and 9-31.

Effect of Curb Radii on Pedestrians

For arterial street design, adequate radii for vehicle operation should be balanced against the needs of pedestrians and the difficulty of acquiring additional right-of-way or corner setbacks. Because the corner radius is often a compromise, its effect on both pedestrians and vehicular movements should be examined.

Metric

Angle of Intersection (Δ)	Design Vehicle	d₂ for cases A and B where:									
		R=4.5 m		R=6 m		R=7.5 m		R=9 m		R=12 m	
		A m	B m	A m	B m	A m	B m	A m	B m	A m	B m
30°	SU	4.3	4.0	4.3	4.0	4.0	4.0	4.0	4.0	4.0	4.0
	BUS	6.7	5.2	5.8	5.2	5.8	5.2	5.8	5.2	5.5	5.2
	WB-12	4.3	4.3	4.3	4.3	4.3	4.3	4.3	4.3	4.3	4.3
	WB-15	6.1	5.2	6.1	5.2	6.1	5.2	5.8	4.9	5.5	4.9
	WB-19	–	–	–	–	–	–	–	–	8.2	5.2
	WB-20	–	–	–	–	–	–	–	–	8.5	5.5
60°	SU	5.8	4.9	5.8	4.9	5.2	4.6	4.9	4.6	4.3	4.3
	BUS	8.5	6.4	7.9	6.1	7.3	6.1	7.0	5.8	6.7	5.5
	WB-12	7.3	5.8	6.7	5.8	6.4	5.8	5.8	5.5	5.2	4.9
	WB-15	9.4	6.7	8.2	6.4	8.5	6.1	7.6	5.8	6.7	5.5
	WB-19	–	–	–	–	–	–	–	–	9.1	6.7
	WB-20	–	–	–	–	–	–	–	–	11.3	7.3
90°	SU	7.9	6.1	7.0	5.5	5.8	4.9	5.2	4.6	4.0	4.0
	BUS	11.6	7.0	10.0	6.7	9.1	6.7	7.6	6.4	6.7	5.5
	WB-12	9.4	6.7	8.2	6.4	7.0	6.4	5.8	5.5	5.2	4.9
	WB-15	12.8	6.7	11.3	7.3	9.8	6.7	8.8	6.4	6.7	5.5
	WB-19	–	–	–	–	–	–	–	–	11.9	7.0
	WB-20	–	–	–	–	–	–	–	–	11.9	7.6

US Customary

Angle of Intersection (Δ)	Design Vehicle	d₂ for cases A and B where:									
		R=15 ft		R=20 ft		R=25 ft		R=30 ft		R=40 ft	
		A ft	B ft	A ft	B ft	A ft	B ft	A ft	B ft	A ft	B ft
30°	SU	14	13	14	13	13	13	13	13	13	13
	BUS	22	17	19	17	19	17	19	17	18	17
	WB-40	14	14	14	14	14	14	14	14	14	14
	WB-50	20	17	20	17	20	17	19	16	18	16
	WB-62	–	–	–	–	–	–	–	–	27	17
	WB-67	–	–	–	–	–	–	–	–	28	18
60°	SU	19	16	19	16	17	15	16	15	14	14
	BUS	28	21	26	20	24	20	23	19	22	18
	WB-40	24	19	22	19	21	19	19	18	17	16
	WB-50	31	22	27	21	28	20	25	19	22	18
	WB-62	–	–	–	–	–	–	–	–	30	22
	WB-67	–	–	–	–	–	–	–	–	37	24
90°	SU	26	20	23	18	19	16	17	15	13	13
	BUS	38	23	33	22	30	22	25	21	22	18
	WB-40	31	22	27	21	23	21	19	18	17	16
	WB-50	42	22	37	24	32	22	29	21	22	18
	WB-62	–	–	–	–	–	–	–	–	39	23
	WB-67	–	–	–	–	–	–	–	–	39	25

Exhibit 9-31. Cross Street Width Occupied by Turning Vehicle for Various Angles of Intersection and Curb Radii

Metric

Angle of Inter-section (Δ)	Design vehicle	R=4.5 m A (m)	R=4.5 m B (m)	R=6 m A (m)	R=6 m B (m)	R=7.5 m A (m)	R=7.5 m B (m)	R=9 m A (m)	R=9 m B (m)	R=12 m A (m)	R=12 m B (m)
		d₂ for cases A and B where:									
120°	SU	10.4	6.7	8.2	5.8	6.4	5.5	5.2	4.9	4.0	4.0
	BUS	14.0	8.5	12.2	7.6	9.8	7.0	7.9	5.8	5.8	5.5
	WB-12	11.3	7.0	8.8	6.7	7.3	6.7	5.8	5.5	5.2	4.9
	WB-15	15.2	8.8	13.1	8.5	11.0	8.2	9.1	7.9	6.7	5.5
	WB-19	–	–	–	–	–	–	–	–	7.9	6.7
	WB-20	–	–	–	–	–	–	–	–	9.1	7.0
150°	SU	12.2	7.6	9.8	6.4	6.7	5.8	5.2	4.9	3.6	3.6
	BUS	14.6	8.5	12.2	7.6	9.8	7.0	6.7	5.5	4.9	4.9
	WB-12	11.9	7.3	8.8	6.7	7.0	6.7	5.8	5.5	5.2	4.9
	WB-15	16.2	9.4	14.0	8.5	11.0	8.2	8.5	7.9	6.7	5.5
	WB-19	–	–	–	–	–	–	–	–	6.1	5.5
	WB-20	–	–	–	–	–	–	–	–	8.2	5.5

US Customary

Angle of Inter-section (Δ)	Design vehicle	R=15 ft A (ft)	R=15 ft B (ft)	R=20 ft A (ft)	R=20 ft B (ft)	R=25 ft A (ft)	R=25 ft B (ft)	R=30 ft A (ft)	R=30 ft B (ft)	R=40 ft A (ft)	R=40 ft B (ft)
		d₂ for cases A and B where:									
120°	SU	34	22	27	19	21	18	17	16	13	13
	BUS	46	28	40	25	32	23	26	19	19	18
	WB-40	37	23	29	22	24	22	19	18	17	16
	WB-50	50	29	43	28	36	27	30	26	22	18
	WB-62	–	–	–	–	–	–	–	–	26	22
	WB-67	–	–	–	–	–	–	–	–	30	23
150°	SU	40	25	32	21	22	19	17	16	12	12
	BUS	48	28	40	25	32	23	22	18	17	16
	WB-40	39	24	29	22	23	22	19	18	17	16
	WB-50	53	31	46	28	36	27	28	26	22	18
	WB-62	–	–	–	–	–	–	–	–	20	18
	WB-67	–	–	–	–	–	–	–	–	27	18

NOTE: P DESIGN VEHICLE TURNS WITHIN 3.6 m [12 FT] WIDTH WHERE R=4.5 m [15 FT] OR MORE. NO PARKING ON EITHER STREET.

CASE A
VEHICLE TURNS FROM PROPER LANE AND SWINGS WIDE ON CROSS STREET
d₁=3.6 m [12 FT] d₂ IS VARIABLE

CASE B
TURNING VEHICLE SWINGS EQUALLY WIDE ON BOTH STREETS
d₁=d₂ BOTH VARIABLE

Note: P design vehicle turns within 3.6 m [12 ft] where R=4.5 m [15 ft] or more. No parking on either street.

Exhibit 9-31. Cross Street Width Occupied by Turning Vehicle for Various Angles of Intersection and Curb Radii (Continued)

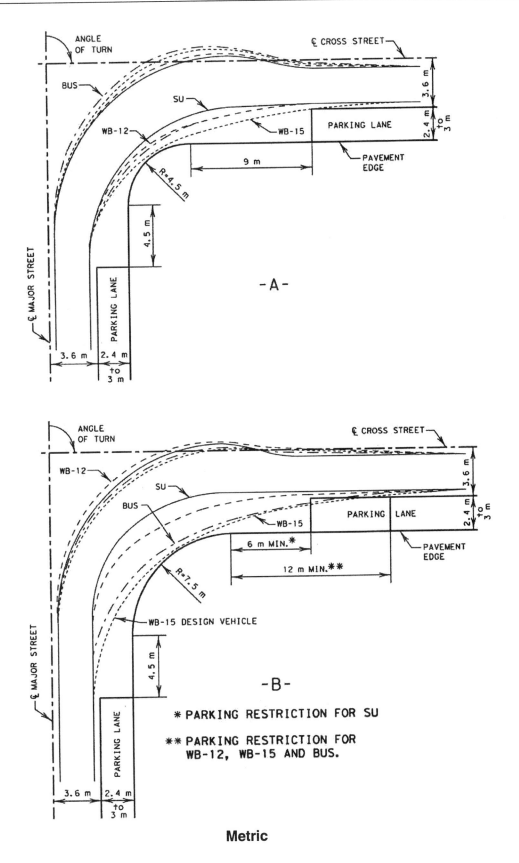

Metric

Exhibit 9-32. Effect of Curb Radii and Parking on Right Turning Paths

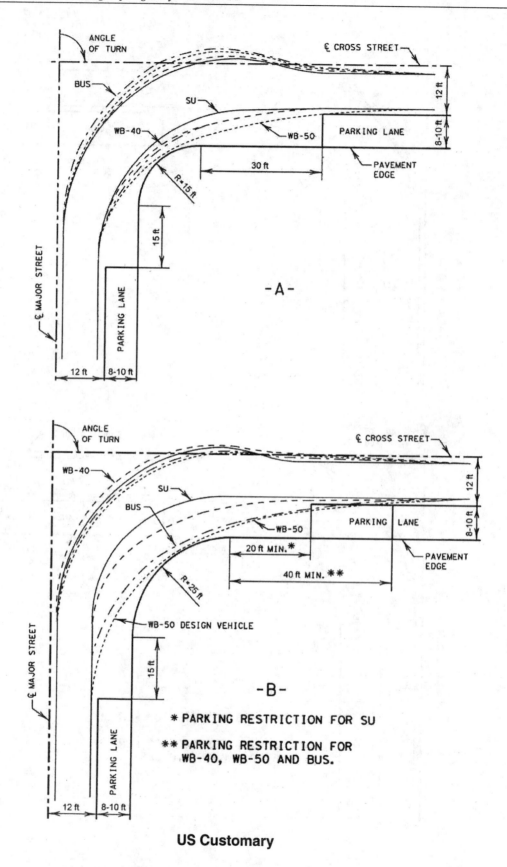

US Customary

Exhibit 9–32. Effect of Curb Radii and Parking on Right Turning Paths (Continued)

Crosswalk distances and right-of-way or corner setback needs increase with the curb return radius. The added crosswalk distances between curbs as compared with the normal curb-to-curb street widths are shown in Exhibit 9-33 based on the assumptions that the sidewalk centerline at a right-angle intersection is in line with the middle of a border and that the same curb radius is used on all four corners.

The additional right-of-way or corner setback resulting from various curb radii for border widths of 3 and 6 m [10 and 20 ft] is shown in Exhibit 9-34. The dimensions shown in Exhibits 9-33 and 9-34 vary somewhat with intersection angles that differ from 90 degrees.

The dimensions presented in Exhibits 9-33 and 9-34 demonstrate why curb radii of only 3 to 4.5 m [10 to 15 ft] have been used in most cities. Where larger radii are used, an intermediate refuge or median island is desirable or crosswalks may need to be offset so that crosswalk distances are not objectionable. In summary, the corner radii proposed at an intersection on urban arterial streets should satisfy the needs of the drivers using them, the amount of right-of-way available, the angle of turn between the intersection legs, the number of pedestrians using the crosswalk, the width and number of lanes on the intersecting street, and the posted speeds on each street. The following is offered as a guide:

- Radii of 4.5 to 7.5 m [15 to 25 ft] are adequate for passenger vehicles. These radii may be provided at minor cross streets where there is little occasion for trucks to turn or at major intersections where there are parking lanes. Where the street has sufficient capacity to retain the curb lane as a parking lane for the foreseeable future, parking should be restricted for appropriate distances from the crossing as shown in Exhibit 9-32.
- Radii of 7.5 m [25 ft] or more should be provided at minor cross streets, on new construction and on reconstruction projects where space permits.
- Radii of 9 m [30 ft] or more should be provided at minor cross streets where practical so that an occasional truck can turn without too much encroachment.
- Radii of 12 m [40 ft] or more, or preferably three-centered curves, or simple curves with tapers to fit the paths of large truck combinations, should be provided where such combinations or buses turn frequently. Where speed reductions would cause problems, longer radii should be considered.
- Curb radii should be coordinated with crosswalk distances or special designs should be used to make crosswalks efficient for all pedestrians (see Chapter 4).

Curb radii at corners on two-way streets have little effect on left-turning movements. Where the width of an arterial street is equivalent to four or more lanes, generally there is no problem of encroachment by left-turning vehicles.

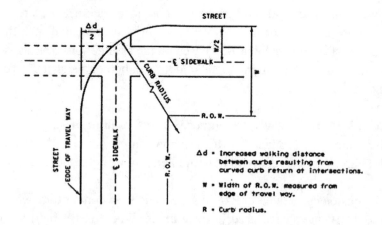

CURB RADIUS, R	ADDED CROSSWALK DISTANCE Δd	
	W = 3m [10 Ft]	W = 6m [20 Ft]
METERS [FEET]	METERS [FEET]	METERS [FEET]
3 [10]	0.8 [3]	0.0 [0]
6 [20]	4.0 [14]	1.6 [5]
9 [30]	8.0 [27]	4.6 [15]
12 [40]	12.4 [42]	8.1 [27]
15 [50]	16.9 [57]	12.0 [40]

Exhibit 9-33. Variations in Length of Crosswalk With Different Curb Radii and Width of Borders

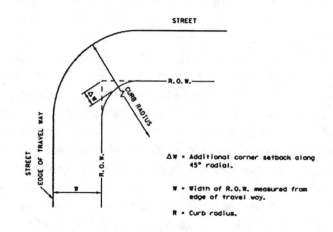

CURB RADIUS, R	ADDITIONAL CORNER SETBACK ΔW	
	W = 3m [10 Ft]	W = 6m [20 Ft]
METERS [FEET]	METERS [FEET]	METERS [FEET]
3 [10]	0.0 [0]	0.0 [0]
6 [20]	1.3 [4]	0.0 [0]
9 [30]	2.5 [8]	1.3 [4]
12 [40]	4.0 [13]	2.5 [8]
15 [50]	5.0 [17]	4.0 [13]

Exhibit 9-34. Corner Setbacks with Different Curb Radii and Width of Borders

Corner Radii Into Local Urban Streets

Because of space limitations, presence of pedestrians, and generally lower operating speed in urban areas, curve radii for turning movements may be smaller than those normally used in rural areas. Corner radii to accommodate right-turning movements depend largely on the number and type of turning vehicles and the volume of pedestrians. Minimum turning paths for passenger vehicles and all other design vehicles are included in Chapter 2.

Guidelines for right-turning radii into minor side streets in urban areas usually range from 1.5 to 9 m [5 to 30 ft] and most are between 3 and 4.5 m [10 to 15 ft]. Where a substantial number of pedestrians are present, the lower end of the ranges described below may be appropriate. Most passenger cars operating at very low speed on lanes 3 m [10 ft] or more in width are able to make a right turn with a curb radius of about 4.5 m [15 ft] with little encroachment on other lanes. However, operation of these vehicles at increased speeds or of larger vehicles even at a very low speed generally results in substantial encroachment on adjacent lanes at either the beginning or the end of the turn, or both.

Where there are curb parking lanes on both of the intersecting streets and parking is restricted for some distance from the corner, the extra width provided by the restriction serves to increase the usable radius. On most streets, curb radii of 3 to 4.5 m [10 to 15 ft] are reasonable because streets and sidewalks are generally confined within the public right-of-way, and larger radii can be obtained only by narrowing sidewalks at corners and increasing the length of pedestrian crosswalks. However, to ensure efficient traffic operation on arterial streets carrying heavy traffic volumes, it is desirable to provide corner radii of 4.5 to 7.5 m [15 to 25 ft] for passenger vehicles and 9 to 15 m [30 to 50 ft] for most trucks and buses, provided there are no significant pedestrian conflicts. Where large truck combinations turn frequently, somewhat larger radii should be provided for turns.

The WB-19 [WB-62] and larger trucks generally are not engaged in local travel destinations but are used principally for "over-the-road" transportation between trucking terminals or industrial or commercial areas. Ideally, such destinations are located near major highway facilities that are designed to accommodate the larger combination units.

If trucks are routed over local streets to reach their destinations, careful consideration should be given to the network to be used. Generally, this network should not include narrow streets, streets with relatively small right-turning radii at intersections, or streets with parking and significant pedestrian crossing volumes.

ISLANDS

General Characteristics

An island is a defined area between traffic lanes used for control of vehicle movements. Islands also provide an area for pedestrian refuge and traffic control devices. Within an intersection, a median or an outer separation is also considered an island. This definition makes evident that an island is no single physical type. It may range from an area delineated by a raised curb to a pavement area marked out by paint or thermoplastic markings. Where traffic entering an intersection is directed into definite paths by islands, this design feature is termed a channelized intersection.

Channelizing islands generally are included in intersection design for one or more of the following purposes:

- Separation of conflicts
- Control of angle of conflict
- Reduction in excessive pavement areas
- Regulation of traffic and indication of proper use of intersection
- Arrangements to favor a predominant turning movement
- Protection of pedestrians
- Protection and storage of turning and crossing vehicles
- Location of traffic control devices

Islands serve three primary functions: (1) channelization—to control and direct traffic movement, usually turning; (2) division—to divide opposing or same direction traffic streams, usually through movements; and (3) refuge—to provide refuge for pedestrians. Most islands combine two or all of these functions.

Islands generally are either elongated or triangular in shape and are normally situated in areas unused for vehicle paths. Islands should be located and designed to offer little obstruction to vehicles, be relatively inexpensive to build and maintain, and occupy a minimum of roadway space; however, they should be commanding enough that motorists will not drive over them. The dimensions and details depend on the particular intersection design and should conform to the general principles that follow.

Curbed islands are sometimes difficult to see at night because of the glare from oncoming headlights or from distant luminaires or roadside businesses. Accordingly, where curbed islands are used, the intersection should have fixed-source lighting or appropriate delineation such as curb-top reflectors.

Where various intersections are involved along a route and the warrants are sufficiently similar to enhance driver expectancy, it is desirable to provide a common geometric design for each intersection. Reference can also be made to the *Manual on Uniform Traffic Control Devices* (MUTCD) (**9**) for design guidance.

Under certain conditions, painted, flush medians and islands or traversable type medians may be preferable to the raised curb type islands. Such conditions are: in lightly developed areas that will not be considered for access management, at intersections where approach speeds are relatively high; where there is little pedestrian traffic; where fixed-source lighting is not provided; where signals, signs, or luminaire supports are not needed in the median or corner islands; in areas of significant snow plowing; and where extensive development exists along a street and there is a need for left-turn areas into many entrances.

Painted islands may be used at the traveled way edge. At some intersections, both curbed and painted islands may be desirable. All pavement markings should be reflectorized. The use of thermoplastic striping, raised dots, spaced and raised retroreflective markers, and other forms of long-life markings also may be desirable. This subject is discussed in the MUTCD (**9**).

Channelizing Islands

Channelizing islands that control and direct traffic movements into the proper paths for their intended use are an important part of intersection design. Confusing traffic movements resulting from spacious areas may be eliminated by the conversion of unused areas into islands that leave little to driver discretion. Channelizing islands may be of many shapes and sizes, depending on the conditions and dimensions of the intersection. Some of those conditions are illustrated in Exhibit 9-35. A common form is the corner triangular shape that separates right-turning traffic from through traffic. Central islands may serve as a guide around which turning vehicles operate.

Channelizing islands should be placed so that the proper course of travel is immediately obvious, easy to follow, and of unquestionable continuity. When designing an island, attention should be given to the fact that the driver's eye view is different from the plan view. Particular care should be taken where the channelization is on or beyond a crest of a vertical curve, however slight, or where there is substantial horizontal curvature on the approach to or through the channelized area. The outlines of islands should be easily flowing curved or straight lines nearly parallel to the line of travel.

Where islands separate turning traffic from through traffic, the radii of curved portions should equal or exceed the minimum for the turning speeds expected. Drivers should not be confronted suddenly with an unusable area in the normal vehicle path. Islands first approached by traffic should be indicated by a gradually widening and marking or a conspicuously rumble strip that directs traffic to each side.

Intersections with multiple turning lanes may need three or more islands to channelize the various movements. There is a practical limitation to the use of multiple islands for channelizing traffic. A group of islands outlining several one-way lanes may cause confusion and result in wrong-way movements into opposing traffic lanes. Such layouts may be confusing to drivers using them for the first time, and with additional trips, the same drivers may use them as desired.

However, with the possibility of confusion, this suggests that a few large islands are preferable to a greater number of smaller islands. At intersections where the area for multiple-lane channelization is restricted, it may be advisable to try temporary layouts of movable stanchions or sandbags and observe traffic flow with several variations of sizes or shapes of islands before designing and constructing the permanent islands.

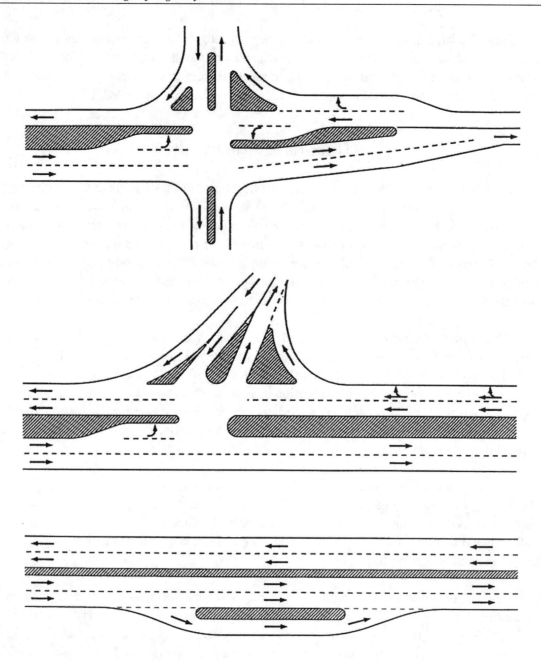

Exhibit 9-35. General Types and Shapes of Islands and Medians

Properly placed islands are advantageous where through and turning movements are heavy. However, at minor intersections on two-lane highways, channelization may be of questionable value, especially in rural areas where small curbed islands were provided. The use of curbed islands generally should be reserved for multilane highways or streets and for the more important intersections on two-lane highways. In or near urban areas where speeds are low and drivers are accustomed to confined facilities, channelization can be expected to work well. Curbed islands generally should not be used in rural areas and at isolated locations unless the intersection is lighted and curbs are delineated.

Marked channelization (painting or striping) can be made to increase efficiency and has the advantage of easy modification when warranted by driver behavior. If a more positive barrier is needed, curbed islands may be constructed, but the marked channelization may well serve initially to establish the best layout arrangement before permanent construction is established. However, it should be noted that inclement weather decreases the effectiveness of flush channelization.

Divisional Islands

Divisional islands often are introduced on undivided highways at intersections. They alert the drivers to the crossroad ahead and regulate traffic through the intersection. These islands are particularly advantageous in controlling left turns at skewed intersections and at locations where separate channels are provided for right-turning traffic. A variety of divisional islands that separate opposing traffic are illustrated in Exhibit 9-36.

Where an island is introduced at an intersection to separate opposing traffic on a four-lane road or on a major two-lane highway carrying high volumes, particularly where future conversion to a wider highway is likely, two full lanes should be provided on each side of the dividing island. In other instances, narrower roadways may be used. For moderate volumes, roadway widths shown under Case II (one-lane, one-way operation with provision for passing a stalled vehicle) in Exhibit 3-55 are appropriate. For light volumes and where small islands are needed, widths on each side of the island corresponding to Case I in Exhibit 3-55 may be used.

Widening a roadway to include a divisional island (Exhibit 9-36) should be done in such a manner that the proper paths to follow are unmistakably evident to drivers. The alignment should require no appreciable conscious effort in vehicle steering. Often the highway is on a tangent, and to introduce dividing islands, reverse curve alignment would be needed. Tapers can be used, but should be consistent with lane shifts at the design speed. In rural areas, where speeds are generally high, reversals in curvature should preferably be with radii of 1,165 m [3,825 ft] or greater. Sharper curves may be used on intermediate-speed roads (up to 70 km/h [45 mph]) with radii of 620 m [2,035 ft] or greater. Usually, the roadway in each direction of travel is bowed out, more or less symmetrically about the centerline as shown in Exhibit 9-36. Widening may also be affected on one side only with one of the roadways continuing through the intersection on a straight course. When this arrangement is used for a two-lane road that is planned for future conversion to a divided highway, the traveled way on tangent alignment will become a permanent part of the ultimate development.

Widening on tangent alignment, even with flat curves, may produce some appearance of distorted alignment. Where the road is on a curve or on widening alignment, advantage should be taken of the curvature in spreading the traffic lanes without using reverse curves, as illustrated in number C and D in Exhibit 9-36.

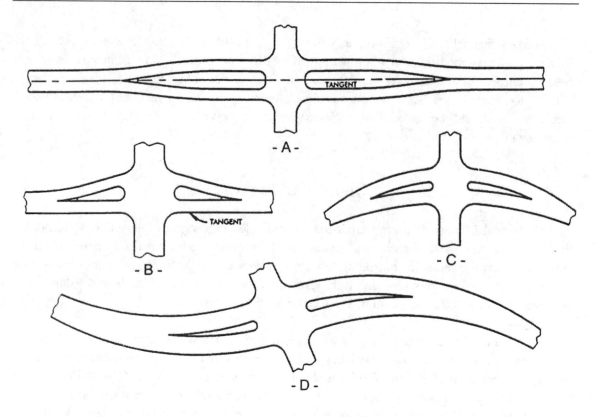

Exhibit 9-36. Alignment for Addition of Divisional Islands at Intersections

Refuge Islands

A refuge island for pedestrians is one at or near a crosswalk or bicycle path that aids and protects pedestrians and bicyclists who cross the roadway. Raised-curb corner islands and center channelizing or divisional islands can be used as refuge areas. Refuge islands for pedestrians and bicyclists crossing a wide street, for loading or unloading transit riders, or for wheelchair ramps are used primarily in urban areas.

The location and width of crosswalks, the location and size of transit loading zones, and the provision of wheelchair ramps influence the size and location of refuge islands. Refuge islands should be a minimum of 1.8 m [6 ft] wide when they will be used by bicyclists. Pedestrians and bicyclists should have a clear path through the island and should not be obstructed by poles, sign posts, utility boxes, etc. Chapter 4 contains details of wheelchair ramp design that affect the minimum size of the small islands.

In rural, as well as in urban, areas many of the islands designed for the function of channelization are also of the type and location to serve as refuge for pedestrians. The islands in Exhibit 9-35 are examples. The general principles for island design also apply directly to providing refuge islands.

Island Size and Designation

Island sizes and shapes vary materially from one intersection to another, as shown in Exhibit 9-35. Further variations, not illustrated, occur at multiple and acute-angle intersections. Islands should be sufficiently large to command attention. The smallest curbed corner island normally should have an area of approximately 5 m^2 [50 ft^2] for urban and 7 m^2 [75 ft^2] for rural intersections. However, 9 m^2 [100 ft^2] is preferable for both. Accordingly, corner triangular islands should not be less than about 3.5 m [12 ft], and preferably 4.5 m [15 ft], on a side after the rounding of corners.

Elongated or divisional islands should be not less than 1 m [4 ft] wide and 6 to 8 m [20 to 25 ft] long. In special cases where space is limited, elongated islands may be reduced to a minimum width of 0.5 m [2 ft]. In general, introducing curbed divisional islands at isolated intersections on high-speed highways is undesirable unless special attention is directed to providing high visibility for the islands. Curbed divisional islands introduced at isolated intersections on high-speed highways should be 30 m [100 ft] or more in length. When situated in the vicinity of a high point in the roadway profile or at or near the beginning of a horizontal curve, the approach end of the curbed island should be extended to be clearly visible to approaching drivers.

Islands should be delineated or outlined by a variety of treatments, depending on their size, location, and function. The type of area in which the intersection is located, rural versus urban, also governs the design. In a physical sense, islands can be divided into three groups: (1) raised-curb islands, (2) islands delineated by pavement markings or reflectorized markers placed on paved areas, and (3) islands formed by the pavement edges and possibly supplemented by delineators on posts or other guideposts, or mounded-earth treatment beyond and adjacent to the pavement edges.

The curbed island treatment is universal and provides the greatest positive guidance. In rural areas where curbs are uncommon, this treatment often is limited to corner islands of small to intermediate size. Conversely, in urban areas, the use of this type of island is common.

Island delineation of unused paved areas, by pavement markings, is common in urban districts where speeds are low and space is limited. In rural areas, this type may be used to minimize maintenance problems on high approach speeds or where snow removal is more difficult with curbed islands. Group 2 islands also are applicable on low-volume highways where the added expense of curbs may not be warranted and where the islands are not large enough for delineation by pavement edges alone.

The Group 3 treatment by its nature applies to other than small channelizing islands and is primarily used at rural intersections where there is space for large-radius intersection curves and wide medians.

The central area of large channelizing islands in most cases has a turf or other vegetative cover. As space and the overall character of the highway determine, low plant material may be included, but it should not obstruct sight distance. Ground cover or plant growth, such as turf, vines, and shrubs, can be used for channelizing island and provides excellent contrast with the paved areas, assuming that the ground cover is cost-effective and can be properly maintained. Small curbed islands may be mounded, but where pavement cross slopes are outward, large islands should be depressed to avoid draining water and snow melt across the pavement. This feature is especially desirable where alternate freezing and thawing occurs. For small curbed islands and in areas where growing conditions are not favorable, some type of paved surface is used on the island. In many respects, the curbed-island cross-section design is similar to that discussed in the section on "Medians" in Chapter 4.

Island Delineation and Approach Treatment

Delineation of small islands is effected primarily by curbs and curb-top reflectors. Large curbed islands may be sufficiently delineated by color and texture contrast of vegetative cover, mounded earth, shrubs, reflector posts, signs, or any combination of these. In rural areas, island curbs should usually be a sloping type.

Chapter 4 indicates different curb types. The most commonly used height of curb is 150 mm [6 in]. Vertical or sloping curbs could be appropriate in urban areas, depending on the conditions. In addition, high-visibility sloping curbs may be advantageous at critical locations.

The outline of a curbed island is determined by the edge of through-traffic lanes or turning roadways. Lateral clearance is provided to the face of the curbed island. The points of intersection of the sides of a curbed island are rounded or beveled for visibility and construction simplicity. The amount that a curbed island is offset from the through-traffic lane is influenced by the type of edge treatment and other factors such as island contrast, length of taper, or auxiliary pavement preceding the curbed island. Since curbs influence the lateral placement of a vehicle in a lane, they should be offset from the edge of through-traffic lanes even if they are sloping. Curbs need not be offset from the edge of a turning roadway, except to reduce their vulnerability to turning trucks.

Details of curbed corner island designs used in conjunction with turning roadways are shown in Exhibits 9-37 and 9-38. The approach corner of each curbed island is designed with an approach nose treatment. Three curbed triangular island sizes, small, intermediate, and large, are shown for two general cases of through-traffic lanes edges: (1) The curbed corner island is located along an urban street with curb and gutter, or (2) the curbed corner island is located on a highway with shoulders.

Small curbed corner islands are those of minimum or near-minimum size, as previously discussed. Large curbed corner islands are those with side dimensions of at least 30 m [100 ft]. All curbed islands in Exhibits 9-37 and 9-38 are shown with approach noses and merging ends rounded with appropriate radii of 0.6 to 1 m [2 to 3 ft]. The approach corner is rounded with a radius of 0.6 to 1.5 m [2 to 5 ft].

Exhibit 9-37 shows curbed corner islands adjacent to through-traffic lanes on an urban street. Where the approach roadway has a curb and gutter, the curbed island may be located at the edge of the through lane with a gradual taper to the nose offset. Where the large-size island is uncurbed, the indicated offsets of the curbed island are desirable but not essential. However, any fixed objects within the island areas should be offset an appropriate distance from the through lanes.

The approach nose of a curbed island should be conspicuous to approaching drivers and should be definitely clear of vehicle paths, physically and visually, so that drivers will not shy away from the island. Reflectorized markers may be used on the approach nose of the curbed island. The offset from the travel lane to the approach nose should be greater than that to the face of the curbed island, normally about 0.6 m [2 ft]. For curbed median islands, the face of curb at the approach island nose should be offset at least 0.6 m [2 ft] and preferably 1.0 m [3 ft] from the normal median edge of the traveled way. The island should then be gradually widened to its full width. For other curbed islands, the total nose offset should be 1 to 2 m [3 to 6 ft] from the normal edge of through lanes and 0.6 to 1 m [2 to 3 ft] edge of the traveled way of a turning roadway. Large offsets should be provided where the curbed corner island is preceded by a right-turn deceleration lane.

Where a curbed corner island is proposed on an approach roadway with shoulders, the face of curb on the corner island should be offset by an amount equal to the shoulder width. If the corner island is preceded by a right-turn deceleration lane, the shoulder offset should be at least 2.4 m [8 ft].

Curbed corner islands and median noses should be ramped down as shown in Exhibit 9-39 and provided with devices to give advance warning to approaching drivers and especially for nighttime driving. Pavement markings in front of the approach nose are particularly advantageous on the areas shown as stippled in Exhibit 9-37. To the extent practical, other high-visibility indications should be used, such as reflectorized curb-top markers mounted on the curb or median surface. The curbs of all islands located in the line of traffic flow should be marked in accordance with the MUTCD (**9**).

Delineation is especially pertinent at the approach nose of a divisional island. In rural areas, the approach should consist of a gradually widening of the divisional island as indicated in Exhibit 9-40. Although not as frequently obtainable, this same design also should be striven for in urban areas. Preferably, the approach should gradually change to a raised surface with texture or to jiggle bars that may be crossed readily even at considerable speed. This transition section should be as long as practical. The cross sections in Exhibit 9-40 demonstrate the transition. The face of curb at the approach island nose should be offset at least 0.5 m [2 ft] and preferably 1 m [3 ft] from the normal edge of traveled way, and the widened pavement gradually should be transitioned to the normal width toward the crossroad.

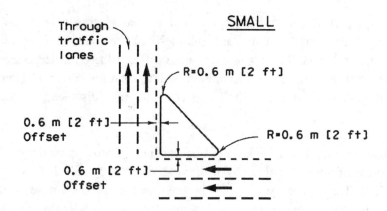

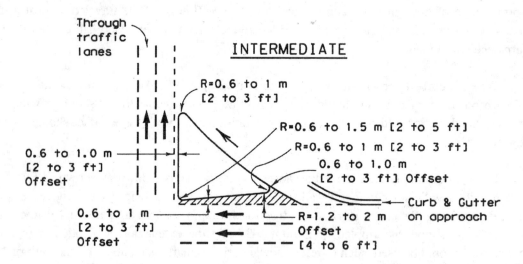

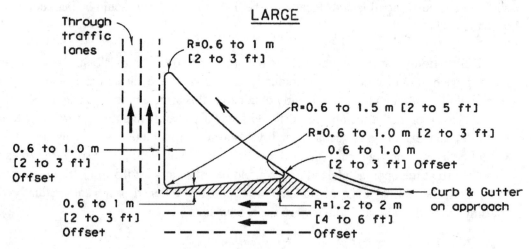

TRIANGULAR CURBED ISLAND ON URBAN STREETS

Exhibit 9-37. Details of Corner Island Designs for Turning Roadways (Urban Location)

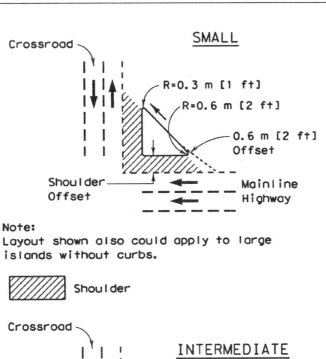

SMALL

Note:
Layout shown also could apply to large islands without curbs.

▨ Shoulder

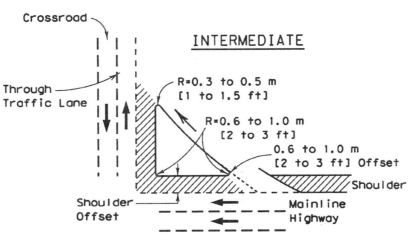

INTERMEDIATE

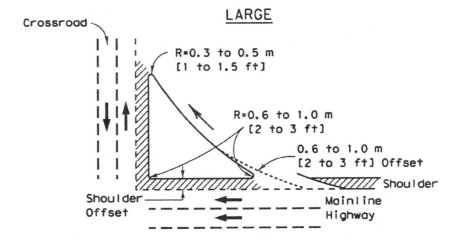

LARGE

TRIANGULAR CURBED ISLAND WITH SHOULDERS

Exhibit 9-38. Details of Corner Island Designs for Turning Roadways (Rural Cross Section on Approach)

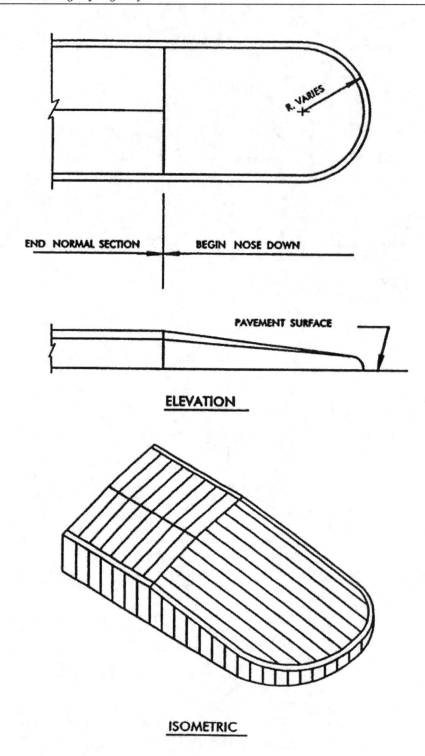

END NORMAL SECTION   BEGIN NOSE DOWN

R. VARIES

PAVEMENT SURFACE

ELEVATION

ISOMETRIC

Exhibit 9-39. Nose Ramping at Approach End of Median or Corner Island

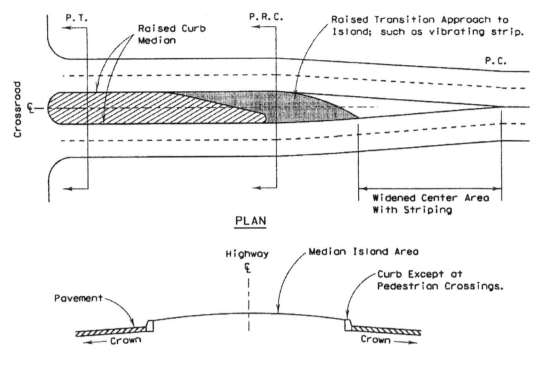

PLAN

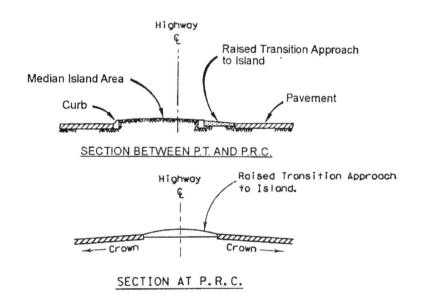

SECTION AT P.T.

SECTION BETWEEN P.T. AND P.R.C.

SECTION AT P.R.C.

Exhibit 9-40. Details of Divisional Island Design

TURNING ROADWAYS WITH CORNER ISLANDS

Where the inner edges of the traveled way for right turns are designed to accommodate semitrailer combinations or where the design permits passenger vehicles to turn at speeds of 15 km/h [10 mph] or more, the pavement area within the intersection may become excessively large and consequently does not provide for the proper control of traffic. To avoid this condition, a corner island can be provided to form a separate turning roadway between the two intersection legs.

Intersections that have large paved areas, such as those with large corner radii or those with oblique angle crossings, permit and encourage uncontrolled vehicle movements, involve long pedestrian crossings, and have unused pavement areas. Even at a simple intersection, appreciable areas may exist on which some motorists can wander from their natural and expected paths. Therefore, conflicts may be reduced by use of corner triangular islands.

Right-Angle Turns With Corner Islands

The principal controls for the design of turning roadways are the alignment of the traveled way edge and the turning roadway width. These design features ensure that a vehicle can be accommodated while turning at the selected turning roadway speed. With radii greater than the minimum edge of traveled way, controls result in an area large enough for a triangular island to be designed between the left edge of the turning roadway and the traveled way edges of the two through highways. Such an island is desirable for delineating the path of through and turning traffic, for the placement of signs, and for providing a refuge for pedestrians and bicycles. Larger islands may be needed to locate signs and to facilitate snow-removal operations.

A turning roadway should be designed to provide at least the minimum size island and the minimum width of roadway. The turning roadway should be wide enough to permit the right and left wheel tracks of a selected vehicle to be within the edges of the traveled way by about 0.6 m [2 ft] on each side. Generally, the turning roadway width should not be less than 4.2 m [14 ft]. When the turning roadway is designed for a semitrailer combination, a much wider roadway is needed. To discourage passenger vehicles from using this wider roadway as two lanes, the roadway may be reduced in size by marking out part of the roadway with paint or thermoplastic markings.

Exhibit 9-41 shows minimum turning roadway designs for a 90-degree right turn. A design based on a minimum size island and a minimum turning roadway width of 4.2 m [14 ft] (Exhibit 9-41A) results in a circular arc of 18-m [60-ft] radius (not shown) for the right edge of the traveled way for the turning roadway or in a three-centered curve (as shown) with radii of 45, 15, and 45 m [150, 50, and 150 ft] with the middle curve being offset 1 m [3 ft] from the tangent edges extended. This design not only permits passenger vehicles to turn at a speed of about 25 km/h [15 mph] but also enables SU design vehicles to turn on a radius (right front wheel) of approximately 20 m [65 ft] and still clear the turning roadway by about 0.3 m [1 ft] on each side.

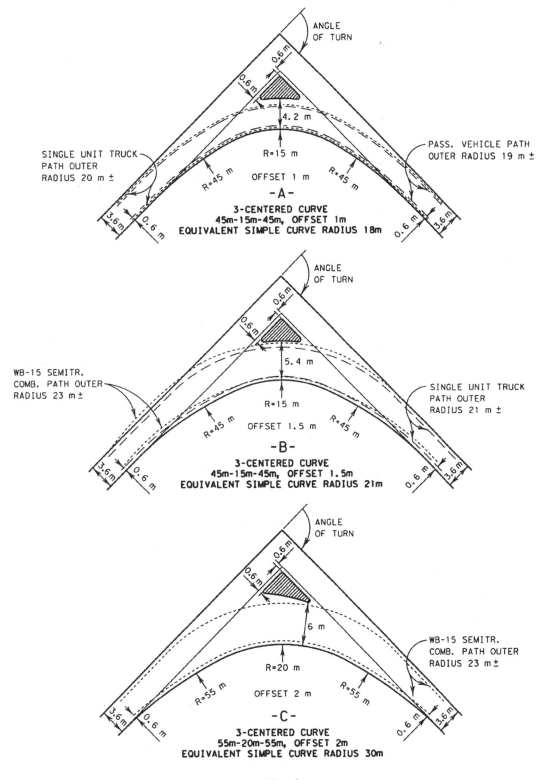

ANGLE
OF TURN

0.6 m

0.6 m

4.2 m

SINGLE UNIT TRUCK
PATH OUTER
RADIUS 20 m ±

PASS. VEHICLE PATH
OUTER RADIUS 19 m ±

R=45 m

R=15 m

R=45 m

OFFSET 1 m

3.6 m

0.6 m

0.6 m

3.6 m

- A -
3-CENTERED CURVE
45m-15m-45m, OFFSET 1m
EQUIVALENT SIMPLE CURVE RADIUS 18m

ANGLE
OF TURN

0.6 m

0.6 m

5.4 m

WB-15 SEMITR.
COMB. PATH OUTER
RADIUS 23 m±

SINGLE UNIT TRUCK
PATH OUTER
RADIUS 21 m ±

R=45 m

R=15 m

R=45 m

OFFSET 1.5 m

3.6 m

0.6 m

0.6 m

3.6 m

-B-
3-CENTERED CURVE
45m-15m-45m, OFFSET 1.5m
EQUIVALENT SIMPLE CURVE RADIUS 21m

ANGLE
OF TURN

0.6 m

0.6 m

6 m

WB-15 SEMITR.
COMB. PATH OUTER
RADIUS 23 m±

R=55 m

R=20 m

R=55 m

OFFSET 2 m

3.6 m

0.6 m

0.6 m

3.6 m

-C-
3-CENTERED CURVE
55m-20m-55m, OFFSET 2m
EQUIVALENT SIMPLE CURVE RADIUS 30m

Metric

Exhibit 9-41. Minimum Turning Roadway Designs With Corner Islands at Urban Locations

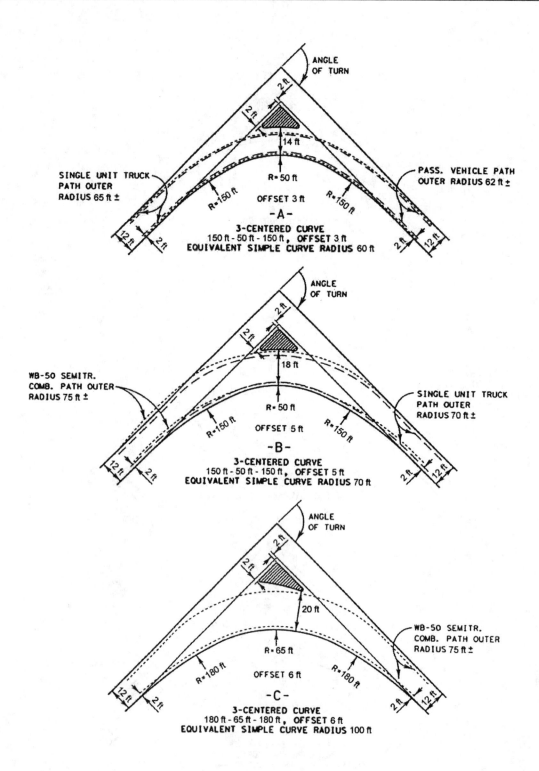

US Customary

Exhibit 9-41. Minimum Turning Roadway Designs With Corner Islands at Urban Locations (Continued)

By increasing the turning roadway width to 5.4 m [18 ft] and using the same combination of curves but with the middles curve being offset 1.5 m [5 ft] from the tangent edges extended, a more desirable arrangement results as shown in Exhibit 9-41B. This design enables the SU design vehicle to use a 21-m [70-ft] turning radius with adequate clearances and makes it possible for the WB-15 [WB-50] vehicle to negotiate the turn with only slight encroachment on adjacent through-traffic lanes.

At locations where a significant number of semitrailer combinations, particularly the longer units, will be turning, the arrangement shown in Exhibit 9-41C should be used. This design, consisting of a minimum curve of 20 m [65 ft] radius, offset 2 m [6 ft], and terminal curves of 55 m [180 ft] radii, generally provides for a WB-15 [WB-50] design vehicle passing through a 6-m [20-ft] turning roadway width and greatly benefits the operation of smaller vehicles.

However, the designer should be aware of larger semitrailer combinations on designated roadways and the effects these vehicles will have on turning roadway designs. The designer should reference the truck turning templates in Chapter 2 to meet his or her design needs. As previously stated, turning roadway widths can be reduced with paint or thermoplastic markings to channelize passenger cars and discourage the usage of the wider roadway as two turning lanes.

In urban areas, the island in all instances should be located about 0.6 m [2 ft] outside the traveled way edges extended, as shown in Exhibit 9-41C. For high-speed highways, the offset from the through lanes to the face of curb normally should be shoulder width. In rural areas, the use of painted corner islands may be considered. When raised corner islands are used in rural locations, they should have a sloping curb face. For more information, reference can be made to Exhibits 9-37 and 9-38 and the accompanying discussion on island types.

For each minimum design shown in Exhibit 9-41, a three-centered compound curve is recommended; however, asymmetric compound or two-centered curves could also be used, particularly where the design provides for the turning of trucks. Although an equivalent simple curve of a given radius is noted in each exhibit, its use in the two latter designs may result in design vehicle encroachments on the shoulder or island.

Oblique-Angle Turns with Corner Islands

The minimum design dimensions for oblique-angle turns are determined on a basis similar to that for right-angle turns, and values are given in Exhibit 9-42. Curve design for the inner edge of the traveled way, turning roadway width, and the approximate island size are indicated for the three chosen design classifications described at the bottom of the table. For a particular intersection, the designer may choose from the three minimum designs shown in accordance with vehicle size, the volume of traffic anticipated, and the physical controls at the site.

In Exhibit 9-42, no design values are given for angles of turn less than 75 degrees. If practical, angles of intersection less than 75 degrees should not be used. For flat angles of turn, the design of turning roadways involve relatively large radii and are not considered in the

Metric						US Customary					
Angle of turn (degrees)	Design classification	Three-centered compound curve Radii (m)	Three-centered compound curve Offset (m)	Width of lane (m)	Approx. island size (m^2)	Angle of turn (degrees)	Design classification	Three-centered compound curve Radii (ft)	Three-centered compound curve Offset (ft)	Width of lane (ft)	Approx. island size (ft^2)
75	A	45-23-45	1.0	4.2	5.5	75	A	150-75-150	3.5	14	60
	B	45-23-45	1.5	5.4	5.0		B	150-75-150	5.0	18	50
	C	55-28-55	1.0	6.0	5.0		C	180-90-180	3.5	20	50
90[a]	A	45-15-45	1.0	4.2	5.0	90[a]	A	150-50-150	3.0	14	50
	B	45-15-45	1.5	5.4	7.5		B	150-50-150	5.0	18	80
	C	55-20-55	2.0	6.0	11.5		C	180-65-180	6.0	20	125
105	A	36-12-36	0.6	4.5	6.5	105	A	120-40-120	2.0	15	70
	B	30-11-30	1.5	6.6	5.0		B	100-35-100	5.0	22	50
	C	55-14-55	2.4	9.0	5.5		C	180-45-180	8.0	30	60
120	A	30-9-30	0.8	4.8	11.0	120	A	100-30-100	2.5	16	120
	B	30-9-30	1.5	7.2	8.5		B	100-30-100	5.0	24	90
	C	55-12-55	2.5	10.2	20.0		C	180-40-180	8.5	34	220
135	A	30-9-30	0.8	4.8	43.0	135	A	100-30-100	2.5	16	460
	B	30-9-30	1.5	7.8	35.0		B	100-30-100	5.0	26	370
	C	48-11-48	2.7	10.5	60.0		C	160-35-160	9.0	35	640
150	A	30-9-30	0.8	4.8	130.0	150	A	100-30-100	2.5	16	1400
	B	30-9-30	2.0	9.0	110.0		B	100-30-100	6.0	30	1170
	C	48-11-48	2.1	11.4	160.0		C	160-35-160	7.1	38	1720

[a] Illustrated in Exhibit 9-41.

Notes: Asymmetric three-centered compound curve and straight tapers with a simple curve can also be used without significantly altering the width of roadway or corner island size. Painted island delineation is recommended for islands less than 7 m^2 [75 ft^2] in size.

Design classification:

A—Primarily passenger vehicles; permits occasional design single-unit truck to turn with restricted clearances.
B—Provides adequately for SU; permits occasional WB-15 [WB-50] to turn with slight encroachment on adjacent traffic lanes.
C—Provides fully for WB-15 [WB-50].

Exhibit 9-42. Typical Designs for Turning Roadways

minimum class. Such turning angles should have individual designs to fit site controls and traffic conditions.

For angles of turn between 75 and 120 degrees, the designs are governed by a minimum island size, which provides for larger turns than minimum turning radii. For angles of turn 120 degrees or more, the sharpest turning path of a design vehicle is selected and the curves on the inner edge of traveled way generally control the design, which results in an island size greater than the minimum. In Exhibit 9-42, the inner edge of traveled way arrangement for designs B and C for turning angles between 120 and 150 degrees are the same as those given in Exhibit 9-20 for single-unit trucks and semitrailer combinations, respectively. The size of islands for the larger turning angles given in the last column of Exhibit 9-42 indicate the areas of unused pavement that are eliminated by the use of islands.

FREE-FLOW TURNING ROADWAYS AT INTERSECTIONS

An important part of the design on some intersections is the design of a free-flow alignment for right turns. Ease and smoothness of operation can result when the free-flow turning roadway is designed with compound curves preceded by a right-turn deceleration lane, as indicated in Exhibits 9-43B and 9-43C. The shape and length of these curves should be such that they: (1) allow drivers to avoid abrupt deceleration, (2) permit development of some superelevation in advance of the maximum curvature, and (3) enable vehicles to follow natural turning paths. The design speed of a free-flow turning roadway for right turns may vary between the end of the right-turn deceleration lanes and the central section. The design speed of the turning roadway may be equal to, or possibly within 20 to 30 km/h [10 to 20 mph] less, than the through roadway design speed. Refer to Exhibit 3-43 for minimum radii for right-turning traffic. Turning roadways at intersections should use the "upper-range" design speeds whenever practical although the "middle-range" speeds may be used in constrained situations.

SUPERELEVATION FOR TURNING ROADWAYS AT INTERSECTIONS

General Design Guidelines

The general factors that control the maximum rates of superelevation for open highway conditions as discussed in Chapter 3 also apply to turning roadways at intersections. Maximum superelevation rates up to 10 percent may be used where climatic conditions are favorable. However, maximum rates up to 8 percent generally should be used where snow and icing conditions prevail.

In intersection design, the free flow of turning roadways is often of limited radii and length. When speed is not affected by other vehicles, drivers on turning roadways anticipate the sharp curves and accept operation with higher side friction than they accept on open highway curves of the same radii. This behavior stems from their desire to maintain their speed through the curve;

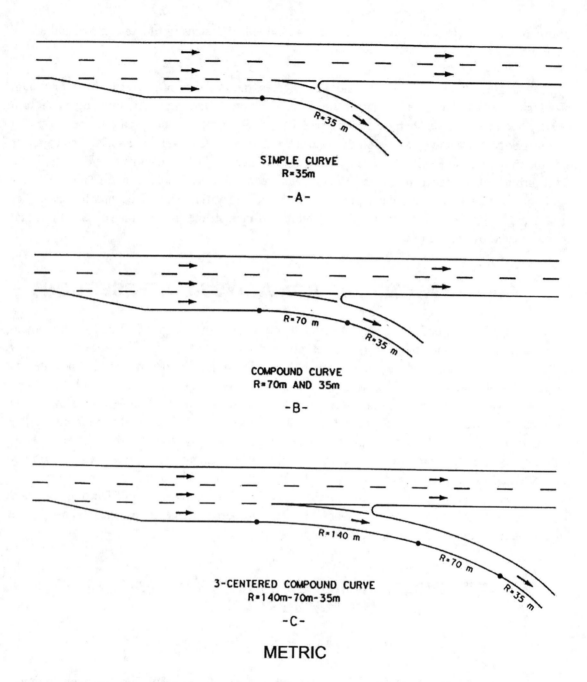

Exhibit 9-43. Use of Simple and Compound Curves at Free Flow Turning Roadways

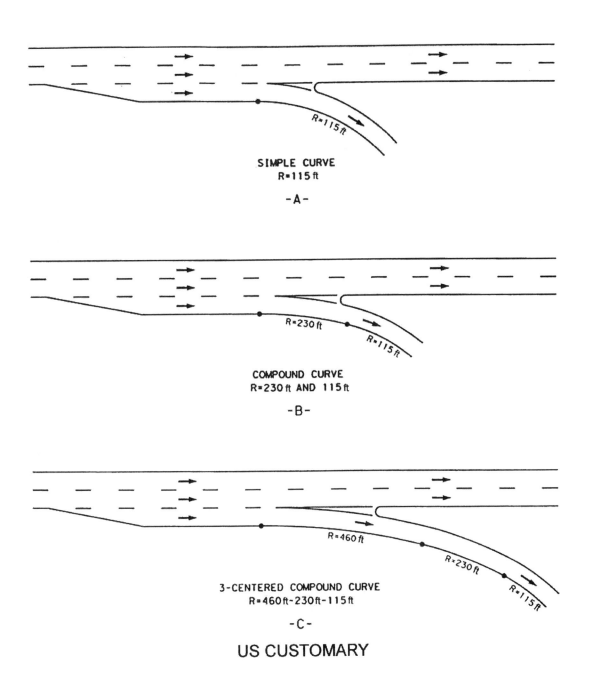

SIMPLE CURVE
R=115 ft

-A-

COMPOUND CURVE
R=230 ft AND 115 ft

-B-

3-CENTERED COMPOUND CURVE
R=460 ft-230 ft-115 ft

-C-

US CUSTOMARY

**Exhibit 9-43. Use of Simple and Compound Curves at Free Flow Turning Roadways
(Continued)**

although some speed reduction typically occurs. When other traffic is present, drivers will travel more slowly on turning roadways than on open highway curves of the same radii because they must diverge from and merge with through traffic. Therefore, in designing for safe operation, periods of light traffic volumes and corresponding speeds will generally control.

Designs with gradually changing curvature, effected by the use of compound curves, spirals, or both, permit desirable development of superelevation. For these designs, the design superelevation rates and corresponding radii listed in Exhibits 3-21 through 3-25 are desirable.

The practical difficulty of attaining superelevation without abrupt cross-slope change at turning roadway terminals, primarily because of sharp curvature and short lengths of turning roadway, most often prevents the development of a generous rate of superelevation. Abrupt changes in cross slope can adversely affect the stability of trucks and other vehicles with high centers of gravity. The design superelevation rates and corresponding radii listed in Exhibit 3-43 can be used when conditions justify the conservative use of superelevation.

Superelevation Runoff

The principles of superelevation runoff design discussed in Chapter 3 generally apply to free flow turning roadways at intersections. In general, the rate of change in cross slope in the runoff section should be based on the maximum relative gradients Δ listed in Exhibit 3-27. The values listed in this table are applicable to a single lane of rotation. The adjustment factors b_w listed in Exhibit 3-28 allow for slight increases in the effective gradient for wider rotated widths. The effective maximum relative gradients (equal to $\Delta \div b_w$) that can be used for a range of turning roadway widths are listed in Exhibit 9-44.

Usually, the profile of one edge of the traveled way is established first, and the profile on the other edge is developed by stepping up or down from the first edge by the amount of desired superelevation at that location. This step is done by plotting a few control points on the second edge by using the maximum relative gradients in Exhibit 9-44 and then plotting a smooth profile for the second edge of traveled way. Drainage may be an additional control, particularly for curbed roadways.

Development of Superelevation at Turning Roadway Terminals

Superelevation commensurate with curvature and speed seldom is practical at terminals where: (1) a flat intersection curve results in little more than a widening of the traveled way, (2) it is desirable to retain the cross slope of the traveled way, and (3) there is a practical limit to the difference between the cross slope on the traveled way and that on the intersection curve. Too great a difference in cross slope may cause vehicles traveling over the cross-over crown line to sway sideways. When vehicles, particularly high-bodied trucks, cross the crown line at other than low speed and at an angle of about 10 to 40 degrees, the body throw may make vehicle control difficult.

Metric				US Customary			
Design speed (km/h)	Effective maximum relative gradient (%)			Design speed (mph)	Effective maximum relative gradient (%)		
	Rotated width (m)				Rotated width (ft)		
	3.6 m	5.4 m	7.2 m		12 ft	18 ft	24 ft
20	0.80	0.96	1.00	15	0.78	0.94	1.00
30	0.75	0.90	1.00	20	0.74	0.89	0.99
40	0.70	0.84	0.93	25	0.70	0.84	0.93
50	0.65	0.78	0.87	30	0.66	0.80	0.88
60	0.60	0.72	0.80	35	0.62	0.75	0.83
70	0.55	0.66	0.73	40	0.58	0.70	0.77
80	0.50	0.60	0.67	45	0.54	0.65	0.72
90	0.47	0.57	0.63	50	0.50	0.60	0.67
100	0.44	0.53	0.59	55	0.47	0.57	0.63
110	0.41	0.49	0.55	60	0.45	0.54	0.60
120	0.38	0.46	0.51	65	0.43	0.52	0.57
130	0.35	0.42	0.47	70	0.40	0.48	0.53
				75	0.38	0.46	0.51
				80	0.35	0.42	0.47

Note: Based on maximum relative gradients listed in Exhibit 3-27 and the adjustment factors in Exhibit 3-28. One lane is assumed to equal 3.6 m [12 ft]. Gradients for speeds of 80 km/h [50 mph] and above are applicable to turning roadways at interchanges (i.e., ramps).

Exhibit 9-44. Effective Maximum Relative Gradients

General Procedure

For design of a highway, the through traffic lanes may be considered fixed in profile and cross slope. As the exit curve diverges from the through traveled way, the curved (or tangent) edge of the widening section can only gradually vary in elevation from the edge of through lane. Shortly beyond the point where the full width of the turning roadway is attained, an approach nose separates the two pavements. Where the exit curve is relatively sharp and without taper or transition, little superelevation in advance of the nose can be developed in the short distance available. Beyond the nose substantial superelevation usually can be attained, the amount depending on the length of the turning roadway curve. Where this curve deviates gradually from a traveled way, a desirable treatment of superelevation may be effected.

The method of developing superelevation at turning roadway terminals is illustrated diagrammatically in Exhibits 9-45 through 9-48. Exhibit 9-45 illustrates the variation in cross slope where a turning roadway leaves a through road that is on tangent. From point A to B, the normal cross slope on the through-traffic lane is extended to the outer edge of auxiliary lane. The additional width at B is nominal, less than 1 m [3 ft], and projecting the cross slope across this width simplifies construction. Beyond point B, the width is sufficient that the cross slope on the auxiliary lane can be the same or begin to be steeper than the cross slope on the adjacent through-traffic lane, as at C. At D where the full width of the turning roadway is attained, a still greater slope can be used. Superelevation is further increased adjacent to the nose at E and is facilitated

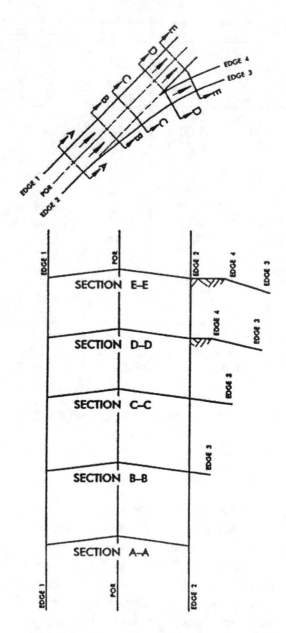

Exhibit 9-45. Development of Superelevation at Turning Roadway Terminals

somewhat by sloping downward the pavement wedge formed between the right edge of the travel way and the extended left traveled way edge of the turning roadway. Beyond the nose, as at E, the traveled way is transitioned as rapidly as conditions permit until the full desired superelevation is attained.

Exhibit 9-46 is a similar illustration for the condition where the through lanes and the turning roadway curve in the same direction. The desired superelevation on the exit roadway, which generally is steeper than that on the through lanes, can be attained in a relatively short distance. At C the cross slope of the through lane is extended over the widened traveled way. At

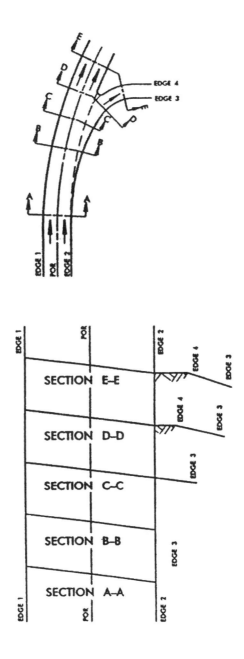

Exhibit 9-46. Development of Superelevation at Turning Roadway Terminals

D somewhat variable cross sections are introduced, the full superelevation being reached in the vicinity of E.

A less favorable situation occurs when the joining facilities curve in opposite directions, as in Exhibit 9-47. Because of the rate of superelevation on the through roadway, it may be impractical to slope the auxiliary lane in a direction opposite to that of the through lanes for reasons of appearance and riding quality. In a typical treatment for a moderate rate of

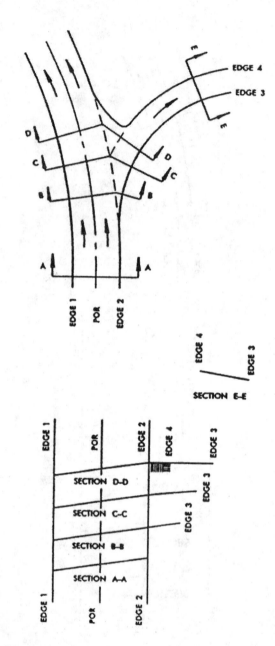

Exhibit 9-47. Development of Superelevation at Turning Roadway Terminals

superelevation, the rate of cross slope on the through roadway is extended onto the auxiliary lane, as at B. At C it may still continue upward, but at a lesser rate. The break between the two slopes becomes more pronounced at D, the added pavement being nearly horizontal. Some superelevation is introduced at the nose, either by a single crown line centering on the nose or by a double break in the cross slope over the pavement wedge in front of the nose. Most of the superelevation should be gained beyond the nose.

On designs with a parallel speed-change lane, as in Exhibit 9-48, part of the cross slope change may be made over the length of this lane. Usually, more than half of the total

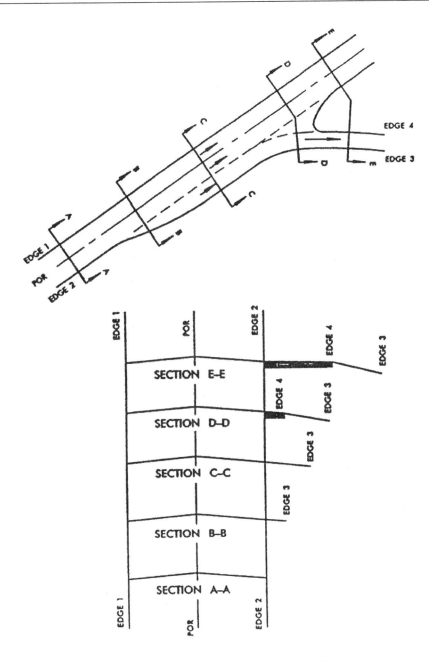

Exhibit 9-48. Development of Superelevation at Turning Roadway Terminals

superelevation rate can be attained at D, and the full desired superelevation can be reached at or just beyond the nose.

The discussion and arrangements illustrated in Exhibits 9-45 through 9-48 for exit terminals are also directly applicable to entrance terminals, except that the details at the merging end are different from those of an approach nose. The merging end of an entrance terminal would be located in proximity of D.

Turn-Lane Cross-Slope Rollover

The design control at the crossover line (not to be confused with the crown line normally provided at the centerline of a roadway) is the algebraic difference in cross slope rates of the two adjacent lanes. Where both roadways slope down and away from the crossover crown line, the algebraic difference is the sum of their cross slope rates; where they slope in the same direction, it is the difference of their cross slope rates. A desirable maximum algebraic difference at a crossover crown line is 4 or 5 percent, but it may be as high as 8 percent at low speeds and where there are few trucks. The suggested maximum differences in cross slope rates at a crown line, related to the speed of turning traffic, are given in Exhibit 9-49.

Metric		US Customary	
Design speed of exit or entrance curve (km/h)	Maximum algebraic difference in cross slope at crossover line (%)	Design speed of exit or entrance curve (mph)	Maximum algebraic difference in cross slope at crossover line (%)
30 and under	5.0 to 8.0	20 and under	5.0 to 8.0
40 and 50	5.0 to 6.0	25 and 30	5.0 to 6.0
60 and over	4.0 to 5.0	35 and over	4.0 to 5.0

Exhibit 9-49. Maximum Algebraic Difference in Cross Slope at Turning Roadway Terminals

Superelevation Transition and Gradeline Control

The attainment of superelevation over the gradually widening auxiliary lane and over the whole of the turning roadway terminals should not be abrupt. The design should be in keeping with the cross-slope controls, given in Exhibit 9-49.

As an example, consider an arrangement as in Exhibit 9-45, in which the limiting curve of the turning roadway has a radius of 70 m [230 ft], corresponding to a design speed of 50 km/h [30 mph]. From Exhibit 3-31, the limiting superelevation rate would be 11 percent or less. Because the roadway width is variable, the transition of cross-slope change should be developed by using the method of traveled way edge change in grade with respect to the point of rotation for a full-width auxiliary lane. Elevations developed by this method should then be converted to a change in elevation between the edge of the traveled way of the through lane and the edge of the full-width pavement of the auxiliary lane. They then should be prorated for the actual partial widths of the auxiliary lane. In this example, the traveled way edge change in grade should be no greater than 0.65 percent [0.66 percent].

An alternate method, which has been noted with respect to rideability, comfort, and appearance of the roadway in cross-slope transition areas, is to establish a rate of change in the roadway cross slope. The rate of cross slope is a function of traveled way width and the change in grade of the edge of traveled way with respect to the point of roadway rotation. This method results in the edge grade being equal to the roadway width, which is rotated, times the rate of change in cross slope. Thus, if the edge of traveled way grade change is 0.65 percent

[0.66 percent] and the width of roadway being rotated (the assumption being that the full width of the auxiliary lane is applied for calculating the grade change of the edge of traveled way) is 3.6 m [12 ft], the rate of change in cross slope is 5.41 percent [5.58 percent] per 30 m [100 ft] length.

In Exhibit 9-45, if the cross slope on the through roadway is 1 percent and the distance from A to B, and also from B to C, is 15 m [50 ft], trial cross-slope rates would be 1 percent [1 percent] at A, 3.71 percent [3.79 percent] at B, and 6.41 percent [6.58 percent] at C. Here the cross-over crown line control (Exhibit 9-48) is barely satisfied, because at the critical section C, the algebraic difference in cross-slope rates is 5.41 percent. If the remaining lengths of C to D and D to E are 7 m [25 ft] apart, the cross-slope rate would be 7.67 percent [7.97 percent] at D and 8.93 percent [9.35 percent] at E.

The cross slope of the edge of traveled way in front of the nose at E could be some intermediate rate, such as 4 percent. On the second trial a better adjustment of superelevation transition results by using a lower change in the cross slope rate for the turning roadway, such as 4 percent per 30 m [100 ft] length.

This procedure of establishing superelevation cross slopes at given points is a preliminary step in design. Elevations on the roadway edges resolved from these cross slopes serve as control points for drawing the edge of traveled way profiles on the turning roadway. Excellent practical results are obtained by plotting to large vertical scale the profiles for both edges of the turning roadway and the edge and centerline of the through roadway in juxtaposition on a single profile drawing. Important points such as approach noses or merging ends also are located. Only one profile, either the stationed centerline or edge of traveled way, is depicted in true length, but the inaccuracy in length of the other profiles is small and it is easy to locate points thereon in the field by radial measurement from the stationed line. The three-dimensional condition can be readily visualized.

Mathematically derived vertical curves, as used for open highways, are not always practical at intersections, but the profile curves can be developed readily with a spline or irregular curve templates. All needed elevations can be read directly from the profiles when they are drawn to large enough vertical scale. The final profile may not precisely produce the selected cross slope at all of the control points, but this problem is not serious as long as the cross-slope change is progressive and within the design control limits. The principal criterion is the development of smooth-edge profiles that do not appear distorted to the driver. Another method of obtaining a three-dimensional presentation is to plot contour lines on a layout of the intersection area. A scale drawing will provide an accurate picture with the additional advantage of showing drainage patterns, sumps, and irregular slope conditions.

TRAFFIC CONTROL DEVICES

Traffic control devices are used to regulate, warn, and guide traffic and are a primary determinant in the efficient operation of intersections. It is essential that intersection design be accomplished simultaneously with the development of signal, signing, and pavement marking plans to ensure that sufficient space is provided for proper installation of traffic control devices.

Geometric design should not be considered complete nor should it be implemented until it has been determined that needed traffic devices will have the desired effect in controlling traffic.

Most of the intersection types illustrated and described in the following discussions are adaptable to either signing control, signal control, or a combination of both. At intersections that do not need signal control, the normal roadway widths of the approach highways are carried through the intersection with the possible addition of speed-change lanes, median lanes, auxiliary lanes, or pavement tapers. Where volumes are sufficient to indicate signal control, the number of lanes for through movements may also need to be increased. Where the volume approaches the uninterrupted flow capacity of the intersection leg, the number of lanes in each direction may have to be doubled at the intersection to accommodate the volume under stop-and-go control. Other geometric features that may be affected by signalization are length and width of storage areas, location and position of turning roadways, spacing of other subsidiary intersections, access connections, and the possible location and size of islands to accommodate signal posts or supports.

At high-volume intersections at grade, the design of the signals should be sophisticated enough to respond to the varying traffic demands, the objective being to keep the vehicles moving through the intersection. Factors affecting capacity and computation procedures for signalized intersections are covered in the HCM (**6**).

An intersection that needs traffic signal control is best designed by considering jointly the geometric design, capacity analysis, design hour volumes, and physical controls. Details on the design and location of most forms of traffic control signals, including the general warrants, are given in the MUTCD (**9**).

INTERSECTION SIGHT DISTANCE

General Considerations

Each intersection has the potential for several different types of vehicular conflicts. The possibility of these conflicts actually occurring can be greatly reduced through the provision of proper sight distances and appropriate traffic controls. The avoidance of conflicts and the efficiency of traffic operations still depend on the judgment, capabilities, and response of each individual driver.

Stopping sight distance is provided continuously along each highway or street so that drivers have a view of the roadway ahead that is sufficient to allow drivers to stop. The provision of stopping sight distance at all locations along each highway or street, including intersection approaches, is fundamental to intersection operation.

Vehicles are assigned the right-of-way at intersections by traffic-control devices or, where no traffic-control devices are present, by the rules of the road. A basic rule of the road, at an intersection where no traffic-control devices are present, requires the vehicle on the left to yield to the vehicle on the right if they arrive at approximately the same time. Sight distance is provided at intersections to allow drivers to perceive the presence of potentially conflicting vehicles. This should occur in sufficient time for a motorist to stop or adjust their speed, as appropriate, to avoid colliding in the intersection. The methods for determining the sight distances needed by drivers approaching intersections are based on the same principles as stopping sight distance, but incorporate modified assumptions based on observed driver behavior at intersections.

The driver of a vehicle approaching an intersection should have an unobstructed view of the entire intersection, including any traffic-control devices, and sufficient lengths along the intersecting highway to permit the driver to anticipate and avoid potential collisions. The sight distance needed under various assumptions of physical conditions and driver behavior is directly related to vehicle speeds and to the resultant distances traversed during perception-reaction time and braking.

Sight distance is also provided at intersections to allow the drivers of stopped vehicles a sufficient view of the intersecting highway to decide when to enter the intersecting highway or to cross it. If the available sight distance for an entering or crossing vehicle is at least equal to the appropriate stopping sight distance for the major road, then drivers have sufficient sight distance to anticipate and avoid collisions. However, in some cases, this may require a major-road vehicle to stop or slow to accommodate the maneuver by a minor-road vehicle. To enhance traffic operations, intersection sight distances that exceed stopping sight distances are desirable along the major road.

Sight Triangles

Specified areas along intersection approach legs and across their included corners should be clear of obstructions that might block a driver's view of potentially conflicting vehicles. These specified areas are known as clear sight triangles. The dimensions of the legs of the sight triangles depend on the design speeds of the intersecting roadways and the type of traffic control used at the intersection. These dimensions are based on observed driver behavior and are documented by space-time profiles and speed choices of drivers on intersection approaches (**10**). Two types of clear sight triangles are considered in intersection design, approach sight triangles, and departure sight triangles.

Approach Sight Triangles

Each quadrant of an intersection should contain a triangular area free of obstructions that might block an approaching driver's view of potentially conflicting vehicles. The length of the legs of this triangular area, along both intersecting roadways, should be such that the drivers can see any potentially conflicting vehicles in sufficient time to slow or stop before colliding within the intersection. Exhibit 9-50A shows typical clear sight triangles to the left and to the right for a vehicle approaching an uncontrolled or yield-controlled intersection.

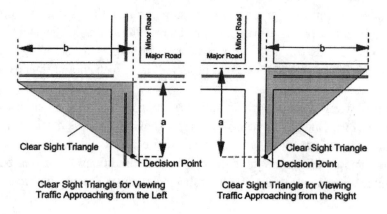

Clear Sight Triangle for Viewing
Traffic Approaching from the Left

Clear Sight Triangle for Viewing
Traffic Approaching from the Right

A – Approach Sight Triangles

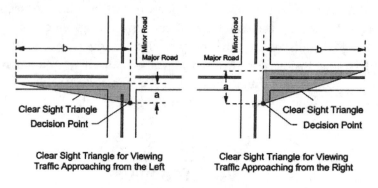

Clear Sight Triangle for Viewing
Traffic Approaching from the Left

Clear Sight Triangle for Viewing
Traffic Approaching from the Right

B – Departure Sight Triangles

Exhibit 9-50. Intersection Sight Triangles

The vertex of the sight triangle on a minor-road approach (or an uncontrolled approach) represents the decision point for the minor-road driver (see Exhibit 9-50A). This decision point is the location at which the minor-road driver should begin to brake to a stop if another vehicle is present on an intersecting approach. The distance from the major road, along the minor road, is illustrated by the dimension "a" in Exhibit 9-50A.

The geometry of a clear sight triangle is such that when the driver of a vehicle without the right of way sees a vehicle that has the right of way on an intersecting approach, the driver of that potentially conflicting vehicle can also see the first vehicle. Dimension "b" illustrates the length of this leg of the sight triangle. Thus, the provision of a clear sight triangle for vehicles without the right-of-way also permits the drivers of vehicles with the right-of-way to slow, stop, or avoid other vehicles, should it become necessary.

Although desirable at higher volume intersections, approach sight triangles like those shown in Exhibit 9-50A are not needed for intersection approaches controlled by stop signs or traffic signals. In that case, the need for approaching vehicles to stop at the intersection is determined by

656

the traffic control devices and not by the presence or absence of vehicles on the intersecting approaches.

Departure Sight Triangles

A second type of clear sight triangle provides sight distance sufficient for a stopped driver on a minor-road approach to depart from the intersection and enter or cross the major road. Exhibit 9-50B shows typical departure sight triangles to the left and to the right of the location of a stopped vehicle on the minor road. Departure sight triangles should be provided in each quadrant of each intersection approach controlled by stop or yield signs. Departure sight triangles should also be provided for some signalized intersection approaches (see Case D in the section on "Intersection Control").

The recommended dimensions of the clear sight triangle for desirable traffic operations where stopped vehicles enter or cross a major road are based on assumptions derived from field observations of driver gap-acceptance behavior (**10**). The provision of clear sight triangles like those shown in Exhibit 9-50B also allows the drivers of vehicles on the major road to see any vehicles stopped on the minor-road approach and to be prepared to slow or stop, if necessary.

Identification of Sight Obstructions Within Sight Triangles

The profiles of the intersecting roadways should be designed to provide the recommended sight distances for drivers on the intersection approaches. Within a sight triangle, any object at a height above the elevation of the adjacent roadways that would obstruct the driver's view should be removed or lowered, if practical. Such objects may include: buildings, parked vehicles, highway structures, roadside hardware, hedges, trees, bushes, unmowed grass, tall crops, walls, fences, and the terrain itself. Particular attention should be given to the evaluation of clear sight triangles at interchange ramp/crossroad intersections where features such as bridge railings, piers, and abutments are potential sight obstructions.

The determination of whether an object constitutes a sight obstruction should consider both the horizontal and vertical alignment of both intersecting roadways, as well as the height and position of the object. In making this determination, it should be assumed that the driver's eye is 1,080 mm [3.5 ft] above the roadway surface and that the object to be seen is 1,080 mm [3.5 ft] above the surface of the intersecting road.

This object height is based on a vehicle height of 1,330 mm [4.35 ft], which represents the 15th percentile of vehicle heights in the current passenger car population less an allowance of 250 mm [10 in]. This allowance represents a near-maximum value for the portion of a passenger car height that needs to be visible for another driver to recognize it as the object. The use of an object height equal to the driver eye height makes intersection sight distances reciprocal (i.e., if one driver can see another vehicle, then the driver of that vehicle can also see the first vehicle).

Where the sight-distance value used in design is based on a single-unit or combination truck as the design vehicle, it is also appropriate to use the eye height of a truck driver in checking sight obstructions. The recommended value of a truck driver's eye height is 2,330 mm [7.6 ft] above the roadway surface.

Intersection Control

The recommended dimensions of the sight triangles vary with the type of traffic control used at an intersection because different types of control impose different legal constraints on drivers and, therefore, result in different driver behavior. Procedures to determine sight distances at intersections are presented below according to different types of traffic control, as follows:

Case A—Intersections with no control
Case B—Intersections with stop control on the minor road
    Case B1—Left turn from the minor road
    Case B2—Right turn from the minor road
    Case B3—Crossing maneuver from the minor road
Case C—Intersections with yield control on the minor road
    Case C1—Crossing maneuver from the minor road
    Case C2—Left or right turn from the minor road
Case D—Intersections with traffic signal control
Case E—Intersections with all-way stop control
Case F—Left turns from the major road

Case A—Intersections With No Control

For intersections not controlled by yield signs, stop signs, or traffic signals, the driver of a vehicle approaching an intersection should be able to see potentially conflicting vehicles in sufficient time to stop before reaching the intersection. The location of the decision point (driver's eye) of the sight triangles on each approach is determined from a model that is analogous to the stopping sight distance model, with slightly different assumptions.

While some perceptual tasks at intersections may need substantially less time, the detection and recognition of a vehicle that is a substantial distance away on an intersecting approach, and is near the limits of the driver's peripheral vision, may take up to 2.5 s. The distance to brake to a stop can be determined from the same braking coefficients used to determine stopping sight distance in Exhibit 3-1.

Field observations indicate that vehicles approaching uncontrolled intersections typically slow to approximately 50 percent of their midblock running speed. This occurs even when no potentially conflicting vehicles are present (**10**). This initial slowing typically occurs at deceleration rates up to 1.5 m/s^2 [5 ft/s^2]. Deceleration at this gradual rate has been observed to begin even before a potentially conflicting vehicle comes into view. Braking at greater deceleration rates, which can approach those assumed in stopping sight distance, can begin up to 2.5 s after a vehicle on the intersecting approach comes into view. Thus, approaching vehicles may be traveling at less than their midblock running speed during all or part of the perception-reaction time and can, therefore, where necessary, brake to a stop from a speed less than the midblock running speed.

Exhibit 9-51 shows the distance traveled by an approaching vehicle during perception-reaction and braking time as a function of the design speed of the roadway on which the intersection approach is located. These distances should be used as the legs of the sight triangles shown in Exhibit 9-50A. Referring to Exhibit 9-50A, highway A with an assumed design speed of 80 km/h [50 mph] and highway B with an assumed design speed of 50 km/h [30 mph] require a clear sight triangle with legs extending at least 75 m and 45 m [245 and 140 ft] along highways A and B, respectively. Exhibit 9-52 indicates the length of the legs of the sight triangle from Exhibit 9-51.

Metric		US Customary	
Design speed (km/h)	Length of leg (m)	Design speed (mph)	Length of leg (ft)
20	20	15	70
30	25	20	90
40	35	25	115
50	45	30	140
60	55	35	165
70	65	40	195
80	75	45	220
90	90	50	245
100	105	55	285
110	120	60	325
120	135	65	365
130	150	70	405
		75	445
		80	485

Note: For approach grades greater than 3%, multiply the sight distance values in this exhibit by the appropriate adjustment factor from Exhibit 9-53.

Exhibit 9-51. Length of Sight Triangle Leg—Case A—No Traffic Control

This clear triangular area will permit the vehicles on either road to stop, if necessary, before reaching the intersection. If the design speed of any approach is not known, it can be estimated by using the 85th percentile of the midblock running speeds for that approach.

The distances shown in Exhibit 9-51 are generally less than the corresponding values of stopping sight distance for the same design speed. This relationship is illustrated in Exhibit 9-52. Where a clear sight triangle has legs that correspond to the stopping sight distances on their respective approaches, an even greater margin of efficient operation is provided. However, since field observations show that motorists slow down to some extent on approaches to uncontrolled intersections, the provision of a clear sight triangle with legs equal to the full stopping sight distance is not essential.

Where the grade along an intersection approach exceeds 3 percent, the leg of the clear sight triangle along that approach should be adjusted by multiplying the appropriate sight distance from Exhibit 9-51 by the appropriate adjustment factor from Exhibit 9-53.

If the sight distances given in Exhibit 9-51, as adjusted for grades, cannot be provided, consideration should be given to installing regulatory speed signing to reduce speeds or installing stop signs on one or more approaches.

No departure sight triangle like that shown in Exhibit 9-50B is needed at an uncontrolled intersection because such intersections typically have very low traffic volumes. If a motorist finds it necessary to stop at an uncontrolled intersection because of the presence of a conflicting vehicle on an intersecting approach, it is very unlikely another potentially conflicting vehicle will be encountered as the first vehicle departs the intersection.

Case B—Intersections with Stop Control on the Minor Road

Departure sight triangles for intersections with stop control on the minor road should be considered for three situations:

Case B1—Left turns from the minor road;
Case B2—Right turns from the minor road; and
Case B3—Crossing the major road from a minor-road approach.

Intersection sight distance criteria for stop-controlled intersections are longer than stopping sight distance to ensure that the intersection operates smoothly. Minor-road vehicle operators can wait until they can proceed safely without forcing a major-road vehicle to stop.

Case B1—Left Turn From the Minor Road

Departure sight triangles for traffic approaching from either the right or the left, like those shown in Exhibit 9-50B, should be provided for left turns from the minor road onto the major road for all stop-controlled approaches. The length of the leg of the departure sight triangle along the major road in both directions is the recommended intersection sight distance for Case B1.

The vertex (decision point) of the departure sight triangle on the minor road should be 4.4 m [14.5 ft] from the edge of the major-road traveled way. This represents the typical position of the minor-road driver's eye when a vehicle is stopped relatively close to the major road. Field observations of vehicle stopping positions found that, where necessary, drivers will stop with the front of their vehicle 2.0 m [6.5 ft] or less from the edge of the major-road traveled way. Measurements of passenger cars indicate that the distance from the front of the vehicle to the driver's eye for the current U.S. passenger car population is nearly always 2.4 m [8 ft] or less (**10**). Where practical, it is desirable to increase the distance from the edge of the major-road traveled way to the vertex of the clear sight triangle from 4.4 m to 5.4 m [14.5 to 18 ft]. This increase allows 3.0 m [10 ft] from the edge of the major-road traveled way to the front of the

METRIC

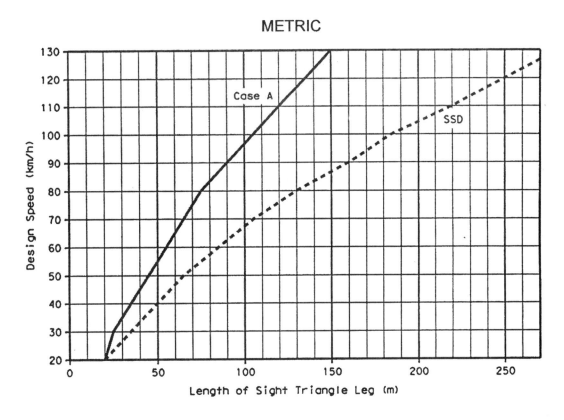

US CUSTOMARY

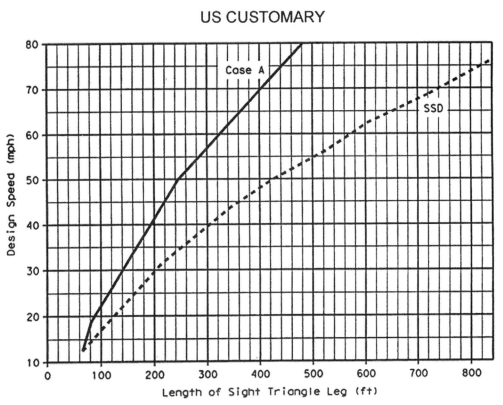

Exhibit 9-52. Length of Sight Triangle Leg—Case A—No Traffic Control

Metric

Approach grade (%)	Design speed (km/h)											
	20	30	40	50	60	70	80	90	100	110	120	130
−6	1.1	1.1	1.1	1.1	1.1	1.1	1.2	1.2	1.2	1.2	1.2	1.2
−5	1.0	1.0	1.1	1.1	1.1	1.1	1.1	1.1	1.2	1.2	1.2	1.2
−4	1.0	1.0	1.0	1.1	1.1	1.1	1.1	1.1	1.1	1.1	1.1	1.1
−3 to +3	1.0	1.0	1.0	1.0	1.0	1.0	1.0	1.0	1.0	1.0	1.0	1.0
+4	1.0	1.0	1.0	1.0	0.9	0.9	0.9	0.9	0.9	0.9	0.9	0.9
+5	1.0	1.0	1.0	0.9	0.9	0.9	0.9	0.9	0.9	0.9	0.9	0.9
+6	1.0	1.0	0.9	0.9	0.9	0.9	0.9	0.9	0.9	0.9	0.9	0.9

US Customary

Approach grade (%)	Design speed (mph)													
	15	20	25	30	35	40	45	50	55	60	65	70	75	80
−6	1.1	1.1	1.1	1.1	1.1	1.1	1.1	1.2	1.2	1.2	1.2	1.2	1.2	1.2
−5	1.0	1.0	1.1	1.1	1.1	1.1	1.1	1.1	1.1	1.2	1.2	1.2	1.2	1.2
−4	1.0	1.0	1.0	1.1	1.1	1.1	1.1	1.1	1.1	1.1	1.1	1.1	1.1	1.1
−3 to +3	1.0	1.0	1.0	1.0	1.0	1.0	1.0	1.0	1.0	1.0	1.0	1.0	1.0	1.0
+4	1.0	1.0	1.0	1.0	1.0	0.9	0.9	0.9	0.9	0.9	0.9	0.9	0.9	0.9
+5	1.0	1.0	1.0	1.0	0.9	0.9	0.9	0.9	0.9	0.9	0.9	0.9	0.9	0.9
+6	1.0	1.0	1.0	0.9	0.9	0.9	0.9	0.9	0.9	0.9	0.9	0.9	0.9	0.9

Exhibit 9-53. Adjustment Factors for Sight Distance Based on Approach Grade

Note: Based on ratio of stopping sight distance on specified approach grade to stopping sight distance on level terrain.

stopped vehicle, providing a larger sight triangle. The length of the sight triangle along the minor road (distance a in Exhibit 9-50B) is the sum of the distance from the major road plus 1/2 lane width for vehicles approaching from the left, or 1-1/2 lane width for vehicles approaching from the right.

Field observations of the gaps in major-road traffic actually accepted by drivers turning onto the major road have shown that the values in Exhibit 9-54 provide sufficient time for the minor-road vehicle to accelerate from a stop and complete a left turn without unduly interfering with major-road traffic operations. The time gap acceptance time does not vary with approach speed on the major road. Studies have indicated that a constant value of time gap, independent of approach speed, can be used as a basis for intersection sight distance determinations. Observations have also shown that major-road drivers will reduce their speed to some extent when minor-road vehicles turn onto the major road. Where the time gap acceptance values in Exhibit 9-54 are used to determine the length of the leg of the departure sight triangle, most major-road drivers should not need to reduce speed to less than 70 percent of their initial speed (**10**).

The intersection sight distance in both directions should be equal to the distance traveled at the design speed of the major road during a period of time equal to the time gap. In applying Exhibit 9-54, it can usually be assumed that the minor-road vehicle is a passenger car. However, where substantial volumes of heavy vehicles enter the major road, such as from a ramp terminal, the use of tabulated values for single-unit or combination trucks should be considered.

Exhibit 9-54 includes appropriate adjustments to the gap times for the number of lanes on the major road and for the approach grade of the minor road. The adjustment for the grade of the minor-road approach is needed only if the rear wheels of the design vehicle would be on an upgrade that exceeds 3 percent when the vehicle is at the stop line of the minor-road approach.

The intersection sight distance along the major road (dimension b in Exhibit 9-50B) is determined by:

Metric	US Customary
$$ISD = 0.278\, V_{major}\, t_g$$	$$ISD = 1.47\, V_{major}\, t_g \qquad (\,9\text{-}1\,)$$
where:	where:
ISD = intersection sight distance (length of the leg of sight triangle along the major road) (m)	ISD = intersection sight distance (length of the leg of sight triangle along the major road) (ft)
V_{major} = design speed of major road (km/h)	V_{major} = design speed of major road (mph)
t_g = time gap for minor road vehicle to enter the major road (s)	t_g = time gap for minor road vehicle to enter the major road (s)

Design vehicle	Time gap (s) at design speed of major road (t_g)
Passenger car	7.5
Single-unit truck	9.5
Combination truck	11.5

Note: Time gaps are for a stopped vehicle to turn left onto a two-lane highway with no median and grades 3 percent or less. The table values require adjustment as follows:

For multilane highways:
For left turns onto two-way highways with more than two lanes, add 0.5 seconds for passenger cars or 0.7 seconds for trucks for each additional lane, from the left, in excess of one, to be crossed by the turning vehicle.

For minor road approach grades:
If the approach grade is an upgrade that exceeds 3 percent; add 0.2 seconds for each percent grade for left turns

Exhibit 9-54. Time Gap for Case B1—Left Turn from Stop

For example, a passenger car turning left onto a two-lane major road should be provided sight distance equivalent to a time gap of 7.5 s in major-road traffic. If the design speed of the major road is 100 km/h [60 mph], this corresponds to a sight distance of 0.278(100)(7.5) = 208.5 or 210 m [1.47(60)(7.5) = 661.5 or 665 ft], rounded for design.

A passenger car turning left onto a four-lane undivided roadway will need to cross two near lanes, rather than one. This increases the recommended gap in major-road traffic from 7.5 to 8.0 s. The corresponding value of sight distance for this example would be 223 m [706 ft]. If the minor-road approach to such an intersection is located on a 4 percent upgrade, then the time gap selected for intersection sight distance design for left turns should be increased from 8.0 to 8.8 s, equivalent to an increase of 0.2 s for each percent grade.

The design values for intersection sight distance for passenger cars are shown in Exhibit 9-55. Exhibit 9-56 includes design values, based on the time gaps for the design vehicles included in Exhibit 9-54.

No adjustment of the recommended sight distance values for the major-road grade is generally needed because both the major- and minor-road vehicle will be on the same grade when departing from the intersection. However, if the minor-road design vehicle is a heavy truck and the intersection is located near a sag vertical curve with grades over 3 percent, then an adjustment to extend the recommended sight distance based on the major-road grade should be considered.

Metric				US Customary			
Design speed (km/h)	Stopping sight distance (m)	Intersection sight distance for passenger cars		Design speed (mph)	Stopping sight distance (ft)	Intersection sight distance for passenger cars	
		Calculated (m)	Design (m)			Calculated (ft)	Design (ft)
20	20	41.7	45	15	80	165.4	170
30	35	62.6	65	20	115	220.5	225
40	50	83.4	85	25	155	275.6	280
50	65	104.3	105	30	200	330.8	335
60	85	125.1	130	35	250	385.9	390
70	105	146.0	150	40	305	441.0	445
80	130	166.8	170	45	360	496.1	500
90	160	187.7	190	50	425	551.3	555
100	185	208.5	210	55	495	606.4	610
110	220	229.4	230	60	570	661.5	665
120	250	250.2	255	65	645	716.6	720
130	285	271.1	275	70	730	771.8	775
				75	820	826.9	830
				80	910	882.0	885

Note: Intersection sight distance shown is for a stopped passenger car to turn left onto a two-lane highway with no median and grades 3 percent or less. For other conditions, the time gap must be adjusted and required sight distance recalculated.

Exhibit 9-55. Design Intersection Sight Distance—Case B1—Left Turn From Stop

Sight distance design for left turns at divided-highway intersections should consider multiple design vehicles and median width. If the design vehicle used to determine sight distance for a divided-highway intersection is larger than a passenger car, then sight distance for left turns will need to be checked for that selected design vehicle and for smaller design vehicles as well. If the divided-highway median is wide enough to store the design vehicle with a clearance to the through lanes of approximately 1 m [3 ft] at both ends of the vehicle, no separate analysis for the departure sight triangle for left turns is needed on the minor-road approach for the near roadway to the left. In most cases, the departure sight triangle for right turns (Case B2) will provide sufficient sight distance for a passenger car to cross the near roadway to reach the median. Possible exceptions are addressed in the discussion of Case B3.

If the design vehicle can be stored in the median with adequate clearance to the through lanes, a departure sight triangle to the right for left turns should be provided for that design vehicle turning left from the median roadway. Where the median is not wide enough to store the design vehicle, a departure sight triangle should be provided for that design vehicle to turn left from the minor-road approach.

The median width should be considered in determining the number of lanes to be crossed. The median width should be converted to equivalent lanes. For example, a 7.2-m [24-ft] median should be considered as two additional lanes to be crossed in applying the multilane highway adjustment for time gaps in Exhibit 9-54. Furthermore, a departure sight triangle for left turns from the median roadway should be provided for the largest design vehicle that can be stored on

METRIC

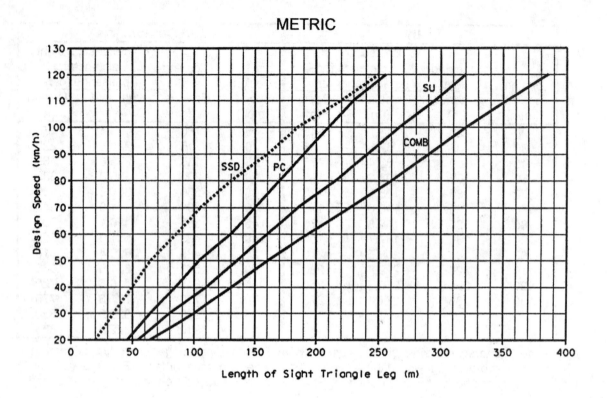

US CUSTOMARY

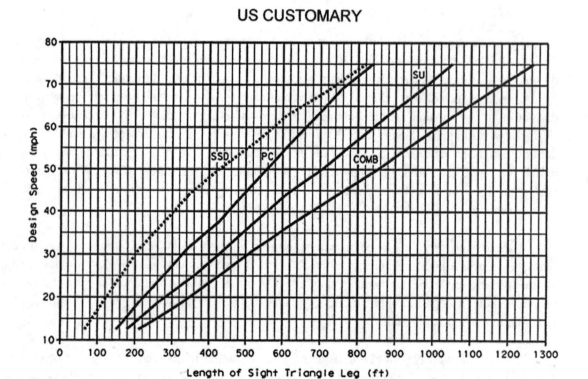

Exhibit 9-56. Intersection Sight Distance—Case B1—Left Turn from Stop

the median roadway with adequate clearance to the through lanes. If a divided highway intersection has a 12-m [40-ft] median width and the design vehicle for sight distance is a 22-m [74-ft] combination truck, departure sight triangles should be provided for the combination truck turning left from the minor-road approach and through the median. In addition, a departure sight triangle should also be provided to the right for a 9-m [30-ft] single unit truck turning left from a stopped position in the median.

If the sight distance along the major road shown in Exhibit 9-55, including any appropriate adjustments, cannot be provided, then consideration should be given to installing regulatory speed signing on the major-road approaches.

Case B2—Right Turn from the Minor Road

A departure sight triangle for traffic approaching from the left like that shown in Exhibit 9-50B should be provided for right turns from the minor road onto the major road. The intersection sight distance for right turns is determined in the same manner as for Case B1, except that the time gaps (t_g) in Exhibit 9-54 should be adjusted. Field observations indicate that, in making right turns, drivers generally accept gaps that are slightly shorter than those accepted in making left turns (**10**). The time gaps in Exhibit 9-54 can be decreased by 1.0 s for right-turn maneuvers without undue interference with major-road traffic. These adjusted time gaps for the right turn from the minor road are shown in Exhibit 9-57. Design values based on these adjusted time gaps are shown in Exhibit 9-58 for passenger cars. Exhibit 9-59 includes the design values for the design vehicles for each of the time gaps in Exhibit 9-57. When the minimum recommended sight distance for a right-turn maneuver cannot be provided, even with the reduction of 1.0 s from the values in Exhibit 9-54, consideration should be given to installing regulatory speed signing or other traffic control devices on the major-road approaches.

Case B3—Crossing Maneuver from the Minor Road

In most cases, the departure sight triangles for left and right turns onto the major road, as described for Cases B1 and B2, will also provide more than adequate sight distance for minor-road vehicles to cross the major road. However, in the following situations, it is advisable to check the availability of sight distance for crossing maneuvers:

- where left and/or right turns are not permitted from a particular approach and the crossing maneuver is the only legal maneuver;
- where the crossing vehicle would cross the equivalent width of more than six lanes; or
- where substantial volumes of heavy vehicles cross the highway and steep grades that might slow the vehicle while its back portion is still in the intersection are present on the departure roadway on the far side of the intersection.

Design vehicle	Time gap (s) at design speed of major road (t_g)
Passenger car	6.5
Single-unit truck	8.5
Combination truck	10.5

Note: Time gaps are for a stopped vehicle to turn right onto or cross a two-lane highway with no median and grades 3 percent or less. The table values require adjustment as follows:

For multilane highways:
For crossing a major road with more than two lanes, add 0.5 seconds for passenger cars and 0.7 seconds for trucks for each additional lane to be crossed and for narrow medians that cannot store the design vehicle.

For minor road approach grades:
If the approach grade is an upgrade that exceeds 3 percent, add 0.1 seconds for each percent grade.

Exhibit 9-57. Time Gap for Case B2—Right Turn from Stop and Case B3—Crossing Maneuver

Metric				US Customary			
Design speed (km/h)	Stopping sight distance (m)	Intersection sight distance for passenger cars		Design speed (mph)	Stopping sight distance (ft)	Intersection sight distance for passenger cars	
		Calculated (m)	Design (m)			Calculated (ft)	Design (ft)
20	20	36.1	40	15	80	143.3	145
30	35	54.2	55	20	115	191.1	195
40	50	72.3	75	25	155	238.9	240
50	65	90.4	95	30	200	286.7	290
60	85	108.4	110	35	250	334.4	335
70	105	126.5	130	40	305	382.2	385
80	130	144.6	145	45	360	430.0	430
90	160	162.6	165	50	425	477.8	480
100	185	180.7	185	55	495	525.5	530
110	220	198.8	200	60	570	573.3	575
120	250	216.8	220	65	645	621.1	625
130	285	234.9	235	70	730	668.9	670
				75	820	716.6	720
				80	910	764.4	765

Note: Intersection sight distance shown is for a stopped passenger car to turn right onto or cross a two-lane highway with no median and grades 3 percent or less. For other conditions, the time gap must be adjusted and required sight distance recalculated.

Exhibit 9-58. Design Intersection Sight Distance—Case B2—Right Turn from Stop and Case B3—Crossing Maneuver

668

METRIC

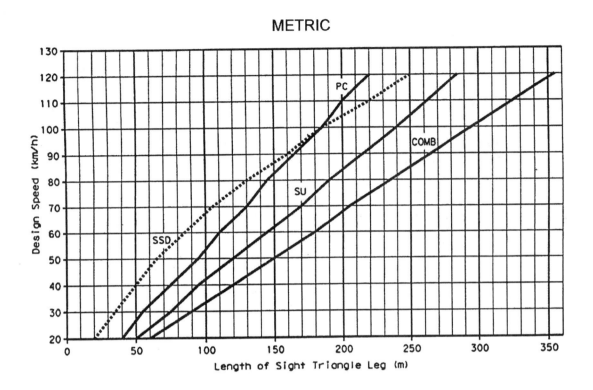

US CUSTOMARY

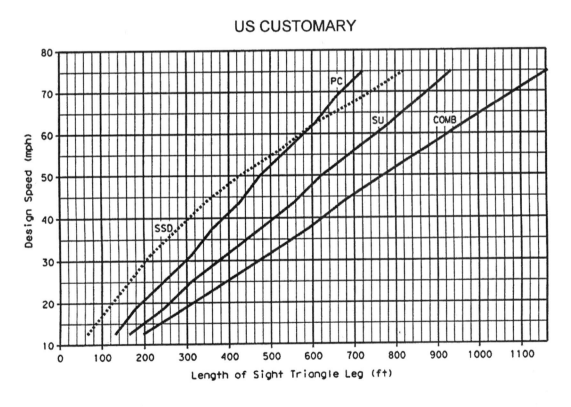

Exhibit 9-59. Intersection Sight Distance—Case B2—Right Turn from Stop and Case B3—Crossing Maneuver

The formula for intersection sight distance in Case B1 is used again for the crossing maneuver except that time gaps (t_g) are obtained from Exhibit 9-57. Exhibit 9-57 presents time gaps and appropriate adjustment factors to determine the intersection sight distance along the major road to accommodate crossing maneuvers. At divided highway intersections, depending on the relative magnitudes of the median width and the length of the design vehicle, intersection sight distance may need to be considered for crossing both roadways of the divided highway or for crossing the near lanes only and stopping in the median before proceeding. The application of adjustment factors for median width and grade is discussed under Case B1.

Exhibit 9-58 shows the design values for passenger cars for the crossing maneuver based on the unadjusted time gaps in Exhibit 9-57. Exhibit 9-59 includes the design values based on the time gaps for the design vehicles in Exhibit 9-57.

Case C—Intersections With Yield Control on the Minor Road

Drivers approaching yield signs are permitted to enter or cross the major road without stopping, if there are no potentially conflicting vehicles on the major road. The sight distances needed by drivers on yield-controlled approaches exceed those for stop-controlled approaches.

For four-leg intersections with yield control on the minor road, two separate pairs of approach sight triangles like those shown in Exhibit 9-50A should be provided. One set of approach sight triangles is needed to accommodate crossing the major road and a separate set of sight triangles is needed to accommodate left and right turns onto the major road. Both sets of sight triangles should be checked for potential sight obstructions.

For three-leg intersections with yield control on the minor road, only the approach sight triangles to accommodate left- and right-turn maneuvers need be considered, because the crossing maneuver does not exist.

Case C1—Crossing Maneuver From the Minor Road

The length of the leg of the approach sight triangle along the minor road to accommodate the crossing maneuver from a yield-controlled approach (distance a in Exhibit 9-50A) is given in Exhibit 9-60. The distances in Exhibit 9-60 are based on the same assumptions as those for Case A except that, based on field observations, minor-road vehicles that do not stop are assumed to decelerate to 60 percent of the minor-road design speed, rather than 50 percent.

Sufficient travel time for the major road vehicle should be provided to allow the minor-road vehicle: (1) to travel from the decision point to the intersection, while decelerating at the rate of 1.5 m/s^2 [5 ft/s^2] to 60 percent of the minor-road design speed; and then (2) to cross and clear the intersection at that same speed. The intersection sight distance along the major road to accommodate the crossing maneuver (distance b in Exhibit 9-50A) should be computed with the following equations:

Metric	**US Customary**
$$t_g = t_a + \frac{w + L_a}{0.167V_{minor}}$$ $$b = 0.278V_{major}\,t_g$$	$$t_g = t_a + \frac{w + L_a}{0.88V_{minor}}$$ $$b = 1.47V_{major}\,t_g \qquad (\,9\text{-}2\,)$$
where: t_g = travel time to reach and clear the major road (s) b = length of leg of sight triangle along the major road (m) t_a = travel time to reach the major road from the decision point for a vehicle that does not stop (s) (use appropriate value for the minor-road design speed from Exhibit 9-60 adjusted for approach grade, where appropriate) w = width of intersection to be crossed (m) L_a = length of design vehicle (m) V_{minor} = design speed of minor road (km/h) V_{major} = design speed of major road (km/h)	where: t_g = travel time to reach and clear the major road (s) b = length of leg of sight triangle along the major road (ft) t_a = travel time to reach the major road from the decision point for a vehicle that does not stop (s) (use appropriate value for the minor-road design speed from Exhibit 9-60 adjusted for approach grade, where appropriate) w = width of intersection to be crossed (ft) L_a = length of design vehicle (ft) V_{minor} = design speed of minor road (mph) V_{major} = design speed of major road (mph)

The value of t_g should equal or exceed the appropriate travel time for crossing the major road from a stop-controlled approach, as shown in Exhibit 9-57. The design values for the time gap (t_g) shown in Exhibit 9-60 incorporate these crossing times for two-lane highways and are used to develop the length of the leg of the sight triangle along the major road in Exhibit 9-61. These basic unadjusted lengths are illustrated in Exhibit 9-62 for passenger cars and should be calculated separately for other design vehicle types.

The distances and times in Exhibit 9-60 should be adjusted for the grade of the minor-road approach using the factors in Exhibit 9-53. If the major road is a divided highway with a median wide enough to store the design vehicle for the crossing maneuver, then only crossing of the near lanes needs to be considered and a departure sight triangle for accelerating from a stopped position in the median should be provided based on Case B3. For median widths not wide enough to store the design vehicle, the crossing width should be adjusted as discussed in Case B1.

Metric					US Customary				
Design speed (km/h)	Minor-road approach		Travel time (t_g) (seconds)		Design speed (mph)	Minor-road approach		Travel time (t_g) (seconds)	
	Length of leg[1] (m)	Travel time t_a[1,2] (seconds)	Calculated value	Design value[3,4]		Length of leg[1] (ft)	Travel time t_a[1,2] (seconds)	Calculated value	Design value[3,4]
20	20	3.2	7.1	7.1	15	75	3.4	6.7	6.7
30	30	3.6	6.2	6.5	20	100	3.7	6.1	6.5
40	40	4.0	6.0	6.5	25	130	4.0	6.0	6.5
50	55	4.4	6.0	6.5	30	160	4.3	5.9	6.5
60	65	4.8	6.1	6.5	35	195	4.6	6.0	6.5
70	80	5.1	6.2	6.5	40	235	4.9	6.1	6.5
80	100	5.5	6.5	6.5	45	275	5.2	6.3	6.5
90	115	5.9	6.8	6.8	50	320	5.5	6.5	6.5
100	135	6.3	7.1	7.1	55	370	5.8	6.7	6.7
110	155	6.7	7.4	7.4	60	420	6.1	6.9	6.9
120	180	7.0	7.7	7.7	65	470	6.4	7.2	7.2
130	205	7.4	8.0	8.0	70	530	6.7	7.4	7.4
					75	590	7.0	7.7	7.7
					80	660	7.3	7.9	7.9

[1] For minor-road approach grades that exceed 3 percent, multiply the distance or the time in this table by the appropriate adjustment factor from Exhibit 9-53.

[2] Travel time applies to a vehicle that slows before crossing the intersection but does not stop.

[3] The value of t_g should equal or exceed the appropriate time gap for crossing the major road from a stop-controlled approach.

[4] Values shown are for a passenger car crossing a two-lane highway with no median and grades 3 percent or less.

Exhibit 9-60. Case C1—Crossing Maneuvers From Yield-Controlled Approaches—Length of Minor Road Leg and Travel Times

Metric

Major road design speed (km/h)	Stopping sight distance (m)	Minor-road design speed (km/h) Design values (m)						
		20	30-80	90	100	110	120	130
20	20	40	40	40	40	45	45	45
30	35	60	55	60	60	65	65	70
40	50	80	75	80	80	85	90	90
50	65	100	95	95	100	105	110	115
60	85	120	110	115	120	125	130	135
70	105	140	130	135	140	145	150	160
80	130	160	145	155	160	165	175	180
90	160	180	165	175	180	190	195	205
100	185	200	185	190	200	210	215	225
110	220	220	200	210	220	230	240	245
120	250	240	220	230	240	250	260	270
130	285	260	235	250	260	270	280	290

US Customary

Major road design speed (mph)	Stopping sight distance (ft)	Minor-road design speed (mph) Design values (ft)							
		15	20-50	55	60	65	70	75	80
15	80	150	145	150	155	160	165	170	175
20	115	200	195	200	205	215	220	230	235
25	155	250	240	250	255	265	275	285	295
30	200	300	290	300	305	320	330	340	350
35	250	345	335	345	360	375	385	400	410
40	305	395	385	395	410	425	440	455	465
45	360	445	430	445	460	480	490	510	525
50	425	495	480	495	510	530	545	570	585
55	495	545	530	545	560	585	600	625	640
60	570	595	575	595	610	640	655	680	700
65	645	645	625	645	660	690	710	740	755
70	730	690	670	690	715	745	765	795	815
75	820	740	720	740	765	795	820	850	875
80	910	790	765	790	815	850	875	910	930

Note: Values in the table are for passenger cars and are based on the unadjusted distances and times in Exhibit 9-60. The distances and times in Exhibit 9-60 need to be adjusted using the factors in Exhibit 9-53.

Exhibit 9-61. Length of Sight Triangle Leg Along Major Road—Case C1—Crossing Maneuver at Yield Controlled Intersections

METRIC

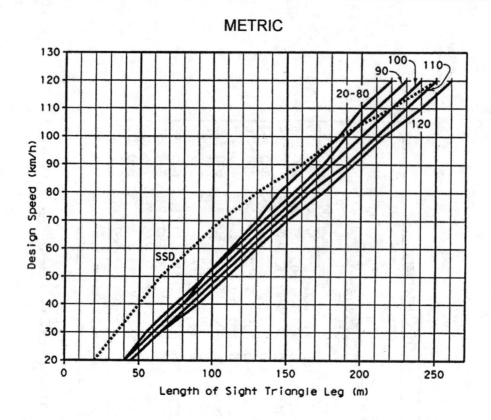

US CUSTOMARY

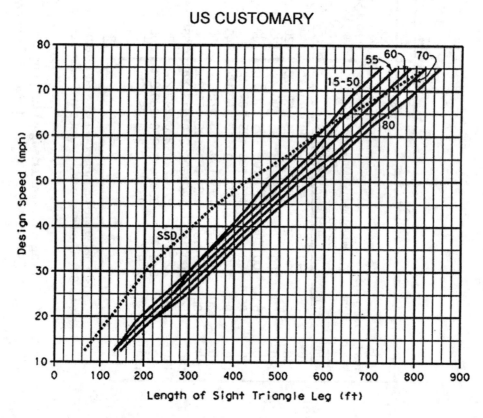

Exhibit 9-62. Length of Sight Triangle Leg Along Major Road for Passenger Cars—Case C1—Crossing Maneuver

Case C2—Left- and Right-Turn Maneuvers

The length of the leg of the approach sight triangle along the minor road to accommodate left and right turns without stopping (distance a in Exhibit 9-50A) should be 25 m [82 ft]. This distance is based on the assumption that drivers making left and right turns without stopping will slow to a turning speed of 16 km/h [10 mph].

The leg of the approach sight triangle along the major road (distance b in Exhibit 9-50A) is similar to the major-road leg of the departure sight triangle for a stop-controlled intersections in Cases B1 and B2. However, the time gaps in Exhibit 9-54 should be increased by 0.5 s to the values shown in Exhibit 9-63. The appropriate lengths of the sight triangle leg are shown in Exhibit 9-64 for passenger cars and in Exhibit 9-65 for the general design vehicle categories. The minor-road vehicle needs 3.5 s to travel from the decision point to the intersection. This represents additional travel time that is needed at a yield-controlled intersection, but is not needed at a stop-controlled intersection (Case B). However, the acceleration time after entering the major road is 3.0 s less for a yield sign than for a stop sign because the turning vehicle accelerates from 16 km/h [10 mph] rather than from a stop condition. The net 0.5-s increase in travel time for a vehicle turning from a yield-controlled approach is the difference between the 3.5-s increase in travel time and the 3.0-s reduction in travel time.

Departure sight triangles like those provided for stop-controlled approaches (see Cases B1, B2, and B3) should also be provided for yield-controlled approaches to accommodate minor-road vehicles that stop at the yield sign to avoid conflicts with major-road vehicles. However, since approach sight triangles for turning maneuvers at yield-controlled approaches are larger than the departure sight triangles used at stop-controlled intersections, no specific check of departure sight triangles at yield-controlled intersection should be needed.

Yield-controlled approaches generally need greater sight distance than stop-controlled approaches, especially at four-leg yield-controlled intersections where the sight distance needs of the crossing maneuver should be considered. If sight distance sufficient for yield control is not available, use of a stop sign instead of a yield sign should be considered. In addition, at locations where the recommended sight distance cannot be provided, consideration should be given to installing regulatory speed signing or other traffic control devices at the intersection on the major road to reduce the speeds of approaching vehicles.

Case D—Intersections With Traffic Signal Control

At signalized intersections, the first vehicle stopped on one approach should be visible to the driver of the first vehicle stopped on each of the other approaches. Left-turning vehicles should have sufficient sight distance to select gaps in oncoming traffic and complete left turns. Apart from these sight conditions, there are generally no other approach or departure sight triangles needed for signalized intersections. Signalization may be an appropriate crash countermeasure for higher volume intersections with restricted sight distance that have experienced a pattern of sight-distance related crashes.

Design vehicle	Time gap (t_g) seconds
Passenger car	8.0
Single-unit truck	10.0
Combination truck	12.0

Note: Time gaps are for a vehicle to turn right or left onto a two-lane highway with no median. The table values require adjustments for multilane highways as follows:

For left turns onto two-way highways with more than two lanes, add 0.5 seconds for passenger cars or 0.7 seconds for trucks for each additional lane, from the left, in excess of one, to be crossed by the turning vehicle.

For right turns, no adjustment is necessary.

Exhibit 9-63. Time Gap for Case C2—Left or Right Turn

Metric				US Customary			
Design speed (km/h)	Stopping sight distance (m)	Length of leg Passenger cars		Design speed (mph)	Stopping sight distance (ft)	Length of leg Passenger cars	
		Calculated (m)	Design (m)			Calculated (ft)	Design (ft)
20	20	44.5	45	15	80	176.4	180
30	35	66.7	70	20	115	235.2	240
40	50	89.0	90	25	155	294.0	295
50	65	111.2	115	30	200	352.8	355
60	85	133.4	135	35	250	411.6	415
70	105	155.7	160	40	305	470.4	475
80	130	177.9	180	45	360	529.2	530
90	160	200.2	205	50	425	588.0	590
100	185	222.4	225	55	495	646.8	650
110	220	244.6	245	60	570	705.6	710
120	250	266.9	270	65	645	764.4	765
130	285	289.1	290	70	730	823.2	825
				75	820	882.0	885
				80	910	940.8	945

Note: Intersection sight distance shown is for a passenger car making a right or left turn without stopping onto a two-lane road.

Exhibit 9-64. Design Intersection Sight Distance—Case C2—Left or Right Turn at Yield Controlled Intersections

METRIC

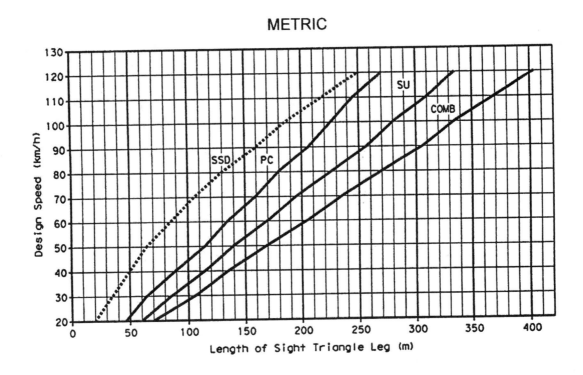

US CUSTOMARY

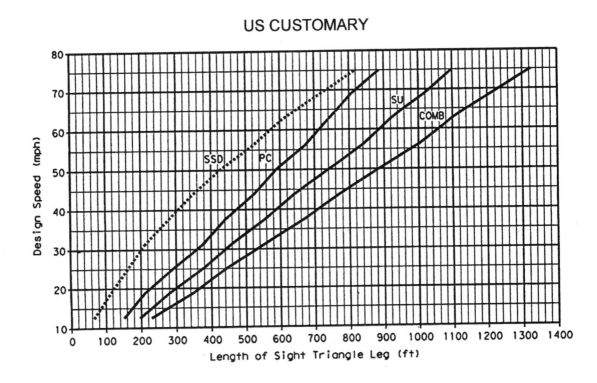

Exhibit 9-65. Intersection Sight Distance—Case C2—Yield Controlled Left or Right Turn

However, if the traffic signal is to be placed on two-way flashing operation (i.e., flashing yellow on the major-road approaches and flashing red on the minor-road approaches) under off-peak or nighttime conditions, then the appropriate departure sight triangles for Case B, both to the left and to the right, should be provided for the minor-road approaches. In addition, if right turns on a red signal are to be permitted from any approach, then the appropriate departure sight triangle to the left for Case B2 should be provided to accommodate right turns from that approach.

Case E—Intersections With All-Way Stop Control

At intersections with all-way stop control, the first stopped vehicle on one approach should be visible to the drivers of the first stopped vehicles on each of the other approaches. There are no other sight distance criteria applicable to intersections with all-way stop control and, indeed, all-way stop control may be the best option at a limited number of intersections where sight distance for other control types cannot be attained.

Case F—Left Turns From the Major Road

All locations along a major highway from which vehicles are permitted to turn left across opposing traffic, including intersections and driveways, should have sufficient sight distance to accommodate the left-turn maneuver. Left-turning drivers need sufficient sight distance to decide when it is safe to turn left across the lane(s) used by opposing traffic. Sight distance design should be based on a left turn by a stopped vehicle, since a vehicle that turns left without stopping would need less sight distance. The sight distance along the major road to accommodate left turns is the distance traversed at the design speed of the major-road in the travel time for the design vehicle given in Exhibit 9-66.

Design vehicle	Time gap (s) at design speed of major road (t_g)
Passenger car	5.5
Single-unit truck	6.5
Combination truck	7.5

Adjustment for multilane highways:
    For left-turning vehicles that cross more than one opposing lane, add 0.5 seconds for passenger cars and 0.7 seconds for trucks for each additional lane to be crossed.

Exhibit 9-66. Time Gap for Case F—Left Turns From the Major Road

The table also contains appropriate adjustment factors for the number of major-road lanes to be crossed by the turning vehicle. The unadjusted time gap in Exhibit 9-66 for passenger cars was used to develop the sight distances in Exhibit 9-67 and illustrated in Exhibit 9-68.

Metric				US Customary			
		Intersection sight distance				Intersection sight distance	
		Passenger cars				Passenger cars	
Design speed (km/h)	Stopping sight distance (m)	Calculated (m)	Design (m)	Design speed (mph)	Stopping sight distance (ft)	Calculated (ft)	Design (ft)
20	20	30.6	35	15	80	121.3	125
30	35	45.9	50	20	115	161.7	165
40	50	61.2	65	25	155	202.1	205
50	65	76.5	80	30	200	242.6	245
60	85	91.7	95	35	250	283.0	285
70	105	107.0	110	40	305	323.4	325
80	130	122.3	125	45	360	363.8	365
90	160	137.6	140	50	425	404.3	405
100	185	152.9	155	55	495	444.7	445
110	220	168.2	170	60	570	485.1	490
120	250	183.5	185	65	645	525.5	530
130	285	198.8	200	70	730	566.0	570
				75	820	606.4	610
				80	910	646.8	650

Note: Intersection sight distance shown is for a passenger car making a left turn from an undivided highway. For other conditions and design vehicles, the time gap should be adjusted and the sight distance recalculated.

Exhibit 9-67. Intersection Sight Distance—Case F—Left Turn From Major Road

If stopping sight distance has been provided continuously along the major road and if sight distance for Case B (stop control) or Case C (yield control) has been provided for each minor-road approach, sight distance will generally be adequate for left turns from the major road. Therefore, no separate check of sight distance for Case F may be needed.

However, at three-leg intersections or driveways located on or near a horizontal curve or crest vertical curve on the major road, the availability of adequate sight distance for left turns from the major road should be checked. In addition, the availability of sight distance for left turns from divided highways should be checked because of the possibility of sight obstructions in the median.

At four-leg intersections on divided highways, opposing vehicles turning left can block a driver's view of oncoming traffic. Exhibit 9-98, presented later in this chapter, illustrates intersection designs that can be used to offset the opposing left-turn lanes and provide left-turning drivers with a better view of oncoming traffic.

METRIC

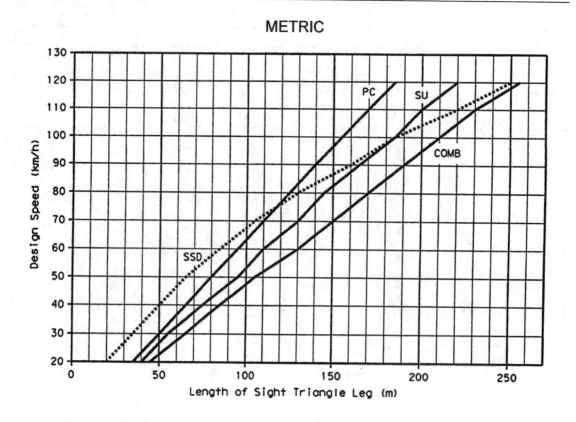

US CUSTOMARY

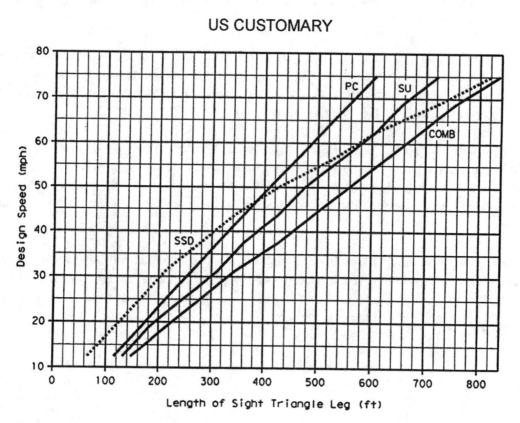

Exhibit 9-68. Intersection Sight Distance—Case F—Left Turn from Major Road

Effect of Skew

When two highways intersect at an angle less than 60 degrees, and when realignment to increase the angle of intersection is not justified, some of the factors for determination of intersection sight distance may need adjustment.

Each of the clear sight triangles described above are applicable to oblique-angle intersections. As shown in Exhibit 9-69, the legs of the sight triangle will lie along the intersection approaches and each sight triangle will be larger or smaller than the corresponding sight triangle would be at a right-angle intersection. The area within each sight triangle should be clear of potential sight obstructions as described previously.

At an oblique-angle intersection, the length of the travel paths for some turning and crossing maneuvers will be increased. The actual path length for a turning or crossing maneuver can be computed by dividing the total widths of the lanes (plus the median width, where appropriate) to be crossed by the sine of the intersection angle. If the actual path length exceeds the total widths of the lanes to be crossed by 3.6 m [12 ft] or more, then an appropriate number of additional lanes should be considered in applying the adjustment for the number of lanes to be crossed shown in Exhibit 9-54 for Case B1 and in Exhibit 9-57 for Cases B2 and B3. For Case C1, the w term in the equation for the major-road leg of the sight triangle to accommodate the crossing maneuver should also be divided by the sine of the intersection angle to obtain the actual path length. In the obtuse-angle quadrant of an oblique-angle intersection, the angle between the approach leg and the sight line is often so small that drivers can look across the full sight triangle with only a small head movement. However, in the acute-angle quadrant, drivers are often required to turn their heads considerably to see across the entire clear sight triangle. For this reason, it is recommended that the sight distance criteria for Case A not be applied to oblique-angle intersections and that sight distances at least equal to those for Case B should be provided, whenever practical.

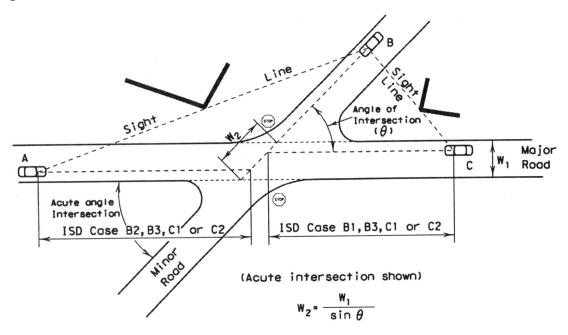

Exhibit 9-69. Sight Triangles at Skewed Intersections

STOPPING SIGHT DISTANCE AT INTERSECTIONS FOR TURNING ROADWAYS

General Considerations

The values for stopping sight distance as computed in Chapter 3 for open highway conditions are applicable to turning roadway intersections of the same design speed. The values from Chapter 3, together with the value for a design speed of 15 km/h [10 mph], are shown in Exhibit 9-70. These distances have been rounded upward to provide an increased factor of safety.

Metric								US Customary								
Design speed (km/h)	15	20	30	40	50	60	70	Design speed (mph)	10	15	20	25	30	35	40	45
Stopping sight distance (m)	15	20	35	50	65	85	105	Stopping sight distance (ft)	50	80	115	155	200	250	305	360

Exhibit 9-70. Stopping Sight Distance for Turning Roadways

These sight distances should be available at all points along a turning roadway; wherever practical, longer sight distances should be provided. They apply as controls in design of both vertical and horizontal alignment.

Vertical Control

The length of vertical curve is predicated, as it is for open highway conditions, on sight distance measured from the height of eye of 1,080 mm [3.5 ft] to the height of object of 600 mm [2 ft]. Formulas shown in the section on "Crest Vertical Curves" in Chapter 3 apply directly. Exhibits 3-75 and 3-76 show the relation between design speed, algebraic difference in gradient, and length of crest vertical curve to provide stopping distance. The factor K is constant for each design speed and the length of vertical curve is found by multiplying A, the algebraic differences in percent of grades, by K.

For design speeds of less than 60 km/h [40 mph], sag vertical curves, as governed by headlight sight distances, theoretically should be longer than crest vertical curves. Lengths of sag vertical curves are found by substituting the stopping sight distances from Exhibit 9-70 in the formulas in the section on "Sag Vertical Curves" in Chapter 3. Because the design speed of most turning roadways is governed by the horizontal curvature and the curvature is relatively sharp, a headlight beam parallel to the longitudinal axis of the vehicle ceases to be a control. Where practical, longer lengths for both crest and sag vertical curves should be used.

Horizontal Control

The sight distance control as applied to horizontal alignment has an equal, if not greater, effect on design of turning roadways than the vertical control. The sight line along the center line of the inside lane around the curve, clear of obstructions, should be such that the sight distance measured on an arc along the vehicle path equals or exceeds the stopping sight distance given in Exhibit 9-70. A likely obstruction may be a bridge abutment or line of columns, wall, cut sideslope, or a side or corner of a building.

The lateral clearance, centerline of inside lane to sight obstruction, for various radii and design speeds, is shown in Exhibit 3-57. The lateral clearances shown in this exhibit apply to the conditions where the horizontal curve is longer than the stopping sight distance. Where the curve length is shorter than the sight distance control, the lateral clearance of Exhibit 3-57 results in greater sight distance. In this case the lateral clearance is best determined by scaling on a plan layout of the turning roadway in a manner indicated by the sketch in Exhibit 3-8 or 3-58. The lateral clearance, so determined, should be tested at several points.

DESIGN TO DISCOURAGE WRONG-WAY ENTRY

An inherent problem of interchanges is the possibility of a driver entering one of the exit terminals from the crossroad and proceeding along the major highway in the wrong direction in spite of signing. This wrong-way entrance maneuver is becoming more of a problem with the increased number of interchanges. However, attention to several details of design at the intersection can discourage this maneuver.

As shown in Exhibits 9-71 and 9-72, a sharp or angular intersection is provided at the junction of the left edge of the ramp entering the crossroad and the right edge of the traveled way. The control radius should be tangent to the crossroad centerline, not the edge. This type of design discourages the improper right turn onto the one-way ramp.

As shown in the same figures, islands can be used in the terminal areas where ramps intersect the crossroads. The islands provide a means of channelizing the traffic into proper paths and can be effectively used for sign placement. Design of the islands should take into consideration initial or future signal installations at the ramp terminals.

Provision of a median as a deterrent to wrong-way movement, as illustrated in Exhibit 9-72, is a very effective treatment. The median makes the left-turn movement onto the exit ramp terminal very difficult, and a short-radius curve or angular break is provided at the intersection of the left edge of the exit ramp and the crossroad to discourage wrong-way right turns from the crossroads.

Additional design techniques to reduce wrong-way movements are (1) providing for all movements to and from the freeway to reduce intentional wrong-way entry, (2) using conventional, easily recognized interchange patterns to reduce driver confusion and hence wrong-

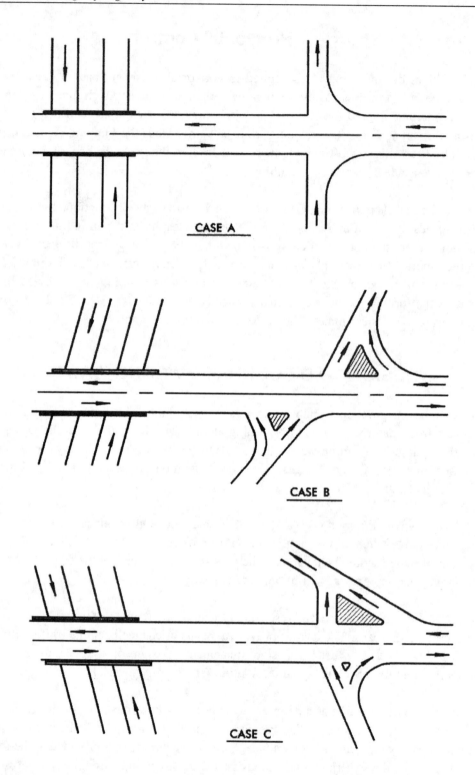

Exhibit 9-71. Two-Lane Crossroad Designs to Discourage Wrong-Way Entry

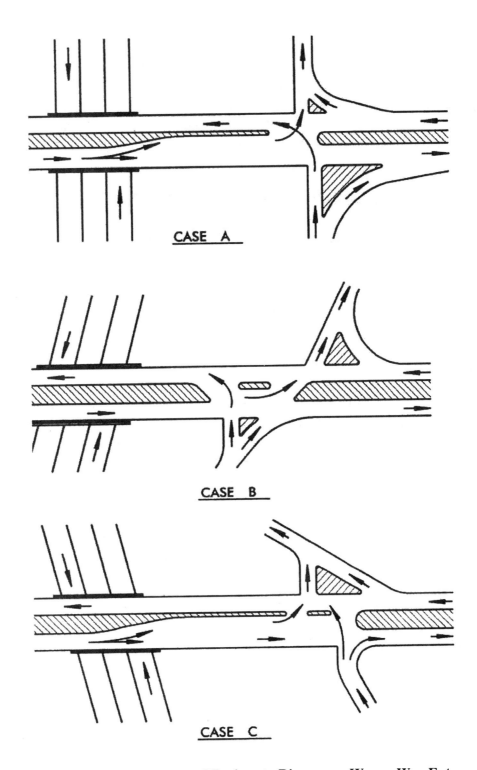

CASE A

CASE B

CASE C

Exhibit 9-72. Divided Crossroad Designs to Discourage Wrong-Way Entry

way entry, and (3) narrowing the arterial highway median opening to prevent left-turn movements onto freeway off-ramps.

Signs and supplementary pavement markings are among the most important devices to discourage wrong-way turns. Signing layouts are fully discussed in the MUTCD (**9**). Other devices such as pavement markings or flashing lights have been used effectively to prevent wrong-way movements. These devices are discussed elsewhere in this book and in the MUTCD.

GENERAL INTERSECTION TYPES

General Design Considerations

General types of intersections and terminology are indicated in Exhibits 9-73 and 9-74. The geometric forms are the three-leg, four-leg, and multileg intersections. Further classification includes such variations as unchannelized, flared, and channelized intersections. Details and specific adaptations of each general type are demonstrated in the section of this chapter on "Types and Examples of Intersections."

Many factors enter into the choice of type of intersection and the extent of design of a given type, but the principal controls are the design-hour traffic volume, the character or composition of traffic, and the design speed. The character of traffic and design speed affect many details of design, but in choosing the type of intersection they are not as significant as the traffic volume. Of particular significance are the actual and relative volumes of traffic involved in various turning and through movements.

When designing an intersection, left-turning traffic should be removed from the through lanes, whenever practical. Therefore, provisions for left turns (i.e., left-turn lanes) have widespread application. Ideally, left-turn lanes should be provided at driveways and street intersections along major arterial and collector roads wherever left turns are permitted. In some cases or at certain locations, providing for indirect left turns (jughandles, U-turn lanes, and diagonal roadways) may be appropriate to improve safety and preserve capacity. The provision of left-turn lanes has been found to reduce crash rates anywhere from 20 to 65 percent (**1**). Left-turn facilities should be established on roadways where traffic volumes are high enough or safety considerations are sufficient to warrant them. They are often needed to ensure adequate service levels for the intersections and the various turning movements.

Guidelines for when left-turn lanes should be provided are set forth in several documents for both signalized and unsignalized intersections (**11, 12, 13**). These guidelines key the need for left-turn lanes to (a) the number of arterial lanes, (b) design, and operating speeds, (c) left-turn volumes, and (d) opposing traffic volumes.

The HCM (**6**) indicates that exclusive left-turn lanes at signalized intersections should be installed as follows:

- Where fully protected, left-turn phasing is to be provided;

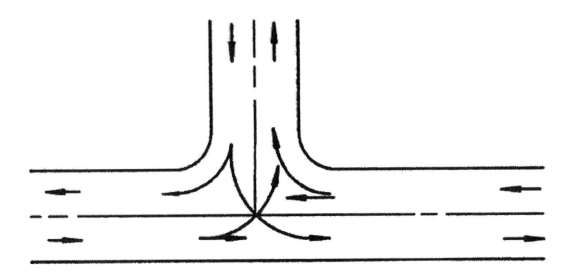

UNCHANNELIZED-T

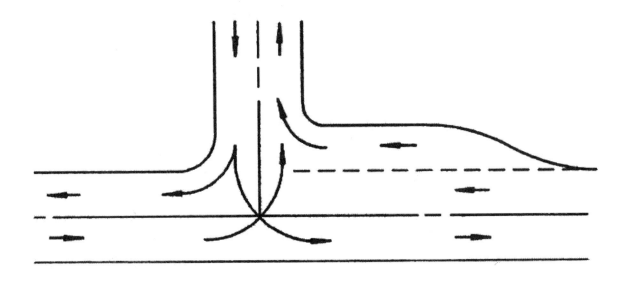

T-INTERSECTION WITH RIGHT TURN LANE

Exhibit 9-73. General Types of Intersections

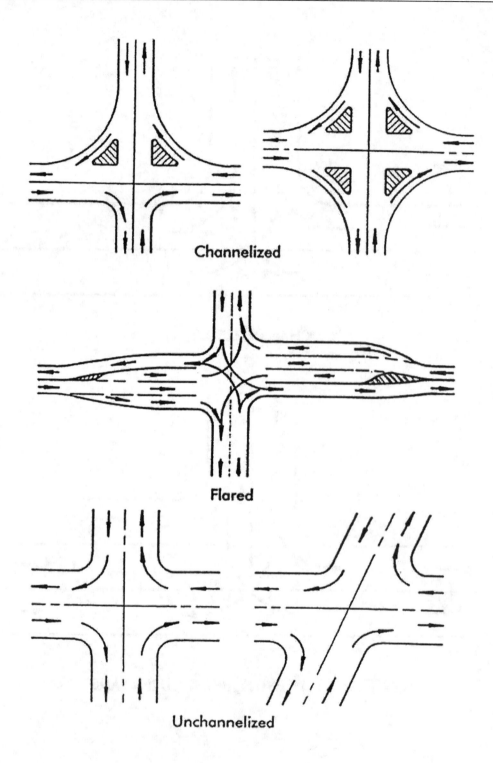

Exhibit 9-74. General Types of Intersections

- Where space permits, left-turn lanes should be considered when left-turn volumes exceed 100 vph (left-turn lanes may be provided for lower volumes as well on the basis of the judged need and state of local practice, or both); and
- Where left-turn volumes exceed 300 vph, a double left-turn lane should be considered.

Exhibit 9-75 is a guide to traffic volumes where left-turn facilities should be considered on two-lane highways. For the volumes shown, left turns and right turns from the minor street can be equal to, but not greater than, the left turns from the major street.

Metric					US Customary				
Opposing volume (veh/h)	Advancing volume (veh/h)				Opposing volume (veh/h)	Advancing volume (veh/h)			
	5% left turns	10% left turns	20% left turns	30% left turns		5% left turns	10% left turns	20% left turns	30% left turns
60-km/h operating speed					40-mph operating speed				
800	330	240	180	160	800	330	240	180	160
600	410	305	225	200	600	410	305	225	200
400	510	380	275	245	400	510	380	275	245
200	640	470	350	305	200	640	470	350	305
100	720	515	390	340	100	720	515	390	340
80-km/h operating speed					50-mph operating speed				
800	280	210	165	135	800	280	210	165	135
600	350	260	195	170	600	350	260	195	170
400	430	320	240	210	400	430	320	240	210
200	550	400	300	270	200	550	400	300	270
100	615	445	335	295	100	615	445	335	295
100-km/h operating speed					60-mph operating speed				
800	230	170	125	115	800	230	170	125	115
600	290	210	160	140	600	290	210	160	140
400	365	270	200	175	400	365	270	200	175
200	450	330	250	215	200	450	330	250	215
100	505	370	275	240	100	505	370	275	240

Exhibit 9-75. Guide for Left-Turn Lanes on Two-Lane Highways (6)

Additional information on left-turn lanes, including their suggested lengths, can be found in published sources (**2, 11, 13**). In the case of double left-turn lanes, a capacity analysis of the intersection should be performed to determine what traffic controls are needed in order for it to function properly.

Local conditions and the cost of right-of-way often influence the type of intersection selected as well as many of the design details. Limited sight distance, for example, may make it desirable to control traffic by yield signs, stop signs, or traffic signals when the traffic densities are less than those ordinarily considered appropriate for such control. The alignment and grade of the intersecting roads and the angle of intersection may make it advisable to channelize or use auxiliary pavement areas, regardless of the traffic densities. In general, traffic service, highway design designation, physical conditions, and cost of right-of-way are considered jointly in choosing the type of intersection.

For the general benefit of through-traffic movements, the number of crossroads, intersecting roads, or intersecting streets should be minimized. Where intersections are closely spaced on a two-way facility, it is seldom practical to provide signals for completely coordinated traffic movements at reasonable speeds in opposing directions on that facility. At the same time the resultant road or street patterns should permit travel on roadways other than the predominant highway without too much inconvenience. Traffic analysis is needed to determine whether the road or street pattern, left open across the predominate highway, is adequate to serve normal traffic plus the traffic diverted from any terminated road or street.

The functional classification of the road, the patterns of traffic movement at the intersections and the volume of traffic on each approach, including pedestrians, during one or more peak periods of the day, are indicative of the type of traffic control devices necessary, the roadway widths needed (including auxiliary lanes), and where applicable, the degree of channelization needed to expedite the movement of all traffic. The differing arrangement of islands and the shape and length of auxiliary lanes depend on whether signal control is provided.

The composition and character of traffic are a design control. Movements involving large trucks need larger intersection areas and flatter approach grades than those needed at intersections where traffic consists predominantly of passenger cars. Bus stops located near an intersection may further modify the arrangement. Approach speeds of traffic also have a bearing on the geometric design as well as on control devices and markings.

The number and locations of the approach roadways and their angles of intersection are major controls for the intersection geometric pattern, the location of islands, and the types of control devices. Intersections preferably should be limited to no more than four approach legs. Two or more crossroads intersecting an arterial highway in close proximity should be combined into a single crossing.

The distances between intersections influence the degree of channelization at any one particular intersection. For example, where intersections are closely spaced, turn restrictions may be imposed at some intersections and pedestrian crossings may be prohibited at others. This makes some channelizing islands and auxiliary pavement areas unnecessary, or it may be appropriate to introduce continuous auxiliary lanes between two or more intersecting roads or streets to handle a buildup and weaving of traffic. Where crossroads are widely spaced, each intersection should accommodate all crossing, turning, and pedestrian movements.

CHANNELIZATION

Channelization is the separation or regulation of conflicting traffic movements into definite paths of travel by traffic islands or pavement marking to facilitate the orderly movements of both vehicles and pedestrians. Proper channelization increases capacity, provides maximum convenience, and instills driver confidence. Improper channelization has the opposite effect and may be worse than none at all. Over channelization should be avoided because it could create confusion and worsen operations. In some cases a simple channelization improvement can result in dramatic operational efficiencies. Most of these cases involve some type of left-turn treatment.

Left-turn lanes at intersections reduce rear-end exposure and provide a comfortable means for making a left turn.

Channelization of intersections is generally considered for one or more of the following factors:

- The paths of vehicles are confined by channelization so that not more than two paths cross at any one point.
- The angle and location at which vehicles merge, diverge, or cross are controlled.
- The amount of paved area is reduced and thereby decreases vehicle wander and narrows the area of conflict between vehicles.
- Clearer indications are provided for the proper path in which movements are to be made.
- The predominant movements are given priority.
- Areas are provided for pedestrian refuge.
- Separate storage lanes permit turning vehicles to wait clear of through-traffic lanes.
- Space is provided for traffic control devices so that they can be more readily perceived.
- Prohibited turns are controlled.
- The speeds of vehicles are restricted to some extent.

Design of a channelized intersection usually involves the following significant controls: the type of design vehicle, the cross sections on the crossroads, the projected traffic volumes in relation to capacity, the number of pedestrians, the speed of vehicles, the location of any needed bus stops, and the type and location of traffic control devices. Furthermore, the physical controls such as right-of-way and terrain have an effect on the extent of channelization that is economically practical.

Certain principles should be followed in the design of a channelized intersection, but the extent to which they are applied will depend on the characteristics of the total design plan. These principles are:

- Motorists should not be confronted with more than one decision at a time.
- Unnatural paths that involve turns greater than 90 degrees or sudden and sharp reverse curves should be avoided.
- Areas of vehicle conflict should be reduced as much as practical. However, merging and weaving areas should be as long as conditions permit. Channelization should be used to keep vehicles within well-defined paths that minimize the area of conflict.
- Traffic streams that cross without merging and weaving should intersect desirably at right angles with a range of 60 to 120 degrees acceptable.
- The angle of intersection between merging streams of traffic should be appropriate to provide adequate sight distance.
- The points of crossing or conflict should be studied carefully to determine if such conditions would be better separated or consolidated to simplify design with appropriate control devices added to ensure efficient operation.
- Refuge areas for turning vehicles should be provided clear of through traffic.

- Islands used for channelization should not interfere with or obstruct bicycle lanes at intersections.
- Prohibited turns should be blocked wherever practical.
- Location of essential control devices should be established as a part of the design of a channelized intersection.
- Channelization may be desirable to separate the various traffic movements where multiple phase signals are used.

SPEED-CHANGE LANES AT INTERSECTIONS

Drivers leaving a highway at an intersection are usually required to reduce speed before turning. Drivers entering a highway from a turning roadway accelerate until the desired open-road speed is reached. When undue deceleration or acceleration by leaving or entering traffic takes place directly on the highway traveled way, it disrupts the flow of through traffic. To preclude or minimize these undesirable aspects of operation at intersections, speed-change lanes are provided on highways having expressway characteristics and are frequently used on other main highway intersections.

A speed-change lane is an auxiliary lane, including tapered areas, primarily for the acceleration or deceleration of vehicles entering or leaving the through-traffic lanes. The terms "speed-change lane," "deceleration lane," or "acceleration lane," as used here, apply broadly to the added pavement joining the traveled way of the highway or street with that of the turning roadway and do not necessarily imply a definite lane of uniform width. A speed-change lane should be of sufficient width and length to enable a driver to maneuver a vehicle into it properly, and once into it, to make the necessary change between the speed of operation on the highway or street and the lower speed on the turning roadway. Deceleration and acceleration lanes may be designed in conjunction with each other, the relationship depending on the arrangement of the intersection and traffic needs. They may be designed as parts of intersections but are especially important at ramp junctions where turning roadways meet high-speed traffic lanes.

Warrants for the use of speed-change lanes cannot be stated definitely. Many factors should be considered, such as speeds, traffic volumes, percentage of trucks, capacity, type of highway, service provided, and the arrangement and frequency of intersections. Observations and considerable experience with speed-change lanes have led to the following general conclusions:

- Speed-change lanes are warranted on high-speed and on high-volume highways where a change in speed is necessary for vehicles entering or leaving the through-traffic lanes.
- All drivers do not use speed-change lanes in the same manner; some use little of the available facility. As a whole, however, these lanes are used sufficiently to improve highway operation.
- Use of speed-change lanes varies with volume, the majority of drivers using them at high volumes.
- The directional type of speed-change lane consisting of a long taper fits the behavior of most drivers and does not require maneuvering on a reverse-curve path.

- Deceleration lanes on the approaches to intersections that also function as storage lanes for turning traffic are particularly advantageous, and experience with them generally has been favorable.

A median lane provides refuge for vehicles awaiting an opportunity to turn, and thereby keeps the highway traveled way clear for through traffic. The width, length, and general design of median lanes are similar to those of any other deceleration lane, but their design includes some additional features discussed in the section on "Auxiliary Lanes" later in this chapter.

Deceleration lanes are always advantageous, particularly on high speed roads, because the driver of a vehicle leaving the highway has no choice but to slow down on the through-traffic lane if a deceleration lane is not provided. The failure to brake by the following drivers because of a lack of alertness causes many rear-end collisions. Acceleration lanes are not always desirable at stop-controlled intersections where entering drivers can wait for an opportunity to merge without disrupting through traffic. Acceleration lanes are advantageous on roads without stop control and on all high-volume roads even with stop control where openings between vehicles in the peak-hour traffic streams are infrequent and short. (For additional design guidance relative to lengths of deceleration and acceleration auxiliary lanes, refer to Chapter 10.)

MEDIAN OPENINGS

General Design Considerations

Medians are discussed in Chapter 4 chiefly as an element of the cross section. General ranges in width are given, and median width at intersections is treated briefly. For intersection conditions the median width, the location and length of the opening, and the design of the median end are developed in combination to fit the character and volume of through and turning traffic. Median openings should reflect street or block spacing and the access classification of the roadway. In addition, full median openings should be consistent with traffic signal spacing criteria. In some situations, median openings should be eliminated or directionalized.

Spacing of openings should be consistent with access management classifications or criteria. Where the traffic pattern at an intersection shows that nearly all traffic travels through on the divided highway and the volume is well below capacity, a median opening of the simplest and least costly design may be sufficient. This type of opening permits vehicles to make cross and turning movements, but in doing so they may encroach on adjacent lanes and usually will not have a protected space clear of other traffic. Where a traffic pattern shows appreciable cross and turning movements or through traffic of high speed and high volume, the shape and width of the median opening should provide for turning movements to be made without encroachment on adjacent lanes and with little or no interference between traffic movements.

The design of a median opening and median ends should be based on traffic volumes, urban/rural area characteristics, and type of turning vehicles, as discussed in Chapter 2. Crossing and turning traffic should operate in conjunction with the through traffic on the divided highway. Design should be based on the volume and composition of all movements occurring simultaneously during the design hours. The design of a median opening becomes a matter of considering what traffic is to be accommodated, choosing the design vehicle to use for layout controls for each cross and turning movement, investigating whether larger vehicles can turn without undue encroachment on adjacent lanes, and finally checking the intersection for capacity. If the capacity is exceeded by the traffic demand, the design must be expanded, possibly by widening or otherwise adjusting widths for certain movements. Urban/rural characteristics may influence the median width selected. Intersections in urban/suburban areas have been found to operate more safely with narrow medians, while unsignalized intersections in rural areas have been found to operate more safely with wider medians. Traffic control devices such as yield signs, stop signs, or traffic signals may be needed to regulate the various movements effectively and improve the effectiveness of operations. However, wide medians may lead to inefficient traffic signal operation.

Control Radii for Minimum Turning Paths

An important factor in designing median openings is the path of each design vehicle making a minimum left turn at 15 to 25 km/h [10 to 15 mph]. Where the volume and type of vehicles making the left-turn movement call for higher than minimum speed, the design may be made by using a radius of turn corresponding to the speed deemed appropriate. However, the minimum turning path at low speed is needed for minimum design and for testing layouts developed for one design vehicle for use by an occasional larger vehicle.

The paths of design vehicles making right turns are given in Chapter 2 and are discussed in this chapter in the section on "Types of Turning Roadways." Any differences between the minimum turning radii for left turns and those for right turns are small and are insignificant in highway design. Minimum 90-degree left-turn paths for design vehicles are shown in Exhibit 9-76. Exhibit 9-76A shows these paths positioned as they would govern median end design for vehicles leaving a divided highway. Exhibit 9-76B shows them positioned for left turns to enter a divided highway. In both cases it is assumed that the inner wheel of each design vehicle clears the median edge and centerline of the crossroad by 0.6 m [2 ft] at the beginning and end of the turn. For comparison, circular arcs of 12-, 15-, and 23-m [40-, 50-, and 75-ft] radii and tangent to the crossroad centerline and the median edge are also shown. The transition paths of the inner rear wheels are long, particularly for the semitrailer vehicles when completing the turn. Where the controlling circular arc is sharper than these long transition paths, drivers can, and habitually do, swing wide and turn on a reverse or S-curve path instead of turning directly to traverse the minimum paths shown.

The traveled way edges that most closely fit the paths of turning vehicles are transitional; however, for sharp turns at intersections, designs closely fitting these paths are three-centered curves. Design guidance for three-centered curves is discussed in the section on "Types of Turning Roadways" in this chapter. The same curves are applicable to left turns and should be used where there is a physical edge of traveled way for left turns, as in a channelized intersection and on ramps for the predominant highway.

The customary intersection on a divided highway does not have a continuous physical edge of traveled way delineating the left-turn path. Instead, the driver has guides at the beginning and

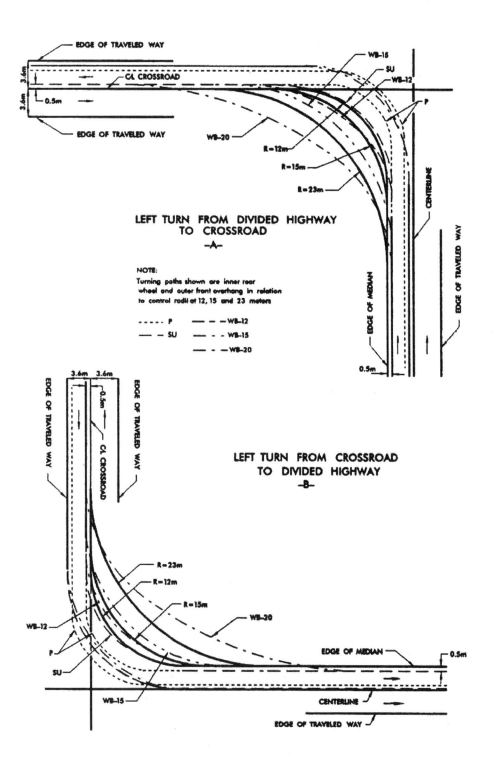

LEFT TURN FROM DIVIDED HIGHWAY
TO CROSSROAD
-A-

NOTE:
Turning paths shown are inner rear
wheel and outer front overhang in relation
to control radii at 12, 15 and 23 meters

LEFT TURN FROM CROSSROAD
TO DIVIDED HIGHWAY
-B-

Metric

Exhibit 9-76. Control Radii at Intersections for 90-Degree Left Turns

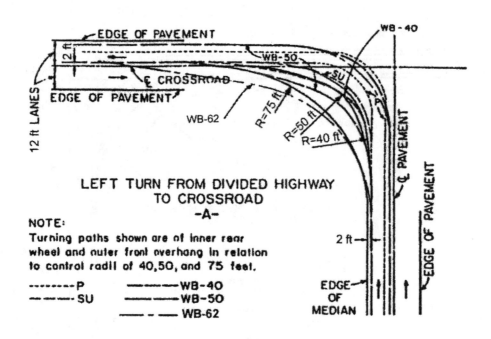

LEFT TURN FROM DIVIDED HIGHWAY TO CROSSROAD
-A-

NOTE:
Turning paths shown are of inner rear wheel and outer front overhang in relation to control radii of 40, 50, and 75 feet.

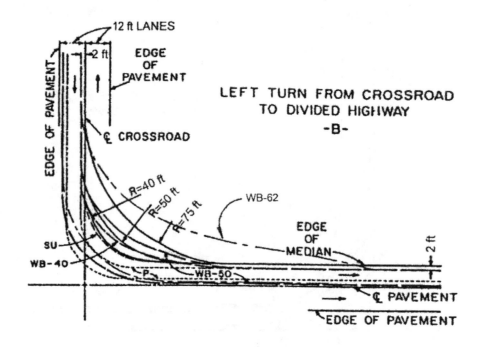

LEFT TURN FROM CROSSROAD TO DIVIDED HIGHWAY
-B-

US Customary

Exhibit 9-76. Control Radii at Intersections for 90-Degree Left Turns (Continued)

at the end of the left-turn operation: (1) the centerline of an undivided crossroad or the median edge of a divided crossroad, and (2) the curved median end. For the central part of the turn the driver has the open central intersection area in which to maneuver. Under these circumstances for minimum design of the median end, the precision of compound curves does not appear necessary, and simple curves for the minimum assumed edge of left turn have been found satisfactory. The larger the simple curve radius used, the better it will accommodate a given design vehicle, but the resulting layout for the larger curve radius will have a greater length of median opening and greater paved areas than one for a minimum radius. These areas may be sufficiently large to result in erratic maneuvering by small vehicles, which may interfere with other traffic. To reduce the effective size of the intersection for most motorists, consideration should be given to providing an edge marking corresponding to the desired turning path for passenger cars, while providing sufficient paved area to accommodate the turning path of an occasional large vehicle.

By considering the range of radii for minimum right turns and the need for accommodation of more than one type of vehicle at the usual intersections, the following control radii can be used for minimum practical design of median ends: a control radius of 12 m [40 ft] accommodates P vehicles suitably and occasional SU vehicles with some swinging wide; one of 15 m [50 ft] accommodates SU vehicles and occasional WB-12 [WB-40] vehicles with some swinging wide; and one of 23 m [75 ft] accommodates WB-12 [WB-40] and WB-15 [WB-50] vehicles with only minor swinging wide at the end of the turn.

These relations are shown generally in Exhibits 9-77 through 9-83. In the following explanation, each control radius design is tested for use by larger vehicles and for occasional movements other than those for which the design is developed. The figures indicate how the design may be tested for protection of cross traffic. This test is followed by development of median opening designs for traffic conditions in which the volumes of through and turning movements are such that it is desirable to provide space in the median for turning vehicles clear of through traffic.

Metric			US Customary		
M Width of median (m)	L = Minimum length of median opening (m)		M Width of median (ft)	L = Minimum length of median opening (ft)	
	Semicircular	Bullet nose		Semicircular	Bullet nose
1.2	22.8	22.8	4	76	76
1.8	22.2	18.0	6	74	60
2.4	21.6	15.9	8	72	53
3.0	21.0	14.1	10	70	47
3.6	20.4	12.9	12	68	43
4.2	19.8	12.0 min	14	66	40 min
4.8	19.2	12.0 min	16	64	40 min
6.0	18.0	12.0 min	20	60	40 min
7.2	16.8	12.0 min	24	56	40 min
8.4	15.6	12.0 min	28	52	40 min
9.6	14.4	12.0 min	32	48	40 min
10.8	13.2	12.0 min	36	44	40 min
12.0	12.0 min	12.0 min	40	40 min	40 min
15.0	12.0 min	12.0 min	50	40 min	40 min
18.0	12.0 min	12.0 min	60	40 min	40 min

Exhibit 9-77. Minimum Design of Median Openings (P Design Vehicle, Control Radius of 12 m [40 ft])

697

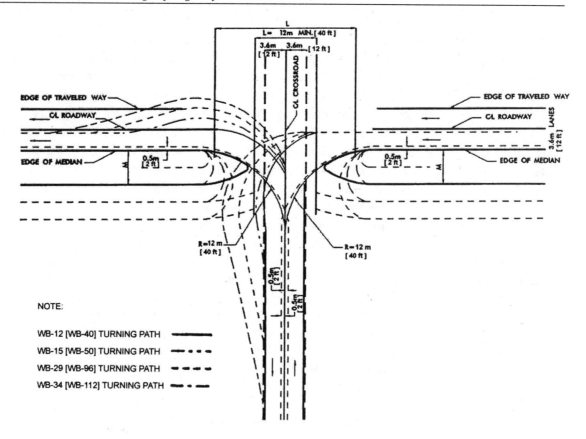

Exhibit 9-78. Minimum Design of Median Openings (P Design Vehicle, Control Radius of 12 m [40 ft])

Metric			US Customary		
M Width of median (m)	L = Minimum length of median opening (m)		M Width of median (ft)	L = Minimum length of median opening (ft)	
	Semicircular	Bullet nose		Semicircular	Bullet nose
1.2	28.8	28.8	4	96	96
1.8	28.2	22.8	6	94	76
2.4	27.6	20.4	8	92	68
3.0	27.0	18.6	10	90	62
3.6	26.4	17.4	12	88	58
4.2	25.8	15.9	14	86	53
4.8	25.2	15.0	16	84	50
6.0	24.0	13.2	20	80	44
7.2	22.8	12.0 min	24	76	40 min
8.4	21.6	12.0 min	28	72	40 min
9.6	20.4	12.0 min	32	68	40 min
10.8	19.2	12.0 min	36	64	40 min
12.0	18.0	12.0 min	40	60	40 min
15.0	15.0	12.0 min	50	50	40 min
18.0	12.0 min	12.0 min	60	40 min	40 min
21.0	12.0 min	12.0 min	70	40 min	40 min

Exhibit 9-79. Minimum Design of Median Openings (SU Design Vehicle, Control Radius of 15 m [50 ft])

Metric			US Customary		
M Width of median (m)	L = Minimum length of median opening (m)		M Width of median (ft)	L = Minimum length of median opening (ft)	
	Semicircular	Bullet nose		Semicircular	Bullet nose
1.2	43.8	36.6	4	146	122
1.8	43.2	34.5	6	144	115
2.4	42.6	33.0	8	142	110
3.0	42.0	31.5	10	140	105
3.6	41.4	30.0	12	138	100
4.2	40.8	28.8	14	136	96
4.8	40.2	27.6	16	134	92
6.0	39.0	25.5	20	130	85
7.2	37.8	23.4	24	126	78
8.4	36.6	21.9	28	122	73
9.6	35.4	20.1	32	118	67
10.8	34.2	18.6	36	114	62
12.0	30.0	17.1	40	100	57
18.0	27.0	12.0 min	60	90	40 min
24.0	21.0	12.0 min	80	70	40 min
30.0	15.0	12.0 min	100	50	40 min
33.0	12.0 min	12.0 min	110	40 min	40 min
36.0	12.0 min	12.0 min	120	40 min	40 min

Exhibit 9-80. Minimum Design of Median Openings (WB-12 [WB-40] Design Vehicle, Control Radius of 23 m [75 ft])

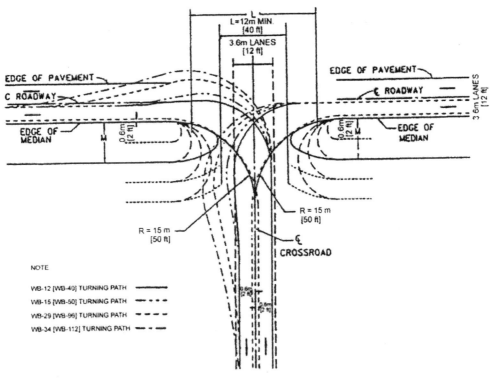

Exhibit 9-81. Minimum Design of Median Openings (SU Design Vehicle, Control Radius of 15 m [50 ft])

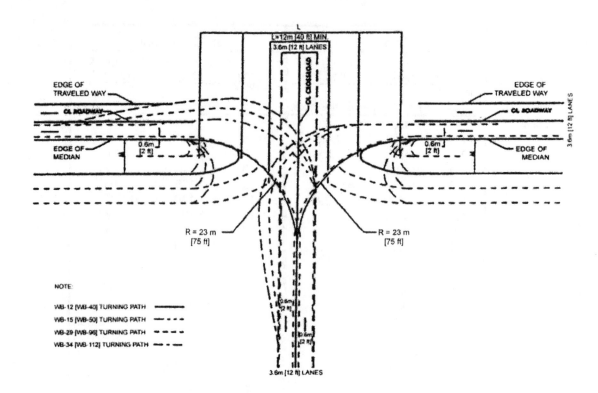

Exhibit 9-82. Minimum Design of Median Openings (WB-12 [WB-40] Design Vehicle, Control Radius of 23 m [75 ft])

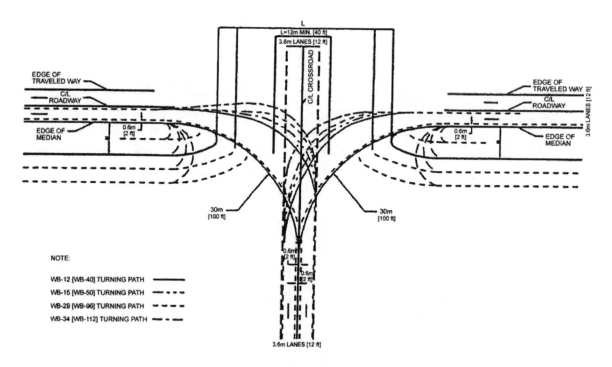

Exhibit 9-83. Minimum Design of Median Openings (Radius of 30 m [100 ft])

Shape of Median End

One form of a median end at an opening is a semicircle, which is a simple design that is satisfactory for narrow medians. However, the several disadvantages of semicircular ends for medians greater than about 3.0 m [10 ft] in width are widely recognized, and other more desirable shapes are generally used.

Alternate minimum designs for median ends to fit the design control radii of 12, 15, 23, and 30 m [40, 50, 75, and 100 ft] are shown in Exhibits 9-78, 9-81, 9-82, and 9-83. The paths of design vehicles are shown on the basis of the inner rear wheel beginning and ending the left-turn maneuver 0.6 m [2 ft] from the edge of the median and the centerline of the undivided crossroad. The alternate minimum designs are a semicircular end and a bullet nose form. The indicated PC of the control radius on the median edge is a common PC for both forms of median end. The bullet nose is formed by two symmetrical portions of control radius arcs and an assumed small radius (e.g., 0.6 m [2 ft] is used, to round the nose). The bullet nose design closely fits the path of the inner rear wheel and results in less intersection pavement and a shorter length of opening than the semicircular end. These advantages are operational in that the driver of the left-turning vehicle channelized for a greater portion of the path has a better guide for the maneuver, and the elongated median is better positioned to serve as a refuge for pedestrians crossing the divided highway.

For medians about 1.2 m [4 ft] wide there is little or no difference between the two forms of median end. For a median width of 3.0 m [10 ft] or more the bullet nose is superior to the semicircular end and preferably should be used in design. On successively wider medians the bullet nose end results in shorter lengths of openings. For median widths greater than 4.2 m [14 ft] and a 12-m [40-ft] control radius (Exhibit 9-78), the minimum length of opening to provide for cross traffic becomes a positive control. The minimum ends for medians about 4.2 m [14 ft] or more wide also take the shape of squared or flattened bullet ends, the flat end being parallel to the crossroad centerline. This shape retains the advantages over semicircular median ends regardless of the median width because of the channelizing control. The bullet nose curves are such as to position the left-turning vehicles to turn to or from the crossroad centerline, whereas the semicircular end tends to direct the left off movement onto the opposing traffic lane of the crossroad.

Minimum Length of Median Opening

For any three- or four-leg intersection on a divided highway the length of median opening should be as great as the width of crossroad traveled way plus shoulders. Where the crossroad is a divided highway, the length of opening should be at least equal to the width of the crossroad traveled ways plus that of the median.

The use of a minimum length of opening without regard to the width of median or the control radius should not be considered except at very minor crossroads. Care should be taken not to make the median opening longer than necessary at rural unsignalized intersections. The minimum length of opening for U-turns is discussed later in this chapter in the section, "Indirect Left Turns and U-Turns."

Median Openings Based on Control Radii for Design Vehicles

Passenger Vehicles

Exhibit 9-78 shows minimum median opening designs based on a control radius of 12 m [40 ft] for a 90-degree intersection. The control radius is made tangent to the upper median edge and to the centerline of the undivided crossroad, thereby locating the semicircular median end or forming a portion of a bullet nose end. The resulting lengths of opening vary with the width of median, as shown in the tabulation on the figure. For each of the median widths indicated, the channelizing and area differences between the semicircular and bullet nose ends are apparent.

The control radius of 12 m [40 ft] accommodates P design vehicles making turns somewhat above minimum, the actual path of which is not shown in Exhibit 9-78, but is shown in Exhibit 9-76 instead.

The paths of the WB-12 [WB-40] and WB-15 [WB-50] design vehicles making minimum left turns both off and onto the divided highway are shown in Exhibit 9-78 to indicate how these large vehicles can turn at an intersection designed for passenger cars. Only the inner wheel track and outer front overhang paths are indicated. The paths are depicted from a position parallel to the median edge or centerline of the crossroad and at the beginning of the turn, and they indicate that swinging wide and reversing are needed at the end of the turn. Drivers of large vehicles making sharp left turns also may swing right before turning left. However, paths might be a combination of these two extremes, swinging out before beginning the left turn with infringement on the outer lane of the divided highway and also swinging wide and reversing at the end of the turn. The path with the parallel movement at the beginning of the turn is shown because it indicates the maximum encroachment.

In Exhibit 9-78, the WB-12 [WB-40] design vehicle turning from the divided highway encroaches about 1 m [3 ft] beyond the two-lane (projected) crossroad edge of traveled way, and the WB-15 [WB-50] vehicle encroaches about 3.3 m [11 ft]. With wide crossroads this encroachment is within the median opening, but with two-lane crossroads, as shown in Exhibit 9-78, the encroachment may be beyond the median end, particularly with wide medians having a minimum length of opening. As the left turn is completed, the encroachment may be beyond the edge of traveled way for right turns located diagonally opposite the beginning of the left-turn movement off the divided highway. With wide crossroads this encroachment does not extend beyond the right-turn edge of traveled way (not shown in Exhibit 9-78), but with two-lane crossroads and narrow medians it may extend beyond. By swinging over a short distance on the divided highway before beginning the turn, most drivers could pass through these openings and remain on the paved areas. Although this procedure is used extensively, it should be discouraged through a more expansive design where practical.

For turns onto the divided highway the paths show varying degrees of encroachment on the right lane of the divided highway. The SU design vehicle encroaches about 0.3 m [1 ft] on the right lane of a four-lane divided highway, the WB-12 [WB-40] vehicle about 1.5 m [5 ft], and the WB-15 [WB-50] vehicle about 3.0 m [10 ft]. These distances can be lessened by the drivers anticipating the turn and swinging right before turning left, if space is available. This space depends on the median width, the length of opening as governed by the number of lanes on the crossroad, and other limitations such as triangular islands for channelizing right-turn movements.

Exhibit 9-78 indicates that minimum median openings based on a control radius of 12 m [40 ft] are not well suited for lengths of opening for two-lane crossroads because trucks cannot turn left without difficult maneuvering and encroachment on median ends or outer shoulders, or both, depending on the median width. It may be suitable for wide crossroad traveled ways, but for these cases it is advantageous to use a control radius greater than 12 m [40 ft], which enables all vehicles to turn at a little greater speed and enables trucks to maneuver and turn with less encroachment. Exhibits 9-78, 9-81, and 9-82 show the squared or truncated bullet nose design in conjunction with the 12-m [40-ft] minimum length of opening. Provision of longer tapers not only avoids this somewhat awkward-looking design but also provides for other important objectives as well. This topic is discussed further in the section of this chapter on "Above-Minimum Designs for Direct Left Turns."

Single-Unit Trucks or Buses

Exhibit 9-81 shows minimum median opening designs for a 90-degree intersection, based on a control radius of 15 m [50 ft]. The basis of development, median ends, and turning paths shown are similar to those of Exhibit 9-78. As indicated in Exhibit 9-76, the control radius of 15 m [50 ft] accommodates the SU design vehicle making minimum left turns without encroachment on adjacent lanes. The paths of the WB-12 [WB-40] and WB-15 [WB-50] design vehicles making left turns both off and onto the divided highway are shown in Exhibit 9-81 to indicate how these large vehicles can turn at an intersection designed for the SU vehicles.

The WB-15 [WB-50] design vehicle would encroach about 1 m [4 ft] beyond a 7.2-m [24-ft] crossroad in turning off the divided highway, but encroachment could be reduced by swinging wide at the beginning of the turn. In turning onto the divided highway it would encroach about 2 m [7 ft] on the adjacent lane, a distance that could be reduced but not eliminated by swinging wide at the beginning of the turn, but, to do so, the length of opening would have to be greater than the 12-m [40-ft] minimum.

Exhibit 9-81 indicates that minimum lengths of median openings based on a control radius of 15 m [50 ft] are suited for truck operation, except that WB-15 [WB-50] vehicles will encroach on adjacent lanes. For these cases, additional advantage is gained by using a control radius greater than 15 m [50 ft] where WB-15 [WB-50] semitrailers are expected to turn.

Semitrailer Combinations

Exhibit 9-82 shows minimum median opening designs for a 90-degree intersection, which are based on a control radius of 23 m [75 ft] while Exhibit 9-83 is based on a control radius of 30 m [100 ft]. The 23-m [75-ft] control radius is sufficiently large to accommodate a WB-12 [WB-40] design vehicle, and the minimum path of the WB-15 [WB-50] vehicle indicates that it can also use this design without undue encroachments. The left turn to leave the divided highway can be made within a two-lane crossroad. In the left turn to enter the divided highway, the WB-15 [WB-50] vehicle would encroach on the adjacent lane about 0.5 m [2 ft]. This encroachment can be avoided by swinging wide at the beginning of the turn. The minimum median opening length to accommodate a WB-12 [WB-40] design vehicle with a control radius of 23 m [75 ft] is shown in Exhibit 9-80.

Effect of Skew

A control radius for design vehicles as the basis for minimum design of median openings results in lengths of openings that increase with the skew angle of the intersection. Although the bullet nose end remains preferable, the skew introduces other variations in the shape of median end. At a skewed crossing, the control radius R should be used in the acute angle to locate the PT on the median edge, point 1 as shown in Exhibit 9-84. The arc for this radius is the equivalent of

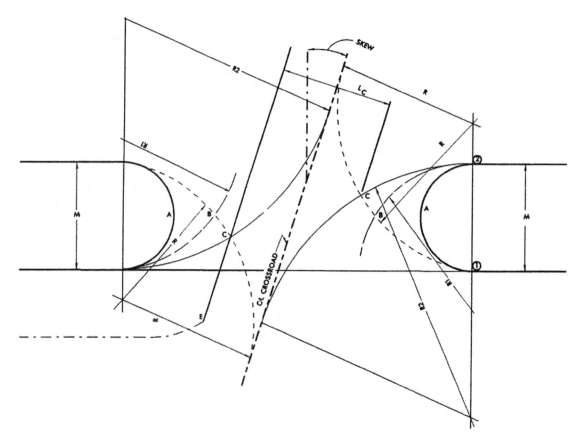

Exhibit 9-84. Minimum Design of Median Openings (Effect of Skew)

the minimum inner path for the vehicle in turning more than 90 degrees. With this PT as a design control, several alternate designs that depend on the skew angle, median width, and control radius may be considered.

Semicircular ends (A in Exhibit 9-84) result in very long openings and minor channelizing control for vehicles making a left turn with less than 90 degrees in the turning angle.

A symmetrical bullet nose B with curved sides determined by the control radius and tangent at points 1 and 2 is a layout similar to those in Exhibits 9-78, 9-81, 9-82, and 9-83. This design also has little channelizing control for vehicles turning left less than 90 degrees from the divided highway. An asymmetrical bullet nose (C in Exhibit 9-84) with radii R and R_2 has the most positive control and less paved area than designs A and B. The second radius R_2, which exceeds R, is that determined to be tangent at point 2 and also to the crossroad centerline. In this design the nose is offset from the median center paths.

The length of the opening of these alternates for a given median width decreases in the order discussed, A to C. For wide medians and a large skew, the length of openings may not be sufficient to accommodate the crossroad and flattened ends, or an above-minimum design should be used. With design ends B or C, a flattened bullet nose treatment for adequate length of opening would be made parallel to the crossroad, as shown at the lower left, C-E, in Exhibit 9-84 for the length of opening L. In such cases, the use of a control radius larger than the minimum should be considered.

Exhibit 9-86 shows typical values obtained for the minimum median ends designed with a control radius of 15 m [50 ft] (the same as in Exhibit 9-81) for a range of skew angles and median widths. Lengths of openings, measured normal to the crossroad, are shown for median ends A, B, and C, as shown in Exhibit 9-84. For any one median width with semicircular ends and skew of 20 degrees and 40 degrees, the openings are about one-third and two-thirds longer, respectively, than that for a 90-degree crossing. Likewise, the symmetrical bullet nose ends give lengths of openings for a 20-degree skew that are about one and one-half times, and for a 40-degree skew about two times, the lengths for a 90-degree crossing.

Metric			US Customary		
Design vehicles accommodated	Control radius (m)		Design vehicles accommodated	Control radius (ft)	
	12	15	23		40	50	75
Predominant	P	SU	WB-12	Predominant	P	SU	WB-40
Occasional	SU	WB-12	WB-15	Occasional	SU	WB-40	WB-50

Exhibit 9-85. Design Controls for Minimum Median Openings

In general, median openings longer than 25 m [80 ft] should be avoided, regardless of skew. This plan may call for special channelization, left-turn lanes, or adjustment to reduce the crossroad skew, all of which result in above-minimum designs.

Preferably, each skew crossing should be studied separately with trial graphical solutions on a suitable scale to permit the designer to make comparisons and choose the preferred layout. In general, the asymmetrical bullet nose end (C in Exhibit 9-84) is preferable. Where end B is not greatly different, the practical aspects of symmetry may make it preferable. In some cases end C-E may well be considered in the alternate studies.

For the preceding discussion the design controls for minimum median openings for left turns are summarized in Exhibit 9-85.

Above-Minimum Designs for Direct Left Turns

Median openings that enable vehicles to turn on minimum paths, and at 15 to 25 km/h [10 to 15 mph], are adequate for intersections where traffic for the most part proceeds straight through the intersection. Where through-traffic volumes and speeds are high and left-turning movements are important, undue interference with through traffic should be avoided by providing median openings that permit turns without encroachment on adjacent lanes. This arrangement would enable turns to be made at speeds above that for the minimum vehicle paths and provide space for vehicle protection while turning or stopping. The general pattern for minimum design can be used with larger dimensions.

A variety of median-opening arrangements may be considered that depend on the control dimensions (width of median and width of crossroad or street, or other) and the size of vehicle to be used for design control.

Median openings having above-minimum control radii and bullet nose median ends are shown in Exhibit 9-87. The design controls are the three radii R, R_1, and R_2. Radius R is the control radius for the sharpest portion of the turn, R_1 defines the turnoff curve at the median edge, and R_2 is the radius of the tip. When a sufficiently large R_1 is used, an acceptable turning speed for vehicles leaving the major road is ensured and a sizable area inside the inner edge of through-traffic lane between points 1 and 2 may be available for speed change and protection from turning vehicles. Radius R_1 may vary from about 25 to 120 m [80 to 400 ft], or more.

The tabulated values shown, 30, 50, and 70 m [90, 170, and 230 ft], are established minimum radii for turning speeds of 30, 40, and 50 km/h [20, 25, and 30 mph], respectively. The radii will vary depending on the maximum superelevation rate selected. In this case, the ease of turning probably is more significant than the turning speeds because the vehicle will need to slow down to about 15 to 25 km/h [10 to 15 mph] at the sharp part of the turn or may need to stop at the crossroad. Radius R_2 can vary considerably, but is pleasing in proportion and appearance when it is about one-fifth of the median width. Radius R is tangent to the crossroad centerline (or edge of crossroad median). Radii R and R_1 comprise the two-centered curve between the terminals of the left turn. For simplicity, the PC is established at point 2. Radius R cannot be smaller than the minimum control radius for the design vehicle, or these vehicles will be unable to turn to or from the intended lane even at low speed. To avoid a large opening, R should be held to a reasonable minimum (e.g., 15 m [50 ft]), as used in Exhibit 9-87.

Metric

Skew angle (degrees)	Width of median (m)	Semi-circular A	Symmetrical B	Asymmetrical C	R for Design C (m)
0	3	27	19	–	–
	6	24	13	–	–
	9	21	12 Min.	–	–
	12	18	12 Min.	–	–
	15	15	12 Min.	–	–
	18	13	12 Min.	–	–
10	3	32	24	23	21
	6	28	17	16	20
	9	25	14	12 Min.	20
	12	21	12 Min.	12 Min.	19
	15	18	12 Min.	12 Min.	18
	18	14	12 Min.	12 Min.	18
20	3	36	29	27	29
	6	32	22	20	28
	9	28	18	14	26
	12	24	14	12 Min.	25
	15	20	12 Min.	12 Min.	23
	18	16	12 Min.	12 Min.	21
30	3	41	34	32	42
	6	36	27	23	39
	9	31	23	17	36
	12	27	19	13	33
	15	23	15	12 Min.	30
	18	18	12	12 Min.	27
40	3	44	38	35	63
	6	39	32	27	58
	9	35	27	20	53
	12	29	23	15	47
	15	24	19	12 Min.	42
	18	19	15	12 Min.	36

US Customary

Skew angle (degrees)	Width of median (ft)	Semi-circular A	Symmetrical B	Asymmetrical C	R for Design C (ft)
0	10	90	62	–	–
	20	80	44	–	–
	30	70	40 Min.	–	–
	40	60	40 Min.	–	–
	50	50	40 Min.	–	–
	60	40	40 Min.	–	–
10	10	106	80	77	70
	20	94	58	54	68
	30	82	45	40 Min.	65
	40	71	40 Min.	40 Min.	63
	50	60	40 Min.	40 Min.	61
	60	47	40 Min.	40 Min.	59
20	10	121	97	90	97
	20	107	74	65	92
	30	94	59	48	86
	40	81	48	40 Min.	82
	50	68	40 Min.	40 Min.	76
	60	54	40 Min.	40 Min.	71
30	10	135	114	105	140
	20	120	91	77	130
	30	104	75	58	120
	40	90	62	42	110
	50	76	51	40 Min.	100
	60	60	40	40 Min.	90
40	10	148	127	118	210
	20	131	106	90	193
	30	115	90	68	174
	40	98	77	51	156
	50	81	64	40 Min.	139
	60	64	51	40 Min.	121

Note: A, B, and C in the headings for this exhibit refer to the types of median ends shown in Exhibit 9-84.

Exhibit 9-86. Effect of Skew on Minimum Design for Median Openings (Typical Values Based on Control Radius of 15 m [50 ft])

Metric						US Customary					
M width of median (m)	Dimensions in meters when					M width of median (ft)	Dimensions in feet when				
	R_1 = 30 m		R_1 = 50 m		R_1 = 70 m		R_1 = 90 ft		R_1 = 170 ft		R_1 = 230 ft
	L	B	L	B	L	B		L	B	L	B	L	B
6.0	18.0	20.2	20.2	24.4	21.3	27.6	20	58	65	66	78	71	90
9.0	15.1	21.4	17.7	26.5	19.0	30.4	30	48	68	57	85	63	101
12.0	12.8	22.4	15.6	28.3	17.1	32.7	40	40	71	50	90	57	109
15.0	–	–	13.8	29.9	15.4	34.7	50	–	–	44	95	51	115
18.0	–	–	–	–	13.8	36.7	60	–	–	–	–	46	122
21.0	–	–	–	–	12.4	38.4	70	–	–	–	–	41	128

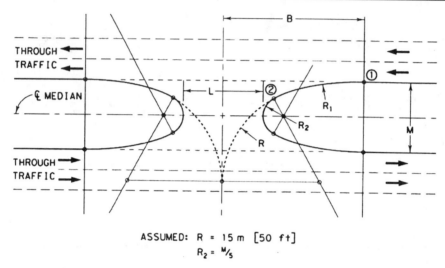

ASSUMED: R = 15 m [50 ft]
$R_2 = M/5$

**Exhibit 9-87. Above Minimum Design of Median Openings
(Typical Bullet-Nose Ends)**

The length of median opening is governed by the radii. For medians wider than about 9 m [30 ft] coupled with a crossroad of four or more lanes, the control radius R generally will need to be greater than 15 m [50 ft] or the median opening will be too short. A rounded value can be chosen for the length of opening (e.g., 15 or 18 m [50 or 60 ft]) and that dimension can be used to locate the center for R. Then R becomes a check dimension to ensure the workability of the layout. The tabulation of values in Exhibit 9-87 shows the resultant lengths of median openings over a range of median widths for three assumed values of R_1 and for R assumed to be 15 m [50 ft]. Dimension "B" is included as a general design control and for comparison with other above-minimum designs.

The median end designs in Exhibit 9-87 do not positively provide protection areas within the limits of the median width. A design using R_1 = 30 m [100 ft] or more provides space for at least a single passenger vehicle to pause in an area clear of both the through-traffic lanes and the crossroad lanes with wide medians; such radii may provide enough protection space for larger design vehicles. At skewed intersections, above-minimum designs with bullet nose median ends can be applied directly. Where the skew is 10 degrees or more, adjustments in R and R_2 from the values shown are needed to provide the appropriate length of opening.

INDIRECT LEFT TURNS AND U-TURNS

General Design Considerations

Divided highways need median openings to provide access for crossing traffic in addition to left-turning and U-turning movements. The discussions to follow deal with the various design methods that accommodate these movements predicated on median width.

At intersections where the median is too narrow to provide a lane for left-turning vehicles and the traffic volumes or speeds, or both, are relatively high, safe, efficient operation is particularly troublesome. Vehicles that slow down or stop in a lane primarily used by through traffic to turn left greatly increase the potential for rear-end collision.

Other factors that should receive special consideration in design for left- and U-turning movements are the turning paths of the various design vehicles in conjunction with narrow medians. The demands for left- or U-turn maneuvers in the urban or heavily developed residential or commercial sectors also present problems with respect to efficient operation.

The design plans shown in Exhibits 9-88 and 9-89 offer two options with respect to indirect left turns and also provide for indirect U-turning movements. Exhibit 9-88 involves a jug-handle-type ramp or diagonal roadway that intersects a secondary crossing roadway. The motorist exits via the jug-handle-type ramp and makes a left turn onto the crossroad. For a U-turn maneuver, the motorist makes an additional left turn onto the divided highway.

Exhibit 9-89 shows an at-grade loop that can serve as an alternate to the jug-handle-type ramp. The loop design might be considered when the jug-handle-type ramps would need costly right-of-way, the opposite quadrant being less costly. There might be other justifications in selecting the loop instead of the ramp, such as improved vertical alignment and comparative grading costs.

Exhibit 9-90 illustrates a design that provides for indirect left turns to be made from the right, via separate turning roadways connected to a crossroad. Such arrangements have the advantage of eliminating left turns from the through lanes and providing storage for left-turning vehicles not available on the highway itself. The left-turning vehicles with little extra travel distance are able to cross the main highway with appropriate traffic control devices. Exhibit 9-90 illustrates three design options that might be adaptable to various roadway patterns. The turn from bottom to left is accomplished via the added left-turn slip ramp at the lower right (similar to previous discussions). This arrangement permits left turns onto the minor road under traffic signal protection and prevents cars making left turns from blocking the lane adjacent to the medians. Where there is a parallel roadway nearby, the added ramp may connect to it, as shown in the upper left or alternately as shown by the dashed-line connection. However, this design is less desirable because the vehicles must pass through the intersection twice and create delays by reducing speed in turning right. This delay might be overcome by the introduction of auxiliary lanes if space is available.

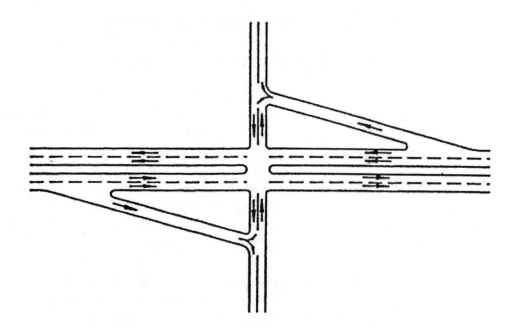

Exhibit 9-88. Jughandle-Type Ramp with Crossroad

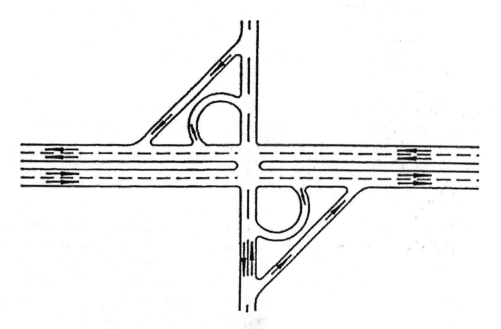

Exhibit 9-89. At-Grade Loop (Surface Loop) with Crossroad

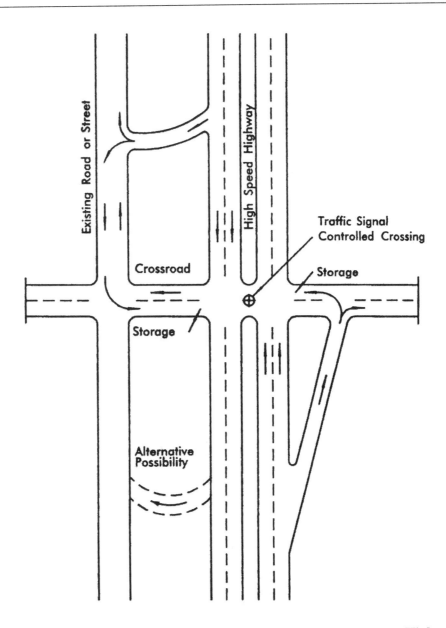

Exhibit 9-90. Special Indirect Left-Turn Designs for Traffic Leaving Highway with Narrow Median

Indirect Left Turn or Indirect U-Turn—Using Local Streets

Highways without control of access that involve narrow nontraversable medians and where the adjacent property owners enter the divided highway by right turn only must gain access to the opposite traveled way by one of three types of operation and control. The first option is to use the interconnecting street patterns. This operation involves making the initial right turn, proceeding by continuous right turns around the block to the median opening that services the secondary crossroads, and then turning left. Variations of access to the divided highway would also prevail for the property owners on the adjacent street patterns. However, the around-the-block principle would still control movements with respect to exit and return trips. The around-the-block option needs careful examination of existing turning radii to accommodate SU vehicles and estimation of the number of WB vehicles that might use this method of indirect left turns or indirect U-turns. This approach needs careful design attention with respect to restrictive parking, regulatory signs, and signal control devices in the proximity of each intersection.

The second alternative, which would only benefit the property owners abutting the divided highway, is to provide median openings for the individual properties. This option would defeat a major purpose of the median and would lead to complete erosion of this control feature.

The around-the-block alternative is not always practical, especially where radial routes traverse areas that are only partially developed and where there is no established pattern of adjacent roadways, often with no existing roads or streets running parallel to the through highway. Even where there is a suitable network of adjacent roadways, the adverse travel is objectionable. The increased traffic volumes passing through four intersections and the left turn to the arterial might be as much of a hindrance to the free flow of traffic as a direct left-turn maneuver.

The third alternative is use of the design principles previously described with respect to constructing jug-handle-type ramps or at-grade intersecting loops.

From the preceding discussion it is apparent that, wherever practical, a newly designed divided highway should have a median width that can accommodate normal left turns and U-turns by using a median storage lane that will protect and store the design-hour turning volume.

Indirect Left Turn or Indirect U-Turn—Wide Medians

Exhibit 9-91 illustrates an indirect left turn for two arterials where left turns are heavy on both roads. The north-south roadway is undivided and the east-west roadway is divided with a wide median. Because left turns from the north-south road would cause congestion because of the lack of storage, left turns from the north-south road are prohibited at the main intersection. Left-turning traffic turns right onto the divided road and then makes a U-turn at a one-way crossover located in the median of the divided road. Auxiliary lanes are highly desirable on each side of the median between the crossovers for storage of turning vehicles.

The crossover should be 120 to 180 m [400 to 600 ft] away from the intersection to allow the left-turn traffic to approach the intersection on a green signal. This scheme provides a slight increase in capacity at very little cost with no additional acquisition of right-of-way. The main disadvantage is that the left-turn traffic has to pass through the same intersection twice. This maneuver also may be confusing to motorists unfamiliar with the design and thus needs special signing.

Special left-turn considerations may also be needed at major crossroads, particularly at isolated intersections in suburban areas, where one or more of the left-turning movements are so

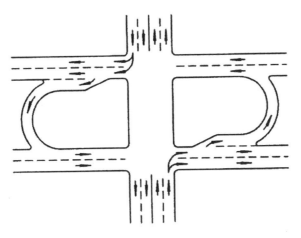

Exhibit 9-91. Indirect Left Turn Through a Crossover

large that they cannot be handled by the conventional median lanes and where there is insufficient width to install two median lanes. In such instances, the jug-handle-type ramp or the at-grade loop pattern that provides an exit or entrance ramp in two of the crossing quadrants would be appropriate.

Location and Design of U-Turn Median Openings

Median openings designed to accommodate vehicles making U-turns only are needed on some divided highways in addition to openings provided for cross and left-turning movements. Separate U-turn median openings may fit at the following locations:

- Locations beyond intersections to accommodate minor turning movements not otherwise provided in the intersection or interchange area. The major intersection area is kept free for the important turning movements, in some cases obviating expensive ramps or additional structures.
- Locations just ahead of an intersection to accommodate U-turn movements that would interfere with through and other turning movements at the intersection. Where a fairly wide median on the approach highway has few openings, U-turns are necessary for motorists to reach roadside areas. Advance separate openings to accommodate them outside the intersection proper will reduce interference.
- Locations occurring in conjunction with minor crossroads where traffic is not permitted to cross the major highway but instead is required to turn right, enter the through traffic stream, weave to the left, U-turn, and then return. On high-speed or high-volume highways, the difficulty of weaving and the long lengths involved usually make this design pattern undesirable unless the volumes intercepted are light and the median is of adequate width. This condition may occur where a crossroad with high-volume traffic, a shopping area, or other traffic generator that needs a median opening nearby and additional median openings would not be practical.
- Locations occurring where regularly spaced openings facilitate maintenance operations, policing, repair service of stalled vehicles, or other highway-related activities. Openings

for this purpose may be needed on controlled-access highways and on divided highways through undeveloped areas.

- Locations occurring on highways without control of access where median openings at optimum spacing are provided to serve existing frontage developments and at the same time minimize pressure for future median openings. A preferred spacing at 400 to 800 m [0.25 to 0.50 mi] is suitable in most instances. Fixed spacing is not necessary, nor is it fitting in all cases because of variations in terrain and local service needs.

For a satisfactory design for U-turn maneuvers, the width of the highway, including the median, should be sufficient to permit the design vehicle to turn from an auxiliary left-turn lane in the median into the lane next to the outside shoulder or outside curb and gutter on the roadway of the opposing traffic lanes.

Medians of 5.0 m [16 ft] and 15 m [50 ft] or wider are needed to permit passenger and single-unit truck traffic, respectively, to turn from the inner lane (next to the median) on one roadway to the outer lane of a two-lane opposing roadway. Also, a median left-turn lane is highly desirable in advance of the U-turn opening to eliminate stopping on the through lanes. This scheme would increase the median width by approximately 3.6 m [12 ft].

Wide medians are uncommon in highly developed areas. Consequently, special U-turn designs should be considered where right-of-way is restricted, speeds are low, and signal control is used downstream to provide sufficient gaps in the traffic stream. Median widths of 2 to 12 m [7 to 40 ft] may be used for U-turn openings to permit passenger vehicles or single-unit trucks to turn from the inner lane in one direction onto the shoulder of a four-lane divided highway in the other direction. This special U-turn feature can be incorporated into the design of an urban roadway section by constructing a short segment of shoulder area along the outside edge of the traveled way across from the U-turn opening. The outside curb and gutter section would then be carried behind the shoulder area and the shoulder would be designed as a pavement.

Where U-turn openings are proposed for access to the opposite side of a multilane divided street, they should be located 15 to 30 m [50 to 100 ft] in advance of the next downstream left-turn lane. For U-turn openings designed specifically for the purpose of eliminating left-turn movement at a major intersection, they should be located downstream of the intersection, preferably midblock between adjacent cross road intersections. This type of U-turn opening should be designed with a median left-turn lane for storage.

Normally, U-turns should not be permitted from the through lanes. However, where medians have adequate width to shield a vehicle stored in the median opening, through volumes are low and left-turn/U-turns are infrequent, this type of design may be permissible. Minimum widths of median to accommodate U-turns by different design vehicles turning from the lane adjacent to the median are given in Exhibit 9-92. These dimensions are for a four-lane divided facility. If the U-turn is made from a median left-turn/U-turn lane, the width needed is the separator width; the total median width needed would include an additional 3.6 m [12 ft] for a single median turn lane. At major intersections, many jurisdictions allow both left turns and U-turns to be made around the curbed nose at the end of a left-turn lane. Where dual left-turn lanes are needed along a street

METRIC

TYPE OF MANEUVER		M - MIN. WIDTH OF MEDIAN (m) FOR DESIGN VEHICLE						
		P	WB-12	SU	BUS	WB-15	WB-18	TDT
		LENGTH OF DESIGN VEHICLE (m)						
		5.7	15.0	9.0	12.0	16.5	19.5	35.4
INNER LANE TO INNER LANE	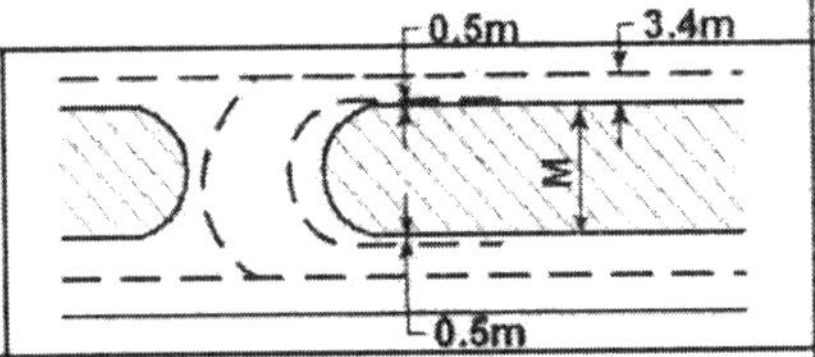	9	18	19	19	21	21	30
INNER LANE TO OUTER LANE	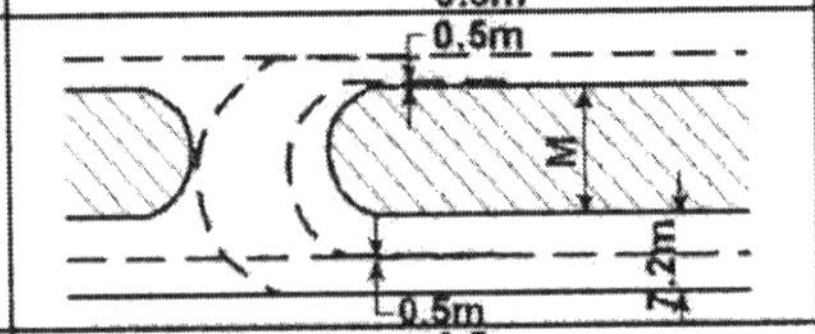	5	15	15	15	18	18	27
INNER LANE TO SHOULDER	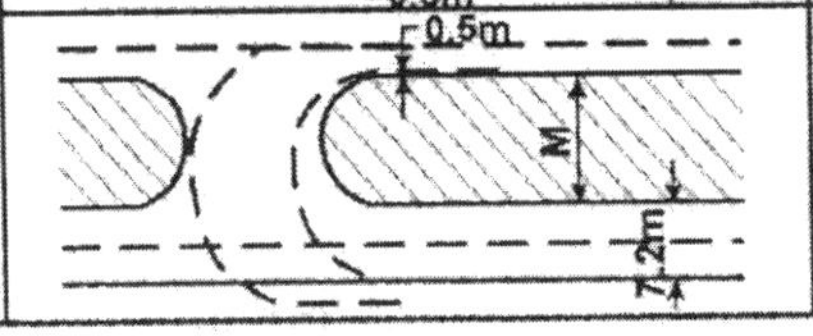	2	12	12	12	15	15	24

US CUSTOMARY

TYPE OF MANEUVER		M - MIN. WIDTH OF MEDIAN (ft) FOR DESIGN VEHICLE						
		P	WB-40	SU	BUS	WB-50	WB-60	TDT
		LENGTH OF DESIGN VEHICLE (ft)						
		19	50	30	40	55	65	118
INNER LANE TO INNER LANE	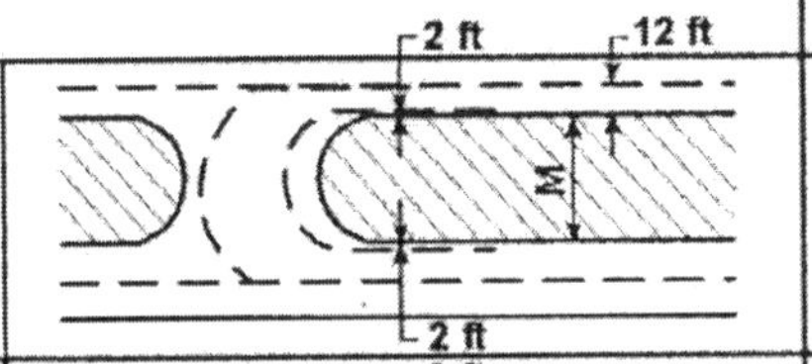	30	61	63	63	71	71	101
INNER LANE TO OUTER LANE	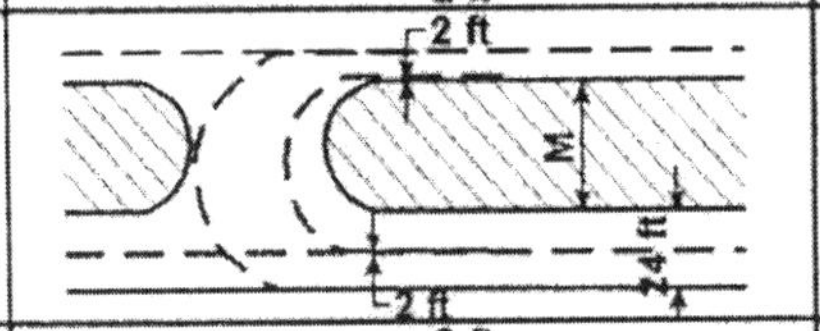	18	49	51	51	59	59	89
INNER LANE TO SHOULDER	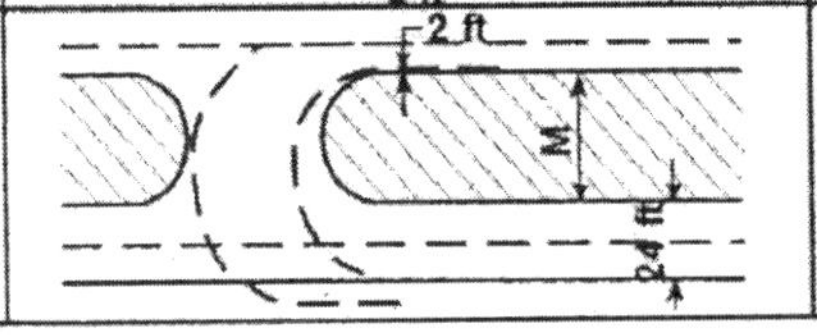	8	39	41	41	49	49	79

Exhibit 9-92. Minimum Designs for U-turns

with a raised-curb median, left turns and U-turns may be permitted from the inside lane and left turns only may be allowed from the outside turn lane.

Exhibit 9-93 illustrates special U-turn designs with narrow medians. In Exhibit 9-93A, the U-turning vehicle swings right from the outer lane, loops around to the left, stops clear of the divided highway until a suitable gap in the traffic stream develops, and then makes a normal left turn onto the divided highway. In Exhibit 9-93B, the U-turning vehicle begins on the inner lane of the divided highway, crosses the through-traffic lanes, loops around to the left, and then merges with the traffic. To deter vehicles from stopping on through lanes, a left-turn lane with proper storage capacity should be provided to accommodate turning vehicles.

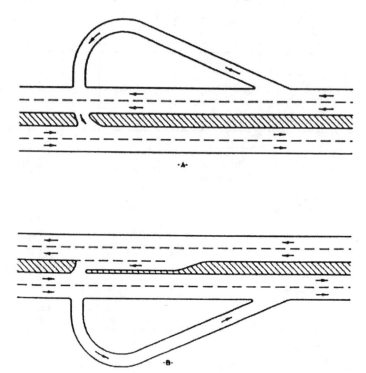

Exhibit 9-93. Special Indirect U-Turn with Narrow Medians

FLUSH OR TRAVERSABLE MEDIANS

The foregoing discussion of design for indirect left turns and indirect U-turns with raised curb medians brings into focus the difficulties involved in providing access to abutting property, especially where such access is by commercial vehicles. These conditions are very common in commercial and industrial areas where property values are high and rights-of-way for wide medians are difficult to acquire. Under such conditions, paved flush or traversable-type medians 3.0 to 4.8 m [10 to 16 ft] wide may be the optimum type of design for left-turning vehicles.

Exhibit 9-94 illustrates two kinds of typical marking applications. Exhibit 9-94A shows a flush median with separate left-turn lanes marked for turns into cross streets and Exhibit 9-94B illustrates a flush or traversable-type median with both separate left-turn lanes into cross streets and two-way left-turn lanes marked at midblock.

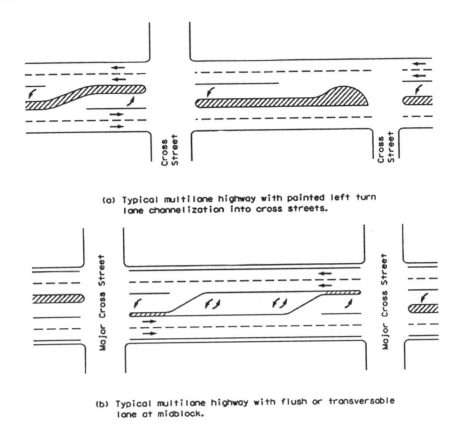

(a) Typical multilane highway with painted left turn
lane channelization into cross streets.

(b) Typical multilane highway with flush or transversable
lane at midblock.

Exhibit 9-94. Flush or Traversable Median Lane Markings

In general, two-way left-turn lanes should be used only in an urban setting where operating speeds are relatively low and where there are no more than two through lanes in each direction. The operational characteristics of two-way left-turn lanes with more than two through lanes in each direction is the subject of ongoing research, and caution is recommended at this time when considering more than a five-lane cross section. This subject is discussed in Chapter 4 and in the MUTCD (**9**). Additional research is available on the effects of midblock left turns (**14**).

AUXILIARY LANES

General Design Considerations

From the foregoing discussions it is appropriate to deal with the design elements of auxiliary lanes as they relate to median openings with left-turning movements. In general, auxiliary lanes are used preceding median openings and are also used at intersections preceding right-turning movements. Auxiliary lanes may also be added to increase capacity and improve safety at an intersection. In many cases, an auxiliary lane may be desirable after completing a right-turn movement to provide for acceleration, maneuvering, and weaving.

Auxiliary lanes should be at least 3 m [10 ft] wide and desirably should equal that of the through lanes. Where curbing is to be used adjacent to the auxiliary lane, an appropriate curb offset should be provided. The length of the auxiliary lanes for turning vehicles consists of three components: (1) entering taper, (2) deceleration length, and (3) storage length.

Desirably, the total length of the auxiliary lane should be the sum of the length for these three components. Common practice, however, is to accept a moderate amount of deceleration within the through lanes and to consider the taper length as a part of the deceleration within the through lanes. Each component of the auxiliary length is discussed in the following section.

Deceleration Length

Provision for deceleration clear of the through-traffic lanes is a desirable objective on arterial roads and streets and should be incorporated into design, whenever practical. The approximate total lengths needed for a comfortable deceleration to a stop from the full design speed of the highway are as follows: for design speeds of 50, 60, 70, 80, and 90 km/h [30, 40, 45, 50, and 55 mph], the desirable deceleration lengths of the auxiliary lane are 70, 100, 130, 165, and 205 m [230, 330, 430, 550, and 680 ft], respectively (**15**). These approximate lengths are based on grades of less than 3 percent.

On many urban facilities, it is not practical to provide full length of auxiliary lane for deceleration and, in many cases, the storage length overrides the deceleration length. In such cases, at least part of the deceleration must be accomplished before entering the auxiliary lane. Inclusion of the taper length as part of the deceleration distance for an auxiliary lane assumes that an approaching turning vehicle can decelerate comfortably up to 15 km/h [10 mph] in a through lane before entering the auxiliary lane. Shorter auxiliary lane lengths will increase the speed differential between turning vehicles and through traffic. A 15 km/h [10 mph] differential is commonly considered acceptable on arterial roadways. Higher speed differentials may be acceptable on collector highways and streets due to higher levels of driver tolerance for vehicles leaving or entering the roadway due to slow speeds or high volumes. Therefore, the lengths given above should be accepted as a desirable goal and should be provided where practical. The deceleration lengths stated above are applicable to both left- and right-turning lanes, but the approach speed is usually lower in the right lane than in the left lane.

Storage Length

The auxiliary lane should be sufficiently long to store the number of vehicles likely to accumulate during a critical period. The storage length should be sufficient to avoid the possibility of left-turning vehicles stopping in the through lanes waiting for a signal change or for a gap in the opposing traffic flow.

At unsignalized intersections, the storage length, exclusive of taper, may be based on the number of turning vehicles likely to arrive in an average two-minute period within the peak hour. Space for at least two passenger cars should be provided; with over 10 percent truck traffic, provisions should be made for at least one car and one truck. The two-minute waiting time may need to be changed to some other interval that depends largely on the opportunities for completing the left-turn maneuver. These intervals, in turn, depend on the volume of opposing traffic. Where the volume of turning traffic is high, a traffic signal will often be needed.

At signalized intersections, the storage length needed depends on the signal cycle length, the signal phasing arrangement, and the rate of arrivals and departures of left-turning vehicles. The storage length is a function of the probability of occurrence of events and should usually be based on one and one-half to two times the average number of vehicles that would store per cycle, which is predicated on the design volume. This length will be sufficient to serve heavy surges that occur from time to time. As in the case of unsignalized intersections, provision should be made for storing at least two vehicles. Traffic signal design fundamentals are discussed further in MUTCD (**9**).

Where turning lanes are designed for two-lane operation, the storage length is reduced to approximately one-half of that needed for single-lane operation. For further information, refer to the HCM (**6**).

Taper

On high-speed highways it is common practice to use a taper rate that is between 8:1 and 15:1 (longitudinal:transverse or L:T). Long tapers approximate the path drivers follow when entering an auxiliary lane from a high-speed through lane. However, long tapers tend to entice some through drivers into the deceleration lane—especially when the taper is on a horizontal curve. Long tapers constrain the lateral movement of a driver desiring to enter the auxiliary lanes. This problem primarily occurs on urban curbed roadways.

For urbanized areas, short tapers appear to produce better "targets" for the approaching drivers and to give more positive identification to an added auxiliary lane. Short tapers are preferred for deceleration lanes at urban intersections because of slow speeds during peak periods. The total length of taper and deceleration length should be the same as if a longer taper was used. This results in a longer length of full-width pavement for the auxiliary lane. This type of design may reduce the likelihood that entry into the auxiliary lane may spill back into the through lane. Municipalities and urban counties are increasingly adopting the use of taper lengths such as 30 m [100 ft] for a single-turn lane and 45 m [150 ft] for a dual-turn lane for urban streets.

Some agencies permit the tapered section of deceleration auxiliary lanes to be constructed in a "squared-off" section at full paving width and depth. This configuration involves a painted delineation of the taper. The abrupt squared-off beginning of deceleration exits offers improved driver commitment to the exit maneuver and also contributes to driver security because of the elimination of the unused portion of long tapers. The design involves transition of the outer or median shoulders around the squared-off beginning of the deceleration lane.

The squared-off design principle can be applied to median deceleration lanes, and it can also be used at the beginning of deceleration right-turn exit terminals when there is a single exit lane.

When two or more exit lanes are used, the tapered designs discussed in the section on "Speed-Change Lanes" in Chapter 10 are recommended. Additional guidance for lengths of tapers may be found in the MUTCD (**9**).

The longitudinal location along the highway, where a vehicle will move from the through lane to a full-width deceleration lane, will vary depending on many factors. These factors include the type of vehicle, the driving characteristics of the vehicle operator, the speed of the vehicle, weather conditions, and lighting conditions.

Straight-line tapers are frequently used, as shown in Exhibit 9-95A. The taper rate may be 8:1 [L:T] for design speeds up to 50 km/h [30 mph] and 15:1 [L:T] for design speeds of 80 km/h [50 mph]. Straight-line tapers are particularly applicable where a paved shoulder is striped to delineate the auxiliary lane. Short, straight-line tapers should not be used on curbed urban streets because of the probability of vehicles hitting the leading end of the taper with the resulting potential for a driver losing control. A short curve is desirable at either end of long tapers as shown in Exhibit 9-95B, but may be omitted for ease of construction. Where curves are used at the ends, the tangent section should be about one-third to one-half of the total length.

Symmetrical reverse curve tapers are commonly used on curbed urban streets. Exhibit 9-95C shows a design taper with symmetrical reverse curves.

A more desirable reverse-curve taper is shown in Exhibit 9-95D where the turnoff curve radius is about twice that of the second curve. When 30 m [100 ft] or more in length is provided for the tapers in Exhibit 9-95D, tapers 1 and 2 would be suitable for low-speed operations. All the dimensions and configurations shown in Exhibit 9-95 are applicable to right-turn lanes as well as left-turn lanes.

Median Left-Turn Lanes

A median left-turn lane is an auxiliary lane for storage or speed change of left-turning vehicles located at the left of one-directional roadway within a median or divisional island. Inefficiencies in operations may be evident on divided highways where such lanes are not provided. Median lanes, therefore, should be provided at intersections and at other median openings where there is a high volume of left turns or where the vehicular speeds are high. Minimum designs of median openings are shown in Exhibits 9-77 through 9-84. Median lane designs for various widths of median are shown in Exhibits 9-96 and 9-97.

Median widths of 6 m [20 ft] or more are desirable at intersections with single median lanes, but widths of 4.8 to 5.4 m [16 to 18 ft] permit reasonably adequate arrangements. Where two median lanes are used, a median width of at least 8.4 m [28 ft] is desirable to permit the installation of two 3.6-m [12-ft] lanes and a 1.2-m [4-ft] separator. Although not equal in width to a normal traveled lane, a 3.0-m [10-ft] lane with a 0.6-m [2-ft] curbed separator or with traffic buttons or paint lines, or both, separating the median lane from the opposing through lane may be acceptable where speeds are low and the intersection is controlled by traffic signals.

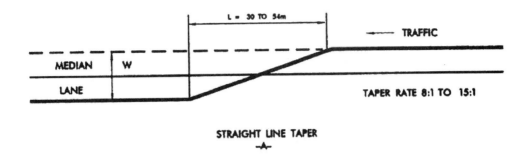

STRAIGHT LINE TAPER
-A-

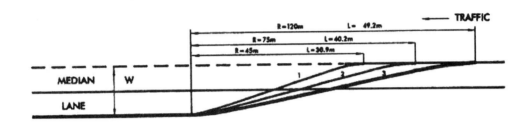

PARTIAL TANGENT TAPER
-B-

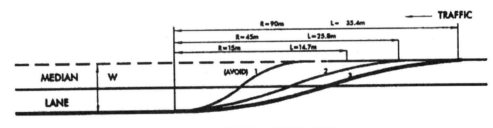

SYMMETRICAL REVERSE CURVE
-C-

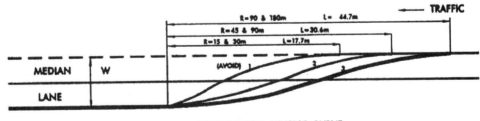

ASYMMETRICAL REVERSE CURVE
-D-

DIMENSIONS ARE ALSO APPLICABLE FOR RIGHT-TURN FLARES.

Exhibit 9-95. Taper Design for Auxiliary Lanes (Metric)

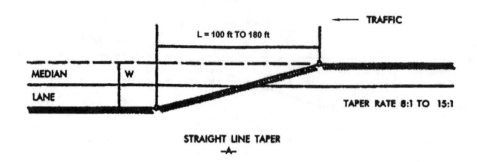

STRAIGHT LINE TAPER
-A-

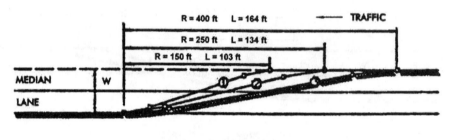

PARTIAL TANGENT TAPER
-B-

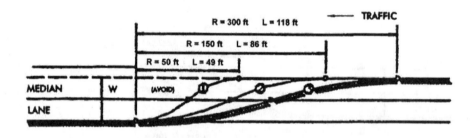

SYMMETRICAL REVERSE CURVE
-C-

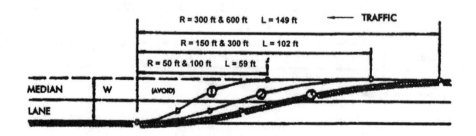

ASYMMETRICAL REVERSE CURVE
-D-

DIMENSIONS ARE ALSO APPLICABLE FOR RIGHT-TURN FLARES.

Exhibit 9-95. Taper Design for Auxiliary Lanes (U.S. Customary) (Continued)

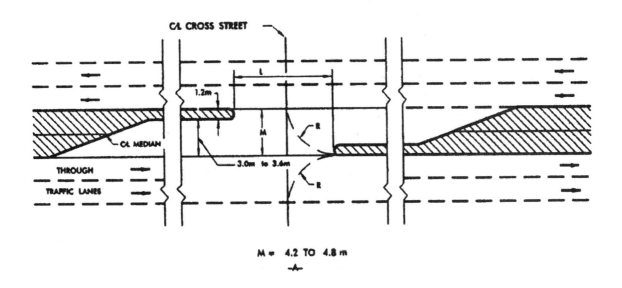

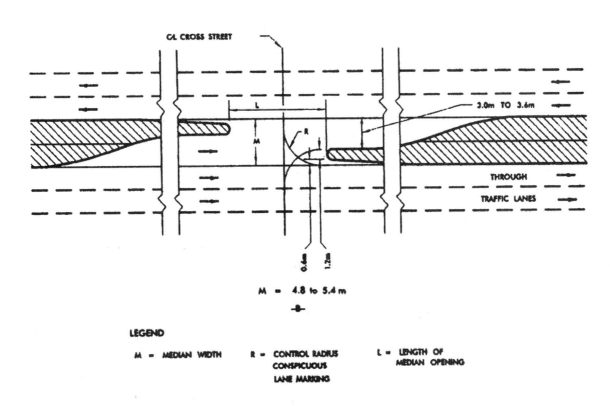

Exhibit 9-96. 4.2 to 5.4 m [14 to 18 ft] Median Width Left-Turn Design (Metric)

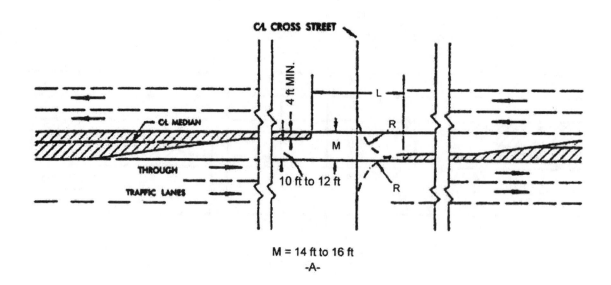

M = 14 ft to 16 ft
-A-

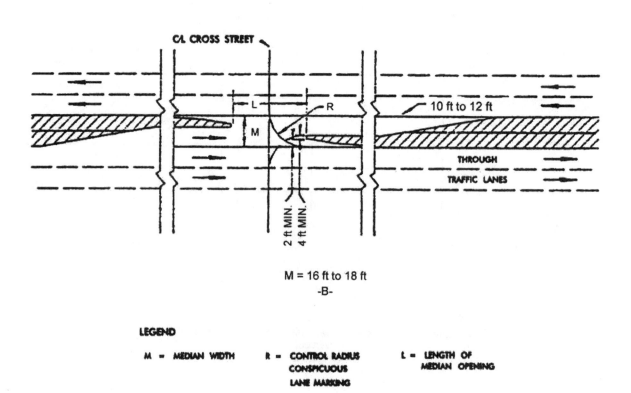

M = 16 ft to 18 ft
-B-

LEGEND

M = MEDIAN WIDTH   R = CONTROL RADIUS   L = LENGTH OF
                        CONSPICUOUS   MEDIAN OPENING
                        LANE MARKING

Exhibit 9-96. 4.2 to 5.4 m [14 to 18 ft] Median Width Left-Turn Design (U.S. Customary)
(Continued)

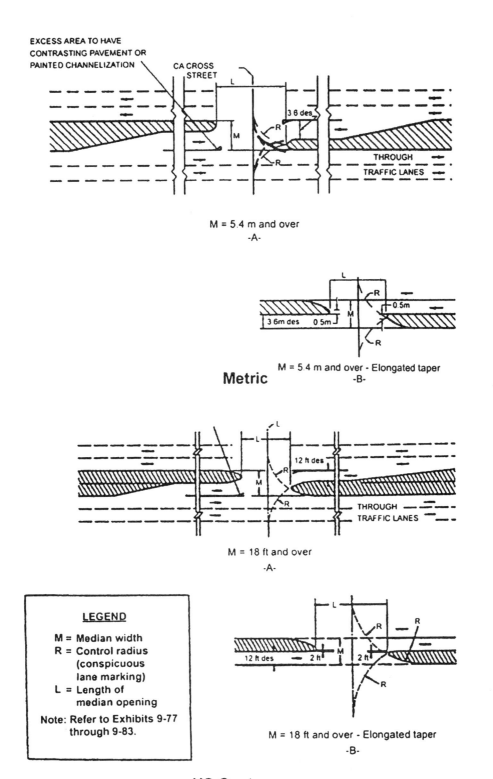

Exhibit 9-97. Median Left-Turn Design for Median Width in Excess of 5.4 m [18 ft]

Exhibit 9-96A shows a minimum design for a median left-turn lane within a median 4.2 to 4.8 m [14 to 16 ft] wide. A curbed divider width of 1.2 m [4 ft] is recommended, and the median left-turn lane should be 3.0 to 3.6 m [10 to 12 ft] wide. Exhibit 9-96B shows a typical median left-turn design within a median width of 4.8 to 5.4 m [16 to 18 ft]. The only change in this design from that in Exhibit 9-96A is a 0.6-m [2-ft] minimum offset to the approach nose. Exhibit 9-97 illustrates a more liberal median left-turn design within a median width of 5.4 m [18 ft] or more. On these medians the elongated tapers may be desirable. For medians 5.4 m [18 ft] wide or more, a flush, color-contrasted divider is recommended to delineate the area between the turning lane and the adjacent through lane in the same direction of travel.

Pavement markings, contrasting pavement texture, signs, and physical separators may be used to discourage the through driver from inadvertently entering the wrong lane.

Where two median lanes are used, left-turning vehicles leave the through lanes to enter the median lanes in single file, but once within the median lanes, the vehicles are stored in two lanes. On receiving the green indication, the left-turning vehicles turn simultaneously from both lanes.

With three-phase signal control, such an arrangement results in an increase in capacity of approximately 180 percent of that of a single median lane. Occasionally, there are operational problems as a result of the two-abreast turns, especially the problems of sideswipe crashes. These usually result from too sharp a turning radius or a roadway that is too narrow. The receiving leg of the intersection should have adequate width to accommodate two lanes of turning traffic. A width of 9 m [30 ft] is used by several highway agencies. Double turning lanes should only be used with signalization providing a separate turning phase.

Median End Treatment

The form of treatment given the end of the narrowed median adjacent to lanes of opposing traffic depends largely on the available width. The narrowed median may be curbed to delineate the lane edge; to separate opposing movements; to provide space for signs, markers, and luminaire supports; and to protect pedestrians. For a discussion on "ramped down" approaches to curb medians, reference can be made to the earlier section in this chapter, "Islands." To serve these purposes satisfactorily, the minimum narrowed median width of no less than 1.2 m [4 ft] is recommended and is preferably 1.8 to 2.4 m [6 to 8 ft] wide. These dimensions can be provided within a median 4.8 to 5.4 m [16 to 18 ft] wide and a turning lane width of 3.6 m [12 ft].

For medians wider than about 5.4 m [18 ft], as shown in Exhibit 9-97, it is usually preferable to provide some offset between the left-turn lanes in the opposing directions of travel. Offset left-turn lanes of this type are discussed in the next section.

For curbed dividers 1.2 m [4 ft] or more in width at the narrowed end, the curbed nose can be offset from the opposing through-traffic lane 0.6 m [2 ft] or more, with gradual taper beyond to make it less vulnerable to contact by through traffic, as shown in Exhibit 9-96B. The shape of the nose for curbed dividers 1.2 m [4 ft] wide usually is semicircular, but for a wider width the ends are normally shaped to a bullet nose pattern to conform better with the paths of turning vehicles.

Offset Left-Turn Lanes

For medians wider than about 5.4 m [18 ft], it is desirable to offset the left-turn lane so that it will reduce the width of the divider to 1.8 to 2.4 m [6 to 8 ft] immediately before the intersection, rather than to align it exactly parallel with and adjacent to the through lane. This alignment will place the vehicle waiting to make the turn as far to the left as practical, maximizing the offset between the opposing left-turn lanes, and thus providing improved visibility of opposing through traffic. The advantages of offsetting the left-turn lanes are (1) better visibility of opposing through traffic; (2) decreased possibility of conflict between opposing left-turn movements within the intersection; and (3) more left-turn vehicles served in a given period of time, particularly at a signalized intersection (**16**). Parallel offset left-turn lanes may be used at both signalized and unsignalized intersections. This left-turn lane configuration is referred to as a parallel offset left-turn lane and is illustrated in Exhibit 9-98A.

An offset between opposing left-turn vehicles can also be achieved with a left-turn lane that diverges from the through lanes and crosses the median at a slight angle. Exhibit 9-98B illustrates a tapered offset left-turn lane of this type. Tapered offset left-turn lanes provide the same advantages as parallel offset left-turn lanes in reducing sight distance obstructions and potential conflicts between opposing left-turn vehicles and in increasing the efficiency of signal operations. Tapered offset left-turn lanes are normally constructed with a 1.2-m [4-ft] nose between the left-turn lane and the opposing through lanes. Tapered offset left-turn lanes have been used primarily at signalized intersections.

This type offset is especially effective for turning radii allowance where trucks with long rear overhangs, such as logging trucks, are turning from the mainline roadway. This same type of offset geometry may also be used for trucks turning right with long rear overhangs.

Parallel and tapered offset left-turn lanes should be separated from the adjacent through traffic lanes by painted or raised channelization.

SIMULTANEOUS LEFT TURNS

Simultaneous left turns may be considered at an intersection of two major highways. Exhibit 9-99 indicates traffic patterns that should be considered in the design. Marking details are given in the MUTCD (**9**).

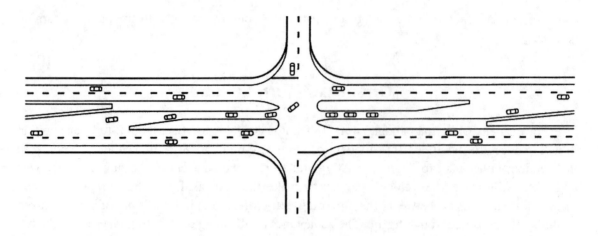

-A- PARALLEL

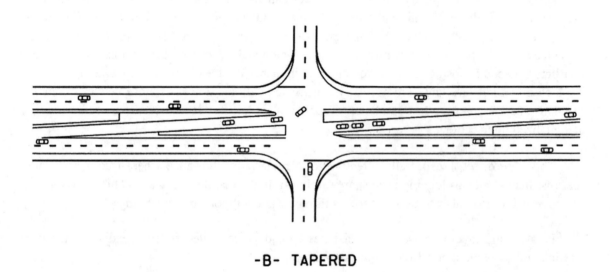

-B- TAPERED

Exhibit 9-98. Parallel and Tapered Offset Left-Turn Lane

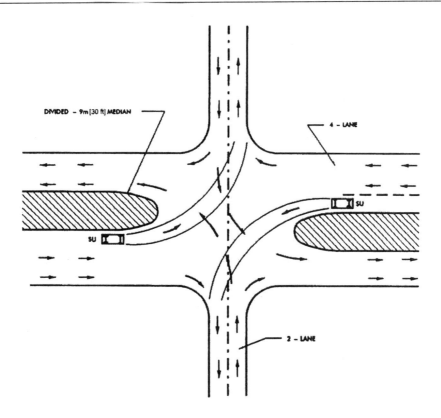

Exhibit 9-99. Four-Leg Intersection Providing Simultaneous Left Turns

INTERSECTION DESIGN ELEMENTS WITH FRONTAGE ROADS

Frontage road cross-sectional elements, functional characteristics, and service value as collectors are discussed in Chapters 4, 6, and 7. The discussion to follow concerns frontage road design elements with respect to the operational features where the frontage road intersects the major highway. Frontage roads are generally needed adjacent to arterials or freeways where adjacent property owners are not permitted direct access to the major facility. Short lengths of frontage roads may be desirable along urban arterials to preserve the capacity of the arterial through control of access. Much of the improvement in capacity may be offset by the added conflicts introduced where the frontage road and arterial intersect the crossroad. Not only is there an increase in the number of conflicting movements, but the confusing pattern of roadways and separations can lead to wrong-way entry. Inevitably, where an arterial is flanked by frontage roads, the problems of design and traffic control at intersections are far more complex than where the arterial consists of a single roadway. Three intersections (two, if there is only one frontage road) actually exist at each cross street.

In lightly developed areas, such as through single-family residential neighborhoods, an intersection designed to fit minimum turning paths of passenger vehicles may operate satisfactorily. In heavily developed areas, however, particularly through commercial districts where frontage roads receive heavy use, an intersection designed with restricted geometrics will seldom operate satisfactorily unless certain traffic movements are prohibited. Separate signal indications can be used to relieve some of the conflicts between the various movements but only at the expense of delay to most of the traffic.

The preferred alternative to restricting turns is to design the intersection with expanded dimensions, particularly the width of outer separation. This design permits the intersections between the crossroad and frontage roads to be well removed from the crossroad intersection with the main lanes.

For satisfactory operation with moderate-to-heavy traffic volumes on the frontage roads, the outer separation should be 50 m [150 ft] or more in width at the intersection. The 50-m [150-ft] dimension is derived on the basis of the following considerations:

- This dimension is about the shortest acceptable length needed for placing signs and other traffic control devices to provide proper direction to traffic on the crossroad.
- It usually affords acceptable storage space on the crossroad in advance of the main intersection to avoid blocking the frontage road.
- It enables turning movements to be made from the main lanes onto frontage roads without seriously disrupting the orderly movement of traffic.
- It facilitates U-turns between the main lanes and two-way frontage roads. (Such a maneuver is geometrically possible with a somewhat narrower separation but is extremely difficult with commercial vehicles).
- It alleviates the potential of wrong-way entry onto through lanes of the predominant highway.

However, wider separations can enhance operations significantly. Outer separations of 100 m [300 ft] allow for overlapping left turn lanes and provide a minimal amount of vehicle storage. The design year traffic volumes, turning movements, signal phasing, and storage requirements should determine the ultimate outer separation distance.

Narrower separations are acceptable where frontage-road traffic is very light, where the frontage road operates one-way only, or where some movements can be prohibited. Turning movements that are affected most by the width of outer separation are (1) left turns from the frontage road onto the crossroad, (2) U-turns from the through lanes of the predominant highway onto a two-way frontage road, and (3) right turns from the through lanes of the predominant highway onto the crossroad. By imposing the restrictions, as may be appropriate on some or all of these movements, outer separations as narrow as 2.4 m [8 ft] may operate satisfactorily. With such narrow separations caution should be exercised in assessing the risk of wrong-way entry onto the through lanes.

Except for the width of the outer separation, the design elements for intersections involving frontage roads are much the same as those for conventional intersections. Exhibit 9-100 shows two arrangements of highways with frontage roads intersecting cross streets. Turning movements are shown on the assumption that frontage road volumes are very light and that all movements will be under traffic signal control.

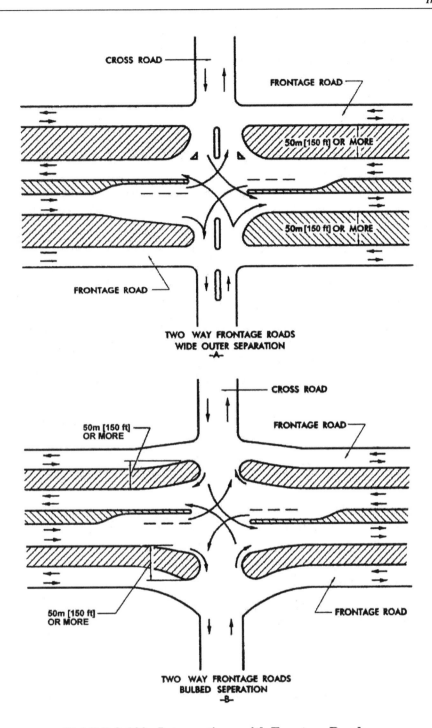

Exhibit 9-100. Intersections with Frontage Roads

As illustrated, deceleration or storage lanes may be provided for the right-turn movements adjacent to the through roadway. Because traffic turning right must cross the path of traffic on the frontage road, the need for such storage lanes is usually greater in this case than in the case of conventional intersections. Storage and speed-change lanes are clearly delineated by pavement markings. Contrasting surfaces are also desirable.

Exhibit 9-100A shows a simple intersection design with an outer separation of 50 m [150 ft] or more in width. The intersections of the two-way frontage roads and the crossroad are sufficiently removed from the through roadways that they might operate as separate intersections. The major elements in the design of the outer intersections are adequate width, adequate radii for right turns, and divisional islands on the crossroad.

Exhibit 9-100B shows a design that would be adaptable for two-way frontage roads in areas where right-of-way considerations would preclude the design shown in Exhibit 9-100A. With narrow outer separations between intersections, a bulb treatment of the outer separations, as shown, formed by a reverse-curve alignment of the frontage road on each side of the crossroad is needed to widen the outer separation to a desirable width at the crossroad. The length of the reverse curve is a matter of frontage road design, governed by design speeds and right-of-way controls. The widths of the outer separation bulbs should be based on the pattern and volumes of traffic, but the right-of-way controls may also govern because additional area is needed at the intersection. The width of outer separation at the crossroad opening should be at least 18 m [60 ft], which might be acceptable for light-to-moderate frontage road traffic, but preferably it should be 50 m [150 ft] or more. A width of 9.6 m [32 ft] for the outer separation is the minimum that will permit a U-turn by a passenger car from the through lanes onto the frontage road. Widths of 22 m [74 ft] or more are needed for trucks and buses. Where such movements are likely to occur frequently, the width of separations should be considerably greater, desirably 50 m [150 ft] or more.

BICYCLES AT INTERSECTIONS

When on-street bicycle lanes and/or off-street bicycle paths enter an intersection, the design of the intersection should be modified accordingly. These modifications may include special sight distance considerations, wider roadways to accommodate on-street lanes, special lane markings to channelize and separate bicycles from right-turning vehicles, provisions for left-turn bicycle movements, or special traffic signal designs (such as conveniently located push buttons at actuated signals or even separate signal indications for bicyclists). Further guidance in providing for bicycles at intersections can be found in the AASHTO *Guide for the Development of Bicycle Facilities* (**5**).

WHEELCHAIR RAMPS AT INTERSECTIONS

When designing a project that involves curbs and adjacent sidewalks to accommodate pedestrian traffic, proper attention should be given to the needs of persons with disabilities, whose mobility depends on wheelchairs and other devices. Related design criteria and illustrations are given in Chapter 4.

LIGHTING AT INTERSECTIONS

Lighting may affect the safety of highway and street intersections, as well as efficiency of traffic operations. Statistics indicate that the nighttime crash rates are higher than that during daylight hours. This fact, to a large degree, may be attributed to impaired visibility. In urban and suburban areas where there are concentrations of pedestrians and roadside and intersectional interferences, fixed-source lighting tends to reduce crashes. Whether or not rural intersections should be lighted depends on the planned geometrics and the turning volumes involved. Intersections that are not channelized are seldom lighted. However, for the benefit of nonlocal highway users, lighting at rural intersections is desirable to aid the driver in ascertaining sign messages during non-daylight periods.

Intersections with channelization, particularly multiple-road geometrics, should include lighting. Large channelized intersections especially need illumination because of the higher range of turning radii that are not within the lateral range of vehicular headlight beams. Vehicles approaching the intersection should also reduce speed. The indication of this need should be definite and visible at a distance from the intersection that may be beyond the range of headlights. Illumination of intersections with fixed-source lighting accomplishes this need.

The planned location of intersection luminaire supports should be designed in accordance with current roadside safety concepts. Additional discussions and design guidance can be found in NCHRP Report 152 (**17**) and the AASHTO *Roadside Design Guide* (**18**).

DRIVEWAYS

Driveways are, in effect, intersections and should be designed consistent with their intended use. For further discussion of driveways, refer to Chapter 4. The number of crashes is disproportionately higher at driveways than at other intersections; thus their design and location merit special consideration.

Ideally, driveways should not be located within the functional area of an intersection or in the influence area of an adjacent driveway. The functional area extends both upstream and downstream from the physical intersection area and includes the longitudinal limits of auxiliary lanes. The influence area associated with a driveway includes (1) the impact length (the distance back from a driveway that cars begin to be affected), (2) the perception-reaction distance, and (3) the car length.

The spacing of driveways should reflect the impact lengths and influence areas associated with motorists entering or leaving a driveway. The impact length represents the distance upstream when the brake lights of through vehicles are activated or there is a lane change due to a turning vehicle.

The impact lengths associated with motorists entering or leaving a driveway should be considered in establishing driveway separation distances. Exhibit 9-101 shows impact lengths based on recent research relating to vehicles making right turns into driveways (**1**). For example, 20 percent of the right-lane through vehicles were impacted at an approximate distance of 52 m

[172 ft] or more in advance of a driveway at 50 km/h [30 mph] speed. At 80 km/h [50 mph], 20 percent of the right-lane through vehicles were impacted a distance of 105 m [345 ft] or more in advance of a right-turn location. Influence areas can be obtained by adding the perception-reaction distance and car length to the distance shown in Exhibit 9-101. The functional area of an intersection should reflect these influence areas.

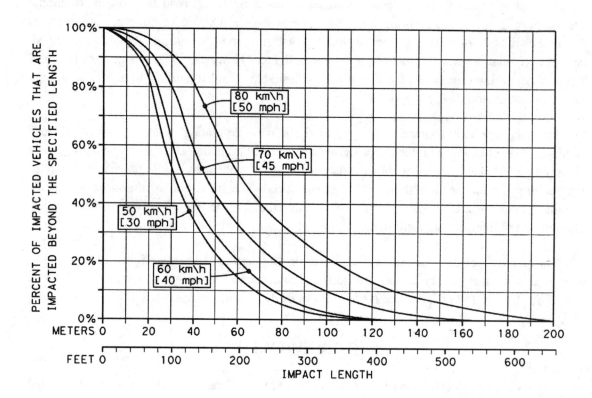

Exhibit 9-101. Cumulative Frequency Distribution of Impact Lengths

When access points are signalized, their location should fit in the time-space pattern of adjacent major intersections to the maximum extent practical.

The regulation and design of driveways are intimately linked with the type of road and zoning of the roadside. On new highways, right-of-way can be obtained to provide the desired degree of driveway regulation and control. In some cases, additional right-of-way can be acquired with the reconstruction of an existing highway or agreements can be made to improve existing undesirable access conditions. Often the desired degree of driveway control should be effected through the use of police powers to require permits for all new driveways, through adjustments of existing driveways, or through access-management regulations.

The main objectives of driveway regulation are to provide desirable spacing of driveways and to ensure that a proper internal layout is being proposed. Achieving these objectives depends on the type and extent of legislative authority granted the highway agency. Many states and cities have developed policies for driveways and have separate units to handle the design details that are incidental to checking requests and issuing permits for new driveways or requested changes to existing driveway connections. Major controls and design features are discussed in other reference sources (**19, 20, 21**).

RAILROAD-HIGHWAY GRADE CROSSINGS

A railroad-highway crossing, like any highway-highway intersection, involves either a separation of grades or a crossing at-grade. The geometrics of a highway and structure that involves the overcrossing or undercrossing of a railroad are substantially the same as those for a highway grade separation without ramps.

The horizontal and vertical geometrics of a highway approaching a railroad grade crossing should be constructed in a manner that does divert driver attention to roadway conditions.

Horizontal Alignment

If practical, the highway should intersect the tracks at a right angle with no nearby intersections or driveways. This layout enhances the driver's view of the crossing and tracks, reduces conflicting vehicular movements from crossroads and driveways, and is preferred for bicyclists. To the extent practical, crossings should not be located on either highway or railroad curves. Roadway curvature inhibits a driver's view of a crossing ahead, and a driver's attention may be directed toward negotiating the curve rather than looking for a train. Railroad curvature may inhibit a driver's view down the tracks from both a stopped position at the crossing and on the approach to the crossings. Those crossings that are located on both highway and railroad curves present maintenance problems and poor rideability for highway traffic due to conflicting superelevations.

Where highways that are parallel with main tracks intersect highways that cross the main tracks, there should be sufficient distance between the tracks and the highway intersections to enable highway traffic in all directions to move expeditiously. Where physically restricted areas make it impossible to obtain adequate storage distance between the main track and a highway intersection, the following should be considered:

- Interconnection of the highway traffic signals with the grade crossing signals to enable vehicles to clear the grade crossing when a train approaches.
- Placement of a "Do Not Stop on Track" sign on the roadway approach to the grade crossing.

Vertical Alignment

It is desirable from the standpoint of sight distance, rideability, braking, and acceleration distances that the intersection of highway and railroad be made as level as practical. Vertical curves should be of sufficient length to ensure an adequate view of the crossing.

In some instances, the roadway vertical alignment may not meet acceptable geometrics for a given design speed because of restrictive topography or limitations of right-of-way. To prevent drivers of low-clearance vehicles from becoming caught on the tracks, the crossing surface should be at the same plane as the top of the rails for a distance of 0.6 m [2 ft] outside the rails. The surface of the highway should also not be more than 75 mm [3 in] higher or lower than the top of nearest rail at a point 9 m [30 ft] from the rail unless track superelevation makes a different level appropriate, as shown in Exhibit 9-102. Vertical curves should be used to traverse from the highway grade to a level plane at the elevation of the rails. Rails that are superelevated, or a roadway approach section that is not level, will necessitate a site specific analysis for rail clearances.

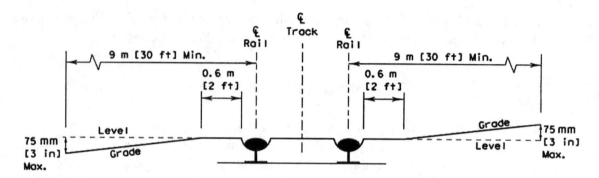

Exhibit 9-102. Railroad-Highway Grade Crossing

General

The geometric design of railroad-highway grade crossings should be made jointly when determining the warning devices to be used. When only passive warning devices such as signs and pavement markings are used, the highway drivers are warned of the crossing location but must determine whether or not there are train movements for which they should stop. On the other hand, when active warning devices such as flashing light signals or automatic gates are used, the driver is given a positive indication of the presence or the approach of a train at the crossing. A large number of significant variables should be considered in determining the type of warning device to be installed at a railroad grade crossing. For certain low-volume highway crossings where adequate sight distance is not available, additional signing may be needed.

Traffic control devices for railroad-highway grade crossings consist primarily of signs, pavement markings, flashing light signals, and automatic gates. Criteria for design, placement, installment, and operation of these devices are covered in the MUTCD (9), as well as the use of various passive warning devices. Some of the considerations for evaluating the need for active warning devices at a grade crossing include the type of highway, volume of vehicular traffic, volume of railroad traffic, maximum speed of the railroad trains, permissible speed of vehicular traffic, volume of pedestrian traffic, crash history, sight distance, and geometrics of the crossing. The potential for complete elimination of grade crossings without active traffic control devices (e.g., closing lightly used crossings and installing active devices at other more heavily used crossings) should be given prime consideration.

These guidelines are not all inclusive. Situations not covered by these guidelines should be evaluated using good engineering judgment. Additional information on railroad-highway grade crossings can be found in various published sources (**22, 23, 24, 25, 26, 27**).

Numerous index formulas have been developed to assess the relative conflict potential at railroad-highway grade crossings on the basis of various combinations of its characteristics. Although no single formula has universal acceptance, each has its own values in establishing an index; when used with sound engineering judgment, each formula provides a basis for a selection of the type of warning devices to be installed at a given crossing.

The geometric design of a railroad-highway grade crossing involves the elements of alignment, profile, sight distance, and cross section. The appropriate design may vary with the type of warning device used. Where signs and pavement markings are the only means of warning, the highway should cross the railroad at or nearly at right angles. Even when flashing lights or automatic gates are used, small intersection angles should be avoided. Regardless of the type of control, the roadway gradient should be flat at and adjacent to the railroad crossing to permit vehicles to stop, when necessary, and then proceed across the tracks without difficulty.

Sight distance is a primary consideration at crossings without train-activated warning devices. A complete discussion of sight distance at grade crossings can be found in two published sources (**24, 27**).

As in the case of a highway intersection, there are several events that can occur at a railroad-highway grade intersection without train-activated warning devices. Two of these events related to determining the sight distance are:

- The vehicle operator can observe the approaching train in a sight line that will allow the vehicle to pass through the grade crossing prior to the train's arrival at the crossing.
- The vehicle operator can observe the approaching train in a sight line that will permit the vehicle to be brought to a stop prior to encroachment in the crossing area.

Both of these maneuvers are shown as Case A of Exhibit 9-103. The sight triangle consists of the two major legs (i.e., the sight distance, d_H, along the highway and the sight distance, d_T, along the railroad tracks). Case A of Exhibit 9-104 indicates values of the sight distances for various speeds of the vehicle and the train. These distances are developed from two basic formulas:

Metric	US Customary
$$d_H = AV_v t + \frac{BV_v^2}{a} + D + d_e$$ $$d_T = \frac{V_T}{V_V}\left[(A)V_v t + \frac{BV_v^2}{a} + 2D + L + W\right]$$	$$d_H = AV_v t + \frac{BV_v^2}{a} + D + d_e$$ $$d_T = \frac{V_T}{V_V}\left[(A)V_v t + \frac{BV_v^2}{a} + 2D + L + W\right]$$ **(9-3)**

where:

Metric	US Customary
A = constant = 0.278 B = constant = 0.039 d_H = sight-distance leg along the highway allows a vehicle proceeding to speed V_v to cross tracks even though a train is observed at a distance d_T from the crossing or to stop the vehicle without encroachment of the crossing area (m) d_T = sight-distance leg along the railroad tracks to permit the maneuvers described as for d_H (m) V_v = speed of the vehicle (km/h) V_T = speed of the train (km/h) t = perception/reaction time, which is assumed to be 2.5 s (This is the same value used in Chapter 3 to determine the stopping sight distance.) a = driver deceleration, which is assumed to be 3.4 m/s² (This is the same value used in Chapter 3 to determine stopping sight distance.) D = distance from the stop line or front of the vehicle to the nearest rail, which is assumed to be 4.5 m d_e = distance from the driver to the front of the vehicle, which is assumed to be 2.4 m L = length of vehicle, which is assumed to be 20 m W = distance between outer rails(for a single track, this value is 1.5 m)	A = constant = 1.47 B = constant = 1.075 d_H = sight-distance leg along the highway allows a vehicle proceeding to speed V_v to cross tracks even though a train is observed at a distance d_T from the crossing or to stop the vehicle without encroachment of the crossing area (ft) d_T = sight-distance leg along the railroad tracks to permit the maneuvers described as for d_H (ft) V_v = speed of the vehicle (mph) V_T = speed of the train (mph) t = perception/reaction time, which is assumed to be 2.5 s (This is the same value used in Chapter 3 to determine the stopping sight distance.) a = driver deceleration, which is assumed to be 11.2 ft/s². (This is the same value used in Chapter 3 to determine stopping sight distance.) D = distance from the stop line or front of the vehicle to the nearest rail, which is assumed to be 15 ft d_e = distance from the driver to the front of the vehicle, which is assumed to be 8 ft L = length of vehicle, which is assumed to be 65 ft W = distance between outer rails(for a single track, this value is 5 ft)

Corrections should be made for skew crossings and highway grades that are other than flat.

METRIC

$$d_H = AV_v t + \frac{BV_v^2}{a} + D + d_e$$

$$d_T = \frac{V_T}{V_v}\left[(A)V_v t + \frac{BV_v^2}{a} + 2D + L + W\right]$$

d_H	=	Sight distance along highway
d_T	=	Sight distance along railroad tracks
V_v	=	Velocity of vehicle
t	=	Perception/reaction time (assumed 2.5 s)
f	=	Coefficient of friction time (see Exhibit 3-1)
D	=	Distance from stop line to near rail (assumed 4.5 m)
W	=	Distance between outer rails (single track W = 1.5 m)
L	=	Length of vehicle (assumed 20 m)
V_T	=	Velocity of train
d_e	=	Distance from driver to front of vehicle (assumed 2 4 m)

Adjustments must be made for skewed crossings.
Assumed flat highway grades adjacent to and at crossings.

US CUSTOMARY

$$d_H = AV_v t + \frac{BV_v^2}{a} + D + d_e$$

$$d_T = \frac{V_T}{V_v}\left[(A)V_v t + \frac{BV_v^2}{a} + 2D + L + W\right]$$

d_H	=	Sight distance along highway
d_T	=	Sight distance along railroad tracks
V_v	=	Velocity of vehicle
t	=	Perception/reaction time (assumed 2.5 s)
f	=	Coefficient of friction (see Exhibit 3-1)
D	=	Distance form stop line to near rail (assumed 15 ft)
W	=	Distance between outer rails (single track W = 5 ft)
L	=	Length of vehicle (assumed 65 ft)
V_T	=	Velocity of train
d_e	=	Distance from driver to front of vehicle (assumed 8 ft)

Adjustments must be made for skewed crossings.
Assumed flat highway grades adjacent to and at crossings.

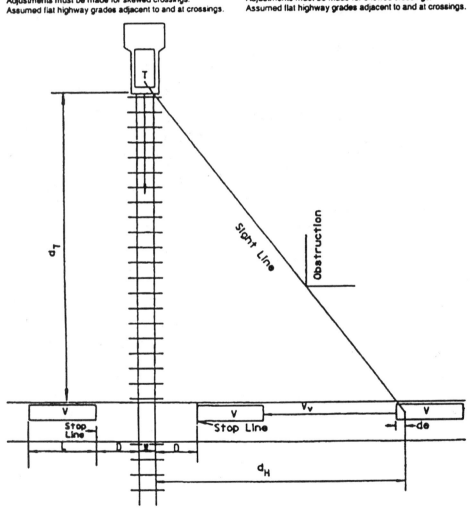

Exhibit 9-103. Case A: Moving Vehicle to Safely Cross or Stop at Railroad Crossing

Case B of Exhibit 9-104 contains various values of departure sight distance for a range of train speeds. When a vehicle has stopped at a railroad crossing, the next maneuver is to depart from the stopped position. The vehicle operator should have sufficient sight distance along the tracks to accelerate the vehicle and clear the crossing prior to the arrival of a train, even if the train comes into view just as the vehicle starts, as shown in Exhibit 9-105. These values are obtained from the formula:

Metric	US Customary
$$d_T = AV_T \left[\dfrac{V_G}{a_1} + \dfrac{L+2D+W-d_a}{V_G} + J \right]$$	$$d_T = AV_T \left[\dfrac{V_G}{a_1} + \dfrac{L+2D+W-d_a}{V_G} + J \right] \quad (9\text{-}4)$$
where: A = constant = 0.278 d_T = sight distance leg along railroad tracks to permit the maneuvers described as for d_H (m) V_T = speed of train (km/h) V_G = maximum speed of vehicle in first gear, which is assumed to be 2.7 m/s a_1 = acceleration of vehicle in first gear, which is assumed to be 0.45 m/s^2 L = length of vehicle, which is assumed to be 20 m D = distance from stop line to nearest rail, which is assumed to be 4.5 m J = sum of perception and time to activate clutch or automatic shift, which is assumed to be 2.0 s W = distance between outer rails for a single track, this value is 1.5 m $$d_a = \frac{V_G^2}{2a_1}$$ or distance vehicle travels while accelerating to maximum speed in first gear $$\left[\frac{V_G^2}{2a_1} = \frac{(2.7)^2}{(2)(0.45)} = 8.1\,m \right]$$	where: A = constant = 1.47 d_T = sight distance leg along railroad tracks to permit the maneuvers described as for d_H (ft) V_T = speed of train (mph) V_G = maximum speed of vehicle in first gear, which is assumed to be 8.8 fps a_1 = acceleration of vehicle in first gear, which is assumed to be 1.47 ft/s^2 L = length of vehicle, which is assumed to be 65 ft D = distance from stop line to nearest rail, which is assumed to be 15 ft J = sum of perception and time to activate clutch or automatic shift, which is assumed to be 2.0 s W = distance between outer rails for a single track, this value is 5 ft $$d_a = \frac{V_G^2}{2a_1}$$ or distance vehicle travels while accelerating to maximum speed in first gear $$\left[\frac{V_G^2}{2a_1} = \frac{(8.8)^2}{(2)(1.47)} = 26.3\,ft \right]$$

Adjustments should be made for skewed crossings and for highway grades other than flat.

Metric

Train speed (km/h)	Case B Departure from stop	Case A Moving vehicle — Vehicle speed (km/h) — Distance along railroad from crossing, d$_T$(m)											
	0	10	20	30	40	50	60	70	80	90	100	110	120	130
10	45	39	24	21	19	19	19	19	20	21	21	22	23	24
20	91	77	49	41	38	38	38	39	40	41	43	45	47	48
30	136	116	73	62	57	56	57	58	60	62	64	67	70	73
40	181	154	98	82	77	75	76	77	80	83	86	89	93	97
50	227	193	122	103	96	94	95	97	100	103	107	112	116	121
60	272	232	147	123	115	113	113	116	120	124	129	134	140	145
70	317	270	171	144	134	131	132	135	140	145	150	156	163	169
80	362	309	196	164	153	150	151	155	160	165	172	179	186	194
90	408	347	220	185	172	169	170	174	179	186	193	201	209	218
100	453	386	245	206	192	188	189	193	199	207	215	223	233	242
110	498	425	269	226	211	207	208	213	219	227	236	246	256	266
120	544	463	294	247	230	225	227	232	239	248	258	268	279	290
130	589	502	318	267	249	244	246	251	259	269	279	290	302	315
140	634	540	343	288	268	263	265	271	279	289	301	313	326	339
Distance along highway from crossing, d$_H$ (m)		15	25	38	53	70	90	112	136	162	191	222	255	291

US Customary

Train speed (mph)	Case B Departure from stop	Case A Moving vehicle — Vehicle speed (mph) — Distance along railroad from crossing, d$_T$(ft)							
	0	10	20	30	40	50	60	70	80
10	240	146	106	99	100	105	111	118	126
20	480	293	212	198	200	209	222	236	252
30	721	439	318	297	300	314	333	355	378
40	961	585	424	396	401	419	444	473	504
50	1201	732	530	494	501	524	555	591	630
60	1441	878	636	593	601	628	666	709	756
70	1681	1024	742	692	701	733	777	828	882
80	1921	1171	848	791	801	838	888	946	1008
90	2162	1317	954	890	901	943	999	1064	1134
Distance along highway from crossing, d$_H$ (ft)		69	135	220	324	447	589	751	931

Exhibit 9-104. Required Design Sight Distance for Combination of Highway and Train Vehicle Speeds; 20-m [65-ft] Truck Crossing a Single Set of Tracks at 90°

METRIC

$$d_T = AV_T \left[\frac{V_G}{a_1} + \frac{L + 2D + W - d_a}{V_G} + J \right]$$

d_T	=	Sight distance along railroad tracks to allow a stopped vehicle to depart and safely cross the railroad tracks
V_T	=	Velocity of train
V_G	=	Maximum speed of vehicle in first gear (assumed 2.7 m/s)
a_1	=	Acceleration of vehicle in first gear (assumed 0.45 m/s²)
$d_a = \dfrac{V_G{}^2}{2a}$	=	Or distance vehicle travels while accelerating to maximum speed in first gear
D	=	Distance form stop line to near rail (assumed 4.5 m)
W	=	Distance between outer rails (single track W = 1.5 m)
L	=	Length of vehicle (assumed 20 m)
J	=	Perception/reaction time (assumed 2.0 s)

Adjustments must be made for skewed crossings.
Assumed flat highway grades adjacent to and at crossings.

US CUSTOMARY

$$d_T = AV_T \left[\frac{V_G}{a_1} + \frac{L + 2D + W - d_a}{V_G} + J \right]$$

d_T	=	Sight distance along railroad tracks to allow a stopped vehicle to depart and safely cross the railroad tracks
V_T	=	Velocity of train
V_G	=	Maximum speed of vehicle in first gear (assumed 8.8 fps)
a_1	=	Acceleration of vehicle in first gear (assumed 1.47 ft)
$d_a = \dfrac{V_G{}^2}{2a}$	=	Or distance vehicle travels while accelerating to maximum speed in first gear
D	=	Distance form stop line to near rail (assumed 15 ft)
W	=	Distance between outer rails (single track W = 5 ft)
L	=	Length of vehicle (assumed 65 ft)
J	=	Perception/reaction time (assumed 2.0 s)

Adjustments must be made for skewed crossings.
Assumed flat highway grades adjacent to and at crossings.

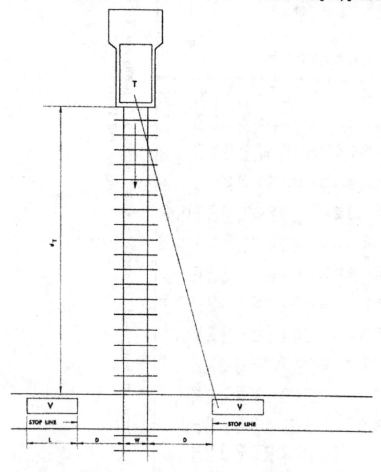

Exhibit 9-105. Case B: Departure of Vehicle From Stopped Position to Cross Single Railroad Track

Sight distances of the order shown in Exhibit 9-104 are desirable at any railroad grade crossing not controlled by active warning devices. Their attainment, however, is difficult and often impractical, except in flat, open terrain.

In other than flat terrain, it may be appropriate to rely on speed control signs and devices and to predicate sight distance on a reduced vehicle speed of operation. Where sight obstructions are present, it may be appropriate to install active traffic control devices that will bring all highway traffic to a stop before crossing the tracks and will warn drivers automatically in time for an approaching train.

The driver of a stopped vehicle at a crossing should see enough of the railroad track to be able to cross it before a train reaches the crossing, even though the train may come into view immediately after the vehicle starts to cross. The length of the railroad track in view on each side of the crossing should be greater than the product of the train speed and the time needed for the stopped vehicle to start and cross the railroad. The sight distance along the railroad track may be determined in the same manner as it is for a stopped vehicle crossing a preference highway, which is covered previously in this chapter. In order for vehicles to cross two tracks from a stopped position, with the front of the vehicle 4.5 m [15 ft] from the closest rail, sight distances along the railroad, in meters [feet], should be determined by the formula with a proper adjustment for the W value.

The highway traveled way at a railroad crossing should be constructed for a suitable length with all-weather surfacing. A roadway section equivalent to the current or proposed cross section of the approach roadway should be carried across the crossing. The crossing surface itself should have a riding quality equivalent to that of the approach roadway. If the crossing surface is in poor condition, the driver's attention may be devoted to choosing the smoothest path over the crossing. This effort may well reduce the attention given to observance of the warning devices or even the approaching train. Information concerning various surface types that may be used can be found in *Railroad-Highway Grade Crossing Surfaces* (**26**).

REFERENCES

1. Gluck, J. S., H. S. Levinson, and V. Stover. *Impacts of Access Management Techniques*, NCHRP Report 420, Washington, D.C.: Transportation Research Board, 1999.
2. Neuman, T. R. *Intersection Channelization Design Guide*, NCHRP Report 279, Washington, D.C.: Transportation Research Board, November 1985.
3. Jacquemart, G. *Modern Roundabout Practice in the United States*, NCHRP Synthesis of Highway Practice 264, Washington, D.C.: Transportation Research Board, 1998.
4. Robinson, B. W., et al. *Roundabouts: An Informational Guide*, Report No. FHWA-RD-00-067, McLean, Virginia: U.S. Department of Transportation, Federal Highway Administration, June 2000.
5. AASHTO. *Guide for the Development of Bicycle Facilities*, Washington, D.C.: AASHTO, 1999.
6. Transportation Research Board. *Highway Capacity Manual*, Special Report 209. Washington, D.C.: Transportation Research Board, 2000 or most current edition.

7. *Symposium on Geometric Design for Large Trucks, Transportation Research Record 1052,* Transportation Research Board, 1986.

8. Transportation Research Board. *Twin-Trailer Trucks,* Special Report 211, Washington, D.C.: Transportation Research Board, 1986.

9. U.S. Department of Transportation, Federal Highway Administration. *Manual on Uniform Traffic Control Devices for Streets and Highways,* Washington, D.C., 1988 or most current edition.

10. Harwood, D. W., J. M. Mason, R. E. Brydia, M. T. Pietrucha, and G. L. Gittings. *Intersection Sight Distance,* NCHRP Report 383, Washington, D.C.: Transportation Research Board, 1996.

11. Harmelink, M. D. "Volume Warrants for Left-Turn Storage Lanes at Unsignalized Grade Intersections," *Highway Research Record 211,* Highway Research Board, 1967.

12. Koepke, F. J., and H. S. Levinson. *Access Management Guidelines for Activity Centers,* NCHRP Report 348, Washington, D.C.: Transportation Research Board, 1992.

13. Pline, J. E. *Left-Turn Treatments at Intersections,* NCHRP Synthesis of Highway Practice 225, Washington, D.C.: Transportation Research Board, 1996.

14. Bonneson, J. A., and P. T. McCoy. *Capacity and Operational Effects of Midblock Left-Turn Lanes,* NCHRP Report 395, Washington, D.C.: Transportation Research Board, 1997.

15. U.S. Department of Transportation, Federal Highway Administration. "Access Management, Location and Design," National Highway Institute Course No. 15255, June 1998.

16. Harwood, D. W., M. T. Pietrucha, M. D. Wooldridge, R. E. Brydia, and K. Fitzpatrick. *Median Intersection Design,* NCHRP Report 375, Washington, D.C.: Transportation Research Board, 1995.

17. Walton, N. E., and N. J. Rowan. *Warrants for Highway Lighting,* NCHRP Report 152, Washington, D.C.: Transportation Research Board, 1974.

18. AASHTO. *Roadside Design Guide,* Washington, D.C.: AASHTO, 1996.

19. Glennon, J. C., J. J. Valenta, B. A. Thorson, and J. A. Azzeh. *General Framework for Implementing Access Control Techniques. Volume I: Technical Guidelines for the Control of Direct Access to Arterial Highways,* Report No. FHWA-RD-76-87, McLean, Virginia: U.S. Department of Transportation, Federal Highway Administration, August 1975.

20. Glennon, J. C., J. J. Valenta, B. A. Thorson, and J. A. Azzeh. *Detailed Description of Access Control Techniques. Volume II: Technical Guidelines for the Control of Direct Access to Arterial Highways,* Report No. FHWA-RD-76-87, McLean, Virginia: U.S. Department of Transportation, Federal Highway Administration, August 1975.

21. Institute of Transportation Engineers. *Guidelines for Driveway Design and Location,* Washington, D.C.: Institute of Transportation Engineers, 1986.

22. *Traffic Control Devices and Rail-Highway Grade Crossings, Transportation Research Record 1114,* Transportation Research Board, 1987.

23. Taggart, R. C., P. Lauria, G. Groat, C. Rees, and A. Brick-Turin. *Evaluating Grade-Separated Rail and Highway Crossing Alternatives,* NCHRP Report 288, Washington, D.C.: Transportation Research Board, 1987.

24. U.S. Department of Transportation, Federal Highway Administration. *Railroad-Highway Grade Crossing Handbook,* Report No. FHWA-TS-86-215, Washington, D.C.: Federal Highway Administration, September 1986.

25. Richards, H. A., and G. S. Bridges. "Railroad Grade Crossings," Chapter 1 in *Traffic Control and Roadway Elements—Their Relationship to Highway Safety*, Washington, D.C.: Automotive Safety Foundation, 1968.

26. Headley, W. J. *Railroad-Highway Grade Crossing Surfaces*, Implementation Package 79-8, Federal Highway Administration, August 1979.

27. Nichelson, G. R. "Sight Distance and Approach Speed," presented at the 1987 National Conference on Highway-Rail Safety, Association of American Railroads, September 1987.

CHAPTER 10
GRADE SEPARATIONS AND INTERCHANGES
INTRODUCTION AND GENERAL TYPES OF INTERCHANGES

The ability to accommodate high volumes of traffic safely and efficiently through intersections depends largely on the arrangements provided for handling intersecting traffic. The greatest efficiency, safety, and capacity are attained when the intersecting traveled ways are grade separated. An interchange is a system of interconnecting roadways in conjunction with one or more grade separations that provides for the movement of traffic between two or more roadways or highways on different levels.

The selection of the appropriate type of grade separation and interchange, along with its design, is influenced by many factors, such as highway classification, character and composition of traffic, design speed, and degree of access control. In addition to these controls, signing needs, economics, terrain, and right-of-way are of great importance in designing facilities with adequate capacity to safely accommodate traffic demands.

To avoid conflicts between vehicles, pedestrians, and/or bicycles within interchanges, it is preferable to separate their movements. When separation cannot be provided, each interchange site should be studied and alternate designs considered to determine the most appropriate arrangement of structures and ramps to accommodate bicycle and pedestrian traffic through the interchange area.

Interchanges vary from single ramps connecting local streets to complex and comprehensive layouts involving two or more highways. The basic interchange configurations are shown in Exhibit 10-1. Any one configuration can vary extensively in shape and scope, and there are numerous combinations of interchange types that are difficult to designate by separate names. An important element of interchange design is the assembly of one or more of the basic types of ramps, which are discussed later in this chapter. The layout for any specific ramp and type of traffic movement will reflect surrounding topography and culture, cost, and degree of flexibility in desired traffic operation. The practical aspects of topography, culture, and cost may be determining factors in the configuration and nature of ramps, but the desired traffic operation should predominate in design.

Exhibit 10-1A and 10-1B illustrate typical three-leg interchanges. Exhibit 10-1A is a trumpet interchange, named for the trumpet or jug-handle ramp configuration. Exhibit 10-1B is a three-level, directional, three-leg interchange. With ramps in one quadrant, the interchange in Exhibit 10-1C is not suitable for freeway systems but becomes very practical for an interchange between a major highway and a parkway. This design is appropriate for parkways because design speeds are usually lower, large trucks are prohibited, and turning movements are light. A typical diamond interchange is illustrated in Exhibit 10-1D. Diamond interchanges have numerous other configurations incorporating frontage roads and continuous collector or distributor roads. Exhibit 10-1E is a single-point urban interchange (SPUI). The SPUI is a form of a diamond interchange with a single signalized intersection through which all left turns utilizing the interchange must travel. All right turns into and out of ramp approaches are generally free flow. Exhibit 10-1F presents a partial cloverleaf that contains two cloverleaf loops and two diagonal ramps. Varying configurations favor heavier traffic movements. A full cloverleaf, as shown in Exhibit 10-1G, gives each interchanging movement an independent ramp; however, it generates weaving maneuvers that occur either in the area adjacent to the through lanes or on collector-distributor roads. Exhibit 10-1H illustrates a fully directional interchange.

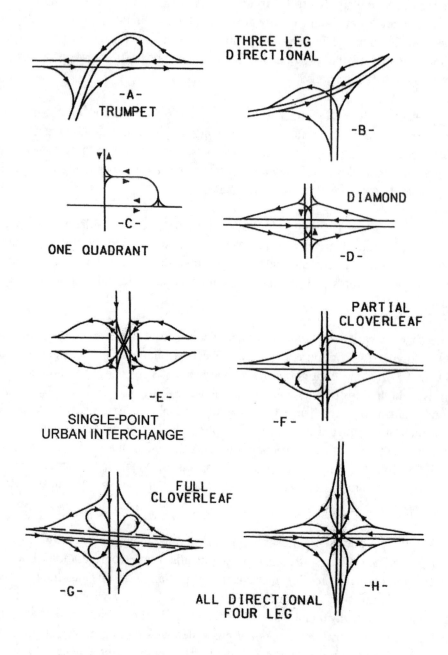

Exhibit 10-1. Interchange Configurations

WARRANTS FOR INTERCHANGES AND GRADE SEPARATIONS

An interchange can be a useful and an adaptable solution for many intersection problems either by reducing existing traffic bottlenecks or by improving safety. However, the high cost of constructing an interchange limits its use to those cases where the additional expenditure can be justified. An enumeration of the specific conditions or warrants justifying an interchange at a given intersection is difficult and, in some instances, cannot be conclusively stated. Because of the wide variety of site conditions, traffic volumes, highway types, and interchange layouts, the warrants that justify an interchange may differ at each location. The following six conditions, or warrants, should be considered when determining if an interchange is justified at a particular site:

1. **Design designation.** The determination to develop a highway with full control of access between selected terminals becomes the warrant for providing highway grade separations or interchanges for all intersecting roadways crossing the highway. Although access control, provision of medians, and elimination of parking and pedestrian traffic are important, the separation of grades on freeways provides the greatest increment of safety. Once it has been decided to develop a route as a freeway, it should be determined whether each intersecting highway will be terminated, rerouted, or provided with a grade separation or interchange. The chief concern is the continuous flow on the major road. If traffic on the minor road will cross the freeway, a grade separation or interchange is provided. Thus, an intersection that might warrant only traffic signal control, if considered as an isolated case, will warrant a grade separation or interchange when considered as a part of a freeway.

2. **Reduction of bottlenecks or spot congestion.** Insufficient capacity at the intersection of heavily traveled routes results in intolerable congestion on one or all approaches. Inability to provide essential capacity with an at-grade facility provides a warrant for an interchange where development and available right-of-way permit. Even on facilities with partial control of access, the elimination of random signalization contributes greatly to improvement of free-flow characteristics.

3. **Safety improvement.** Some at-grade intersections have a disproportionate rate of serious crashes. If inexpensive methods of eliminating crashes are likely to be ineffective or impractical, a highway grade separation or interchange may be warranted. Crash-prone intersections are frequently found at the junction of comparatively light-traveled highways in sparsely settled rural areas where speeds are high. In such areas, structures can usually be constructed at little cost compared with urban areas, right-of-way is not expensive, and lower cost improvements can be justified by the elimination of only a few serious crashes. Serious crashes at heavily traveled intersections, of course, also provide a warrant for interchange facilities. In addition to greater safety, the operational efficiency for all traffic movements is also improved at the interchange.

4. **Site topography.** At some sites, grade-separation designs are the only type of intersection that can be constructed economically. The topography at the site may be such that, to satisfy appropriate design criteria, any other type of intersection is physically impossible to develop or is equal to or greater than the cost of a grade-separated design.

5. Road-user benefits. The road-user costs due to delays at congested at-grade intersections are large. Road-user costs, such as fuel and oil usage, wear on tires, repairs, delay to motorists, and crashes that result from speed changes, stops, and waiting, are well in excess of those for intersections permitting uninterrupted or continuous operation. In general, interchanges involve somewhat more total travel distance than direct crossings at grade, but the added cost of the extra travel distance is less than the cost savings resulting from the reduction in stopping and delay. The relation of road-user benefits to the cost of improvement indicates an economic warrant for that improvement. For convenience, the relation is expressed as a ratio and represents the annual benefit divided by the annual capital cost of the improvement. Annual benefit is the difference in road-user costs between the existing and the improved condition. Annual capital cost is the sum of interest and amortization for the cost of the improvement. The larger the ratio, the greater the justification insofar as road-user benefits are concerned.

Comparison of these ratios for design alternatives is an important factor in determining the type and extent of improvement to be made. If used for justifying a single project or design, a ratio in excess of one is appropriate for minimum economic justification. Furthermore, interchanges usually are adaptable to stage construction, and initial stages may produce incremental benefits that compare even more favorably with incremental costs.

6. Traffic volume warrant. A traffic volume warrant for interchange treatment may be the most tangible of any interchange warrant. Although a specific volume of traffic at an intersection cannot be completely rationalized as the warrant for an interchange, it is an important guide, particularly when combined with the traffic distribution pattern and the effect of traffic behavior. However, volumes in excess of the capacity of an at-grade intersection would certainly be a warrant. Interchanges are desirable at cross streets with heavy traffic volumes because the elimination of conflicts due to high crossing volume greatly improves the movement of traffic.

Not all warrants for grade separations are included in the warrants for interchanges. Additional warrants for grade separations include grade separations that would:

- Serve local roads or streets that cannot practically be terminated outside the right-of-way limits of freeways.
- Provide access to areas not served by frontage roads or other means of access.
- Eliminate a railroad-highway grade crossing.
- Serve unusual concentrations of pedestrian traffic (for instance, a city park developed on both sides of a major arterial).
- Serve bikeways and routine pedestrian crossings.
- Provide access to mass transit stations within the confines of a major arterial.
- Assure free-flow operation of certain ramp configurations and serve as part of an interchange.

ADAPTABILITY OF HIGHWAY GRADE SEPARATIONS AND INTERCHANGES

The three general types of intersections are: at-grade intersections, highway grade separations without ramps, and interchanges. For each type of intersection, there is a range of situations for which the intersection is practical, but the limits of that range are not sharply defined. Furthermore, there is much overlapping between these ranges, and the final selection of intersection type is frequently a compromise after joint consideration of design traffic volume and pattern, cost, topography, and availability of right-of-way.

Traffic and Operation

Each intersection type accommodates through traffic to varying degrees of efficiency. Where traffic on the minor crossroad is considerably less than on the major road, through traffic on the major road is minimally inconvenienced on at-grade intersections, particularly where topography is flat. Where the minor crossroad traffic volume is sufficient to justify a traffic signal, delay is experienced by all through traffic. Where through and crossroad volumes are nearly equal, about 50 percent of the traffic on each approach is required to stop.

Through traffic has no delays at highway grade separations except where approach gradients are long and steep and many heavy trucks are included in the traffic stream. Ramps at interchanges have no severe effect on through traffic except where the capacity is not adequate, the merging or speed-change lanes are not of adequate length, or a full complement of turning roadways is not provided.

Turning movements can affect traffic operations at an intersection and is accommodated to varying degrees, depending on the type of at-grade intersection or interchange. At interchanges, ramps are provided for turning movements. Where turning movements are light and some provision is made for all turning movements, a one-quadrant ramp design may suffice. However, left-turning movements on both highways may be no better accommodated than at an intersection at grade. Ramps provided in two quadrants may be situated such that crossings of through movements occur only at the crossroad and the major highway is free of such interference. An interchange with a ramp for every turning movement is suitable for heavy volumes of through traffic and for any volume of turning traffic, provided the ramps and terminals are designed with sufficient capacity.

Right-turning movements at interchanges follow simple direct or nearly direct paths on which there is little possibility of driver confusion. Cloverleaf interchanges involve loop paths for the left-turning movements, which may confuse drivers, and which involve added travel distance, and in some cases induce weaving movements. The diamond pattern of ramps is simple and more adaptable than a cloverleaf in cases where direct left turns are fitting on the minor road. However, where traffic on the minor road is sufficient to justify the expenditure to eliminate the at-grade left turns, a cloverleaf or higher type interchange should be considered.

Except on freeways, interchanges usually are provided only where crossing and turning traffic cannot readily be accommodated by an at-grade intersection. Some driver confusion may be unavoidable on interchanges, but such difficulties are minor in comparison to the benefits gained by the reduction of delays, stops, and crashes. Furthermore, confusion is minimized as interchanges become more frequent, drivers gain experience in operating through them, interchange designs are improved, and the quality and use of signing and other control devices are increased. Where interchanges are infrequent, publicity, education, and enforcement regarding proper usage of ramp patterns provided are valuable in ensuring efficient operation.

Interchanges are adaptable to various traffic mixes. The presence of a high proportion of heavy trucks in the traffic stream makes interchanges especially desirable. Interchanges help to maintain the capacity of the intersecting highways by minimizing vehicle delays caused by heavy trucks that do not have the accelerating ability that passenger cars have.

Site Conditions

In rolling or hilly topography, interchanges usually can be well fitted to the existing ground, and the through roads often can be designed more generously than if an at-grade intersection were provided. Such terrain may also simplify the design of some ramps. Other ramps, however, may involve steep grades, substantial length, or both, depending on the site terrain. Interchange design is simple in flat terrain, but grades may be introduced that do not favor vehicle operation. However, interchanges in flat terrain generally are not as pleasing in appearance as those fitted to rolling terrain. When it is practical to regrade the whole of the interchange area and to landscape it properly, most of the deficiency in appearance can be overcome.

The right-of-way needed for an interchange is largely dependent on the number of turning movements for which separate ramps need to be provided. The actual area needed for any particular interchange also depends on the highway type, topography, overall criteria of interchange development, and the impact on property access that may occur with provision of an interchange. The construction of an interchange may require adjustment in the existing highway profiles, complicate local access, or involve circuitous travel paths.

Type of Highway and Intersecting Facility

Interchanges are practical for all types of intersecting highways and for any range of design speeds. Conflicts from vehicles stopping and turning at an intersection increase with the design speed such that high-design-speed highways warrant interchange treatment earlier than low-design-speed roads with similar traffic volumes. The ramps on a high-design-speed highway should permit suitably high-turning speeds and include sufficiently long speed-change lanes.

Interchanges provide areas suitable for landscape development. For some conditions, the two-level nature of an interchange is a disadvantage with respect to appearance and may block a driver's view of the landscape. On the other hand, consideration of the architectural features in the structural design, the flattening and rounding of slopes for erosion control, and landscape treatment can provide a development that is aesthetically pleasing. Landscape development may involve above-minimum layouts rather than less costly structures or ramps with minimal grading.

Interchanges are essential components of freeways. With full control of access, grade separations are provided at all crossroads of sufficient importance to prohibit their termination. The interchange configuration will vary with the terrain, development along the highway, and right-of-way conditions, but in general it will be based on ramp layouts to expedite entrance to or exit from the freeway. In addition, ramp connections may involve frontage roads.

The extent to which local service should be maintained or provided is also a consideration in selecting the intersection type. Whereas local service can be provided readily on certain types of at-grade intersections, it may be difficult to provide for some types of interchanges.

ACCESS SEPARATIONS AND CONTROL ON THE CROSSROAD AT INTERCHANGES

As one of the most critical elements in the design of freeways and other high-volume highways, interchanges are expensive to build and equally expensive to upgrade. Therefore, it is essential that they be designed and operated as efficiently as practical. To preserve their intended function, adequate geometry at ramp termini and appropriate access control along crossroads are essential.

Many older interchanges have been designed with only limited access control on the intersecting crossroad. As a result, considerable development may occur in close proximity to the intersection of the ramp terminus and the crossroad. Over time, such ramp terminals, as well as several nearby access connections, require signalization, causing traffic operational problems.

In urbanized areas, high turning volumes and close spacing between adjacent ramp terminals and access connections create operational problems on the crossroad that affect traffic on the ramp and may spill back onto the mainline freeway. These problems consist of queue spillback, stop-and-go travel, heavy weaving volumes, and poor traffic signal progression.

Access control should be an integral part of the design of highways whose primary function is mobility, and it is a highly desirable feature along the crossroad at an interchange to provide efficient traffic operations and safety. Access control can be accomplished by purchasing access rights or by establishing access-control policies along the crossroad.

To ensure efficient operations along the crossroad at an interchange, adequate lengths of access control should be part of the overall design of an interchange. This minimizes spillback on the ramp and crossroad approaches to the ramp terminal, provides adequate distances for crossroad weaving, provides space for merging maneuvers, and provides space for storage of turning vehicles at access connections on the crossroad (**1, 2**).

Exhibit 10-2A illustrates the elements to be considered in determining access separation and access-control distances in the vicinity of free-flowing ramp entrances and exits. These include the distances needed to enter and weave across the through-traffic lanes, move into the left-turn lane, store left turns with a low likelihood of failure, and extend from the stop line to the centerline of the intersecting road or driveway. In addition, driver perception-reaction distance may be included in the computation. Where only right-turn access is involved and there are no left turns or median breaks, the weaving distance governs.

Exhibit 10-2B illustrates factors affecting access separation and control distances along a crossroad where there is a diamond interchange and the ramp termini are controlled by either a traffic signal or stop sign (**1**).

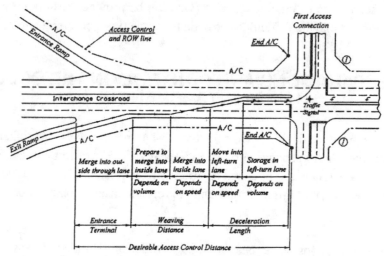

-A- Free-Flow Ramps Entering and Exiting From Crossroad

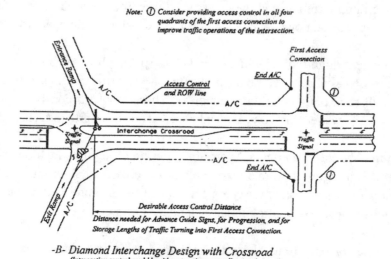

-B- Diamond Interchange Design with Crossroad
(Intersection control could be either stop signs or traffic signals.)

Exhibit 10-2. Factors Influencing Length of Access Control Along an Interchange Crossroad

Safety

Elimination or minimization of crossing and turning conflicts can be very effective in improving safety, especially at intersections. Regardless of design, signing, and signalization, at-grade intersections have a potential for crashes resulting from vehicle-vehicle conflicts. This is due, in part, to conflicting crossing and turning movements that occur within a limited area.

By separating the grades of the intersecting roadways, crashes caused by crossing and turning movements can be reduced. The grade separation structure itself may be a roadside obstruction; however, this can be minimized by the use of adequate clear roadside widths and protective devices at bridge abutments and piers. Where access between intersecting roadways will be provided, an interchange is appropriate for providing the maximum degree of safety. Depending on the interchange configuration used, left turns may be entirely eliminated or confined to the crossroad. Right-turning traffic can be accommodated on ramps that provide operation approaching the equivalent of free flow. Thus, conflicts caused by crossing traffic can be eliminated or minimized.

Stage Development

Where the ultimate development consists of a single grade-separation structure, stage construction may be uneconomical unless provisions are made in the original design for a future stage of construction. Ramps, however, are well adapted to stage development.

Economic Factors

Initial Costs

An interchange is the most costly type of intersection. The combined cost of the structure, ramps, through roadways, grading and landscaping of large areas, and possible adjustments in existing roadways and utilities generally exceeds the cost of an at-grade intersection. Directional interchanges involve more than one structure, and their cost is usually greater than any simple interchange.

Maintenance Costs

Each type of intersection has appreciable and distinct maintenance costs. Interchanges have large paved and variable slope areas, the maintenance of which, together with that of the structure, signs, and landscaping, exceeds that of an at-grade intersection. In addition, interchanges often involve maintenance and operation costs for lighting.

Vehicular Operating Costs

In a complete analysis of the adaptability of an interchange, vehicular operating costs should be compared between the interchange and other intersections. The values are so dependent on traffic, site, and design that a general comparison cannot be cited. Through traffic at an interchange usually follows a direct path with only a minor speed reduction. The added vehicular costs related to the change in grade when passing over or under the structure may need to be considered only when grades are steep, a condition that is usually limited to the minor intersecting roads. Right-turning traffic is subject to added vehicular costs of deceleration and acceleration and may also be subject to the costs of operation on a grade; however, travel distance is usually shorter than that on an at-grade intersection. Left-turning traffic is subject to added costs of acceleration and deceleration and usually to added travel distance compared with direct left turns at grade. Directional ramps may eliminate large speed changes and save travel distance, as compared with at-grade intersections. For any one vehicle, these differences in operating costs may appear insignificant, but when considered in cumulative totals, they indicate a pronounced benefit to traffic as a whole at the intersection. For intermediate-to-heavy traffic, the total vehicle-operating costs at an intersection usually will be lower with an interchange than with an at-grade design, especially if the through movements predominate.

GRADE SEPARATION STRUCTURES

Introduction

Various types of structures are employed to separate the grades of two intersecting roadways or a highway and a railroad. Although many phases of structural design should also be considered, this discussion is confined to the geometric features of grade separation structures. Some phases of structural design are referred to because of their effect on geometric design. This discussion largely concerns highway grade separations, but most of the geometric design features also apply to railroad grade separations.

Types of Separation Structures

Grade-separation structures are identified by three general types: deck type, through, and partial through. The deck type is most common for grade separations. However, the through and partial through types are appropriate for railroad structures. In special cases where the spans are long and the difference in elevation between the roadways is to be severely limited, truss bridges may be used.

Through girder bridges, in comparison to through deck-type bridges, will decrease vertical restrictions. In the case where the upper roadway extends from hilltop to hilltop and vertical clearance is not a problem, deck-type structures, such as trusses, arches, girders, etc., may be appropriate. A through plate girder bridge is often used for railroad separations when the railroad overpasses the highway or street. The through plate girder and through truss bridges produce a greater sense of visual restriction than deck-type structures; therefore, lateral clearances from the edge of lane should be as great as practical.

In any single separation structure, care should be exercised in maintaining a constant clear roadway width and a uniform protective railing or parapet. The type of structure best suited to grade separations is one that gives drivers little sense of restriction. Where drivers take practically no notice of a structure over which they are crossing, their behavior is the same or nearly the same as at other points on the highway, and sudden, erratic changes in speed and direction are unlikely. On the other hand, it is virtually impossible not to notice a structure overpassing the roadway being used. For this reason, every effort should be made to design the structure so that it fits the environment in a pleasing and functional manner without drawing unnecessary or distracting attention. Collaboration between the bridge and highway engineers throughout the various stages of planning and design can provide excellent results in this regard. Overpass structures should have liberal lateral clearances on the roadways at each level. All piers and abutment walls should be suitably offset from the traveled way. The finished underpass roadway median and off-shoulder slopes should be rounded, and there should be a transition to backslopes to redirect errant vehicles away from protected or unprotected structural elements.

A grade-separation structure should conform to the natural lines of the highway approaches in alignment, profile, and cross section. Fitting structures to the highway may result in variable structural widths, flared roadways, flared parapets or bridge railing, and non-symmetrical substructure units. Such dimensional variations are recognized as essential by both highway and bridge engineers and result in individual designs for each separate structure. In addition to the above geometric considerations, other conditions such as span lengths, depths of structure, foundation material at the site, aesthetics, safety, and especially skew may substantially influence the engineering and cost feasibility of the structure being considered. The bridge engineer should be consulted during alignment (horizontal and vertical) studies, and close coordination should be maintained throughout the design phase so that the most prudent design from the standpoint of safety, functionality, and economics of the total highway (including the bridge) can be selected. Many times a minor adjustment in alignment can substantially reduce serious structural problems, especially with wide structures.

For the overpass highway, the deck-type structure is most suitable. Although the supports may present both lateral and vertical clearance problems on the lower roadway, they are out of sight for motorists on the upper roadway. The deck-type bridge at the upper roadway has unlimited vertical clearance; lateral clearance is controlled only by location of the protective barrier. The parapet system should provide a freedom of view from the passing vehicles insofar as practical; however, capability to redirect errant vehicles should have precedence over preserving the motorist's view. The parapet and railing should have an appearance of strength and the ability to safely redirect the design vehicle(s) under the design impact conditions. Consideration should also be given to containing and redirecting larger vehicles crossing the structure. The end posts of through trusses should be protected by a suitable approach traffic barrier and transition section. Spans at highway grade separations should not be long enough to require through trusses. In special cases where spans are long and the difference in elevation between the two roadways is to be limited, all practical designs should be compared by the bridge engineer for suitability, including economic and aesthetic considerations.

For the underpass highway, the most desirable structure from the standpoint of vehicular operation is one that will span the entire highway cross section and provide a lateral clearance of structural supports from the edge of roadway that is consistent with good roadside design. The lateral clearance between the edge of roadway and the structural supports should be as wide and flat as practical to provide usable recovery space for errant vehicles and to prevent distraction in the motorist's peripheral field of vision. In the case of depressed roadways, lateral clearances may be reduced, as discussed in the section "Lateral Clearances" later in this chapter. On divided highways, center supports should be used only where the median is wide enough to provide sufficient lateral clearance or narrow enough to need protective barriers. The usual lateral clearance of an underpass at piers or abutments may allow sufficient room to construct additional lanes under the structure in the future, but at a sacrifice of recovery space. In anticipation of future widening, the piers or abutment design should provide footings with sufficient cover after widening. The bridge engineer should be advised when future widening is contemplated. A greater sense of openness results with end spans than with full-depth abutments. Perched stub or semistub abutments can also provide appropriate visual clearance.

In urban areas, although not all cross streets are important enough to warrant interchange ramps with the main line, a sufficient number of cross streets should be separated in grade to preserve the continuity of traffic flow on the local street system. As a matter of economics, however, it is seldom practical to continue all cross streets across the main line. Most streets that cross the major roadway, whether or not they connect with it, experience a rapid increase in traffic following construction of the major roadway as a result of intensified land development and local street closures within the main-line corridor. Terminated and through streets may be intercepted by one-way frontage roads on each side of the main facility. Access between the main roadway and frontage roads can be provided by slip ramps at prescribed intervals to serve traffic demands.

On elevated facilities with viaduct construction, cross streets are relatively undisturbed; however, on all other types of roadways, considerable savings can be achieved by terminating some of the less important cross streets. Special consideration is needed relative to the spacing and treatment of cross streets on these roadways. Arterials and other major cross streets should continue across the main line without interruption or deviation. Grade separations should be of sufficient number and capacity to accommodate not only the normal cross traffic but also the traffic diverted from the other streets terminated by the main facility and the traffic generated by access connections to and from the main line. Thus, determination of the number and location of cross streets to be separated in grade needs a thorough analysis of traffic on the street system, in addition to that on the main line and its interchanges.

Insofar as throughway operation is concerned, there is no minimum spacing or limit to the number of grade-separated cross streets. The number and their location along one corridor are governed by the local street system, existing or planned. Depending on features of the city street network such as block length, the presence or absence of frontage roads, and degree of adjacent urban development, it may be appropriate to provide more crossings than otherwise needed for the principal cross streets. Where frontage roads are not provided or where they are used only intermittently, more crossings may be needed to provide convenient access to all areas. Other factors that may affect the number and spacing of cross streets are the location of schools, recreational areas, other public facilities, school bus routes, and fire-fighting equipment routes. Whereas in and near downtown districts, cross streets continuing across the throughway may be located at intervals of two or three blocks, and sometimes every block, in intermediate areas they are likely to be three to five blocks apart, and in residential or outlying districts they should be at greater intervals.

Cross streets should also fit the existing, revised, or expected pattern of transit operation and the needs of pedestrians and bicyclists. For the most part, pedestrians and bicyclists are accommodated on structures that also serve vehicular traffic. Because extra travel distance is more acceptable for vehicular travel and bicyclists than for pedestrians, it is appropriate to add separate pedestrian crossings, particularly where there are large numbers of pedestrians, such as near schools, churches, and factories.

Although the streets that are to cross the major roadway should be selected during the planning stage, all crossings need not necessarily be constructed initially. Normally, structures carrying the major roadway should be constructed initially, as it is impractical to disrupt the main line after it is open to traffic. However, some of the planned structures to carry cross streets over the major roadway may be deferred until fully justified by traffic growth or other planned developments. The system of overcrossing streets should be coordinated with and shown in the design of the major roadway, and a plan should be developed showing those that are to be constructed initially and those that are to be provided later. Such a plan should show the traffic circulation scheme at initial and later stages, and it should be checked periodically against traffic needs of the major roadway, the interchanges, and the street system.

The new cross-street structure and approaches are usually designed for projected traffic 10 to 20 years in the future. In many cases, the existing cross street is not as generously designed at either side of the separation structure as the newly designed separation. Improvement of the cross street may not be scheduled for several years. Therefore, there should be a suitable transition of the new work to the existing facility in a manner that will promote the safe, orderly movement of traffic.

In many instances, the existing street approaching the major roadway needs some improvements to increase capacity and facilitate traffic more efficiently to and from the major roadway. Typical improvements include lane and shoulder widening, control of parking and pedestrian movement, improvements of intersections with traffic signals, marking, channelization, and one-way operation where appropriate.

Where a city street underpasses a major roadway, the underside of the structure is a design feature that deserves special treatment for aesthetic reasons. Because of numerous pedestrians and slower moving traffic, the underside of a structure as viewed from the cross street is especially noticeable to local citizens. It should therefore be as open as practical to allow the maximum amount of light and air below. An open-type structural design is also needed to improve the sight distance, especially if there are intersections adjacent to the structure.

On sections of roadway that are elevated on a viaduct, the local street system may be left relatively undisturbed unless there is a need to realign the cross street or widen it for additional capacity. Structural openings should allow for future expansion of approach width and vertical clearance.

Cross-street overcrossings and undercrossings have many features in common such as lane and shoulder widths, corner curb radii, storage for turning vehicles, horizontal clearances, curbs, and sidewalks.

Typical highway separation structures are depicted in Exhibits 10-3 and 10-4. The bridge span arrangement is determined principally by the need for a clear roadside recovery area, although sight distance is an important design element for all roadways and diamond interchanges.

Exhibit 10-3. Typical Grade Separation Structures With Closed Abutments

Exhibit 10-4. Typical Grade Separation Structure With Open-End Span

A single simple-span girder bridge may be used with spans of up to approximately 45 m [150 ft] and can accommodate conditions of severe skew and horizontal curvature. Spans of greater length need greater structure depth and higher approach embankments. The structure depth for single-span girder bridges is approximately 1/15 to 1/30 of the span.

The conventional type of overpass structure over divided highways is currently a two-span, deck-type bridge. When bridging with two or more spans, the deck-girder-type bridge, either steel or concrete, is usually continuous in design for reasons of economy, providing some saving in structure depth and elimination of deck joints over the piers.

As an alternative to the girder bridge, a deck-type, single-span rigid frame or a three-span rigid-frame, slant leg bridge may be used for aesthetic purposes where appropriate. At special geographic locations, where excess vertical clearance is available and the skew is not severe, a spandrel arch bridge may be economically and aesthetically desirable when foundation support is adequate. This type of bridge is also inherently pleasing in appearance.

Two or more structures are not uncommon at interchanges with direct connections for left-turning movements. In special cases, several structures may be combined to form one multilevel structure. Two variations of roadways crossing at three or four levels are shown in Exhibit 10-5. Designs that include three- and four-level structures may not be more costly than the equivalent number of conventional structures to provide the same traffic service, particularly in urban areas where right-of-way costs are high.

—A—

—B—

Exhibit 10-5. Multilevel Grade Separation Structures

Overpass Versus Underpass Roadways

General Design Considerations

A detailed study should be made at each proposed highway grade separation to determine whether the main road should be carried over or under the crossroad. Often this decision is based on features such as topography or highway classification. It may be appropriate to make several nearly complete preliminary layout plans before an appropriate decision can be reached. General guidelines for over-versus-under preference follow, but such guidelines should be used in combination with detailed studies of the grade separation as a whole.

At any site, the issues governing whether a road should be carried over or under usually fall into one of three groups: (1) the influence of topography predominates and, therefore, the design should be closely fitted to it; (2) the topography does not favor any one arrangement; and (3) the alignment and gradeline controls of one highway predominate and, therefore, the design should accommodate that highway's alignment instead of the site topography.

As a rule, a design that best fits the existing topography is the most pleasing and economical to construct and maintain, and this factor becomes the first consideration in design. Where topography does not govern, as is common in the case of flat topography, it may be appropriate to study secondary factors, and the following general guidelines should be examined:

- For the most part, designers are governed by the need for economy, which is obtained by designs that fit existing topography, not only along the intersecting highways but also for the whole of the area to be used for ramps and slopes. Thus, it is appropriate to consider alternatives in the interchange area as a whole to decide whether the major road should go over or under the crossroad.
- An undercrossing highway has a general advantage in that an approaching interchange may be easily seen by drivers. As a driver approaches, the structure appears ahead, making the presence of the upper-level crossroad obvious, and providing advance warning of the likely presence of interchange ramps.
- Through traffic is given aesthetic preference by a layout in which the more important road is the overpass. A wide overlook can be provided from the structure and its approaches, giving drivers a minimum feeling of restriction.
- Where turning traffic is significant, the ramp profiles are best fitted when the major road is at the lower level. The ramp grades then assist turning vehicles to decelerate as they leave the major highway and to accelerate as they approach it, rather than the reverse. In addition, for diamond interchanges, the ramp terminal is visible to drivers as they leave the major highway.
- In rolling topography or in rugged terrain, major-road overcrossings may be attainable only by a forced alignment and rolling gradeline. Where there otherwise is no pronounced advantage to the selection of either underpass or an overpass, the design that provides the better sight distance on the major road (desirably passing distance if the road is two-lane) should be preferred.
- An overpass offers the best possibility for stage construction, both in the highway and structure, with minimum impairment of the original investment. The initial development of only part of the ultimate width is a complete structure and roadway in itself. By lateral extension of both or construction of a separate structure and roadway for a divided highway, the ultimate development is reached without loss of the initial facility.
- Troublesome drainage problems may be reduced by carrying the major highway over without altering the crossroad grade. In some cases, the drainage problem alone may be

sufficient reason for choosing to carry the major highway over rather than under the crossroad.

- Where topography control is secondary, the cost of bridges and approaches may determine whether the major roadway underpasses or overpasses the minor facility. A cost analysis that takes into account the bridge type, span length, roadway cross section, angle of skew, soil conditions, and cost of approaches will determine which of the two intersecting roadways should be placed on structure.

- An underpass may be more advantageous where the major road can be built close to the existing ground, with continuous gradient and with no pronounced grade changes. Where the widths of the roads differ greatly, the quantity of earthwork makes this arrangement more economical. Because the minor road usually is built to less generous design criteria than the major road, grades on it may be steeper and sight distances shorter, with resultant economy in grading volume and pavement area on the shorter length of road to be rebuilt above the general level of the surrounding country.

- Frequently, the choice of an underpass at a particular location is determined not by conditions at that location, but by the design of the highway as a whole. Grade separations near urban areas constructed as parts of a depressed expressway, or as one raised above the general level of adjoining streets, are good examples of cases where decisions regarding individual grade separations are subordinated to the general development.

- Where a new highway crosses an existing route carrying a large volume of traffic, an overcrossing by the new highway causes less disturbance to the existing route and a detour is usually not needed.

- The overcrossing structure has no limitation as to vertical clearance, which can be a significant advantage in the case of oversized loads requiring special permits on a major highway or route.

- Desirably, the roadway carrying the highest traffic volume should have the fewest number of bridges for better rideability and fewer conflicts when repair and reconstruction are needed.

- In some instances, it may be appropriate to have the higher volume facility depressed and crossing under the lower volume facility to reduce noise impact.

- In some instances, the lower volume facility should be carried over if there is a pronounced economic advantage.

Structure Widths

Roadways with wide shoulders, wide gutters, and flat slopes are the safest and give the driver a sense of freedom. Poles, walks, bridge columns, bridge railing, and parapets located close to the traveled way are potential obstructions and cause drivers to shy away from them. For this reason, the clear width on bridges should preferably be as wide as the approach roadway in order to give drivers a sense of openness and continuity.

On long bridges, particularly on long-span structures where cost per square meter [yard] is greater than the cost on short-span structures, widths that are less than ideal may be acceptable; however, economy alone should not be the governing factor in determining structure widths. The analysis of traffic characteristics, safety features, emergency contingencies, and benefit/cost ratios should be fully considered before the desirable structure width is compromised.

When determining the appropriate width of the roadway over or under a grade separation, in determining the dimensions, location, and design of the structure as a whole, and in detailing features adjacent to the road, the designer should aim to provide a facility on which driver reaction and vehicle placement will be essentially the same as elsewhere on the intersecting roads. However, the width should not be so great as to result in the high cost of structure without proportionate value in usefulness and safety.

Underpass Roadways

For each underpass, the type of structure used should be determined by the dimensional, load, foundation, and general site needs for that particular location. Only the dimensional details are reviewed herein.

Although it is an expensive element, an underpass is only one component of the total facility and should, therefore, be consistent with the design standards of the rest of the facility to the extent practical. It is desirable that the entire roadway cross section, including the median, traveled way, shoulders, and clear roadside areas, be carried through the structure without change. However, several possible limitations may require some reduction in the basic roadway cross section: structural design limitations; vertical clearance limitations; controls on grades and vertical clearance; limitations due to skewed crossings, appearance, or aesthetic dimension relations; and cost factors, such as those encountered in lengthy depressed sections of roadway. On the other hand, where conditions permit a substantial length of freeway to be developed with desirable lateral dimensions, an isolated overpass along the section should not be designed as a restrictive element. In such cases, the additional structural costs are strongly encouraged to ensure consistency through the facility.

Lateral Clearances

Minimum lateral clearances at underpasses are illustrated in Exhibit 10-6. For a two-lane roadway or an undivided multilane roadway, the cross-section width at underpasses will vary, depending on the design criteria appropriate for the particular functional classification and traffic volume. The minimum lateral clearance from the edge of the traveled way to the face of the protective barrier should be the normal shoulder width.

On divided highways, the clearances on the left side of each roadway are usually governed by the median width. A minimum median width of 3.0 m [10 ft] may be used on a four-lane roadway to provide 1.2-m [4-ft] shoulders and a rigid median barrier. For a roadway with six or more lanes, the minimum median width should be 6.6 m [22 ft] to provide 3.0-m [10-ft] shoulders and a rigid median barrier. Exhibit 10-6A shows the minimum lateral clearances to a continuous median barrier, either concrete or metal, for the basic roadway section and for an underpass where there is no center support. The same clearance dimensions are applicable for a continuous wall on the left. Where a concrete median barrier is used, its base should be aligned with respect to the traveled way, as shown in Exhibit 10-6A.

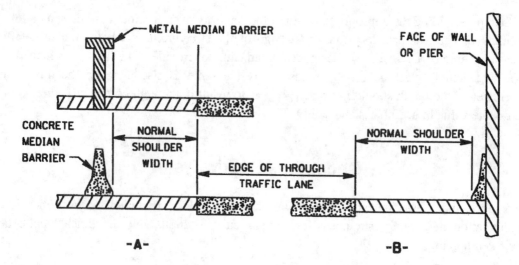

CONTINUOUS WALL OR BARRIER

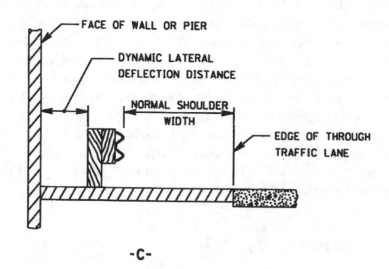

WITH GUARDRAIL RIGHT OR LEFT

Exhibit 10-6. Lateral Clearances for Major Roadway Underpasses

Exhibit 10-6B shows the minimum lateral clearance on the right side of the roadway as applicable to a continuous wall section. A concrete barrier is constructed integrally with the wall. For this situation, the lateral clearance on the right should be measured to the base of the barrier. For design with a continuous concrete barrier on the right, usually a section similar to a median barrier, Exhibit 10-6B is applicable. The same type of barrier may be used as an introduced feature where conditions lead to structure design with full-depth abutments.

As discussed in earlier chapters, the shoulder on high-speed highways should be flush with the traveled way. Continuous curbs on high-speed highways should be limited to special situations, such as drainage systems on the outside of shoulders. Such curbs should be carried through the underpass. Where walkways are provided, the full shoulder section should be maintained through the underpass and the span increased by the width of the walk. Where a curb is needed along solid abutments or walls, a concrete barrier may be used. See Chapter 4 for a discussion on other types of barriers, their warrants, and issues concerning their placement.

Where conditions preclude the clear roadside design concept, all abutments, piers, and columns should be shielded with suitable protective devices unless they are so situated that they cannot be hit by out-of-control vehicles. Protective devices are usually not needed along continuously walled sections.

Guardrail installed along the face of an exposed pier or abutment should have a clearance appropriate to the dynamic lateral deflection of the particular rail type. The rail cannot cushion and deflect an errant vehicle unless there is sufficient lateral space clear of the bridge support. Exhibit 10-6C shows the limits of the dynamic lateral deflection distance between the face of bridge support and the back of the rail system. Guardrail attached flush with the exposed faces of piers, abutments, and bridge railings should be stiffened preceding the obstruction to avoid snagging an errant vehicle. This may be accomplished by reducing the post spacing, increasing the post imbedment, increasing the rail section modulus, and/or transitioning to a different, stiffer barrier (i.e., metal to concrete). The rail should be fastened securely enough to develop its full strength longitudinally. For further details, see the AASHTO *Roadside Design Guide* (**3**).

Where structural design and cost features make it necessary to reduce the horizontal clearance through an underpass, the change in lateral width should be accomplished through gradual adjustments in the cross section of the approach roadway rather than abruptly at the structure. Such transitions in width should have a gradual rate of 50:1 or more (longitudinal:lateral).

Vertical Clearance

Vertical clearance is typically determined for an entire route and may be governed by the established policies of the highway system. Although State laws vary somewhat, most States permit the vehicle height, including load, to be between 4.1 m [13.5 ft] and 4.4 m [14.5 ft]. The vertical clearance of all structures above the traveled way and shoulders should be at least 0.3 m [1 ft] greater than the legal vehicle height, and allowance should be made for future resurfacing.

Additional vertical clearance is desirable to compensate for several resurfacings, for snow or ice accumulation, and for an occasional slightly overheight load. The recommended minimum vertical clearance is 4.4 m [14.5 ft], and the desirable vertical clearance is 5.0 m [16.5 ft].

Some roadways are parts of systems or routes for which a minimum vertical clearance of 4.9 m [16 ft], plus an allowance for future resurfacing, has been established. Freeway and arterial systems are generally provided with such clearance, but for other routes a lower minimum vertical clearance is acceptable.

To permit the movement of exceptionally high loads through an urban area, it is desirable to have at least one route with structures designed so that the movement can be easily accommodated. This design could entail the use of deck-type bridges, street lights mounted higher than normal, underground utilities, mast-arm-supported traffic signals, which can be swung to one side, etc.

Where a depressed facility is a parkway with traffic restricted to passenger vehicles, the vertical clearance at structures should be 4.6 m [15 ft], and in no case should it be less than 3.8 m [12.5 ft]. The minimum clearance should be obtained within all portions of the roadway.

Overpass Roadways

The roadway dimensional design of an overpass or other bridge should be the same as that of the basic roadway. The bridge is a small part of the continuous roadway and should be designed without change in cross-section dimensions, unless the cost becomes prohibitive.

This section covers the general dimensional features for single structures typically used at a grade separation, a stream crossing, or a single-structure interchange. Overpasses usually are deck structures. Their major dimensional features are the parapet rail system, lateral clearances, and the median treatment, where applicable. Typical overpass structures are shown in Exhibit 10-7. For further discussion see also the sections under "Curbs" and "Traffic Barriers" in Chapter 4.

Bridge Railings

The typical bridge railing has some form of concrete base or parapet on which metal or concrete rail or rails are mounted on structurally adequate posts. The bridge railing should be designed to accommodate the design vehicle(s) on the structure under the design impact conditions. That is, the design vehicle should be safely redirected, without penetration or vaulting over the railing. Likewise, the railing should not pocket or snag the design vehicle, causing abrupt deceleration or spinout, and it should not cause the design vehicle to roll over.

Most bridge railings in service are of a rigid, non-yielding design. Several railings incorporate energy-absorbing features in their design to reduce vehicle impact severity. Where noise is a factor, solid rails may be considered for their added value in noise attenuation.

At certain locations, there may be a need to provide a pedestrian walkway or bicycle path on the freeway overpass. In these situations, a barrier-type bridge rail of adequate height should be installed between the pedestrian walkway and the roadway. Also, a pedestrian rail or screen should be provided on the outer edge of the walkway.

Exhibit 10-7. Typical Overpass Structures

Bridge railings located on the inside of horizontal curves may restrict stopping sight distance. Adjustment of the horizontal alignment or the offset to the bridge railing may be needed to provide adequate stopping sight distance.

Lateral Clearances

On overpass structures, it is desirable to carry the full width of the approach roadway across all structures. For facilities other than freeways, exception may be made on major structures with a high unit cost. The selection of cross-section dimensions that are different from those on the approach roadway should be subject to individual economic studies. Refer to previous chapters on arterials, collectors, and local roads and streets for permissible deviations from providing full approach roadway width across bridges. In the case of a curbed roadway, the minimum structure width should match the curbed approach roadway.

When the full approach roadway width is continued across the structure, the parapet rail, both left and right, should align with the guardrail on the approach roadway. For example, where the typical practice of the highway agency is to place the longitudinal barrier 0.6 m [2 ft] from the outer edge of the surfaced shoulder, the bridge rail should be placed 0.6 m [2 ft] outside the effective edge of the shoulder. This provides additional clearance for high-speed operation and door-opening space for vehicles stopped on the shoulder of the structure. Some agencies prefer to place the roadway longitudinal barrier 0.6 m [2 ft] from the outer edge of the shoulder and the bridge rail at the shoulder edge. In this case, a transition rate of about 20:1 is appropriate to taper the longitudinal barrier into the bridge rail.

At some interchanges, additional width for speed-change lanes or weaving sections is needed across overpass structures. Where the auxiliary lane is a continuation of a ramp, the horizontal clearance to the bridge rail should be at least equal to the width of shoulder on the approach ramp. Where the auxiliary lane is a weaving lane connecting entrance and exit ramps or is a parallel-type speed-change lane across the entire structure, the clearance to the parapet should be of uniform width and be at least equal to the shoulder width on the ramp.

Medians

On a divided highway with a wide median or one being developed in stages, the overpass will likely be built as two parallel structures. The approach width of each roadway should be carried across each individual structure. If separate parallel structures are used, the width of opening between structures is unimportant.

Where the approach is a multilane, undivided roadway or one with a flush median less than 1.2-m [4-ft] wide, a raised median is considered unnecessary on short bridges of about 30 m [100 ft] in length but is desirable on bridges of 120 m [400 ft] or more in length. On bridges between 30 m [100 ft] and 120 m [400 ft] in length, local conditions such as traffic volume, speed, sight distance, need for luminaire supports, future improvement, approach cross section, number of lanes, and whether the roadway is to be divided determine whether or not medians are warranted.

Where there are medians of narrow or moderate width on approaches to long single structures, the structure should be wide enough to accommodate the same type of median barrier as is used in the median of the approach roadway.

Longitudinal Distance to Attain Grade Separation

The longitudinal distance needed for adequate design of a grade separation depends on the design speed, the roadway gradient, and the amount of rise or fall needed to achieve the separation. Exhibit 10-8 shows the horizontal distances needed in flat terrain. It may be used as a guide for preliminary design to determine quickly whether or not a grade separation is practical for given conditions, what gradients may be involved, and what profile adjustments, if any, may be needed on the cross street. These data also may serve as a general guide in other than flat terrain, and adjustments can be made in the length of the terminal vertical curves. The chart is useful where the profile is rolled to overpass some cross streets and to underpass others, and it is useful for design of an occasional grade separation on a facility located at ground level, such as a major street or at-grade expressway.

The distance needed to achieve a grade separation can be determined from Exhibit 10-8 for gradients ranging from 2 to 7 percent and for design speeds (V_o) ranging from 50 km/h to 110 km/h [30 mph to 70 mph]. Design speeds (V_o) of 80 km/h to 110 km/h [50 mph to 70 mph] are applicable to urban freeways, and 60 km/h (50 km/h in special cases) [40 mph (30 mph in special cases)] is used on major arterials. The curves are derived with the same approach gradient on each side of the structure. However, values of D from Exhibit 10-8 also are applicable to combinations of unequal gradients. Distance D is equal to the length of the initial vertical curve, plus one-half the central vertical curve, plus the length of tangent between the curves. Lengths of vertical curves, both sag and crest, are minimums based on the minimum stopping sight distance. Longer curves are desirable. Length D applies equally to an overpass and an underpass, despite the fact that the central crest vertical curve may be longer than the central sag vertical curve for comparable values of H and G.

Certain characteristics and relations in Exhibit 10-8 are worthy of note:

- For the usual profile rise (or fall) needed for a grade separation (H of 7.5 m [25 ft] or less), gradients greater than 3 percent for a design speed of 110 km/h [70 mph], 4 percent for 100 km/h [60 mph], 5 percent for 80 km/h [50 mph], and 6 percent for 60 km/h [40 mph] cannot be used. For values of H less than 7.5 m [25 ft], flatter gradients than cited above should generally be used. The lower terminal of the gradient lines on the chart, marked by a small circle, indicates the point where the tangent between curves is 0 and below which a design for the given grade is not feasible (i.e., a profile condition where the minimum central and end curves for the gradient would overlap).

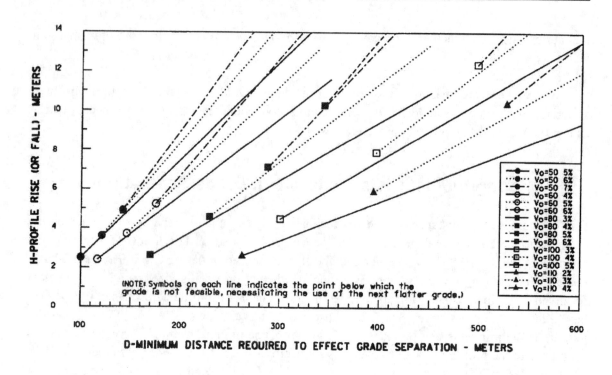

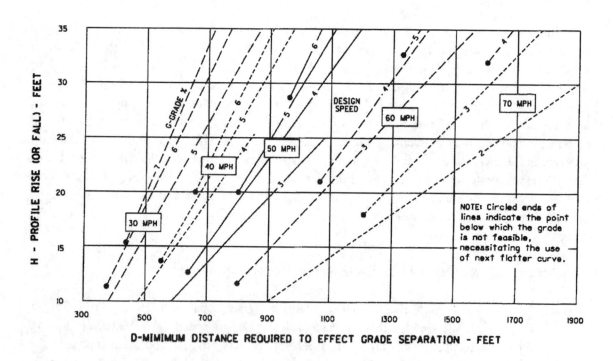

Exhibit 10-8. Flat Terrain, Distance Required to Effect Grade Separation

NOTE:"MINIMUM VERTICAL CLEARANCE SHOULD BE CHECKED UNDER THE OUTSIDE EDGE OF THE OVERCROSSING STRUCTURE."

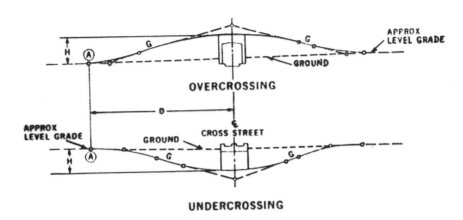

Exhibit 10-8. Flat Terrain, Distance Required to Effect Grade Separation (Continued)

- For given H and design speed, distance D is shortened a negligible amount by increasing the gradient above 4 percent for a design speed of 80 km/h [50 mph] and above 5 percent for 60 and 50 km/h [40 and 30 mph]. Distance D varies to a greater extent, for given H and G, with changes in design speed.

A 6.0- to 6.6-m [20- to 22-ft] difference in elevation is usually needed at a grade separation of two highways for essential vertical clearance and structural thickness. The same dimension generally applies to a highway undercrossing a railroad, but about 8.4 m [28 ft] is needed for a highway overcrossing a main line railroad. In level terrain, these vertical dimensions correspond to H, the rise or fall needed to achieve a grade separation. In practice, however, H may vary over a wide range because of topography. Where a relatively short distance is available for a grade separation, it may be appropriate to reduce H to keep D within the distance available. This reduction is accomplished by raising or lowering the intersecting street or railroad.

Grade Separations Without Ramps

There are many situations where grade separations are constructed without the provision of ramps. For example, some major arterials intersecting the existing highway must be kept open for access but carry only low traffic volumes. Lacking a suitable relocation plan for the crossroad, a highway grade separation without ramps may be provided. All drivers desiring to turn to or from that road are required to use other existing routes and enter or leave the highway at other locations. In some instances, these vehicles may have to travel a considerable extra distance, particularly in rural areas.

In other situations, despite sufficient traffic demand, ramps may be omitted (1) to avoid having interchanges so close to each other that signing and operation would be difficult, (2) to eliminate interference with large highway traffic volumes, and (3) to increase safety and mobility by concentrating turning traffic where it is practical to provide adequate ramp systems. On the

773

other hand, undue concentration of turning movements at one location should be avoided where it would be better to provide several interchanges.

In rugged topography, the site conditions at an intersection may be more favorable for provision of a grade separation than an at-grade intersection. If ramp connections are difficult or costly, it may be practical to omit them and accommodate turning movements at other intersecting roads.

INTERCHANGES

General Considerations

There are several basic interchange configurations to accommodate turning movements at a grade separation. The type of configuration used at a particular site is determined by the number of intersection legs, expected volumes of through and turning movements, type of truck traffic, topography, culture, design controls, and proper signing. The designer's initiative also plays an important role.

While interchanges are custom designed to fit specific site conditions, it is desirable that the overall pattern of exits along the freeway have some degree of uniformity. Furthermore, from the standpoint of driver expectancy, it is desirable that all interchanges have one point of exit located in advance of the crossroad wherever practical.

Signing and operations are major considerations in the design of the interchanges. The signing of each design should be tested to determine if it can provide for the smooth, safe flow of traffic. The need to simplify interchange design from the standpoint of signing and driver understanding cannot be overstated.

To prevent wrong-way movements, all freeway interchanges with non-access-controlled highways should provide ramps to serve all basic directions. Drivers expect freeway-to-freeway interchanges to provide all directional movements. As a special case treatment, a freeway-to-freeway movement may be omitted if the turning traffic is minor and can be accommodated by and given the same route signing over other freeway facilities.

The accommodation of pedestrians and bicyclists also should be considered in the selection of an interchange configuration.

For convenience, examples of interchange configurations are illustrated in the following discussion in general terms for three- and four-leg intersections and for special designs involving two or more structures. The general interchange configurations are shown either schematically or as examples of existing facilities.

Three-Leg Designs

An interchange with three intersecting legs consists of one or more highway grade separations and one-way roadways for all traffic movements. When two of the three intersection legs form a through road and the angle of intersection is not acute, the term T interchange applies. When all three intersection legs have a through character or the intersection angle with the third intersection leg is small, the interchange may be considered a Y configuration. A clear distinction between the T and Y configurations is not important. Regardless of the intersection angle and through-road character, any basic interchange pattern may apply for a wide variety of conditions. Three-leg interchanges should only be considered when future expansion to the unused quadrant is either impossible or highly unlikely. This is due in part to the fact that three-leg interchanges are very difficult to expand or modify in the future.

Exhibit 10-9 illustrates patterns of three-leg interchanges with one grade separation. Exhibits 10-9A and 10-9B show the widely used trumpet pattern. Through-traffic movements, from points a to c, are on direct alignment. A criterion for selection of either design is the relative volumes of the left-turning movements, the more direct alignment favoring the heavier volume and the loop favoring the lesser volume. Skewed crossings are more desirable than right-angle crossings because the skewed crossing has a somewhat shorter travel distance and flatter turning radius for the heavier left-turning volume, and there is less angle of turn for both left turns. In Exhibit 10-9A, the curvature of the loop b-a begins before the structure, warning the driver to anticipate a major break in curvature. The transition spirals provide for a smooth speed change and steering maneuver both into the loop and onto the high-speed facility. The oblong shape of the loop allows the curvature of the high-volume left turn, c to b, to be flattened, allowing higher operating speeds to be attained. The exit to the loop ramp of Exhibit 10-9B is placed well in advance of the structure to provide sufficient deceleration length in the approach to the break in curvature. Curves with spiral transitions are effective in developing the desired shape of ramps. The curvature of the left turn, b-a, is initiated in advance of the structure for driver anticipation.

The other type of three-leg single-structure interchange shown in Exhibit 10-9C is less common, with loops for both left-turning movements. The interchange in Exhibit 10-9C has an excellent field of usage as the initial stage of an ultimate cloverleaf. A collector-distributor road is provided to eliminate weaving on the main road. In the second stage, the roadway forming the fourth leg opposite the stem of the T is developed, and the remaining ramps are added. With respect to traffic, this type of interchange is inferior to those in Exhibits 10-9A and 10-9B because both left-turns movements use loops and weave across each other. Furthermore, the small-radius loop ramps are not considered to be an appropriate method of terminating a freeway. Although the pattern is appropriate for interchanges where the left-turning volumes are not great, the configurations in Exhibits 10-9A and 10-9B are preferable if they are equally adaptable to the site conditions. For comparable conditions, construction costs for Exhibits 10-9A and 10-9B should be about the same.

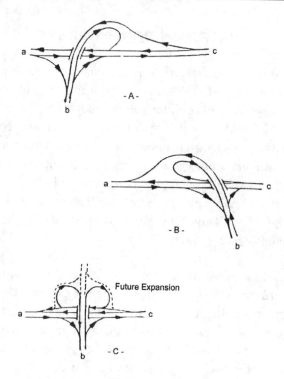

Exhibit 10-9. Three-Leg Interchanges With Single Structures

Exhibit 10-10 illustrates high-type T and Y interchanges, each with more than one structure or with one three-level structure, that provide for all of the movements without loops. These configurations are more costly than single-structure configurations and are justified only where all movements are large.

In Exhibit 10-10A, all movements are directional, three structures are needed, and weaving is avoided. This plan is suitable for the intersection of a through freeway with the terminal of another major freeway. Some or all of the interchanging movements will need at least two-lane roadways. All entrances and exits are designed as branch connections or major forks, as discussed later in this chapter. The alignment of this interchange may be adjusted to reduce the right-of-way needs, forming an interchange with only one three-level structure, as illustrated in Exhibit 10-10B.

Operationally, the configuration in Exhibit 10-10A might be superior to the configuration in Exhibit 10-10B because of the inherent sharp curvature on movement c-b in Exhibit 10-10B. While complete cost comparison involves a special analysis, there usually is little difference in cost. In some cases, the more complex three-level structure has been found to be less costly.

Exhibit 10-10C illustrates a three-leg interchange with a double jug-handle pattern. This pattern applies where it is appropriate to carry one of the freeways through the interchange with minimal deviation in alignment but where the intersecting radius is also considerably important. Interchanging traffic enters and exits the freeway on the right and ramps are usually only single-lane roadways. This pattern involves the use of three structures, at least two of which span double roadways. As shown in Exhibit 10-10D, the basic pattern can be arranged so that the two left-turn ramps and the through roadway meet at a common point where a three-level structure replaces the three structures shown.

Exhibit 10-10E is another variation of the configuration in Exhibits 10-10C and 10-10D. Separate roadways are provided for each left-turning movement with two, two-level structures separating the ramps from the through movements. The grade separation structures should be spaced sufficiently far apart to permit the placement of the separate ramp, b-a, between them, thus avoiding the third structure of Exhibit 10-10C. This design may be altered, as shown in Exhibit 10-10F. This arrangement provides smoother alignment on the ramps, but successful operation depends on provision of a weaving section that is suitably long for these two movements.

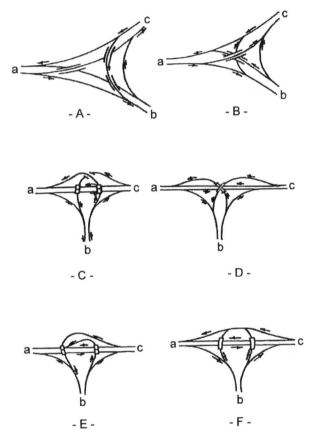

Exhibit 10-10. Three-Leg Interchanges With Multiple Structures

Exhibit 10-11 shows a trumpet interchange at the junction of a freeway and a major local road in a rural area. A unique feature of this configuration is that the local road overpasses one roadway of the freeway and underpasses the other because of the steep slope on the terrain. This pattern also explains the relatively sharp radius on the loop. The design favors the heavier traffic movement that is provided by the semidirect connection, and the loop handles the lighter volume.

Exhibit 10-12 shows an interchange between two freeways in a rural area. The directional design with large radii permits high-speed operation for all movements. The frontage roads

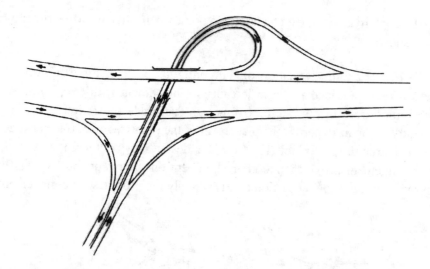

Exhibit 10-11. Three-Leg Interchange (T-Type or Trumpet)

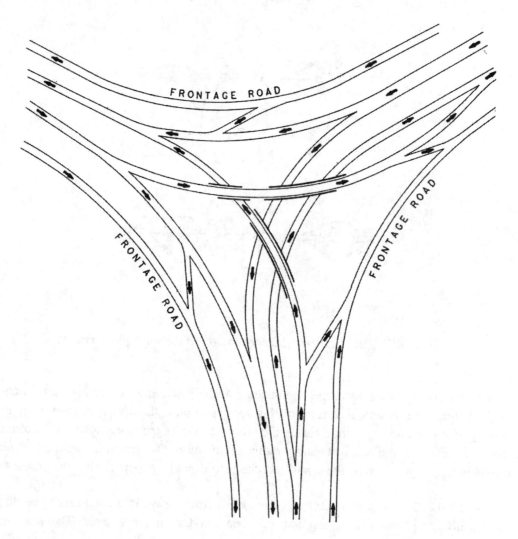

Exhibit 10-12. Three-Leg Interchange Semidirectional Design

provide desirable one-way operations with connections from the interchange roadways being as long as practical. The separation distance between major forks and the ramp terminals that follow should be sufficient to provide for smooth traffic operations. There are three separate structures in this configuration.

Exhibit 10-13 shows a directional, three-leg interchange between two freeways at a river crossing. The turning roadways are liberally designed to permit high-speed operation. Note the major fork and branch connection on the river crossings. A much more expansive gore area is provided on the divergence than on the convergence for a recovery area and possible installation of an attenuator.

Exhibit 10-13. Directional Three-Leg Interchange of a River Crossing

Exhibit 10-14 illustrates a trumpet-type interchange. The two-lane exit and entrance at the bridge in the foreground of the exhibit serve both the local street system and the freeway-to-freeway movements. All interchange movements are usually provided in this type of interchange, and the exits on the curves are properly designed to discourage inadvertent exits. The exit in the bottom of the foreground of the exhibit is placed so that it commences in advance of the main-line curve. The gores are liberally designed with good delineation.

Exhibit 10-14. Trumpet Freeway-to-Freeway Interchange

Four-Leg Designs

Interchanges with four intersection legs may be grouped under five general configurations: (1) ramps in one quadrant, (2) diamond interchanges, (3) single-point urban interchanges (SPUIs), (4) full or partial cloverleafs (including ramps in two or three quadrants), and (5) interchanges with direct and semidirect connections. Operational characteristics and adaptations of each configuration are discussed separately. Actual examples of existing or planned interchanges are presented for each type.

Ramps in One Quadrant

Interchanges with ramps in only one quadrant have application for an intersection of roadways with low traffic volumes. Where a grade separation is provided at an intersection because of topography, even though volumes do not justify the structure, a single two-way ramp of near-minimum design usually will suffice for all turning traffic. The ramp terminals may be simple T intersections.

Appropriate locations for this type of interchange are very limited. A typical location would be at the intersection of a scenic parkway and a State or county two-lane highway where turning movements are light, there is minimal truck traffic, and the terrain and preservation of natural environment typically take precedence over providing additional ramps.

At some interchanges it may be appropriate to limit ramp development to one quadrant because of topography, culture, or other controls, even though the traffic volumes justify more extensive turning facilities. With ramps in only one quadrant, a high degree of channelization at the ramp terminals, at the median, and at the left-turn lanes on the through facilities is normally needed to control turning movements properly.

In some instances, a one-quadrant interchange may be constructed as the first step in a stage construction program. In this case, the initial ramps should be designed as a part of the ultimate development.

Exhibit 10-15A illustrates a one-quadrant interchange at the intersection of a State highway and a scenic parkway located in a rural mountainous area. The elongated shape of the ramp was determined largely by topography. Traffic entering both through roadways is under stop-sign control. Although traffic volumes are low, the turning traffic consists of a substantial proportion of the total volume.

Exhibit 10-15B is a one-quadrant interchange designed to function as an early phase of stage construction. On future construction, it is readily adaptable to become a part of a full or partial cloverleaf interchange without major renovation. The channelization, although elaborate, is conducive to safety and attractive landscaping.

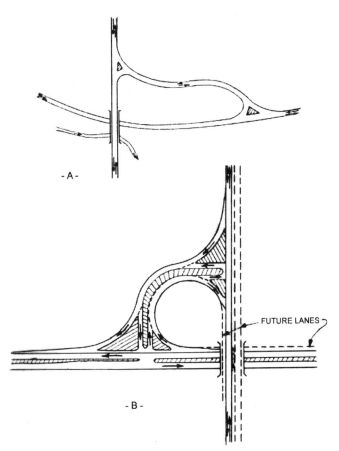

- A -

- B -

- FUTURE LANES

Exhibit 10-15. Four-Leg Interchanges, Ramps in One Quadrant

Diamond Interchanges

The simplest and perhaps most common interchange configuration is the diamond. A full diamond interchange is formed when a one-way diagonal ramp is provided in each quadrant. The ramps are aligned with free-flow terminals on the major highway, and the left turns at grade are confined to the crossroad. The diamond interchange has several advantages over a comparable partial cloverleaf: all traffic can enter and leave the major road at relatively high speeds, left-turning maneuvers entail little extra travel, and a relatively narrow band of right-of-way is needed, sometimes no more than that needed for the highway alone.

Diamond interchanges have application in both rural and urban areas. They are particularly adaptable to major-minor crossings where left turns at grade on the minor road are fitting and can be handled with minimal interference to traffic approaching the intersection from either direction. The intersection on the crossroad formed by the terminals functions as any other T intersection at grade and should be designed as outlined in Chapter 9. However, because these intersections have four legs, two of which are one-way, they present a problem in traffic control to prevent wrong-way entry from the crossroad. For this reason, a median should be provided on the crossroad to facilitate proper channelization. While this median can be a painted median, a depressed or raised median with a sloping curb is preferred. In most cases, additional signing to help prevent improper use of the ramps should be incorporated in the interchange design. Wrong-way entry problems are further discussed in Chapter 9 in the section on "Design to Discourage Wrong-Way Entry."

Diamond interchanges usually need signalization where the cross street carries moderate-to-large traffic volumes. The capacity of the ramps and that of the cross street may be determined by the signal-controlled ramp terminals. In such a case, roadway widening may be needed on the ramps or on the cross street through the interchange area, or both. While a single-lane ramp may adequately serve traffic from the freeway, it may have to either be widened to two or three lanes or be channelized for storage at the cross street, or both, in order to provide the capacity needed for the at-grade condition. This design would prevent stored vehicles from extending too far along the ramps or onto the freeway. Left-turning movements in the most common diamond interchange configurations, as shown in Exhibit 10-16, usually need multiphase control.

Exhibits 10-16 through 10-18 illustrate a variety of diamond interchange configurations. These interchanges may be designed with or without frontage roads. Designs with frontage roads are common in built-up areas, often as part of a series of such interchanges along a freeway. Ramps should connect to the frontage road at a minimum distance of 100 m [350 ft] from the crossroad. Greater distances are desirable to provide adequate weaving length, space for vehicle storage, and turn lanes at the crossroad. Exhibit 10-16C is a spread diamond rural interchange with the potential for conversion to a cloverleaf.

In a diamond interchange, the greatest impediment to smooth operations is left-turning traffic at the crossroad terminal. Arrangements that may be suitable to reducing traffic conflicts are shown in Exhibits 10-17 and 10-18. By using a split diamond (i.e., each pair of ramps connected to a separate crossroad about a block apart), as shown in Exhibit 10-17A, conflicts are

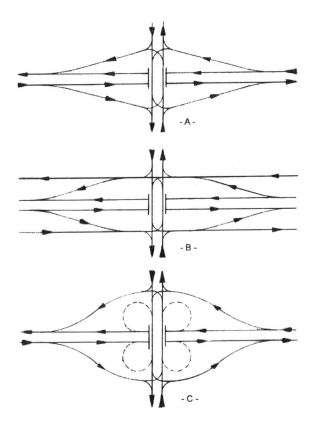

Exhibit 10-16. Diamond Interchanges, Conventional Arrangements

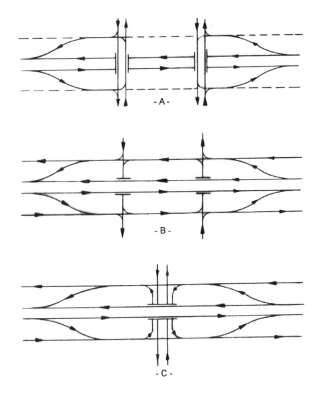

Exhibit 10-17. Diamond Interchange Arrangements to Reduce Traffic Conflicts

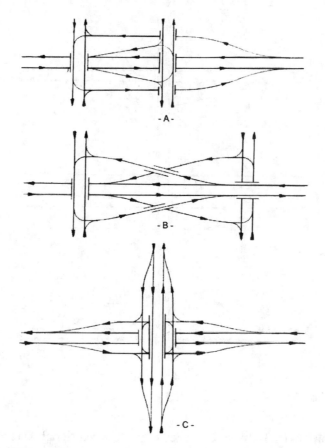

Exhibit 10-18. Diamond Interchanges with Additional Structures

minimized by handling the same traffic at four rather than two crossroad intersections, reducing the left-turn movements at each intersection from two to one. A disadvantage with this arrangement is that traffic leaving the freeway cannot return to the freeway at the same interchange. Frontage roads (shown as dashed lines) are optional.

Exhibit 10-17B shows a split diamond in conjunction with a pair of one-way cross streets and one-way frontage roads. Simplicity of layout and operation of both the crossroad and the at-grade terminals result. Traffic leaving the freeway is afforded easy access to return to the freeway and continue the journey in the same direction.

Exhibit 10-17C shows a diamond interchange with frontage roads and separate turnaround provisions. These are highly desirable if the cross street has heavy traffic volumes and there is considerable demand for the U-turning movement. The turnaround roadways are adjacent to the cross street with additional width provided beneath the structure or, if the cross street overpasses the freeway, on top of the structure. As an alternative, separate structures may be provided for the U-turn movements.

Exhibit 10-18 shows diamond interchanges with more than one structure. The layout in Exhibit 10-18A and the "criss-cross" arrangement in Exhibit 10-18B are sometimes dictated by topographic conditions or right-of-way restrictions. The operational performance of the interchanges in Exhibits 10-18A and 10-18B are the same as those shown in Exhibit 10-17A. The layout of Exhibit 10-18B also may be used to eliminate weaving between two closely spaced interchanges. These layouts may be further modified by the use of one-way operation on the cross streets. The deficiency of both layouts in Exhibits 10-18A and 10-18B is that traffic that has left the freeway cannot return directly to it and continue in the same direction. The spacing of the crossroads is determined primarily by grade constraints and acceleration and deceleration lengths.

The double or three-level diamond in Exhibit 10-18C, which has a third-level structure and four pairs of ramps, provides uninterrupted flow of through traffic on both of the intersecting highways. Only the left-turning movements cross at grade. This arrangement is applicable where the cross street carries large traffic volumes and topography is favorable. The right-of-way needed is much less than that for other layouts having comparable capacity. Although large through and turning volumes can be handled, it is disadvantageous for intersections of two freeways in that some of the turning movements must either stop or slow down substantially. Signals are used in high-volume situations, and their efficiency is dependent on the relative balance in left-turn volumes. They are normally synchronized to provide continuous movement through a series of left turns once the area is entered.

Exhibits 10-19 and 10-20 present examples of diamond interchange configurations that are somewhat different from the conventional application. Exhibit 10-19 shows a freeway that features a three-level diamond interchange at the crossing of another controlled-access facility. In urban areas, where a crossing street carries a high volume of traffic, the three-level diamond interchange may be appropriate. Exhibit 10-20 shows a spread diamond interchange with provision in all four quadrants for future loop ramps. When there is sufficient traffic to justify the additional loop ramps, the diamond ramps would also be modified to convert the interchange to a full cloverleaf. The interchange in the exhibit is currently located in a rural area, but the area is rapidly developing. The ramps are all 4.2 m [14 ft] wide with grades less than 1 percent. The minor road is a four-lane divided highway with left-turn lanes at the diamond ramp terminals. Traffic control consists only of stop signs for traffic on the off-ramps from the freeway.

Exhibit 10-19. Freeway With a Three-Level Diamond Interchange

Exhibit 10-20. Existing Four-Leg Interchange With Diamond Stage Construction

It may be beneficial to consider the use of "X" pattern ramps at diamond interchanges in urban areas. With this ramp pattern, the entrance occurs prior to the intersection while the exit occurs after the cross street. This configuration, as shown in Exhibit 10-21, can improve traffic flow characteristics for the through roadways around diamond interchanges. However, driver expectancy should be considered.

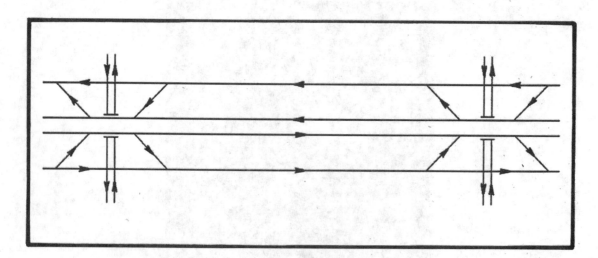

Exhibit 10-21. X-Pattern Ramp Arrangement

Single-Point Urban Interchanges

The single-point urban interchange (SPUI) is a relatively recent development in interchange design with the first SPUIs being constructed in the early 1970s. The SPUI is also known as an urban interchange or a single-point diamond interchange. The primary features of a SPUI are that all four turning moves are controlled by a single traffic signal and opposing left turns operate to the left of each other.

SPUIs are typically characterized by narrow right-of-way, high construction costs, and greater capacity than conventional tight diamond interchanges. These interchanges can be constructed either with or without frontage roads. They are primarily suited for urban areas where right-of-way is restricted but may also be applicable to rural settings where it is undesirable to utilize adjacent right-of-way due to environmental, geographical, or other constraints.

SPUIs offer several advantages. These include construction in a relatively narrow right-of-way, resulting in potentially significant cost reductions. The primary operational advantage of this interchange configuration is that vehicles making opposing left turns pass to the left of each other rather than to the right, so their paths do not intersect. In addition, the right-turn movements are typically free-flow movements and only the left turns must pass through the signalized intersection. As a result, a major source of traffic conflict is eliminated, increasing overall intersection efficiency and reducing the traffic signal needed from four-phase to three-phase operation. Since the SPUI has only one intersection, as opposed to two for a diamond interchange, the operation of the single traffic signal on the crossroad may result in reduced delay through the intersection area when compared to a diamond interchange. Curve radii for left-turn movements through the intersection are significantly flatter than at conventional intersections and, therefore, the left turns move at higher speeds. The above-mentioned operational improvements result in a higher capacity than a conventional tight diamond interchange.

The primary disadvantage of SPUIs is high construction costs associated with bridges. Overpass SPUIs need long bridges to span the large intersection below. A two-span structure is not a design option because a center column would conflict with traffic movements. Single-span overpass bridges are typically 65 m [220 ft] in length, while three-span bridges often exceed 120 m [400 ft]. As shown in Exhibit 10-22, the SPUI underpass tends to be wide and often is "butterfly" in shape, resulting in high costs. Where right-of-way is tightly constrained, SPUIs typically utilize extensive retaining walls, further adding to the cost. However, the higher construction cost of SPUIs is often offset by the reduced right-of-way cost. Exhibit 10-23 shows an underpass SPUI in restricted right-of-way.

A second potential problem encountered with SPUIs is the length and geometry of the path for left-turning vehicles through the intersection. Like most typical intersections, left-turning vehicles pass to the left of opposing left-turning vehicles. However, due to the size and distance between opposing approaches, the path of left-turning vehicles does not resemble a quarter of a circle found at typical intersections, but rather resembles a quarter of an ellipse. To provide positive guidance for this non-traditional path, various features have been developed. At a minimum, 0.6-m [2-ft] dashed lane lines should be painted through the intersection. Another

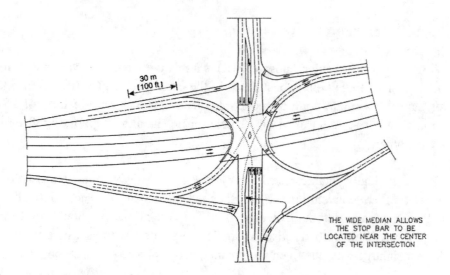

Exhibit 10-22. Underpass Single Point Urban Interchange

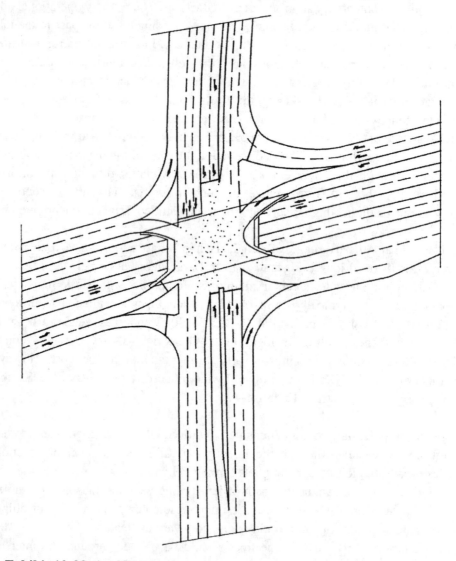

Exhibit 10-23. An SPUI Underpass in Restricted Right-of-Way

option is to install directional guide lights, such as airport runway lights, flush with the pavement surface. These lights can be sequenced in coordination with the traffic signal at the intersection.

On underpass SPUIs, airport runway lights placed in the bridge deck may conflict with the reinforcing steel and therefore may not be practical.

A skew angle between the two roadway alignments has an adverse effect on SPUIs because it increases clearance distances and adversely affects sight distance. Severe skew in alignments may also increase the length of the bridge and widen the distance between the stop bars on the local streets. Extreme care should be exercised in planning SPUIs when the skew angle approaches 30 degrees.

Several basic design considerations can optimize the geometrics and operation of an SPUI. First, it is desirable that the left-turn curve be a single radius. This will, however, typically result in additional right-of-way and/or a larger bridge structure. Where it is not practical to provide a single radius, and curves are compounded from a larger to a smaller radii, the second curve should be at least half the radius of the first. Another important design feature is to provide stopping sight distance on the left-turn movements equal to or exceeding the design speed for the curve radius involved. A third design feature that can improve intersection operation is to provide additional median width on the cross street. The stop bar location on the cross street is dependent on the wheel tracks from the opposing ramp left-turn movement (See Exhibit 10-22). By widening the median, the stop bar on the cross street can be moved forward, thus reducing the size of the intersection and the distance each vehicle travels through the intersection. The results include greater available green time and less potential driver confusion due to an expansive intersection area.

An SPUI with frontage roads, as illustrated in Exhibit 10-24, introduces additional considerations into the design. Frontage roads should be one way in the direction of the ramp traffic. A slip ramp from the mainline to the frontage road provides access to and from the intersection. This ramp should connect to the frontage road at least 200 m [650 ft], and preferably greater than 300 m [1,000 ft], from the crossroad. The traffic signal needs a fourth phase to provide through movements on the frontage roads. A free-flow, U-turn movement may be desirable to expedite movements from one direction on the frontage road to the other.

Because of the size, shape, and operational characteristics of SPUIs, pedestrian movement through the intersection should be given careful consideration. Pedestrian crossing of the local street at ramp terminals typically adds a signal phase and involves considerable green time, resulting in reduced operational efficiency. Therefore, the overall design should include provision for pedestrian crossings at adjacent intersections instead of at the ramp terminal intersection. Pedestrian movements parallel to the local street are more readily handled. If, however, crosswalks are provided at ramps, they should be perpendicular to the ramp direction of travel and near to the local street. Perpendicular crosswalks minimize the length of the crossing and therefore minimize conflicting movements. Crosswalks located near the local street meet driver expectation and allow good sight distance to the pedestrian crossing.

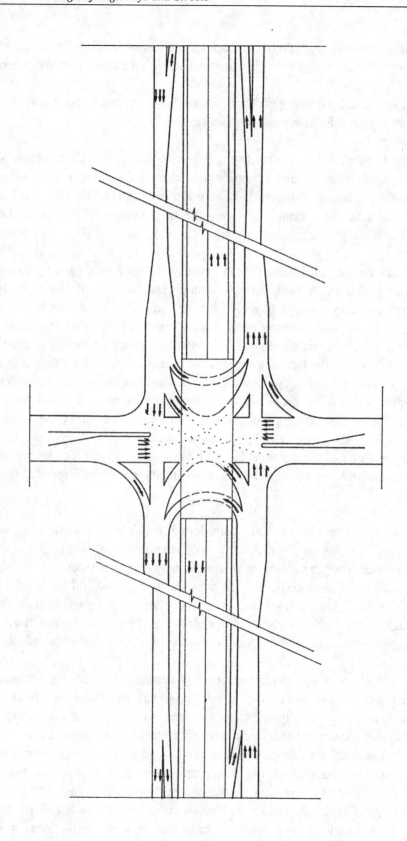

Exhibit 10-24. Overpass Layout With a Frontage Road and a Separate U-Turn Movement

Right-turn lanes at SPUIs are typically separated from the left-turn lanes, often by a considerable distance. The exit ramp right turn can be a free or controlled movement. The design of free right turns should include an additional lane on the cross street beginning at the free right-turn lane for at least 60 m [200 ft] before being merged. Free-flow right turns from the exit ramp to an arterial crossroad are not desirable when the nearest intersection on the crossroad is within 150 m [500 ft] because there may be inadequate weaving distance between the exit ramp and the adjacent intersection. Heavy pedestrian traffic also can diminish the desirability of free right-turn lanes by adding a potential conflict with non-controlled vehicular traffic. Where the right-turn move is controlled by stop sign or traffic signal, adequate right-turn storage on the exit ramp should be provided to prevent blockage of vehicles turning left or traveling straight. Free-flow right turns on entrance ramps pose little operational problem, assuming adequate merge length is provided on the entrance ramp. As shown at the upper left portion of Exhibit 10-22, the right-turn lane should extend at least 30 m [100 ft] beyond the convergence point before beginning the merge.

Exhibit 10-25 illustrates both an underpass and an overpass SPUI.

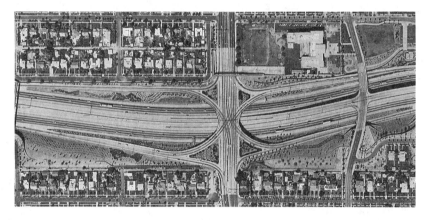

Underpass SPUI

Overpass SPUI

Exhibit 10-25. Underpass SPUI and Overpass SPUI

Cloverleafs

Cloverleafs are four-leg interchanges that employ loop ramps to accommodate left-turning movements. Interchanges with loops in all four quadrants are referred to as "full cloverleafs" and all others are referred to as "partial cloverleafs." A full cloverleaf may not be warranted at major-minor crossings where, with the provision of only two loops, freedom of movement for traffic on the major road can be maintained by confining the direct at-grade left turns to the minor road. The principal disadvantages of the cloverleaf are the additional travel distance for left-turning traffic, the weaving maneuver generated, the very short weaving length typically available, and the relatively large right-of-way areas needed. When collector-distributor roads are not used, further disadvantages include weaving on the main line, the double exit on the main line, and problems associated with signing for the second exit. Because cloverleafs are considerably more expansive than diamond interchanges, they are less common in urban areas and are better adapted to suburban or rural areas where space is available.

The travel distance on a loop, as compared with that of a direct left turn at grade, increases rapidly with an increase in design speed. On a loop designed for 30 km/h (30-m radius [20 mph (110-ft radius)], the extra travel distance is approximately 200 m [650 ft]; for a design speed of 40 km/h (50-m radius) [25 mph (170-ft radius)], it is about 300 m [1,000 ft], and for a design speed of 50 km/h (80-m radius) [30 mph (250-ft radius)], it is around 500 m [1,500 ft]. Thus, for a 10-km/h [5-mph] increase in design speed, extra travel distance increases 50 percent; the right-of-way area needed increases by about 130 percent.

Travel time on loops varies almost directly with the design speed, the increased speed being more than balanced by increased distance. For an increase of 10 km/h [5 mph] in loop design speed, travel time increases 20 to 30 percent or approximately 7 s. This increase in travel time is actually somewhat less when the overall maneuver is considered because of deceleration and acceleration outside the limits of the loop proper. (In any case, the travel time via a loop may be much less than the travel time when making a direct left turn.)

The advantages of increased speed should be weighed against the disadvantages of increased travel time, distance, and right-of-way. It should also be noted that large trucks may not be able to operate as efficiently on smaller radii curves. Considering all factors, experience shows that the practical size of loops resolves into approximate radii of 30 to 50 m [100 to 170 ft] for minor movements on highways with design speeds of 80 km/h [50 mph] or less and 50 to 75 m [170 to 250 ft] for more important movements on highways with higher design speeds. A continuous additional lane is needed for deceleration, acceleration, and weaving between the on- and off-loop ramps. Additional structure width or length is usually needed for this lane. The lateral placement of bridge abutments or columns, or main-line bridge widths, may have to be reviewed due to the effects of offtracking at the exit end of the ramp curve.

The cloverleaf involves weaving maneuvers as discussed in the section of this chapter on "Weaving Sections." This is not objectionable when the left-turning movements are relatively light, but when the sum of traffic on two adjoining loops approaches about 1,000 vph, interference mounts rapidly, which results in a reduction in speed of through traffic. The weaving lengths presented in Exhibit 10-68 later in this chapter should be provided on low-volume cloverleaf interchanges. When the weaving volume in a particular weaving section exceeds 1,000 vph, the quality of service on the main facility deteriorates rapidly, thus generating a need to transfer the weaving section from the through lanes to a collector-distributor road. A loop rarely operates with more than a single line of vehicles, regardless of the roadway width, and thus has a design capacity limit of 800 to 1,200 vph, the higher figure being applicable only where there are no trucks and where the design speed for the ramp is 50 km/h [30 mph] or higher. Loop ramp capacity is, therefore, a major control in cloverleaf designs. Loops may be made to operate with two lanes abreast, but only by careful attention to design of the terminals and design for weaving, which would need widening by at least two additional lanes through the separation structure. To accomplish this type of design, the terminals should be separated by such great distances and the loop radii should be made so large that cloverleafs with two-lane loops are not generally economical from the standpoint of right-of-way, construction, cost, and amount of out-of-direction travel. Loops that operate with two lanes of traffic, therefore, are considered exceptional cases.

Where no direct left turns are permitted on either the main facility and the crossroad, but all turning movements are to be accommodated, a four-quadrant cloverleaf interchange is the minimum interchange configuration that will suffice. When a full cloverleaf interchange is used in conjunction with a freeway and the sum of the traffic on two adjoining cloverleaf loops approaches about 1,000 vph, collector-distributor roads should be considered. Collector-distributor roads are generally not cost effective where the ramp volumes are low and are not expected to increase significantly. The use of acceleration or deceleration lanes with cloverleaf interchanges is one possible alternative to collector-distributor roads.

Exhibit 10-26 shows an existing cloverleaf interchange between a freeway and an expressway with partial control of access, located on the edge of a rapidly expanding suburban area. Collector-distributor roads have been provided along the freeway in expectation of heavy weaving volumes. Because of the high unit cost of right-of-way, this design was more economical with the collector-distributor roads and loop ramps with smaller radii. A design with ramps of larger radii would have been used if the collector-distributor roads had not been provided. The grades are relatively flat, with 3 percent being the maximum.

Exhibit 10-26. Four-Leg Interchange, Full Cloverleaf With Collector-Distributor Roads

Exhibit 10-27 shows a cloverleaf interchange between a freeway and a divided arterial street. Collector-distributor roads serve all the ramp movements on the freeway.

Exhibit 10-27. Cloverleaf Interchange With Collector-Distributor Roads

Partial Cloverleaf Ramp Arrangements. In the design of partial cloverleafs, the site conditions may offer a choice of quadrants to use. However, at a particular interchange site, topography and culture may be the factors that determine the quadrants in which the ramps and loops can be developed. There is considerable operational advantage in certain arrangements of ramps. These are discussed and summarized in the following analysis.

Ramps should be arranged so that the entrance and exit turns create the least impediment to the traffic flow on the major highway. The following guidelines should be considered in the arrangement of the ramps at partial cloverleafs:

- The ramp arrangement should enable major turning movements to be made by right-turn exits and entrances.
- Where through-traffic volume on major highways is decidedly greater than that on the intersecting minor road, preference should be given to an arrangement placing the right

turns, either exit or entrance, on the major highway even though it results in a direct left turn off the crossroad.

These controls do not always lead to the most direct turning movements. Instead, drivers frequently are required to first turn away from or drive beyond the road of their intended direction. Such arrangements cannot be avoided if the through-traffic movements, for which the separation is provided, are to be facilitated to the extent practical.

Exhibit 10-28 illustrates the manner in which the turning movements are made for various two- and three-quadrant cloverleaf arrangements. When ramps in two quadrants are adjacent and on the same side of the minor road, as shown in Exhibits 10-28A and 10-28B, or diagonally opposite each other, as shown in Exhibits 10-28E and 10-28F, all turning movements to and from the major road are accomplished by right turns. Any choice between the arrangement in Exhibit 10-28A and its alternate arrangement (ramps in the other two quadrants) will depend on the predominant turning movements or the availability of right-of-way, or both. When the ramps in two quadrants are adjacent but on the same side of the major road (Exhibits 10-28B and 10-28D), four direct left turns fall on the major road. This arrangement and its alternate are the least desirable of the six possible arrangements, and their use should be avoided.

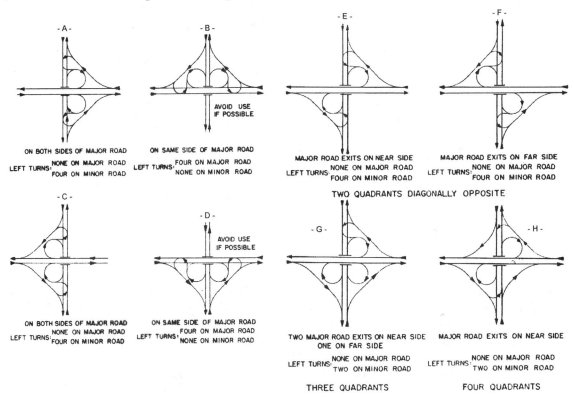

Exhibit 10-28. Schematic of Partial Cloverleaf Ramp Arrangements, Exit and Entrance Turns

The arrangement with ramps in diagonally opposite quadrants is advantageous in that the turning movements in both directions in the quadrants in which the ramps are located are made by desirable right-turn exits and entrances. At interchanges where turning movements in one quadrant predominate, the best two-quadrant arrangement has ramps in that quadrant and in the quadrant diagonally opposite. Where turning movements in two adjacent quadrants are of nearly the same importance, arrangements shown in Exhibits 10-28A, 10-28E, and 10-28F are applicable in that all turns to and from the major road are on the right. However, the arrangement in Exhibit 10-28E is preferable because the ramps are on the near side of the structure as drivers approach on the major road. With this plan, it may be practical to provide for high-speed turns from the major road, and drivers desiring to turn are not confused by ramps that may be hidden by the structure, as shown in Exhibit 10-28F.

There are four possible arrangements for ramps in three quadrants, including the arrangement in Exhibit 10-28G and the alternate arrangements in which each of the other three quadrants has no ramps. In an arrangement with ramps in three quadrants, six of the eight turning movements can be made by right-turn exits and entrances, and the other two are made by right turns on the major road and corresponding left turns on the minor road. The determination of which quadrant is to be without ramps is usually dependent on the availability of right-of-way and the predominant turning movements to be handled.

In some cases, it is desirable to provide diagonal ramps in all four quadrants, but with loops in one, two, or three of the quadrants. Exhibit 10-28H shows a design with loops in diagonally opposite quadrants. This design has the advantage of providing all right exits. Storage of vehicles waiting to make the left turn at the at-grade intersections occurs on the ramp and not on either of the through highways. In addition, there is no weaving on the major highway.

Exhibit 10-29 shows an existing partial cloverleaf interchange where a two-lane highway overpasses a six-lane freeway in a suburban area. The design consists of ramps in diagonally opposite quadrants, arranged to needed minimum frontage on the crossroad. Directional islands and merge lanes at the ramp terminals permit free-moving right turns to and from the crossroad. The only traffic control needed is stop signs at the crossroad for the left turns from the off-ramps. Protected left-turn bays on the crossroad are desirable.

Exhibit 10-29. Four-Leg Interchange (Partial or Two-Quadrant Cloverleaf with Ramps Before Main Structure)

Exhibit 10-30 shows an existing partial cloverleaf with ramps in diagonally opposite quadrants. In relation to the major highway, the ramps are in opposite quadrants from those of the two previous examples. A major highway crosses over a four-lane freeway. Ramps are located to avoid heavy commercial and residential development in the other two quadrants. Direct left turns are confined to the minor road where the terminals are channelized by divisional islands. In the right half of the exhibit, the intersection of the ramps with the minor highway is signalized. The outer connections are designed to encourage high-speed merging with the freeway traffic. The loops have slightly larger radii than the previous example and are designed for a speed of 50 km/h [30 mph].

Exhibit 10-30. Four-Leg Interchange (Partial or Two-Quadrant Cloverleaf with Ramps Beyond Main Structure)

Any other arrangement of two loop ramps and four diagonal ramps does involve direct left turns from the minor road onto a ramp. Note the triangular island channelization at the ends of the two ramps from which left turns are made onto the two-lane high-volume crossroad. This design provides adequate left-turn storage and a free-flow right-turn movement from the freeway to the crossroad. The design of free-flow right turns should include an extension of the right turn-lane for at least 60 m [200 ft] along the crossroad to allow adequate space for merging. Free right turns are not desirable when the adjacent intersection is within 150 m [500 ft] because there may be insufficient weaving area for vehicles making a right turn onto the crossroad and then turning left at the adjacent intersection.

The loop ramps are carried on separate roadways under structure spans, so the loop traffic merges with the outer connection traffic before entering the freeway. This design entails a considerable extra cost but reduces the points of conflict on the freeway.

Directional and Semidirectional Interchanges

Direct or semidirect connections are used for important turning movements to reduce travel distance, increase speed and capacity, eliminate weaving, and avoid the need for out-of-direction travel in driving on a loop. Higher levels of service can be realized on direct connections and, in some instances, on semidirect ramps because of relatively high speeds and the likelihood of better terminal design. Often a direct connection is designed with two lanes. In such cases, the ramp capacity may approach the capacity of an equivalent number of lanes on the through highway.

In rural areas, there rarely is a volume justification for provision of direct connections in more than one or two quadrants. The remaining left-turning movements usually are handled satisfactorily by loops or at-grade intersections. At least two structures are needed for such an interchange. There are many possible arrangements with direct and semidirect connections, but only the more basic arrangements are discussed herein.

A direct connection is defined as a one-way roadway that does not deviate greatly from the intended direction of travel. Interchanges that use direct connections for the major left-turn movements are directional interchanges. Direct connections for one or all left-turn movements would qualify an interchange to be also considered directional even if the minor left-turn movements are accommodated on loops.

When one or more interchange connections are indirect in alignment yet more direct than loops, the interchange is described as semidirectional. All left-turn connections or only those that accommodate major left-turn movements may be semidirect in alignment. On direct or semidirect interchanges usually more than one highway grade separation is involved.

Semidirect or direct connections for one or more left-turning movements are often appropriate at major interchanges in urban areas. In fact, interchanges involving two freeways almost always require directional layouts. In such cases, turning movements in one or two quadrants often are comparable in volume to through movements. In comparison to loops, direct or semidirect connections have shorter travel distance, higher speeds of operation, and a higher level of service, and they often avoid the need for weaving.

There are many configurations for directional interchanges that use various combinations of directional, semidirectional, and loop ramps. Any one of them may be appropriate for a certain set of conditions, but only a limited number of patterns are generally used. The most common configurations fill the least space, have the fewest or least complex structures, minimize internal weaving, and fit the common terrain and traffic conditions. Basic patterns of selected semidirectional and all directional interchanges are illustrated in Exhibits 10-31 through 10-33, with distinctions made as to configurations with and without weaving.

With Loops and Weaving. Common arrangements where turning movements in one quadrant predominate are shown in Exhibits 10-31A and 10-31B. The predominant turning movement bypasses the central portion of the interchange via two-lane terminals. The minor turning movements pass through weaving sections between loops on each highway. In both

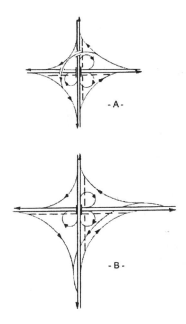

NOTE: WEAVING ADJACENT TO THE THROUGH LANES IS
ELIMINATED BY PROVIDING COLLECTOR-
DISTRIBUTOR ROADS AS SHOWN BY DOTTED LINES.

Exhibit 10-31. Semidirect Interchanges With Weaving

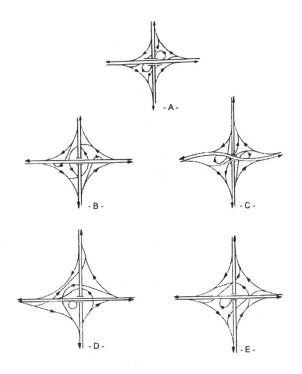

Exhibit 10-32. Semidirect Interchanges With No Weaving

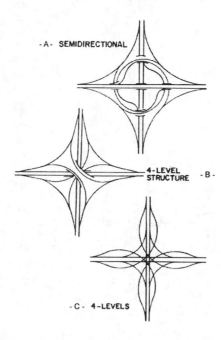

Exhibit 10-33. Semidirectional and Directional Interchanges—Multilevel Structures

figures, a semidirect connection is used without affecting the alignment of the intersecting highways. Both arrangements involve three structures, and the area occupied is about the same as or somewhat greater than a full cloverleaf.

The efficiency and capacity of all the layouts shown in Exhibit 10-31 may be improved by eliminating weaving on the main roadways through the use of a collector-distributor road, as shown dashed in Exhibit 10-31.

With Loops and No Weaving. Semidirectional interchanges that do not involve weaving but include loops are shown in Exhibit 10-32. The through lanes do not need to be spread apart for any of these configurations; however, four or more structures are needed. Single exits on the right side along with right-hand entrances enhance the operational characteristics of these designs.

Fully Directional. Fully directional interchanges are generally preferred where two high-volume freeways intersect. Since traffic movements between the two freeways are free-flow with this interchange configuration, there are no at-grade intersections, only direct ramp connections from one freeway to the other. Fully directional interchanges are costly to construct due to the increased number and length of ramps and the increased number of bridge crossings, but offer high capacity movements for both through and turning traffic with comparatively little additional area needed for construction. The configuration and design of each interchange is uniquely based on the traffic volumes and patterns, environmental considerations, costs, etc. As a result, detailed and time-consuming studies are usually needed for each interchange and should include a study of all likely alternatives. A detailed discussion is, therefore, not within the purview of this policy; however, Exhibits 10-33A through 10-33C are included to show diagrammatic layouts.

Weaving, left-side exits, and left-side entrances are undesirable within directional interchanges; however, there may be instances where they cannot be reasonably avoided because of site restrictions or other considerations. With heavy left-turn movements, the terminals should be designed as major forks and branch connections, as covered later in this chapter.

The most widely used directional interchange configuration is the four-level layout shown in Exhibit 10-33B. A variation of this configuration is the four-level interchange with two exits from both major roadways, as shown in Exhibit 10-33C. Exhibit 10-34 shows a diagram of an existing interchange between two high-volume freeways in a suburban area. Other examples of semidirectional or directional interchanges are shown in Exhibits 10-35 through 10-37.

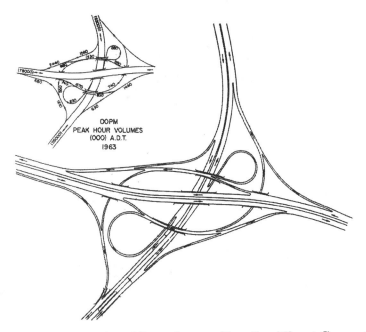

Exhibit 10-34. Directional Interchange, Two Semidirect Connections

Exhibit 10-35. Four-Level Directional Interchange

Exhibit 10-36. Four-Level Directional Interchange

Exhibit 10-37. Semidirectional Interchange With Loops

Other Interchange Configurations

Offset Interchanges

Exhibit 10-38 illustrates an offset interchange arrangement between freeways that may be suitable where there are major buildings or other developments near the crossing of the freeways. This arrangement consists of a pair of trumpet interchanges, one on each highway, which is connected to each other with a ramp highway. The length of the connecting roadway depends on the distances between each trumpet interchange and the crossing of the freeways. As illustrated in Exhibit 10-38, the ramp highway may include local service connections, in this case accommodated by a diamond interchange.

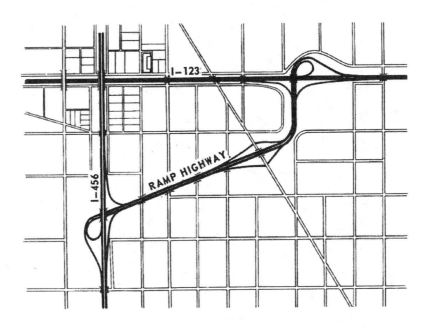

Exhibit 10-38. Offset Interchange via Ramp Highway

A disadvantage of this interchange configuration is the substantial out-of-direction travel for six of the eight turning movements between the freeways. However, when one pair of these movements is predominant, the ramp highway may be located in such a way that favors these movements. When considered from a city street system, the overall configuration may be confusing to unfamiliar drivers. With adequate signing, however, most motorists should be unaware of the unusual pattern.

Combination Interchanges

When one or two turning movements have very high volumes with respect to the other turning movements, analysis may indicate the need for a combination of two or more of the previously discussed interchanges.

Exhibit 10-39 shows an existing diamond interchange in which a semidirectional ramp has been added to accommodate the high-volume, left-turning traffic. The complementary high-volume, right-turning movement in the opposite direction of travel is provided with a liberal radius to facilitate high speeds. Because the cross street connects a city on the left with a four-lane freeway, relatively high volumes result in that direction. This design needs two more structures than a diamond interchange. Three of the crossroad terminals are channelized with separate right- and left-turning roadways.

Exhibit 10-39. Four-Leg Interchange, Diamond With a Semidirect Connection

Exhibit 10-40 presents an existing cloverleaf interchange between two freeways in which a semidirect connection has been substituted for the loop ramp in the upper left quadrant. The interchange is located at the edge of a suburban area that is rapidly developing both industrially and residentially, and where considerably higher volumes are expected in the future. The ramps along the left-to-right roadway have been combined so that the traffic merges with the freeway traffic at a single point at the upper left. The semidirectional turning roadway permits traffic to travel at operating speeds approaching that on the main roadways. The complement of this movement is provided with a high-type two-lane ramp with more liberal radii than provided for the remaining movements. The design was complicated by an adjacent railroad and two adjacent important local roads, one of which was relocated from the center of the interchange area. An additional structure and spans were needed for the local road crossings to provide the two high-speed directional ramps.

Exhibit 10-40. Four-Leg Interchange, Cloverleaf With a Semidirect Connection

Unusual ramp configurations are shown in Exhibits 10-41 and 10-42. Exhibit 10-41 shows a complex interchange arrangement at a crossing of two major routes in an urban area. The interchange design chosen minimizes disruption of existing development.

Exhibit 10-41. Complex Interchange Arrangement

Exhibit 10-42. Freeway with a Three-Level Cloverleaf Interchange

Exhibit 10-42 shows the unusual arrangement of a three-level cloverleaf. In this case, environmental constraints and other site restrictions made the use of this configuration appropriate. Advance traffic studies were carefully prepared to ensure that the loop ramps would continue to function properly as the traffic volumes increased. Signing was also critical to the proper operation of the facility. Another interesting aspect of this interchange is the park-and-ride public transportation facility provided underneath the structures on the lowest level. While this interchange does not have widespread applicability, it exemplifies the type of innovative design that is sometimes needed to provide a functional facility under restrictive conditions.

General Design Considerations

Determination of Interchange Configuration

The need to use interchanges may occur in the design of all functionally classified roadways, as discussed previously in the section on "Warrants for Interchanges and Grade Separations" in this chapter. Interchange configurations are covered in two categories, "system interchanges" and "service interchanges." The term "system interchanges" is used to identify interchanges that connect two or more freeways, whereas the term service interchange applies to interchanges that connect a freeway to lesser facilities.

In rural areas, interchange configurations are selected primarily on the basis of service demand. When the intersecting roadways are freeways, directional interchanges may be needed for high-turning volumes.

A combination of directional, semidirectional, and loop ramps may be appropriate where turning volumes are high for some movements and low for others. When loop ramps are used in combination with direct and semidirect ramp designs, it is desirable that the loops be arranged in such a way that weaving sections are avoided.

A cloverleaf interchange is the minimum design that can be used at the intersection of two fully controlled access facilities or where left turns at grade are prohibited. A cloverleaf interchange is adaptable in a rural environment where right-of-way is not prohibitive and weaving is minimal. When designing a cloverleaf interchange, careful attention should be given to the potential improvement in operational quality that would be realized if the design included collector-distributor roads on the major roadway.

A simple diamond interchange is the most common interchange configuration for the intersection of a major roadway with a minor facility. The capacity of a diamond interchange is limited by the capacity of the at-grade terminals of the ramps at the crossroad. High through and turning volumes could preclude the use of a simple diamond unless signalization is used.

Partial cloverleaf designs with loops in opposite quadrants are very desirable because they eliminate the weaving problem associated with the full cloverleaf designs. They may also provide superior capacity to other interchange configurations. Partial cloverleaf designs are especially appropriate where rights-of-way are not available (or expensive) in one or more quadrants or some of the movements are disproportionate to the others. This is especially true for heavy left-turn volumes where loop ramps may be utilized to accommodate the left-turn movements.

Generally, interchanges in rural areas are widely spaced and can be designed on an individual basis without any appreciable effect from other interchanges within the system. However, the final configuration of an interchange may be determined by the need for route continuity, uniformity of exit patterns, single exits in advance of the separation structure, elimination of weaving on the main facility, signing potential, and availability of right-of-way. Sight distance on the highways through a grade separation should be at least as long as that needed for stopping and preferably longer. Where exits are involved, decision sight distance is preferred, although not always practical.

Selecting an appropriate interchange configuration in an urban environment involves considerable analysis of prevailing conditions so that the most practical interchange configuration alternatives can be developed. At a new location, it is desirable that the interchange be planned into the location study so that the final alignment is compatible, both horizontally and vertically, with the interchange site. Generally, in urban areas, interchanges are so closely spaced that each interchange may be influenced directly by the preceding or following interchange to the extent that additional traffic lanes may be needed to satisfy capacity, weaving, and lane balance.

On a continuous urban route, all the interchanges should be integrated into a system design rather than considered on an individual basis. Line sketches for the entire urban corridor can be prepared, and several alternate interchange combinations developed for analysis and comparisons.

During the analysis procedure, a thorough study of the crossroad should be made to determine its potential for handling the heavier volume of traffic that an interchange would discharge. The ability of the crossroad to receive traffic from and discharge traffic to the main roadway has considerable bearing on the interchange geometry. For example, loop ramps may be needed to eliminate heavy left turns on a conventional diamond interchange.

In the process of developing preliminary line-sketch studies, systems interchanges may be inserted at freeway-to-freeway crossings and varying combinations of service interchanges developed for lesser crossroads. Generally, cloverleaf interchanges with or without collector-distributor roads are not practical for urban construction because of the excessive right-of-way needs.

Once several alternates have been prepared for the system design, they can be compared on the following principles: (1) capacity, (2) route continuity, (3) uniformity of exit patterns, (4) single exits in advance of the separation structure, (5) with or without weaving, (6) potential for signing, (7) cost, (8) availability of right-of-way, (9) potential for stage construction, and (10) compatibility with the environment. The most desirable alternatives can be retained for plan development.

In the case of an isolated interchange well removed from the influence of other interchanges, the criteria set forth for rural interchange determination apply. Exhibit 10-43 depicts interchanges that are adaptable on freeways as related to classifications of intersecting facilities in rural, suburban, and urban environments.

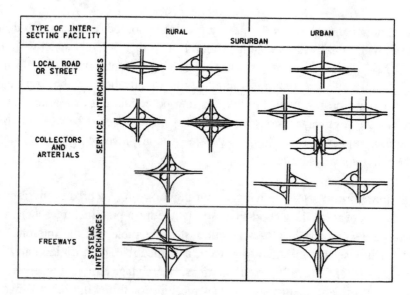

Exhibit 10-43. Adaptability of Interchanges on Freeways as Related to Types of Intersecting Facilities

Approaches to the Structure

Alignment, profile, and cross section. Traffic passing through an interchange should be afforded the same degree of utility and safety as that given on the approaching highways. The design speed, alignment, profile, and cross section in the intersection area, therefore, should be consistent with those on the approaching highways, even though this may be difficult to attain. The presence of the structure itself is somewhat of an obstruction, which should not be augmented by inconsistent designs that might encourage undesirable driver behavior. Preferably, the geometric design at the highway grade separation should be better than that for the approaching highways to counterbalance any possible sense of restriction caused by abutments, piers, curbs, and rails. Desirably, the alignment and profile of the through highways at an interchange should be relatively flat with high visibility. Sometimes it will be practical to design only one of the intersecting roadways on a tangent with flat grades. Preferably, the major highway should be so treated.

The general controls for horizontal and vertical alignment and their combination, as stated in Chapter 3, should be adhered to closely. In particular, any relatively sharp horizontal or vertical curves should be avoided. Horizontal curves that begin at or near a pronounced crest or sag should be kept to a minimum and should satisfy the design criteria established for open-highway conditions. Gradients that may slow down commercial vehicles or that may be difficult to negotiate under icy conditions should be avoided. Reduction of vehicle speeds by long upgrades encourages passing, which is undesirable in the vicinity of ramp terminals. Slow-moving through vehicles also encourage abrupt cutting in by vehicles leaving and entering the highways.

For a grade separation without ramps, the alignment and cross section of the approaches do not present a problem except where the median is widened to accommodate a middle pier or where the median is narrowed for structure economy. With ramps, changes in alignment and cross section may be needed to ensure proper operation and to develop the capacity needed at the ramp terminals, particularly where there is not a full complement of ramps and where some left turns at grade are provided. On a divided highway, the provision of direct left turns may involve widening of the cross section to ensure a suitably wide median for a combined speed-change and storage lane. On an undivided multilane highway, the introduction of a median is usually appropriate to ensure that the direct left turn is made to the proper ramp. Where a two-lane highway is carried through an interchange, wrong-way left turns are likely to occur, even with the provision of a full complement of ramps. For high-speed or high-volume conditions, this factor may warrant a divided section through the interchange area to prevent such turns.

A four-lane highway should be divided at interchanges. Since four-lane highways may carry enough traffic to justify the elimination of at-grade left turns, a nontraversable median should be provided to ensure that drivers use the proper ramps for left-turning maneuvers. At-grade left turns preferably should be accommodated within a suitably wide median.

Widening a roadway cross section to gain the desired width for a divisional island in an interchange area is done in the same manner as that done at any other intersection. Some of the more typical widening situations are illustrated in Exhibit 10-44. Exhibit 10-44A shows the customary symmetrical development of a divisional island on a four-lane undivided highway.

Traffic in each direction is required to traverse two reverse curves. Exhibit 10-44B shows a divisional island developed on a four-lane undivided highway in which the centerline is offset through the interchange area. Traffic in each direction enters the area without traversing any curvature, but traverses one reverse curve beyond the structure and ramp terminals. The scenario in Exhibit 10-44B is not usable on existing four-lane highways unless the approaches are reconstructed to obtain the center line offset.

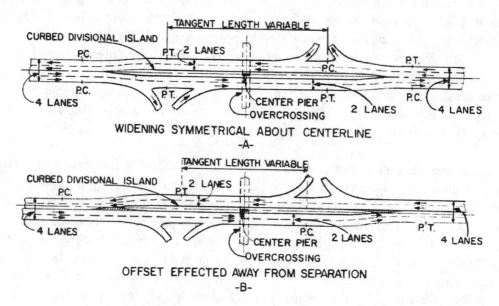

Exhibit 10-44. Widening for Divisional Island at Interchanges

Sight distance. Sight distance on the highways through a grade separation should be at least as long as that needed for stopping and preferably longer. Where exits are involved, decision sight distance is preferred, although not always practical. Design of the vertical alignment is the same as that at any other point on the highway.

The horizontal sight distance limitations of piers and abutments at curves usually present a more difficult problem than that of vertical limitations. With the minimum radius for a given design speed (see Chapter 3), the normal lateral clearance at piers and abutments of underpasses does not provide the minimum stopping sight distance. Similarly, on overpasses with the sharpest curvature for the design speed, sight distance deficiencies result from the usual offset to bridge rails. Thus, above-minimum radii should be used for curvature on highways through interchanges. If sufficiently flat curvature cannot be used, the clearances to abutments, piers, or rails should be increased to obtain the proper sight distance, even though this involves increasing structure spans or widths.

Ramp terminals at crossroads should be treated as at-grade intersections and should be designed in accordance with Chapter 9.

Interchange Spacing

Interchange spacing has a pronounced effect on freeway operations. In areas of concentrated urban development, proper spacing usually is difficult to attain because of traffic demand for frequent access. Minimum spacing of arterial interchanges (distance between intersecting streets with ramps) is determined by weaving volumes, ability to sign, signal progression, and lengths of speed-change lanes. A general rule of thumb for minimum interchange spacing is 1.5 km [1 mi] in urban areas and 3.0 km [2 mi] in rural areas. In urban areas, spacing of less than 1.5 km [1 mi] may be developed by grade-separated ramps or by adding collector-distributor roads.

Uniformity of Interchange Patterns

When a series of interchanges are being designed, attention should be given to the group of interchanges as a whole, as well as to each individual interchange. Interchange uniformity and route continuity are interrelated concepts, and both can be obtained under ideal conditions. Considering the need for high capacity, appropriate level of service, and maximum safety in conjunction with freeway operations, it is desirable to provide uniformity in exit and entrance patterns. Because interchanges are closely spaced in urban areas, shorter distances are available in which to inform drivers of the course to follow when exiting a freeway. An inconsistent arrangement of exits between successive interchanges causes driver confusion, resulting in drivers slowing down on high-speed lanes and making unexpected maneuvers. Examples of inconsistent exit arrangements are illustrated in Exhibit 10-45A and include inconsistency of exit ramp locations with respect to the structure (near and far side of structure) and exit ramps on the left side of the traveled way. The difficulty of left-entrance merging with high-speed through traffic and the requisite lane changing for left-exit ramps make these layouts undesirable. Except in highly special cases, all entrance and exit ramps should be on the right. To the extent practical, all interchanges along a freeway should be reasonably uniform in geometric layout and general appearance, as shown in Exhibit 10-45B.

Route Continuity

Route continuity refers to the provision of a directional path along and throughout the length of a designated route. The designation pertains to a route number or a name of a major highway. Route continuity is an extension of the principle of operational uniformity coupled with the application of proper lane balance and the principle of maintaining a basic number of lanes.

The principle of route continuity simplifies the driving task in that it reduces lane changes, simplifies signing, delineates the through route, and reduces the driver's search for directional signing.

Desirably, the through driver, especially the unfamiliar driver, should be provided a continuous through route on which changing lanes is not necessary to continue on the through route.

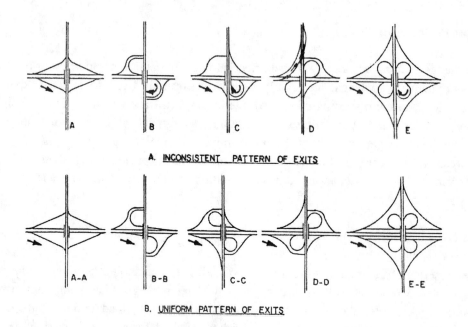

A. INCONSISTENT PATTERN OF EXITS

B. UNIFORM PATTERN OF EXITS

Exhibit 10-45. Arrangement of Exits Between Successive Interchanges

In the process of maintaining route continuity, particularly through cities and bypasses, interchange configurations need not always favor the heavy movement but rather the through route. In this situation, heavy movements can be designed on flat curves with reasonably direct connections and auxiliary lanes, equivalent operationally to through movements.

Exhibit 10-46 illustrates the principle of route continuity as applied to a hypothetical route, Interstate 15, as it intersects other major high-volume routes (service interchanges not shown). In Exhibit 10-46A route continuity is maintained on the designated route by keeping it on the left of all other entering or exiting routes. In Exhibit 10-46B, route continuity is disrupted by other routes exiting or entering on the left, except for the northbound direction of the last interchange.

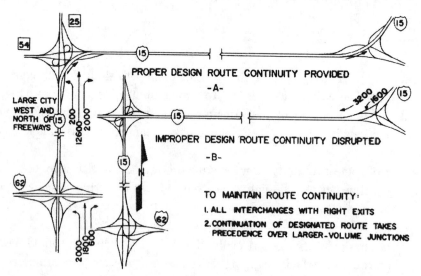

Exhibit 10-46. Interchange Forms to Maintain Route Continuity

Overlapping Routes

In some situations, two or more routes occupy a single alignment within a corridor. In rural areas, the problem of handling overlapping routes is generally limited to providing adequate signing and maintaining route continuity. In urban areas, the complexity of the problem increases with the probability of weaving and the need for additional capacity and lane balance.

In urban areas, it is preferable not to have overlapping routes, especially for only short distances. When routes overlap, signing is more complicated, and the decision process for the driver is more demanding.

The provision for route continuity through overlapping sections is essential. However, in some instances, this provision poses a problem in determining which route should have precedence, and the problem is especially acute when both routes have the same classification. Through a process of subclassification (i.e., U.S., State, city, or county route), a priority may be established for one of the overlapping roadways. All other factors being equal, priority should be assigned to the route that handles the highest volume of through traffic.

Once priority for one of the overlapping roadways has been established, basic lanes, lane balance, and other principles of interchange design can be applied to the design of the overlapping section. The lower classified facility should enter and exit on the right, thus conforming to the concept of route continuity.

On overlapping roadways, weaving is usually involved. However, on longer overlaps, the problem of weaving is minimized. Where the overlap is short, such as between successive interchanges, careful attention should be given to the design of weaving sections and lane balance.

In a situation where a major arterial would be overlapped by a lesser roadway, the minor facility may be designed as a collector-distributor road with transfer roads connecting the two facilities, as shown in Exhibit 10-47. This design removes weaving from the major roadway and transfers it to the minor facility. (See the discussion of collector-distributor roads in this chapter.)

Signing and Marking

The safety, efficiency, and clarity of paths to be followed at interchanges depend largely on their relative spacing, geometric layout, and effective signing. The location of and minimum distances between ramp junctions depend to a large degree on whether or not effective signing can be provided to inform, warn, and control drivers. Location and design of interchanges, individually and as a group, should be evaluated for proper signing. Signs should conform to the *Manual on Uniform Traffic Control Devices* (MUTCD) (**4**).

Pavement striping, delineators, and other markings are also important elements of driver communication at interchanges. These should be uniform and consistent with the MUTCD (**4**).

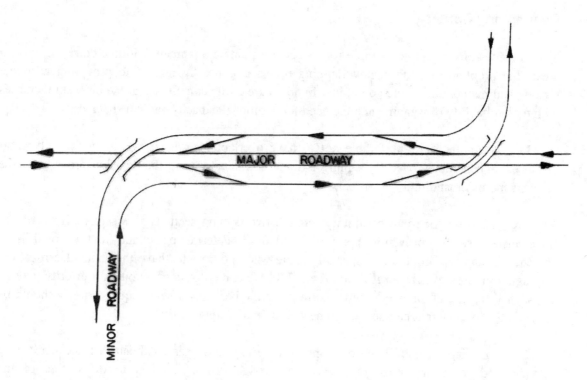

Exhibit 10-47. Collector-Distributor Road on Major-Minor Roadway Overlap

Basic Number of Lanes

Fundamental to establishing the number and arrangement of lanes on a freeway is the designation of the basic number of lanes. A certain consistency should be maintained in the number of lanes provided along any route of arterial character. Thus, the basic number of lanes is defined as a minimum number of lanes designated and maintained over a significant length of a route, irrespective of changes in traffic volume and lane-balance needs. Stating it another way, the basic number of lanes is a constant number of lanes assigned to a route, exclusive of auxiliary lanes.

As illustrated in Exhibit 10-48, the basic number of lanes on freeways is maintained over significant lengths of the routes, as A to B or C to D. The number of lanes is predicated on the general volume level of traffic over a substantial length of the facility. The volume considered here is the DHV (normally, representative of the morning or evening weekday peak).

Localized variations are ignored, so short sections of roadway carrying lower volumes would theoretically have reserve capacity, and short sections of roadway carrying somewhat higher volumes would be compensated for by the addition of auxiliary lanes introduced within these sections.

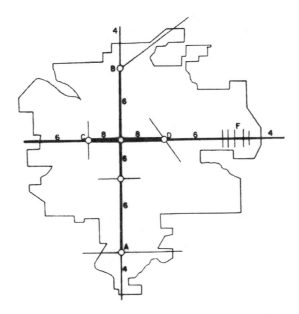

Exhibit 10-48. Schematic of Basic Number of Lanes

An increase in the basic number of lanes is needed where traffic builds up sufficiently to justify an extra lane and where such buildup raises the volume level over a substantial length of the facility.

The basic number of lanes may be decreased where traffic volumes are significantly reduced for a substantial length of highway. Lane reductions are discussed later in this chapter.

Coordination of Lane Balance and Basic Number of Lanes

To realize efficient traffic operation through and beyond an interchange, there should be a balance in the number of traffic lanes on the freeway and ramps. Design traffic volumes and a capacity analysis determine the basic number of lanes to be used on the highway and the minimum number of lanes on the ramps. The basic number of lanes should be established for a substantial length of freeway and should not be changed through pairs of interchanges, simply because there are substantial volumes of traffic entering and leaving the freeway. In other words, there should be continuity in the basic number of lanes. As described later in this section, variations in traffic demand should be accommodated by means of auxiliary lanes where needed.

After the basic number of lanes is determined for each roadway, the balance in the number of lanes should be checked on the basis of the following principles:

1. At entrances, the number of lanes beyond the merging of two traffic streams should not be less than the sum of all traffic lanes on the merging roadways minus one, but may be equal to the sum of all traffic lanes on the merging roadway (See Exhibit 10-49).
2. At exits, the number of approach lanes on the highway should be equal to the number of lanes on the highway beyond the exit, plus the number of lanes on the exit, minus one.

Exceptions to this principle occur at cloverleaf loop-ramp exits that follow a loop-ramp entrance and at exits between closely spaced interchanges (i.e., interchanges where the distance between the end of the taper of the entrance terminal and the beginning of the taper of the exit terminal is less than 450 m [1,500 ft], and a continuous auxiliary lane between the terminals is being used). In these cases, the auxiliary lane may be dropped in a single-lane exit with the number of lanes on the approach roadway being equal to the number of through lanes beyond the exit plus the lane on the exit.

3. The traveled way of the highway should be reduced by not more than one traffic lane at a time.

Typical examples of lane balance are shown in Exhibit 10-49.

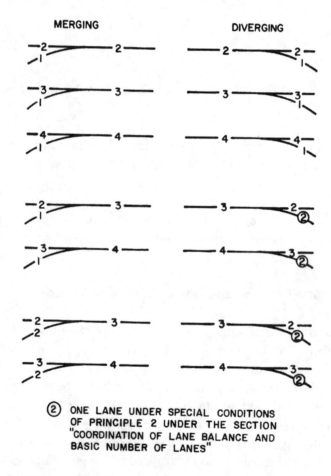

MERGING   DIVERGING

② ONE LANE UNDER SPECIAL CONDITIONS OF PRINCIPLE 2 UNDER THE SECTION "COORDINATION OF LANE BALANCE AND BASIC NUMBER OF LANES"

Exhibit 10-49. Typical Examples of Lane Balance

The principles of lane balance seem to conflict with the concept of continuity in the basic number of lanes, as illustrated in Exhibit 10-50. The exhibit shows three different arrangements where a four-lane freeway in one direction of travel has a two-lane exit followed by a two-lane entrance.

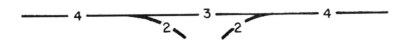

LANE BALANCE BUT NO COMPLIANCE WITH BASIC NUMBER OF LANES

- A -

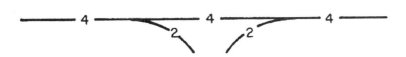

NO LANE BALANCE BUT COMPLIANCE WITH BASIC NUMBER OF LANES

- B -

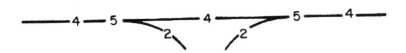

COMPLIANCE WITH BOTH LANE BALANCE AND BASIC NUMBER OF LANES

- C -

Exhibit 10-50. Coordination of Lane Balance and Basic Number of Lanes

In Exhibit 10-50A, lane balance is maintained, but there is no compliance with the basic number of lanes. This pattern may cause confusion and erratic operations for through traffic on the freeway. Even though traffic volumes are reduced through the interchange, there is no assurance that traffic demand will not increase under certain circumstances. Unduly large concentrations of through traffic may be caused by special events or by closures or reduction in capacity of other parallel facilities because of crashes or maintenance operations. Under such circumstances, any lanes that have been dropped on a freeway between interchanges (based on capacity and lane-balance needs as dictated by the normal DHV) produce definite bottlenecks.

The arrangement shown in Exhibit 10-50B provides continuity in the basic number of lanes but does not conform with the principles of lane balance. With this arrangement, the large exiting or entering traffic volume needing two lanes would have difficulty in diverging from or merging with the main line flow.

Exhibit 10-50C illustrates an arrangement in which the concepts of lane balance and basic number of lanes are brought into harmony by means of building on the basic number of lanes (i.e., by adding auxiliary lanes or removing them from the basic width of the traveled way). Auxiliary lanes may be added to satisfy capacity and weaving needs between interchanges, to accommodate traffic pattern variations at interchanges, and for simplification of operations (as reducing lane changing). The principles of lane balance should be applied in the use of auxiliary lanes. In this manner, the appropriate balance between traffic load and capacity is provided, and lane balance and operational flexibility are realized.

Design details of multilane ramp terminals with auxiliary lanes are covered below in the section on "Auxiliary Lanes."

Auxiliary Lanes

An auxiliary lane is defined as the portion of the roadway adjoining the traveled way for speed change, turning, storage for turning, weaving, truck climbing, and other purposes supplementary to through-traffic movement. The width of an auxiliary lane should be equal to the through lanes. An auxiliary lane may be provided to comply with the concept of lane balance, to comply with capacity needs, or to accommodate speed changes, weaving, and maneuvering of entering and leaving traffic. Where auxiliary lanes are provided along freeway main lanes, the adjacent shoulder should desirably be 2.4 to 3.6 m [8 to 12 ft] in width, with a minimum 1.8 m [6 ft] wide shoulder considered.

Operational efficiency may be improved by using a continuous auxiliary lane between the entrance and exit terminals where (1) interchanges are closely spaced, (2) the distance between the end of the taper on the entrance terminal and the beginning of the taper on the exit terminal is short, and/or (3) local frontage roads do not exist. An auxiliary lane may be introduced as a single exclusive lane or in conjunction with a two-lane entrance. The termination of the auxiliary lane may be accomplished by several methods. The auxiliary lane may be dropped in a two-lane exit, as illustrated in Exhibit 10-51A. This treatment complies with the principles of lane balance. Some agencies prefer to drop the auxiliary lane in a single-lane exit, as illustrated in Exhibit 10-51B. This treatment is in accordance with the exceptions listed under Principle 2 of lane balance as presented in the earlier section on "Coordination of Lane Balance and Basic Number of Lanes." Another method is to carry the full-width auxiliary lane to the physical nose before it is tapered into the through roadway. This design provides a recovery lane for drivers who inadvertently remain in the discontinued lane (see Exhibit 10-51C). When these methods of terminating the auxiliary lane (Exhibits 10-51B and Exhibit 10-51C) are used, the exit gore should be visible throughout the length of the auxiliary lane.

If local experience with single-exit design indicates problems with turbulence in the traffic flow caused by vehicles attempting to recover and proceed on the through lanes, the recovery lane should be extended 150 to 300 m [500 to 1,000 ft] before being tapered into the through lanes (see Exhibit 10-51D). Within large interchanges, this distance should be increased to 450 m [1,500 ft]. When an auxiliary lane is carried through one or more interchanges, it may be dropped as indicated above, or it may be merged into the through roadway approximately 750 m [2,500 ft] beyond the influence of the last interchange (see Exhibit 10-51E).

When interchanges are widely spaced, it might not be practical or necessary to extend the auxiliary lane from one interchange to the next. In such cases, the auxiliary lane originating at a two-lane entrance should be carried along the freeway for an effective distance beyond the merging point, as shown in Exhibit 10-52A1 and 10-52A2. An auxiliary lane introduced for a two-lane exit should be carried along the freeway for an effective distance in advance of the exit and extended onto the ramp, as shown in Exhibit 10-52B1 and 10-52B2. Exhibits 10-52A1 and 10-52B1 utilize taper designs, whereas Exhibits 10-52A2 and 10-52B2 show parallel designs.

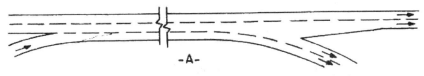

AUXILIARY LANE DROPPED ON EXIT RAMP

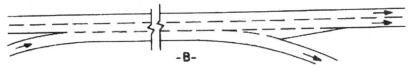

AUXILIARY LANE BETWEEN CLOVERLEAF LOOPS OR CLOSELY SPACED INTERCHANGES DROPPED ON SINGLE EXIT LANE.

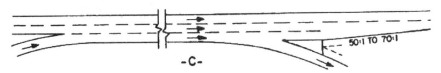

AUXILIARY LANE DROPPED AT PHYSICAL NOSE

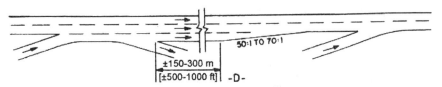

AUXILIARY LANE DROPPED WITHIN AN INTERCHANGE

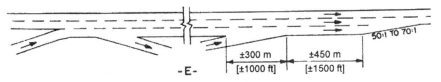

AUXILIARY LANE DROPPED BEYOND AN INTERCHANGE

Exhibit 10-51. Alternative Methods of Dropping Auxiliary Lanes

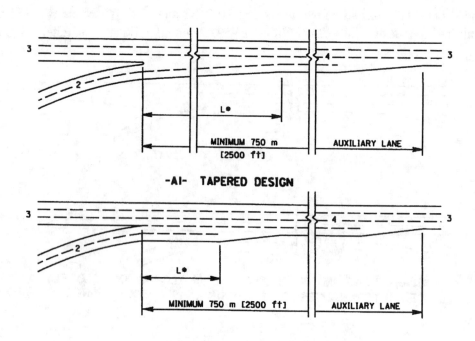

-A1- TAPERED DESIGN

-A2- PARALLEL DESIGN (PREFERRED)

AUXILIARY LANE EXTENDED FOR EFFECTIVE DISTANCE BEYOND ENTRANCE
* Refer to Exhibit 10-76 for minimum length criteria.

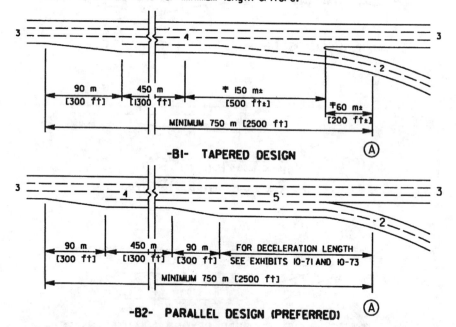

-B1- TAPERED DESIGN

-B2- PARALLEL DESIGN (PREFERRED)

AUXILIARY LANE INTRODUCED FOR EFFECTIVE DISTANCE IN ADVANCE OF EXIT
∓ Varies with angle of divergence
Ⓐ Point controlling speed on ramp

Exhibit 10-52. Coordination of Lane Balance and Basic Number of Lanes Through Application of Auxiliary Lanes

Generally, parallel designs are preferred. While tapered designs are acceptable, some agencies are concerned about the inside merge on the tapered entrance ramps. Auxiliary lanes should not be shorter than those shown later in this chapter for single-lane ramps in Exhibit 10-70 and Exhibit 10-73 with adjustments for grades as suggested in Exhibit 10-71. It is not precisely known what the effective length of the introduced auxiliary lane should be under these circumstances. Experience indicates that minimum distances of about 750 m [2,500 ft] produce the desired operational effect and enables development of the full capacity of two-lane entrances and exits.

For those instances where an auxiliary lane extends for a long distance from an entrance at one interchange to an exit at the next interchange, unfamiliar motorists may perceive the auxiliary lane as an additional through lane. For these situations, an auxiliary lane may be terminated, as discussed in the subsequent section entitled "Lane Reductions" or by providing a two-lane exit.

Auxiliary lanes are used to balance the traffic load and maintain a more uniform level of service on the highway. They facilitate the positioning of drivers at exits and the merging of drivers at entrances. Thus, the concept is very much related to signing and route continuity. Careful consideration should be given to the design treatment of an auxiliary lane because it may have the potential for trapping a driver at its termination point or the point where it is continued onto a ramp or turning roadway.

Exhibit 10-53 illustrates the application of an auxiliary lane that is terminated through a multilane exit terminal. The outside basic lane automatically becomes an interior lane with the addition of the auxiliary lane. From this interior lane a driver may exit right or proceed straight ahead. Although the driver has two choices of direction of travel, the design of multilane exit terminals is not to be confused with the optional lane concept, as discussed in the subsection on "Major Forks and Branch Connections" later in this chapter "Traveled-Way Widths." The example complies with the principles of lane balance and basic number of lanes. The design emphasizes the through route and allows drivers to make their decision to travel through or turn right well in advance of the exit point, or fairly close to it as a result of the additional maneuver area.

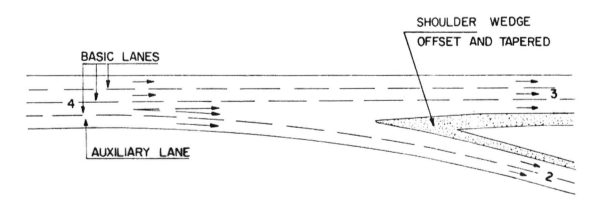

Exhibit 10-53. Auxiliary Lane Dropped at Two-Lane Exit

Lane Reductions

As discussed in the section "Basic Number of Lanes" and the section "Coordination of Lane Balance and Basic Number of Lanes" earlier in this chapter, the basic number of lanes should be maintained over a significant length of freeway. Lane reductions should not be made between and within interchanges simply to accommodate variations in traffic volumes. Instead, auxiliary lanes, as needed, are added or removed from the basic number of lanes, as described in the section on "Auxiliary Lanes" in this chapter.

A reduction in the basic number of lanes may be made beyond a principal interchange involving a major fork or at a point downstream from an interchange with another freeway. This reduction may be made provided the exit volume is sufficiently large to change the basic number of lanes beyond this point on the freeway route as a whole. Another case where the basic number of lanes may be reduced is where a series of exits, as in outlying areas of the city, causes the traffic load on the freeway to drop sufficiently to justify the lesser basic number of lanes. Dropping a basic lane or an auxiliary lane may be accomplished at a two-lane exit ramp or between interchanges.

If a basic lane or an auxiliary lane is to be dropped between interchanges, it should be accomplished at a distance of 600 to 900 m [2,000 to 3,000 ft] from the previous interchange to allow for adequate signing.

The reduction should not be made so far downstream that motorists become accustomed to a number of lanes and are surprised by the reduction (see Exhibit 10-51E). Desirably, the lane-drop transition should be located on tangent horizontal alignment and on the approach side of any crest vertical curve. A sag vertical curve is also a good location for a lane drop because it provides good visibility. Preferably, the lane reduction should be made on the right side following an exit ramp because there is likely to be less traffic in that lane. A right-side lane reduction has advantages in that speeds are generally lower and the merging maneuver from the right is more familiar to most motorists, because it is similar to a merge at an entrance ramp. Left-side lane reductions may not function as well because of generally higher speeds and the less familiar left-side merge.

The end of the lane drop should be tapered into the highway in a manner similar to that at a ramp entrance. Preferably, the rate of taper should be longer than that for a ramp. The minimum taper rate should be 50:1, and the desirable taper rate is 70:1.

If there is a lane reduction of a basic lane or an auxiliary lane within an interchange, it should be made in conjunction with a two-lane exit, as shown in Exhibit 10-51A, or in a single-lane exit with an adequate recovery lane, as discussed in the section of this chapter on "Auxiliary Lanes."

Weaving Sections

Weaving sections are highway segments where the pattern of traffic entering and leaving at contiguous points of access results in vehicle paths crossing each other. Weaving sections may occur within an interchange, between entrance ramps, followed by exit ramps of successive interchanges, and on segments of overlapping roadways.

Because considerable turbulence occurs throughout weaving sections, interchange designs that eliminate weaving entirely or at least remove it from the main facility are desirable. Weaving sections may be eliminated from the main facility by the selection of interchange forms that do not have weaving or by the incorporation of collector-distributor roads. Interchanges that provide all exit movements before any entrance movements will eliminate weaving.

Although interchanges that do not involve weaving operate better than those that do, interchanges with weaving areas generally are less costly than those without. Designs that avoid weaving movements may need a greater number of structures or larger and more complex structures, with some direct connections. Joint evaluation of the total interchange cost and the specific volumes to be handled is needed to reach a sound decision between design alternatives. The partial cloverleaf design with loops in opposite quadrants eliminates the weaving sections, does not involve direct connections or additional structures, and has been found by some States to operate superior to all other interchanges with a single separation structure.

Where cloverleaf interchanges are used, consideration should be given to the inclusion of collector-distributor roads on the main facility, or possibly both facilities where warranted.

The capacity of weaving sections may be seriously restricted unless the weaving section has adequate length, adequate width, and lane balance (See Chapter 2 for procedures for determining weaving lengths and widths). Refer to the *Highway Capacity Manual* (HCM) (**5**) for capacity analysis of weaving sections.

Collector-Distributor Roads

Collector-distributor roads between two interchanges and continuous collector-distributor roads are discussed in Chapter 8. Collector-distributor roads within an interchange are discussed in this section.

A full cloverleaf interchange in an urban or suburban area is a typical example of a single interchange that should be analyzed for the need for collector-distributor roads within the interchange. Collector-distributor roads may be one or two lanes in width, depending on capacity needs. Lane balance should be maintained at entrances and exits to and from the main line, but strict adherence is not mandatory on the collector-distributor road proper because weaving is handled at reduced speed. The design speed usually ranges from 60 to 80 km/h [40 to 50 mph], but should not be less than 20 km/h [10 mph] below the design speed of the main roadway. Operational problems will occur if collector-distributor roads are not properly signed, especially those servicing more than one interchange.

Outer separations between the main line and the collector-distributor roads should be as wide as practical; however, minimum widths are tolerable. The minimum width should allow for shoulder widths equal to that on the main line and for a suitable barrier to prevent indiscriminate crossovers.

The advantages of using collector-distributor roads within an interchange are that weaving is transferred from the main roadway, single entrances and exits are developed, all main-line exits occur in advance of the structure, and a uniform pattern of exits can be maintained (See the following section on "Two-Exit Versus Single-Exit Interchange Design.").

Two-Exit Versus Single-Exit Interchange Design

In general, interchanges that are designed with single exits are superior to those with two exits, especially if one of the exits is a loop ramp or if the second exit is a loop ramp preceded by an entrance loop ramp. Whether used in conjunction with a full cloverleaf or with a partial cloverleaf interchange, the single-exit design may improve operational efficiency of the entire facility.

The purposes for developing single exits, where applicable, are as follows:

- To remove weaving from the main facility and transfer it to a slower speed facility.
- To provide a high-speed exit from the main roadway for all exiting traffic.
- To simplify signing and the decision process.
- To satisfy driver expectancy by placing the exit in advance of the separation structure.
- To provide uniformity of exit patterns.
- To provide decision sight distance for all traffic exiting from the main roadway.

The full cloverleaf interchange, where a weaving section exceeds 1,000 vph, is an example where operational efficiency may be improved by the development of single exits and entrances.

The loop ramps of a full cloverleaf interchange create a weaving section adjacent to the outside through lane, and considerable deceleration-acceleration must occur in the through lane. By using collector-distributor roads, as shown in Exhibit 10-26, a single exit is provided and weaving is transferred to the collector-distributor road. Without a collector-distributor road, the second exit of a cloverleaf interchange occurs beyond the separation structure and, in many cases, is hidden behind a crest vertical curve. The single-exit design places the exit from the main line in advance of the structure and is conducive to a uniform pattern of exits. Where the through roadway overpasses the crossroad in a vertical curve, it may be more difficult to develop full-decision sight distance for the loop ramp exit of a conventional cloverleaf interchange. The use of the single-exit design may make it easier to obtain the desired decision sight distance due to the exit occurring on the upgrade.

Some arrangements of partial cloverleaf loop ramps may feature single exits, as shown in Exhibit 10-28F, and still be inferior because they do not provide any of the desirable purposes previously discussed.

On a full cloverleaf interchange, the single exit is developed by using a collector-distributor road for the full length of the interchange. On certain partial cloverleaf arrangements, the single exit can be developed by elongating the loop ramp in the upstream direction to the point where it diverges from the right-turn movement well in advance of the separation structure. The elongation of the loop ramp may be done with a spiral, simple curve, tangent, or a combination of these.

Full diamond and SPUIs feature single exits and entrances and accomplish the desired intent of single exits as well as entrances. There are some cases where a single exit does not work as well as two exits, such as at high-volume, high-speed directional interchanges. The problem usually occurs at the fork following the single exit from the freeway, particularly when the traffic volume is great enough to warrant a two-lane exit and the distance from the exit terminal to the fork is insufficient for weaving and proper signing. There is often some confusion at this second decision point, resulting in poor operation and a high crash potential. Because of this, it may be advantageous on some directional interchanges to provide two exits on each freeway leg.

Generally, the provision for single exits is more costly because of the added roadway, longer bridges and, in some cases, additional separation structures. The overall efficiency of a cloverleaf interchange with collector-distributor roads should be taken into consideration. Where ramp volumes are low and not expected to increase significantly, or where a particular cloverleaf weave does not exceed about 1,000 vph, it will often be impractical to use collector-distributor roads. These conditions can be expected in rural areas or on low-volume freeways.

Collector-distributor roads may still be an option if significant future turning volumes are expected or site investigations reveal a definitive need for such a configuration. Exhibit 10-54 shows various interchange configurations that are compatible with the concepts of uniform exit patterns and exits in advance of the separation structure.

Wrong-Way Entrances

Wrong-way entrance onto freeways and arterial streets is not a frequent occurrence, but it should be regarded as a serious problem wherever the likelihood exists because each occurrence has such a high potential for culminating in a serious crash. This problem should be given special consideration at all stages of design. Most wrong-way entrances occur at freeway offramps, at intersections at grade along divided arterial streets, and at transitions from undivided to divided highways. Several factors that contribute to wrong-way entrances are related to interchange design. These factors concern the interchange configuration and, more particularly, the crossroad terminal of the exit ramps, which are discussed below.

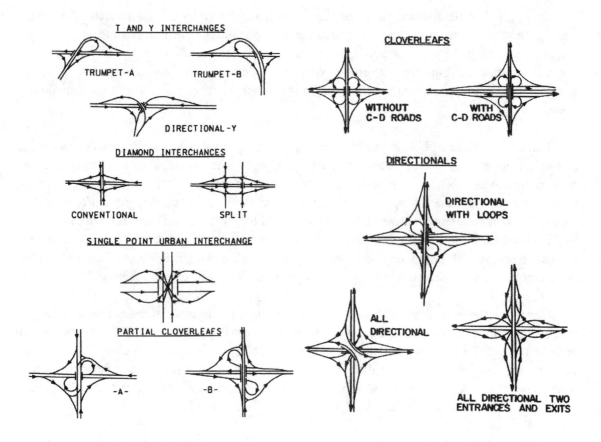

Exhibit 10-54. Interchange Forms with One and Two Exits

Partial interchanges are particularly troublesome. Where provision is not made for any one or more of the movements at an interchange, wrong-way entry may occur. Exit ramps that connect to two-way frontage roads are also conducive to wrong-way entry. Without channelization on the frontage road, they appear as open entries. Some of the "scissors" channelization has proved to be confusing, resulting in wrong-way use.

Exit ramps with a sweeping connection to the street (e.g., outer connection, loop, and some diamond ramps) have a low rate of wrong-way entry. However, one-way ramps that connect as an unchannelized "T" intersection are most likely to lead to wrong-way entry.

Unusual or odd arrangements of exit ramps are confusing and conducive to wrong-way entry. An example is the button-hook or J-shaped ramp that connects to a parallel or diagonal street or frontage road, often well-removed from the interchange structure and other ramps. Another example is a pair of right-turn connections to a lateral or parallel street (frontage road) offset from the separation structure.

On undivided crossroads, a non-traversable median (except at turn points) introduced within the interchange limits helps prevent wrong-way entry on diamond, partial cloverleaf, and full cloverleaf interchanges. Where adjacent off-and-on ramps join a minor road, the ramp roadways should be separated. The ramp-crossroad intersection at a diamond interchange should be well-removed from any other nearby intersection, such as a frontage road-crossroad intersection. Local road connections within the length of any exit ramp should be avoided. Temporary ramp terminals warrant special attention in layout details to avoid wrong-way entry paths.

Left-side main roadway exit ramps should be avoided because they may appear to be a right-side entrance ramp to a confused motorist. Many wrong-way entries occur at night when volumes are low and control devices are less effective. Lighting should be considered to minimize such conditions.

Open sight distances throughout the entire length of the ramp help prevent wrong-way use. Especially important is the driver's view of the ramp terminal when approaching from the cross street.

In the design of any interchange, consideration should be given to the likelihood of wrong-way travel and to the practical measures that may be taken in the design and traffic control for preventing or discouraging such usage. Signing to prevent wrong-way entry should be in accordance with the MUTCD (**4**).

Ramps

Types and Examples

The term "ramp" includes all types, arrangements, and sizes of turning roadways that connect two or more legs at an interchange. The components of a ramp are a terminal at each leg and a connecting road. The geometry of the connecting road usually involves some curvature and a grade. Generally, the horizontal and vertical alignment of ramps is based on lower design speeds than the intersecting highways, but in some cases it may be equal.

Exhibit 10-55 illustrates several types of ramps and their characteristic shapes. Various configurations are used; however, each can be broadly classified as one of the types shown. Each ramp generally is a one-way roadway. Diagonal ramps (Exhibit 10-55A) are almost always one-way but usually have both a left- and right-turning movement as the terminal on the minor intersecting road. A diagonal ramp may be largely tangent or wishbone in shape with a reverse curve. Diamond interchanges generally have four diagonal ramps.

A loop ramp may have single turning movements (left or right) or double turning movements (left and right) at either or both ends. Exhibit 10-55D shows the case where there are only single turns made at both ends of the ramp. With this loop pattern, a left-turning movement is made without an at-grade crossing of the opposing through traffic. Instead, drivers making a left-turn travel beyond the highway separation, turn to the right through approximately 270 degrees to enter the other highway. The loop usually involves more indirect travel distance than any other type of ramp.

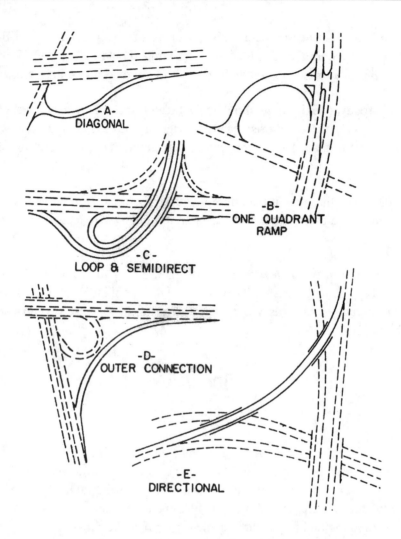

Exhibit 10-55. General Types of Ramps

With a semidirect connection (Exhibit 10-55C), the driver makes a right turn first, heading away from the intended direction, gradually reversing, and then completing the movement by following directly around and entering the other road. This semidirect connection may also be used for right turns, but there is little reason for its use if the conventional diagonal can be provided. A descriptive term frequently associated with this type of ramp is "jug-handle," the obvious plan shape. Travel distance on this ramp is less than that for a comparable loop and more than that for a direct connection. At least three structures or a three-level structure is needed. Exhibit 10-55D is termed an outer connection, while Exhibit 10-55E is referred to as a direct connection.

The different ramp patterns of an interchange (i.e., the different types of interchange configurations) are made up of various combinations of these types of ramps. For example, the trumpet configuration has one loop, one semidirectional ramp, and two right-turn directional or diagonal ramps.

General Ramp Design Considerations

Design speed. Desirably, ramp design speeds should approximate the low-volume running speed on the intersecting highways. This design speed is not always practical, and lower design speeds may be selected, but they should not be less than the low range presented in Exhibit 10-56. Only those values for highway design speeds of at least 80 km/h [50 mph] apply to freeway and expressway exits. The application of values in Exhibit 10-56 to various conditions and ramp types is discussed as follows:

Portion of ramp to which design speed is applicable. Values in Exhibit 10-56 apply to the sharpest, or controlling, ramp curve, usually on the ramp proper. These speeds do not pertain to the ramp terminals, which should be properly transitioned and provided with speed-change facilities adequate for the highway speed involved.

Ramps for right turns. An upper-range value of design speed is often attainable on ramps for right turns, and a value between the upper and lower range is usually practical. The diagonal ramp of a diamond interchange may also be used for right turns. For these diagonal ramps, a value in the middle range is usually practical.

Loops. Upper-range values of design speed generally are not attainable on loop ramps. Ramp design speeds above 50 km/h [30 mph] for loops involve large areas, rarely available in urban areas, and long loops, which are costly and require left-turning drivers to travel a considerable extra distance. Minimum values usually control, but for highway design speeds of more than 80 km/h [50 mph], the loop design speed preferably should be no less than 40 km/h (50-m radius) [25 mph (150-ft radius]. If less restrictive conditions exist, the loop design speed and the radius may be increased.

Semidirect connections. Design speeds between the middle and upper ranges shown in Exhibit 10-56 should be used. A design speed less than 50 km/h [30 mph] should not be used. Generally, for short single-lane ramps, a design speed greater than 80 km/h [50 mph] is not practical. For two-lane ramps, values in the middle and upper ranges are appropriate.

Direct connections. Design speeds between the middle and upper ranges shown in Exhibit 10-56 should be used. The minimum design speed preferably should be 60 km/h [40 mph].

Different design speeds on intersecting highways. The highway with the greater design speed should be the control in selecting the design speed for the ramp as a whole. However, the ramp design speed may vary, the portion of the ramp closer to the lower speed highway being designed for the lower speed. This variation in ramp design speed is particularly applicable where the ramp is on an upgrade from the higher speed highway to the lower speed highway.

At-grade terminals. Where a ramp joins a major crossroad or street, forming an intersection at grade, Exhibit 10-56 is not applicable to that portion of the ramp near the intersection because a stop sign or signal control is normally employed. This terminal design should be predicated on near-minimum turning conditions, as given in Chapter 9. In urban areas, where the land adjacent to the interchange is developed commercially, provisions for pedestrian movements through the interchange area should also be considered.

Metric								
Highway design speed (km/h)	50	60	70	80	90	100	110	120
Ramp design speed (km/h)								
Upper range (85%)	40	50	60	70	80	90	100	110
Middle range (70%)	30	40	50	60	60	70	80	90
Lower range (50%)	20	30	40	40	50	50	60	70
Corresponding minimum radius (m)	see Exhibit 3-43							

US Customary									
Highway design speed (mph)	30	35	40	45	50	55	60	65	70	75
Ramp design speed (mph)									
Upper range (85%)	25	30	35	40	45	48	50	55	60	65
Middle range (70%)	20	25	30	33	35	40	45	45	50	55
Lower range (50%)	15	18	20	23	25	28	30	30	35	40
Corresponding minimum radius (ft)	see Exhibit 3-43								

Exhibit 10-56. Guide Values for Ramp Design Speed as Related to Highway Design Speed

Curvature. The design guidelines for turning roadways at interchanges are discussed in Chapter 3. They apply directly to the design of ramp curves. Compound or spiral curve transitions are desirable to: (1) obtain the desired alignment of ramps, (2) provide for a comfortable transition between the design speeds of the through and turning roadways, and (3) fit the natural paths of vehicles. Caution should be exercised in the use of compound curvature to prevent unexpected and abrupt speed adjustments. Additional design information regarding the use of compound curves is contained in Chapter 3.

The general shape of a ramp evolves from the type of ramp selected, as previously described and shown in Exhibit 10-55. The specific shape, or curvature, of a ramp may be influenced by such factors as traffic pattern, traffic volume, design speed, topography, culture, intersection angle, and type of ramp terminal.

Several ramp shapes may be used for the loop and outer connection of a semidirectional interchange, as shown in Exhibit 10-57A. Except for its terminals, the loop may be a circular arc or some other symmetrical or asymmetrical curve, formed with spiral transitions. The asymmetrical arrangement may fit where the intersecting roads are not of the same importance and the ramp terminals are designed for different speeds, the ramp in part functioning as a speed-change area. Similar shapes may be dictated by right-of-way controls, profile and sight distance conditions, and terminal location. The freeway terminal should normally be placed in advance of the structure. The most desirable alignment for an outer connection is one on a continuous curve (line A). This arrangement, however, may involve questionably extensive right-of-way. Another desirable arrangement has a central tangent and terminal curves (lines B-B and C-C). Where the loop is more important than the outer connection, reverse alignment on the outer connection may be used to reduce the area of right-of-way, as shown by line D-D. Any combination of lines B, C, and D may be used for a practical shape.

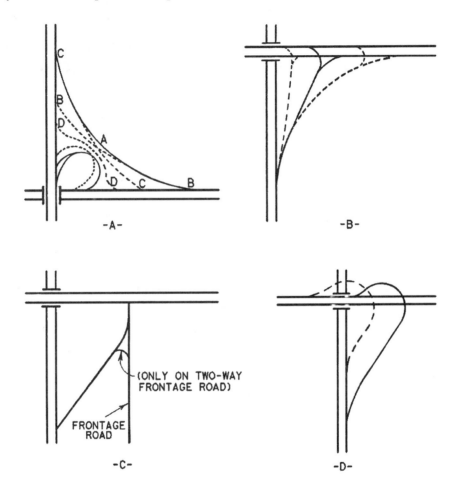

Exhibit 10-57. Ramp Shapes

In Exhibit 10-57A, the loop and the outer connection are separated, as is generally desirable. However, where the movements are minor and economy is desired, a portion of the two ramps may be combined into a single two-way roadway. Where this design is used, a barrier should separate the traffic in two directions. This design is generally discouraged.

Diagonal ramps may assume a variety of shapes, depending on the pattern of turning traffic and right-of-way limitations. As shown in Exhibit 10-57B, the ramp may be a diagonal tangent with connecting curves (solid line). To favor a right-turning movement, the ramp may be on a continuous curve to the right with a spur to the left for left turns. On restricted right-of-way along the major highway, it may be appropriate to use reverse alignment with a portion of the ramp being parallel to the through roadway.

Another variation of diagonal ramps, usually called "slip ramps," connects with a parallel frontage road, as shown in Exhibit 10-57C. Where this design is used, it is desirable to have one-way frontage roads. Ramps to two-way frontage roads introduce the possibility of wrong-way entry onto the through lanes. If two-way frontage roads are used, special attention should be given in the design and signing of ramps to discourage the possibility of wrong-way entry.

The shape of a semidirect connection (Exhibit 10-57D) is influenced by the location of the terminals with respect to the structures, the extent to which the structure is widened, and the curve radii needed to maintain a desired turning speed for an important left-turning movement. The angular position or the curvature may be dictated somewhat by the relative design speeds of the intersection legs and by the proximity of other roadways.

Sight distance. Sight distance along a ramp should be at least as great as the design stopping sight distance. Sight distance for passing is not needed. There should be a clear view of the entire exit terminal, including the exit nose and a section of the ramp roadway beyond the gore.

The sight distance on a freeway preceding the approach nose of an exit ramp should exceed the minimum stopping sight distance for the through traffic design speed, desirably by 25 percent or more. Decision sight distance, as discussed in Chapter 3, is desired where feasible. There should be a clear view of the entire exit terminal, including the exit nose. See Chapter 3 for ranges in design values for stopping sight distance on horizontal and vertical curves for turning roadways and open road conditions.

Grade and profile design. The profile of a typical ramp usually consists of a central portion on an appreciable grade, coupled with terminal vertical curves and connections to the profiles of the intersection legs. The following references to ramp gradient pertain largely to the central portion of the ramp profile. Profiles at the terminals largely are determined by through-road profiles and are seldom tangent grades.

Ramp grades should be as flat as practical to minimize the driving effort needed in maneuvering from one road to another. Most ramps are curved, and steep ramp grades in combination with curves hamper traffic flow. The slowing down of vehicles on an ascending ramp is not as serious as on a through road, provided the speed is not decreased sufficiently to result in a peak-hour backup onto the through road. Most diamond ramps are only 120 to 360 m [400 to 1,200 ft] long, and the short central portion with the steepest gradient has only moderate operational effect. Accordingly, gradients on ramps may be steeper than those on the intersecting highways. For any one ramp, the gradient to be used is dependent on a number of factors unique to that site and quadrant. The flatter the gradient on a ramp, the longer it will be, but the effect of gradient on ramp length is not substantial. The conditions and designs at ramp terminals frequently have an effect equally as great as the effect of the gradient. For example, when the ramp profile is opposite in direction to that of the through highway, a fairly long vertical curve is needed because of the large algebraic difference in grade; this adds considerably to the length of ramp. As another example, additional length may be needed to warp the ramp profile to attain superelevation or to provide drainage.

In general, adequate sight distance is more important than a specific gradient control and should be favored in design. Usually, these two controls are compatible. On one-way ramps, a distinction should be made between ascending and descending gradients. For high-speed ramp designs, the values cited in the next paragraph apply. However, with proper ramp terminal facilities, short upgrades of 7 to 8 percent permit good operation without unduly slowing down passenger cars. Short upgrades of as much as 5 percent do not unduly interfere with truck and bus operation. On one-way downgrade ramps, gradients of up to 8 percent do not cause undesirable operation due to excessive acceleration of passenger vehicles. However, there is a greater potential for heavy trucks to increase their speeds on downgrades. Therefore, downgrades should desirably be limited to 3 or 4 percent on ramps with sharp horizontal curvature and significant heavy truck or bus traffic. In many areas, consideration of snow and ice conditions may limit the choice of gradient regardless of the direction of the grade.

From the foregoing discussion, it can be seen that ramp grades are not directly related to design speed; however, design speed is a general indication of the quality of design being used, and the gradient for a ramp with a high design speed should be flatter than for one with a low design speed. As general criteria, it is desirable that upgrades on ramps with a design speed of 70 to 80 km/h [45 to 50 mph] be limited to 3 to 5 percent; those for a 60-km/h [40-mph] design speed, to 4 to 6 percent; those for a 40- to 50-km/h [25- to 30-mph] design speed, to 5 to 7 percent; and those for a 30- to 40-km/h [15- to 25-mph] design speed, to 6 to 8 percent. Where appropriate for topographic conditions, grades steeper than desirable may be used. One-way downgrades on ramps should be held to the same general maximums, but in special cases they may be 2 percent greater. Where ramp terminals are properly located and fit other design needs and where the curvature conforms to a reasonable design speed, ramps are generally long enough to attain the difference in levels with grades that are level or, at least, not too steep. The cases in which grade is a determining factor in the length of the ramp are as follows: (1) for intersection angles of 70 degrees or less, the ramp may need to be located farther from the structure to provide a ramp of sufficient length with reasonable grade; (2) where the intersection legs are on appreciable grade, with the upper road ascending and the lower road descending from the structure, the ramp will have to attain a large difference in elevation that increases with the distance from the structure; (3) where a ramp leaves the lower road on a downgrade and meets the higher road on a downgrade, longer-than-usual vertical curves at the terminals may need a long ramp to meet grade limitations. For these reasons, alignment and grade of a ramp should be determined jointly.

Vertical curves. Usually, ramp profiles assume the shape of the letter "S" with a sag vertical curve at the lower end and a crest vertical curve at the upper end. Additional vertical curves may be needed, particularly on ramps that overpass or underpass other roadways. Where a crest or sag vertical curve extends onto the ramp terminal, the length of curve should be determined by using a design speed between those on the ramp and the highway. See Chapter 3 for design values for open and turning roadway conditions.

Superelevation and cross slope. The following guidelines should be used for cross-slope design on ramps:

1. Superelevation rates, as related to curvature and design speed on ramps, are given in Exhibit 3-21 through 3-25. Where drainage impacts to adjacent property or the frequency of slow-moving vehicles are important considerations, the superelevation rates and corresponding radii in Exhibit 3-40 can be used.

2. The cross slope on portions of ramps on tangent normally should be sloped one way at a practical rate ranging from 1.5 to 2 percent for high-type pavements.

3. In general, the rate of change in cross slope in the superelevation runoff section should be based on the maximum relative gradients Δ listed in Exhibit 3-27. The values listed in this table are applicable to single-lane rotation. The adjustment factors b_w listed in Exhibit 3-28 allow for slight increases in the effective gradient for wider rotated widths. The effective maximum relative gradients (equal to $\Delta \div b_w$) applicable to a range of roadway widths are listed in Exhibit 9-43. The superelevation development is started or ended along the auxiliary lane of the ramp terminal. Alternate profile lines for both edges should be studied to ensure that all profiles match the control points and that no unsightly bumps and dips are inadvertently developed. Spline profiles are very useful in developing smooth lane/shoulder edges.

4. Another important control in developing superelevation along the ramp terminal is that of the crossover crown line at the edge of the through-traffic lane. The maximum algebraic difference in cross slope between the auxiliary lane and the adjacent through lane is shown in Exhibit 9-49.

5. Three segments of a ramp should be analyzed to determine superelevation rates that would be compatible with the design speed and the configuration of the ramp. The exit terminal, the ramp proper, and the entrance terminal should be studied in combination to ascertain the appropriate design speed and superelevation rates.

The guidelines in Item 5 can vary by the type of ramp configuration used. Three ramp configurations are described in the following paragraphs. The diamond ramp usually consists of a high-speed exit terminal, tangent or curved alignment on the ramp proper, and stop or yield conditions at the entrance terminal. Deceleration to the first controlling curve speed should occur on the auxiliary lane of the exit terminal and continued deceleration to stop or yield conditions should occur on the ramp proper. As a result, superelevation rate and radii used should reflect a decreasing sequence of design speeds for the exit terminal, ramp proper, and entrance terminal.

The loop ramp consists of a moderate-speed exit terminal connecting to a slow-speed ramp proper, which in turn connects to a moderate-speed acceleration lane. The curvature of the ramp proper may be a simple curve or a combination of curves, and is determined by the design speed and superelevation rate used. Superelevation should be gradually developed into and out of the curves for the ramp proper, as detailed later in this discussion.

Direct and semi-direct ramps generally are designed with a high-speed exit, a moderate- or high-speed ramp proper, and a high-speed entrance. As a result, the design speed and superelevation rates used are comparable to open-road conditions.

On ramps designed for speeds of 70 km/h [45 mph] or less, superelevation ranges given in Exhibit 3-40 are appropriate for design of the ramp proper.

The method of developing superelevation at free-flow ramp terminals is illustrated in Exhibit 10-58.

Exhibit 10-58A shows a tapered exit from a tangent section with the first ramp curve falling beyond the design deceleration length. The normal cross slope is projected onto the auxiliary lane, and no superelevation is needed until the first ramp proper curve is reached.

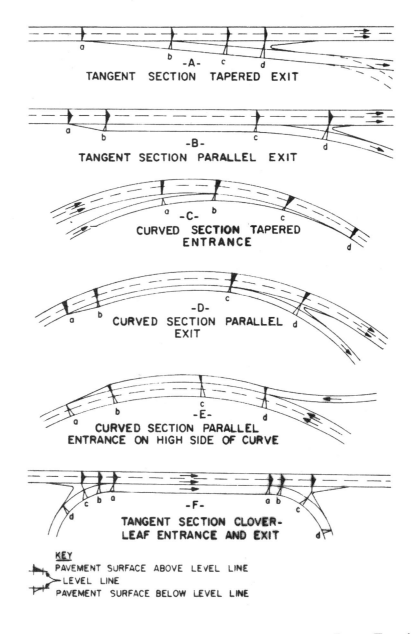

Exhibit 10-58. Development of Superelevation at Free-Flow Ramp Terminals

Exhibit 10-58B shows a parallel-type exit from a tangent section that leads into a flat exiting curve. At point b, the normal cross slope of the through roadway is projected onto the auxiliary lane. At point c, the cross slope can be gradually changed to start the development of superelevation for the exiting curve. At point d, two breaks in the crossover crown line would be conducive to developing a full superelevation in the vicinity of the physical nose.

Exhibits 10-58C and 10-58D show ramp terminals on which the superelevation of the through roadway would be projected onto the auxiliary lane. Exhibit 10-58E shows a parallel entrance terminal on the high side of a curve. At point e, the superelevation on the ramp begins to decrease and is gradually decreased through the tangent section to point d. At point d, the cross slope is gradually rotated to eventually meet the superelevation rate of the mainline at point c.

Exhibit 10-58F shows a parallel exit from a tangent section with sharp curvature developing in advance of the physical nose. This design is typical for cloverleaf loops. Part of the cross-slope transition can be accomplished over the length of the parallel lane with about half of the total superelevation being developed at point b. Full superelevation of the ramp proper is reached beyond the physical nose.

Care should be exercised to see that the rate of change in cross slope in the runoff section is based on the maximum relative gradients listed in Exhibit 9-44 and that the algebraic difference in cross slope does not exceed the values presented in Exhibit 9-49.

Gores. The term "gore" indicates an area downstream from the shoulder intersection points as illustrated in Exhibit 10-59. The physical nose is a point upstream from the gore, having some dimensional width that separates the roadways. The painted nose is a point, having no dimensional width, occurring at the separation of the roadways. The neutral area refers to the triangular area between the painted and the gore nose and incorporates the physical nose. The geometric layout of these is an important part of exit ramp terminal design. It is the decision point area that should be clearly seen and understood by approaching drivers. Furthermore, the separating ramp roadway not only should be clearly evident but should also have a geometric shape appropriate for the likely speeds at that point. In a series of interchanges along a freeway, the gores should be uniform and have the same appearance to drivers.

As a general rule, the width at the gore nose is typically between 6.0 to 9.0 m [20 to 30 ft], including paved shoulders, measured between the traveled way of the main line and that of the ramp. This dimension may be increased if the ramp roadway curves away from the freeway immediately beyond the gore nose or if speeds in excess of 100 km/h [60 mph] are expected to be common.

The entire triangular area, or neutral area, should be striped to delineate the proper paths on each side and to assist the driver in identifying the gore area. The MUTCD (**4**) may be referenced for guidance on channelization. Standard or snow-plowable raised reflective markers can be employed for additional delineation.

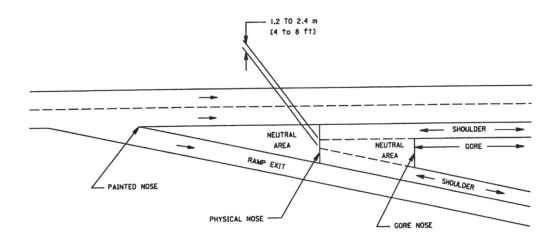

Exhibit 10-59. Typical Gore Area Characteristics

Rumble strips may be placed in the gore area but should not be located too close to the gore nose because such placement renders them ineffective for warning high-speed vehicles. In all cases, supplemental devices of this type should be placed to provide the driver with ample time to correct the vehicle's path safely.

The rate of crashes in gore areas is typically greater than the rate of run-off-the-road crashes at other locations. For this reason, the gore area, and the unpaved area beyond, should be kept as free of obstructions as practical to provide a clear recovery area. The unpaved area beyond the gore nose should be graded as nearly level with the roadways as practical so that vehicles inadvertently entering will not be overturned or abruptly stopped by steep slopes. Heavy sign supports, luminaire supports, and roadway structure supports should be kept well out of the graded gore area. In addition, yielding or breakaway supports should be employed for the exit sign, and concrete footings, where used, should be kept flush with the ground level.

Unfortunately, there will be situations where placement of a major obstruction in a gore is unavoidable. Gores that occur at exit ramp terminals on elevated structures are a prime example. Also, there are occasions when a bridge pier in a gore cannot be avoided. Guardrails and bridge rails are designed to handle angular impacts but are not effective in handling the kind of near head-on impacts that occur at these gores.

In recognition of this problem, a considerable effort has been directed toward the development of cushioning or energy-dissipating devices for use in front of fixed objects. At present, several types of crash cushions are being used. These devices substantially reduce the severity of fixed-object collisions. Thus, adequate space should be provided for the installation of a crash-cushion device whenever a major obstruction is present in a gore on a high-speed highway. Reference may be made to Chapter 4 and to the *Roadside Design Guide* (**3**) for details on the installation of crash-cushion devices.

Although the term "gore" generally refers to the area between a through roadway and an exit ramp, the term may also be used to refer to the similar area between a through roadway and a converging entrance ramp. At an entrance terminal, the point of convergence (beginning of all paved area) is defined as the "merging end." In shape, layout, and extent, the triangular maneuver area at an entrance terminal is much like that at an exit. However, it points downstream and separates traffic streams already in lanes; thus, it is less of a decision area. The width at the base of the paved triangular area is narrower, however, and is usually limited to the sum of the shoulder widths on the ramp and freeway plus a narrow physical nose 1.2 to 2.4 m [4 to 8 ft] wide.

Exhibit 10-60 illustrates typical gore designs for free-flow exit ramps. Exhibits 10-60A and 10-60B depict a recovery area adjacent to the outside through lane and moderate offset of the ramp traveled way to the left.

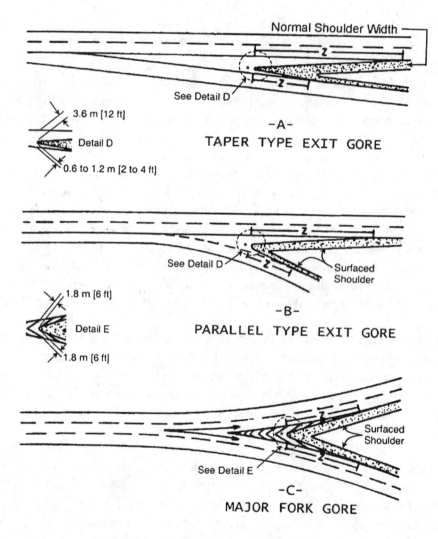

Exhibit 10-60. Typical Gore Details

Exhibit 10-60C presents a major fork, with neither diverging roadway having priority. The offset is equal for each roadway, and striping or rumble strips are placed upstream from the physical nose. Desirably, curbs, utility poles, and sign supports should be omitted from the gore area, especially on high-speed facilities. When curbs are used, they should be low-profile, sloping designs and the geometry of the gore area intersection points is usually curved. When curbs are not used, the geometry of the gore area intersection points can be squared or truncated.

Exhibit 10-61 gives the minimum lengths for tapers beyond the offset nose (shown as length Z in Exhibit 10-60). However, another alternative for providing a recovery area is the use of the paved shoulder of the through lane.

Metric		US Customary	
Design speed of approach highway (km/h)	Length of nose taper (Z) per unit width of nose offset	Design speed of approach highway (mph)	Length of nose taper (Z) per unit width of nose offset
50	15.0	30	15.0
60	20.0	35	17.5
70	22.5	40	20.0
80	25.0	45	22.5
90	27.5	50	25.0
100	30.0	55	27.5
110	35.0	60	30.0
120	40.0	65	32.5
		70	35.0
		75	37.5

Exhibit 10-61. Minimum Length of Taper Beyond an Offset Nose

Exhibit 10-62 shows an entrance ramp, as at a cloverleaf loop, where a reduction in the ramp lane width is appropriate in order to maintain a single-lane entrance. Another option is to begin the reduction in the ramp lane width at the end of the ramp curvature.

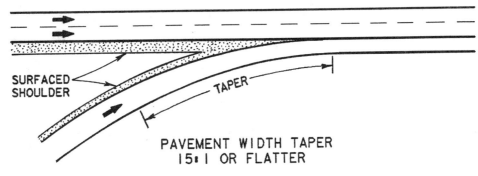

Exhibit 10-62. Traveled-Way Narrowing on Entrance Ramps

Exhibit 10-63 presents a photograph of a single-lane exit. The striping, pavement reflectors, delineators, and fixed-source lighting help guide the exiting motorist.

Exhibit 10-63. Gore Area, Single-Lane Exit

Exhibit 10-64 shows a gore at a major fork between two freeways with three lanes on each roadway. The small angle of divergence results in the long, gradual split with a clear recovery area. Overhead signs are provided.

Exhibit 10-64. Gore Area, Major Fork

Whereas Exhibit 10-65 shows a gore at a two-lane exit from a freeway, Exhibit 10-66 shows a typical gore and ramp terminal for a ramp entering a freeway.

Exhibit 10-65. Gore Area, Two-Lane Exit

Exhibit 10-66. Entrance Terminal

Ramp Traveled-Way Widths

Width and cross section. Ramp traveled-way widths are governed by the type of operation, curvature, and volume and type of traffic. It should be noted that the roadway width for a turning roadway includes the traveled-way width plus the shoulder width or equivalent clearance outside the edges of the traveled way. The section "Widths for Turning Roadways" in Chapter 3 may be referenced for additional discussion on the treatments at the edge of traveled way. Design widths of ramp traveled ways for various conditions are given in Exhibit 10-67. Values are shown for three general design traffic conditions, as follows:

Traffic Condition A—predominantly P vehicles, but some consideration for SU trucks.

Traffic Condition B—sufficient SU vehicles to govern design, but some consideration for semitrailer vehicles.

Traffic Condition C—sufficient buses and combination trucks to govern design.

Traffic conditions A, B, and C are described in broad terms because design traffic volume data for each type of vehicle are not available to define these traffic conditions with precision in relation to traveled-way width. In general, traffic condition A has a small volume of trucks or only an occasional large truck, traffic condition B has a moderate volume of trucks (in the range of 5 to 10 percent of the total traffic), and traffic condition C has more and larger trucks.

Shoulders and lateral clearances. Design values for shoulders and lateral clearances on the ramps are as follows:

- When paved shoulders are provided on ramps, they should have a uniform width for the full length of ramp. For one-way operation, the sum of the right and left shoulder widths should not exceed 3.0 to 3.6 m [10 to 12 ft]. A paved shoulder width of 0.6 to 1.2 m [2 to 4 ft] is desirable on the left with the remaining width of 2.4 to 3.0 m [8 to 10 ft] used for the paved right shoulder.
- The ramp traveled-way widths from Exhibit 10-67 for Case II and Case III should be modified when paved shoulders are provided on the ramp. The ramp traveled-way width for Case II should be reduced by the total width of both right and left shoulders. However, in no case should the ramp traveled-way width be less than needed for Case I. For example, with condition C and a 125-m [400-ft] radius, the Case II ramp traveled-way width without shoulders is 6.6 m [22 ft]. If a 0.6-m [2-ft] left shoulder and a 2.4-m [8-ft] right shoulder are provided, the minimum ramp traveled-way width should be 4.8 m [16 ft].
- Directional ramps with a design speed over 60 km/h [40 mph] should have a paved right shoulder width of 2.4 to 3.0 m [8 to 10 ft] and a paved left shoulder width of 0.3 to 1.8 m [1 to 6 ft].
- For freeway ramp terminals where the ramp shoulder is narrower than the freeway shoulder, the paved shoulder width of the through lane should be carried into the exit terminal. It should also begin within the entrance terminal, with the transition to the

US Customary

Pavement width (ft) — Design traffic conditions

Radius on inner edge of pavement R (ft)	Case I — One-lane, one-way operation—no provision for passing a stalled vehicle			Case II — One-lane, one-way operation—with provision for passing a stalled vehicle			Case III — Two-lane operation—either one-way or two-way		
	A	B	C	A	B	C	A	B	C
50	18	18	23	20	26	30	31	36	45
75	16	17	20	19	23	27	29	33	38
100	15	16	18	18	22	25	28	31	35
150	14	15	17	18	21	23	26	29	32
200	13	15	16	17	20	22	26	28	30
300	13	15	15	17	20	22	25	28	29
400	13	15	15	17	19	21	25	27	28
500	12	15	15	17	19	21	25	27	28
Tangent	12	14	14	17	18	20	24	26	26

Width modification regarding edge treatment

	Case I	Case II	Case III
No stabilized shoulder	None	None	None
Sloping curb	None	None	None
Vertical curb: one side	Add 1 ft	None	Add 1 ft
two sides	Add 2 ft	Add 1 ft	Add 2 ft
Stabilized shoulder, one or both sides	Lane width for conditions B & C on tangent may be reduced to 12 ft where shoulder is 4 ft or wider	Deduct shoulder width; minimum pavement width as under Case I	Deduct 2 ft where shoulder is 4 ft or wider

Metric

Pavement width (m) — Design traffic conditions

Radius on inner edge of pavement R (m)	Case I — One-lane, one-way operation—no provision for passing a stalled vehicle			Case II — One-lane, one-way operation—with provision for passing a stalled vehicle			Case III — Two-lane operation—either one-way or two-way		
	A	B	C	A	B	C	A	B	C
15	5.4	5.5	7.0	6.0	7.8	9.2	9.4	11.0	13.6
25	4.8	5.0	5.8	5.6	6.9	7.9	8.6	9.7	11.1
30	4.5	4.9	5.5	5.5	6.7	7.6	8.4	9.4	10.6
50	4.2	4.6	5.0	5.3	6.3	7.0	7.9	8.8	9.5
75	3.9	4.5	4.8	5.2	6.1	6.7	7.7	8.5	8.9
100	3.9	4.5	4.8	5.2	5.9	6.5	7.6	8.3	8.7
125	3.9	4.5	4.8	5.1	5.9	6.4	7.6	8.2	8.5
150	3.6	4.5	4.5	5.1	5.8	6.4	7.5	8.2	8.4
Tangent	3.6	4.2	4.2	5.0	5.5	6.1	7.3	7.9	7.9

Width modification regarding edge treatment

	Case I	Case II	Case III
No stabilized shoulder	None	None	None
Sloping curb	None	None	None
Vertical curb: one side	Add 0.3 m	None	Add 0.3 m
two sides	Add 0.6 m	Add 0.3 m	Add 0.6 m
Stabilized shoulder, one or both sides	Lane width for conditions B & C on tangent may be reduced to 3.6 m where shoulder is 1.2 m or wider	Deduct shoulder width; minimum width as under Case I	Deduct 0.6 where shoulder is 1.2 m or wider

Note: A = predominantly P vehicles, but some consideration for SU trucks.
B = sufficient SU vehicles to govern design, but some consideration for semitrailer combination trucks.
C = sufficient bus and combination-trucks to govern design.

Exhibit 10-67. Design Widths for Turning Roadways

narrower ramp shoulder accomplished gracefully on the ramp end of the terminal. Abrupt changes should be avoided.

- Ramps should have a lateral clearance on the right outside of the edge of the traveled way of at least 1.8 m [6 ft], and preferably 2.4 to 3.0 m [8 to 10 ft], and a lateral clearance on the left of at least 1.2 m [4 ft] beyond the edge of traveled way.

- Where ramps pass under structures, the total roadway width should be carried through the structure. Desirably, structural supports should be located beyond the clear zone. As a minimum, structural supports should be at least 1.2 m [4 ft] beyond the edge of paved shoulder. The AASHTO *Roadside Design Guide* (3) provides guidance on clear zone and the use of roadside barriers.

- Ramps on overpasses should have the full approach roadway width carried over the structure.

- Edge lines or some type of color or texture differentiation between the traveled way and shoulder is desirable.

Shoulders and curbs. Shoulders should be provided on ramps and ramp terminals in interchange areas to provide a space that is clear of the traveled way for emergency stopping, to minimize the effect of breakdowns, and to aid drivers who may be confused.

Ramps at interchanges should be designed without curbs. Curbs should be considered only to facilitate particularly difficult drainage situations, such as in urban areas where restrictive right-of-way favors enclosed drainage. In some cases, curbs are used at the ramp terminals but are omitted along the central ramp portions. Where curbs are not used, full-depth paving should be provided on shoulders because of the frequent use of shoulders for turning movements.

On low-speed facilities, curbs may be placed at the edge of roadway. Vertical curbs are seldom used in conjunction with shoulders, except where pedestrian protection is needed. Where curbs are used on high-speed facilities, sloping curbs should be placed at the outer edge of the shoulder. Because of fewer restrictions and more liberal designs in rural areas, the need for curbs seldom arises. See Chapter 4 for a full discussion of shoulder cross-section elements.

Ramp Terminals

The terminal of a ramp is that portion adjacent to the through traveled way, including speed-change lanes, tapers, and islands. Ramp terminals may be the at-grade type, as at the crossroad terminal of diamond or partial cloverleaf interchanges, or the free-flow type where ramp traffic merges with or diverges from high-speed through traffic at flat angles. Design elements for the at-grade type are discussed in Chapter 9, and those for the free-flow type are discussed in the following sections.

Terminals are further classified as either single or multilane, according to the number of lanes on the ramp at the terminal and as either a taper or parallel type, according to the configuration of the speed-change lane.

Left-hand entrances and exits. Left-hand entrances and exits are contrary to the concept of driver expectancy when intermixed with right-hand entrances and exits. Therefore, extreme care should be exercised to avoid left-hand entrances and exits in the design of interchanges.

Left-side ramp terminals break up the uniformity of interchange patterns and generally create uncertain operation on through roadways. Left-hand entrances and exits are considered satisfactory for collector-distributor roads; however, their use on high-speed, free-flow ramp terminals is not recommended. Because left-hand entrances and exits are contrary to driver expectancy, special attention should be given to signing and the provision for decision sight distance in order to alert the driver that an unusual situation exists.

Terminal location and sight distance. Where diamond ramps and partial cloverleaf arrangements intersect the crossroad at grade, an at-grade intersection is formed. Desirably, this intersection should be located an adequate distance from the separation structure to provide adequate sight distance for all approaches. Sight distance criteria are detailed in Chapter 3.

Drivers prefer and expect to exit in advance of the separation structure. The use of collector-distributor roads and single exits on partial cloverleafs and other types of interchange configurations automatically positions the main line exit in advance of the separation structure.

Designs that result in an exit concealed behind a crest vertical curve should be avoided, especially on high-speed facilities. Desirably, high-speed entrance ramp terminals should be located on descending grades to aid truck acceleration. Adequate sight distance at entrance terminals should be available so that merging traffic on the ramp can adjust speed to merge into gaps on the main facility.

Loop ramps that are located beyond the structure, as in the conventional cloverleaf or in certain arrangements of partial cloverleafs, usually need a parallel deceleration lane. The actual exit from the auxiliary lane is difficult for drivers to locate even when sight distance is not restricted by a vertical curve. Placing the exit in advance of the structure via a single exit alleviates this problem. See the section on "Two-Exit Versus Single-Exit Interchange Design" earlier in this chapter.

Ramp terminal design. Profiles of ramp terminals should be designed in association with horizontal curves to avoid sight restrictions that will adversely affect operations. At an exit into a ramp on a descending grade, a horizontal curve ahead should not appear suddenly to a driver using the ramp. Instead, the initial crest vertical curve should be made longer and sight distance over it should be increased so that the location and direction of the horizontal curve are apparent to the driver in sufficient time for the driver to respond appropriately. At an entrance terminal from a ramp on an ascending grade, the portion of the ramp intended for acceleration and the ramp terminal should closely parallel the through-lane profile to permit entering drivers to have a clear view of the through road ahead, to the side, and to the rear.

It is desirable that profiles of highway ramp terminals be designed with a platform on the ramp side of the approach nose or merging end. This platform should be at least 60 m [200 ft] in length and should have a profile that does not greatly differ from that of the adjacent through-traffic lane.

A platform area should also be provided at the at-grade terminal of a ramp. The length of this platform should be determined from the type of traffic control and the capacity at the terminal. For further discussion, see Chapter 9.

Traffic control. On major highways, ramps are arranged to facilitate all turning movements by merging or diverging maneuvers. On minor highways, some of the left-turning movements often are made at grade. The left-turning movements leaving the crossing highway preferably should have median left-turn lanes. For low-volume crossroads the left-turning movements from ramps normally should be controlled by stop signs. The right-turning movements from ramps into multilane crossroads should be provided with an acceleration lane or generous taper, or should be controlled by stop or yield signs. Ramps approaching stop signs should be nearly perpendicular to the crossroad and be nearly level for storage of several vehicles.

Traffic signal controls may be needed at ramp terminals on the minor road where there is sufficient volume of through and turning traffic. In such cases, the intersections formed at the terminals should be designed and operated in the same manner as any other traffic-signal-controlled intersection at grade. Signal controls should be avoided on express-type highways and confined to the minor highways on which other intersections are at grade and some of which are signalized. In or near urban areas, signal control is especially appropriate at ramp terminals on streets crossing over or under an expressway. Here the turning movements usually are sizable, and the cost of right-of-way and improvements is high. As a result, appreciable savings may be realized by the use of diamond ramps with high-type terminals on the expressway and signalized terminals on the cross streets. Warrants for the installation of traffic signals that can be applied to diamond ramp terminals are given in Part 4 of the MUTCD (**4**).

Distance between a free-flow terminal and structure. The terminal of a ramp should not be near the grade-separation structure. If it is not practical to place the exit terminal in advance of the structure, the exiting terminal on the far side of the structure should be well removed in order that, when leaving, drivers have some distance after passing the structure in which to see the turnout and begin the turnoff maneuver. Decision sight distance is recommended where practical. The distance between the structure and the approach nose at the ramp terminal should be sufficient for exiting drivers to leave the through lanes without undue hindrance to through traffic. Such distance also aids drivers entering from a ramp terminal on the far side of the structure to have a clear view well back on the through road behind or to the left. Such drivers may be able to see back along the road beyond the limits of the structure, but as a general rule, the crest of the profile at an overpass and the columns, abutments, and approach walls at an underpass obstruct or impair their view of the traffic stream into which they merge.

The conditions for determining the distance between a structure and the far side approach nose are similar to those discussed for speed-change lanes. A minimum distance between the structure and an exit nose of about the same length as a speed-change taper is suggested. Decision sight distances are desirable but are not rigid controls for ramp design. Topographic or right-of-way controls may govern the overall shape of the ramp.

While safety and convenience are enhanced by a long separation distance between a structure and an exit ramp terminal, this distance can be too long for certain ramp arrangements such as cloverleaf loop ramps. Unusually large right-of-way needs as well as increased travel time and length on the loops may result. Where only one loop is needed and it falls on the far side of the structure, a speed-change lane should be developed on the near side of the structure and carried across the structure if sight distance is a problem.

The separation distance between a structure and a ramp terminal does not need to be as long for ramp terminals on the near side of a structure as for those beyond the structure. Both the view of the terminal ahead for drivers approaching on the through road and the view back along the road for drivers on an entrance ramp are not affected by the structure. Where an entrance ramp curve on the near side of the structure needs an acceleration lane, the ramp terminal should be located to provide sufficient length for it between the terminal and the structure, or the acceleration lane maybe continued through or over the structure. Where ramp terminals on the far side of a structure are located close to it, the horizontal sight line may be limited by the abutment or parapet; available sight distance should, therefore, be checked.

Distance between successive ramp terminals. On urban freeways, two or more ramp terminals are often located in close succession. To provide sufficient weaving length and adequate space for signing, a reasonable distance should be provided between successive ramp terminals. Spacing between successive outer ramp terminals is dependent on the classification of the interchanges involved, the function of the ramp pairs (entrance or exit), and weaving potential.

The five possible ramp-pair combinations are: (1) an entrance followed by an entrance (EN-EN), (2) an exit followed by an exit (EX-EX), (3) an exit followed by an entrance (EX-EN), (4) an entrance followed by an exit (EN-EX) (weaving), and (5) turning roadways.

Exhibit 10-68 presents recommended minimum ramp terminal spacing for the various ramp-pair combinations as they are applicable to interchange classifications.

Where an entrance ramp is followed by an exit ramp, the absolute minimum distance between the successive noses is governed by weaving considerations. The spacing policy for EN-EX ramp combinations is not applicable to cloverleaf loop ramps. For these interchanges, the distance between EN-EX ramp noses is primarily dependent on loop ramp radii and roadway and median widths. A recovery lane beyond the nose of the loop ramp exit is desirable.

When the distance between the successive noses is less than 450 m [1,500 ft], the speed-change lanes should be connected to provide an auxiliary lane. This auxiliary lane improves traffic operation over relatively short sections of the freeway route and is not considered an addition to the basic number of lanes. See the section "Auxiliary Lanes" in this chapter for alternate methods of dropping these lanes.

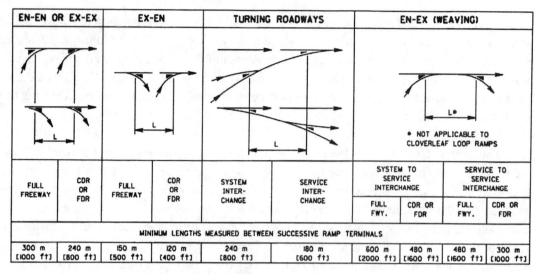

EN-EN OR EX-EX		EX-EN		TURNING ROADWAYS		EN-EX (WEAVING)			
FULL FREEWAY	CDR OR FDR	FULL FREEWAY	CDR OR FDR	SYSTEM INTER-CHANGE	SERVICE INTER-CHANGE	SYSTEM TO SERVICE INTERCHANGE		SERVICE TO SERVICE INTERCHANGE	
						FULL FWY.	CDR OR FDR	FULL FWY.	CDR OR FDR
MINIMUM LENGTHS MEASURED BETWEEN SUCCESSIVE RAMP TERMINALS									
300 m [1000 ft]	240 m [800 ft]	150 m [500 ft]	120 m [400 ft]	240 m [800 ft]	180 m [600 ft]	600 m [2000 ft]	480 m [1600 ft]	480 m [1600 ft]	300 m [1000 ft]

NOTES: FDR - FREEWAY DISTRIBUTOR ROAD EN - ENTRANCE
CDR - COLLECTOR DISTRIBUTOR ROAD EX - EXIT

THE RECOMMENDATIONS ARE BASED ON OPERATIONAL EXPERIENCE AND NEED FOR FLEXIBILITY AND ADEQUATE SIGNING. THEY SHOULD BE CHECKED IN ACCORDANCE WITH THE PROCEDURE OUTLINED IN THE HIGHWAY CAPACITY MANUAL (4) AND THE LARGER OF THE VALUES IS SUGGESTED FOR USE. ALSO, A PROCEDURE FOR MEASURING THE LENGTH OF THE WEAVING SECTION IS GIVEN IN CHAPTER 24 OF THE 2000 HIGHWAY CAPACITY MANUAL (4). THE "L" DISTANCES NOTED IN THE FIGURES ABOVE ARE BETWEEN LIKE POINTS, NOT NECESSARILY "PHYSICAL" GORES. A MINIMUM DISTANCE OF 90 m [270 ft] IS RECOMMENDED BETWEEN THE END OF THE TAPER FOR THE FIRST ON RAMP AND THE THEROETICAL GORE FOR THE SUCCEDING ON RAMP FOR THE EN-EN (SIMILIAR FOR EX-EN).

Exhibit 10-68. Recommended Minimum Ramp Terminal Spacing

Speed-change lanes. Drivers leaving a highway at an interchange are required to reduce speed as they exit on a ramp. Drivers entering a highway from a turning roadway accelerate until the desired highway speed is reached. Because the change in speed is usually substantial, provision should be made for acceleration and deceleration to be accomplished on auxiliary lanes to minimize interference with through traffic and to reduce crash potential. Such an auxiliary lane, including tapered areas, may be referred to as a speed-change lane. The terms "speed-change lane," "deceleration lane," or "acceleration lane" as used herein apply broadly to the added lane joining the traveled way of the highway with that of the turning roadway and do not necessarily imply a definite lane of uniform width. This additional lane is a part of the elongated ramp terminal area.

A speed-change lane should have sufficient length to enable a driver to make the appropriate change in speed between the highway and the turning roadway in a safe and comfortable manner. Moreover, in the case of an acceleration lane, there should be additional length to permit adjustments in speeds of both through and entering vehicles so that the driver of the entering vehicle can position himself opposite a gap in the through-traffic stream and maneuver into it before reaching the end of the acceleration lane. This latter consideration also influences both the configuration and length of an acceleration lane.

Two general forms of speed-change lanes are: (1) the taper and (2) the parallel type. The taper type provides a direct entry or exit at a flat angle, whereas the parallel type has an added lane for changing speed. Either type, when properly designed, will operate satisfactorily. However, the parallel type is still favored in certain areas. Furthermore, some agencies use the taper type for exits and the parallel type for entrances.

See Chapter 9 for discussion of speed-change lanes as applicable to at-grade intersections.

Single-Lane Free-Flow Terminals, Entrances

Taper-type entrance. When properly designed, the taper-type entrance usually operates smoothly at all volumes up to and including the design capacity of merging areas. By relatively minor speed adjustment, the entering driver can see and use an available gap in the through-traffic stream. A typical single-lane, taper-type entrance terminal is shown in Exhibit 10-69A.

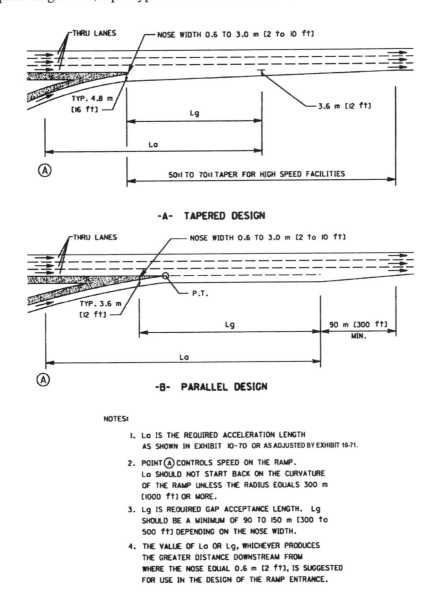

NOTES:

1. Lo IS THE REQUIRED ACCELERATION LENGTH AS SHOWN IN EXHIBIT 10-70 OR AS ADJUSTED BY EXHIBIT 10-71.

2. POINT (A) CONTROLS SPEED ON THE RAMP. Lo SHOULD NOT START BACK ON THE CURVATURE OF THE RAMP UNLESS THE RADIUS EQUALS 300 m [1000 ft] OR MORE.

3. Lg IS REQUIRED GAP ACCEPTANCE LENGTH. Lg SHOULD BE A MINIMUM OF 90 TO 150 m [300 to 500 ft] DEPENDING ON THE NOSE WIDTH.

4. THE VALUE OF Lo OR Lg, WHICHEVER PRODUCES THE GREATER DISTANCE DOWNSTREAM FROM WHERE THE NOSE EQUAL 0.6 m [2 ft], IS SUGGESTED FOR USE IN THE DESIGN OF THE RAMP ENTRANCE.

Exhibit 10-69. Typical Single-Lane Entrance Ramps

The entrance is merged into the freeway with a long, uniform taper. Operational studies show a desirable rate of taper of about 50:1 to 70:1 (longitudinal to lateral) between the outer edge of the acceleration lane and the edge of the through-traffic lane. The gap acceptance length, L_g is also a consideration in the design of taper-type entrances, as illustrated in Exhibit 10-69A.

The geometrics of the ramp proper should be such that motorists may attain a speed that is within 10 km/h [5 mph] of the operating speed of the freeway by the time they reach the point where the left edge of the ramp joins the traveled way of the freeway. For consistency of application, this point of convergence of the left edge of the ramp and the right edge of the through lane may be assumed to occur where the right edge of the ramp traveled way is 3.6 m [12 ft] from the right edge of the through lane of the freeway.

The distance needed for acceleration in advance of this point of convergence is governed by the speed differential between the operating speed on the entrance curve of the ramp and the operating speed of the highway. Exhibit 10-70 shows minimum lengths of acceleration distances for entrance terminals. Exhibit 10-69 shows the minimum lengths for gap acceptance. Referring to Exhibit 10-69, the larger value of the acceleration length (L_a) or the gap acceptance (L_g) length is suggested for use in the design of the ramp entrance. Where the minimum values for nose width (0.6 m [2 ft]), lane width 4.8 m [16 ft]), and taper rate (50:1) are used with high traffic volumes, taper lengths longer than the larger of L_a or L_g may be needed to avoid inferior operation and to reduce fairly sharp moves into the mainline traffic stream. Where grades are present on ramps, speed-change lengths should be adjusted in accordance with Exhibit 10-71.

Parallel-type entrances. The parallel-type entrance provides an added lane of sufficient length to enable a vehicle to accelerate to near-freeway speed prior to merging. A taper is provided at the end of the added lane. The process of entering the freeway is similar to a lane change to the left. The driver is able to use the side-view and rear-view mirrors to monitor surrounding traffic.

A typical design of a parallel-type entrance is shown in Exhibit 10-69B. Desirably, a curve with a radius of 300 m [1,000 ft] or more and a length of at least 60 m [200 ft] should be provided in advance of the added lane. If this curve has a short radius, motorists tend to drive directly onto the freeway without using the acceleration lane. This behavior results in undesirable merging operations.

The taper at the downstream end of a parallel-type acceleration lane should be a suitable length to guide the vehicle gradually onto the through lane of the freeway. A taper length of approximately 90 m [300 ft] is suitable for design speeds up to 110 km/h [70 mph].

The length of a parallel-type acceleration lane is generally measured from the point where the left edge of the traveled way of the ramp joins the traveled way of the freeway to the beginning of the downstream taper. Whereas, in the case of the taper type entrance, acceleration is accomplished on the ramp upstream of the point of convergence of the two roadways, acceleration usually takes place downstream from this point in the case of the parallel type. However, a part of the ramp proper may also be considered in the acceleration length, provided the curve approaching the acceleration lane has a long radius of approximately 300 m [1,000 ft] or more, and the motorist on the ramp has an unobstructed view of traffic on the freeway to his or her left. The minimum acceleration lengths for entrance terminals are given in Exhibit 10-70, and the adjustments for grades are given in Exhibit 10-71.

Metric									
Acceleration length, L (m) for entrance curve design speed (km/h)									
Highway	Stop condition	20	30	40	50	60	70	80	
	Speed reached,	and initial speed, V'$_a$ (km/h)							
Design speed, V (km/h)	V$_a$ (km/h)	0	20	28	35	42	51	63	70
50	37	60	50	30	–	–	–	–	–
60	45	95	80	65	45	–	–	–	–
70	53	150	130	110	90	65	–	–	–
80	60	200	180	165	145	115	65	–	–
90	67	260	245	225	205	175	125	35	–
100	74	345	325	305	285	255	205	110	40
110	81	430	410	390	370	340	290	200	125
120	88	545	530	515	490	460	410	325	245

Note: Uniform 50:1 to 70:1 tapers are recommended where lengths of acceleration lanes exceed 400 m.

US Customary									
Acceleration length, L (ft) for entrance curve design speed (mph)									
Highway	Stop condition	15	20	25	30	35	40	45	50
	Speed reached,	and initial speed, V'$_a$ (mph)							
Design speed, V (mph)	V$_a$ (mph)	0	14	18	22	26	30	36	40	44
30	23	180	140	–	–	–	–	–	–
35	27	280	220	160	–	–	–	–	–
40	31	360	300	270	210	120	–	–	–
45	35	560	490	440	380	280	160	–	–
50	39	720	660	610	550	450	350	130	–
55	43	960	900	810	780	670	550	320	150	–
60	47	1200	1140	1100	1020	910	800	550	420	180
65	50	1410	1350	1310	1220	1120	1000	770	600	370
70	53	1620	1560	1520	1420	1350	1230	1000	820	580
75	55	1790	1730	1630	1580	1510	1420	1160	1040	780

Note: Uniform 50:1 to 70:1 tapers are recommended where lengths of acceleration lanes exceed 1,300 ft.

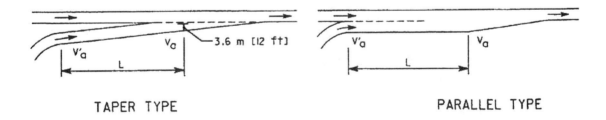

TAPER TYPE   PARALLEL TYPE

Exhibit 10-70. Minimum Acceleration Lengths for Entrance Terminals With Flat Grades of 2 Percent or Less

Metric

Deceleration lanes

Ratio of length on grade to length on level for design speed of turning curve (km/h)[a]

Design speed of highway (km/h)		Upgrade	Downgrade
All speeds	3 to 4%	0.9	1.2
All speeds	5 to 6%	0.8	1.35

Acceleration lanes

Ratio of length on grade to length of level for design speed of turning curve (km/h)[a]

Design speed of highway (km/h)	40	50	60	70	80	All speeds
	3 to 4% upgrade					**3 to 4% downgrade**
60	1.3	1.4	1.4	–	–	0.7
70	1.3	1.4	1.4	1.5	–	0.65
80	1.4	1.5	1.5	1.5	1.6	0.65
90	1.4	1.5	1.5	1.5	1.6	0.6
100	1.5	1.6	1.7	1.7	1.8	0.6
110	1.5	1.6	1.7	1.7	1.8	0.6
120	1.5	1.6	1.7	1.7	1.8	0.6
	5 to 6% upgrade					**5 to 6% downgrade**
60	1.5	1.5	–	–	–	0.6
70	1.5	1.6	1.7	–	–	0.6
80	1.5	1.7	1.9	1.8	–	0.55
90	1.6	1.8	2.0	2.1	2.2	0.55
100	1.7	1.9	2.2	2.4	2.5	0.5
110	2.0	2.2	2.6	2.8	3.0	0.5
120	2.3	2.5	3.0	3.2	3.5	0.5

US Customary

Deceleration lanes

Ratio of length on grade to length on level for design speed of turning curve (mph)[a]

Design speed of highway (mph)		Upgrade	Downgrade
All speeds	3 to 4%	0.9	1.2
All speeds	5 to 6%	0.8	1.35

Acceleration lanes

Ratio of length on grade to length of level for design speed of turning curve (mph)[a]

Design speed of highway (mph)	20	30	40	50	All speeds
	3 to 4% upgrade				**3 to 4% downgrade**
40	1.3	1.3	–	–	0.7
45	1.3	1.35	–	–	0.675
50	1.3	1.4	1.4	–	0.65
55	1.35	1.45	1.45	–	0.625
60	1.4	1.5	1.5	1.6	0.6
65	1.45	1.55	1.6	1.7	0.6
70	1.5	1.6	1.7	1.8	0.6
	5 to 6% upgrade				**5 to 6% downgrade**
40	1.5	1.5	–	–	0.6
45	1.5	1.6	–	–	0.575
50	1.5	1.7	1.9	–	0.55
55	1.6	1.8	2.05	–	0.525
60	1.7	1.9	2.2	2.5	0.5
65	1.85	2.05	2.4	2.75	0.5
70	2.0	2.2	2.6	3.0	0.5

[a] Ratio from this table multiplied by the length in Exhibit 10-70 or Exhibit 10-73 gives length of speed change lane on grade.

Exhibit 10-71. Speed Change Lane Adjustment Factors as a Function of Grade

The operational and safety benefits of long acceleration lanes provided by parallel type entrances are well recognized. A long acceleration lane provides more time for the merging vehicles to find an opening in the through-traffic stream. An acceleration lane length of at least 360 m [1,200 ft], plus the taper, is desirable wherever it is anticipated that the ramp and freeway will frequently carry traffic volumes approximately equal to the design capacity of the merging area.

Single-Lane Free-Flow Terminals, Exits

Taper-type exits. The taper-type exit fits the direct path preferred by most drivers, permitting them to follow an easy path within the diverging area. The taper-type exit terminal beginning with an outer edge alignment break usually provides a clear indication of the point of departure from the through lane and has generally been found to operate smoothly on high-volume freeways. The divergence angle is usually between 2 and 5 degrees.

Studies of this type of terminal show that most vehicles leave the through lane at relatively high speeds, thereby reducing the potential for rear-end collisions as a result of deceleration on the through lane. The speed change can be achieved off the traveled way as the exiting vehicle moves along the taper onto the ramp proper. Exhibit 10-72A shows a typical design for a taper-type exit.

Vehicles should decelerate after clearing the through-traffic lane and before reaching the point limiting design speed for the ramp proper. The length available for deceleration may be assumed to extend from a point where the right edge of the tapered wedge is about 3.6 m [12 ft] from the right edge of the right through lane, to the point of initial curvature of the exit ramp (i.e., the first horizontal curve on the ramp). The length provided between these points should be at least as great as the distance needed to accomplish the appropriate deceleration, which is governed by the speed of traffic on the through lane and the speed to be attained on the ramp. Deceleration may end in a complete stop, as at a crossroad terminal for a diamond interchange, or the critical speed may be governed by the curvature of the ramp roadway. Minimum deceleration lengths for various combinations of design speeds for the highway and for the ramp roadway are given in Exhibit 10-73. Grade adjustments are given in Exhibit 10-71.

The taper-type exit terminal design can be used advantageously in developing the desired long, narrow, triangular emergency maneuver area just upstream from the exit nose located at a proper offset from both the through lane and separate ramp lane. The taper configuration also works well in the length-width superelevation adjustments to obtain a ramp cross slope different from that of the through lane.

The width of the recovery area or the distance between the inner edges of the diverging lanes at the ramp nose is usually 6.0 to 9.0 m [20 to 30 ft]. This entire area should be paved to provide a maneuver and recovery area, but the desired travel path for the ramp roadway should be clearly delineated by pavement markings.

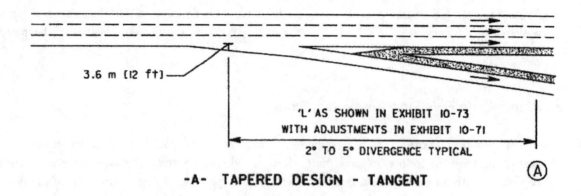

3.6 m [12 ft]

'L' AS SHOWN IN EXHIBIT 10-73
WITH ADJUSTMENTS IN EXHIBIT 10-71

2° TO 5° DIVERGENCE TYPICAL

-A- TAPERED DESIGN - TANGENT

Ⓐ

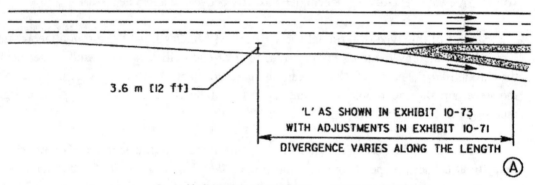

3.6 m [12 ft]

'L' AS SHOWN IN EXHIBIT 10-73
WITH ADJUSTMENTS IN EXHIBIT 10-71

DIVERGENCE VARIES ALONG THE LENGTH

-B- TAPERED DESIGN - CURVILINEAR

Ⓐ

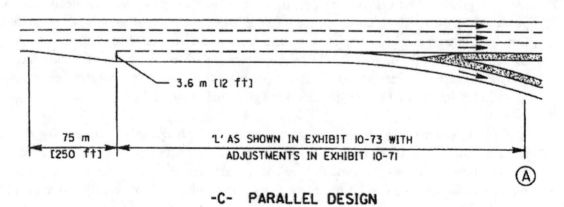

3.6 m [12 ft]

75 m
[250 ft]

'L' AS SHOWN IN EXHIBIT 10-73 WITH
ADJUSTMENTS IN EXHIBIT 10-71

-C- PARALLEL DESIGN

Ⓐ

Ⓐ POINT CONTRLLING SPEED AT RAMP

Exhibit 10-72. Exit Ramps—Single Lane

Metric									
Deceleration length, L (m) for design speed of exit curve V' (km/h)									
Highway design speed, V (km/h)	Speed reached, V$_a$ (km/h)	Stop condition	20	30	40	50	60	70	80
		For average running speed on exit curve V'$_a$ (km/h)							
		0	20	28	35	42	51	63	70
50	47	75	70	60	45	–	–	–	–
60	55	95	90	80	65	55	–	–	–
70	63	110	105	95	85	70	55	–	–
80	70	130	125	115	100	90	80	55	–
90	77	145	140	135	120	110	100	75	60
100	85	170	165	155	145	135	120	100	85
110	91	180	180	170	160	150	140	120	105
120	98	200	195	185	175	170	155	140	120

V = design speed of highway (km/h)
V$_a$ = average running speed on highway (km/h)
V' = design speed of exit curve (km/h)
V'$_a$ = average running speed on exit curve (km/h)

US Customary									
Deceleration length, L (ft) for design speed of exit curve, V' (mph)									
		Stop condition	15	20	25	30	35	40	45	50
		For average running speed on exit curve, V'$_a$ (mph)							
Highway design speed, V (mph)	Speed reached, V$_a$ (mph)	0	14	18	22	26	30	36	40	44
30	28	235	200	170	140	–	–	–	–	–
35	32	280	250	210	185	150	–	–	–	–
40	36	320	295	265	235	185	155	–	–	–
45	40	385	350	325	295	250	220	–	–	–
50	44	435	405	385	355	315	285	225	175	–
55	48	480	455	440	410	380	350	285	235	–
60	52	530	500	480	460	430	405	350	300	240
65	55	570	540	520	500	470	440	390	340	280
70	58	615	590	570	550	520	490	440	390	340
75	61	660	635	620	600	575	535	490	440	390

V = design speed of highway (mph)
V$_a$ = average running speed on highway (mph)
V' = design speed of exit curve (mph)
V'$_a$ = average running speed on exit curve (mph)

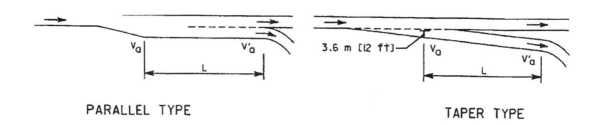

PARALLEL TYPE TAPER TYPE

Exhibit 10-73. Minimum Deceleration Lengths for Exit Terminals With Flat Grades of 2 Percent or Less

Parallel-type exits. A parallel-type exit terminal usually begins with a taper, followed by an added lane that is parallel to the traveled way. A typical parallel-type exit terminal is shown in Exhibit 10-72C. This type of terminal provides an inviting exit area, because the foreshortened view of the taper and the added width are very apparent. Parallel-type exits operate best when drivers choose to exit the through lane sufficiently in advance of the exit nose to permit deceleration to occur on the added lane (deceleration lane) and allows them to follow a path similar to that encouraged by a taper design. Drivers who do not exit the through lane sufficiently in advance of the exit nose will likely utilize a more abrupt reverse-curve maneuver, which is somewhat unnatural and can sometimes result in the driver slowing in the through lane. In locations where both the mainline and ramp carry high volumes of traffic, the deceleration lane provided by the parallel-type exit provides storage for vehicles that would otherwise undesirably queue up on the through lane or on a shoulder, if available.

The length of a parallel-type deceleration lane is usually measured from the point where the added lane attains a 3.6 m [12 ft] width to the point where the alignment of the ramp roadway departs from the alignment of the freeway. Where the ramp proper is curved, it is desirable to provide a transition at the end of the deceleration lane. A compound curve may be used with the initial curve desirably and should have a long radius of about 300 m [1,000 ft] or more. A transition or a long radius curve is also desirable if the deceleration lane connects with a relatively straight ramp. In such cases, a portion of the ramp may be considered as a part of the deceleration length, thus shortening to some extent the appropriate length of contiguous parallel lane. Minimum lengths are given in Exhibit 10-73, and adjustments for grades are given in Exhibit 10-71. Longer parallel-type deceleration lanes are more likely than shorter lanes to be used properly. Lengths of at least 240 m [800 ft] are desirable.

The taper portion of a parallel-type deceleration lane should have a taper of approximately 15:1 to 25:1 [longitudinal:transverse]. A long taper indicates the general path to be followed and reduces the unused portion of the deceleration lane. However, a long taper tends to entice the through driver into the deceleration lane. A short taper produces a better "target" to the approaching driver, giving a positive indication of the added lane ahead.

Free-flow terminals on curves. The previous discussion was based on highways with a tangent alignment. Because the curvature on most freeways is slight, there is usually no need to make any appreciable adjustments at ramp terminals on curves. However, where the curves on a freeway are relatively sharp and there are exits and entrances located on these curves, some adjustments in design may be desirable to avoid operational difficulties.

On freeways having design speeds of 100 km/h [60 mph] or more, the curves are sufficiently gentle so that either the parallel type or the taper type of speed-change lane is suitable. With the parallel type, the design is about the same as that on tangent and the added lane is usually on the same curvature as the main line. With the taper type, the dimensions applicable to terminals located on tangent alignment are also suitable for use on curves. A method for developing the alignment of tapered speed-change lanes on curves is illustrated in Exhibit 10-74. On curved sections, the ramp is tapered at the same rate relative to the through-traffic lanes as on tangent sections.

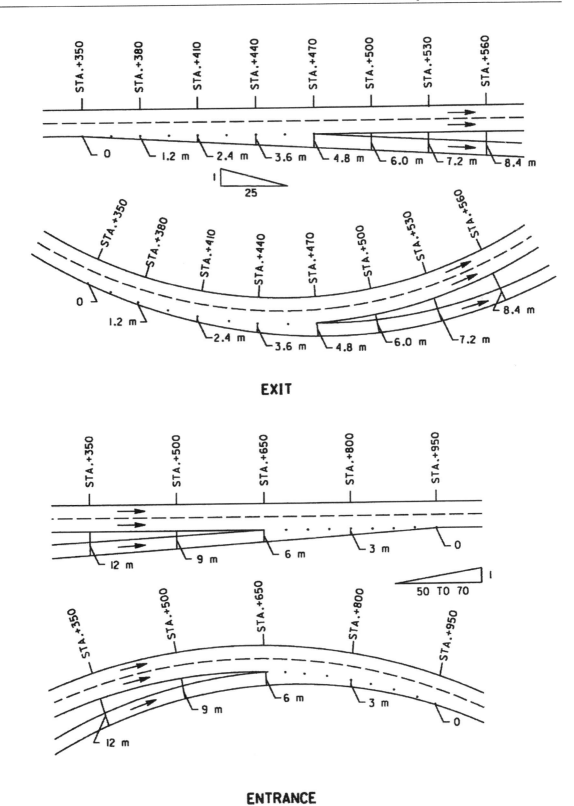

Exhibit 10-74. Layout of Taper-Type Terminals on Curves (Metric)

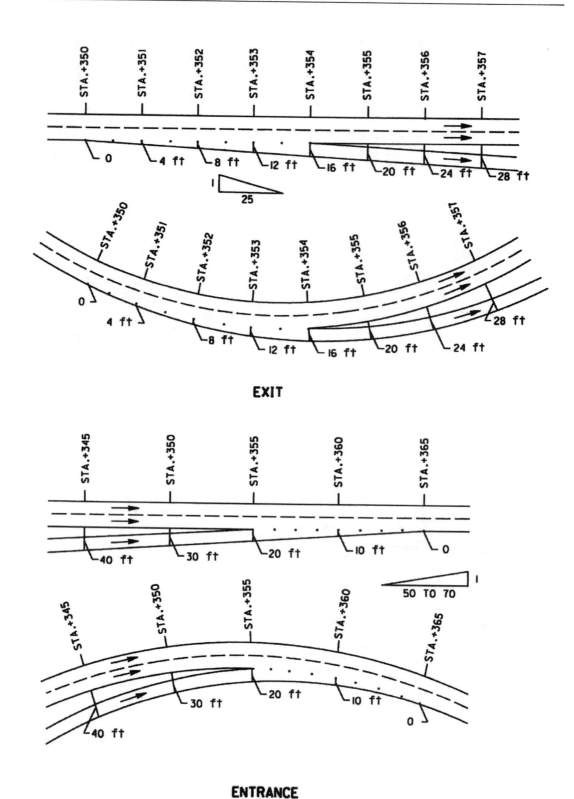

EXIT

ENTRANCE

Exhibit 10-74. Layout of Taper-Type Terminals on Curves (U.S. Customary)

Wherever a part of a tapered speed-change lane falls on curved alignment, it is desirable that the entire length be within the limits of the curve. Where the taper is introduced on tangent alignment just upstream from the beginning of the curve, the outer edge of the taper will appear as a kink at the point of curvature.

At ramp terminals on relatively sharp curves, such as those that may occur on freeways having a design speed of 80 km/h [50 mph], the parallel type of speed-change lanes has an advantage over the taper type. At exits the parallel type is less likely to confuse through traffic, and at entrances this type will usually result in smoother merging operations. Parallel-type speed-change lanes at ramp terminals on curves are illustrated in Exhibit 10-75.

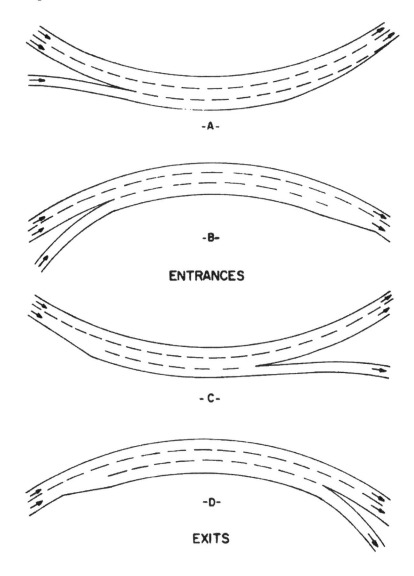

Exhibit 10-75. Parallel-Type Ramp Terminals on Curves

Entrances on curved sections of highway are generally less of a problem than exits. Exhibits 10-75A and Exhibit 10-75B show entrances with the highway curving to the left and right, respectively. It is important that the approach curve on the ramp has a very long radius as it joins the acceleration lane. This aligns the entering vehicle with the acceleration lane and lessens the chances of motorists entering directly onto the through lanes. The taper at the end of the acceleration lane should be long, preferably about 90 m [300 ft] in length. When reverse-curve alignment occurs between the ramp and speed-change lane, an intervening tangent should be used to aid in superelevation transition.

An exit may be particularly troublesome where the highway curves to the left (Exhibit 10-75C) because traffic on the outside lane tends to follow the ramp. Exits on left-turning curves should be avoided, if practical. Caution should be used in positioning a taper-type deceleration lane on the outside of a left-turning main line curve. The design should provide a definite break in the right edge of traveled way to provide a visual cue to the through driver so that the driver is not inadvertently led off the through roadway. To make the deceleration lane more apparent to approaching motorists, the taper should be shorter, preferably no more than 30 m [100 ft] in length. The deceleration lane should begin either upstream or downstream from the PC. It should not begin right at the PC, as the deceleration lane appears to be an extension of the tangent, and motorists are more likely to be confused. The ramp proper should begin with a section of tangent or a long-radius curve to permit a long and gradual reversing of the superelevation.

An alternate design, which will usually avoid operational problems, is to locate the exit terminal a considerable distance upstream from the PC. In this design, separate and parallel ramp roadway is provided to connect with the ramp proper.

With the highway curving to the right and the exit located on the right (Exhibit 10-75D, there is a tendency for vehicles to exit inadvertently. Again, the taper should be short to provide additional "target" value for the deceleration lane. With this configuration, the superelevation of the deceleration lane is readily achieved by continuing the rate from the traveled way and generally increasing it to the appropriate rate for the ramp curve.

Multilane free-flow terminals. Multilane terminals are appropriate where traffic is too great for single-lane operation. Other considerations that may call for multilane terminals are through-route continuity, queuing on long ramps, lane balance, and design flexibility. The most common multilane terminals consist of two-lane entrances and exits at freeways. Other multilane terminals are sometimes termed "major forks" and "branch connections." The latter terms denote a separating and joining of two major routes.

Two-lane entrances. Two-lane entrances are warranted for two situations: either as branch connections or because of capacity needs for the on-ramp. To satisfy lane-balance needs, at least one additional lane should be provided downstream. This addition may be a basic lane, if needed for capacity, or an auxiliary lane that may be dropped 750 to 900 m [2,500 to 3,000 ft] downstream from the entrance or at the next interchange. In some instances, two additional lanes may be needed because of capacity considerations.

If the two-lane entrance is preceded by a two-lane exit, there is probably no need to increase the basic number of lanes on the freeway from a capacity standpoint. In this case, the added lane that results from the two-lane entrance is considered an auxiliary lane, and it may be dropped approximately 750 m [2,500 ft] or more downstream from the entrance. Details of lane drops were discussed earlier in this chapter.

Exhibit 10-76 illustrates simple two-lane entrance terminals where a lane has been added to the freeway. The number of lanes on the freeway has little or no effect on terminal design. Exhibit 10-76A presents a taper-type entrance and Exhibit 10-76B shows a parallel-type entrance. Intermixing of the two designs is not recommended within a system route or an urban-area system.

The basic form or layout of a two-lane taper-type entrance, as shown in Exhibit 10-76A, is the same as a single-lane taper, as described earlier in this chapter, with a second lane added to the right or outer side and continued as an added or auxiliary lane on the freeway. Exhibit 10-70 shows minimum lengths of acceleration distances for entrance ramps. The gap acceptance length is also a consideration as illustrated in Exhibit 10-76A. Where ramp grades are involved, the lengths should be adjusted as shown in Exhibit 10-71. As in the case of a single-lane entrance, it is most desirable that the geometrics of the ramp proper permit motorists to attain the approximate running speed of the freeway before reaching the tapered section.

With the parallel type of two-lane entrance, as shown in Exhibit 10-76B, the left lane of the ramp is continued onto the freeway as an added lane. The right lane of the ramp is carried as a parallel lane for at least 90 to 150 m [300 to 500 ft] and terminated by a tapered section at least 90 m [300 ft] in length. The length of the right lane should, as a minimum, be determined from the acceleration length or gap acceptance length, as shown in Exhibit 10-76B. Major factors in determining the appropriate length are the traffic volume on the ramp and the traffic volume on the freeway.

When the volume of the two-lane ramp, either the taper or parallel type, exceeds the capacity of a through lane as specified in the HCM (5), it is suggested that the value for L_g (Exhibit 10-76) be in the range of 300 to 665 m [900 to 2,000 ft] to allow sufficient time and distance for vehicles in the left ramp lane to move into the mainline lanes, thereby opening space and providing time for vehicles in the right ramp lane to move into the left ramp lane. Following the termination of the left ramp lane, an additional distance in the range of 300 to 665 m [900 to 2,000 ft] should be provided, plus a taper before terminating the right ramp lane.

Although both the taper type and the parallel type of two-lane entrances will operate efficiently when properly designed, some designers prefer the parallel type. This is based on the premise that the taper type involves an "inside merge" with traffic traveling on both sides of the merging lanes. If either vehicle involved with the merging movement abandons the merge, traffic in the adjacent lanes could prevent the merging vehicles from escaping to the adjacent lanes. By contrast, the parallel type allows the merging vehicle to escape to the right shoulder without any interference.

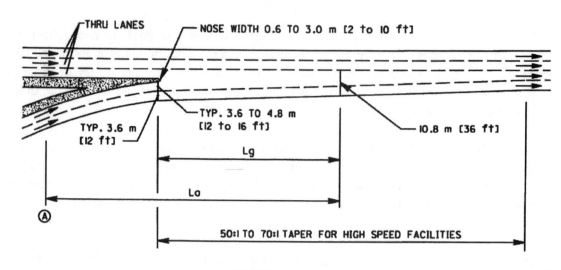

-A- TAPERED DESIGN

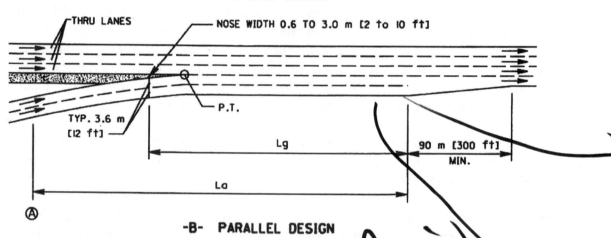

-B- PARALLEL DESIGN

NOTES:

1. La IS THE REQUIRED ACCELERATION LENGTH
   AS SHOWN IN EXHIBIT 10-70 OR AS ADJUSTED BY EXHIBIT 10-71.

2. POINT (A) CONTROLS SPEED ON THE RAMP.
   La SHOULD NOT START BACK ON THE CURVATURE
   OF THE RAMP UNLESS THE RADIUS EQUALS 300 m
   [1000 ft] OR MORE.

3. Lg IS REQUIRED GAP ACCEPTANCE LENGTH. Lg
   SHOULD BE A MINIMUM OF 90 TO 150 m [300 to
   500 ft] DEPENDING ON THE NOSE WIDTH.

4. THE VALUE OF La OR Lg, WHICHEVER PRODUCES
   THE GREATER DISTANCE DOWNSTREAM FROM
   WHERE THE NOSE EQUAL 0.6 m [2 ft], IS SUGGESTED
   FOR USE IN THE DESIGN OF THE RAMP ENTRANCE.

Exhibit 10-76. Typical Two-Lane Entrance Ramps

Where the predominant two-lane entrances in a particular State or locality are of the parallel type and, therefore, drivers are accustomed to that type of entrance, a taper-type entrance would violate driver expectancy, and vice versa. Thus, a particular type of entrance terminal is sometimes condemned as being unsatisfactory when in fact the difficulty may be lack of uniformity. Either form of two-lane entrance is satisfactory if used exclusively within an area or a region, but they should not be intermixed along a given route.

Two-lane exits. Where the traffic volume leaving the freeway at an exit terminal exceeds the design capacity of a single lane, a two-lane exit terminal should be provided. To satisfy lane-balance needs and not to reduce the basic number of through lanes, it is usually appropriate to add an auxiliary lane upstream from the exit. A distance of approximately 450 m [1,500 ft] is recommended to develop the full capacity of a two-lane exit. Typical designs for two-lane exit terminals are shown in Exhibit 10-77; the taper is illustrated in Exhibit 10-77A and the parallel type in Exhibit 10-77B.

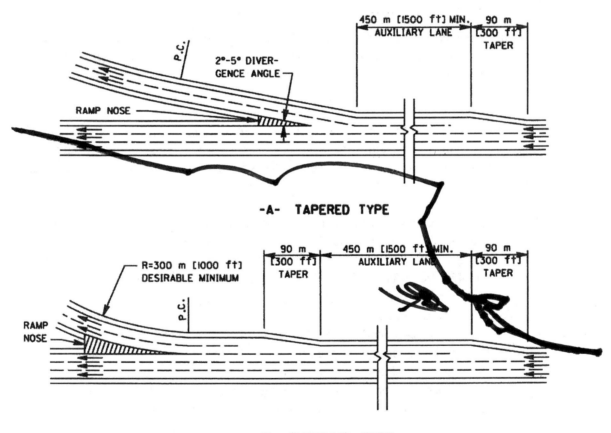

Exhibit 10-77. Two-Lane Exit Terminals

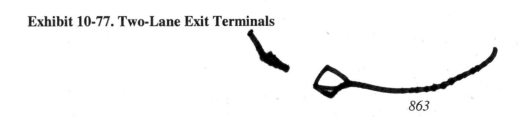

In cases where the basic number of lanes is to be reduced beyond a two-lane exit, the basic number of lanes should be carried beyond the exit before the outer lane is dropped. This design provides a recovery area for any through vehicles that remain in that lane. This was discussed in the section "Lane Reductions" earlier in this chapter.

With the parallel type of two-lane exit, as shown in Exhibit 10-77B, the operation is different from the taper type in that traffic in the outer through lane of the freeway must change lanes in order to exit. In fact, an exiting motorist is required to move two lanes to the right in order to use the right lane of the ramp. Thus, considerable lane changing is needed in order for the exit to operate efficiently. This entire operation takes place over a substantial length of highway, which is dependent in part on the total traffic volume on the freeway and especially on the volume using the exit ramp. The total length from the beginning of the first taper to the point where the ramp traveled way departs from the right-hand through lane of the freeway should range from 750 m [2,500 ft] for turning volumes of 1,500 vph or less upward to 1,000 m [3,500 ft] for turning volumes of 3,000 vph.

Two-lane terminals on curved alignment. The design of ramp terminals where the freeway is on curved alignment is discussed under single-lane terminals. The same principles of design, in which offsets from the edge of roadway are used, may be used in the layout of two-lane terminals.

Major forks and branch connections. A major fork is defined as the bifurcation of a directional roadway of a terminating freeway route into two directional multilane ramps that connect to another freeway, or of a freeway route into two separate freeway routes of about equal importance.

The design of major forks is subject to the same principles of lane balance as any other diverging area. The total number of lanes in the two roadways beyond the divergence should exceed the number of lanes approaching the diverging area by at least one. Operational difficulties invariably develop unless traffic in one of the interior lanes has an option of taking either of the diverging roadways. Accordingly, the nose should be placed in direct alignment with the center line of one of the interior lanes, as illustrated in Exhibits 10-78A, B, or C, where the horizontal alignment of the two departing roadways are in curves. This interior lane is continued as a full-width lane, both left and right of the gore. Thus, the width of this interior lane will be at least 7.2 m [24 ft] at the painted nose (prolongation of pavement-edge stripes) and preferably not over 8.4 m [28 ft]. The length over which the widening from 3.6 to 7.2 m [12 to 24 ft] takes place should be within the range of 300 or 540 m [1,000 or 1,800 ft]. However, in the case where at least one of the approaches is on a tangent alignment and continues on a tangent, a true optional interior lane cannot be physically developed. As such, the principles of the two-lane exit facility should be used as shown in Exhibit 10-78D.

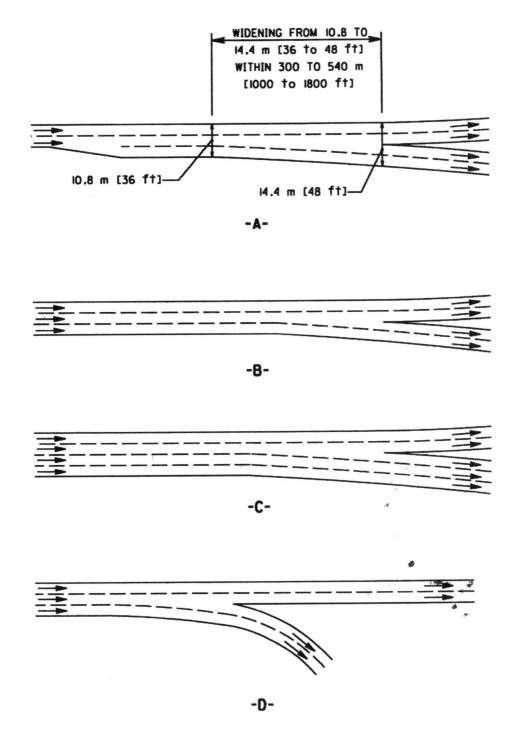

Exhibit 10-78. Major Forks

In the case of a two-lane roadway separating into two, two-lane routes, there is no interior lane. In such cases, it is advisable to widen the approach roadway to three lanes, thus creating an interior lane. The lane is added on the side of the fork that serves the lesser traffic volume. In Exhibit 10-78A, the right (lower) fork would be the more lightly traveled of the two. The

865

widening from 10.8 m [36 ft] for the approach roadway to about 14.4 or 15.0 m [48 or 50 ft] at the painted nose should be accomplished in a continuous sweeping curve with no reverse curvature in the alignment of the roadway edges.

A branch connection is defined as the beginning of a directional roadway of a freeway formed by the convergence of two directional multilane ramps from another freeway or by the convergence of two freeway routes to form a single freeway route.

The number of lanes downstream from the point of convergence may be one lane fewer than the combined total on the two approach roadways. In some cases, the traffic demand may indicate that the number of lanes going away from the merging area be equal to the sum of the number of lanes on the two roadways approaching it, and a design of this type will pose no operational problem. Such a design is illustrated in Exhibit 10-79A.

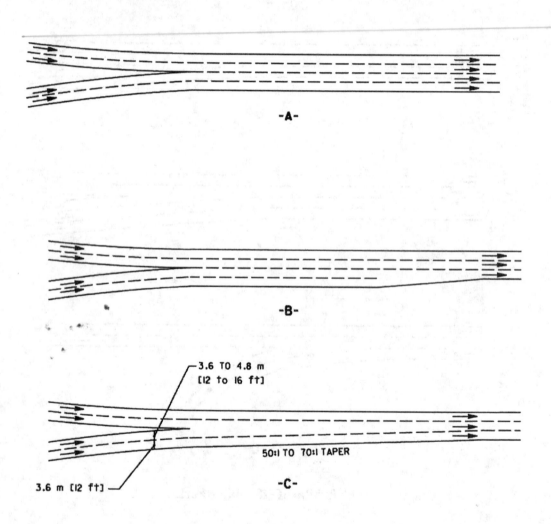

Exhibit 10-79. Branch Connections

Where a lane is to be dropped, which is the more common case, a means for accomplishing the reduction is discussed in the section "Lane Reductions" in this chapter. The lane that is terminated will ordinarily be the exterior lane from the roadway serving the lowest volume per lane. However, some considerations should also be given to the fact that the outer lane from the roadway entering from the right is the slow-speed lane for that roadway, whereas the opposite is true for the roadway entering from the left. If the traffic volumes per lane are about equal, it would be proper to terminate the lane on the right, as shown in Exhibit 10-79B. In any case, consistency within an area or region is often more important than volume per lane since the latter may change with the specific design or with traffic demand changing over time. The lane being terminated should be carried at full lane width for a distance of approximately 300 m [1,000 ft] before being tapered out.

Another consideration is the possibility of a high-speed inside merge, as in Exhibit 10-79C. This merge should be treated as any other high-speed merging situation; see the discussion of the advantages of parallel-type entrance in the earlier section on "Two-Lane Entrances."

Other Interchange Design Features

Testing for Ease of Operation

Each section of freeway that includes a series of interchanges or a succession of exits and entrances should be tested for various operational characteristics of the route including adaptability, capacity, and operational features. The evaluation tests for ease of operation and for route continuity from a driver's point of view, both of which are affected by the location, proximity, and sequence of exits and entrances; the merging, diverging, and weaving movements involved; practicality of signing; and clarity of paths to be followed. This test should be completed after the preliminary design and before each interchange is completed.

A route may be tested by isolating those parts of the plan that will affect drivers on individual paths through the interchange. Viewing an entire plan, as it might be seen from the air, may give an impression of complexity because of the number of exits, entrances, ramps, and structures. Actually, it is not as complex to the drivers, who see only the path they are driving on. On the other hand, certain weaknesses of operation not evident on the overall plan will be revealed in testing a single path of travel.

The plan should be tested by drawing or tracing the individual path for each principal origin and destination and studying thereon those physical features that will be encountered by a driver. The test can also be made on an overall plan on which the path to be studied and the stubs of connecting roads are colored or shaded. The plan should show the peak-hour volumes, number of traffic lanes, and peak-hour and off-peak-hour running speed. Thus, the designer can visualize exactly what the driver sees—only the road being traveled, with the various points of ingress and egress and the directional signs along it—and have a sense of the accompanying traffic.

Such an analysis indicates whether or not confusion is likely because of exits and entrances too close together or interference is likely because of successive weaving sections. It should also show whether or not the path is clearly defined, if it is practical to sign the facility properly, and if major or overhead signs are needed and where they may be placed. The test may show that the path is easy to travel, direct in character, and free from sections that might confuse drivers, or it may show that the path is sufficiently complex and confronted with disturbing elements so that an adjustment in design is appropriate. As a result, it may be appropriate to move or eliminate certain ramps. In an extreme case, the test may show that it is appropriate to change the overall pattern by eliminating an interchange, introducing collector-distributor roads in order to prevent interference with through traffic, or making some other radical change in design.

Exhibit 10-80 is a simple diagrammatic solution to a typical freeway operational problem. The freeway joins a principal arterial at a branch connection and diverges at a major fork in a distance of approximately 1.5 to 5.0 km [1 to 3 mi]. There may be other connections to and from the freeway between these points. The through freeway merges on the left at point X and diverges on the right at point Y. The desirable solution, as shown in this figure, does not involve any lane changes on the through lanes of the freeway. Traffic on the local arterial enters and exits on the right, and there is no disruption of route continuity on either facility.

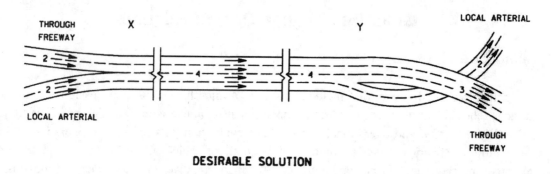

DESIRABLE SOLUTION

NOTES:

1. THE DISTANCE BETWEEN POINTS X AND Y MAY
   BE APPROXIMATELY 1.5 TO 5.0 Km [1 TO 3 MILES]

2. NUMBER OF LANES ARE SHOWN ON EACH ROADWAY

Exhibit 10-80. Diagram of Freeway Operational Problem and Solution

Pedestrians

The accommodation of pedestrians through urban interchanges should be considered early in the development of interchange configurations. High-density land use in the vicinity of an interchange can generate heavy pedestrian movements, resulting in conflicts between vehicles and pedestrians.

The movement of pedestrians through interchanges can be enhanced by providing sidewalks to separate the pedestrian from the vehicular traffic. When sidewalks are provided, they should be placed as far from the roadway as practical and be wide enough to handle the anticipated pedestrian volumes. To maximize pedestrian usage, the sidewalk should provide the most direct route through the interchange with minimal change in vertical alignment. Through complex interchange configurations, the use of informational signing may be appropriate to direct pedestrians to appropriate alternate routes.

Where pedestrians will be crossing an interchange ramp, adequate sight distance should be provided to ensure that drivers detect the presence of pedestrians and pedestrians can perceive gaps in the traffic flow. To provide increased visibility at night, pedestrian ramp crossings should have overhead illumination. Where there are insufficient gaps in the traffic flow to allow pedestrians to cross the ramp, pedestrian-actuated signals or a pedestrian overpass/underpass should be considered.

Ramp Metering

The purpose of ramp metering is to reduce congestion or improve merge operations on urban freeways. The metering may be limited to only one ramp or integrated into a series of entrance ramps.

Ramp metering consists of traffic signals installed on entrance ramps in advance of the entrance terminal to control the number of vehicles entering the freeway. The traffic signals may be pretimed or traffic actuated to release the entering vehicles individually or in platoons.

Pretimed signals release vehicles at regular intervals that have been determined by traffic volume studies. Traffic-actuated signals involve detectors placed on the freeway, upstream of the entrance terminal, to measure the approach traffic volumes. The metering rate is adjusted by comparing the upstream volumes to the downstream capacity, which has been determined either by traffic volume studies or measured by detectors in the pavement.

Ramp metering to improve merge operations involves detectors on the upstream approach of the freeway to determine acceptable gaps in the traffic flow. The traffic on the entrance ramp is released to coincide with the gap detected in the traffic on the freeway. For further information on ramp metering, see the *Highway Capacity Manual* (**5**). In addition, the AASHTO *Guide for the Design of High-Occupancy Vehicle Facilities* (**6**) provides treatments for ramp metering in conjunction with HOV lanes.

Grading and Landscape Development

Grading at an interchange is determined chiefly by the alignments, profiles, cross sections, and drainage needs for the intersecting highways and ramps. Each through roadway or ramp should not be treated as a separate unit and graded to a specified cross section without regard to its relation with adjacent roads and to the surrounding topography. Instead, the whole construction area should be designed as a single unit to keep construction and maintenance costs to a minimum, obtain maximum visibility, and enhance the area's appearance. In some parts, such as at narrow sections between converging roadways, the slopes and grading controls may affect the alignment and profile design.

Contour grading design. An important and early step in interchange design is the initial bridge control study in which the preliminary alignment and profiles of the intersecting roads are developed to determine the controls for bridge design. Alternative treatments of such elements as clearances, curbs, walks, and position and extent of walls should be examined in regard to general grading before conclusions are drawn for the bridge design, particularly for lengths of wing walls. Minor modifications in alignment and profile, in abutments and walls, and in related earthwork may produce a more desirable solution as a whole.

Steep roadside earth slopes should be avoided for all roads and ramps at interchanges. Flat slopes should be used where practical, for economical construction and maintenance, to increase safety and enhance the appearance of the area. Broad rounded drainageways or swale-like depressions should be used, where practical, to encourage healthy turf and easy mowing. V ditches and small ditches with steep side slopes should be avoided. Drainage channels and related structures should be as inconspicuous and maintenance-free as practical. They should not be an eyesore or become an obstacle to an errant vehicle. Transition grading between cut and fill slopes should be long and natural in appearance. The slopes should be well rounded and smooth to blend the highway into the adjacent terrain. The contours should have flowing continuity and be congruous with the form of the roadway and with the adjacent topography as well.

The contour grading and drainage plan should be designed to protect existing trees and preserve other desirable features, as practical. This effort, however, should be consistent with the objectives stated above.

Plantings. Proposed plantings should be selected with regard to their ultimate growth. Improperly located shrubs or trees may decrease horizontal sight distance on curves and seriously interfere with lateral sight distance between adjacent roadways. Even low-lying ground covers may shorten vertical sight distance on curving ramps.

Trees or shrubs may be used to outline travel paths or to give drivers a sense of an obstruction ahead. For example, the ends of a directional island or approach nose may be planted with low-growing shrubs that will be seen from a considerable distance and direct the driver's attention to the need for a turn. Shrubs that could cause vehicle damage on impact or obscure signs or warning devices should be avoided.

The AASHTO *Roadside Design Guide* (3) should be referenced for guidance on minimum clear zones prior to planting trees that will mature to greater than 100 mm [4 in] in diameter. Distances greater than the minimum are often appropriate because overhanging branches create a distraction, and leaves on the roadway reduce the pavement surface friction, especially when wet. In areas where ice and snow are a problem, all trees should be planted an adequate distance from the traveled way to allow for snow drifting and to prevent icing in shaded areas.

Models

Three-dimensional models are helpful in the design of interchanges. Models are particularly useful in communicating the designer's ideas to lay groups and others who are not trained to visualize three dimensions from the plans. Design concept teams and other officials find models helpful in analyzing proposed designs.

Highway models fall into two basic categories—design models and presentation models. Design models are simple and easily adjusted, thus permitting the designer to experiment with different concepts. Presentation models are more permanent and are valuable to highway officials when making a presentation to others who are not familiar with engineering terms and methods.

REFERENCES

1. Gluck, J., H. S. Levinson, and V. Stover. *Impact of Access Management Techniques*, NCHRP Report 420, Washington, D.C.: Transportation Research Board, 1999.

2. *Interchange Operations on the Local Street Side: State of the Art*, Transportation Research Circular 430, Washington, D.C.: Transportation Research Board, July 1999.

3. AASHTO. *Roadside Design Guide*, Washington, D.C.: AASHTO, 1989.

4. U.S. Department of Transportation, Federal Highway Administration. *Manual on Uniform Traffic Control Devices for Streets and Highways*, Washington, D.C.: 1988 or most current edition.

5. Transportation Research Board. *Highway Capacity Manual*, Special Report 209, Washington, D.C.: Transportation Research Board, 2000 or most current edition.

6. AASHTO. *Guide for the Design of High-Occupancy Vehicle Facilities*, Washington, D.C.: AASHTO, 1992.

7. Messer, C. J., and J. A. Bonneson. *Single Point Urban Interchange Design and Operations Analysis*, NCHRP Report 345, Washington, D.C.: Transportation Research Board, 1991.

8. Barnes, M., R. Ervin, C. MacAdam, and R. Scott. *Impact of Specific Geometric Features on Truck Operations and Safety at Interchanges*, Report No. FHWA/RD 86/057, McLean, Virginia: U.S. Department of Transportation, Federal Highway Administration, August 1985.

9. Ferlis, R. A., and L. S. Kagin. *Planning for Pedestrian Movement at Interchanges*, Report No. FHWA/RD 74/65, McLean, Virginia: U.S. Department of Transportation, Federal Highway Administration, July 1974.

INDEX